Plant Pathology

Concepts and Laboratory Exercises

THIRD EDITION

Plant Pathology

Concepts and Laboratory Exercises

THIRD EDITION

Edited by

Bonnie H. Ownley

Robert N. Trigiano

CRC Press
Taylor & Francis Group
Boca Raton London New York

CRC Press is an imprint of the
Taylor & Francis Group, an **informa** business

CRC Press
Taylor & Francis Group
6000 Broken Sound Parkway NW, Suite 300
Boca Raton, FL 33487-2742

First issued in hardback 2017

ISBN 13: 978-1-138-40760-2 (hbk)
ISBN 13: 978-1-4665-0081-5 (pbk)

Library of Congress Cataloging-in-Publication Data

Names: Ownley, Bonnie H., editor. | Trigiano, R. N. (Robert Nicholas), 1953- editor.
Title: Plant pathology concepts and laboratory exercises / editors: Bonnie H. Ownley and Robert N. Trigiano.
Description: Third edition. | Boca Raton : Taylor & Francis, 2016. | Includes bibliographical references and index.
Identifiers: LCCN 2016013150 | ISBN 9781466500815 (alk. paper)
Subjects: LCSH: Plant diseases--Laboratory manuals.
Classification: LCC SB732.56 .P63 2016 | DDC 632/.3--dc23
LC record available at https://lccn.loc.gov/2016013150

Visit the Taylor & Francis Web site at
http://www.taylorandfrancis.com

and the CRC Press Web site at
http://www.crcpress.com

Contents

Preface

We thank those instructors who have adopted the first two editions of *Plant Pathology Concepts and Laboratory Exercises* as a guide for their classes. We also are grateful to them and their students and colleagues for providing invaluable feedback and criticism of the previous editions. We have incorporated many of their ideas into this new, third edition, which includes combining concept and laboratory chapters into one presentation, almost all figures in color, more technical presentations of some topics, a chapter on safety in the laboratory, treatment of organic agriculture and disease, and more extensive chapters about disease diagnostics. We have also improved the binding of the book, which is now spiral bound, allowing the students to access any page easily.

This edition of *Plant Pathology Concepts and Laboratory Exercises* is intended to serve as a primary text for introductory courses and furnishes instructors and students alike with a broad consideration of this important and growing field. It presents many useful protocols and procedures and thus serves as a valuable reference to researchers as well as students in beginning and advanced plant pathology and allied biological sciences courses. The book is intentionally written informally to some extent as it provides the reader with a minimum number of references, but does not lose any essential information or accuracy. Broad topic chapters are authored by specialists with considerable experience in the field and are supported by one or more laboratory exercises illustrating the central concepts of the topic. Each chapter begins with a "Concept Box" highlighting some of the more important ideas contained within the chapter and signals students to read carefully for these primary topics. There is an extensive glossary, which appear as bolded words in each chapter. Collectively, the laboratory exercises are exceptionally diverse in nature, providing something for beginning to advanced students. Most importantly, the authors have successfully completed the exercises/experiments many times, often with either plant pathology or biology classes or in their own research laboratories. All the laboratory protocols are written in procedure boxes that provide step-by-step, easy-to-follow instructions. A unique feature of this text is that the authors have provided the expected results of each of the experiments in general terms. At the end of each exercise, there are a series of questions designed to provoke individual thought and critical examination of the experiment and results. Our intention is that instructors will not attempt to do all the experiments in each chapter, but rather select one or two for each concept that serves the needs and interests of their particular class. For an advanced class, other experiments may be assigned to resourceful students. We caution instructors and students to obtain the proper documents for transport and use of plant pathogenic organisms and to properly dispose of cultures and plant materials at the end of the laboratory exercises. We also support mandatory safety training that is typically available online at many institutions.

This book is divided into five primary sections: Introductory Concepts, Groups of Plant Pathogens and Abiotic Disorders, Plant–Pathogen Interactions, Epidemiology and Disease Control, and Special Topics. Chapter 1 in Part I introduces students to the basic concepts of plant pathology including some historical perspectives, fundamental ideas of what is disease, how disease relates to environment, the host, and time, and provides a very broad overview of organisms that cause disease. Chapter 2 is a new topic in the third edition and describes laboratory safety, media preparation, and solutions. Chapter 3 introduces students to the fundamentals of microscopy, which is a topic often omitted in biological textbooks. Part II includes chapters that detail various disease-causing organisms, plant parasitic plants, and the causes of abiotic diseases. This section begins with a consideration of viruses (Chapter 4), prokaryotic organisms (Chapter 5), and nematodes (Chapter 6). Chapter 7 provides a very broad overview of pathogenic species in the Oomycota (fungus-like organisms) and pathogenic true fungi. The next eight chapters are devoted to species in the Oomycota and various phyla of fungi followed by chapters that focus on soilborne plant pathogens, parasitic seed plants, and disorders caused by abiotic agents. Part III explores plant–pathogen interactions in Chapters 19–21 including treatments of virulence factors, pathogen attack strategies, extracellular enzymes, host defenses, and disruption of plant function. Part IV is anchored with an extensive chapter (Chapter 22) outlining the basic concepts of epidemiology, which is followed in turn by several chapters detailing various strategies for disease control, including host resistance (Chapter 23), plant–fungal interactions (Chapter 24), cultural management of plant disease (Chapter 25), chemical control of disease (Chapter 26), use of microbial control agents (Chapter 27), and integrated pest management (IPM) strategies (Chapter 28). The concluding chapter in this section is an often suggested topic, organic agricultural and plant disease (Chapter 29). Part V is devoted to the treatment of plant disease diagnostics (Chapter 30) and identifying disease-causing organisms using molecular techniques (Chapter 31). Chapter 32 relates fungal and bacterial physiology/nutrition to disease via extracellular enzyme production. This chapter contains many valuable techniques that are applicable to other fields of science. Lastly, Chapter 33 provides explanations and exercises for molecular techniques used in plant pathology and other fields of study.

It is our hope that students and instructors find the format, level, and amount of information contained in the book to be appropriate for an introductory course and some advanced courses. The presentation style has been used very successfully in other books and with the addition of the extensive glossary, useful case studies, and concept boxes, students should find the format stimulating and conducive for learning. We invite and welcome your comments and suggestions for improvements.

B.H. Ownley
R.N. Trigiano
The University of Tennessee

Acknowledgments

We wish to recognize and applaud the extraordinary efforts and talents of all the contributing authors—their creativity, support, advice, understanding, and especially patience throughout the process of developing the third edition of *Plant Pathology Concepts and Laboratory Exercises* were phenomenal. We express our gratitude to the Tennessee Institute of Agriculture for permitting us the time and financial support necessary to complete this project. We thank our colleagues and students who suggested changes to the chapters and the arrangement in the text. We extend very special thanks to Alan S. Windham (The University of Tennessee) and David Shew (North Carolina State University) for the fantastic images donated to this project. We also express our sincere gratitude to our families for always supporting us during the completion of this book.

Editors

Dr. Bonnie H. Ownley is a professor of plant pathology in the Department of Entomology and Plant Pathology at The University of Tennessee in Knoxville. She received her B.S. in biology from the University of North Carolina at Chapel Hill, M.S. in microbiology from Auburn University, Alabama, and PhD in plant pathology, with a minor in soil science, from North Carolina State University, Raleigh. She was a postdoctoral research fellow with the U.S. Department of Agriculture (USDA), Agricultural Research Service, in the Root Disease and Biological Control Research Unit at Pullman, Washington, and a visiting plant pathologist in the Plant Pathology Department at Washington State University before joining the faculty at The University of Tennessee.

Dr. Ownley's research and teaching programs are focused on the etiology, biology, ecology, and environmentally sustainable control of plant pathogens on a variety of food, fiber, and biofuels crops. Her work in biological control of plant diseases is recognized internationally. She has published more than 120 research papers, book chapters, conference proceedings, and popular press articles and has received numerous grants from the USDA, state agencies, private industry, and commodity groups to support her research, teaching, and outreach projects.

Dr. Ownley is the Director of Graduate Studies for the Department of Entomology and Plant Pathology. She is a dedicated teacher and strong proponent of experiential and service learning. Her teaching portfolio includes graduate courses on mycology, phytobacteriology, and soilborne plant pathogens. She has mentored and trained 19 graduate students and served on the research committees of more than 40 additional students. Her teaching has extended beyond the university to include multiple biotechnology workshops for middle and high school teachers across the State of Tennessee and experiential learning summer programs for middle and high school students.

Dr. Ownley has been recognized with numerous awards and honors for her research, teaching, and academic outreach programs, as well as service to the university and community. She has served in leadership roles for the American Phytopathological Society and as Senior Editor for *Phytopathology*. She is currently President of the Faculty Senate of The University of Tennessee, Knoxville. Her service to the university has been wide-ranging, including multiple administrative and faculty search committees, program initiatives, policy development, unit reviews, and strategic planning. Dr. Ownley has worked to improve the workplace and learning environment for faculty, students, and staff through her service to the university, from the department to the system level. Having often been the only woman at the table in the early part of her career, she is committed to eliminating bias and discrimination and educating others that inclusion of underrepresented minorities and women will multiply the possibilities and improve the innovation, creativity, civility, and sense of community of the organization.

Dr. Robert N. Trigiano received his B.S. degree with an emphasis in biology and chemistry from Juniata College, Huntingdon, Pennsylvania, in 1975 and an M.S. in biology (mycology) from the Pennsylvania State University, State College, Pennsylvania, in 1977. He was an associate research agronomist working with mushroom culture and plant pathology for Green Giant Co., Le Sueur, Minnesota, until 1979 and then a mushroom grower for Rol-Land Farms, Ltd., Blenheim, Ontario, Canada, during 1979 and 1980. He completed a PhD degree in botany and plant pathology (comajors) at North Carolina State University at Raleigh in 1983. After concluding postdoctoral work in the Plant and Soil Science Department at The University of Tennessee in Knoxville, he was appointed an assistant professor in the Department of Ornamental Horticulture and Landscape Design at the same university in 1987, promoted to associate professor in 1991 and to professor in 1997. He served as interim head of the department from 1999 to 2001. He then joined the Department of Entomology and Plant Pathology at the University of Tennessee in 2002 and was interim head from 2012 to 2013. In 2015, Dr. Trigiano was selected as an Institute Professor at The University of Tennessee Institute of Agriculture.

Dr. Trigiano is a member of the American Phytopathological Society (APS), the American Society for Horticultural Science (ASHS), and the honorary societies of Gamma Sigma Delta, Sigma Xi, and Phi Kappa Phi. He received the T.J. Whatley Distinguished Young Scientist Award (The University of Tennessee, Institute of Agriculture) and the Gamma Sigma Delta Research individual and team Award of Merit at the University of Tennessee. He is the recipient of the publication awards for the most outstanding educational and ornamental papers in

ASHS and in the Southern region ASHS L. M. Ware distinguished research award. In 2006, he was elected a fellow of the American Society for Horticultural Science. Dr. Trigano was awarded the B. Otto and Kathleen Wheeley Award of Excellence in Technology Transfer and founder and manager of Creative Agricultural Technologies, LLC by the University of Tennessee Research Foundation in 2007. He has been an editor for *Plant Disease*, ASHS journals, *Plant Cell, Tissue and Organ Culture*, *Plant Cell Reports* and is currently the editor-in-chief of *Critical Reviews in Plant Sciences*. Additionally, he has coedited ten books, including *Plant Tissue Culture Concepts and Laboratory Exercises*, *Plant Pathology Concepts and Laboratory Exercises*, and *Plant Development and Biotechnology*.

Dr. Trigiano has received research grants from the United States Department of Agriculture (USDA) and Forest Service, Horticultural Research Institute, and from private industries and foundations. He has published more than 200 research papers, book chapters, patents, and popular press articles. He teaches graduate courses in scientific writing and molecular techniques and has presented numerous workshops on English scientific writing in Germany, the People's Republic of China, and Brazil. His current research interests include molecular markers for breeding ornamental plants, population studies of pathogens and native plants, diseases of ornamental plants, somatic embryogenesis, and micropropagation of ornamental species.

Contributors

Richard E. Baird
Mississippi State University
Starkville, Mississippi

Peter Balint-Kurti
United States Department of Agriculture
Agriculture Research Service
Plant Science Research Laboratory
North Carolina State University
Raleigh, North Carolina

Anton Baudoin
Virginia Polytechnic Institute and State University
Blacksburg, Virginia

D. Michael Benson
North Carolina State University
Raleigh, North Carolina

Ernest C. Bernard
The University of Tennessee
Knoxville, Tennessee

Kira L. Bowen
Auburn University
Auburn, Alabama

Judith K. Brown
The University of Arizona
Tucson, Arizona

Carolee T. Bull
Pennsylvania State University
University Park, Pennsylvania

David M. Butler
The University of Tennessee
Knoxville, Tennessee

Lori M. Carris
Washington State University
Pullman, Washington

Christina Cowger
United States Department of Agriculture
Agriculture Research Service
Plant Science Resesarch Laboratory
North Carolina State University
Raleigh, North Carolina

Bert Cregg
Michigan State University
East Lansing, Michigan

Tom Creswell
Purdue University
West Lafayette, Indiana

Marc A. Cubeta
North Carolina State University
Raleigh, North Carolina

Margery L. Daughtrey
Cornell University
Ithaca, New York

Ricardo Manuel de Seixas Boavida Ferreira
Institute of Agronomy
Technical University of Lisbon
Lisbon, Portugal

Leonardo De La Fuente
Auburn University
Auburn, Alabama

Ann Brooks Gould
Rutgers University
New Brunswick, New Jersey

Kimberly D. Gwinn
The University of Tennessee
Knoxville, Tennessee

Denita Hadziabdic
The University of Tennessee
Knoxville, Tennessee

Jong Hyun Ham
Louisiana State University
Baton Rouge, Louisiana

Kathie T. Hodge
Cornell University
Ithaca, New York

Alejandra I. Huerta
Colorado State University
Fort Collins, Colorado

Teresa M. Jardini
Washington State University
Pullman, Washington

Steven N. Jeffers
Clemson University
Clemson, South Carolina

Steven T. Koike
University of California Cooperative Extension
Salinas, California

Kurt H. Lamour
The University of Tennessee
Knoxville, Tennessee

Blanca B. Landa
Institute for Sustainable Agriculture
Spanish National Research Council
Cordoba, Spain

Marie A.C. Langham
South Dakota State University
Brookings, South Dakota

Larry J. Littlefield
Oklahoma State University
Stillwater, Oklahoma

Stacy J. Mauzey
University of California Cooperative Extension Service
Salinas, California

Dmitri V. Mavrodi
The University of Southern Mississippi
Hattiesburg, Mississippi

Olga V. Mavrodi
Washington State University
Pullman, Washington

Rebecca Ann Melanson
Louisiana State University
Baton Rouge, Louisiana

Sara Alexandra Valadas da Silva Monteiro
Institute of Agronomy
Technical University of Lisbon
Lisbon, Portugal

Sharon E. Mozley-Standridge
Middle Georgia State University
Cochran, Georgia

Lytton J. Musselman
Old Dominion University
Norfolk, Virginia

Daniel L. Nickrent
Southern Illinois University
Carbondale, Illinois

James P. Noe
University of Georgia
Athens, Georgia

Kevin L. Ong
Texas AgriLife Extension Service
Texas A&M University
College Station, Texas

Bonnie H. Ownley
The University of Tennessee
Knoxville, Tennessee

Laura E. Poplawski
The University of Tennessee
Knoxville, Tennessee

David Porter
University of Georgia
Athens, Georgia

Melissa B. Riley
Clemson University
Clemson, South Carolina

Timothy A. Rinehart
United States Department of Agriculture
Agriculture Research Service
Southern Horticultural Laboratory
Poplarville, Mississippi

Erin N. Rosskopf
United States Department of Agriculture
Agriculture Research Service
Horticulture Research Laboratory
Fort Pierce, Florida

Craig S. Rothrock
University of Arkansas
Fayetteville, Arkansas

Isael Rubio
University of Wisconsin
Madison, Wisconsin

Gail Ruhl
Purdue University
West Lafayette, Indiana

Robert E. Schutzki
Michigan State University
East Lansing, Michigan

Barbara B. Shew
North Carolina State University
Raleigh, North Carolina

H. David Shew
North Carolina State University
Raleigh, North Carolina

Nina Shishkoff
United States Department of Agriculture
Agriculture Research Service
Foreign Disease-Weed Science
Fort Detrick, Maryland

Otmar Spring
University of Hohenheim
Stuttgart, Germany

Terry N. Spurlock
University of Arkansas
Monticello, Arkansas

C. Elizabeth Stokes
Mississippi State University
Starkville, Mississippi

Linda S. Thomashow
Root Disease and Biological Control Research Laboratory
Washington State University
Pullman, Washington

Robert N. Trigiano
The University of Tennessee
Knoxville, Tennessee

Phillip A. Wadl
Vegetable Research Laboratory
Charleston, South Carolina

David T. Webb
University of Hawaii
Honolulu, Hawaii

David M. Weller
Root Disease and Biological Control Research Laboratory
Washington State University
Pullman, Washington

Alan S. Windham
The University of Tennessee
Knoxville, Tennessee

Jason E. Woodward
Texas A&M AgriLife Research and Extension Center
Lubbock, Texas

David I. Yates
The University of Tennessee
Knoxville, Tennessee

Ana B. Zacaroni
Federal University of Lavras
Lavras, Brazil

Ning Zhang
Rutgers University
New Brunswick, New Jersey

Part I

Introductory Concepts

1 What Is Plant Pathology?

H. David Shew and Barbara B. Shew

Concept Box

- Most plants are healthy most of their lives; disease is the exception.
- The science of plant pathology had its beginnings in the late blight epidemics of the 1840s in Ireland and Europe.
- The germ theory of disease is the foundation of plant pathology.
- Plant disease is the result of a continuous interaction between a plant and a pathogen in a favorable environment.
- A disease cycle is the series of steps in the interaction of a host and pathogen from inoculation through pathogen reproduction and survival.
- Major pathogens of plants include fungi, fungus-like organisms, bacteria, viruses, nematodes, and parasitic seed plants.
- Diagnosis is the art of identifying disease based on symptoms and signs and associated factors.
- Koch's postulates are a set of rules to establish if a pathogen is the cause of a disease.
- Disease impacts include making plants and plant products scarce, dangerous to consume, and more costly to obtain.

CONCEPT OF PLANT HEALTH

Most plants, for most of their lives, are healthy! This is fortunate, because healthy plants are the foundation of the earth's terrestrial ecosystems. They are the source of the nutrients that sustain the interdependent organisms that together make up a stable ecosystem. Plants capture energy from the sun, and this energy provides food for large and small herbivores, the carnivores that eat the herbivores, and the scavengers that degrade the remains, including those of the plants themselves. Plants also provide energy for a variety of microorganisms that live in and on them, some of which are parasites that cause disease.

The plants that we observe in natural ecosystems are a product of natural selection. They have adapted to the biotic and abiotic environments of the ecosystem that they support by growing and reproducing more efficiently than their competitors. However, evolution is a dynamic process, and ecosystems are subject to changes due to climate change, introduction of new plant and animal species, introduction of exotic pathogens to the undisturbed ecosystem, and adaptation by existing microorganisms. In fact, coevolution with microorganisms, including those that are capable of causing diseases, is an important part of the long-term adaptation of a plant species to its environment. In this dynamic interplay between plants and their microbial companions, the pathogen sometimes gains the advantage and epidemics flare. In agroecosystems, many of the natural checks and balances of natural ecosystems are removed, so epidemics may occur more often and become very severe unless disease management practices are implemented.

In the following chapters, you will be introduced to the broad scientific discipline known as plant pathology. The primary goal of the text and the associated laboratory experiments is to raise your awareness of the importance of plant pathogens and plant diseases. The chapters will introduce you to the vast array of organisms that cause plant diseases and will allow you to experience the dynamic nature of the interactions between microbes and plants and to understand how we have successfully and unsuccessfully attempted to manage the organisms that cause plant diseases.

DEVELOPMENT OF PLANT PATHOLOGY

Plant pathology is a very broad and diverse scientific discipline. It integrates information from all of the core disciplines dealing with plant biology, plant production, microbial biology, and ecology to understand the dynamic interactions that result in disease. The concept that links all of these disciplines together within the science of plant pathology is the concept of disease: how it starts, develops, and spreads, and how it is prevented or managed. Literally, plant pathology is the science that studies plant suffering (pathology: pathos = suffer, and logy = study of). Plant pathologists attempt to improve plant health and crop productivity through the study of plant diseases, so that the severity and impact of diseases (suffering) can be alleviated.

As people moved from foragers to cultivators and began to rely on harvests of food and fiber from cultivated crops, their awareness of plant diseases must have increased. Much like today, the diseases that early agrarians observed no doubt ranged from minor to devastating. There are numerous references to blights and mildews in religious texts, including the Hebrew Bible, and in early Greek and Chinese writings. However, over several thousand years, there was little advancement in the understanding of the causes of disease and in the development of disease management strategies, largely because the biological basis of a disease was unknown. A disease was due to bad weather, toxic air, celestial events, imbalances of the sap, or divine intervention. The fact that microorganisms were only first observed in the seventeenth century, following the invention of the microscope, is hardly surprising. Even then, it would be nearly 200 years before the relationship between diseases and microorganisms was conclusively established. With few exceptions, people, including most scientists, instead believed in **spontaneous generation**. In the case of plant diseases, this belief or theory held that microbes were the result and not the cause of disease or decay.

It took multiple epidemics of the late blight disease of potato in Ireland and other areas of Europe in the 1840s, and the tragic events that followed, to provide the impetus for the founding of the science of plant pathology. The germ theory of disease provided the biological basis of the science. Within 15 years of the epidemics of late blight, Julius Kühn published the first textbook of plant pathology, concluding that both parasitic and nonparasitic factors resulted in plant abnormalities (disease). In the 1860s and 1870s, at least five additional textbooks were written about diseases of different groups of plants. The science of plant pathology continued to develop in response to the need of societies to understand the causes of plant diseases and to find the means to control them. It was the successful demonstration that disease problems could be alleviated by applying this new knowledge that led to the rapid development of plant pathology as a science in the late nineteenth and early twentieth centuries.

Germ Theory of Disease

Germ theory of disease was the single most important discovery in the early development of plant pathology. It states that germs (living organisms) cause diseases. A review of the history of plant pathology reveals that multiple scientists developed early evidence for the role of microbes in disease causality. It is beyond the scope of this chapter to list all those findings, but perhaps Prevost presented the most convincing results in his comprehensive studies of the bunt (smut) disease of wheat in the late eighteenth and early nineteenth centuries. His experimental approach led to a thorough description of the pathogen, its development in the plant, and even the approaches for controlling the disease. His treatise on this disease in 1807 should have provided the evidence needed to establish the germ theory of disease in plants, but his peers rejected the work as unsound (probably meaning too controversial).

It was not until the 1840s that a plant disease drew enough attention from scientists of the day to begin the science of plant pathology. The disease, late blight of potato (Figure 1.1a), ravaged potato crops throughout much of Europe and was especially devastating to the people of Ireland. The suffering that resulted from the consecutive years of epidemics was made worse by the dependence of the Irish population on a single food crop. Ireland was dominated by a landholding arrangement in which poor tenant farmers raised wheat, oats, barley, and other cash crops for export while depending almost exclusively on potatoes for their own sustenance. Even as cash crops continued to be exported, severe epidemics of late blight led to widespread starvation, sickness, and death. The Great Famine resulted in an estimated 1 million deaths and the mass emigration of at least a million more people from Ireland.

The multiple late blight epidemics in the 1840s found a better-prepared scientific community and a more profound need to understand the devastating effects of plant diseases than ever before. Many scientists uncovered important clues in the aftermath of the Great Famine, but the scientific approach used by Anton deBary finally led to the understanding of the true cause of late blight and other plant diseases. In 1861, deBary published his first work on the relationship of the pathogen, *Phytophthora infestans*, to late blight of potato. Much like Prevost, deBary described the development of the disease, from inoculation to symptom development and production of a new generation of spores on inoculated potato tissues. He also demonstrated the survival of the pathogen in potato tubers. He was able to duplicate all the stages of the disease cycle and repeat these stages in controlled

FIGURE 1.1 Symptoms of diseases studied to confirm the germ theory for different pathogen groups. (a) Destruction of a potato crop by late blight of potato caused by the fungus-like organism *Phytophthora infestans.* (b) Blighted terminal of ornamental pear caused by the fire blight bacterium, *Erwinia amylovora.* (c) Alternating light and dark green areas on tobacco leaves caused by tobacco mosaic virus. ([a] Courtesy of Marc Cubeta. With permission. [b, c] Courtesy of H.D. Shew.)

inoculations with the pathogen. In key experiments, deBary inoculated potato plants and compared them under identical conditions with plants that had not been inoculated (controls). When only the inoculated plants became diseased, it was clear that infection by *P. infestans* was the cause of late blight, and the germ theory was validated. By the time deBary published his work in 1861, many other scientists had begun to support the germ theory of disease on other plants, but deBary is credited with providing the conclusive proof for this theory and is often referred to as the founder of plant pathology.

The germ theory for other groups of plant pathogens followed the pivotal work of deBary. Burrill and his student, Arthur, developed the evidence for bacteria as pathogens of plants in the late 1870s and early 1880s in their pioneering work on the fire blight disease of pear (Figure 1.1b). It would be almost 20 additional years before the writings and studies of E.F. Smith helped to garner wide acceptance of the germ theory for bacteria. Acceptance of the germ theory for viruses would soon follow from the findings by three different scientists, Mayer, Ivanowski, and Beijerink, who all worked on tobacco mosaic caused by *tobacco mosaic virus* (Figure 1.1c). The extremely small size of viruses and the fact that they could not be cultured like fungi and bacteria introduced many problems in completing this work. Viruses were thought to be toxins or fluids, and Beijerink referred to them as a *contagium vivum fluidum*, a contagious living fluid. He also used the term virus (Latin for poison) to describe this type of pathogen. The physical nature of viruses was not known until after the invention of the electron microscope in 1931.

WHAT IS PLANT DISEASE?

Communication is based on the assumption that the people who are communicating understand each other. It is important to have a basic understanding of the terms used in any branch of science so that you can communicate effectively to others in the field. This is especially true when there are multiple definitions for a given term, such as disease. If asked to define disease, it is doubtful that any two people in a group or a class will have the exact same definition. However, there would be little disagreement about several key components that ought to be included in any definition of disease. Broadly speaking, disease refers to some type of abnormal condition or something that causes an organism to deviate from a healthy condition. It is important to understand that there is no clear line of demarcation between health and disease. Recognizing that a plant is diseased can be difficult and often requires the artful knowledge of a trained diagnostician. Regardless, most definitions of disease imply that the characteristics of a healthy, normally functioning plant are known. For example, the glossary of the American Phytopathological Society defines disease simply as "the abnormal functioning of an organism," and in 1968, the National Academy of Science defined plant disease as "a harmful alteration of the normal physiological and biochemical development of a plant." These are perfectly acceptable definitions of disease, but for the purposes of this volume, additional components are needed for our working definition. For this volume, we define plant disease as *a condition detrimental to the normal development of a plant resulting from the continuous interaction between the plant and a causal agent leading to the production of symptoms.* There are four key components of this definition, with several separating disease from other detrimental conditions affecting plants.

First, disease is *detrimental* to the development of a plant. Disease may impact any stage of plant development, vegetative or reproductive. Some diseases, known as **damping off**, occur only on seedlings, others occur only on mature or senescing plants, and still others occur throughout the life of a plant. Many things can be detrimental to plant health, so this component of our definition does not separate disease from the negative effects of the physical or chemical environment, nor from damage by organisms such as insects, voles, and other plant-consuming pests.

Second, disease is the result of a *continuous interaction* between the plant and a pathogen. Disease is a dynamic process and takes time to develop; it is not the result of an instantaneous event. When you hear expressions like "the disease happened over night," beware. Although visible symptoms may show up seemingly overnight, the disease process is well underway by the time symptoms become evident. The continuous and progressive nature of disease is one of the components of our definition that separates disease from injury.

Third, disease results from the activities of a *causal agent*. Most people would agree that diseases are caused by something. Specifically, causal agent refers to a **pathogen**, an organism that can cause disease. Usually, a single agent causes a disease, but some diseases are caused by two or more pathogens acting together. As we have seen, the germ theory of disease is a fundamental concept of plant pathology. This is another component of our definition that separates disease from injury and implies that disease is contagious; pathogens can be spread and infect neighboring plants. Injury is not contagious; it is not capable of being moved or spread.

Fourth, disease leads to the expression of *symptoms*. Symptoms are the evidence that something has altered the normal development or appearance of a plant (Figure 1.2). Plants respond to the presence of disease in multiple ways, but symptoms typically fall into groups or categories based on the part or processes of the plant affected, as we will see later. Symptoms may be minor, barely detectable, or severe, up to plant death. Some symptom types are unique to specific diseases, but some are common to many different causes, just as fever in humans can indicate anything from a slight cold to bubonic plague. Furthermore, the presence of symptoms is not unique to disease and thus does not separate disease from injury. By keeping these four components in mind when you think of disease, you can begin to develop a conceptual framework to build upon as we introduce and discuss the many complexities of diseases in plants.

In our narrow sense of disease, biotic organisms cause all plant diseases. In a broader view of disease, abiotic

FIGURE 1.2 Abnormal development in rose caused by a viral infection. Note the proliferation of shoots typical of Rose rosette disease in this shrub rose. (Courtesy of H.D. Shew.)

factors may also cause disease. Most authors consider these abiotic diseases to be *disorders*, because typically either they are not the product of a continuous interaction, or they are not contagious (do not have a pathogen associated with the damage), or both. Disorders can result from nutritional deficiencies or toxicities, exposure to harmful levels of air pollutants, flooding, drought, and many other causes. While this broad sense of disease is helpful to understand and diagnose plant problems, plant pathology is developed with a focus on the nature of diseases caused by living pathogens.

Symptoms and Signs of Plant Diseases

A **symptom** is the visible expression of a disease. Names of symptoms are generally descriptive of the primary abnormality that we see, such as leaf spot, wilting, or stunting. However, some diseases produce a whole **syndrome**, a series of symptoms that are characteristic for that disease. For example, yellowing, wilting, and death of all the plant parts that are above ground typically indicate that the real problem is in the roots, perhaps caused by a root rot disease. Symptoms are often used in the common name of a disease. For example, black root rot is a common name that is used to describe a disease that results in the development of black and rotted roots on a plant. Common names can also be misleading, as many diseases have more than one common name. For example, the disease of peanut caused by the fungus *Sclerotium rolfsii* may be called southern stem rot or white mold.

Plants have a limited ability to express the abnormal or harmful effects of a disease or disorder. Symptoms thus fall into discrete categories based on the type of damage caused, the part of the plant affected, and when it occurs in relation to the development of the plant. A disease may be characterized by one or more symptoms that are diagnostic for that disease, but in other cases additional information is needed to determine which disease (or diseases) is present.

The most common symptom of plant disease is **necrosis**. Necrosis is browning or blackening of host tissues brought about by cell death. This is a very broad category of symptom, and there are numerous necrotic symptom types. A localized area of necrosis is a **lesion**. Perhaps the most common and most easily observed necrotic symptom or lesion is leaf spot. A **leaf spot** is a localized area of necrosis on a leaf. Leaf spots may be very characteristic in some cases, or nondescript in others (Figure 1.3a), and typically have a defined size and shape. Another common necrotic symptom is a **canker**, a sunken area on the main stem or trunk of a plant, sometimes with raised margins (Figure 1.3b). A different type of sunken lesion that occurs on leaves, stems, and fruits is **anthracnose** (Figure 1.3c).

Necrosis may also extend across tissue types. For example, **blight** is a rapid blackening of host tissue and may include leaves, stems, and flowers (Figure 1.1a and b). A general (not localized) type of necrosis is called **rot**. Another descriptive term is often used along with rot to describe a disease, such as root rot, ear rot, stem rot, soft rot, fruit rot, and so on. Finally, necrosis that begins at the top of a plant and progresses downward is a **dieback** (Figure 1.3d). This symptom is typical for root rot and canker diseases, but also is common with abiotic factors that impact root growth.

Symptoms of disease may include various types of color changes. These symptoms may affect leaves, flowers, and fruits. The most common color change is **chlorosis**, which is yellowing due to lack of chlorophyll in leaves. A specific type of chlorosis around a necrotic spot is a **halo** (Figure 1.3a). Leaf spots may have borders of distinct colors as well, such as red or purple. These colors are distinctive symptoms for certain diseases. Loss of color, or **bleaching** of tissue, is also characteristic for some diseases. For example, diseases caused by viruses (Chapter 4) result in many patterns of color variation including **mosaic** (Figure 1.1c), **mottle**, and **ringspot**. A color-breaking virus is famous for driving "Tulipmania" in the seventeenth century. Tulips with variegated flowers were highly prized for their unique color patterns, and collecting them became an obsession with the Dutch. The mania for tulips led speculators to pay much higher prices for even a single blub. Finally, the bubble burst and Tulipmania ended. Today, color breaking in tulips and the variegated flowers and foliage seen in many ornamental plants are the result of selection for genetically inherited traits and not caused by a virus.

A common response of plants to pathogens is the production of overgrowths or galls. A **gall** is a localized swelling or overgrowth of host tissue, which results from cell enlargement (hypertrophy) and cell proliferation

(a) (b) (c) (d)

FIGURE 1.3 Common symptoms of plant diseases. (a) Leaf spot with chlorosis surrounding the spots. (b) Canker with rings of callous tissue around the infected area. (c) Sunken lesion of anthracnose on bell pepper. Note the abundant production of tan-colored spores in the sunken area. (d) Dieback of elm caused by the Dutch elm disease. (Courtesy of H.D. Shew.)

(hyperplasia) (Figure 1.4a and b). Multiple pathogen groups cause galls. Another common growth abnormality is **distortion**, or an abnormal formation or twisting of tissues and organs, especially leaves and fruit.

Symptoms may affect either the entire plants or the localized organs or tissues. **Wilt** is a general response to loss of water brought about by diseases that impede or degrade the vascular system. Wilts may occur as a result of stem cankers and root rots but are most commonly associated with the infection of the vascular system. **Vascular wilt diseases** are caused by pathogens that infect the xylem (Figure 1.5). Wilting may affect all of the plant or occur on one side of the plant or even one side of a leaf as specific areas of the xylem are plugged. One-sided wilting is called unilateral wilting. Another common whole plant symptom is stunting. **Stunting** is a reduction in plant size compared with an uninfected plant growing under the same conditions (Figure 1.6). Stunting is a common symptom in many virus-infected plants and plants that have root rot diseases.

A **sign** is a part of the pathogen on or in the plant that is visible to the unaided eye. Signs may be very large or barely visible without magnification (Figure 1.7). In the broadest sense, a sign could include any visualization of a pathogen on or in the diseased plant, even those pathogens that, like viruses, can only be seen under extremely high magnification. However, we will confine our discussion to signs that can be seen with the unaided eye.

Signs are most easily observed with diseases caused by fungi. During periods of high humidity, fungi often produce visible vegetative growth or spores on infected plants. Some of our most important plant diseases are named for the signs they produce. For example, **rust diseases** (Chapter 14) are named for the rust-colored spores produced in abundance on their hosts (Figure 1.8a), **powdery mildews** (Chapter 12) are named for the powdery appearance of the hyphae and spores that the powdery mildew fungi produce on the surfaces of their hosts (Figure 1.8b), **smut** diseases (Chapter 14) are characterized by the black dusty spores produced in

(a) (b)

FIGURE 1.4 Examples of galls. (a) Small gall of corn smut, with an infected and swollen single kernel. (b) Small elongated gall (bottom) and a cross section through a large gall on pine caused by Fusiform rust disease. Note the very large growth rings in the galled tissue in the cross section. (Courtesy of H.D. Shew.)

FIGURE 1.5 Wilting of snapdragon caused by a plugging of the xylem tissue. The disease is a vascular wilt caused by the fungus *Verticillium dahliae*. (Courtesy of B.B. Shew.)

FIGURE 1.6 Stunting of tomato in the production field due to viral infection. (Courtesy of H.D. Shew.)

FIGURE 1.7 Cluster of basidiocarps (sign) of the root rot pathogen *Armillaria*. (Courtesy of B.B. Shew.)

(a) (b) (c) (d)

FIGURE 1.8 Examples of signs of plant pathogens. (a) Spores of the orange rust pathogen on blackberry. (b) Hyphae and conidia of powdery mildew on cucumber. (c) Abundant conidial production by *Botrytis cinerea*, the cause of gray mold on pansy. (d) Bacterial streaming (arrow) from the cut end of a tobacco stem infected by the vascular wilt bacterium *Ralstonia solanacearum*. (Courtesy of H.D. Shew.)

their hosts (Figure 1.4a), and **downy mildews** (Chapter 8) are named for the downy growth on the undersides of leaves. Pathogens that produce fuzzy masses of spores, or visible clumps or colonies of hyphae, are often referred to as molds (Figure 1.8c). All of these diseases also produce symptoms, but they are named and diagnosed by the signs they produced.

Bacteria (Chapter 5) are extremely small and single-celled, so they are not visible except in mass. Signs of bacteria include streaming and ooze, which is a combination of bacterial cells, extracellular slime produced by bacteria as they cause disease, and host cells (Figure 1.8d). Streaming is most frequently used to diagnose vascular wilts caused by bacteria but can also be used to diagnose leaf spots and other necrotic diseases. The most obvious sign of a nematode disease is the presence of adult females of a certain group of nematodes called cyst nematodes (Chapter 6). The enlarged, globose body of the female can be seen on the surface of infected roots. It changes from white to dark brown as she dies, forming a **cyst** that contains the nematode's eggs. Viral diseases do not have signs, as they are too small to be seen.

THE DISEASE TRIANGLE

As you have seen, disease is a product of complex interactions between host and pathogen. Clearly, if either host or pathogen is absent, there will be no interaction and disease cannot occur. However, host and pathogen will coexist without interacting unless the environment is favorable for infection and disease development. Thus, disease occurs only when host, pathogen, and environment come together at the right time and in the right place. The classic disease triangle (Figure 1.9) illustrates this fundamental concept.

PATHOGENS

With few exceptions, plant diseases are caused by microorganisms that are also parasites on their host. That is, they live in close association with the plant and derive all or most of their nutrients from it. Not all parasites are pathogens; in some cases, a parasite causes little harm to its host. The parasite becomes a pathogen when it impairs normal plant function through its activities, causing disease. Each species of plant pathogen causes

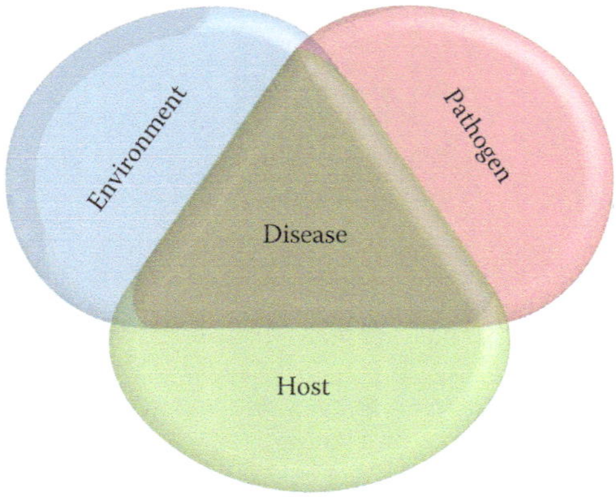

FIGURE 1.9 Disease triangle, illustrating the interaction of host, pathogen, and a favorable environment leading to disease. (Courtesy of Arlene Mendoza-Moran. With permission.)

FIGURE 1.10 Range of resistance responses observed in peanut to *Cylindrocladium* black rot disease. (Courtesy of Jerald Pataky. With permission).

disease on a limited number of plant species. The **host range** of a pathogen consists of all the host species on which it causes disease. Some plant pathogens have a host range of only a single host species, whereas others can attack hundreds of species across many plant families. **Pathogenicity** is the ability of a particular plant pathogen species to cause disease on a particular plant species. Pathogen species and individuals within a species vary in **aggressiveness**, which is the relative ability of the pathogen to inflict damage on its host: pathogens that cause severe symptoms are highly aggressive, whereas those that cause mild symptoms are nonaggressive. Pathogenicity and aggressiveness in the pathogen are inherited traits, but aggressiveness can also vary with the age, life stage, nutrient reserves, or previous exposure to adverse or favorable conditions of the pathogens.

Plants

Most plants are healthy during most of their lives. Just as a given species of pathogen has only a limited number of potential plant hosts, any plant species is host to only a limited number of plant pathogens. The plant species must be **susceptible** to the pathogen for disease to occur. A susceptible host is one that is capable of being attacked by a pathogen. **Susceptibility** (Latin: take up, sustain) is a measure of how well a host can sustain a pathogen's development. Therefore, within a host plant species, susceptibility exists along a continuum and can range from extremely high to low. If the plant actively reduces, delays, or prevents the development of a pathogen or disease, it has some form of **disease resistance** (Chapters 20 and 23). This resistance varies along the same continuum as susceptibility from very low (susceptible) to very high (Figure 1.10). In some cases, resistance is so high among some members of a normally susceptible host species that no disease is apparent at all. Often, however, resistance or susceptibility within a host species is a matter of degree and is described relative to a standard or average type. Like pathogenicity and aggressiveness, susceptibility and resistance are inherited traits. Susceptibility may also vary with age, growth stage, or the condition of the plant before infection. Injuries, nutrient excess or deficiencies, and stresses like drought or cold can **predispose** the plant to infection or make it more vulnerable to disease. Conversely, vigorous plants are usually less likely to become infected or diseased than less healthy ones.

Environment

Often, we think of the environment in terms of physical (e.g., temperature and moisture) and chemical factors (e.g., pH and nutrients). However, the biotic environment also profoundly affects host, pathogen, and disease. The biotic environment includes insects, nitrogen-fixing bacteria, beneficial and antagonistic microbes, competing pathogens, weeds, earthworms, and many other organisms. The physical characteristics and population density of plants may also affect the environment. For example, leaves shade the soil, transpiration increases humidity, and dense plantings reduce wind circulation, creating microclimates that can favor pathogen growth and disease development. Conversely, disease can change the environment near the infected plant, sometimes making that environment more or less favorable for further disease development. For example, defoliation may reduce shading and result in increased air circulation, whereas rotten fruits or vegetables may release moisture and nutrients that promote further infection and decay. Since disease develops over time, these processes may speed up, slow down, or even stop as the environment changes.

Each species of plant and pathogen has a range of environmental conditions that are optimal for growth. Plants are adapted to nearly every environment on earth, but all green plants require sunlight, oxygen, water, nitrogen, and an array of other essential elements. Likewise, plant pathogens can be found wherever there are plants.

In very general terms, plant pathogenic fungi and bacteria tend to thrive in moist (but not necessarily wet) soil and high humidity. The environment that favors a disease will be found within the overlapping range of environments that favor both the host and the pathogen. Sometimes, an environment that is best for the plant is likewise very favorable for the pathogen and thus disease. For example, high soil moisture and fertility may promote rapid plant growth, but at the same time encourage pathogen infection, growth, and reproduction. Conversely, the same conditions may inhibit both plant and pathogen, so that no disease develops. For example, plants usually suffer little disease during dry periods because, like plants, most pathogens need ample moisture to survive and infect. Naturally, some pathogens have evolved to thrive in dry conditions and can take advantage of a drought-stressed host. Disease often develops when the environment is not ideal for the pathogen, but is even more unfavorable for host. For example, seed germination may slow down or even stop in cold, wet soil. This makes seedlings vulnerable to infection by a variety of soil-inhabiting pathogens, even those that prefer or grow optimally at warmer soil temperatures.

Understanding the dynamic balance between host, pathogen, and environment is a key to developing sound disease management strategies. Actions that promote the general health and resistance of the host, remove or impair the pathogen, and shift the physical, chemical, or biotic environment help to reduce or prevent diseases and the losses that they cause.

DISEASE CYCLES

Disease develops when a host and a pathogen interact in a sequential series of events. Because these events recur over time, they can be envisioned as a cycle, paralleling to varying degrees of life cycles of the two organisms. Plant pathologists refer to the continual repeating steps in the interactions between a host and a pathogen as a **disease cycle**. Often, the major stages or steps in disease development are depicted in a diagram that helps us to visualize how a disease progresses from beginning to end (Figure 1.11a and b). Disease cycle diagrams and the concepts they represent provide an excellent framework for understanding how disease develops and even how it can be managed.

In most climatic regions, plants complete a single cycle of growth, development, and reproduction per year. Even in tropical regions, plants grow in cyclic patterns that usually correspond to seasonal changes in the environment. Crop plants are mostly cultivated as annuals, with some notable exceptions, including fruit and tree crops, so they have a single cycle of growth per year. The majority of pathogens have the capacity to complete many life cycles per year, but some are limited to a single cycle. The number of cycles of disease that occur per year (or host growth cycle) is one of the most important characteristics of a particular disease. **Monocyclic** diseases complete one disease cycle per year (Figure 1.11a). **Polycyclic** diseases complete multiple cycles of disease per year (Figure 1.11b). Disease cycles that repeat or extend over more than one year are referred to as **polyetic** diseases.

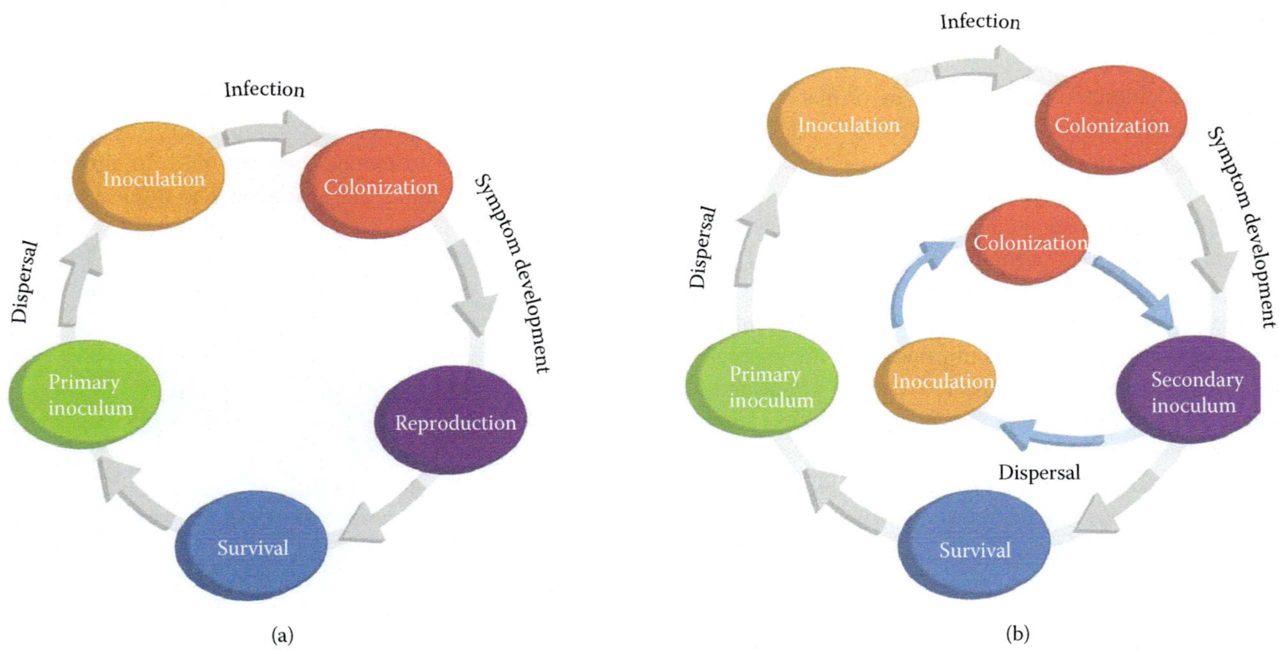

FIGURE 1.11 Graphical representation of the most common types of disease cycles. (a) Monocyclic disease, with only one cycle per plant-growing season. (b) Polycyclic disease, with multiple cycles of disease per plant-growing season. (Courtesy of Arlene Mendoza-Moran. With permission.)

Although the specific features of each disease cycle are unique, some general features are present in all cycles. Plant growth begins from a dormant stage, for example, germination of a seed, growth from tubers, bulbs, or vegetative cuttings, or bud break in perennial species. Similarly, pathogens typically start new growth from a dormant stage. The dormant stage or period of inactivity for a pathogen is referred to as **survival**. Pathogens use a variety of strategies and structures to survive periods when the host is not present or when the environment is not favorable for growth. Specific strategies for survival by specific types of pathogens are discussed in Chapters 4 through 17.

Survival structures are a type of **inoculum**, the part of the pathogen that can infect a plant. The inoculum that begins a disease cycle is referred to as **primary inoculum** (also known as initial inoculum, overseasoning or overwintering inoculum). Germination or regrowth of the pathogen may be dependent on the presence of host exudates or totally independent of the host. For example, soybean cyst nematodes require a specific organic molecule exuded from soybean roots to induce the emergence of juveniles from the survival cysts. In contrast, production of ascospores, the primary inoculum of the Camellia petal blight pathogen *Ciborinia camelliae*, is based solely on environmental conditions. Likewise, the pathogen may actively find the host by swimming or growing toward it or may be carried passively in air, soil, water, or other agents to the host. In all the cases, the success of the primary inoculum in finding, contacting, and infecting a host will determine whether disease will occur. If the primary inoculum is not successful in finding a host, then disease will not develop and no new inoculum will be produced in the next season. Because monocyclic diseases complete only one cycle of infection per year, they depend entirely on infections from primary inoculum. While infection by primary inoculum is also required for polycyclic diseases, it is less important in determining the amount of disease that ultimately develops, because even a few infections from primary inoculum can give rise to many cycles of infections during the year. The chances of finding an **infection court**, a site on a host where the pathogen can infect, are very small for the typical microscopic plant pathogen. **Inoculation** of a host occurs when inoculum finds an infection court. Pathogens compensate for the low chance of success by any one unit of inoculum by producing very high number of individuals. The likelihood that the pathogen will find an infection court may be enhanced by its dispersal mechanisms. Pathogens disseminated by wind and rain may have a very low chance of finding a host, whereas pathogens that are disseminated by a vector have a much higher chance of success. A **vector**, an organism that transmits a pathogen, finds the host for the pathogen. Unfortunately, human activities also inadvertently move pathogens along with host plants. As we will see, this is the case with many of our most devastating epidemics, such as the chestnut blight that killed billions of chestnut trees in the eastern United States during the early twentieth century.

If the environment is favorable following inoculation, infection of the host occurs. **Infection** is the establishment of a food relationship between the host and pathogen and is the second stage of the disease cycle. Pathogens may infect directly through host tissues, but most use natural openings such as stomata or wounds in the plant that are caused by various types of injury. Once infected, all subsequent growth of a pathogen in a host is called **colonization**. Pathogens have specific patterns of colonization within host tissue(s), primarily based on how the pathogen obtains nutrients from host cells during pathogenesis. **Pathogenesis** (meaning: origin of suffering) is the sequence of events that occur during disease development. As the pathogen colonizes the host, it accumulates nutrients that it uses for growth and reproduction. The capture of these nutrients from the host takes place over time. The time between inoculation and reproduction is the **latent period**. This period may be shorter or longer than the **incubation period**, which is defined as the time between inoculation and symptom development. For example, powdery mildew fungi often produce new inoculum before any symptoms are present. Organisms that reproduce on dead tissue or after symptoms appear have latent periods longer than the incubation period. Inoculum that is produced on infected tissue and that is capable of infecting a new host immediately is termed **secondary inoculum**. If the inoculum is dispersed to and infects other plants within the same growing season, it is a polycyclic disease (Figure 1.12a and b). These secondary cycles of infection can result in devastating epidemics such as the potato late blight epidemic discussed earlier. Ultimately, the host will die or the environment will become unfavorable, slowing and then halting disease development on the infected plant. Late in disease development, a pathogen may produce survival structures that will allow it to remain dormant until the time when new host plants are again available. The disease cycle is

(a) (b)

FIGURE 1.12 Examples of splash dispersed pathogens. (a) Black spot of rose, with lesions of various sizes; splashing rain or water disperses spores to other leaves or to neighboring plants. (b) Leaf symptoms of black rot of grape, showing multiple generations of infections on the leaf caused by spores splashing from lesions with active sporulation. (Courtesy of H.D. Shew.)

now completed and will start again when all the factors necessary for the disease occur in the same time and space.

TYPES OF PLANT PATHOGENS

Trophic Levels

Pathogens can be separated into groups based on their trophic lifestyle; that is, how they obtain nutrients. With the exception of some parasitic seed plants, all plant pathogens are **heterotrophic**, which means they must capture their nutrients from another organism. Organisms that use live plant cells as their only source of nutrients are called **biotrophs**. Since they can exist only as parasites, they are also referred to as **obligate parasites**. Parasites that obtain their nutrients from cells that they kill through the production of toxins and enzymes are called **necrotrophs**. Although they feed and reproduce on dead cells, these organisms are not saprotrophs since they killed the host cells prior to using them as nutrients. **Saprotrophs**, on the other hand, have no parasitic phase and derive nutrients from nonliving sources. While some necrotrophs can live as saprotrophs when a susceptible host is not available, others compete poorly with true saprotrophs or even lack the ability to live independently from their hosts. **Hemibiotrophs** are pathogens that begin their relationship with their host as biotrophs, but become necrotrophic in later stages of pathogenesis. Biotrophic pathogens tend to have narrow host ranges, but some hemibiotrophs and necrotrophs also have a limited host range. Some necrotrophs, such as *S. rolfsii*, have a very extensive host range, attacking hundreds of hosts across many plant families. This fungus kills host cells and tissues by producing a very potent toxin, oxalic acid, and copious amounts of cell wall-degrading enzymes. This nonselective strategy of interacting with plants leads to the very wide host range. Not all pathogen groups fit well into these trophic levels. For example, viruses and viroids do not directly absorb and metabolize nutrients, so they do not have a true trophic lifestyle. However, they are obligate parasites in the sense that they replicate only inside living cells.

Fungi

Fungi (singular: fungus) are the most abundant group of plant pathogens. They cause many thousands of different plant diseases, producing a wide range of symptom types, including spots, blights, rots, galls, and wilts. Fungi belong to the Kingdom Fungi and the branch of science that studies fungi is **mycology**. Fungi find and explore new substrates by vegetative growth via microscopic filamentous threads called **hyphae** (singular: **hypha**). Hyphae are composed of individual fungal cells laid end to end. Growth occurs at the tips of the hyphae, or with branching followed by tip growth. When a potential food source is found, they secrete enzymes that break down complex molecules into simple compounds that are readily absorbed and used for growth, reproduction, and survival. In substrates rich in nutrients, the fungus can produce masses of hyphae called a **mycelium**. Mycelium may be visible to the unaided eye and sometimes is diagnostic. When grown in a sterile culture medium, the mycelium that forms is called a **colony**.

Fungi reproduce by the production of **spores**. Spores come in many sizes and shapes and play many roles in the sometimes-complex life cycles of fungi. They allow the fungus to find new sources of food as wind, rain, or vectors disperse them. Spores are the primary survival structure for most fungi and give rise to new growth when conditions become favorable for growth.

Classification of fungi is based on the type of sexual spore produced. For example, the **ascospore** is the sexual spore of the **Ascomycota** (Chapters 11 through 13) and the **basidiospore** is the sexual spore of the **Basidiomycota** (Chapters 14 and 15). A large group of fungal plant pathogens, the **mitosporic fungi** (formerly Deuteromycota) or imperfect fungi, produce only asexual spores called **conidia**. These imperfect or mitosporic (Chapter 13) fungi may lack a sexual cycle entirely or they may reproduce sexually only under very specific and rare conditions so that the sexual stage is unknown. Finally, the **sterile fungi** do not produce spores at all. These fungi propagate and survive either as hyphae or in structures made of masses of hyphae such as **sclerotia** (singular: **sclerotium**).

Many plant pathogenic fungi produce both sexual and asexual spores. Historically, these different spore stages were each given a distinct genus and species name known as a Latin binomial. This dual naming system led to the use of the terms **teleomorph** for the sexual stage, **anamorph** for the asexual stage, and **holomorph** for both the stages. For example, one of the most common plant pathogens is known mostly by its anamorph name, *Rhizoctonia solani*. The teleomorph and holomorph name for the organism is *Thanatephorus cucumeris*, but this name is rarely used because the organism is observed in its anamorph stage most of the time. The dual naming of organisms contradicts the purpose of naming organisms with a unique Latin binomial, which was to eliminate confusing nonstandard common names, thus allowing people to communicate more accurately. The confusion caused by having two names for one organism has led to an effort to establish one scientific name for all organisms (one organism: one name), including plant pathogenic fungi.

Fungi are phylogenetically more closely related to animals than plants. Like plants, fungi have cell walls, but these walls are made primarily of **chitin**, which is also the main component of the exoskeletons of insects and crustaceans. Unlike animals and vascular plants, fungi spend most of their life cycle as **haploid** organisms, meaning they have only one set of chromosomes in their nuclei. Cells may contain one or more nuclei, and cells are separated by the formation of **septations**. These hyphae are thus called septate. The septate hyphae may have characteristic features such as color, branching pattern, or the presence of distinctive structures, but few fungi can be identified by hyphal characteristics alone.

Fungi use multiple strategies to survive. As we have seen, some fungi can survive as saprotrophs in the absence of a host. Others survive as the same spores found during pathogenesis and epidemic development, or they may produce highly specialized survival spores. In some cases, fungi survive in reproductive structures produced on or inside living or dead host tissues. Other fungi survive by producing various kinds of thickened vegetative cells or clumps of cells, including **stromata** (singular: **stroma**), **chlamydospores**, and **microsclerotia**. Sclerotia are very common vegetative survival structures that may survive for years in the absence of a host. Under favorable conditions, they may germinate directly, producing mycelium that acts as primary inoculum, or in some species they may germinate indirectly to give rise to sexual or asexual spores that are primary inoculum. Pathogenic fungi may survive as hyphae in tissues of perennial hosts. The pathogen begins growth as the host begins a new life cycle. For example, in Fusiform rust of pine, the fungus survives vegetatively in perennial galls and produces a new set of spores on the galls each spring (Figure 1.4b).

Oomycetes

A group of organisms very similar to fungi in appearance and function are the Oomycota (Chapter 8). These fungus-like organisms include many important pathogens, including the late blight pathogen, *P. infestans*. Important pathogens in this group include the downy mildews and members of the genera *Phytophthora* and *Pythium*. Appearance can be deceiving, because this group of pathogens may look like fungi, but they are more closely related to plants than animals. Like plants, these organisms have cellulose cell walls, but unlike plants, they are heterotrophic and obtain their nutrients by absorption. These organisms belong to an entirely different kingdom of organisms, the Stramenopila. Most organisms in this group produce characteristic biflagellate swimming spores called **zoospores**, so they are sometimes called water molds. The sexual spore is the **oospore**. Also, unlike fungi, hyphae of the Oomycota do not generally have septations; however, septations are found at the base of reproductive structures. Because the hyphae are mostly **nonseptate**, cells are multinucleate (**coenocytic**), and streaming of the cytoplasm and other cellular contents may be evident upon microscopic examination. Individual nuclei are diploid (2N), with haploid cells produced only inside sexual structures called **oogonia** and **antheridia**. Oospores germinate to produce mycelium or they may produce **sporangia**. These sporangia may in turn bear zoospores or produce mycelium, depending on environmental conditions. Some species of oomycetes produce sexual spores very rarely and some species produce sporangia but never zoospores. These pathogens survive using similar structures and strategies as fungi. Survival structures include oospores, chlamydospores, and hyphae in infected or infested plant debris.

Bacteria

Plant pathogenic bacteria cause a wide range of symptoms and diseases in plants (Chapter 5). Common symptoms of bacterial infection include leaf and fruit spots, soft rots, cankers, galls, and vascular wilt. Plant pathogenic bacteria are unicellular **prokaryotes**; that is, they are organisms that lack a nucleus and other membrane-bound organelles such as mitochondria and chloroplasts. Plant pathogenic bacteria are heterotrophic organisms that attain nutrients by absorption. They secrete enzymes and toxins to kill cells and break down potential sources of nutrients. Bacteria are much smaller than fungi, typically less than 3 μm in length (a human hair is about 75 μm in diameter), and they have relatively simple morphologies, especially compared with fungi and nematodes. With few exceptions, plant pathogenic bacteria have cell walls and are rod-shaped; many also have one or more flagella arranged in various patterns on the cell. A mass of bacteria is called a colony, with color, shape, and morphology of the colony being important characteristics. Cells are often coated with a slime layer, and some bacteria produce colonies that appear distinctly slimy in culture. However, bacteria are very difficult to identify from their appearance under the microscope or in culture. Typically, a range of laboratory tests are needed to identify bacterial species based on their ability to break down different types of carbohydrates and other nutrients under different cultural conditions. Sequences of specific genes are also useful for identification. In addition to the bacteria that have cell walls, the phytoplasmas and spiroplasmas are wall-less bacteria that cause important plant diseases. These wall-less forms are the only obligate parasites among the plant pathogenic bacteria. Most bacteria are present on host surfaces and are intercellular when inside the host. They do not penetrate host cells.

Bacteria multiply by **binary fission** (splitting into two), with cells dividing rapidly when nutrients are available. Bacteria do not reproduce sexually, but bacterial cells sometimes exchange genetic materials through **conjugation** and the transfer of **plasmids**, which are small pieces of DNA found inside the bacterial cell. Bacteria also obtain new DNA and new variability via absorption of free DNA, **transformation**, or as a result of movement from cell to cell via infection by viruses called bacteriophages, **transduction**.

Bacteria require wounds or natural openings to penetrate their hosts. Once inside a host, they reside in the intercellular spaces and produce many of the same types of weapons in pathogenesis as fungi. Unlike many other types of bacteria, most plant pathogenic bacteria do not produce survival spores that are highly resistant to heat and other adverse environments. It is important to note, however, that spore-forming bacteria living on and in plants are a major cause of contaminated produce and that their control is critical for maintaining food safety. Because of their small size and lack of specialized survival structures, plant pathogenic bacteria have evolved somewhat different survival strategies than eukaryotic plant pathogens. Plant pathogenic bacteria typically survive in groups of cells in a **biofilm**. The biofilm is secreted by the bacteria and protects them from harmful external factors. It also allows the bacterial cells to communicate effectively and sense when it is appropriate to initiate life cycle events such as attempting to infect a host. Bacteria survive in vectors, or in biofilms as epiphytes, in host plant debris, in and on seeds, or in soils.

Nematodes

Symptoms caused by plant parasitic nematodes include wilting, yellowing, stunting of entire plants or organs, root or leaf lesions, and galling (Chapter 6). Nematodes are unsegmented roundworms of the phylum Nematoda. Most nematodes are free-living saprotrophs or predators, but many are also important animal or plant parasites. All plant parasitic nematodes are biotrophs; they must feed on living plant cells in order to obtain nutrients. Nematodes have specific feeding habits, feeding either as **ectoparasites**, from outside the host, or as **endoparasites**, from inside the host. Ectoparasites and endoparasites may be either **migratory**, moving from cell to cell, or **sedentary**, staying in one place once a feeding site is established. Most plant parasitic nematodes feed on the roots of plants, but there are some genera that feed on leaves and at least one genus feeds in the xylem of trees.

Nematodes are identified based on their morphological features, as they are readily observed with a low-power microscope. They range in size from 300 µm to almost 4000 µm (4 mm) in length. Many nematodes are long and slender, but some are sausage-shaped and others take on a globose appearance as they mature. Typically, male and female nematodes mate to produce eggs, referred to as **amphimitic** reproduction, but in some species females can produce viable eggs without mating, a process known as **parthenogenesis**. All nematodes produce eggs, and the juveniles differentiate and undergo a molt before hatching. The **second-stage juvenile**, which emerges from the egg, is the first infective stage. Three additional molts occur as the nematode grows and matures. All plant parasitic nematodes have a hollow protrusible stylet that they use to penetrate host cells and obtain nutrients. Various enzymes, toxins, and growth regulators are injected into the host cell during feeding. Although they are biotrophic, migratory nematodes may kill host cells during feeding. In contrast, sedentary feeders must repeatedly feed from the same cells over extended periods without killing them.

Nematodes may have specialized survival structures or stages. Nematodes may survive as eggs, as quiescent or resting larvae, or as adults. Cyst nematodes survive as eggs inside the hardened body of adult female. The wheat seed gall nematode can survive for many years living in a desiccated state inside a wheat seed.

Viruses

Viruses are noncellular entities that do not directly absorb and metabolize nutrients (Chapter 4). However, they are obligate parasites in the sense that they depend entirely on living host cells for the basic materials and cellular mechanisms necessary for their replication. Viruses can replicate only inside living cells. Viruses cause some of the most devastating diseases of plants. Many symptoms caused by viruses, including wilting, yellowing, necrosis, stunting, and lesions, are similar to those caused by other pathogens. Viruses also cause symptoms that are distinctive, including color breaking, mottles, ringspots, vein clearing, and unusual patterns of plant growth.

Viruses are very small (measured in nanometers) in comparison with other pathogens discussed earlier. Virus particles are only visible by electron microscopy and may be rod-shaped, filamentous, or bacteria-like, or have a geometric three-dimensional form (**isometric**). Individual virus particles are called **virions** and are composed of either RNA or DNA surrounded by a protein coat (**capsid**). Viruses are classified based on the organization of their genome, the shape of the particle, and their natural means of transmission and host range. Depending on the virus species, the DNA or RNA can be single- or double-stranded and the synthesis of new nucleic acid may proceed in either a positive or negative direction. Single-stranded RNA species are the most common viruses in plants. In addition, the virus genome may be organized as a single strand or it may be **multipartite**, that is, broken up into two to four separate pieces. Virus species are given descriptive names rather than Latin binomials. Species are named for the host on which they are originally described, then by the most common symptoms associated with the virus, and then the word virus. For example, tomato spotted wilt virus was first observed on tomato, the primary symptoms are spots and wilt, and it is a virus. Virus names are commonly made into acronyms, for example TSWV for tomato spotted wilt virus. The host range of viruses varies greatly from a few species to many hundreds, and the host for which the species is named may not be the host most severely affected. Dispersal of viruses, with the notable exception of tobacco mosaic virus, is highly dependent on a specific relationship with a vector. The types of vectors vary with the virus, but insects are the most common vectors. Among insects, the most common vectors are aphids and whiteflies. Noninsect vectors include mites and plant parasitic nematodes, fungi, and plants. Viruses have very limited survival capacity once outside of its host or vector, so survival is typically in their host, in alternative perennial or annual hosts (including weeds), or in their vectors. These alternative hosts or vectors serve as primary sources of inoculum for epidemics caused by viruses, as do propagative materials like tubers and cuttings.

Parasitic Seed Plants

The vast majority of plant species are **autotrophs**, that is, they produce their own food by photosynthesis, but some species have evolved as parasites of other plants (Chapter 17). In addition to the presence of the plant itself (signs), symptoms of infection with parasitic seed plants include yellowing, poor growth, stunting, wilting, swellings or galls, and excessive branching known as **witches' brooms**. Parasitic plants either lack roots entirely or produce highly modified roots that penetrate the host in order to obtain its nutrients. The common true mistletoes infect the stems of trees but produce their own chlorophyll. They rely on the host plant only for mineral nutrition and water. On the other hand, dwarf mistletoes absorb minerals, plant nutrients, and water from the stems of their conifer hosts, greatly reducing the tree's productivity. They also stimulate the formation of witches' brooms and other growth abnormalities. Species of the vine dodder are very common parasites on stems of both herbaceous and woody plants. Most cause little harm, but others can result in extensive overgrowths and damage to plants. Dodders can also vector certain plant viruses. Some parasitic plants, including witchweed and broomrapes, attach to host roots and use the host's nutrients for their own development. These plants have wide host ranges, which include important crop species such as corn, sorghum, beans, cowpeas, and other legumes. These parasites can be devastating to yield and productivity and can be particularly difficult to control.

Parasitic seed plants reproduce by producing flowers and seeds. As with other higher plant species, they are separated into groups based on their morphology and phylogeny. Parasitic seed plants generally survive as seeds. The seeds are adapted to be dispersed by a variety of mechanisms, including forcible discharge in dwarf mistletoes, and dispersal by birds and other animals in the case of true mistletoes. Other species of parasitic plants are very prolific producers of seeds that survive in soil until a susceptible host is grown.

DISEASE DIAGNOSIS

Earlier in this chapter, we defined disease as *a condition detrimental to the normal development of a plant resulting from the continuous interaction between the plant and a causal agent leading to the production of symptoms*. We can use this definition to look at how diseases are diagnosed (Chapter 30).

Diagnosis begins with observation of the entire plant, including the roots. The diagnostician must first identify the plant and know the characteristics of a normal plant of the same species and variety. This can be a daunting task, considering the thousands of plant species and varieties under cultivation around the world. As we saw with color breaking in tulips, an unusual plant type may be inherited genetically or may be the symptom of a disease. On the other hand, symptoms such as stunting can be subtle and hard to distinguish without the knowledge of normal plant appearance and development.

As the diagnostician examines the plant, they look for symptoms and signs. Signs are usually highly diagnostic of a particular disease because they immediately point to the pathogen. Some symptoms are so diagnostic that no further investigation is needed. More commonly, however, individual symptoms are only clues. For example, wilting is a symptom of many diseases, which can only be distinguished by checking for additional symptoms like root rotting or vascular discoloration or by attempting to culture the pathogen.

Observations or information about the location and history of the problem are equally important. This information will help the diagnostician to separate diseases (resulting from a continuous interaction) from injuries and disorders. The diagnostician must consider the setting: where was the plant growing and how was it cultivated? How quickly did the symptoms appear and where are they distributed in the field, greenhouse, or landscape? As diseases are contagious, we expect them to increase over time and appear in clusters or clumps. In contrast, disorders often develop very quickly and tend to affect all of the plants in an area. Recent weather, for example, rainfall, temperature, cloudy periods, humidity, and frosts, must be considered. Warm, humid periods are ideal for many blight diseases, but rapid death of tender leaves could also indicate a recent frost. Taken together, the recognizable associations of signs, symptoms, timing, location, and history constitute a syndrome or signature that is typical of a particular disease.

Because any plant species is host to only a limited number of pathogens, it is possible to narrow the range of possible causal agents to a fairly manageable few once the signs, symptoms, and history have been investigated. A **host index**, such as the U.S. Department of Agriculture Fungal Database (http://nt.ars-grin.gov/fungaldatabases/), lists all the hosts on which a pathogen has been identified, and most are searchable by host or pathogen. The host index can help to guide the diagnostician as they consider whether additional culturing or diagnostic tests are necessary to identify the causal agent.

When signs are not visible, the affected parts of the plants may be checked microscopically for evidence of the pathogen. Often, the diagnostician can observe the mycelium of fungal or bacterial cells but cannot find spores or other identifying characteristics. In those cases, the diagnostician can attempt to obtain pure cultures of the organism in order to identify it or perform additional diagnostic tests.

Although signs and other direct evidence of a pathogen strongly indicate it as the causal agent, it is important to consider whether the pathogen identified agrees with all the other observations that have been made. The diagnostician should consult reliable references that describe the disease and verify that the observations fit published descriptions before making the final diagnosis.

Often, the exact cause of a disease cannot be determined with certainty. Sometimes, critical information is lacking. It may not be possible to culture the pathogen, either because it is an obligate parasite or because it has died. In other cases, a potential pathogen can be identified, but the symptoms or other observations do not match those described as being caused by the organism in question. Finally, an organism may be positively identified, but not previously described as a pathogen on the host. In that case, the diagnostician may decide to perform Koch's postulates in order to confirm the organism as the cause of the disease.

Koch's Postulates—How We Determine the Cause of a Disease

In the 1870s, the scientist Robert Koch began a series of experiments that would change the science of pathology. From the 1850s to the 1870s, many scientists conducted scientifically sound experiments to prove the germ theory of disease in plants. In the 1860s, Louis Pasteur completed his famous experiments showing that microbes developed on boiled substrates only if they were exposed to air after boiling. This work conclusively disproved the theory of spontaneous generation. However, there was still no widely accepted standard of proof that a specific microbe caused a specific disease.

Koch worked with anthrax, a lethal disease of sheep. He was able to see the bacteria in the blood of infected animals and not in healthy animals. He was also able to initiate disease in healthy animals by injecting them with blood from infected animals. But the question remained: how could he prove that there was not something else in the blood of the infected animals that caused anthrax? He needed reliable methods to **isolate** the bacterium from the blood of the infected animals by growing it in pure culture, free of all other organisms. He could then introduce the purified pathogen into healthy animals and see if disease developed. As momentum increased for studying the importance of the microbial world, Koch and others developed improved techniques for isolating and growing organisms in pure culture. In 1876, Koch published his results and developed the germ theory of disease for anthrax. In the early 1880s, he developed a set of rules known as **Koch's postulates** (Chapter 31). These postulates are now used as a guideline for establishing proof of pathogenicity; that is, proof that a specific pathogen is the cause of a specific disease (Figure 1.13).

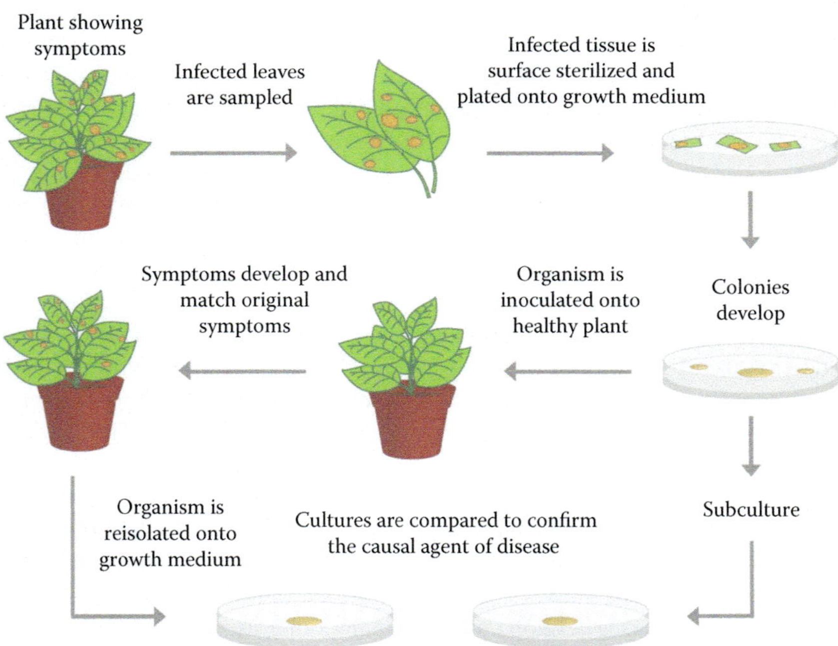

FIGURE 1.13 Diagram of the steps involved in Koch's postulates. These four steps are used to demonstrate that a specific organism causes a disease. Modifications of these steps are needed for pathogens that cannot be cultured. (Courtesy of Arlene Mendoza-Moran. With permission.)

The postulates are as follows:

1. There must be a constant association between the presence of the pathogen and the disease.
2. The pathogen must be isolated into pure culture from the symptomatic host.
3. The pathogen must be inoculated into a healthy host of the same species or variety and the original symptoms reproduced.
4. The pathogen must be reisolated from the inoculated host and be identical to the original pathogen.

It is important to note that Koch's postulates are guidelines for the proof of pathogenicity and must be modified somewhat when applied to those pathogens that cannot be grown in pure culture. In those cases, the pathogen is grown on host plants and not in sterile medium. Interestingly, this is the approach taken by deBary in confirming *P. infestans*, which is very difficult to culture, as the cause of late blight of potato.

IMPACTS OF PLANT DISEASE

Plant diseases have wide-ranging impacts on plants and the organisms that depend on them for their existence. There are many examples of disease in natural and agroecosystems that have had devastating consequences for humans. In general, plant diseases matter to humans because they (i) make food scarce, sometimes to the point of malnutrition or even famine; (ii) increase the cost of food and other products due to yield losses or increased production costs; (iii) make products dangerous for consumption by producing toxins; (iv) damage natural habitats; (v) reduce the esthetic value of produce, flowers, trees, and other plants and plant products; (vi) degrade environmental quality due to loss of key species or some methods of disease control; and (vii) inhibit trade in plant products due to the threat of pathogen movement across geographic or political boundaries. Although not all diseases are dramatic in their effects on plants, most cultivated plants have important diseases that must be managed each year to prevent some of the problems described earlier.

Occasionally, plant diseases are beneficial to humans. For example, corn smut produces edible galls that are highly prized in Mexican cuisine, while certain fungi produce beautiful grains and staining patterns in wood products. In some cases, plants are deliberately inoculated with plant pathogens or managed so that the desirable or beneficial product will be produced.

As we have already seen, plant diseases develop only when all elements of the disease triangle, host, pathogen, and environment, come together at the same time and place. New diseases may occur when host or pathogen is introduced in a new place for the first time. On the other hand, existing diseases can become important or reemerge when one element of the existing disease triangle changes or shifts. What factors lead to these changes and subsequent impacts? Why do some diseases cause extensive losses and others remain minor or unimportant? If we focus on the disease triangle, it is clear that some of our most important epidemics have come about by a significant and often abrupt change in the balance between pathogen, host, and environment.

Exotic Pathogens

Worldwide, many of the most devastating epidemics come about after the introduction of an exotic pathogen. An **exotic pathogen** is one that is introduced into a new geographic area. Of course, the presence of this pathogen becomes noticeable only when it finds a susceptible host, growing in an environment also favorable for the pathogen. This new host is often a relative of one of the pathogen's established hosts; likewise, the plant may already be host to a closely related species of the pathogen. In these cases, host and pathogen have coevolved through a kind of trench warfare, in which each has developed defensive and offensive strategies that keep each partner partially at bay. However, the exotic pathogen may be extremely aggressive on its new host and the host highly susceptible to the exotic pathogen because these partners have not evolved together. The list of epidemics caused by exotic pathogens is a long one, with examples present on all continents and across all pathogen types. In most cases, people inadvertently introduce these pathogens. Well-known examples of epidemics caused by exotic pathogens include late blight of potato, chestnut blight, white pine blister rust, Dutch elm disease, Jarrah dieback in eucalyptus and other species, downy mildew of grape, fire blight of apple and pear, citrus canker, plum pox virus, dogwood anthracnose, Camellia petal blight, and many others. These exotic pathogens, along with numerous insect pests, are the principal justification for establishing regulations and plant quarantine laws by all countries.

An excellent example of an exotic pathogen that resulted in total devastation of a host is the chestnut blight pathogen, *Cryphonectria parasitica*, which was brought into the United States. The American chestnut was the dominant tree in the eastern half of the country and was very important in many aspects of the economy. It has been said that the chestnut tree was used in rural mountain areas from the cradle to the grave. The wood was used not only for building houses and furniture but had many other uses because of its resistance to decay. The same tannins that made the wood resistant to decay were extracted and used in leather tanning. Chestnut trees were an abundant source of nuts for humans and wildlife. The chestnut blight epidemic destroyed billions of chestnut trees from Georgia to Maine and changed Eastern forests forever. What happened?

Cankers from chestnut blight were first observed on trees in the New York Zoological Gardens in 1904. Within just 4 years of the original observation, the disease was present in multiple states in the Northeast, and within 10 years, it was widespread throughout the northeastern states and as far south as North Carolina. All efforts to stop the spread were ineffective. Over 4 billion trees were lost in less than 30 years, and the American chestnut population was destroyed by the early 1950s. The pathogen does not kill roots, so young trees that develop from root sprouts can still be seen occasionally, only to be killed by the pathogen that is now endemic in eastern U.S. forests.

The pathogen was most likely introduced in many locations in small cankers on Japanese and Chinese chestnuts, which have coevolved with the pathogen and are not severely damaged by it. The disease was not reported on Chinese chestnuts until 1913, well after the disease had become epidemic in the United States. In contrast, *C. parasitica* found a highly susceptible host in the fast-growing American chestnut, which had natural growth cracks that served as ideal infection courts for both wind- and rain-dispersed spores. Casual vectors, such as birds that picked up spores on their feet, also moved the pathogen. The pathogen thrived in a highly favorable environment, uninhibited by host resistance or natural antagonists.

Chestnut blight is one of three major diseases that led to the passing of the Plant Quarantine Act of 1912, which was intended to prevent further catastrophes like chestnut blight. The disease changed the composition of eastern hardwood forests and changed the way of life for many people who depended on the tree. There are extensive efforts to breed resistance to the blight pathogen into American chestnut, with seedlings now planted in natural forests to determine whether they will be able to survive.

Exotic Hosts

Sometimes new diseases result when a new host is introduced into an area where it finds a pathogen in wait. This **exotic host** completes the disease triangle, but its impact is usually not as devastating as observed with exotic pathogens. For example, when the Dutch introduced a New World (Americas) crop, tobacco, into Indonesia in the 1860s, the crop thrived until it was planted in an area on the island of Sumatra where a new disease developed. This root and crown rot disease, caused by the oomycete *Phytophthora nicotianae*, was described in 1896 and named black shank. The black shank pathogen was eventually spread on tobacco back to the Western Hemisphere and then to most tobacco-producing areas of the world. If tobacco had never been introduced to Sumatra, it is possible that millions of dollars of annual losses to black shank would have been avoided. A change in the host itself may result in a drastic change in its response to a pathogen. An excellent example of this occurred in the hybrid corn crop in the United States during the late 1960s and early 1970s. Hybrid corn is grown because yields are higher and hybrid plants are more vigorous than their inbred parents, a phenomenon known as hybrid vigor. Production of hybrid seed requires a male and female parent, with the female inbred line being pollinated with pollen from the desired male line. In corn, this meant that the pollen-bearing tassels of female (seed-bearing) plants had to be removed to prevent self-pollination. This was accomplished by hand detasseling, a very labor-intensive job. Plant breeders took advantage of the discovery of a cytoplasmically inherited trait that caused male sterility. A seed parent with male sterile cytoplasm did not have to be detasseled. The Texas cytoplasmic male sterile (Tcms) trait was soon used in almost all seed corn produced in the United States. While the nuclear genes varied in the corn varieties planted from Florida to Wisconsin, their cytoplasm was identical. This genetic uniformity in the cytoplasm was about to spell disaster for the corn crop in 1970. What happened?

Southern corn leaf blight, caused by *Bipolaris maydis*, was a minor disease of corn in the 1960s. Year-to-year losses varied slightly, but the disease was not considered to be a severe threat to corn. In the late 1960s, localized areas of unusually severe disease caused by this organism were observed. When an early season epidemic began in Florida in 1970, one of the worst epidemics in the history of U.S. agriculture was afoot. All the elements of the disease triangle came together to cause a severe epidemic that year. The environment was highly favorable for southern corn leaf blight throughout the entire summer, and several hurricanes helped to spread inoculum up from the South through the Corn Belt. The epidemic spread throughout most of the U.S. corn crop, with losses exceeding a billion dollars (about $6 billion in 2013 dollars). The epidemic sent the seed corn industry into a crash program to breed inbred corn lines without the Tcms trait before the next season. Within 2 years, Tcms was gone from corn germplasm sold in the United States.

The genetically uniform host had been selected for a variant in the pathogen population that was highly aggressive on corn that carried the Tcms trait. This pathogen variant, race T, produced a toxin that specifically recognized and acted on a cytoplasmic trait associated with Tcms. This type of recognition is common with host-specific toxins produced by some fungi. The southern corn leaf blight epidemic uncovered the genetic vulnerability of a major food crop and forever changed our perceptions of the hazards of genetic uniformity. The epidemic demonstrated the importance of cytoplasmic inheritance of traits and clearly showed how a minor pathogen could cause a devastating epidemic when a susceptible host selects for variants in the pathogen population. As discussed in Chapters 23 and 26, strong selection pressures such as host resistance genes and site-specific fungicides result in significant shifts in pathogen populations and increased disease, but this epidemic illustrated that host factors other than disease resistance genes could also have significant impacts on the pathogen and thus disease.

Environment

Even when a host and pathogen are well established in the same place and time, disease will not develop without a favorable environment. People have long recognized that environment plays a role in disease outbreaks, even attributing disease itself to bad air (miasmas) or to excess rain in the days before the germ theory was established. As already discussed, late blight of potato became especially severe in the cloudy, wet summers of the mid-1840s in Ireland and Northern Europe. Likewise, the wet summer of 1970 contributed to the severity of the southern corn leaf blight epidemic of that year. Because of the importance of environment on disease, epidemics of some diseases are very sporadic, with no or minimal disease observed in a season following a severe epidemic. Sclerotinia blight of rapeseed in the Northern Great Plains provides an excellent example of how the environment impacts host and pathogen in disease dynamics. The fungal pathogen *Sclerotinia sclerotiorum* survives in soil as sclerotia, and when soil is cold and saturated for extended periods of time, the sclerotia germinate to produce tiny mushroom-like structures (apothecia) that produce millions of ascospores. Ascospores infect flowers, so disease is severe when the environment favors simultaneous production of ascospores of the pathogen and blooming of the host. Disease problems are minimal if the environment is unfavorable, and the infection court is not present when inoculum is produced. The reliance of this and many other epidemics on specific environmental conditions often allows plant pathologists to forecast or advise growers of potential disease outbreaks based on weather. Growers can then take steps to prevent disease losses through spray programs and other management strategies.

Although we often think of environment in terms of weather, the cultural environment also plays a very important role in disease development. Rotation, tillage, fertility, irrigation, planting date, pruning, and harvesting are examples of cultural practices that may influence disease development. A change in cultural practices can lead to unexpected changes in the balance between host, pathogen, and environment, leading to serious outbreaks of diseases that were of lesser importance previously.

An excellent example of how favorable weather in combination with a change in cultural practices led to major epidemics is seen with Fusarium head blight (FHB), or scab, of wheat and barley in the Upper Midwest of the United States in the 1990s. Although epidemics of this disease had been somewhat common earlier in the twentieth century, scab was considered a minor disease and attracted little interest in the years leading up to the 1990s. All of this changed with a devastating epidemic in 1993 and outbreaks that continued through 1997. Direct losses of over a billion dollars (more than $1.6 billion in 2013 dollars) and additional indirect losses of $4.8 billion were attributed to FHB in 1993, one of the greatest losses caused by a plant disease in North America in a single year. An estimated 70 million tons of barley were lost in the 1993 epidemic alone, and an estimated 100 million acres of wheat and barley were affected. In addition to yield losses, FHB results in contamination of the grain with a toxin called deoxynivalenol (DON). Contaminated grain is hazardous to humans and animals and leads to feed refusal, particularly in swine. Therefore, animal feed, grain, and flour are monitored for DON contamination, and rejection of contaminated products results in losses in addition to gross yield losses. Infected grain can continue to produce DON during malting, posing a health hazard and causing excess foaming (gushing) in beer. Epidemics were especially severe in the Red River Valley of the Dakotas and Minnesota, and many farmers faced foreclosure and bankruptcy.

FHB is commonly caused by the fungal pathogen *Fusarium graminearum*, although other species of *Fusarium* can also cause head blight. The host range of *F. graminearum* includes wheat, barley, corn, and other grain and grass species. The fungus overwinters on the debris of previous crops and produces asexual spores (macroconidia) on the overwintered debris. These macroconidia are one source of primary inoculum. Under relatively warm, wet, humid conditions, the overwintered fungus also produces sexual structures (perithecia) that bear ascospores. The conditions that favor ascospore production also favor infection and colonization of the host. Both the splash-borne macroconidia and the wind-blown ascospores can infect the grain head during flowering. Under favorable conditions, colonization continues, and depending on the time of infection, infected florets may produce no seed at all, produce a shriveled grain called a tombstone, or produce normal-looking seeds contaminated with DON.

The spring and summer of 1993 were extremely wet in the Red River Valley. Rainfall was 250%–600% of normal in that year. In addition, in typical years, rainfall accumulation is greatest in June, before plants produce heads and florets, and July is fairly dry. In 1993, however, rain continued in July, during the entire period of flowering and grain fill. Although less extreme, above normal rainfall patterns continued during the mid-1900s along with continued outbreaks of FHB. Cultural changes also played a major role in the scab epidemics of the 1990s. Traditionally, grains were produced with clean tillage, meaning that weeds and crop debris were buried by use of plowing, disking, and other tillage operations. However, with the advent of better herbicides and soil conservation programs, growers began to adapt conservation or minimum tillage practices. These practices are beneficial to soil, but they also result in increased amounts of debris on the soil surface, which may be colonized by the pathogen. In addition, production of corn, which is also a host of *F. graminearum*, began to expand into wheat production areas. Corn debris is more resistant to breakdown than debris from wheat and other small grains, also resulting in an increase in overwintering inoculum due to a change in the cultural environment.

FHB continues to cause losses to grain crops in the United States, Canada, China, and other countries. As a result of the epidemics of the 1990s, a major effort to develop management strategies and tactics was undertaken in the United States. This scab initiative emphasizes breeding for resistance to both FHB and DON accumulation. In addition, risk assessment tools and weather-based FHB alerts are available to growers throughout the affected areas, so that they can identify periods when the environment favors outbreaks of FHB.

SUGGESTED READINGS

There are many books and articles on plant pathology-related topics. We list a few of those that will allow you to read further on the topics that we have briefly described in this introduction. By having an understanding of the basic concepts in plant pathology, you will be better prepared to critically evaluate both scientific and popular articles on plant disease and how they affect our lives every day.

Agrios, G. N. 2005. *Plant Pathology* (5th edition). San Diego, CA: Academic Press.

Ainsworth, G. C. 1981. *An Introduction to the History of Plant Pathology*. Cambridge, UK: Cambridge University Press.

Horst, R. K. 2013. *Westcott's Plant Disease Handbook* (8th edition). Dordrecht, the Netherlands: Springer.

Large, E. C. 2003. *The Advance of the Fungi*. St. Paul, MN: American Phytopathological Society.

Lucas, J. A. 1998. *Plant Pathology and Plant Pathogens*. Oxford, UK: Blackwell Science.

Maloy, O. C., and T. D. Murray. 2001. *Encyclopedia of Plant Pathology* (Vols. 1 and 2). New York, NY: John Wiley and Sons.

Matossian, M. K. 1989. *Poisons of the Past. Molds, Epidemics, and History*. New Haven, CT: Yale University Press.

Schumann, G. L., and C. J. D'Arcy. 2010. *Essential Plant Pathology* (2nd edition). St. Paul, MN: American Phytopathological Society.

Schumann, G. L., and C. J. D'Arcy. 2012. *Hungry Planet. Stories of Plant Disease*. St. Paul, MN: American Phytopathological Society.

2 Laboratory Skills

Safety and Preparation of Culture Media and Solutions

Robert N. Trigiano and Bonnie H. Ownley

Concept Box

- Laboratory safety is very important. Specific rules are in place to prevent most accidents and should be practiced at all times.
- Preparation of culture media requires the following three steps: combining the ingredients, sterilizing the medium (typically by autoclaving), and dispensing the media into suitable vessels such as Petri dishes or flasks.
- Heat-liable materials in solutions, such as antibiotics and amino acids, may be sterilized by filtration.
- Knowledge about how to make molar, normal, and buffer solutions are necessary for proper experimentation.

Laboratories are an integral part of any science-based learning experience as they are wonderful opportunities to reinforce the ideas presented in lectures and to develop research and critical thinking skills. If done properly, laboratory exercises can be used as occasions to improve organizational, collaborative, and written skills as well. The lessons learned by participating and completing laboratory experiments provide great benefits to students as the exercises help to prepare them to undertake advanced undergraduate and graduate studies. Obviously, we strongly believe that experiential learning is an important facet of instruction in plant pathology as most of the chapters in this book include laboratory exercises. Fortunately, for many students, most introductory plant pathology courses have incorporated some form of laboratory-based learning.

In this chapter, we will focus on three primary areas in laboratory work, namely safety, preparation of culture media, and preparation of solutions. We will not explore all possibilities and specific research techniques as these are adequately represented in the individual chapters of this book and do not warrant additional explanations. However, there are certain common elements to all experiments that bear discussion. We have assumed that most students have had limited exposure to laboratory work, and hence, we will start from the beginning.

The first consideration to any laboratory experience is *safety*. Each experiment presents its own safety concerns, but there are a few general precautions that apply to all situations. By following these basic rules, students may have a safe and productive experience. We suggest that the safety officer of the unit be asked to conduct a safety presentation for the class and/or that, if available, students should be strongly encouraged to take training in chemical and biological safety offered by their institution.

Before the first laboratory exercise, students should be instructed to

- Wear appropriate clothing and shoes. Open-toed shoes are not permitted in the laboratory. Long-sleeved shirts and long pants provide some protection for your skin and should be worn if working with chemical hazards. Lab coats also provide protection and are strongly recommended.
- Remove all jewelry before lab starts and store them in a secure place.
- Eating and drinking are not permitted in the laboratory. Food for human consumption should never be brought into the laboratory.
- Do not apply cosmetics or handle contact lenses while in the laboratory. Chewing gum or use of any tobacco products is strictly prohibited. Leave all books (except your laboratory notebook), coats or jackets, and hats outside of the laboratory.

- Consider constraining long hair, so that it does not get in your way.
- Never pipette anything, even distilled water, by mouth. Mechanical pipetting devices must be used.
- When leaving the lab for the day, wash hands with soap and water.

The instructor should complete the following on the first day of laboratory work:

- Give a tour and demonstrate the safety features of the laboratory, including eye protection (against hazardous solutions and UV light), emergency showers and eyewashes, fire extinguishers and alarms, and emergency phone numbers. The instructor should also outline the evacuation plan for the laboratory in case of an emergency, such as a chemical spill.
- Provide the location of the emergency first aid kit.
- Reinforce the idea that students should not work alone in the laboratory.
- Explain and demonstrate the use of acetonitrile gloves and under which situations they are needed. Never reuse gloves and never wear your gloves outside of the laboratory.
- Explain and demonstrate how to safely dispose of hazardous waste materials and sharp objects (needles, scalpels, etc.).
- Show students how to prepare a label for secondary storage (not the original container) of chemicals and solutions. The label should contain the following information: name, date, contents, and any specific hazard associated with the contents of the bottle.
- Provide an explanation of the "Right to Know" law in regard to any hazards associated with materials and procedures in the laboratory.
- Provide an explanation of Material Safety Data Sheets (MSDS) and how to use this information, including how to read the hazard codes on chemical bottles. A chemical hygiene plan should also be available for the laboratory.
- Explain the purpose and demonstrate the proper use of chemical fume hoods, biosafety cabinets, and laminar flow hoods.
- Demonstrate how to clean up liquid and solid chemical spills or contamination.
- Explain how to work with open flames (alcohol lamps and Bunsen burners) in the laboratory.
- Explain the hazards of handling items stored at low temperatures (–20°C and –80°C).
- Explain why it is necessary to report all accidents immediately to the instructor and seek medical attention if necessary.

We suggest that after completion of the safety presentation, each student sign and date a document that states that they have completed safety training for the laboratory. This should be filed as part of the safety training of students for the laboratory class.

Now that you are familiar with some basic safety materials and procedures, we will discuss preparation of culture media for growing microorganisms. Most of the organisms that will be used in the laboratory can be cultured on various artificial solidified (agar) or liquid nutrient media. There are generally four steps in the preparation of media: (1) measuring and combining the constituents, (2) sterilization, (3) dispensing culture medium into suitable containers, and (4) short- or long-term storage of culture media.

PREPARATION OF CULTURE MEDIUM—STEP 1: COMBINING INGREDIENTS

Many media are used in plant pathology, and several of them are commercially available as dehydrated agar- (a solidifying agent) or broth-based preparations. A commonly used base medium (notice that the singular form uses "ium" instead of "ia") for fungi is potato dextrose (PD) (= glucose, a simple sugar). Potato dextrose agar (PDA) is very simple to prepare from a purchased powdered product and typically contains 4 g potato infusion (primarily starch), 20 g dextrose, and 15 g agar.

MATERIALS

Each student or team of students will need the following items:

- Commercially prepared PDA and PD broth
- Agar
- Top loading balance, spatula, and plastic weigh boats
- 1-L graduated cylinder
- Several 1-L Pyrex bottles with caps
- Distilled or deionized water
- Stir plate with magnetic stir bars
- Autoclave tape
- Permanent marking pen
- 125 cc (mL) rice or oat grains
- 250-mL graduated cylinder
- Aluminum foil

Follow the instructions in Procedure 2.1 to prepare media.

Procedure 2.1
Preparation of Culture Media

Step	Instructions and Comments
	A. Solid Culture Medium
1	Weigh 39 g of powdered PDA preparation into a large plastic weigh boat using a top loading balance. Clean up any spilled materials and never return unused or spilled materials to the stock bottle.
2	Combine the powder with 1 L of distilled water in a glass Pyrex bottle and stir for few minutes to dissolve the sugar and potato infusion—the agar will not dissolve.
3	Label the bottle with "PDA," your name, and date, and place a small strip of autoclave tape on the bottle (Figure 2.1). Lastly, place the cap on the bottle loosely—do NOT tighten the cap completely.
	B. Liquid or PD Broth Medium
1	Weigh 24 g of PD broth (without agar) in a large weigh boat and dispense into 1 L of distilled water. Stir the preparation until all of the powder is dissolved.
2	Label the bottle as before.
	C. Dilute Formulation of PDA, Such as 0.1 PDA
1	Weigh 2.4 g of PD broth (without agar) in a small weigh boat, dissolve completely in 1 L of distilled water, and then add 15 g of agar.
2	Clean up the balance and label the bottle with the same information as with the PDA.
	D. Water Agar
1	Dispense 15 g of agar into 1 L of distilled water and label the bottle as before. Mix, but note that the agar will not dissolve at room temperature.
	E. Rice or Oat Grain Culture Medium
1	Dispense about 150 cc or mL of rice or oat grains into a wide-mouthed 1-L flask. Add 125 mL of distilled water. Cap the flask with aluminum foil. Other grains can be used and the amount of water may need to be adjusted.

FIGURE 2.1 Label on medium bottle including type of medium (PDA), name, and date. (Courtesy of R.N. Trigiano.)

PREPARATION OF CULURE MEDIA—STEP 2: STERILIZATION

The media are now ready to be sterilized. Water and other ingredients that are used to prepare media always contain living microbes in the form of either bacterial cells or fungal hyphae and spores (bacteria or fungi), and these will grow unabated in media if not destroyed. In plant pathology, we need to have axenic (without a stranger) or pure cultures of organisms used in our studies. In other words, we only want our organism of interest to be present. However, there are instances where there is a need to have two or more organisms in the same culture, but that will not be discussed here.

The most frequently used option for sterilizing agar media, liquids, glassware, and instruments is by autoclaving at high temperature and pressure. There are times when certain heat-liable compounds (antibiotics, amino acids, plant growth regulators, fungicides, etc.) will need to be sterilized using a filter with small pores (0.22- or 0.45-μm diameter) that exclude contaminating organisms. Heat sterilization of instruments and glassware in an oven at 204.4°C (400°F) for 2 h can be used, but typically is not. This discussion will focus on autoclaving and filtering for sterilization.

Materials

Each student or team should have the following items:

- Agar, liquid, and grain media prepared from Procedure 2.1
- Access to autoclave gloves
- 50-mL beakers covered with aluminum foil
- Twenty-five 125-mL screw-top Erlenmeyer flasks
- Large metal or "autoclave-safe" pan for holding flasks and bottles
- Empty 1-L storage bottles with screw tops
- Access to an autoclave

Follow the instructions in Procedure 2.2 to sterilize the media that was prepared using Procedure 2.1.

Procedure 2.2

Sterilization of Culture Medium

Step	Instructions and Comments
1	Place the bottles of media into a large pan. You may choose to dispense 25 mL of liquid PB medium into the 125-mL Erlenmeyer flasks at this time. If you choose to dispense after autoclaving, then the Erlenmeyer flasks must be autoclaved. First, add two drops of water to each flask and lightly tighten the cap.
2	Add a drop of water to each 50-mL beaker and cover with a square of aluminum foil. Add a few drops of water to each empty 1000-mL bottle and loosely tighten the cap. Caution: Do not tighten the cap on any bottle containing liquids to be autoclaved.
3	Securely tighten the door of the autoclave. Most recently purchased autoclaves have a selectable program. If this is the type of autoclave in use, select "liquid cycle" that provides at least 121°C and 15 lb. pressure per square inch (psi) for a minimum of 20 min and engage. If the autoclave is older, close and tighten the door, and open the valve that admits steam to the chamber. Begin timing (20 min) when the temperature reaches 121°C and pressure 15 psi.
4	Most modern autoclaves will automatically time the heat/pressure duration. When "liquid cycle" is selected, the autoclave is programmed to release the pressure in the chamber slowly to avoid super heating of the liquid media. If an older autoclave is in use, select slow exhaust. This may take some time to exhaust the pressure completely.
5	After the pressure in the chamber becomes "zero" and the temperature is below 90°C, it is safe to open the autoclave door. Be careful, use autoclave gloves, and keep your face away from the door. Open the door slowly and watch out for hot steam escaping from the chamber.
6	Remove the pans and place the bottles containing PDA on a magnetic stir plate and stir gently. Alternatively, media may be placed in a 60°C water bath until the temperature of the media equilibrates with the temperature of the water. Hot agar medium cannot be dispensed into plastic Petri dishes immediately as the dishes will warp and excess water will condense on the lids.
	Sterilization of Rice and Oat Grains
1	Place the flask containing the water and grain in the autoclave and set for liquid cycle for 90 min. Remove from the autoclave and place at room temperature overnight.
2	The first autoclave cycle will kill actively growing microorganisms, but not heat-resistant bacterial spores. The heat will activate spores, and they will "germinate" and reproduce vegetatively. Reautoclave the grain preparation with a liquid setting for 90 min the next day. "Double autoclaving" should produce sterile grain suitable for growing pure cultures of many fungi. The sterile grains may be aseptically transferred to test tubes or flasks and used to culture fungi.

PREPARATION OF CULTURE MEDIA—STEP 3: DISPENSING MEDIUM

Solidified and liquid media prepared in Procedures 2.1 and 2.2 may be stored for an extended time if desired (see Step 4), but if needed soon, they may be dispensed after the media has been "cooled." A "rule of thumb" is if the bottle containing the autoclaved medium can be held with a "bare" hand without discomfort, it is sufficiently cooled (50°C–55°C) to be poured and for any amendments, such as antibiotics that would have been destroyed or have diminished activity if autoclaved, to be added, and then dispensed. Agar medium may be dispensed into a number of containers, but Petri dishes of various sizes are typically used in the laboratory. In this exercise, agar medium (with and without augmentation) will be poured into Petri dishes and liquid medium dispensed into flasks.

MATERIALS

Each team of students should have the following materials:

- Two sleeves of 10-cm diameter Petri dishes
- Four sleeves of 6-cm diameter Petri dishes
- 1000-μL pipette and sterile tips
- 10-mL syringe and 0.22-μm syringe-driven filter unit (Millipore Corp., Billerica, MA)
- Lab supply vacuum source or vacuum pump with rubber tubing and a bottle top filter (Millipore Corp., Billerica, MA)
- 10 mL water-based food dye solution (substitute for antibiotic solution that needs to be filter-sterilized)
- Disposable acetonitrile gloves
- 25-mL disposable pipettes and manual pipette pump or battery-operated pipette pump
- Two 50-mL sterile beakers covered with aluminum foil

Follow the instructions in Procedure 2.3 to complete this section.

After the medium in the Petri dishes has been solidified, the dishes may be placed in the original plastic sleeve for storage at 4°C. Be sure to write the name of the medium, date, and your name on the sleeve. If you have excess medium that was not poured, the bottle may also be stored at 4°C with identification of the date, type of medium, and who prepared it written on the bottle.

Procedure 2.3

Augmenting Autoclaved Media and Dispensing Medium into Vessels

Step	Instructions and Comments
1	Wear acetonitrile gloves during this operation. Dissolve 100 mg of powdered dye in 10 mL of distilled water. Depending on the size of the class, these solutions may be shared with different groups.
2	Load the 10-mL syringe with the colored dye solution (this is a substitute for water-soluble antibiotics for demonstration purposes only) by drawing the solution into the barrel by pulling the plunger in the syringe upward (Figure 2.2a). Unwrap the sterile 0.22-μm filter and aseptically attach to the syringe—the tapered end of the filter unit should face outward (Figure 2.2b). Remove the aluminum foil from the sterilized beaker and express the dye solution through the filter into the beaker (Figure 2.2c). Adjust the 1000-μL pipette to "1000," place a tip on the end of the pipette and measure 1000 μL or 1 mL of the dye solution into the tip (Figure 2.2d). Dispense into the liter of PDA and stir. The effective concentration of dye solution is 10 mg/L.
3	Steps 3–5 can be completed using a laminar flow hood or on the lab bench using proper aseptic technique. A liter of medium is sufficient to pour 25 mL into each of forty 10-cm diameter Petri dishes or to pour about 15 mL into each of sixty-five 6-cm diameter dishes. Most investigators do not actually measure 25 mL, but approximate the volume poured into each dish. Lift the lid of the first dish, leaving an opening that allows you to pour the medium into the bottom of the Petri dish while holding the bottle of the medium in your other hand. Do not place the lid of the Petri dish on the lab bench. Immediately replace the Petri dish lid after pouring the medium (Figure 2.3a), then place an empty dish on top and repeat the process to pour medium into this dish. Do this in stacks of 10 dishes (Figure 2.3b). The bottle of medium should be swirled a number of times when pouring dishes because agar will settle at the bottom of the bottle. By pouring the dishes in this fashion, water condensate on the lids of the dishes is minimized. Save the original plastic sleeves that housed the sterile plastic dishes and slip the sleeves over a stack of about 20 Petri dishes after the agar has been solidified.

(*Continued*)

4 Typically, if the liquid medium does not require amendments, the medium is dispensed into the flasks and then autoclaved. However, if heat-sensitive materials must be added, follow Procedure 2.3 Step 2 and mix thoroughly. Aseptically remove a sterile 25-mL pipette (has a cotton plug in the barrel near the top) from the wrapping and mount with the milliliter gradations, so that you can easily see them. Draw 25 mL into the pipette and dispense into each 125-mL flask. The same pipette may be used to fill all of the flasks since all flasks are filled with the same medium. Completely tighten the screw cap of the flask.

5 There may be occasions when large volumes of liquid medium may need to be filter-sterilized. For this, a bottle-top filter and a sterile bottle are used. Unwrap the bottle top 0.22-μm filter unit and place the membrane side of the filter on the opening of a sterile bottle (Figure 2.4a). Attach rubber tubing to the nipple on the filter and the other end to a vacuum source. Pour the medium in the reservoir and gently apply vacuum. The medium will be drawn through the membrane and into the sterile bottle (Figure 2.4b). The sterile medium may be dispensed into vessels.

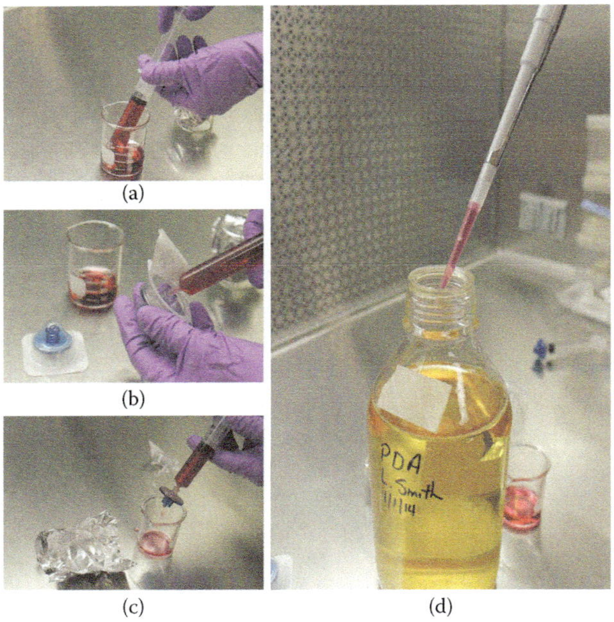

FIGURE 2.2 Syringe and filter sterilization of heat-liable compound. (a) Load solution into a 10-mL disposable plastic syringe. (b) Attach a 0.22-μm filter unit to the nipple of the syringe without touching the filter. (c) Remove aluminum foil cover from a small sterile glass beaker and express the fluid through the filter into the beaker. (d) Using a sterile pipette tip, transfer 1 mL of the sterile liquid into the medium, which has been autoclaved and cooled to about 50°C–55°C. Dispose of the filter and syringe according to the hazard level of the filtered solution. (Courtesy of R.N. Trigiano.)

To use the medium, melt the agar in a hot water bath or steamer. Agar remelts at 85°C. Do NOT autoclave the medium again as this may "caramelize" (change the structure of) the sugar component, which can alter the growth of microorganisms. Liquid medium in either bottles or flasks may also be stored at 4°C. Remember to mark the identification of the medium with type of medium, date, and name.

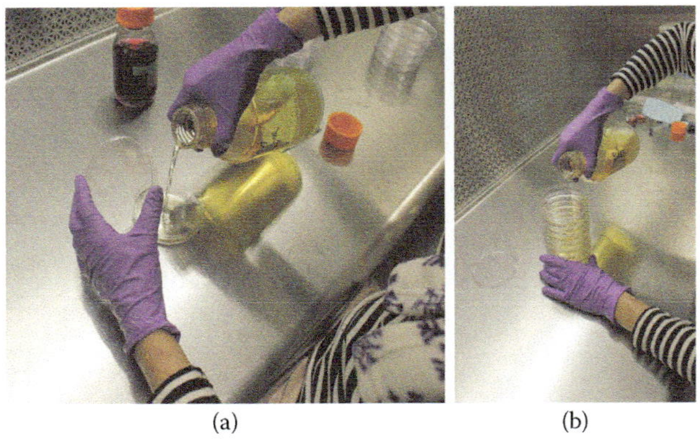

FIGURE 2.3 Pouring agar medium into Petri dishes. (a) Be sure that the medium has been cooled (50°C–55°C) before pouring about 25 mL into each 10-cm diameter Petri dish. This operation can be performed in a laminar flow hood or on the lab bench using aseptic technique. (b) Stack dishes when pouring the medium. This will prevent excess water condensation on the inner surface of the Petri dishes. Fill the empty medium bottle with hot water immediately after all of the medium is dispensed. After the agar has been solidified, the Petri dishes may be stored in the original plastic sleeve. Label the sleeve with the type of medium, date, and name. (Courtesy of R.N. Trigiano.)

PREPARATION OF SOLUTIONS

Occasionally, protocols will require that chemical solutions be made. The composition of most of the solutions generally falls into the following two categories: weight or volume/volume or percentage and molar and normal.

Materials

- Magnesium chloride (easy to dissolve and safe) (any suitable compound may be substituted)
- Disposable acetonitrile gloves
- Glacial acetic acid
- 100- and 1000-mL graduated cylinders

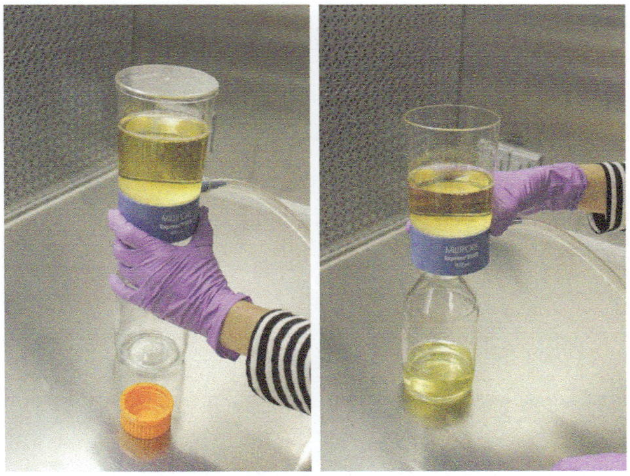
(a) (b)

FIGURE 2.4 Sterilizing a large volume of liquid medium using a bottle top filter. (a) Mount the bottle top filter onto the sterile storage bottle and fill the reservoir with the medium. Attach a vacuum source to the unit. (b) Gently pull the medium through the 0.22-µm filter membrane into the sterile medium bottle with the vacuum. Sterile medium may be either stored or dispensed to other vessels. (Courtesy of R.N. Trigiano.)

- Several glass 1-L storage bottles
- Chemical fume hood
- Milligram balance and small plastic weigh boats
- Top loading balance and large plastic weigh boats
- Magnetic stir bars and stir plate
- Succinic acid anhydrous (without associated water molecules); sodium succinate dibasic·$6H_2O$
- pH meter and electrode (recently calibrated with pH 4.0 and pH 7.0 standards)
- Several 1-L beakers
- Concentrated sulfuric acid

A. Making weight/volume (w/v) solutions. For this example, magnesium chloride will be used to make a 1% (w/v) solution. One milliliter [or 1 cubic centimeter (cc)] of water at room temperature weighs 1 g or 1000 mg. Follow the instructions in Procedure 2.4A to complete this exercise.

B. Making volume/volume (v/v) solutions. Complete this exercise using a fume hood and protective eyewear. Follow the protocol in Procedure 2.4B to complete this exercise.

C. Making molar solutions. A mole of a substance, either an element or a compound, is its atomic weight (in grams) and contains 6.02×10^{23} or Avogadro's number of atoms or molecules. A one molar (1 M) solution is defined as this amount of the substance dissolved in approximately 1 L of water. Follow the instructions in Procedure 2.4C to complete this exercise.

D. Making normal solutions. Normality is based on molar equivalence of either H^+ or OH^- ions and is most often used for strong acids and bases. For example, hydrochloric acid (HCl) completely disassociates into H^+ and Cl^- ions. So, 1 mole of HCl (36.5 g) will yield 1 mole of H^+. In this case, 1 M = 1 Normal (1 N). However, sulfuric acid, another strong acid (H_2SO_4; 98 g/mole), completely disassociates into 2 moles of H^+ and 1 mole of SO_4^{-2}. Therefore, if 98 g sulfuric acid is dissolved in water, it yields 2 M or 2 N of H^+.

Known concentrations or normalities of acids (and bases) may be purchased from manufacturers, but these can be expensive. More typically, concentrated (almost pure) acid is usually obtained from the same sources. However, sometimes the manufacturer does not print the normality of the acid/base on the label. If this is the case, look for the specific gravity (s.g. or g/mL) of the liquid. The s.g. of most concentrated sulfuric acid preparations is about 1.84 g/mL or 1840 g/L. Because the contents of the bottle are 96% acid (see label), the amount of acid is 1840 g $\times$ 0.96 = 1766.4 g of acid per liter. This translates into about 18 moles (1766.4 g/98 g/mole) of sulfuric acid. Remember that each mole of sulfuric acid produces 2 moles of H^+; therefore, the normality of the solution is 2 $\times$ 18 or 36 N. This is a very caustic solution. Always wear protective clothing, eyewear, and rubber gloves while working in a chemical fume hood.

Follow the protocol outlined in Procedure 2.4D to make normal solutions.

E. Making buffer solutions—titration method. Buffers are used to control changes in pH and maintain the original pH of solutions when acids and bases are added to the solution. They have various applications in studies of the growth, enzyme activity, histology of tissues, and so on involving plant pathogens and their hosts. A simple definition of a buffer is the combination of a weak organic acid (e.g., succinic acid) with the salt of that acid (e.g., sodium succinate dibasic). In solution, the acid and salt of the acid are in equilibrium, so that when an acid or base is added, the equilibrium shifts to either the acid or base and pH remains relatively stable. The highest buffering capacity (can add the highest amount of acid or base without appreciably changing the original pH) of the solution is near the mid-range of the buffering system. Most physiological buffers are

Procedure 2.4
Making Solutions and Buffers

Step	Instructions and Comments
	A. Making Weight/Volume (w/v) Solutions
1	To make a 1% (w/v) magnesium chloride solution, weigh 100 mg of the compound and dissolve in 9900 mg or 9.9 mL of water. Typically, this accuracy is not demanded, so 100 mg in 10 mL of water is sufficient. This yields a 1% solution.
	B. Making Volume/Volume (v/v) Solutions
1	While wearing acetonitrile gloves, measure 70 mL of glacial acetic acid in a graduated cylinder and pour into a 1-L storage bottle.
2	Slowly add 930 mL of distilled water to the bottle; be sure that the mouth of the bottle is pointed to the interior of the chemical fume hood. Cap the bottle, mix, and label with name, date, and contents. This makes a 7% solution of glacial acetic acid.
	C. Making Molar Solutions
1	In this example, 0.05 M solutions of succinic acid and sodium succinate dibasic will be made and used when buffers are discussed (Procedure 2.4E). A 0.05 M solution of succinic acid requires 0.05 M × 118.1 g/L = 5.91 g dissolved in 800 mL of distilled water contained in a beaker.
2	Pour the contents of the beaker into a 1000-mL graduated cylinder and bring the volume up to 1000 mL using distilled water. Label the bottle with contents, date, and name.
3	To make a 0.05 M solution of sodium succinate dibasic, weigh 0.05 M × 270.1 g/L = 13.51 g, dissolve as before and bring to 1 L with distilled water. Label this bottle also.
	D. Making Normal Solutions
1	To make a 1 N solution of sulfuric acid, dilute the concentrated acid with water. Use this simple formula to determine how to dilute concentrated sulfuric acid to 1 N. $V1 \times N1 = V2 \times N2$ where V1 = the volume in milliliter of concentrated acid to use N1 = the normality of concentrated acid (36 N) (mL) V2 = the desired volume of 1 N acid (1000 mL) N2 = the desired normality (1 N)
2	Solve the equation for V1: $V1 = (V2 \times N2)/N1$ or $V1 = (1000\ \text{mL} \times 1\ \text{N})/36\ \text{N}$ $V1 = 27.77$ mL of concentrated sulfuric acid
3	Add 27.77 mL of concentrated sulfuric acid to a 1000-mL graduated cylinder and bring to 1000 mL with distilled water to make a 1 N solution of sulfuric acid. It is best to hold the graduated cylinder at a 30° angle with the mouth pointed toward the interior of the chemical fume hold and slowly pour the water down the side of the cylinder. Care should be exercised when adding the water as heat is liberated (exothermic). Label the bottle with contents, date, and name.
	E. Making Buffer Solutions—Titration Method
1	Make a pH 4.5, 0.05 M succinate buffer. Pour 300 mL of 0.05 M succinic acid solution (prepared in Procedure 2.4C) into a 1-L beaker containing a magnetic stir bar. Place on a stir plate and determine the pH of this solution, which should be below 4.5. Caution: stir slowly and be sure the pH electrode is not contacted by the stir bar.
2	Slowly add 0.05 M sodium succinate dibasic (made in Procedure 2.4C) to the succinic acid solution and monitor the change in pH. When the pH meter reads 4.5, the buffer is ready for use.

in the range of 0.05–0.10 M of both the acid and the base. The Henderson–Hasselbalch equation will predict how much of the acid and the salt should be added to achieve a buffer of a specific pH. However, this equation can be difficult to use. An alternative to making a mathematical calculation is to use a titration method to achieve the desired initial pH of the buffer.

Follow the protocol in Procedure 2.4E to make a succinate buffer.

Laboratory exercises are excellent learning experiences, but like anything else, there are real dangers. Follow and practice the safety guidelines presented in the beginning of this chapter. However, emergencies and accidents do occur in laboratories; try to prevent them, but if they happen, be prepared to respond appropriately to them.

3 Proper Use of Compound and Stereo Microscopes

David T. Webb

Concept Box

- Microscopes are used to magnify and examine small objects such as spores, hyphae, and fruiting bodies.
- A compound microscope is used to examine very small objects and typically has 10X ocular (eyepiece) lenses and 4X, 10X, 40X, and 100X (oil immersion) objective lenses. The total magnification can be approximated by multiplying the power of the ocular lens by the power of the objective lens, e.g., 10X × 40X = 400X.
- Stereo or dissecting microscopes are used to examine larger objects and typically have 10X ocular lenses and a variable (0.5X – 4.0X) objective lens.
- When using a compound microscope, begin with the lowest power objective lens (4X), focus, and then progress to higher magnification objective lenses. Never begin your examination with higher magnification (40X or 100X) objective lenses.
- Only lens paper should be used to clean lenses. Do NOT use paper toweling or laboratory wipes as these will scratch the lenses, which are very expensive.
- Measurements of objects with a compound microscope can be made using an ocular micrometer that has been calibrated with a stage micrometer. Dimensions are usually reported in micrometers.
- Commercially prepared slides of plant host and plant pathogen are usually stained with fast green and safranin O, whereas fresh sections of materials are typically stained with phloroglucinol, IKI, or toluidine blue O.

This chapter is written as if the reader is a microscope novice. Experience has taught us that it is best to assume that most students will know next to nothing about using a microscope correctly and that it is best to start from scratch. In some cases, you may have learned some bad practices that need to be corrected. This chapter also covers compound microscopes that have a field diaphragm and a condenser that can be centered and focused to achieve Koehler illumination. Many student scopes do not have these features as their condensers and field diaphragms are fixed or of limited flexibility. In the course of your career, you will encounter microscopes that have the ability to achieve Koehler illumination. At that point, this tutorial will be even more useful.

Although the compound microscope is the most commonly used biological instrument, it is often used improperly. This may not matter with very thin commercial slides at low to medium magnifications. However, proper alignment of the illumination system is essential for viewing thick sections, whole mounts, and highly magnified samples of fungi and bacteria. It is also crucial for studying unstained specimens and for photomicroscopy.

You will be using microscopes throughout your career. If you learn the simple lessons contained within this chapter, you will do much better work and see the exciting world of microscopy in a new light. The modified procedure that we present was developed by the German scientist August Köhler (1866–1948) and bears his name (Köhler, 1893). Recently, his ideas were used to make the EM 910 Electron Microscope by Zeiss. Thus, this procedure, which was introduced in 1893, has been of lasting value.

MONOCULAR, BINOCULAR, AND TRINOCULAR MICROSCOPES

Microscopes are partly categorized by the number of oculars they contain. The first microscopes had one ocular and therefore were monocular. Binocular scopes have two oculars, whereas trinocular scopes have three. The third ocular is modified typically for the use of a camera. This chapter explains the uses of binocular microscopes. The same principles apply to all of the preceding types. However, stereo or dissecting microscopes differ significantly from the typical compound microscope.

THE COMPOUND MICROSCOPE

Because the optical systems in a microscope are composed of many lenses, the term compound microscope is used. This is applied specifically to microscopes that are used to study thin sections with high-power objectives (also known as objective lens; Figure 3.1). Dissecting or stereo microscopes are used to examine larger, three-dimensional specimens at lower magnifications (Figure 3.2). They also have compound lenses, but they are not generally called compound microscopes. Both the types of microscope use **transillumination** (illumination through) in which light passes from the microscope base through the specimen and travels to your eyes through oculars. This requires a special transillumination base for stereo microscopes. **Epiillumination** (illumination from above) is typically used with stereo scopes but is not typically used with compound scopes. This chapter illustrates Zeiss and American Optical microscopes. Your microscopes may be somewhat different, but you should be able to transfer the terminology and procedures described herein to your instrument. First, we will examine compound microscopes and will discuss stereo scopes later.

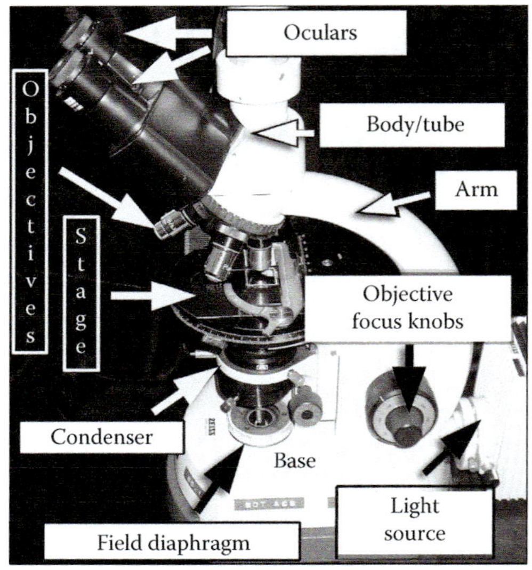

FIGURE 3.1 Zeiss Standard microscope showing major parts of a typical compound microscope. (Courtesy of D.T. Webb.)

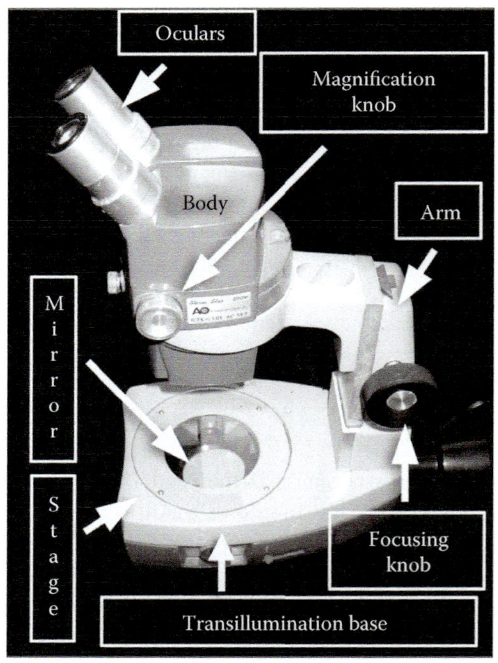

FIGURE 3.2 Typical stereo or dissecting microscope with transillumination base. This is an American Optical Stereo Zoom microscope. (Courtesy of D.T. Webb.)

Microscope Care and Handling

Please treat these instruments with great care—they are expensive and somewhat fragile.

- Value what they can do and handle them with respect.
- Always use two hands to carry microscopes. Place one hand on the arm, the curved area that connects the body to the stage and base, and the other hand under the base of the microscope (Figures 3.1 and 3.2).
- Do NOT carry scopes sideways or upside down, as the oculars and other parts will fall out.
- Use lens paper to clean all lenses on the compound scope before each lab and especially after using immersion oil. Do not use any other kind of paper except lens paper to clean microscope lenses.
- Do not use liquids (except where specified) when cleaning the lenses.
- Always use the correct focusing technique to avoid contact between any objective and your slide.
- Turn off the light when not using microscopes for long time periods.
- Carefully place the power cord or any other cords out of the way at your workspace.

- Always replace the cover on the microscope when you put it away.
- Deal with any problems immediately. Do not use the microscope if you cannot see your specimens clearly.

The Pathway of Light in a Typical Microscope

The light in a typical microscope traverses the following pathway:

Light Source → Mirror → Field Diaphragm →
Condenser → Stage → Specimen → Objective →
Body / Tube → Ocular → Eye or Camera

- Locate the major parts of your microscope by referring to Figure 3.1.

Light is provided by a bulb and is reflected through the field diaphragm, condenser, specimen, objective, tube, and ocular. There are various control knobs on the microscope that affect the light path. In addition, there are knobs for coarse and fine focus, as well as knobs to move the stage.

FOCUSING THE OBJECTIVES

The objectives are focused on the specimen by two sets of knobs that are located on both the sides of your microscope (Figure 3.1). The large outer knob is for coarse focusing. This should be used at the lowest magnification when you first place a specimen on the stage. It should be used with caution at higher magnifications. The smaller, central knob is for fine focusing and is used more at the higher magnifications, but is also used with low-magnification objectives.

- Locate the coarse and fine focusing knobs on each side of your scope.

Moving the Mechanical Stage

Light passes through the stage opening so that it can illuminate the specimen. The knobs that control the mechanical stage (stage transport knobs) are usually on the left side of the microscope as it faces you. One of these moves the stage from side to side (x-axis), whereas the other moves the stage in and out (y-axis). Most stages have x and y scales. These will allow you to record the precise location of the objects that you may want to relocate without searching the entire specimen. Slides are held in place by a mechanical slide holder.

- Locate the stage transport knobs on your microscope.
- Locate the x and y scales on your stage.
- Locate the mechanical slide holder and explore its mode of action.

USING THE CONDENSER

The condenser aligns and focuses light on the specimen. It may be equipped with a high-power condenser lens (HPCL) (Figure 3.3). This is used with 10× to 100× objectives, but is removed from the light path with low-magnification objectives that are typically 4× to 5×. The position of the HPCL may be controlled by a rotating knob. In some microscopes, it is moved in and out of the light path by a push–pull plunger or a lever. Failure to use this lens properly is the most common mistake that people make. The lens is typically left out of the light path at low magnification because it limits the **field** of illumination. A fully illuminated field is achieved with the HPCL out of the light path with low-power objectives. There may not be a large penalty for examining commercial slides at higher magnification with the HPCL out, but there is a severe visual penalty at higher magnifications and with fresh mounts.

- Locate the HPCL on your microscope and determine how to move it into and out of the light path.

Condenser-Centering Screws

A pair of screws, set apart by a 45° angle, are used to center the condenser. These are located along the back

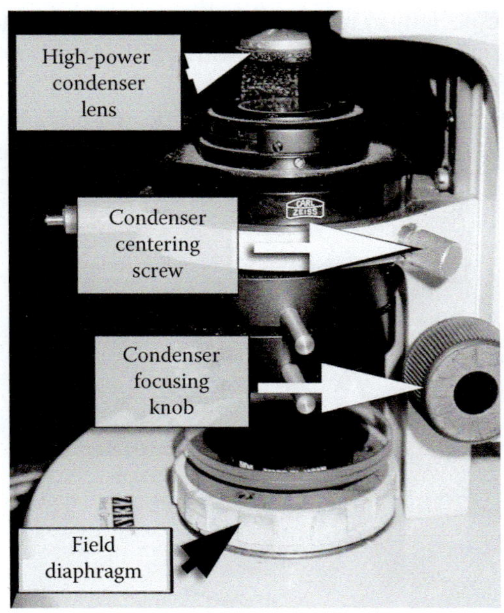

FIGURE 3.3 Side view of a Zeiss Standard microscope showing the high-power condenser lens, the condenser focusing screws, the condenser focusing knob as well as the field diaphragm. (Courtesy of D.T. Webb.)

right and left sides of the condenser on the Zeiss Standard (Figure 3.3). However, they may be found near the front of the condenser with other scopes.

- Locate the condenser-centering screws on your microscope.

The condenser is used to focus light onto the specimen from below. This is an extremely important, but poorly understood, function of the condenser. It is obvious that you must focus the objectives onto the specimen to see it clearly, but it is less obvious that you need to focus the condenser on the specimen so that it is properly illuminated. Imagine that the condenser is a magnifying glass and you want to start a fire. You need to move the magnifying glass to a position that produces the smallest focused beam of sunlight in order to start fire. That is exactly the same concept you need to keep in mind when you focus the condenser. This is done by rotating the condenser-focusing knob that is found on the right side of the condenser with the Zeiss Standard (Figure 3.3).

- Locate the condenser-focusing knob on your microscope.

Condenser (Aperture) Diaphragm

Finally, there is a lever that controls the aperture of the condenser or aperture diaphragm (Figure 3.4). I like to refer to this as the condenser diaphragm to prevent confusion with the field diaphragm. However, it has typically been called the aperture diaphragm. Partially closing this iris improves **contrast** (the difference between light and dark) especially at intermediate and high magnifications. It also increases the **depth of field**, which is very small at high magnifications.

Do NOT use this to increase or decrease brightness! This is the second most frequent mistake that people make. It is best to leave this completely open (rotated to the left on the Zeiss Standard) at the outset. Later, you will experiment with this to see its effects.

- Locate the condenser (aperture) diaphragm lever and manipulate it.
- Leave it in the open position for now.

USING THE FIELD DIAPHRAGM

The light source is usually housed in the base of the microscope. It passes through the field diaphragm that also contains an iris (Figure 3.5). The size of its iris diaphragm is controlled by rotating a knurled ring, which is concentric with it. The field diaphragm controls the area of illumination.

- Locate the field diaphragm on your microscope.
- Manipulate the knurled ring to open and close the iris inside.
- Leave it in the fully open position for now.

Using the Objective Lens

The magnification of an image is regulated by the objectives that are housed in a rotating nosepiece (Figure 3.6). To change objectives, rotate the nosepiece. Ensure that the low-power objective is in place before you start using your scope and when you are finished using it. This prevents damage to the objectives and your specimen.

- Always start viewing with the low-power objective (4× or 5×).
- Do not start by swinging the 10× to 100× objectives into position.
- Be sure that you rotate the nosepiece in the right direction. You do not have to switch from 4×

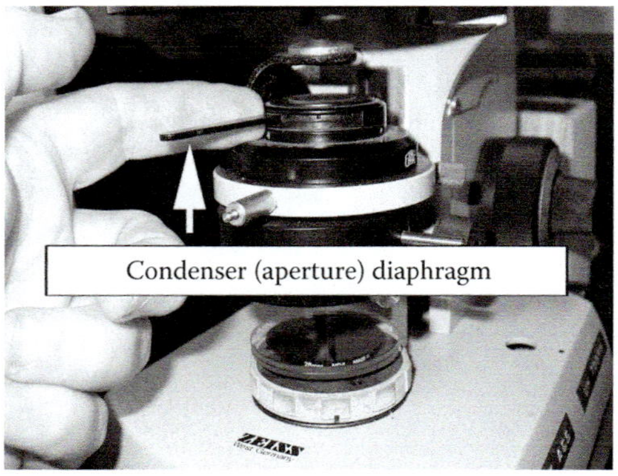

FIGURE 3.4 Lever that controls the condenser (aperture) diaphragm. It is completely open in this position. Rotating the lever to the right closes the iris diaphragm inside. (Courtesy of D.T. Webb.)

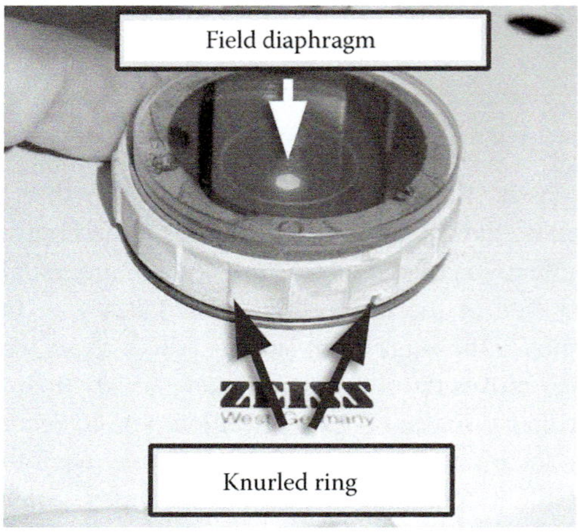

FIGURE 3.5 Field diaphragm that is almost completely closed. The knurled ring is used to open and close the iris inside. (Courtesy of D.T. Webb.)

FIGURE 3.6 Rotating nosepiece with objectives. In this case, the nosepiece is rotated clockwise (left to right) to change objectives to higher magnifications. (Courtesy of D.T. Webb.)

to 100× because objectives may be damaged if they hit the specimen.

- You should focus on the sample with each objective before switching to the next. Focus on the sample using the low-power objective and rotate to the next lens (10×), and refocus and repeat this until you reach the magnification you plan to use. It is vital to focus on your specimen with the 40× objective before switching to 100×.

Higher quality microscopes have objectives that are **parfocal**, the ability to focus with one objective and switch to the next one and the one after that without refocusing. However, this rarely occurs, especially with student scopes. Special care must be used with fresh sections and whole mounts, which can be thick and irregular. Consequently, greater care must be taken when changing objectives. When in doubt, play it safe.

A typical microscope will have a series of objectives like the following 5×, 10×, 20×, 40×, and 100×. The length of the objectives is a rough indication of their relative magnification. However, their magnifications are engraved on them.

- Check the magnification of your objectives.

Most oil immersion objectives have a black line near their tip (Figure 3.7). However, this is not true for all manufacturers. The words "Oil" or "Oel" indicate that it is an oil immersion objective. Oil improves the image because it unites the coverslip and the objective and replaces air with oil. Immersion oil has the same refractive index as glass. Thus, less light scattering and refraction occur. The oil also protects the objective lens from getting scratched.

The markings, homogenous immersion (HI) or multiple immersion (Imm), indicate oil or multiple medium immersion objectives, respectively. Water immersion objectives are marked with the words "water," "waser," "water immersion," or WI. Furthermore, an objective with a **numerical aperture** greater than 1.0 (Figure 3.7) is probably an immersion objective. Oil is the most

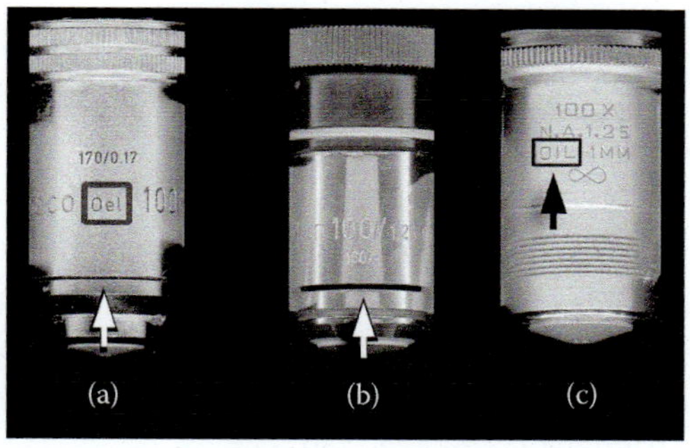

FIGURE 3.7 Oil immersion objectives from (a) Leitz, (b) Zeiss, and (c) American Optical (AO). Note the black lines (indicated by the white arrows) on (a) and (b), which indicate that they are oil immersion objectives. The AO objective (c) lacks this line. However, it has the word OIL inscribed on it as indicated by the black arrow and box. The word Oel (oil) is inscribed on the Leitz objective (a). (Courtesy of D.T. Webb.)

common immersion medium, but water and glycerin can also be used with Imm objectives.

- Check your 100× objective to see its markings.

The instructions for properly using oil immersion objectives can be found in Procedure 3.1.

For optimum results, oil should also be placed between the uppermost condenser lens and the bottom of the slide. However, the condenser needs to have a numerical aperture greater than 1.0 in order for the oil to have a beneficial effect. This may be impractical for routine studies, but should be used for critical examinations and for the most detailed microphotography. However, most oil immersion objectives require you to partially close the condenser aperture. This alleviates the need to add oil to the condenser lens because it effectively lowers the numerical aperture of the condenser, such that adding oil is no longer beneficial.

Using the Oculars

The oculars must be adjusted to suit both of your eyes. You should be able to adjust the interpupillary distance between the two objectives. This means that you can move them to match the distance between your eyes. Grasp the base of the ocular tubes or the plate at their base and gently spread them apart or draw them together. In some cases there is a dial that you can use to move the oculars. Either one or both oculars may be movable. There should be a scale and a reference line or dot that allows you to record the best spacing for your eyes. Thus, you can readily readjust the oculars to your personal setting if they have been moved. Follow the steps in Procedure 3.2 to adjust the interpupillary distance of your oculars.

Head position is very important. You need to find a comfortable distance for your eyes from the oculars

Procedure 3.1

Using an Oil Immersion Objective

Step	Instructions and Comments
1	Locate the area of interest in the center of the field and focus on this with the 40× objective.
2	Raise the objective to its upper limit. Swing the oil immersion objective into viewing position.
3	Place a drop of oil on the 100× oil immersion objective and place a small drop of oil on the coverslip. Do NOT look through the oculars.
4	While observing from the side of the stage with your eyes, focus on the objective and the specimen. Your eyes must be at the same height as the stage. Lower the objective lens carefully (use the coarse-focusing knob) until it just touches the oil on the coverslip. A light flash may be observed when the oil on the objective meets the oil on the slide.
5	Now look through the oculars and use the coarse-focusing knob to bring the specimen into rough focus. Use the fine-focusing knob to complete focusing on the sample.
6	Hereafter, avoid focusing down on the specimen with an oil immersion lens. Change the focus so that the objective is traveling away from the slide. If the image does not come into focus readily, carefully reverse the direction until it does. When in doubt, stop and ask for help. The lens might be dirty or there may be some other problem.
7	Important! Wipe the oil from the objective with lens paper when you are finished. Clean the objective until no more oil is visible on the lens paper. Wipe oil from the coverslip of the slide if it is to be saved.

Procedure 3.2

Adjusting the Interpupillary Distance of the Oculars

Step	Instructions and Comments
1	Position your head so that you can see through the oculars while focusing on a sample. Focus on a specimen using a 10× to 20× objective.
2	Grasp the base of the ocular tubes or use the dial located between the oculars.
3	Move the oculars so that they are as close together as possible. Carefully move the oculars apart until you see only one image.

in order to see things properly. This depends on the type and quality of your oculars and may require some experimentation. Most oculars can be used without eyeglasses that are corrected for nearsightedness or farsightedness. However, they are not compensated for astigmatism. If you have astigmatism, you need to use your eyeglasses or contact lenses. Oculars with eyeglasses engraved on them are suitable for use with glasses (Figure 3.8), but you are not required to wear glasses to use these.

It is important that each ocular is in focus for your eyes when you examine samples. Oculars that are capable of independent focusing will have a scale, a reference line, and a knurled ring on them (Figure 3.8). These markings may be on the ocular tube rather than on the oculars themselves. Follow the instructions in Procedure 3.3 to focus your oculars.

Koehler Illumination

The best resolution occurs when all the elements of the microscope are in perfect alignment and the iris diaphragms are properly adjusted to the best apertures for the objectives you are using. On simple microscopes, you may not be able to focus and align the condenser, but on the Zeiss Standard and many other microscopes, it is possible to do this and achieve "Koehler illumination." This makes a significant difference for viewing unstained and lightly stained samples, especially at high magnifications.

Centering the Lamp Filament

The first step in this process involves centering the lamp filament. This may not be possible with your microscope.

Procedure 3.3
Focusing Oculars for Your Eyes

Step	Instructions and Comments
1	Before you make any adjustments, place a slide on the stage and focus on the central part of the specimen with a 10× to 20× objective.
2	Block one of the oculars. Look through the other ocular with your matching eye (left eye → left ocular or right eye → right ocular) and focus on a fine detail in the center of the specimen with the objective focusing knobs at the rear of the scope.
3	Switch to the other ocular and look through it with the matching eye. Do NOT look through the other ocular while you are doing this!
4	Rotate the knurled ring of the ocular to bring the fine detail into sharp focus. You will need to stabilize this with one hand while you turn it with the other.
5	Check the first ocular to see that the image is still in focus with your other eye.
6	Both the oculars are now focused for your eyes.

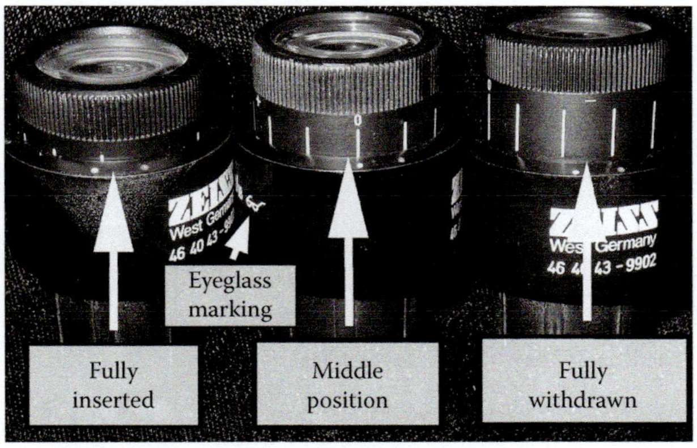

FIGURE 3.8 Zeiss oculars showing three focusing positions. These are focused by grasping the knurled ring at the top of the ocular and rotating it while holding onto the barrel below. The one on the left has been adjusted so that the rotating part of the ocular is fully inserted into the barrel. The one on the right shows the extreme opposite rotation. The one in the center is approximately in the middle. In this case the position of the ocular is indicated by the length of the white lines. In other cases there are numbers that can be used to designate the best focusing position for the ocular. Note the eyeglasses engraved on the middle objective. This indicates that this ocular was designed to be used with your glasses. (Courtesy of D.T. Webb.)

Furthermore, it is best done by someone who is very familiar with this process. Check the illuminator housing at the back of your microscope to see if there are any adjustable screws. If not, you cannot do this. A generic description of this is as follows:

- Turn on the microscope illuminator.
- Place a piece of paper over the field diaphragm.
- If the illumination is uneven, use the lamp-centering screws or rotate the lamp to get uniform illumination, or do both.

Focusing on the Specimen for Koehler Illumination

The recommended procedure for focusing on a specimen as part of achieving Koehler illumination is given in Procedure 3.4. However, when you decide to proceed, it is very important to focus on a specimen before doing anything else. After you have focused on it, you may have to move the specimen out of the light path the first few times you do this.

Focusing and Centering the Field Diaphragm

This is the heart of Koehler illumination. See Procedure 3.5 for the detailed steps in this process. Briefly, completely close the field diaphragm and use the condenser focusing knob to focus the field diaphragm until you see that it is a small polygon of light (Figure 3.9a). Use the condenser centering screws to center the image of the field diaphragm (Figure 3.9b). Partially open the field diaphragm and center it again (Figure 3.9c). Open the field diaphragm until the field is completely illuminated and stop.

Adjusting the Condenser (Aperture) Diaphragm

When working with 10× to 100× objectives, it is important to adjust the condenser diaphragm (Figure 3.4). This is especially true for translucent structures. Closing this iris increases contrast. Thus, something indistinct becomes sharp and something faint becomes dark. It also improves the depth of field, which is critically small at high magnifications. It is usually possible to close the iris and judge its effects subjectively (Delly, 1988). However, there is a "tried and true" procedure (Procedure 3.6) that you should know.

In practice, you can experiment with this while viewing a specimen and adjust it without removing the ocular. This is what I do when I want to take photos. I slowly close

Procedure 3.4

Focusing on a Specimen as Part of Koehler Illumination

Step	Instructions and Comments
1	Adjust the oculars so that they have the correct interpupillary position and are in focus for your eyes.
2	Use a commercial slide with obvious well-stained contents and move the specimen into the light path. Focus on the specimen with your low-power objective. Now, rotate the 10× objective into viewing position.
3	Watching from the side of the stage (not looking through the oculars), lower the 10× objective so that it comes closer to the coverslip.
4	Look through the oculars and rotate the objective focusing knobs so that the objective is retracted from the slide. Make a note on which direction the objectives retract.
5	Stop when the sample comes into focus. Moving the mechanical stage during this process may help.

Procedure 3.5

Focusing and Centering the Condenser

Step	Instructions and Comments
1	Use the 10× objective to focus on the center of a specimen. Reduce the illumination to a moderate level so that you do not hurt your eyes.
2	Check to see that the condenser (aperture) diaphragm is open, and ensure that the HPCL is in the light path.
3	Close the field diaphragm so that the circle of light becomes smaller. Observe the field diaphragm through the oculars when it is being closed (Figure 3.9). When the diaphragm is as small as possible, use the condenser focusing knob (Figure 3.3) to make the "circle" of light as small as possible (Figure 3.9).
4	You should see that the field diaphragm is NOT circular in outline, but has a polygonal shape (Figure 3.9). You may see a red or blue fringe as you bring the field diaphragm into focus. The best position is the one in between the red and blue fringes.
5	Use the condenser-centering screws to center the field diaphragm. Open the field diaphragm by rotating its knurled ring. It may not be perfectly centered. Perform final centering of the field diaphragm when it fills most of the field (Figure 3.10). Expand the field diaphragm just beyond the field of view and stop!
6	Repeat this with the 20× and 40× objectives. For critical work, this should be done for each objective. This is especially important for taking photographs and for examining minute, translucent specimens like fungi and bacteria. This is difficult to do with the 100× objective. However, if you achieve proper alignment with the 40× objective, the 100× will be similar.

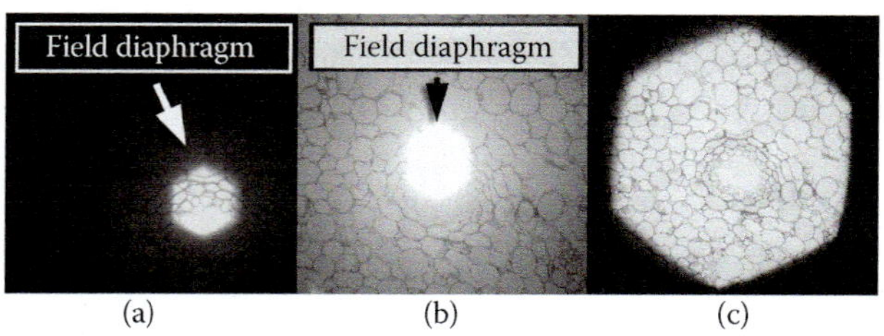

FIGURE 3.9 (a) Focused and uncentered field diaphragm, (b) closed and centered field diaphragm, and (c) partly open and centered field diaphragm. (Courtesy of D.T. Webb.)

this iris until I first see a perceptible change in the specimen and take a photograph. I close it some more and take another photo and repeat this for a third time. In reality, each specimen is different, and strict rules like those in Procedure 3.6 may not give the best results. Closing the condenser diaphragm also increases the depth of field. Thus, more regions of a three-dimensional specimen will be in focus. However, if it is closed too much, a flat indistinct image results.

The examples in Figure 3.10 show how the condenser iris increases contrast and depth of field. It shows

Procedure 3.6

Adjusting the Condenser (Aperture) Diaphragm

Step	Instructions and Comments
1	Complete all of the preceding operations (Procedures 3.1 through 3.5).
2	Place a lightly stained specimen in the light path. Focus on the specimen using a 20× to 40× objective lens.
3	Remove one of the oculars and look directly down the tube at the light field. Close the condenser diaphragm so that it occludes 1/4 to 1/3 of the area. This should give the best contrast.
4	Examine a specimen before and after adjusting this iris.
5	This should be done for each objective for critical viewing.

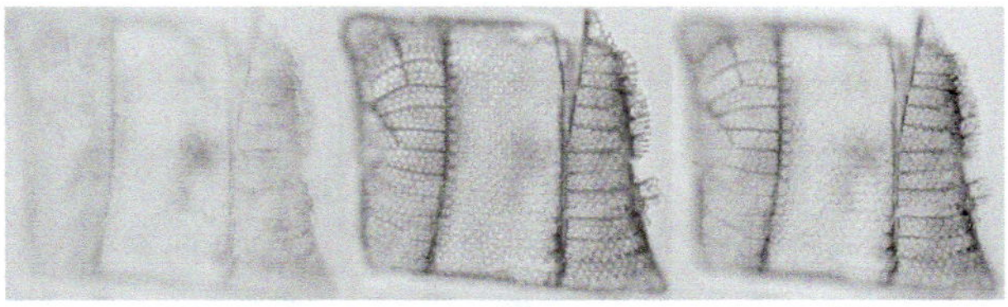

(a) (b) (c)

FIGURE 3.10 View of a diatom with condenser (aperture) diaphragm (a) completely open, (b) completely closed, and (c) partially closed. In (a), it is hard to see any detail and most of the subject is out of focus due to the shallow depth of field. In (b), more details are visible due to increased contrast, and there is a greater depth of field in (b) compared with (a). In (c), more details are visible, and there is a greater depth of field in (c) compared with (a). In this case, (c) should have been the best. However, (b) appears to be the best. (Courtesy of D.T. Webb.)

a diatom frustule that is very translucent. There is little detail when the condenser diaphragm is wide open (Figure 3.10a). When it is fully closed (Figure 3.10b), the contrast and depth of field are greatly increased. When the iris is closed 25%–30%, there is an improved contrast and depth of field with less theoretical potential for aberrations (Figure 3.10c). In this case Figure 3.10c should have been the best image, but Figure 3.10b appears to be the best.

- Experiment with the condenser diaphragm while viewing a lightly stained or unstained slide.
- Once you have achieved what you think gives the best image, remove one of the oculars and see how much of the field is occluded.

Throughout your career, you will be using different stains to study their effects on fresh specimens. Experiment with the aperture diaphragm as you study these. The condenser diaphragm can be used to great effect with this type of material. Although some of these procedures seem to be tedious, they will become routine as you progress in your work. Your results will be superior to others who do not know how to do this.

Simple Measurements with a Compound Microscope

In most cases, you cannot accurately determine the magnification of a compound microscope by multiplying the magnifications of the ocular and the objective lenses. This is a very common misconception. Microscope parts are not manufactured that precisely. Furthermore, the length of the microscope tube/body differs from one type of scope to another. This is especially true in photography because projection lenses of different magnifications are used in place of oculars and the total distance of the light path is different from that used with the oculars. We will work through Procedure 3.7 for calibrating an **ocular micrometer** that can be used to make direct measurements during observations.

Before we proceed, a quick review of the metric system will be helpful. A millimeter (mm) = 10^{-3} m, whereas a micron (μm) = 10^{-6} m. Consequently, 1 mm = 1000 μm, 0.1 mm = 100 μm, and 0.01 mm = 10 μm. A stage micrometer (Figure 3.11) is used to precisely determine the magnification.

Calibrating an Ocular Micrometer

The **stage micrometer** is the "known" micrometer in this process. It has finely etched distance calibrations on its surface. The largest dimensions from one end to the other are millimeters (Figure 3.11a and b). Each millimeter (1000 μm) is divided into 0.1-mm (100-μm) segments (Figure 3.11b). Each 0.1-mm segment is divided into 0.01-mm (10-μm) segments.

Ocular micrometers have precisely etched lines engraved on them. However, because of the differences in the optics of individual microscopes, they must be calibrated with a stage micrometer. Briefly, the stage micrometer and the ocular micrometer are brought together under the microscope at 10× so that

Procedure 3.7
Calibrating an Ocular Micrometer

Step	Instructions and Comments
1	Place the stage micrometer onto the stage of your microscope and move it into the light path. Focus on its scale with the 4× objective. Switch to the 10× objective.
2	Move the stage micrometer so that some of the reference lines on it coincide with reference lines on the ocular micrometer (Figure 3.12).
3	Because the distances between the lines of the stage micrometers are known, you divide this known distance by the number of lines from the ocular micrometer. This gives you the distance measured by the intervals of the ocular micrometer at that magnification.
4	Repeat this for each objective lens on your microscope.

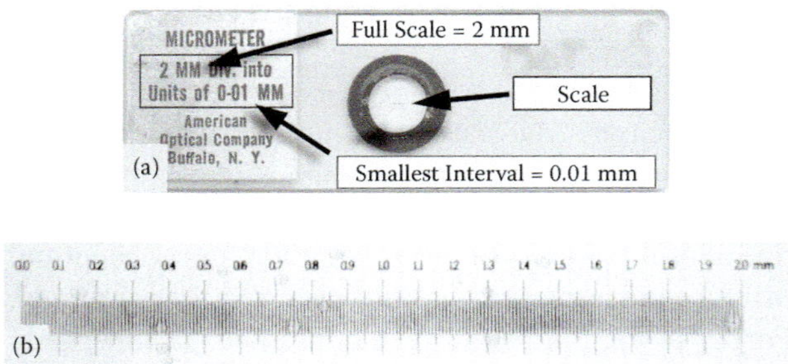

FIGURE 3.11 (a) Stage micrometer and (b) magnified scale of the stage micrometer. The total length of the scale is 2 mm (2000 μm). It is divided into 20 intervals of 0.1 mm or 100 μm (b). Each of these is further divided into intervals of 0.01 mm or 10 μm. (Courtesy of D.T. Webb.)

the intervals of the optical micrometer (unknown) are matched with the intervals of the stage micrometer (known). The actual distance between the units of the ocular micrometer is determined by dividing the known distance (stage micrometer) by the unknown units (ocular micrometer). This gives the actual distance for each unit on the ocular micrometer (Figure 3.12). In the example provided, two large units on the ocular micrometer equal 270 μm, consequently one large unit on the ocular micrometer equals 135 μm and each small unit equals 13.5 μm. Record this value and repeat it for all of the objectives on your scope. If you transfer your ocular micrometer to another microscope, you must repeat this process for that scope. The ocular micrometer can now be used to precisely record the diameter of fungal hyphae in microns.

Typical Stains for Commercial Slides of Plants

Commercial slides are typically stained with Safranin O and Fast Green. Safranin O appears as brilliant red in chromosomes, nuclei, and lignified, suberized, and/or cutinized cell walls. Fast Green appears as brilliant green in cytoplasm and nonlignified cellulosic cell walls. Fast

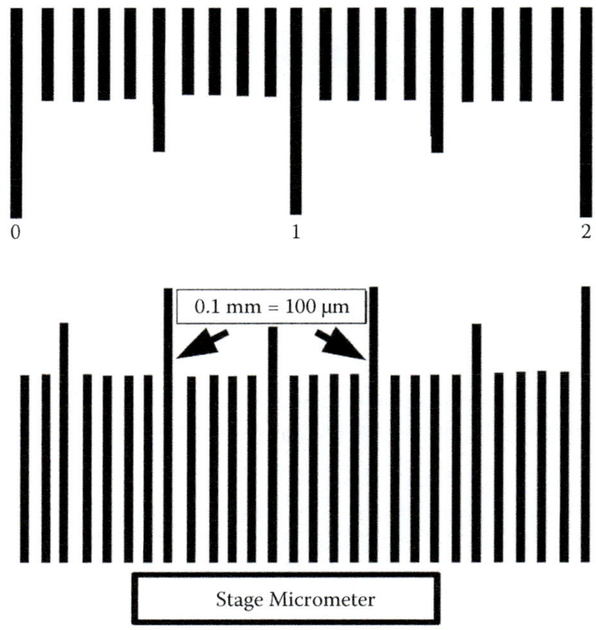

FIGURE 3.12 Diagram of a stage micrometer and ocular micrometer. In this case 2 large units on the ocular micrometer = 270 μm. One small unit on the ocular micrometer = 13.5 μm. (Courtesy of D.T. Webb.)

Green turns blue in basic solutions and may appear blue to bluish-green in the stems and leaves of aquatic plants and most gymnosperms (Johansen, 1940; Ruzin, 1999).

Fresh Sections of Plant Organs

There are several ways to generate fresh sections of plant organs. These include hand sectioning, use of a hand **microtome**, and use of an inexpensive sliding microtome or a traditional sliding microtome. Freezing microtome sections are nearly equivalent to fresh sections in many cases. This is a topic unto itself. However, I will review the process of hand sectioning as this is a quick way to produce useful sections. The details are given in Procedure 3.8.

Procedure 3.8
Making Hand Sections

Step	Instructions and Comments
1	Place a Band-Aid or adhesive tape on the thumb of your left hand. Position the cotton portion at the bottom of your thumb. The thumb is the backstop for this operation. Place another Band-Aid or adhesive tape at the end of your index finger. The index finger will control the height of the specimen and thus its thickness. Grasp the plant structure between your thumb and forefinger so that the top of the specimen extends above the level of your forefinger.
2	Take a single-edge razor blade in your right hand. Ensure that it is wet. Rest the blade on your forefinger and use a slicing motion to cut off the top of the specimen. This is a thick section that is designed to produce a flat surface. Slice away from your thumb in order to avoid cutting your thumb with the blade.
3	Raise the specimen slightly by manipulating it with your fingers and repeat the slicing motion. It is best to make a lot of quick slices rather than a few, slow and careful sections.
4	Thin sections can often be obtained by pressing the blade down with your forefinger and then slicing through the specimen several times. After several sections have been accumulated on the blade, wash them off in a Petri dish of water. Continue slicing until you have several thin sections. These will appear translucent when seen against a dark background. In most cases, the sections will have thin and thick regions. As long as part of the section is thin, you may be able to use it, and thick sections are frequently good for gross anatomy.
5	Sections can be picked up with a forceps or a wet artist's brush. Place these in a drop of water or stain on a microscope slide. Sections will be released from the brush if you rotate it in the water on the slide. The brush is good for delicate specimens.
6	It is a good idea to view unstained sections prior to staining. Proper use of the condenser (aperture) diaphragm is important for viewing unstained samples.

The most important element in this process is the blade that you use. Teflon-coated, stainless steel, safety razor blades or injector blades work fairly well. If you use a double-edged blade, be sure to put tape over one edge to prevent cutting your fingers. Use tape on your fingers as well. Single-edge utility blades give reasonable results, but they lose their edge quickly. Dip the blade in water before you use it. Also, wet the surface of the specimen. Water acts as a lubricant and helps the sections stay on the blade. You should slice quickly rather than try and make one perfect section by cutting slowly.

The ability to make fresh sections allows you to quickly analyze plant organs without resorting to laborious and hazardous procedures. A tremendous amount of information can be derived from these. They have natural colors because they have not been extracted with organic or caustic chemicals. This can be very important in plant pathology if you want to know the exact symptoms or signs of a disease. Sections need not be extremely thin to be of use. In addition, they do not need to be complete or uniform. They provide three-dimensional information, which is not available with extremely thin sections. Your initial attempts at hand sectioning will be frustrating; however, you will become proficient.

Procedures 3.8 through 3.10 work well for stems, petioles, and roots. Leaves are more difficult to section because they are very flexible. One way to overcome this is to make a sandwich of three (5 × 10 mm) pieces of leaf blade attached to the midrib. A reasonable cross-section can be obtained by pressing the leaf pieces together and cutting across the leaf pieces. Another way to do this is to use artificial cork as a support medium. This is far superior to real cork. Hold a 5 × 10 mm leaf piece between two layers of artificial cork to make cross sections. You may want to trim the excess "cork" away from the end so that you do not need to cut through too much of it. Real cork works but is more difficult to use. You can purchase commercial pith, but it is expensive and delicate. Fresh carrot can be used, but there is much debris from it, and it is rather slippery.

Adding Coverslips to Wet Mounts

It is best to use 22 × 40 mm rather than 22 × 22 mm coverslips with wet mounts. It is essential that air bubbles be avoided when adding coverslips to wet mounts. These will interfere with your observations. To avoid air bubbles, follow the steps outlined in Procedure 3.9. Basically, use forceps or a dissecting needle to support one end of the coverslip while it is slowly lowered onto the wet mount. This allows the specimen to be covered by any solution without the formation of large air bubbles. Be sure to wipe excess fluid from the bottom of the slide. Otherwise, the slide will stick to the stage and prevent its transport. Furthermore, excess fluids can damage the stage.

Procedure 3.9
Proper Method for Adding a Coverslip to a Wet Mount

Step	Instructions and Comments
1	Place your samples in 2–3 drops of water or stain in the center of the slide.
2	Use a large (20 × 40 mm) coverslip. Place one end of the coverslip on the slide at a 45° angle without touching the solution containing the specimens. Steady this end with your thumb and index finger.
3	Grasp the other end of the coverslip with fine forceps. Alternatively rest it on a dissecting needle. Slowly lower the forceps or needle until the coverslip touches the solution. Continue until the forceps or needle touches the slide. Release your grip on the forceps. Slowly remove the forceps or needle by sliding them along the slide.
4	Remove excess solution by touching the side of a Kimwipe or paper towel near one of the coverslip edges. Be careful to not sponge out your samples with the excess solution. Slowly remove the Kimwipe so that you do not drag the coverslip over the slide.
5	If you have been using a stain that must be removed, add water to one end of the coverslip. Withdraw the stain at the opposite end by blotting with a Kimwipe or paper towel.
6	Wipe excess fluid from the bottom of the slide or else it will stick on the stage and make slide transport difficult. Excess fluids may damage the stage or other microscope parts.
7	Carefully place the slide into the slide holder on your stage.

Stains for Fresh Sections

It is important to examine unstained samples prior to staining. Plant parts often have natural pigmentation that may not be obvious. This may lead to a misinterpretation of staining reactions. Furthermore, staining often masks natural pigmentation. Consequently, you are losing data if you do not examine the unstained sections.

Safety

All stains have some safety risk associated with them. Wear nitrile gloves in the lab to prevent staining your hands. Never touch your eyes and always wash your hands as soon as possible after using these stains and after any lab in which chemicals have been used.

Stains

Phloroglucinol

Phloroglucinol dissolved in 20% HCl stains lignified, cutinized, and suberized walls a red to pink to an orange color (O'Brien and McCully 1981; Berlyn and Miksche 1976). It is good for staining lignified cells, like sclereids, fibers, and xylem tracheary elements. It also stains the epidermal cuticle. The stain is very easy to use.

Place the sections in the stain, apply a coverslip, and wait for few minutes to observe. You do not need to remove the stain, but you can by applying water to one edge of the coverslip and drawing the stain out the other side with a Kimwipe or paper towel. Because it is dissolved in 20% HCl, it can damage the microscope or your clothes. Be careful to remove the excess stain from your slides and work areas.

IKI

IKI stands for Iodide (I) and Potassium Iodide (KI). It stains starch a blue-black to brown color (O'Brien and McCully, 1981; Berlyn and Miksche, 1976). It can be used exactly as described for Phloroglucinol. It also imparts a golden color to cell walls and nuclei, although this incidental staining is not specific for any substance in them.

Toluidine Blue O

Toluidine Blue O stains lignified walls a blue to blue-green color and pectin-rich walls pink (O'Brien and McCully, 1981; see Procedure 3.10 for details). Some walls, especially those of the phloem, may not stain at all. It is a fast-acting stain, and hence, overstraining may destroy its specificity. You need to act quickly with this stain. It is good to compare results from Toluidine Blue O with those from Phloroglucinol.

Polarizing Filters

A **polarizing filter** causes light to vibrate in one plane and thus produces "plane-polarized light." Light traveling from a source vibrates in all possible planes. Imagine many radii emanating from a common center. These would represent the vibrational planes of the light beam. A polarizer cuts out all but one of these. One can think of polarizers as combs. A comb straightens tangled hair

Procedure 3.10
Staining with Toluidine Blue O

Step	Instructions and Comments
1	Use nitrile gloves to protect your hands from the stain. Add several sections to a drop of water on a slide. Now add 2–3 drops of toluidine blue O to the water. Quickly add a coverslip as in Procedure 3.9. Remove the excess stain by blotting with a Kimwipe. Wipe excess fluid from the bottom of the slide (Procedure 3.9). View right away because this stain fades over the time span of the lab.
2	*Caution*: It is difficult to remove toluidine blue O from clothing; therefore, use it carefully and clean any spills with water. In addition, it is mildly poisonous, so avoid getting it on your skin as much as possible. Be sure to wash your hands well after using the stain.

so that the strands are parallel to one another. Polarizing filters "comb" light so that only one plane passes through.

If two polarizers are parallel to each other, light will pass through because the plane-polarized light that passes through the first filter is parallel to the "teeth" in the second comb. However, if two polarizers are crossed at a 90° angle, no light passes through the second polarizer because the first polarizer eliminates all light that vibrates parallel to the "teeth" in the second polarizer. You can verify this by holding one polarizer while looking at a light source. Take a second polarizer in your other hand and superimpose it on the first. Turn either one until the light is completely blocked. You can do this with polarized sunglasses.

If a crystalline object is placed between crossed polarizers, it will depolarize the light that passes through the first polarizer (Berlyn and Miksche, 1976). This property is known as **birefringence**. The birefringent material will create light that vibrates in the same plane as the second polarizer, and it will be visible while all else will be dark. Cell walls, crystals, and starch grains are birefringent and become apparent using crossed polarizers. This works with unstained and some stained sections. However, staining with IKI may destroy the birefringent properties of starch grains.

Inexpensive polarizing filters can be purchased from ScientificsOnline.com, and also polarizing filters designed for cameras can be used. Use neutral gray polarizing filters. Circular polarizers work better than linear polarizers for microscopy.

- Place one polarizer over the field diaphragm.
- Place another over your ocular or wear polarized sunglasses.
- Focus on your sample.
- Rotate the polarizer over the field diaphragm.

Thick cell walls, starch grains, or crystals will become bright while the background becomes dark. Intermediate effects are also possible and can reveal subtle features that are not visible otherwise. If you want to photograph your results, you will need to place one polarizer between the specimen and your camera lens.

Dissecting or Stereo Microscopes

Dissecting microscopes are also called stereo microscopes because they contain two separate light paths that travel to separate oculars. The specimen is seen from two different angles. This results in three-dimensional images. This is a vital feature for viewing and dissecting three-dimensional subjects. Early dissecting scopes consisted of two monocular scopes bound together. Compound scopes visualize only one light path that goes to both of the oculars. This produces a two-dimensional image. Dissecting scopes have many similarities to compound microscopes (Figures 3.1 and 3.2). The basic parts of a dissecting scope are as follows: base, stage, arm, focus knob, body, magnification knob, and oculars.

- Locate the major parts of the dissecting scope by referring to Figure 3.2.

Some dissecting scopes have a transillumination base. There is a mirror in the base that can direct light through the specimen. This is used to examine translucent specimens. It can provide an overview of a large translucent sample that cannot be seen with a compound scope. The mirror can be rotated using a knob in the base. This yields various angles of illumination. The mirror usually has a white opaque back that can provide diffuse reflected light.

In most cases, epiillumination is used. There may be a "built-in" illuminator or a slot for placing an epiillumination light source above and behind the microscope body (Figure 3.13). This avoids creating shadows as the sample is manipulated on the stage. There may be a way to vary the illumination angle. The American Optical Stereo Zoom scope has two contiguous slots at different angles. In many cases, you will use a separate epiilluminator or a pair of them that can be positioned around the

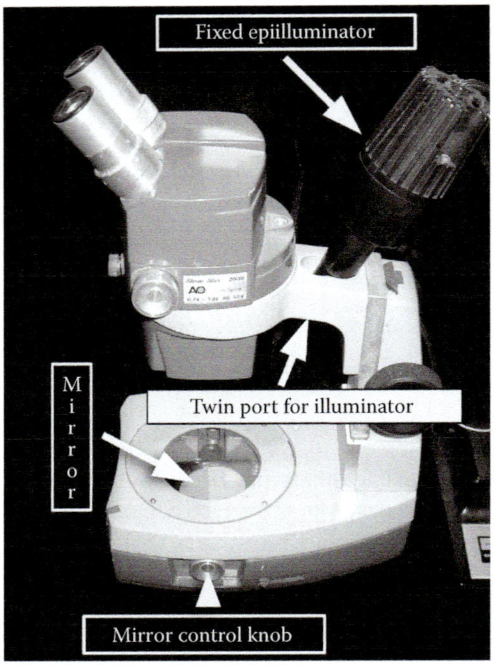

FIGURE 3.13 American Optical stereo zoom microscope with transillumination base showing the mirror and its control knob plus an illuminator placed in one of the twin ports that are designed to provide two angles for epiillumination from behind the body of the microscope. (Courtesy of D.T. Webb.)

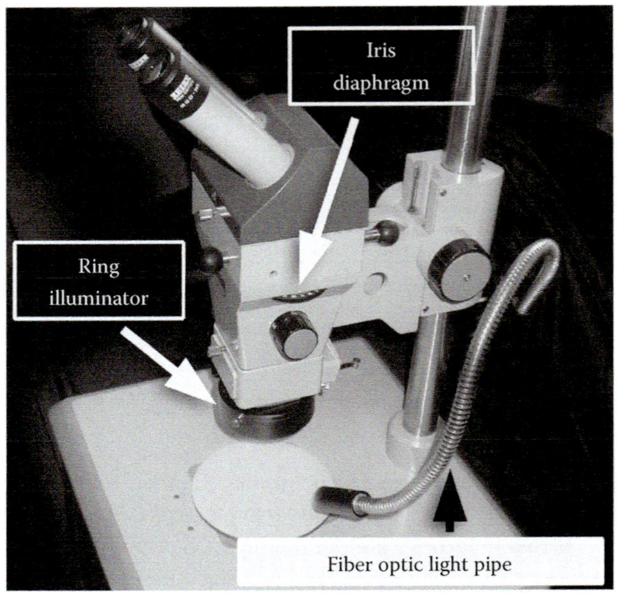

FIGURE 3.14 Zeiss stereo microscope with two types of illuminators. One is a ring illuminator, which provides shadow-free light. The other is a fiber optic illuminator that has two flexible light pipes. These can be adjusted to provide light from various angles. This scope also has an iris diaphragm, which increases the depth of field. This is especially useful for photography. (Courtesy of D.T. Webb.)

scope on your lab bench. There is an adjustable arm that can be used to vary the direction of light. This provides considerable flexibility. Fiber optic illuminators may have one or two flexible light guides that can be bent to provide many angles of illumination (Figure 3.14).

- Identify the types of illuminators available for your scopes.
- Explore their utility with various types of samples.

Uniform, shadow-free illumination can be obtained from ring illuminators mounted just below the objectives (Figure 3.14). Fluorescent ring illuminators are inexpensive and do not produce damaging heat. Light-emitting diode (LED) ring illuminators can produce more light than fluorescent illuminators and are relatively "heat-free." Fiber optic ring illuminators are available, but are significantly more expensive than other light sources. They also produce "cool" light because the heat is dissipated by the transformer box. Shadow-free illuminators are extremely useful for most situations, but the ability to produce shadows in a controlled fashion can be helpful in obtaining three-dimensional relief from your specimen.

The stage has either a removable plate that is translucent or a plate that is white on one side and black on the other. This allows you to vary the background. Stage clips can usually be attached to the base. These are used for samples mounted on glass slides, but can secure some other large samples. There are mechanical stages for stereo scopes. These work like the same devices on compound scopes and are useful when fine adjustments are necessary. There are also clever devices that allow you to tilt your samples. Modeling clay or plasticine can be used to stabilize small samples.

There is typically one focusing knob on each side of the arm, but generally there is no fine-focusing knob. Rotating the focusing knob moves the microscope body up and down. Objectives are located near the base of the body. Magnification is controlled by a knob that is located either on the side or at the top of the body. Older and less expensive microscopes (Figure 3.14) contain a set of fixed objectives that are rotated into the light path by turning the knob. Most modern dissecting scopes have zoom objectives that can achieve continuous magnification over a range, which is typically 1× to 3×. Auxiliary objective lenses can be fitted over the "built in" objectives much like camera filters. These can increase or decrease the magnification.

Some stereo scopes have an iris diaphragm inside the body (Figure 3.14). There is a dial or knurled ring that is used to control the opening of the iris. Closing this increases the depth of field and is useful for photography. It will diminish the amount of light that gets through but can be useful during some dissections. This is usually not present on inexpensive microscopes.

The oculars are inserted into tubes that are attached to the body. These are similar to oculars used on a compound microscope. They must be adjustable for your eyes by following the steps in Procedures 3.2. and 3.3.

There may be an ocular micrometer in one of the oculars. To calibrate this, you may be able to use the same stage micrometer that is used to calibrate a compound scope. Otherwise, use an extremely accurate ruler like the one that can be obtained from Ted Pella Inc. (Product # 54480). Similar calibration aids may be available, but when in doubt, use the best ruler you can obtain.

SUMMARY

In this chapter, we documented the proper way to use compound and stereo microscopes. We emphasized the compound microscope that is more complex in design and more complicated to use. The process of achieving Koehler illumination is presented along with the proper use of the condenser and oculars. Instructions for the calibration of an ocular micrometer with a stage micrometer are provided. The production of fresh sections and the interpretation of various plant stains and polarizing filters are discussed. Various ways to illuminate specimens for stereo microscopes are reviewed.

REFERENCES

Berlyn, G. P., and J. P. Miksche. 1976. *Botanical Microtechnique and Cytochemistry*. Ames, IA: Iowa State University Press.

Delly, J. G. 1988. *Photography through the Microscope*. Rochester, NY: The Eastman Kodak Company.

Johansen, D. A. 1940. *Plant Microtechnique*. New York, NY: McGraw-Hill.

Köhler, A. 1893. A new system of illumination for photomicrographic purposes. In: *Royal Microscopical Society*. Oxford, UK: Koehler Illumination Centenary (1994), pp. 1–5.

O'Brien, T. P., and M. E. McCully. 1981. *The Study of Plant Structure Principles and Selected Methods*. Melbourne, Australia: Termarcarphi Pty. Ltd.

Ruzin, S. E. 1999. *Plant Microtechnique and Microscopy*. New York, NY: Oxford University Press.

ONLINE RESOURCES

A-Z Microscope Glossary of Microscope Terms (http://www.az-microscope.on.ca/glossary-of-terms.html).

Microscopy Primer (http://www.microscopy-uk.org.uk/index.html).

Molecular Expressions Optical Microscopy Primer (http://micro.magnet.fsu.edu/primer/index.html).

Nikon Microscopy U (http://www.microscopyu.com/).

Part II

Groups of Plant Pathogens and Abiotic Disorders

4 Plant Pathogenic Viruses

Marie A.C. Langham and Judith K. Brown

CONCEPT BOX

- Viruses are unique, submicroscopic, and obligate pathogens.
- Viruses are composed of RNA or DNA genomes surrounded by a protein coat (capsid).
- Plant viruses replicate through the assembly of previously formed components, and replication is not separated from the cellular contents by a membrane.
- Plant virus species are named for the host from which they are originally associated and the major symptom that they cause. Virus species are grouped into genera and families.
- Plant viruses are vectored by insects, mites, nematodes, parasitic seed plants, fungi, seed, and pollen.
- Plant viruses can be detected and identified by biological, physical, protein, and nucleic acid properties.

Plant viruses are deceptively simple in their structure and incredibly small. However, this simplicity leads to an intensely interactive relationship and dependency between virus, host, and vector. Comprehending the mechanisms unique to this relationship is vital to understand plant viral diseases. During the past 30 years, our understanding of plant viruses, how they function, and why they cause disease has exceeded the limits imagined by early virologists. Today, new plant viruses are identified rapidly and our awareness of the pathological impact caused by known and newly emerging viruses continues to increase. This impact affects producers and consumers most clearly through economic losses caused by the reduction of yield or quality, adversely affecting plant growth and reproduction, death of host tissues and plants, sterility, crop failure, increased susceptibility to other stresses, loss of aesthetic value, quarantine or eradication of infected plants, and the cost of control and detection programs (Waterworth and Hadidi, 1998).

There is a growing evidence that many viruses exist in the ecosystem that aid in balancing the biodiversity and nutrient recycling, as well as providing protection against invasion of the host by an exotic, possibly more damaging virus. This concept, when applied to most viruses, including plant virology, is still poorly understood, but present and future research that incorporates deep sequencing and community ecological considerations to advance our understanding is expected to create many new and perhaps even surprising principles about virus pathogenicity, virus evolution, and host–parasite interactions. The intricate relationship of virus, vector, and plant host complicates strategies for virus control in cultivated plants and for decreasing the losses that plant viruses cause. Plant viral disease management programs depend on our understanding of the virus–host relationship, and viral disease control remains one of the greatest challenges for the future of plant virology.

WHAT IS A PLANT VIRUS?

Plant viruses are a unique kind of organism that contains, minimally, a nucleic acid genome and a coat protein. They rely on living cellular organisms to contribute diverse biochemical resources (nucleic acids and proteins, and possibly lipids, sugars, and energy) that are needed for multiplication, systemic infection of the host, and departure from that host to infect a new one, deriving from the host for nearly all of their "life-sustaining" functions. The word virus is taken from a Latin word meaning "poison." They are parasitic on their hosts for energy because they cannot produce or store energy in the form of adenosine triphosphate (ATP).

Plant viruses are a diverse group among viruses that infect hosts from unicellular plants to trees. Despite this diversity, plant viruses share a number of characteristics. A good definition of plant viruses focuses on characteristics that all plant viruses have in common. Some plant viruses also infect hosts other than plants, including fungi, insects, and possibly mites and nematodes. These cross-kingdom affiliations are thought to come about through virus "host shifts" in which the virus evolves the ability to enter and/or multiply and systemically infect previously immune species. It is much more difficult to decipher the exact origin of viruses before they specialized in plants.

Genomes and Genome Diversity

Viruses have nucleic acid **genomes**, which may be either ribonucleic acid (RNA) or deoxyribonucleic acid (DNA). No plant virus has been discovered that includes both the types of nucleic acids. However, there are many variations in the structures of the viral genomes. The nucleic acid may be single-stranded (ss) or double-stranded (ds), and it may be linear or circular. The genome may be on a single piece of nucleic acid (**unipartite genome**) or on multiple pieces (**multipartite genome**). These pieces may be **encapsidated** in a single particle or in multiple particles.

There is more biological diversity within viruses than in all the rest of the bacterial, plant, and animal kingdoms combined. This is due to the ability to coevolve with all kinds of living organisms. There are examples of viruses that parasitize all known groups of living organisms. Understanding the ways in which viruses are diverse and the extent of the diversity within and between different viruses are key to comprehend the interactions of viruses with their hosts. Diversification in viruses occurs through multiple mechanisms, including mutation, recombination, and reassortment. The mutation rate is not the same for all kinds of viruses (Figure 4.1), presumably because they coevolve with the host, which also differs with respect to diversification. In addition, each kind of virus has an effective genome size that influences the rates of change that are allowable to maintain genomic stability.

Capsids

Viruses have one or more protein coats or **capsids** surrounding their perimeter. These capsid layers are composed of protein subunits. The subunits may be composed of the same or different types of protein. For example, *Tobamoviruses* have one type of protein subunit in their capsids, whereas *Comoviruses* have two types of protein subunits, and *Phytoreoviruses* have six to seven types of structural protein subunits in their capsids (International Committee on Taxonomy of Viruses [ICTV], 2014). Some viruses may also have a lipoprotein layer associated with them. Viruses that have lipid membranes or layers are expected to interact with cell membranes, and if it is a plant virus, it is likely to infect an insect or other kind of animal host also.

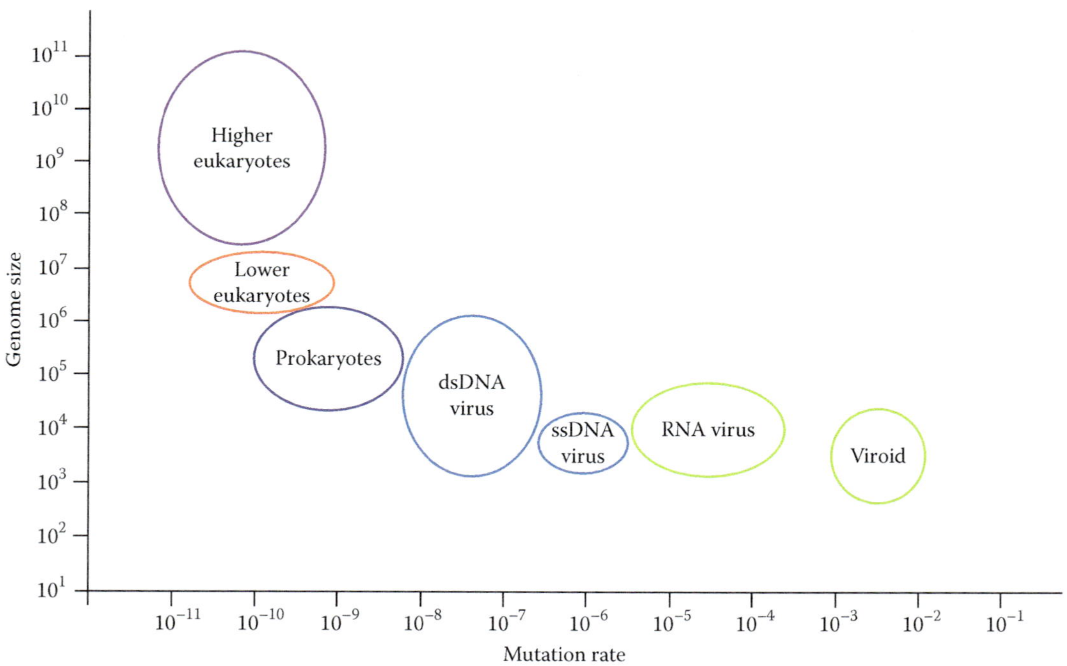

FIGURE 4.1 Relationship between size and mutation rate per nucleotide size of different genomes, with major groups of viruses. (http://viralzone.expasy.org/all_by_species/4136.html [accessed July 1, 2016].

Size and Cellular Association

Viruses are *ultramicroscopic*; that is, they cannot be observed using a light microscope. Their visualization requires an electron microscope, which uses a beam of electrons instead of visible light. The narrower wavelength of the electron beam allows the resolution of smaller objects such as cell organelles and viruses. The relative size comparisons of viruses with other organisms, cells, diverse small molecules, and atoms are illustrated in Figure 4.2. Viruses are measured in units called nanometers (10^{-9} m). More information about viral structure can be learned by applying atomic-level instrumentation, such as x-ray crystallography to reconstruct interactions between molecules and atoms comprising them. Most plant viruses vary in diameter from 20 nm to 250–400 nm (1 nm = 0.0000008 in.); the largest, however, measure about 500 nm in diameter and is about 700–1000 nm in length. Plant viruses are quite small compared with some newly discovered viruses found in the ocean, the mimiviruses and pandoraviruses, which are some of the largest known viruses and whose genomes contain 1–2.5 megabases (Mb) (1 Mb = 1,000,000 base pairs of DNA).

Viruses and their hosts have a more basic relationship than other pathogen systems. Viruses are not separated from their host by a membrane during replication (Hull, 2009). Viruses infect the cellular structure of the host and control a part of the subcellular plant systems. Their dependency on their host for all the basic metabolic systems that are needed to replicate the virus enhances the need for this. You can visualize it as the virus moving into the cell and starting to reprogram some of its functions.

Obligate Parasites

Viruses are obligate parasites. The simplicity of the virus leads to its dependence on the host for many functions. Viruses have no systems for the accumulation of metabolic materials. They have no systems for energy generation (mitochondria), protein synthesis (ribosomes), or capturing light energy (chloroplasts). Thus, viruses are dependent on their host for these functions plus nucleic acid and amino acid synthesis.

Replication by Assembly

Viral replication is dependent on the assembly of new particles from pools of required components (Hull, 2009). These components are synthesized as separate proteins or nucleic acids using the host enzyme systems and the infecting viral genome. New particles are then assembled using these materials. This type of replication contrasts strongly with binary fission or other methods of replication found in prokaryotic and eukaryotic organisms.

HOW ARE VIRUSES NAMED?

Plant viruses are typically named for the host that they are infecting when originally described and for the principal symptom that they cause in this host. The word virus follows these two terms. For example, a virus causing a mosaic symptom in tobacco would be *tobacco mosaic virus* (TMV). This is the species name for the virus.

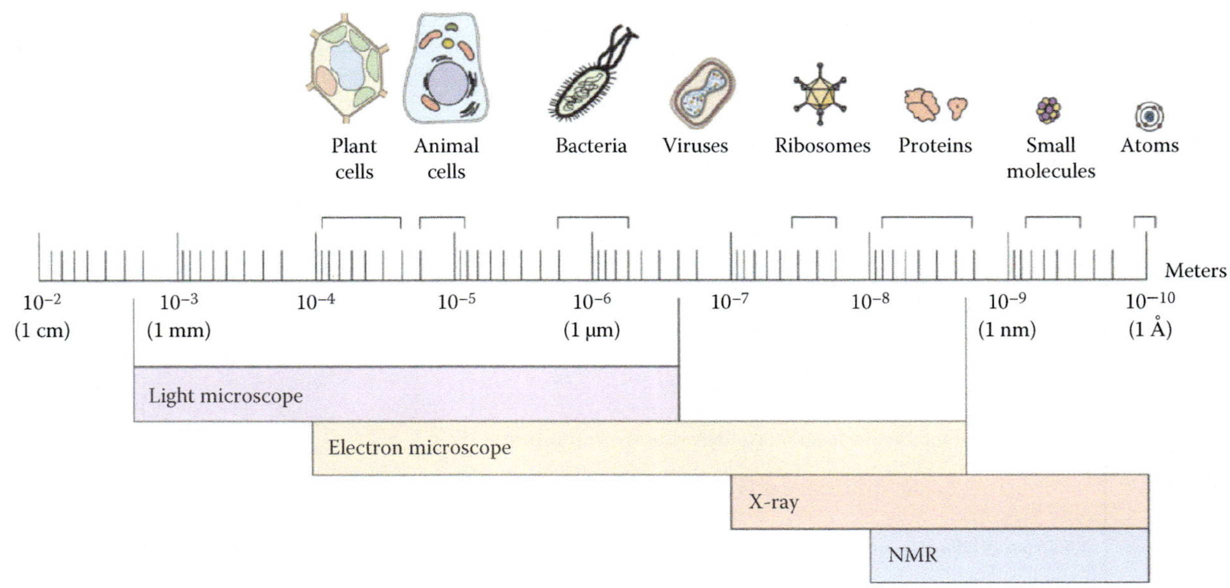

FIGURE 4.2 Relative size comparisons of viruses and types of instrument required to observe them. (Adapted from Levine, A.J., *Viruses*, Scientific American Library, New York, NY, 1991. With permission of Henry Holt and Company, LLC.)

HOW ARE VIRUSES CLASSIFIED?

Taxonomy and Classification

Viruses occupy a special taxonomic position: they are not plants, animals, or prokaryotic bacteria (single-cell organisms without defined nuclei), and they are generally placed in their own kingdom. In fact, viruses should not even be considered organisms, in the strictest sense, because they are not free-living in the classical sense; that is, they cannot reproduce and carry on metabolic processes without a host cell.

The first universally accepted classification system applied to viruses in general was the Baltimore classification system, devised by David Baltimore (1971), which grouped viruses into seven groups (then, families) by the type of genome, strandedness (ss or ds) of the RNA or DNA (nucleic acids), and method of replication. Later, it was observed that viruses classified together in this way could be further separated into groups that shared a similar particle shape and size (Figure 4.3).

The use of the species concept in plant virology began in the recent years following much debate concerning what constitutes a virus species (van Regenmortel et al., 2000). Following the first use of the species name, the virus is referred to by the abbreviation that is given in parenthesis after the first use of the species name. Two levels of taxonomic structure for grouping species are the genus, which is a collection of viruses with similar properties, and the family, which is a collection of related virus genera (Figure 4.4). *Cowpea mosaic virus* (CPMV) is a member of the genus *Comovirus* and the family *Secoviridae* (ICTV, 2014). Virus species may also be subdivided into **strains** and **isolates**. Strains are named when a virus isolate proves to differ from the type of isolate of the species in a definable character, but it does not differ enough to be a new species (Hull, 2002). For example, a virus strain may have altered reactions in an important host, such as producing a systemic reaction in a host that previously had a local lesion reaction or the strain might had an important serological difference. Strains represent mutations or adaptations in the type of virus. Isolates are any propagated culture of a virus with a unique origin or history. Typically, they do not differ sufficiently from the type of isolate of a virus to be a strain.

The International Union of Microbiological Societies (IUMS) has charged the ICTV, a committee of the Virology Division, with the task of developing, refining, and maintaining universal virus taxonomy (Figure 4.5). The goal is to categorize all known viruses using a single classification scheme that reflects their evolutionary relationships through individual phylogenies, similar to a "family tree" that is based on viral genome or gene sequences. Based on the groupings, the ICTV develops an internationally agreed upon taxonomic structure, and the classification and nomenclature follow rules that are established in an "International Code." The universal scheme is applied to all viruses and uses the hierarchical levels of order, family, subfamily, genus, and species. The primary classification, as in all of biology, is the species. The taxonomy and names for all of the different groups of known viruses, and the taxonomic rules and nomenclature, are posted at the ICTV website (http://www.ictvonline.org/index.asp). The purpose of naming a virus (taxon) is to have a way to refer to it, not to indicate its characteristics or the history. If a new kind of virus is found that does not yet adhere to an existing, ratified group, then it is considered "unclassified" until sufficient information becomes available to classify it. The ICTV is not responsible for classifying or naming virus taxa below the species level. This is the responsibility of the international specialist groups, referred to as Study Groups that represent each virus family. The classification and naming of serotypes, genotypes, strains,

Classifications

1.1 Group I: Double-stranded DNA viruses

1.2 Group II: Single-stranded DNA viruses

1.3 Group III: Double-stranded RNA viruses

1.4 Group IV & V: Single-stranded RNA viruses

- 1.4.1 Group IV: Single-stranded RNA viruses—Positive sense
- 1.4.2 Group V: Single-stranded RNA viruses—Negative sense

1.5 Group VI: Positive-sense single-stranded RNA viruses that replicate through a DNA intermediate

1.6 Group VII: Double-stranded DNA viruses that replicate through a single-stranded RNA intermediate

FIGURE 4.3 Baltimore virus classification system. (From Baltimore, D., *Bacteriol Rev.*, 35, 235–241, 1971.)

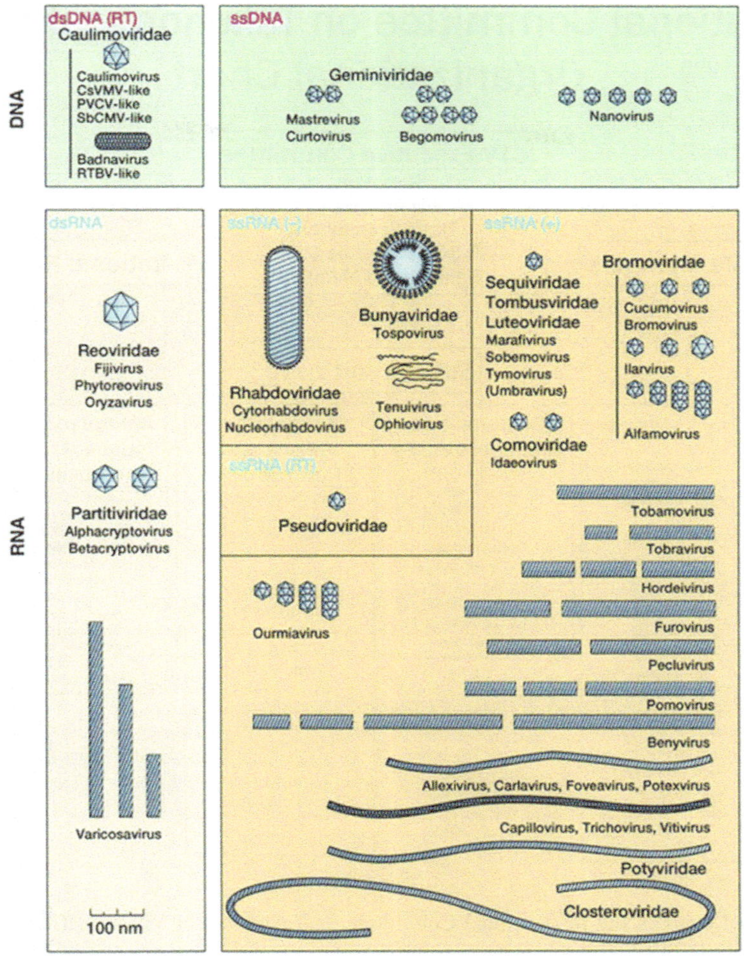

FIGURE 4.4 Examples of recognized plant virus families showing different viral particle morphologies with genome type. (From Van Regenmortel et al., *Seventh Report of the International Committee on Taxonomy of Viruses*, Academic Press, San Diego, CA, 2000. Available from: https://www.researchgate.net/figure/278704382_fig5_Figure-1-1-Families-and-Genera-of-viruses-infecting-plants-Courtesy-MHV-van-Regenmortel [accessed July 1, 2016].)

variants, and isolates of virus species is the responsibility of acknowledged international specialist groups.

WHAT ARE THE IMPORTANT HISTORICAL DEVELOPMENTS IN PLANT VIROLOGY?

One of the first references to a disease caused by a plant virus occurred during Tulipomania in seventeenth-century Holland (1600–1660; Hull, 2009). This is an unusual case of a virus that increased the value of infected plants. It began with the importation of tulips into Holland from Persia. The beautiful flowers quickly became popular and were grown throughout Holland. People began noticing that some flowers were developing streaks and broken color patterns. These tulips were called "bizarres," and people quickly learned that they could produce more bizarres by planting a bizarre tulip in a bed or by rubbing the bizarres bulb onto plain tulip plants. They did not realize that they were transmitting a pathogen. Bizarres became so popular that a single bulb was worth large sums of money, thousands of pounds of cheese, or acres of land. One case is recorded where a man offered his daughter in marriage in exchange for a single bizarre tulip bulb.

In 1886, Adolf Mayer scientifically confirmed a primary principle of plant virology when he transmitted TMV to healthy tobacco plants by rubbing them with sap from infected plants. The newly rubbed plants displayed the same symptoms as the original infected plants (Scholthof et al., 1999). Mayer's research established the contagious nature of plant viruses and the first procedure for mechanical transmission of a virus. Today, mechanical transmission enables virologists to transmit some viruses for experimental purposes and to evaluate plants for resistance and tolerance to viral diseases.

The independent experiments of a Russian scientist, Dmitri Ivanowski, in 1892, and a Dutch scientist, M.W. Beijerinck, in 1898, first indicated the unique nature of viral pathogens. Both scientists extracted plant sap from tobacco

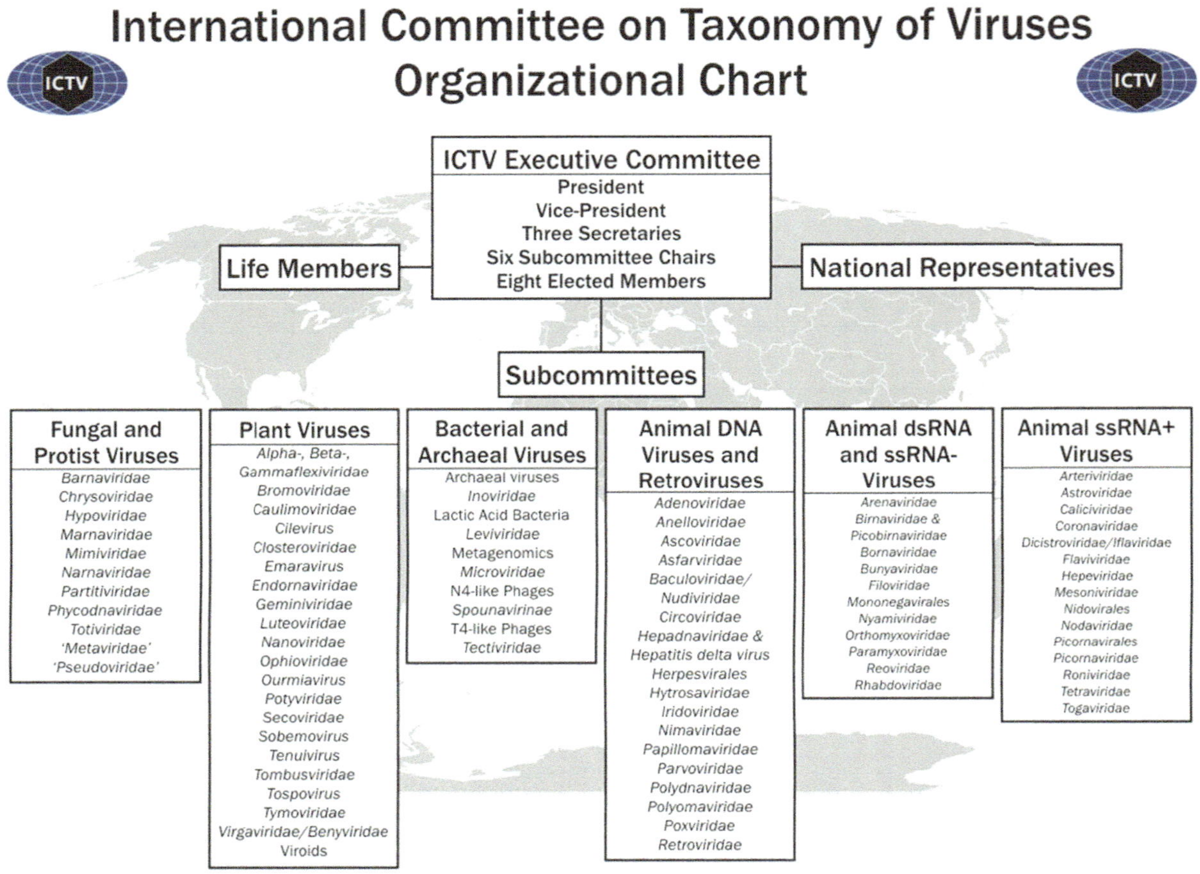

FIGURE 4.5 Organizational chart of the International Committee on Taxonomy of Viruses (ICTV). http://www.ictvonline.org/organization.asp [accessed July 1, 2016].)

plants infected with TMV. The sap was then passed through a porcelain bacterial filter that retained bacteria and larger pathogens. If bacteria or other organisms were the cause of the disease, the filter would have retained the pathogen, and the sap that had been passed through the filter would not transmit the disease. When Ivanowski or Beijerinck filtered sap from TMV-infected plants and inoculated the filtered sap onto healthy tobacco plants, the plants became diseased. Beijerinck recognized that this indicated the unique nature of this disease. Beijerinck declared that TMV was a new type of pathogen that he called a *contagium fluidium vivium* (a contagious living fluid; Scholthof et al., 1999). Later scientists discovered that the ultramicroscopic particle nature of viruses was not truly fluid. However, this was the first indication that viral pathogens represented a new and unique type of pathogen.

Clues to this unique nature would wait until the research of W.M. Stanley in 1935. Working with TMV, Stanley extracted gallons of plant sap and used the newly developed technique of protein fractionation by precipitation with salts and other chemicals. Each fraction was tested by inoculation on susceptible hosts to determine where the infectivity remained. Gradually, Stanley isolated the infective fraction into a pure form in which he crystallized TMV. Stanley was the first person to purify a virus, and in 1946, he was awarded the Nobel Prize for this accomplishment (Scholthof, 2001).

In Stanley's first analyses of purified TMV solution, he found only protein. In 1936, Bawden and Pirie found the presence of phosphorous, which indicated that the solution contained nucleic acid (Bawden et al., 1936). Fraenkel-Conrat was able to isolate the RNA from TMV in 1956 and used it to infect healthy tobacco plants, proving that the RNA was the source of infectivity and establishing that RNA contained the genome of TMV (Fraenkel-Conrat, 1956; Creager et al., 1999).

In this section, we have discussed only a few of the most important historical principles in plant virology. However, the brevity of this section should not be used to judge the importance of historical research. The theory and accomplishments of today's research in plant virology is built upon the accomplishments of many researchers who preceded today's researchers, and their contributions are the foundation for tomorrow's research. A comprehensive history of virology, including animal, human, bacterial, and plant virus examples, is available as an eBook, *Foundations of Virology* (2014).

WHAT ARE THE SYMPTOMS FOUND IN PLANTS INFECTED WITH VIRUS?

Symptoms are the host's response to infection and typically the first signal that attracts the attention of the pathologist, producer, or homeowner. However, viruses are sometimes referred to as great imposters because viral symptoms are often mistaken for other diseases or conditions. Viral diseases may be misidentified as nutritional deficiencies, toxins, genetic abnormalities, mineral toxicities, pesticide damage, environmental stresses, insect feeding, or infection by other plant pathogen groups. Symptomatology can provide a strong first indication of a possible viral infection, but it cannot be utilized as the sole basis for diagnosis, due to possible confusion with other conditions. Classic symptoms of viral infections can be grouped by their similarities. The two primary categories are symptoms caused by localized infection and symptoms caused by systemic infections.

SYMPTOMS CAUSED BY LOCALIZED INFECTION

Local Lesions

Local lesions (Figure 4.6) are often the initial symptom that develops on an inoculated plant and are usually observed within 3–4 days of viral inoculation. They occur when viral infection moves from cell to cell. This symptom can be generated by the host's defense responses triggering cell death in a hypersensitive reaction to the presence of the infecting virus, or by the virus's lack of compatible systemic movement mechanisms. Plant viruses may overcome this initial defensive reaction to spread systemically, or they may never spread beyond the initial infection site. Local lesions may be either necrotic or chlorotic. In virus and host combinations that produce local lesions, they can be utilized to quantify viral infectivity levels by inoculating a host with a specific volume of a solution containing extracted virus and counting the number of local lesions that are produced in the inoculated area.

SYMPTOMS CAUSED BY SYSTEMIC INFECTION

Symptoms Based on Changes in Chlorophyll or Other Pigments

Patterns of lighter and darker pigmentation are usually referred to as **mosaics** or **mottles** (Figure 4.7). Areas of lighter pigmentation may be pale green, yellow, or white and are caused by decrease in chlorophyll, decrease or destruction of chloroplasts, or other damage to the plant's chlorophyll system. Other theories suggest that mosaics and mottles may also be due to increased pigmentation in the areas of darker green. TMV is a classic example of a virus that can produce mosaic in systemically infected hosts (see Case Study 4.1). When mosaics develop in a floral part, this can be referred to as a **color break**. Other types of patterning that can be formed from changes in chlorophyll or other pigments include **stripes**, **streaks** (Figure 4.8), **ringspots**, **line patterns**, **vein banding** (Figure 4.9), **vein clearing**, and **yellows** (Figure 4.10). **Necrosis** or death of cells, tissues, or organs often follows the development of one more symptom types associated with changes in host plant pigments (Figure 4.11; Verbeek et al., 2007).

FIGURE 4.6 Local lesions on *Chenopodium amaranticolor*. A number of plant viruses cause symptoms like these when inoculated to this indicator species. (From Hull, R., *Matthews' Plant Virology* (4th ed.), Academic Press, New York, NY, 2002.)

FIGURE 4.7 Mosaic and distortion are prominent on the leaf of this plant infected with *Squash mosaic virus* (SqMV). (Courtesy of M. Langham.)

CASE STUDY 4.1

Tobacco mosaic virus (TMV)—A "Stable" Problem in Tomato Production

- TMV is a serious problem to commercial and home tomato production.
  - TMV causes a light to dark green mosaic when infecting tomato. Some TMV strains may also produce a bright yellow mosaic in tomato.
  - Deformation of the leaves includes curling, size reduction, and malformed shapes.
  - Tomato fruits ripen unevenly, and they are reduced in both size and number. At times, tomato fruits may develop internal browning, particularly in the earliest clusters.
- TMV is a highly stable virus and has high rates of mechanical transmission.
  - TMV adheres to the drying tomato seed and survives the desiccation during drying due to its stable nature. Thus, it is borne on the exterior of these seed as a contaminant. As the seedling emerges from the germinating seed, any minor wound allows TMV to infect the seedling.
  - Workers, who handle infected plants during transplanting, pruning, or other normal greenhouse activities, contaminate their hands with TMV during these processes.
  - After contaminating their hands, workers may infect any healthy plants they contact and injure. Even a tiny injury can be enough for TMV to infect the plant.
  - Additionally, workers who use tobacco products, such as cigarettes or smokeless tobacco, may contaminate their hands with TMV while handling these products and transmit TMV through mechanical transmission while handling the tomato plants.
- Identification and detection to establish the identity of the virus infecting the crop is an important foundation for viral disease management. Knowing the correct identity aids in selecting appropriate resistance sources, developing vector control plans, and selecting other appropriate tools such as planting dates, sanitation plans, and eradication protocols.
- Host plant resistance provides the first line of defense against viral diseases.
  - Resistance is a management tool that should be utilized whenever possible.
  - In tomatoes, several single dominant genes have been identified for resistance to TMV and are available in a variety of tomato cultivars. However, utilizing host plant resistance depends on careful consideration when purchasing seed and cannot help the current tomato crop if it is susceptible to TMV.
- Exclusion of primary inoculum is another basic step in TMV control.
  - Since TMV is often introduced on contaminated seeds, purchasing seed that have been treated to eliminate TMV is a good option.
  - If virus-free seeds are not available in the variety that you desire, disinfesting the seeds by seed treatment with 10% Na_3PO_4 for 15 min or by other established disinfestation protocols is recommended.
- Sanitation is a major control for TMV, especially in greenhouse production.
  - The stability of TMV is difficult to conceive in comparison with the fragility of other viruses. TMV can persist in contaminated soil; on benches, pots, gravel, and farm machinery; and even in Arctic ice cores.
  - Due to TMV's high stability, all equipment and surfaces in the production area should be cleaned with soap and water, rinsed, and treated with a 10% bleach solution. Some decontamination methods include rinses with 20% (weight to volume) powdered nonfat skim milk solution with 1% Tween 20 as addition of skim milk has been shown to lessen TMV contamination.
  - Workers in tomato production need to clean their hands and tools frequently. Tobacco users can become contaminated with TMV by handling tobacco products and should take special care to disinfest when entering tomato production areas.
- Eradication is also a vital step to TMV control.
  - In greenhouse tomato production, any plant or weed that is a potential host for TMV should be removed from the production area. This includes plants surrounding the exterior of the greenhouse. Even in field

production where complete removal is impossible, limiting these secondary hosts can improve TMV control.
- Remove infected debris from the production area.
- Rogue infected plants. When removing infected plants, it is not enough to simply remove the plant showing symptoms. Whether this plant is touching or rubbing another plant should be checked. TMV is highly mechanically transmissible so that all plants surrounding the symptomatic tomato plant should also be removed.

- Yield losses occur in both field and greenhouse tomato production. However, due to the confined nature and labor-intense processes of greenhouse production, TMV is particularly devastating in greenhouse tomato production, and it remains a constant problem.

FIGURE 4.8 Streaks and stripes in three wheat lines infected with *Wheat streak mosaic virus* (WSMV). (Courtesy of M. Langham.)

FIGURE 4.9 Vein banding in tobacco infected with a TMV strain. (Courtesy of B. Ruden and M. Langham.)

Symptoms Caused by Growth Abnormalities

Symptoms that are due to growth abnormalities can affect any part of the host plant. The most common growth abnormalities are **stunting** and **dwarfing**, which are caused by a reduction in the size of the infected host plant (Figure 4.12). Some viruses may stunt the plant only slightly, whereas others affect the host plant dramatically. Stunting may include changes in the size of all plant parts such as leaves in addition to height. Stunting that includes shortening of the internodes is often referred to as a bushy stunt or a **rosette**. Growth abnormalities can also result from tissue overgrowth. **Tumors**, **galls**, and **enations** are examples of symptoms produced by **hyperplasia** and **hypertrophy**. Other growth abnormalities are caused by a combination of overgrowth and undergrowth. **Distortions**, **epinasty**, **shoe stringing**, and **leaf rolling** are additional examples of abnormalities that are also produced by viruses.

Symptoms Affecting Reproduction

If you question any producer about what is his greatest concern about his crop, one of the most common answers is "How much will my crop produce this year?" Viral symptoms that affect reproduction have some of the most direct effects on the answer to this question. **Sterility** directly affects the infected host plant's ability to produce viable seed. These effects may be through floral abnormalities, decreases in flowering, reduced viability of pollen or ovaries, inhibited seed development, or reduced seed set. The production of sterility is also linked with changes in the plant's metabolism and changes in its biochemical signaling that triggers flowering and seed set. Yield losses (Figure 4.13) can take many forms in addition to sterility. It may be a reduction in the total reproduction of the plant. Seed or fruit produced may be shriveled, reduced in size, and distorted or inferior in

FIGURE 4.10 Foliar chlorosis and green vein banding of melon plants infected with a *Crinivirus* (*Closteroviridae*). (Courtesy of J.K. Brown.)

FIGURE 4.11 Veinal necrosis and interveinal chlorosis of bean plant infected with virus.

quality. *Wheat streak mosaic virus* (WSMV) is a good example of a virus that causes yield loss due to undersized and shriveled grain in addition to reducing the total number of seeds set (see Case Study 4.2). *Plum pox virus* (PPV) is an example of a virus that not only disfigures the fruit but also reduces the carbohydrate level in the fruit of the infected trees.

Other Symptoms Caused by Plant Viruses

Some plant viruses cause their hosts to produce symptoms that are not often associated with viral diseases.

FIGURE 4.12 Symptoms in pepper infected with a whitefly-transmitted *Begmovirus* (*Geminiviridae*), showing shortened internodes, stunting, and reduced leaf size. (Courtesy of J.K. Brown.)

FIGURE 4.13 Mature soybean plants (C, uninoculated; I, inoculated) demonstrating the effect of *Bean pod mottle virus* (BPMV) on soybean yield and productivity. (Courtesy of M. Langham.)

Necrosis is the death of cells, tissues, organs, or the whole plant. Although it seems to be a disadvantage for a virus to kill its host, necrosis is a prominent symptom produced in some infected host plants and may indicate a basic incompatibility between the virus and the host. **Wilting** is another unique symptom produced during some viral infections (Figure 4.14).

CASE STUDY 4.2

Wheat streak mosaic virus (WSMV) in Winter Wheat—Following The Vector

- WSMV is the most severe viral disease affecting winter wheat produced in the Great Plains region of the United States. In addition to winter wheat, WSMV can infect spring wheat, oats, barley, rye, corn, sorghum, millet, and many introduced and native grasses.
- WSMV symptoms:
  - Mosaic and streaking—WSMV symptoms begin as mosaic patterns that develop in the leaf lamina between parallel veins of the wheat leaf (Figure 4.8). As the mosaic intensifies, it spreads as streaks along the leaf following the parallel vein pattern. Depending on the cultivar, environment, and stage of infection, the streaks may be light green or yellow. Typically, the yellowing intensifies as the wheat develops. Leaves may be predominately yellow in severe mosaic infections.
  - Stunting—WSMV also has a significant effect on plant height. Stunting in WSMV can range from less than 5% to more than 75%.
  - Growth and yield effects—WSMV affects the number of secondary tillers developed by the wheat plant. Wheat plants may also tend to flatten and spread toward the ground in a symptom called prostrating. It may cause sterility in the grain, but more commonly, the grain is poorly filled. WSMV delays wheat maturity. This adds to the yield loss in wheat-growing areas when drought conditions begin early in the summer.
- Vector and disease cycle:
  - The wheat curl mite, *Aceria. tosichella* Keifer, is an eriophyid mite that is approximately 0.2 mm. It serves as the vector to transmit WSMV and continues its movement through the disease cycle. In the spring, the mites move WSMV from winter wheat to other hosts (spring wheat, corn, or annual and perennial grasses), and in the fall, the mites infect the emerging winter wheat with WSMV.
- Host plant resistance is a primary management strategy for WSMV.
  - In susceptible cultivars, yield losses of up to 70% can be observed depending on the cultivar and the environmental conditions.
  - No truly resistant cultivar is widely available, but cultivars with improved levels of resistance or tolerance are available and should be selected whenever possible.
- Can controlling a vector always control the virus?
  - In field situations, WSMV can often be identified beginning at the field margin and progressing inward into the field with the prevailing wind. In this situation, producers often think about applying a miticide to the plants that are showing symptoms. However, this typically does not control the virus spread or lessen the WSMV infection rate.
  - How long does it take a plant to display symptoms once it is infected? It is often several days depending on the environment. When the producer sprays only the symptomatic plants, the miticide does not cover the plants that are infected but have no visible symptoms. Thus, the virus has already spread prior to applying the miticide.
  - Additionally, the plants displaying symptoms represent where the wheat curl mites were feeding 10–14 days ago. While it is true that this indicates that mite populations are established in this area, it does not show the current advancing margin of the mite population, which has still been spreading for the 10–14 days those symptoms took to develop.
- Other controls for WSMV:
  - Eradication—Eradicating all green plant materials, particularly volunteer wheat or grasses, for 2–3 weeks before planting will decrease the primary inoculum sources that are present within the field and will lessen the WSMV incidence.

(Continued)

- Cultural control—Delay in winter wheat planting until grass species around the field have died back for the winter will also decrease the incidence of WSMV. As these grass species brown during the fall, wheat curl mites move from dying host plants to plants that are green. Timing planting to miss the peak of this mite movement avoids the emerging winter wheat being infected at its most susceptible stage.
- Eradication and delayed planting are good tools that producers do not always use due to other urgent needs in fall scheduling or difficulty in judging peak mite movement.

FIGURE 4.14 Foliar necrosis and wilting symptoms in tomato plant infected with "apex necrosis" disease transmitted by whitefly (genus, *Torradovirus*; family, *Sesquiviridae*).

Recovery or **symptom suppression** is found with some virus and host combinations. However, environment and plant stage may also affect this reaction. Other viruses have the primary symptom of being **symptomless**. The cryptoviruses were not found for many years due to remaining symptomless, although they are transmitted in pollen and seed. Finally, some viruses, such as *Barley stripe mosaic virus* (BSMV), cause abnormalities in the segregation of phenotypic traits (**aberrant ratio**) or may cause an increase in the mutation rate in the infected host plants.

WHAT DO VIRUSES LOOK LIKE?

Virion shape is one of the most fundamental properties of viruses. Plant viruses are based on four types of architecture or morphology. **Icosahedral** (isometric) viruses are basically spherical in shape. However, on closer examination, they are not simply smooth but are faceted. These viral structures are suggestive of the geodesic domes designed by Buckminster Fuller (Morgan, 2006). Icosahedral viruses have 20 facets or faces. Second, **rigid rod viruses** are all based on protein coats surrounding a helical nucleic acid strand. Rigid rod viruses are typically shorter and have a greater diameter than **flexuous rods**. In addition, the central canal, an open region in the center of the viral helix, is more apparent in rigid rod viruses. Third, flexuous rods are typically very flexible and may bend into many formations. They are narrow in diameter and are longer than rigid rods. Lastly, **bacilliform viruses** are short, thick particles (rods) that are rounded on both ends. When particles are found with only one rounded end, they are referred to as bullet-shaped. Electron micrographic and computer models generated from crystallography can be found at http://www.virology.wisc.edu/virusworld/viruslist.php and http://www.virology.net/Big_Virology/BVHomePage.html.

HOW ARE VIRUSES TRANSMITTED?

Plant viruses are not capable of penetrating the plant cuticle, epidermis, and cell wall. Thus, they cannot disperse from plant to plant without the assistance of a vector. Plant viruses are dependent on **vectors** to breach the epidermal layer of the plant and to place them within a living host cell. This dependency on vectors is so great that sometimes it can be considered to add an additional component to the classic disease pyramid (Figure 4.15). Virus vectors include insects, mites, nematodes, fungi, seed, and dodder (Chapter 17). It also includes humans, animals, and other organisms that transfer viruses through mechanical transmission. For experimental transmission, viruses are often transferred using mechanical methods. This is discussed in

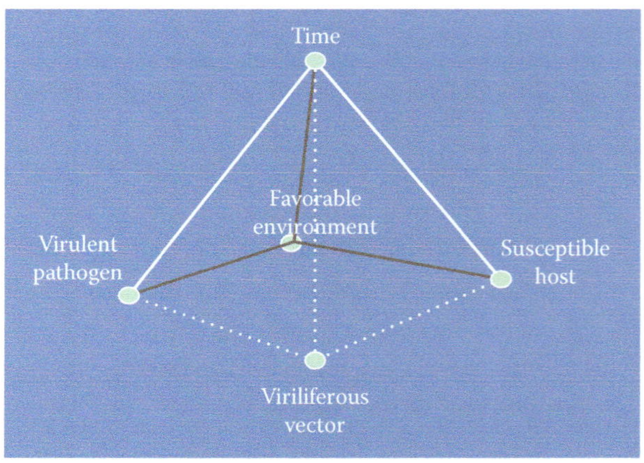

FIGURE 4.15 The dependency of plant viruses on their vectors can be visualized by the addition of a vertex to the disease pyramid to illustrate this vital factor in viral disease development.

- Mechanical transmission allows the transmission of plant viruses without a vector.
- An abrasive is used to make wounds that penetrate the epidermal layers, cell wall, and cell membrane. Silica carbide is the most widely used abrasive, but sand, bentonite, and celite have also been utilized.
- These wounds must not cause the cell to die, because the viruses are obligate parasites and require a living cell. This type of wounding is described as nonlethal.
- Mechanical transmission is effective for viruses that infect epidermal cells. Other viruses that are limited to other tissues such as phloem are not mechanically transmissible.
- Some viruses that are highly stable such as TMV may be mechanically transmitted through accidental contact between healthy and infected plants, wounding the plants to allow infected plant sap to be transmitted to the healthy plants. In addition, contacting healthy plants after handling plants infected with stable viruses may allow you to become the vector by transmitting infected plant sap to the healthy plants.
- Mechanical transmission allows scientists to study the effects of plant viruses without using the vector. It is also used in the evaluation of new plant cultivars for viral disease resistance.

FIGURE 4.16 Mechanical Transmission of Plant Viruses

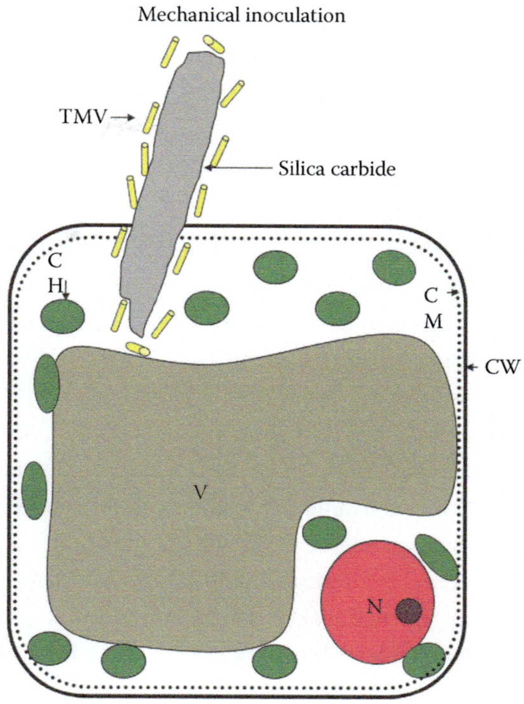

FIGURE 4.17 Mechanical inoculation of plant viruses requires wounding of the plant cell by an abrasive. Silica carbide particles rupture the cell wall (CW) and cell membrane (CM) and allow viruses to enter the cell. Other structures shown in the cell include nucleus (N), chloroplasts (CH), and central vacuole (V). (Courtesy of M. Langham.)

more detail in Figure 4.16 and diagrammatically shown in Figure 4.17 and Experiments 1 (Procedure 4.1) and 2 (Procedure 4.2). Each virus evolves a unique and specific relationship with its vector. Viruses are dependent on this complex interaction and have developed many methods for capitalizing on the biology of their vectors. Thus, understanding the relationship between virus and vector is vital in developing control programs.

Insect Vectors

Approximately 90% of all known plant viruses are transmitted by insects classified in the suborder, Homoptera (order: Hemiptera) (see references in Hogenhout et al., 2008). These insects feed in the vascular system of the plant, either the xylem or the phloem, using small slender stylets that penetrate the cells causing minimal damage. During feeding on phloem or xylem contents, feeding behaviors are utilized to ingest sap, during which virus particles are taken up, "ingested," or delivered to the site with insect saliva when it is delivered during feeding, egested. Most of these viruses do not replicate in their insect vector, but have evolved a relationship by which they either adhere to particular locations in the mouthparts or stylets, the foregut, or circulate in the body and enter the salivary glands. In these instances, the virus–vector relationship ranges from relatively virus–vector specific for nonpersistently transmitted viruses, which may have more than one vector species, to moderately to highly specific as occurs with semipersistently or persistently transmitted viruses that have one, several, or a single vector species, respectively. Viruses that circulate and multiply in their vector are thought to have evolved first as viruses of the insect host and then to have adapted to the plant host, thereby making a "host-shift."

Aphids

Aphids transmit more viruses than any other vector group. Viruses have developed four types of interactions with aphids. These interactions are dependent on the infected plant tissue, the association of the virus with the vector, and virus replication (or lack of replication) in the vector (Table 4.1).

Nonpersistent Transmission

Viruses that are transmitted in a nonpersistent manner infect the epidermal cells of the host plant. This type of transmission is dependent on the sampling behavior of the aphid that quickly probes in and out of the epidermal cells in order to determine host suitability. The virus forms a brief association with two sites. The first site is at the tip of the **stylet**, and the second is found just before the **cibarial pump** at the top of the stylet. This association lasts only until the next probe of the aphid when it is flushed from the stylet during the process of **egestion**. Nonpersistent

Procedure 4.1

Mechanical Transmission of *Tobacco mosaic virus* (TMV)

Step	Instructions and Comments
1	Control plants. Collect 0.1 g of healthy plant tissue. Grind the tissue thoroughly in a mortar and pestle. Add 9.9 mL of phosphate buffer. Grind the mixture again. Remove any plant material that has not been thoroughly macerated. Add 1% silica carbide to the sap extract (silica carbide can also be dusted on the leaves with a sprayer; however, adding it to the sap mixture decreases the amount that may be accidentally inhaled). Stir the mixture well and thoroughly wet the inoculation pad. Silica carbide settles to the bottom of the sap extract and must be stirred each time the inoculation pad is soaked. With the soaked inoculation pad, firmly, but gently, rub the upper surface of the leaves on two tobacco plants. Excessive pressure will result in lethal damage to the epidermal cells from the silica carbide; however, too little pressure will not wound the cells for viral entry. Use paper tags to mark the leaves that are being inoculated. Note: For best results, complete the control plants before you handle any infected plant material and thoroughly wash your hands between each step. In addition, if your virus has a lower or higher concentration in the source plants than usual, the tissue and buffer ratio may need to be adjusted for best lesion development.
2	Experimental plants. Repeat Step 1 utilizing infected plant tissue.
3	Observe plants for the next 2–4 weeks for symptom development.

Procedure 4.2

Mechanical Transmission of Plants for Host Ranges

Step	Instructions and Comments
1	Select a test virus and a compatible set of hosts.
2	Utilize five plants per species (for each student or group) for the experimental group and another group of five plants per species for the control group. If you do not have the greenhouse space for every group of students to do all of the host plants, assign different hosts to each group. An advantage of this format is that students need to compare the symptoms that develop in the different species.
3	Prepare the sap extract from the control plants and inoculate the leaves as detailed in Procedure 4.1.
4	Prepare the sap extract from the infected plants and inoculate the leaves as detailed in Procedure 4.1.
5	Changing inoculation pads between species is a good procedure in case one species is contaminated with a seed-borne virus. Plants with viscous sap, such as *Chenopodium*, should be inoculated last as the sap may contaminate the extract and cause inhibition of the virus.
6	Plants should be observed daily. Record the date of symptom appearance, symptom type, and changes in symptoms as plants mature.

aphid transmission requires only seconds for acquisition and transmission and is increased by preaquisition starvation or by any other condition that increases sampling behavior of the aphid (Table 4.1). Evidence suggests that in some nonpersistently transmitted viruses, such as cucumoviruses, the ability to bind to the aphid is dependent on a property of the coat protein, such as conformation or the binding of metal ions (Ng and Falk, 2006). Nonpersistent transmission in other viruses requires the presence of an additional protein called helper component, helper factor, or aphid transmission factor to assist in the binding of the virus to specific regions in the aphid stylet. This binding is referred to as the bridge hypothesis (Ng and Falk, 2006; Hull, 2009). Potyviruses are among the most important group of viruses transmitted in a nonpersistent manner.

Semipersistent Transmission

Semipersistent viruses form an association with the lining of the aphid foregut. "Semipersistent" emphasizes that these viruses are retained longer than nonpersistent viruses. However, the phrase, "nonpersistently

TABLE 4.1
Comparison of Aphid Transmission Characteristics

	Types of Transmission			
Characteristic	**Nonpersistent**	**Semipersistent**	**Persistent, Circulative, Nonpropagative**	**Persistent, Circulative, Propagative**
Tissue infected by transmitted virus	Epidermis	Mesophyll and phloem	Phloem	Phloem
Virus interaction sites	Stylet tip and preciberium	Foregut	Hindgut and salivary glands	Hindgut and salivary glands; also infects many tissues of the host
Type of feeding behavior associated with transmission	Sampling	Phloem probing	Phloem probing	Phloem probing
Acquisition time	Seconds	Minutes to hours	20 minutes to hours	20 minutes to hours
Inoculation time	Seconds	Minutes to hours	20 minutes to hours	20 minutes to hours
Retention time	Minutes to hours	Hours to days	Days to life	Days to life
Latent period	No	No	Yes (hours)	Yes (days to weeks)
Retained through molt	No	No	Yes	Yes
Found in the hemolymph	No	No	Yes	Yes
Replicates and infects the host tissues	No	No	No	Yes
Transovarial passage—infects offspring	No	No	No	Yes

transmitted foregut-borne virus" is also used to describe this group (Hull, 2009). Acquisition of these viruses requires phloem-feeding leading to longer acquisition times than nonpersistently transmitted viruses, transmission requires minutes, and retention of the virus typically lasts for hours (Table 4.1). Virus retention does not last through the aphids' developmental molts as the lining of the foregut is shed with the rest of the cuticular exoskeleton. Semipersistent transmission may require the presence of a helper component or a helper virus (Hull, 2002). *Cauliflower mosaic virus* (CaMV) is semipersistently transmitted by aphids and depends on coat protein and two nonvirion proteins in transmission (Ng and Falk, 2006).

Persistent, Circulative, Nonpropagative Transmission

All viruses transmitted in this manner must be taken from the phloem of an infected plant and placed in the phloem of a healthy plant. Thus, they are dependent on the phloem-probing behavior of aphids. Approximately 20 min is required for aphids to establish phloem probes, which is considered the minimum time needed for acquisition and this type of transmission. Viruses transmitted in a persistent, circulative, nonpropagative manner form a close association with their aphid vectors. The virus moves through the intestinal tract to the hindgut where it passes into the aphid's **hemolymph** by endocytosis into coated pits and vesicles (Hull, 2009). The virus then circulates throughout the hemocoel and moves from the hemocoel into the **salivary glands** by passing through the basal membrane of the salivary gland. The virus is then injected into the healthy plant with the saliva during egestion. Viruses transmitted in this manner are retained for days to weeks. Retention time is correlated with the amount of virus in the hemocoel and the level of virus in the hemolymph is related to the acquisition time that the vector feeds on the infected host. *Barley yellow dwarf virus* (BYDV) and *Cereal yellow dwarf virus* (CYDV), two of the most widely distributed viruses in the world, are transmitted in this manner.

Persistent, Circulative, Propagative Transmission

Persistent circulative propagative transmission has many characteristics in common with persistent nonpropagative viruses. Viruses transmitted in this manner are phloem-limited viruses. Transmission is dependent on phloem probes and passage of the virus into the hemolymph. However, after entering the hemolymph, the virus infects and replicates in the aphid. Many tissues of the aphid such as brain and ganglions, salivary glands, ovaries, fat body, and muscle can be infected. The virus can also pass through the ovaries into the offspring, which are **viruliferous** when they are born.

This is termed transovarial passage of the virus. Some viruses in the *Rhabdoviridiae*, such as *Sowthistle yellow vein virus* (SYVV), can multiply in their aphid vector (Hull, 2009).

Beetles

Beetle transmission is unique among all other types of virus transmission in that the specificity of virus transmission is not in the ability of the beetle to acquire the virus, but in the interaction of the virus and host after transmission. Beetles can acquire both transmissible and nontransmissible viruses. Both types of viruses may be found in the hemolymph of some beetles. Other viruses are found only in the gut lumen and mid-gut epithelial cells (Wang et al., 1994). However, the specificity of transmission does not depend on these factors. Beetles spread a layer of predigestive material known as **regurgitate** on the leaves as they feed. This layer contains high concentrations of deoxyribonucleases, ribonucleases, and proteases. When viruliferous beetles spread this layer, they also deposit virus particles in the wound at the feeding site. Beetle transmissible viruses are able to move through the vascular system to an area away from the wound site with its high level of ribonuclease in order to establish infection (Gergerich et al., 1984; Gergerich and Scott, 1988a; Gergerich and Scott, 1988b). Nontransmissible viruses are retained at the wound site where the ribonuclease levels inhibit their ability to infect the plant (Field et al., 1994). These unique movement systems of beetle-transmitted viruses within the plant are an important area of current research. *Comovirus* and *Sobemovirus* are among the most important viral genera transmitted by beetles.

Leafhoppers

Leafhopper transmission of viruses highly parallels aphid transmission with one exception, that is, nonpersistent transmission does not occur in leafhoppers. Leafhoppers transmit viruses by semipersistent transmission, persistent circulative nonpropagative transmission, or persistent circulative propagative transmission. Important examples of viruses transmitted by leafhoppers include *Beet curly top virus* (BCTV), *Potato yellow dwarf virus* (PYDV), and *Maize stripe virus* (MSV).

Mealybugs

Mealybug transmission of plant viruses is not very common. However, several ampeloviruses (genus, *Ampelovirus* and *Closterovirus*; family, *Closteroviridae*) transmitted by mealybugs are known to infect cherry trees, grapevines, and pineapple plants worldwide (http://viralzone.expasy.org/all_by_species/285.html), causing yellowing and necrosis of phloem. In West Africa, over 10 species of mealybugs have been reported to transmit cacao swollen shoot virus to *Theobroma cacao* (L.). At least 11 members of the genus *Badnavirus* (family, *Caulimoviridae*) are transmitted by mealybug vectors, including *Banana streak virus*, and viruses of *Citrus*, *Dioscorea* (water yam), and black pepper (*Piper nigrum*). Evidence suggests that mealybug transmission is either nonpersistent or semipersistent, but the mode of transmission is poorly studied for these viruses.

Thrips

Virus transmission by thrips has been a dynamic area of research in the recent years. The rapid increase in the importance of *Tospovirus* and the diseases that they cause in both greenhouses and fields has stimulated much of this research. Members of the genus *Tospovirus* are persistently and propagatively transmitted by thrips. Thrip larvae acquire the virus while feeding on virus-infected tissue and the virus crosses through the midgut barrier and enters the salivary glands. The virus must be acquired by immature thrips (adult thrips cannot acquire the virus) because the virus passes from larvae to adult thrip as it undergoes pupation and the changes associated with maturity. After a thrip acquires the virus, infection is established in the midgut. It first replicates in the midgut epithelium before spreading to the circular and longitudinal muscles. By the time the thrip has matured to an adult, the virus has spread to the visceral muscles. As the virus is not found in the hemolymph, it is hypothesized that it reaches the salivary glands through ligaments that stretch from the salivary gland to the muscle (Hull, 2009; Whitfield et al., 2005). Thrips transmit three additional viral genera, *Ilarvirus*, *Sobemovirus*, and *Carmovirus* through the movement of virus-infected pollen. The virus from the infected pollen is then transmitted to the host plant through wounds caused by thrip feeding (Hull, 2009).

Whiteflies

The mode of transmission of viruses in whiteflies varies with the genus of virus. Members of the genus *Begomovirus* are transmitted in a persistent circulative manner (Brown, 2001), resembling somewhat, aphid transmission of *Luteovirus*. However, this relationship may be more complex than it appears due to the extended retention lengths and the transovarial passage of some species of this viral genus (Hull, 2002). In contrast, members of the genera *Closterovirus* and *Crinivirus* are transmitted in a foregut-borne, semipersistent manner (Hull, 2002). Activities of more than one different coat protein or other viral encoded proteins (like heat shock proteins) may also be necessary for transmission. *Lettuce infectious yellows virus* (LIYV), a closterovirus (genus, *Crinivirus*; family, *Closteroviridae*), has a minor coat protein (CPm) that is necessary for transmission (Ng and Falk, 2006). Regardless of which virus is transmitted or what components are required for transmission, whiteflies present constant challenges as virus vectors due to their dynamic increase in population, resistance to control, and changes in their biotype.

Noninsect Vectors

Mites

Eriophyid mites are tiny **arthropods** (0.2 mm length) that are known to transmit several plant viruses, including WSMV. Mites acquire the virus during larval stages. As in thrips, adult mites cannot acquire the virus, but both the larvae and adults transmit the virus (Slykhuis, 1955). Mites remain infective for over 2 months (Hull, 2002). WSMV particles have been found in the midgut, body cavity, and salivary glands of the mite (Paliwal, 1980). However, there has been no evidence to conclude the replication of virus in the mite. WSMV has been shown to require the potyvirus helper component-proteinase (HC-Pro) for transmission by the wheat curl mite (*Aceria tosichella* [Keifer]). This was the first demonstration of the requirement for HC-Pro in any vector except an aphid (Stenger et al., 2005).

Nematodes

Nematodes that transmit plant viruses are all migratory ectoparasites. Three genera of nematodes, *Longidorus*, *Xiphinema*, and *Trichodorus*, are primarily associated with the transmission of viruses. Nematodes feeding on virus-infected plants retain virus on the stylet, buccal cavity, or esophagus. When the nematodes are feeding on healthy host plants, the retained virus is released into the feeding site to infect the new host. Change in pH associated with saliva movement when the nematode begins to feed is hypothesized to cause release of the absorbed virus particles (Hull, 2009). Viruses in the *Tobravirus* and the *Nepovirus* genera are transmitted by nematodes.

Fungi and Fungus-Like Organisms

The chytridiomycete, *Olpidium*, and the plasmodiophoromycete, fungus-like *Polymyxa* and *Spongospora*, transmit viruses as they infect the root systems of their hosts. Zoospores released from the infected plants may carry virus either externally or internally. Viruses absorbed to the external surface of the zoospore, such as some members of the *Tombusviridae*, are released to infect the new plant. Virus absorbed to the zoospore flagellum can enter the zoospore when its flagellum is retracted to encyst. The process through which viruses are carried internally is undefined, but the coat protein appears to be the basis of the interaction between virus and zoospore. *Bymovirus* and *Furovirus* are examples of viral genera that are transmitted in this manner (Hull, 2009). Rhizomania of sugarbeets, caused by *Beet necrotic yellow vein virus* (BNYVV) transmitted by *Polymyxa*, is a good example of a viral disease transmitted in this manner that is a major economic problem.

Seed and Pollen

Viruses may be transmitted into seed by two methods. In the first method, the virus infects the embryo within the seed, and when the seedling emerges, it is already infected. This is often referred to as true seed transmission. The second transmission method is through contamination of the seed, especially the seed coat. As the germinating seedling emerges from the seed, the virus infects the plant through wounds or through microfissures caused during cell maturation. TMV is transmitted to tomato seedlings by contamination of the seed coat.

Pollen may also transmit viruses. In addition to infecting the ovule during pollination, infected pollen may be moved to uninfected plants and infect them during pollination (Hull, 2009). Pollen may also carry virus into wounds. BSMV is an example of a virus transmitted by pollen.

Dodder

The parasitic seed plant, dodder (*Cuscuta* sp.), sinks haustoria into the phloem of the plants that it parasitizes. This connection allows carbohydrates and other compounds to move into the dodder's phloem. When a dodder plant connects a healthy and virus-infected host plant, viruses can be transmitted from the infected plant through the dodder to the phloem of the noninfected plant.

Vegetative Propagation and Grafting

Viruses that systemically infect plants can be transmitted by vegetative propagation of a portion of the infected plant. This portion can range from leaves, stems, branches, and roots to bulbs, corms, and tubers. Grafting is a form of vegetative transmission, and transmission occurs through the newly established vascular system linking the graft and the scion (Hull, 2002).

HOW ARE PLANT VIRUSES DETECTED AND IDENTIFIED?

Detection and identification of plant viruses are two of the most important procedures in plant virology. Correct identification of viruses is critical to establish control tactics for the disease. Procedures for the identification of plant viruses can be divided into the following categories.

Biological Activity

Infectivity assays, indicator hosts, and host range studies are all types of bioassays that are based on defining the interaction of the viral pathogen and its hosts. Infectivity assays measure the range of viral infectivity by determining the number of host plants infected at different dilutions. Indicator hosts are certain species of plants that have known reactions to a wide range of viruses. *Chenopodium quinoa* and different species and cultivars of *Nicotiana* are widely used as indicator hosts. Host range studies test the ability of the virus to infect different plant species. Some types of viruses have very narrow host range and infect only closely related plants. For example, *Maize dwarf mosaic virus* (MDMV) infects only monocots. TMV infects a wide range of host plants, including plants from Solanaceae, Chenopodiaceae, and Compositae. Although indicator plants and bioassays were once used as the principle method of virus identification, they are currently employed only for primary characterization of new viruses. Another type of biological activity that is an important identification characteristic is the mode of transmission. Identification of the vector association helps indicate the relationship of the unknown virus to characterized groups.

Physical Properties

The most important physical property used in detection and identification is morphology or particle size and shape, which are usually determined by electron microscopy. Viruses can be visualized by negative stains of sap extracts or as purified virus solutions, thin sections of infected tissue to localize the virus within the cellular structure, and immunospecific electron microscopy that combines electron microscopy and **serology** to capture virus particles on coated electron microscope grids. Visualization of the virus provides the virion shape and size. It can also determine the presence of features such as spikes or other features of the capsid surface.

Other physical properties that have been classically utilized for the identification of viruses measure the stability of the particle outside the host. These properties include the following: thermal inactivation point, the temperature at which a virus loses all infectivity; longevity *in vivo*, the length of time a virus can be held in sap before it loses its infectivity; and dilution endpoint, the greatest dilution of sap at which the virus titer is capable of causing infection in a susceptible host. Many of these characteristics take several weeks to obtain and are no longer used for routine diagnostics. However, they continue to be utilized as part of the official virus description.

Serology

Serology is based on the ability of the viral capsid protein to elicit an antigenic response and stimulate the production of antibodies (Figure 4.18) against the antigen in an animal, usually a rabbit, goat, or mouse. Areas of the capsid with unique shape and amino acid composition are capable of inducing immune responses in avians and mammals. These uniquely shaped areas are referred to as epitopes and are only a few amino acids in length. Animals injected with plant viruses produce antibodies (**immunoglobulins**) that can recognize and attach to the epitopes. These antibodies can be isolated from the serum or utilized as the serum fraction (**antiserum**). The antibodies are termed polyclonal due to the presence of many **antibody** types in the serum. Antisera can be exchanged between researchers around the world in order to compare plant viruses from many countries. Antisera can also be frozen and utilized to compare viruses over many years. This allows researchers to make comparisons over time to follow the evolution and epidemiology of the viral disease.

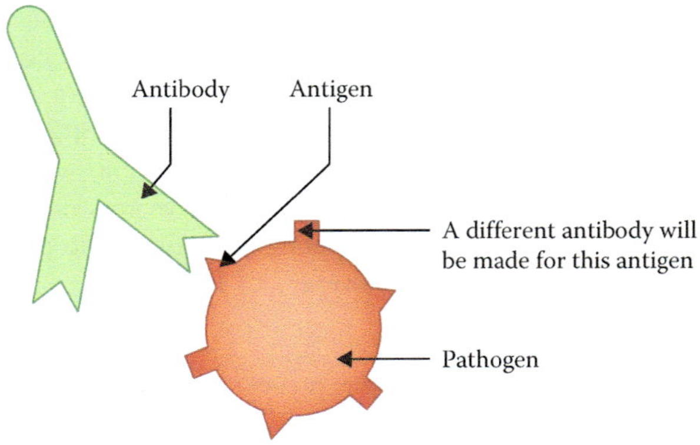

FIGURE 4.18 Cartoon drawing of an antibody and the epitope that it specifically recognizes on the antigen, to which it has been raised. (Retrieved from https//www.leavingbio.net/The%20 Human%20 Defense%20 System-web-2_files/image003.gif [accessed July 1, 2016].)

Monoclonal antibodies are produced by the fusion of an isolated spleen cell from a mouse immunized to the plant virus and a murine myeloma cell. The resulting hybridoma cell line produces only one type of antibody. Advantages to monoclonal antibodies include a single antibody that can be well characterized, identification of the eliciting epitope, production of large amounts of antibody, and the ability of the producing cell line to be multiplied and frozen.

Antisera have allowed the development of many rapid and widely utilized detection assays. The most commonly utilized serological assay is enzyme-linked immunosorbent assay (ELISA). Many variations exist in the ELISA procedure, but the most widely adopted protocol is the double antibody sandwich (Figure 4.19). This procedure starts by trapping a layer of antibodies on the well surfaces of a polystyrene microtiter plate. Attaching the antibodies to a solid surface is important since it allows all reactants that do not attach to the antibody

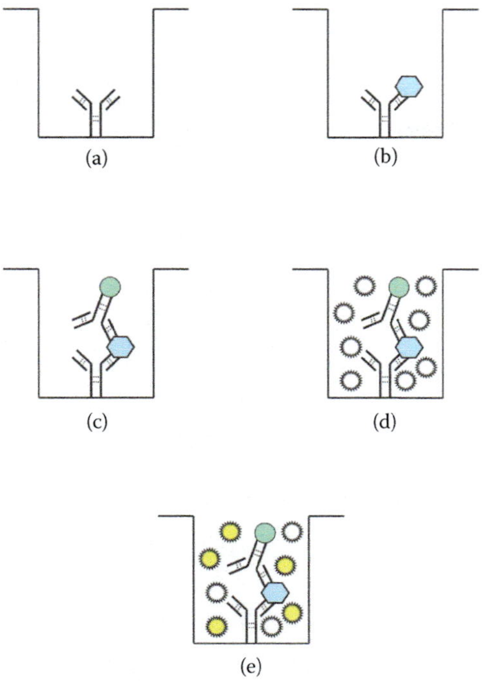

FIGURE 4.19 The principal steps in a double-antibody sandwich (DAS). (a) Antibodies (isolated immunoglobulin-G) of a virus are attached to a polystyrene microtiter plate well by incubation with an alkaline pH 9.0 carbonate buffer. Unattached antibodies are removed from the well by washing with phosphate-buffered saline with Tween 20 added. (b) Sap extracted from plant samples is diluted and added to the well. Virus contained in infected samples binds to the matching antibody. Washing removes unbound materials from the well. (c) A second layer of antibodies that have been conjugated to an enzyme (typically alkaline phosphatase) is added to the well and attached to the virus. If the virus has not been trapped by the primary layer of the antibody, these detecting antibodies are removed by the wash and thus are not present to react in the remaining steps. (d) A substrate solution is added to the wells. (e) The enzyme attached to the detecting antibody causes a color change in the substrate. The intensity of the color is proportional to the amount of enzyme present and can be quantified by spectrophotometry. (Courtesy of M. Langham.)

and well surface to be washed away between steps. The attached antibodies are used to trap virus particles from sap solutions. A second layer of antibodies conjugated to an enzyme is then used to label the virus. Alkaline phosphatase is the most widely utilized enzyme, but other enzymes such as horseradish peroxidase may also be used. After the final washing, a substrate solution is added to the wells. Substrates are chosen because they change color after being acted on by the conjugated enzyme. The color change can be quantified by reading the absorbance of a known wavelength of light when passed through each well. The amount of virus in the test solution is proportional to the light absorbance. ELISA is the basis for the development of many of the rapid diagnostic tests used by growers, producers, and agricultural consultants (Procedure 4.3).

Molecular Properties

The molecular weight and number of proteins forming the capsid layer are important characteristics of plant viruses. These characteristics can be determined by standard gel electrophoresis procedures. Determination of the amino acid content and sequence are also important in understanding particle structure and function.

Nucleic Acid

The ability to compare nucleic acid characteristics rapidly opened a new dimension in the identification of plant viruses. Identification based on the nucleic acid genome has enhanced the ability to identify strains and new viruses with similar characteristics. Comparisons of viral genomes have facilitated organization of plant viruses into the current taxonomic system and have had important implications for concepts of viral evolution. Nucleic acid tests provide some of the best tools for viral identification. The most powerful of these new tools is polymerase chain reaction (PCR).

Polymerase Chain Reaction

PCR is an enzymatically driven chemical reaction by which complete genes or selected fragments can be amplified (to make more), from a portion of a nucleic acid genome. The targeted region is defined by the use of short sequences of matching nucleotides, called "primers" that anneal or "bind" to the denatured, template of choice, or "target" dsDNA strands (forward and reverse) in solution (Figure 4.20). Amplification occurs by sequential round of copying the target enzymatically (Figure 4.21) using a DNA polymerase that catalyzes the synthesis of the DNA strands between the primer annealing sites, using dNTPs and magnesium, also present in the buffer mix. The particular DNA polymerase used in PCR must be stable at high temperatures (up to 95°C) to withstand sequential rounds of denaturation (heating to separate the dsDNA strands), "annealing" or hybridization of primers (55°C) to "known" sequence, followed by enzymatic extension (72°C) of the DNA strand (see Palumbi et al., 1991, for details). Initially, the use of PCR was limited to DNA viruses; however, the use of reverse transcriptase to generate DNA strands (cDNA) from RNA viral genomes has expanded our ability to utilize this technology with RNA viruses.

Procedure 4.3
Serological Detection of Plant Viruses

Step	Instructions and Comments
1	Complete any preliminary step that is required for the assay that has been selected according to the directions that come with the assay.
2	Macerate healthy tissue in the recommended extraction buffer.
3	Macerate series of unknown samples in the recommended extraction buffer.
4	Macerate infected tissue in the recommended extraction buffer.
5	ELISA—load the recommended sample amount into each test well. Be sure to include positive and negative controls as well as buffer for standardized checks. Allow to incubate as directed by the kit.
6	Immunostrip assays—Place the wicking end of the strip into the sap extracts to the depth indicated by the manufacturer. Allow to incubate as directed by the kit.
7	ELISA—After incubation and washing, ELISA plates must be incubated with the secondary antibody, washed, and developed as directed by the instructions in the kit. The intensity of the color in the test wells is dependent on the amount of virus captured on the plate.
8	Immunostrip assays—Immunostrips are ready as soon as their incubation time is completed. Most strips will have one line for a positive control and one for the sample. Be sure and follow the instructions on the interpretation of the kit that you have as these can vary with different virus combinations.
9	Record and compare results.

HOW ARE PLANT VIRUSES CONTROLLED?

Control of plant viruses has a different primary focus than many other pathogens because there are no practical therapeutic or curative treatments for plant viruses. Thus, plant virus control focuses on preventative measures.

Host Plant Resistance

Host plant resistance is the major approach to control viral diseases, but has not been identified in many crop species. It is typically the most economical control measure because it requires low input from the producer. In addition, resistant plants eliminate the need for controlling the vector and are selective against the primary pathogen (Khetarpal et al., 1998). Resistance to plant viruses can be due to the inability to establish infection; inhibited or delayed viral multiplication; blockage of movement; and resistance to the vector and viral transmission from it (Jones, 1998). Development of plant cultivars with viral resistance is the primary goal of many plant virology and breeding projects. The three lines of wheat infected by WSMV in Figure 4.8 are from the evaluation of wheat lines for breeding selection.

In some crops where strong resistance to viral diseases is not available, **tolerance** may be an alternative goal for plant breeding (Hull, 2009). Tolerant plants become infected with the virus, but they do not develop severe reactions or yield losses. One disadvantage to the use of tolerance is that it does not reduce the inoculum available in the environment, as the virus is able to replicate in the tolerant plants.

In addition to traditional breeding, two other methods are available for increasing the resistance of the host plant. In **cross protection**, a viral strain that produces only mild or no symptoms is inoculated as a protecting strain. When a second, more severe strain is used as a challenge inoculation, the presence or effects of the first virus block its infection. This effect was first observed by McKinney (1929), and it has been used successfully in the control of some viruses, such as *citrus tristeza virus* (CTV) in Brazil and South Africa (Lecoq, 1998). Limitations to the usefulness of cross protection include the inability of viruses to cross protect, the yield reductions caused by mild strains, and the possible mutation of the mild strain to a more virulent form.

Genetic engineering provides a second method for enhancing host plant resistance. It is particularly valuable in situations where no natural source of resistance has been identified. Gene coding for coat protein and replicase (Kaniewski and Lawson, 1998), **antisense RNA**, and **ribozymes** (Tabler et al., 1998) have been used to confer resistance to different viruses. This technology holds much promise for the future. However, problems with stability of the inserts, expression of the inserted genes, unanticipated effects on the host plant, and public acceptance of genetic engineering remain as challenges to this technique.

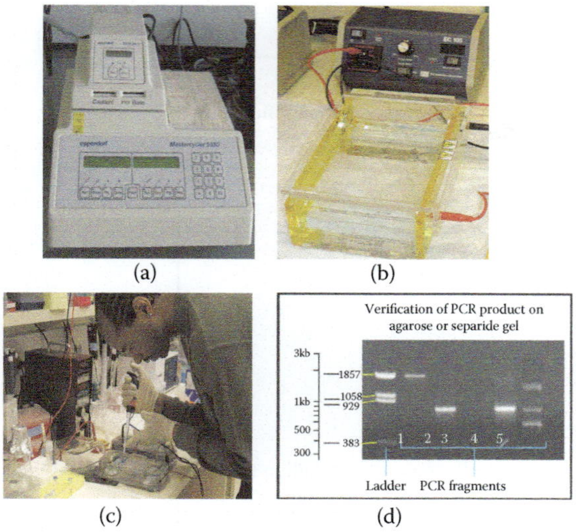

FIGURE 4.20 Polymerase chain reaction (PCR) machine (a) for PCR amplification of nucleic acids targeted by sequence specific primers, and a typical agarose gel electrophoresis apparatus with power supply (b) used to separate PCR products by size in an electrical field. PCR products are loaded into wells made in an agarose (0.7%–2.0%) gel matrix. The sample is first mixed with loading buffer and a blue dye, the latter, to allow for tracking of the sample DNA as it is pushed in the electrical field from the negatively charged (cathode, black) pole, while also being pulled toward the positively charged (anode, red) pole. When the dye reaches the bottom of the gel, the electricity is turned off. The gel is then stained with a fluorescent dye and viewed under ultraviolet light to visualize the PCR products (d). The gel photo (c) shows the "amplicons" or PCR products that have migrated in the buffer, in relation to one another, with the largest fragments running most slowly, and the smallest running the fastest. The size of the dsDNA can be estimated using the bands of known size, referred to as a "ladder," which are loaded onto the gel alongside the samples. ([a] Retrieved from https://commons.wikimedia.org/wiki/File:Pcr_machine.jpg. Author: Magnus Manske, public domain [accessed July 1, 2016]. [b] Retrieved from https://commons.wikimedia.org/wiki/File:Gel_electrophoresis_apparatus.JPG. Author: Jeffrey M. Vinocur. [accessed July 1, 2016]. [c] Retrieved from https://commons.wikimedia.org/wiki/File:Load_DNA_Gel.jpg. Author: Maggie Bartlett, NHGRI, public domain [accessed July 1, 2016]. [d] Retrieved from http://missinglink.ucsf.edu/lm/molecularmethods/images/clip_image004.gif. iROCKET Learning Module: Molecular Methods. Authors: Doris Wang, Arif Hussain, and Nili Sommovilla [accessed July 1, 2016].)

Vector Control

Controlling insect or other vectors to minimize viral transmission differs from simply controlling pests. While controlling pests, the principal goal is to reduce the levels of pests below the levels that cause economic damage. However, low levels of pests may still be significant in the transmission of viral diseases (Satapathy, 1998). To determine the suitability of vector control, it is necessary to determine the vector and virus affecting the host (Hull, 2009). This information establishes the type of relationship involved in transmission. For example, aphids transmitting MDMV require only seconds to accomplish this because the virus is transmitted in a nonpersistent manner. *Oat blue dwarf virus* (OBDV), in contrast, is transmitted by leafhoppers in a persistent, circulative, and propagative manner that requires the leafhopper to feed on the plant for over 20 min. Thus, chemical treatments may be more effective on the leafhoppers transmitting OBDV than on the aphids transmitting MDMV. In addition, effects of the controls on the feeding behaviors of the hosts should be considered. Vectors moving into the field from the surrounding hosts or due to seasonal migration may affect control measures. Each of these affects the probability for successful control of the virus. In the future, understanding the biological interactions between viruses and vectors may allow development of unique controls based on these intricate molecular mechanisms (Andret-Link and Fuchs, 2005)

Quarantine and Eradication

Control of viral movement between fields, states, countries, and continents is often difficult and time consuming. **Quarantine** acts as the first line of defense against the introduction of many foreign viruses. However, the ability to detect viruses and other submicroscopic pathogens for exclusion is limited in comparison with other pathogens (Foster and Hadidi, 1998). Viruses may enter into new areas through importation of seeds, nursery stock, viruliferous vectors, plants, or experimental material (Foster and Hadidi, 1998). When exclusion of these pathogens fails, the cost of control and **eradication** of the virus (if possible) is often overwhelming. A current example of this is the PPV outbreak in the Northeastern United States. Costs for this outbreak included loss of hundreds of nursery trees, cost of the quarantine (detection, identification, and tree removal), and loss of income to the growers.

Cultural Controls

Cultural controls for viruses can be referred to as "wise production practices." These include any production practice or method that eliminates or significantly reduces the threat of viruses. Virus control is based on breaking the bridge of living hosts that is necessary to complete the seasonal movement of a virus. Any cultural practice that modifies the time or spatial interaction of virus and host can be an effective control. One common cultural control is modifying the date of planting. In South Dakota, delaying the planting of winter wheat until the seasonal end of wheat curl mite movement is one of the most effective

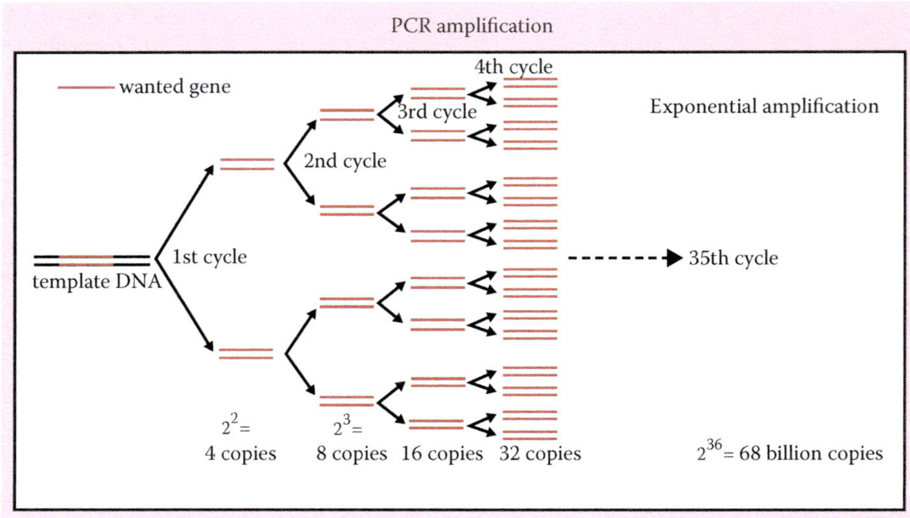

FIGURE 4.21 Illustration of the exponential amplification of the template or "target" dsDNA due to repeated cycles (usually 30–40) of amplification. PCR products can be cloned and the DNA sequence determined, an outcome not possible without prior amplification. This makes PCR such a powerful technique. (Retrieved from http://stratfeed.cra.wallonie.be/img/page/PCR_web_page5.jpg [accessed July 1, 2016].)

controls available. However, other concerns such as early planting of winter wheat to increase the amount of residue cover for the winter conflict with virus control. Thus, producers must decide the wise balance point between these goals for their farm. Other common cultural controls include production of virus-free seeds and plants through indexing and **seed certification** programs, elimination of volunteer plants, elimination of alternate host plants, elimination of overlap between the planting of two crops, and sanitation of equipment.

Plant virology is a diverse area of plant pathology. Research with plant viruses includes determining the basic molecular structure and mechanisms of viruses, studying the cellular interaction, understanding vector relationships, or working with farmers and producers in the field. However, the ultimate goal of assisting farmers and producers across the world in the control of plant viruses remains the same.

SATELLITE AGENTS THAT DEPEND ON PLANT VIRUSES

A **satellite** is a subviral agent that cannot function independently without the assistance of a **helper virus**. Helper viruses are capable of independent replication and can cause infections independently. However, satellites are as dependent on the virus as the virus is on its plant host, and they are sometimes referred to as small molecular parasites. The most common of the satellites are satellite viruses and satellite RNAs; however, small satellite DNAs have been associated with the begomoviruses.

Satellite Viruses

Satellite viruses are viruses that form complete intact virus particles, but they are not capable of independent replication. Satellite viruses possess the genetic code for their own coat proteins, which are used to **encapsidate** their genome. Thus, they are serologically distinct from their helper virus. However, the RNA genome of the satellite virus is dependent on the RNA replication enzymes of the helper virus for replication. Typically, satellite viruses are smaller than their helper virus. Satellite tobacco necrosis virus (STNV) is 17 nm in diameter, whereas *Tobacco necrosis virus* (TNV) is 26 nm in diameter. *Panicum mosaic virus* (PMV) measures approximately 25–30 nm in diameter, and its satellite, satellite panicum mosaic virus (SPMV) is approximately 15–18 nm in diameter. The effects of satellite viruses on the diseases caused by their helper viruses can vary, but satellite viruses have been associated with increased cellular and long-distance transport in the host and in translational stimulation.

Satellite RNAs

Satellite RNAs have been identified in association with several helper viruses. They are distinct from the helper virus because satellite RNAs do not contain any sequence homology with the helper virus. They are dependent on the helper virus for replication. In addition, they do not produce their own coat protein and are encapsidated in the helper virus coat protein. Satellite RNAs are divided into the following three basic groups: (1) large satellite RNAs that are 0.7 kb or larger and contain a single open reading frame (ORF) that can be translated into a protein; (2) small linear RNAs that are less than 0.7 kb and have no ORFs; and (3) small circular satellite RNAs that are about 350–450 nucleotides and occur in circular and linear forms. The first satellite RNA was described in 1972 when a lethal necrotic reaction of tomato plants infected with *Cucumber mosaic virus* (CMV) occurred in France. A fifth segment of RNA (CRNA5) that was not part of the CMV genome was discovered in CMV isolated from the infected plants. This new satellite RNA was responsible for intensifying the symptoms caused by CMV. Many more satellite RNAs have been discovered since this initial discovery, and there are satellite RNAs that lessen or do not change the effects of the helper virus as well as those that intensify them (Hu et al., 2009; Hull, 2009).

FIGURE 4.22 Symptoms in tomato plants of tomato chlorotic dwarf viroid, which is readily transmitted through contact in greenhouses. (Courtesy of J.K. Brown.)

VIROIDS—ANOTHER SUBCELLULAR PATHOGEN

Are viruses the only subcellular pathogen in plants? Viruses are the most commonly described subcellular pathogen group in plants, but they are not the only one. Viroids are subcellular pathogens simply composed of small "naked" RNA genomes. These genomes are circular pieces of ssRNA ranging from approximately 200 to 400 nucleotides. Viroids differ from viruses, because they do not have a protein coat to surround their RNA nor do they produce proteins as part of their infection or replication. Viroids replicate using the host's RNA replication enzymes with an asymmetric or symmetric rolling circle mechanism.

How can a naked RNA strand survive to pass from host to host? Viroid RNA has a number of sequences that are self-complimentary and align to form internal base pairing. These internal "double-stranded areas" transform the viroid RNA from an open circle to a rod-like shape with unpaired areas forming loops. This rod-like shape is very resistant to destruction by enzymes, heat, and ultraviolet light. With this resistance and stability, viroids can be transmitted mechanically at a high rate. In field situations, this high rate of mechanical transmission allows viroids to be readily spread by contaminated farm machinery, hand tools, or people (Figure 4.22). Foliar contact by infected plants is an important transmission pathway with viroids. Viroids are also transmitted by vegetative propagation and grafting, especially in horticultural crops. Seed transmission has also been demonstrated for many viroids. Within the plant cell, viroids utilize the cell's internal movement pathways to translocate rapidly from the entry point to desired locations in the cells throughout the plant. Viroids classified in the Pospiviroidae accumulate and replicate in the cell's nucleus, and those classified in Avsunviroidae accumulate and replicate in the chloroplasts (Hammond and Owens, 2006).

Viroids are pathogenic and cause striking disease symptoms in their plant hosts. The first viroid disease described was potato spindle tuber caused by potato spindle tuber viroid (PSTVd). PSTVd was first described in the early 1920s, but it was assumed to be a virus until its identity as a new pathogen which was described by Theodore O. Diener in 1971 (Owens and Verhoeven, 2009). Today, more than 30 viroid diseases are known to infect a diverse number of agronomic and horticultural crops (ICTV, 2014).

LABORATORY EXERCISES

The ultramicroscopic nature of plant viruses causes many students and producers to wonder about working with plant viruses. "How can something that you cannot see or feel cause a disease?," "How can you determine if a plant is infected by a virus when you cannot even see it with a microscope?," and "How can something so small even exist?" are questions often asked about plant viruses. Masses of mycelial growth and sclerotia produced during the fungal disease and white mold or the erumpent pustules produced by rust fungi are quickly recognizable as pathogenic agents. However, identifying ultramicroscopic pathogens such as viruses requires different concepts than pathogens that are quickly visualized with the naked eye or with a hand lens. In general, plant viruses are most commonly tracked by the effects that they produce in host plants or by assays that detect these ultramicroscopic pathogens.

Experiment 1. Mechanical Transmission of Plant Viruses (TMV)

Mechanical transmission or inoculation is a widely used technique in plant viral experiments and is effective for most viruses that infect the epidermal tissue layers of their host plants. It is utilized for the propagation of plant viruses, infectivity comparisons, host range determinations, and host plant resistance evaluations. Successful mechanical inoculation is dependent on the condition of the virus-infected tissue source, condition of the host plants to be inoculated, inoculum preparation, use of abrasive, and environment where the plants are maintained during symptom development. The "ideal" propagation host for inoculum preparation is a young, systemically infected plant with a high virus concentration (titer) in the tissue. The propagation host should not develop necrotic symptoms. It should contain neither high concentrations of inhibitors that interfere with infection establishment nor tannins or other phenolic compounds that can precipitate virus particles. Unfortunately, it is not always possible to utilize an "ideal" propagation host. Propagation hosts with less desirable characteristics may require modifying inoculation protocols. For example, utilizing a host with a low virus concentration or one that produces local lesions may require modification of buffer to tissue ratios. Additionally, growing an ideal propagation host in poor conditions may cause changes in the virus concentration or in compounds present in the tissue.

Virus release from the propagation host's cellular structure necessitates maceration of infected tissue (Figure 4.23a). Tissue may be ground in a mortar and pestle for small volumes. Large inoculum volumes that require macerating large quantity of infected tissue are typically prepared using blenders. Disrupting plant cells not only releases virus particles but also breaks

FIGURE 4.23 Mechanical inoculation of plant viruses is commonly utilized to inoculate new host plants. (a) Infected plant material is ground in a mortar and pestle to macerate the tissue as the beginning step in the preparation of sap extracts for inoculation. (b) Sap extract is prepared by adding buffer to macerated leaf tissue and grinding. (c) Mechanical inoculation of *Phaseolus vulgaris* cv. provider is accomplished by gently rubbing the surface of the plant leaves with a mixture of sap extract and silica carbide. (d) Mechanical inoculation of tobacco demonstrating the technique for supporting the leaf and coating it with inoculum during inoculation. (Courtesy of M. Langham.)

the central vacuole and other smaller vacuoles releasing their contents into the extracted sap. The first effect of releasing the vacuole contents is to increase the acidity (lowering the pH) of the plant sap extract. As the sap pH decreases, proteins (including viruses with their protein capsid) precipitate out of solution. Thus, buffers (Figure 4.23b) are added to the tissue during grinding to stabilize the extract at a pH favorable to virus infectivity, integrity, and suspension. Although the majority of viruses can be transmitted in the pH 7 (neutral) range, the desired pH varies with the virus being used for inoculation. This is not the only role that buffers serve in preparing inoculum. The ionic composition of the buffer contributes to inoculation success. Addition of phosphate to an inoculation buffer increases the infection rate during inoculation. Thus, phosphate is one of the most commonly utilized buffer components (Hull, 2002). Other ions may be added to the inoculation buffer if required for virus stability. For example, some viruses require the presence of calcium or magnesium to maintain their structure, and these ions must be added to the inoculation buffer.

Reducing agents, **chelating agents**, and **competitors** are other chemicals that may be added to inoculation buffers for viruses with specific requirements. Sodium sulfite, 2-mercaptoethanol, sodium thioglycolate, and cysteine hydrochloride are examples of frequently utilized reducing agents. For example, sodium sulfite is often used in inoculation buffers for *Tomato spotted wilt virus*. Quinones, tannins, and other phenolic compounds in plant sap can cause precipitation of plant viruses. This activity is countered by chelating agents and competitors. Chelating agents such as diethyldithiocarbamate are added to bind the copper ions found in polyphenoloxidase, the enzyme that oxidizes phenolic compounds in the plant. Competitors, such as hide powder, have been used to dilute the effects of phenolic compounds by serving as alternate sites for phenolic activity. Finally, mechanical inoculation is based on our ability to create nonlethal wounds in plant cells for virus entry (Figure 4.23c and d). Wounds in the cell wall and membrane are created by the use of an abrasive and allow the plant virus to enter the host plant. The most commonly used abrasive is silica carbide; however, celite, bentonite, and acid-washed sand have also been utilized (Hull, 2002).

Materials

Each student or team of students will require the following items:

- TMV-infected host plant tissue (*Nicotiana tabacum* is a good systemic host for providing inoculum)
- Healthy host plant tissue (*N. tabacum*)
- Mortars and pestles
- Inoculation pads (sterile cheesecloth rectangles [approximately 2–3 cm × 10–15 cm])
- Silica carbide (600 mesh)
- 0.02 M phosphate buffer, pH 6.8–7.2 (add 3.48 g of K_2HPO_4 and 2.72 g of KH_2PO_4 to 1 L of distilled water to prepare this buffer)
- *N. tabacum* plants with three to four well-expanded leaves (four plants per student or group); *C. quinoa*, *Nicotiana glutinosa*, and *N. tabacum* cv. Samsun N may also be used if desired
- 10-mL pipettes
- Top loading balance
- Markers, paper tags, and string

Follow the protocol outlined in Procedure 4.1 to complete this part of the experiment.

Anticipated Results

Plants should be observed closely for 2–3 weeks. Mechanical damage due to excessive pressure during inoculation (sometimes referred to as "inoculation burns") appears first and can be observed as necrotic damage to the epidermis. It appears on both the plants inoculated with virus and those inoculated with healthy tissue. Local lesion hosts begin to display lesions in 3–5 days on the leaves that were inoculated. Development of local lesions from pinpoint to larger size can be observed. Systemic symptoms become apparent from 10 days to 2 weeks depending on the growth conditions. Systemic symptoms first appear in newly expanding leaves.

Questions

- Why is a propagation host needed that does not have necrotic reactions to the virus?
- If an abrasive was not included in the inoculum or dusted on the plants, would the inoculated plants become infected?
- Why should plants inoculated with healthy plant sap be included in all mechanical inoculation studies?
- How many virus particles are required to initiate an infection in a plant?
- How do local lesion infections differ from systemic infections?
- How can you prevent people from becoming plant virus vectors in field or greenhouse situations?
- How can mechanical inoculation be utilized in developing host plant resistance in crop plants?
- Does a virus whose genome is in multiple particles require a higher concentration of virus in the inoculum than a virus whose genome is in a single particle?
- What is a viral inhibitor, and why does it cause problems when you are inoculating plants?

- Is mechanical transmission of diseases a concern when you are using farm machinery for normal operations such as plowing or spraying in a field?

Experiment 2. Mechanical Transmission of Plant Viruses for Host Ranges

Experiments to study host range can be done by utilizing the inoculation procedure in the transmission exercise to inoculate a number of different host plants, including susceptible and nonsusceptible plant species.

Materials

- Select a test virus and a compatible host set. Table 4.2 provides some examples of plants and compatible viruses. Other combinations can be found in the VIDE database (Brunt et al., 1996).
- The remaining materials are the same as in Experiment 1.
- Follow the protocol outlined in Procedure 4.2 to complete this part of the experiment.

Anticipated Results

A variety of symptoms should be formed. Within 3–5 days, local lesions should be visible. Students should observe the number of lesions and whether or not they are necrotic or chlorotic. Systemic symptoms will form later (approximately 10–14 days). Some hosts may not be susceptible and will not have any symptoms. Control plants should be carefully observed for inoculation damage and compared with the experimental plants.

Questions

- Why would the environment where an inoculated plant is being grown have an effect on the rate and type of symptom expression?
- Why does it require 10 days to 2 weeks for a plant to develop systemic symptoms when it only takes 3–4 days for a plant to develop local lesions?
- Can a plant with a local lesion reaction develop a systemic reaction also?
- How can knowing the host range of a virus assist in predicting the susceptibility of an untested plant species?
- Can a host range help identify plants that might be alternate hosts of the virus and enable it to survive when the affected crop is not in season?
- Would a virus only infect hosts that are closely related to each other?

TABLE 4.2
Examples of Plant Species Utilized in a Host Range Experiment

Virus	Plant Species	Reaction[a]
TMV		
	Chenopodium amaranticolor	Susceptible
	C. quinoa	Susceptible
	Cucumis sativus	Susceptible
	Cucurbita pepo	Susceptible
	Lactuca sativa	Susceptible
	N. glutinosa	Susceptible
	N. tabacum	Susceptible
	Avena sativa	Not susceptible
	Pisum sativum	Not susceptible
	Spinacia oleracea	Not susceptible
	Triticum aestivum	Not susceptible
	Zea mays	Not susceptible
	Zinnia elegans	Not susceptible
BPMV		
	Glycine max	Susceptible
	Lens culinaris	Susceptible
	Phaseolus vulgaris cvs. Black Valentine, Tendergreen, or Pinto	Susceptible
	P. sativum	Susceptible
	Vigna unguiculata	Susceptible
	C. sativus	Not susceptible
	N. glutinosa	Not susceptible
	N. tabacum	Not susceptible
	Petunia × hybrida	Not susceptible
TRSV		
	Beta vulgaris	Susceptible
	C. amaranticolor	Susceptible
	C. sativus	Susceptible
	C. pepo	Susceptible
	N. tabacum	Susceptible
	Petunia × hybrida	Susceptible
	P. vulgaris	Susceptible
	Brassica campestris ssp. *rapa*	Not susceptible
	Helianthus annuus	Not susceptible
	Secale cereale	Not susceptible
	Trifolium hybridum	Not susceptible
	T. aestivum	Not susceptible

[a] Variation in reaction is possible due to virus isolate and inoculation conditions.

Experiment 3. Detection of Plant Viruses

Assays that detect plant viruses are a quick and efficient way to confirm the presence of a viral pathogen and to determine its identity. Detection assays are rapid and do not require the amount of greenhouse space or host plants that mechanical transmission and host range assays do. However, detection assays (depending on the type and the virus being identified) may be unavailable, limited in sample number, or costly. They may also lack the ability to differentiate between closely related virus strains. When developing detection assays, a unique identifier for the virus in question is needed. Viruses are composed only of the capsid (protein coat) and the nucleic acid genome. Thus, the unique identifier for the assay must detect characteristics in either viral protein or nucleic acid.

Serological assays have been the most widely used detection method for plant viruses since their development. Today, PCR techniques (described previously) are becoming more available and affordable; however, serology remains widely used, especially in large-scale studies, first and confirming identifications, and quick industry applications. In order to develop a serological-based assay for a virus, the virus must be able to elicit an immune response in an animal (**immunogenic**). This ability depends on the capsid protein having distinctive clusters of six to eight amino acids (antigenic determinants) called epitopes. When purified virus is injected into an animal, it is recognized as a foreign protein and its immune system builds antibodies that recognize the epitopes found in the viral capsid protein. Antibodies that have been made to a purified virus can be utilized in a number of assays.

Materials

- Virus-infected plant tissue—Selection of the plant and virus to utilize is dependent on many factors. With the exception of TMV, availability of virus-infected materials varies from region to region. Select a virus that is appropriate for your location.
- Healthy plant tissue that is the same plant species and cultivar as the infected material.
- A series of two to six plant tissue samples prepared by the instructor and labeled only by a number or other code per student or student group; the instructor will know which samples are infected and which are not.
- Immunoassay supplies—When selecting an immunological assay to do in class, consider the amount of time that is available for your lab. Immunostrip assays are rapid and will easily fit into a limited time. ELISA typically requires multiple days. Students can do the entire protocol if facilities are available for students to have access through the required assay period. Alternatively, all steps except for the final rinse and development can be done prior to lab. The final development step easily fits into a short lab period. Unless other sources are available, kits that come complete with buffers and required supplies are preferable. ELISA or Immunostrip kits are available from the two following suppliers:

  Agdia
  52642 County Road 1
  Elkhart, IN 46514, USA
  1-800-62-AGDIA
  http://www.agdia.com/
  AC Diagnostics, Inc.
  1131 W Cato Springs Road
  Fayetteville, AR 72701, USA
  (479) 595-0320; (479) 251-1960
  http://www.acdiainc.com

- Grinding tissue samples—Tissue samples can be ground in mesh sample bags such as ACC00930 from Agdia or mortars and pestles. If the tissue is flat and care is taken not to perforate the bag, it can also be done in a sandwich bag that seals shut with a smooth rounded object such as a pestle to gently macerate the tissue.
- Washing—ELISAs require washing between each step with phosphate-buffered saline with Tween 20 added. If a mechanical ELISA washer is not available, this can be done manually by carefully removing the contents of the test well and filling the wells with wash buffer three to five times, preferably allowing 5 min for the plate to soak between each wash. For the washes after the plate has incubated with samples, care must be taken to avoid cross contaminating the wells. Samples should be removed by careful pipetting. Care should be taken with subsequent washes to preserve the unique profile of each sample well. When wells are filled with identical solutions (e.g., wash buffer), the contents can be removed by a rapid flicking motion over the sink.
- Micropipettes and disposable tips.
- Guide to record data.
- Follow the protocol outlined in Procedure 4.3 to complete this part of the experiment.

Anticipated Results

Infected plant samples are expected to produce positive reactions (color changes in the ELISA substrate reagent or lines in the Immunostrip assays). Healthy plant samples should not provide these changes. Based on these reactions, the coded tissue samples should be identified as infected or healthy. Variations may be seen in the intensity of the reactions depending on the virus concentration in each tissue sample.

Questions

- Why is it necessary to add a negative control to the immunoassay tests?
- Why is a positive control also required?
- What would it mean if the wells that contained buffer as a sample reacted positively?
- What advantages does serological identification of the virus provide?
- A virus is seed transmitted. How can immunoassays be used to assist in the control of this virus?
- What would happen if a test sample was infected with more than one virus?
- Why would it be beneficial to a producer to rapidly identify a virus infecting his crop?
- If virus A is transmitted by aphids and virus B is transmitted by nematodes, how would knowing which virus is infecting the plants help control the virus?

REFERENCES

Andret-Link, P., and M. Fucha. 2005. Transmission specificity of plant viruses by vectors. *J. Plant Pathol.* 87:153–165.

Baltimore, D. 1971. Expression of animal virus genomes. *Bacteriol. Rev*. 35:235–41.

Bawden, F. C., N. W. Pirie, J. D. Bernal, and I. Fankuchen. 1936. Liquid crystalline substances from virus-infected plants. *Nature* 138:1051–1055.

Brown, J. K. 2001. *The Molecular Epidemiology of Begomoviruses*. Pages 279–316 in: *Trends in Plant Virology*, J. A. Khan, and J. Dykstra (eds). New York, NY: The Haworth Press, Inc.

Brunt, A. A., K. Crabtree, M. J. Dallwitz, A. J. Gibbs, L. Watson, and E. J. Zurcher. (eds). 1996 onwards. `Plant Viruses Online: Descriptions and Lists from the VIDE Database. Version: 20th August 1996.' URL http://biology.anu.edu.au/Groups/MES/vide/.

Creager, A. N. H., K. B. G. Scholthof, V. Citovsky, and H. B. Scholthof. 1999. Tobacco mosaic virus: Pioneering research for a century. *Plant Cell* 11:301–308.

Field, T. K., C. A. Patterson, R. C. Gergerich, and K. S. Kim. 1994. Fate of virus in bean leaves after deposition by *Epilachana varivestis*, a beetle vector of viruses. *Phytopathology* 84:1346–1350.

Foster, J. A., and A. Hadidi. 1998. Exclusion of plant viruses. Pages 208-229 in: *Plant Virus Disease Control*, A. Hadidi, R. K. Khetarpal, and H. Koganezawa (eds). St. Paul, MN: APS Press.

Fraenkel-Conrat, H. 1956. The role of the nucleic acid in the reconstitution of active tobacco mosaic virus. *J. Amer. Chem. Soc*. 78:882–883.

Gergerich, R. C., and H. A. Scott. 1988a. Evidence that virus translocation and virus infection of non-wounded cells are associated with transmissibility by leaf-feeding beetles. *J. Gen Virol*. 69:2935–2938.

Gergerich, R. C., and H. A. Scott. 1988b. The enzymatic function of ribonuclease determines plant virus transmission by leaf-feeding beetles. *Phytopathology* 78:270–272.

Gergerich, R. C., H. A. Scott, and J. P. Fulton. 1984. Evidence that ribonuclease in beetle regurgitant determines the transmission of plant viruses. *J. Gen. Virol*. 67:367–370.

Hammond, R. W., and R. A. Owens. 2006. Viroids: New and continuing risks for horticultural and agricultural crops. *APSnet Features*. doi:10.1094/APSnetFeature-2006-1106.

Hogenhout, S. A., El-Desouky Ammar, A. E. Whitfield, and M. G. Redinbaugh. 2008. Insect vector interactions with persistently transmitted viruses. *Annu. Rev. Phytopathol.* 46:327–359.

Hu, C. C., Y. H. Hsu, and N. S. Lin. 2009. Satellite RNAs and satellite viruses of plants. *Viruses* 1:1325–1350. doi:10.3390/v1031325.

Hull, R. 2002. *Matthews' Plant Virology* (4th ed). New York, NY: Academic Press.

Hull, R. 2009. *Comparative Plant Virology* (2nd ed). New York, NY: Elsevier Academic Press.

International Committee on Taxonomy of Viruses (ICTV). 2014. Virus taxonomy: 2013 Release. http://www.ictvonline.org/virusTaxonomy.asp.

Jones, A. T. 1998. Control of virus infection in crops through breeding plants for vector resistance. Pages 41–55 in: *Plant Virus Disease Control*, A. Hadidi, R. K. Khetarpal, and H. Koganezawa (eds). St. Paul, MN: APS Press.

Kaniewski, W., and C. Lawson. 1998. Coat protein and replicase-mediated resistance to plant viruses. Pages 65–78 in: *Plant Virus Disease Control*, A. Hadidi, R. K. Khetarpal, and H. Koganezawa (eds). St. Paul, MN: APS Press.

Khetarpal, R. K., B. Maisonneuve, Y. Maury, B. Chalhoub, S. Dinant, H. Lecoq, and A. Varma. 1998. Breeding for resistance to plant viruses. Pages 14–32 in: *Plant Virus Disease Control*, A. Hadidi, R. K. Khetarpal, and H. Koganezawa (eds). St. Paul, MN: APS Press.

Lecoq, H. 1998. Control of plant virus diseases by cross protection. Pages 33-40 in: *Plant Virus Disease Control*, A. Hadidi, R. K. Khetarpal, and H. Koganezawa (eds). St. Paul, MN: APS Press.

McKinney, H. H. 1929. Mosaic diseases in the Canary Islands, West Africa and Gibraltar. *J. Agric. Res*. 39:557–578.

Morgan, G. J. 2006. Virus design, 1955-1962: Science meets art. *Phytopathology* 96:1287–1291.

Ng, J. C. K., and B. W. Fald. 2006. Virus-vector interactions mediating nonpersistent and semipersistent transmission of plant viruses. *Annu. Rev. Phytopathol*. 44:183–212.

Owens, R.A., and J.Th. J. Verhoeven. 2009. Potato spindle tuber. *The Plant Health Instructor.* doi:10.1094/PHI-I-2009-0804-01.

Paliwal, Y. C. 1980. Relationship of wheat streak mosaic and barley stripe mosaic viruses to vector and nonvector eriophyid mites. *Arch. Virol*. 63:123–132.

Palumbi, S. R., A. Martin, S. Romano, W. O. McMillan, L. Stice, and G. Grabowski. 1991. *The Simple Fool's Guide to PCR, Version 2.0*. Privately published, Honolulu, HI: University of Hawaii.

Satapathy, M. K. 1998. Chemical control of insect and nematode vectors of plant viruses. Pages 188–195 in: *Plant Virus Disease Control*, A. Hadidi, R. K. Khetarpal, and H. Koganezawa (eds). St. Paul, MN: APS Press.

Scholthof, K. B, J. G. Shaw, and M. Zaitlin (eds). 1999. *Tobacco Mosaic Virus–One Hundred Years of Contributions to Virology*. St. Paul, MN: APS Press.

Scholthof, K. B. 2001. 1898 - The beginning of Virology... time marches on. *Plant Health Instructor.* doi:10.1094/PHI-I-2001-0129-01.

Slykhuis, J. T. 1955. *Aceria tulipae* Keifer (Acarina: Eriophyidae) in relation to the spread of wheat streak mosaic virus. *Phytopathology* 45:116–128.

Stenger, D. C., G. L. Hein, F. E. Gildow, K. M. Horken, and R. French. 2005. Plant virus HC-PEO is a determinant of eriophyid mite transmission. *J. Virol.* 79:9054–9061.

Tabler, M., M. Tsagris, and J. Hammond. 1998. Antisense RNA and ribozyme-mediated resistance to plant viruses. Pages 79–93 in: *Plant Virus Disease Control*, A. Hadidi, R. K. Khetarpal, and H. Koganezawa (eds). St. Paul, MN: APS Press.

van Regenmortel, M. H. V., C. M. Fauquet, D. H. L. Bishop, E. B. Carstens, M. K. Estes, S. M. Lemon, J. Maniloff, M. A. Mayo, D. J. McGeoch, C. R. Pringle, and R. B. Wickner. 2000. *Virus Taxonomy--Classification and Nomenclature of Viruses. Seventh Report of the International Committee on Taxonomy of Viruses.* San Diego, CA: Academic Press.

Verbeek, M., A. M. Dullemans, J. F. J. M. Van den Heuvel, P. C. Maris, and R. A. A. Van der Vlugt. 2007. Identification and characterisation of tomato torrado virus, a new plant picorna-like virus from tomato. *Arch Virology* 152:881–890.

Wang, R. Y., R. C. Gergerich, and K. S. Kim. 1994. The relationship between feeding and virus retention time in beetle transmission of plant viruses. *Phytopathology* 84:995–999.

Waterworth, H. E., and A. Hadidi. 1998. Economic losses due to plant viruses. Pages 1–13 in: *Plant Virus Disease Control*, A. Hadidi, R. K. Khetarpal, and H. Koganezawa (eds). St. Paul, MN: APS Press.

Whitfield, A. E., D. E. Ullman, and T. L. German. 2005. Tospovirus-thrips interactions. *Annu. Rev. Phytopathol.* 43:459–489.

5 Plant Pathogenic Prokaryotes

Carolee T. Bull, Steven T. Koike, Alejandra I. Huerta, Teresa M. Jardini, Stacy J. Mauzey, Isael Rubio, and Ana B. Zacaroni

Concept Box

- Bacteriology is a fascinating topic of research because the organisms are microscopic and therefore seemingly invisible; finding out where they are and what they do in the environment is like solving a mystery.
- Phytobacteriology can be like a "who done it" because we often try to solve major economic problems and identify "who" attacked the broccoli.
- What we think bacteria are related to and not (classification) changes with the techniques available. Unfortunately, this can lead to changes in the names of our favorite organisms and sometimes it becomes difficult to keep track of all those names.
- Identification of bacteria is based on tests used to separate them into different groups. Identification methods change as the methods used in classifications change.
- Critical steps toward disease diagnosis are made with a good initial description of the crime scene (the host, the symptoms, the environment, and other field details).
- Managing diseases caused by bacteria follows integrated pest management principles that rely on multiple strategies for controlling the disease, minimizing impacts on the environment, and favoring long-term sustainability.

Phytobacteriology is the study of plant pathogenic prokaryotes. **Prokaryotes** (Greek: *pro* meaning before and *karyon* meaning nut or kernel) are single-celled microorganisms that are unified into a group based on how they differ from **eukaryotes** (Greek: *eu* meaning true and *karyon* meaning nut or kernel). Specifically, eukaryotes have chromosomes surrounded by a nuclear membrane (**true nucleus**). The genetics of prokaryotes are easier to study because they have a single circular chromosome located in the cytoplasm of the cells and because they do not have organelles (**chloroplast** and **mitochondria**) with their own chromosomes. However, prokaryotes present their own difficulties in that their microscopic nature makes them difficult to see and study in the environment. A great deal of interesting research involves developing techniques to allow researchers to ask which bacteria are out there, where do these organisms live, and what they do there? The study of microbiology is largely a study dedicated to divulging the secret lives of prokaryotes.

Bacteria cause diseases on a wide range of plants. In 1920, just 40 short years after the isolation of the first bacterial plant pathogen, Erwin F. Smith (an American, U.S. Department of Agriculture [USDA] researcher who is largely thought of as the father of phytobacteriology) catalogued diseases in a quarter of the plant families known at the time. He predicted that bacterial diseases would be found on all plant families and maybe even on all plant species. However, the diseases that attract the most attention are those that cause damage at economically or ecologically important levels. Although the field of phytobacteriology is well over 100 years old, every year new bacterial diseases of important plants are described, creating a great deal of detective work for phytobacteriologists who want to find out "who done it?" (e.g., who ruined the broccoli?).

Like bacteria, most bacterial diseases are ephemeral. Although a wide range of bacteria can cause severe disease and significant losses worldwide, few cause consistent and complete crop loss. Bacterial diseases of cultivated crops are often sporadic and cause minor cosmetic symptoms. Bacterial diseases may not cause complete crop loss, but they can limit the supply of certain commodities, decrease the overall quality, and reduce yield, thus resulting in economic impact. In the worst cases, crops from individual growers or from entire regions can be destroyed in some seasons due to severe disease outbreaks. Additionally, the impact can be a matter of human survival if the severely affected crops are major food staples of populations with few alternative resources.

TAXONOMY OF PLANT PATHOGENIC BACTERIA

In 1920, Smith's list of plant pathogenic bacteria included only the following three genera: *Aplanobacter*, *Bacillus*, and *Bacterium*. The number of genera of bacterial plant pathogens has increased since then and continues to do so. Pathogens described in Smith's era are still causing important diseases and set the stage for modern research that has greatly expanded the frontiers of not only phytobacteriology, but science in general. Smith and a team of phytobacteriologists* at the USDA documented the causes (etiology) of many of the bacterial diseases that remain problematic today. In 1989, there were only nine bacterial genera to which bacterial plant pathogens were assigned. By 1996, the number had increased to 26 genera and the latest catalogues of Names of Plant Pathogenic Bacteria lists 38 genera to which bacterial plant pathogens have been assigned. However, the picture is complex as some of these genera contain named pathogens that are also listed as synonyms from other genera.

Understanding the relationships among organisms (**classification**), ensuring that all scientists use the same unique name for each different organism (**nomenclature**), and recognizing organisms that have already been classified and named (**identification**) are three functions of **taxonomy**. **Taxonomists** often apply new classification techniques to old problems and propose new classifications and corresponding new names based on their findings. Additionally, new names arise due to the reporting of new diseases of novel etiology. Recently, our increasing ability to characterize **unculturable bacteria** has resulted in an increase in the number of named bacterial plant pathogens. Regardless of the cause, phytobacteriologists struggle to know which names are up-to-date because choosing the appropriate name requires that researchers first understand the classification from which those names were derived. If they agree with that classification, then they should use the corresponding names associated with it. It is the responsibility of each researcher to know which classification scheme best suits the organisms that are the subject of their research and to choose the name that corresponds to that classification scheme. There are rules for how to construct names of bacteria for new classifications, and lists are generated to help bacteriologists know which names have been proposed according to the rules (Table 5.1).

* Incidentally, many women were equal partners on this team. Phytobacteriology and microbiology were two fields in which female scientists were more common and more accepted at the start of the twentieth century than they were in other fields of science.

You do not have to study plant pathology long before you notice that the names of some pathogens change. For the seemingly continuous name changing, you can first blame Carl Linnaeus and second the desire of biologists to have names coupled with classification. Linnaeus did a great favor to all by shrinking the species names from the paragraph-long descriptions used as the names of organisms in his time to two words, a **binomial** species name. A binomial species name consists of the genus and a specific epithet. For example, the **species name** for one important plant pathogen is *Pseudomonas syringae*. This pathogen belongs to the **genus** *Pseudomonas* (from the Greek for false unit). The species name is not *syringae*, but is the binomial *Pseudomonas syringae*. The word *syringae* is the **specific epithet** for this pathogen and should never be written without a genus name (or single letter abbreviation: *P. syringae*).

Binomials require name changes if novel classifications place the pathogen in a new or different genus. For example, a pathogen cannot be called *P. solanacearum* if it is a member of the genus *Burkholderia* or *Ralstonia*; thus, *P. solanacearum* was changed to *B. solanacearum* and then to *R. solanacearum* due to the proposed changes in the classification of this pathogen. If a species name was one word and was not made from both the genus name and the specific epithet, names would not change based on the changes in our understanding of which genus the organism belongs. Although other naming systems have been proposed, researchers tend to prefer keeping Linnaeus' binomial system both because of tradition and because this system creates names that reflect relationships and changes in our understanding of those relationships.

Nomenclature for many plant pathogenic bacteria is made even more complicated because a **pathovar** epithet is often needed to distinguish among distinct pathogens within the same species on the basis of **pathogenicity** on one or more plant hosts. Like species names, pathovar names will change if pathogens are transferred to a new or different species or genus. For example, *P. syringae* pv. *alisalensis* was transferred to the species *P. cannabina* and therefore became *P. cannabina* pv. *alisalensis* in the process. Luckily, the International Society of Plant Pathology's Committee on the Taxonomy of Bacterial Plant Pathogens (Table 5.1) is ready to help phytobacteriologists and others understand the issues related to the taxonomy of these organisms.

What is a bacterial species? This is a question of much debate because it is difficult to stretch species concepts that were originally made for eukaryotic species to bacteria. Researchers deal with this conundrum by trying to determine the relationships among bacteria. In the past, **phenotypic** tests were the primary methods used for defining these relationships on the premise that organisms that are more closely related are more similar phenotypically. The most widely known and used phenotypic test is

TABLE 5.1
Publications and Websites Useful for Understanding the Taxonomy of Plant Pathogenic Bacteria

Lists of Validly Published Names of Bacteria

List of prokaryotic names with standing in nomenclature, http://www.bacterio.cict.fr/.

Bull, C. T., S. H. De Boer, T. P. Denny, G. Firrao, M. Fischer-Le Saux, G. S. Saddler, M. Scortichini, D. E. Stead, and Y. Takikawa. 2012. List of new names of plant pathogenic bacteria (2008-2010). *J. Plant Pathol.* 94:21–27. http://www.isppweb.org/about_tppb.asp.

Bull, C. T., S. H. De Boer, T. P. Denny, G. Firrao, M. Fischer-Le Saux, G. S. Saddler, M. Scortichini, D. E. Stead, and Y. Takikawa. 2010. Comprehensive list of names of plant pathogenic bacteria, 1980-2007. *J. Plant Pathol.* 92:551–592. http://www.isppweb.org/about_tppb.asp.

Young, J. M., C. T. Bull, S. H. De Boer, G. Firrao, L. Gardan, G. E. Saddler, D. E. Stead, and Y. Takikawa. 2004. Names of plant pathogenic bacteria published since 1995. Report of the Taxonomy of Bacterial Plant Pathogens Committee of the International Society of Plant Pathology. This publication is no longer up to date but is an online comprehensive list through 2004. http://www.isppweb.org/names_bacterial_new2004.asp.

Codes of Nomenclature

Lapage, S. P., P. H. A. Sneath, E. F. Lessel, V. B. D. Skerman, H. P. R. Seeliger, and W. A. Clark (eds.). 1992. *International Code of Nomenclature of Bacteria. Bacteriological Code, 1990 Revision.* Washington, D.C: American Society for Microbiology. http://www.ncbi.nlm.nih.gov/books/NBK8817/.

Dye, D. W., J. F. Bradbury, M. Goto, A. C. Hayward, R. A. Lelliott, and M. N. Schroth. 1980. International standards for naming pathovars of phytopathogenic bacteria and a list of pathovar names and pathotype strains. *Rev. Plant Pathol.* 59:153–168.

Young, J. M., C. T. Bull, S. H. De Boer, G. Firrao, L. Gardan, G. E. Saddler, D. E. Stead, and Y. Takikawa. 2001. International standards for naming pathovars of phytopathogenic bacteria. http://www.isppweb.org/about_tppb_naming.asp.

Useful References

Bull, C. T., S. H. De Boer, T. P. Denny, G. Firrao, M. Fischer-Le Saux, G. S. Saddler, M. Scortichini, D. E. Stead, and Y. Takikawa. 2008. Demystifying nomenclature of bacterial plant pathogens. *J. Plant Pathol.* 90:403–417.

Cintas, N. A., C. T. Bull, S. T. Koike, and H. Bouzar. 2001. A new bacterial leaf spot disease of broccolini, caused by *Pseudomonas syringae* pv. *maculicola*, in California. *Plant Dis.* 85:1207.

Koike, S. T., P. Gladders, and A. O. Paulus. 1998. *Vegetable Diseases: A Color Handbook.* New York, NY: Academic Press.

Vinatzer, B. A., and C. T. Bull. 2009. The impact of genomic approaches on our understanding of diversity and taxonomy of plant pathogenic bacteria. In: R. W. Jackson (ed). *Plant Pathogenic Bacteria: Genomics and Molecular Biology.* Norfolk, UK: Caister Academic Press, pp. 37–61.

Young, J. M., C. T. Bull, S. H. De Boer, G. Firrao, L. Gardan, G. E. Saddler, D. E. Stead, and Y. Takikawa. 2001. Classification, nomenclature, and plant pathogenic bacteria—A Clarification. *Phytopathology* 91:617–620.

Committee on the Taxonomy of Plant Pathogenic Bacteria (ISPP) http://www.isppweb.org/about_tppb.asp.

the Gram stain technique (or a substitute, the KOH test; Figure 5.1; Procedure 5.1), which is used to separate bacteria into groups based on their cell wall structure. We describe two additional phenotypic tests that not only are used for characterization of bacteria, but can also be the starting points for additional research. Studies on **bacteriocin** (toxins produced by bacteria to which related bacteria are sensitive) production and sensitivity to specific **bacteriophages** (viruses that attack bacteria) could lead to research on the survival of bacteria in the environment as well as being useful for developing phenotypic typing systems (Procedures 5.2 and 5.3).

A particular subgroup of bacteria pose special challenges when it comes to research and nomenclature. Gram-positive **fastidious prokaryotes** are a group of bacteria that cannot be grown in culture and therefore cannot satisfy the rules for naming bacteria. Instead of traditional binomials, the designation *Candidatus* (italicized) is used as a means of naming these unculturable bacteria. In each case, the word *Candidatus* precedes the name of the organism (not italicized) and the entire designation is written between quotation marks (e.g., "*Candidatus* Liberibacter psyllaurous"). There are currently only three plant pathogenic *Candidatus*

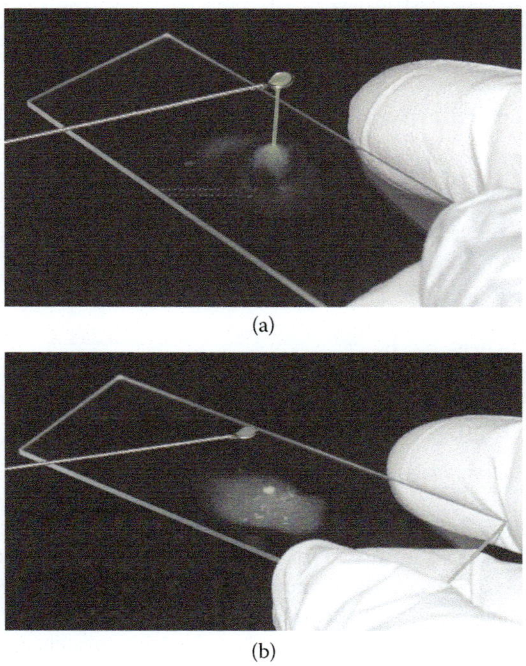

FIGURE 5.1 KOH test to determine Gram character of bacteria. (a) KOH-positive test result and (b) KOH-negative test result. (Courtesy of T.M. Jardini.)

genera: "*Candidatus* Liberibacter," "*Candidatus* Phlomobacter," and "*Candidatus* Phytoplasma."

Phenotypic tests are still used in combination with genotypic tests to define bacterial relationships in polyphasic classifications using all of the phenotypic and genotypic data available. Sequence-based methods are now taking a leading role in classification and identification of plant pathogens. Previously, relatedness and bacterial species were defined by **percent DNA/DNA hybridization** (or percent DNA homology). The now outdated bacterial species definition required that bacteria with ≥70% **DNA homology** and (**ΔTm**) of <5° were the members of the same species. This species definition is being replaced by species concepts that will use sequence-based classification schemes; these concepts have the potential to be more useful to phytobacteriologists than the previous approach.

The cost of sequencing entire genomes of bacteria is becoming quite affordable, and classification and identification will be done using entire genomes. Currently, **whole genome sequencing** is used to answer specific research questions, but has not yet been used for truly taxonomic purposes. Laboratories can, however, afford to regularly sequence individual genes, and so those are used for classification and identification. The 16S rDNA sequences have been used to analyze the relationships among all organisms and place them in the **tree of life**. Other **housekeeping genes** (needed genes) are used to study the relationships among organisms in more detail. **Multilocus sequence analysis** (MLSA; Figure 5.2) uses four to eight housekeeping genes to analyze the relationships and develop new classifications. **Multilocus sequence typing** (MLST) is used to identify the organisms based on which **alleles** of these four to eight genes do the organisms have. The use of multiple genes can identify incidents of **horizontal gene transfer**. One drawback is that different genes or gene fragments are used to identify species and pathovars within different genera. However, MLSA is much better than DNA/DNA hybridization for delineating species and pathovars because the data needed are **transportable**. All the data needed for comparisons are in databases that can be used by anyone in the world. Researchers do not need to obtain the type and pathotype strains nor DNA from the organisms to complete the MLST. The **plant-associated microbes database** (PAMDB) hosts the sequences for many MLST schemes of plant pathogenic bacteria (www.pamdb.org).

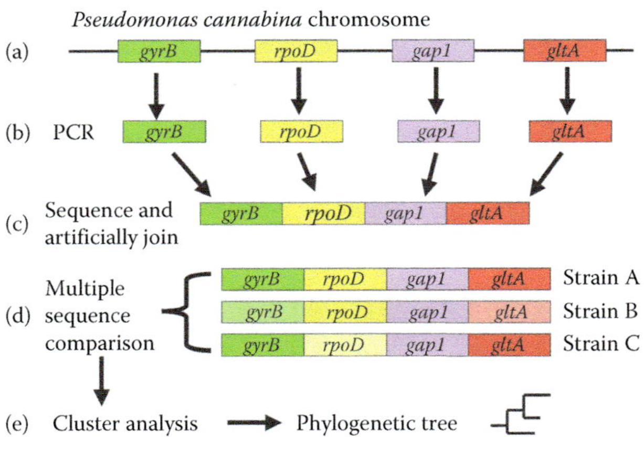

FIGURE 5.2 Multilocus sequence analysis work flow. (a) Genomic DNA is isolated from the target organism. (b) Specific housekeeping genes are amplified in separate PCR reactions. (c) Amplicons are sequenced and, using sequence analysis software, sequences are cut to a standard size and concatenated in a standard order for further analysis. (d) Concatenated sequences are aligned. (e) Alignments are subjected to cluster analysis for building phylogenetic trees. (Courtesy of C.T. Bull.)

BACTERIAL DISEASE DIAGNOSIS AND PATHOGEN IDENTIFICATION

For crops that are produced commercially on a large scale, or for a small number of plants grown in someone's backyard, the occurrence of plant diseases is a significant concern. Therefore, people who grow plants at any scale need to diagnose and identify problems affecting their crops. Plant disease diagnosis is a process involving analysis of **symptoms** and **signs**, compilation of cropping and problem histories, clinical testing of plant materials, and other investigative steps (Table 5.2). This is analogous to collecting evidence at a crime scene. Disease diagnosis can be considered the art of plant pathology, whereas pathogen identification involves the science of taxonomy. The extent of the investigations leading to a diagnosis or identification of a pathogen will always be dependent on the goal of the investigation.

SYMPTOMS AND SIGNS CAUSED BY BACTERIAL PLANT PATHOGENS

Diseases caused by plant pathogenic bacteria result in a wide range of symptoms. The majority of plant pathogenic bacteria can cause one or more common, typical symptoms. For example, **leaf spots** (sometimes called leaf lesions) are discrete, well-defined, and limited spots or sections of leaves that turn yellow, brown, black, or other abnormal color (Figure 5.3a through d). If leaf spots become numerous and combine or coalesce together, the symptom will cause much of the foliage to become diseased and die, causing foliage **blights** (Figure 5.4). Leaf **chlorosis**, or yellowing of leaf tissue, occurs when leaf spots or other infected areas turn yellow due to bacterial infection, colonization, or

TABLE 5.2
Information Collected on Disease Samples That Is Useful in Identification and Further Research

Disease Outbreak Information
Date of collection
Grower/location of outbreak
Type of production (greenhouse, field, etc.)
Cropping history (including previous disease problems)
Irrigation practices
Host Plant Information
Common name
Scientific name
Cultivar
Seed/germplasm source
Plant organ affected (leaves, stem, roots, etc.)
Symptoms
Photograph of symptoms and signs on individual plants
Description of typical symptoms and signs
Percent field affected (number of acres) and description of the types of damage
Isolations Made
What media were used?
ID# from collector/collection #
Degree to which initial isolation cultures appear pure
General description of colonies and initial test results
Each isolate from:
One or multiple plants
Plants from the same field or multiple fields

(a) (b) (c) (d)

FIGURE 5.3 Leaf spot symptoms caused by bacterial diseases. (a) Leaf spot symptoms of bacterial leaf spot (*Pseudomonas syringae* pv. *apii*) of celery. (b) Chlorosis and leaf spot symptoms of bacterial blight (*P. cannabina* pv. *alisalensis*) of cauliflower. (c) Shot hole symptoms of bacterial leaf spot (*Xanthomonas campestris* pv. *vitians*) of lettuce. (d) Water-soaked leaf lesions of bacterial leaf spot (*Xanthomonas campestris* pv. *vitians*) of lettuce. (Courtesy of S.T. Koike.)

FIGURE 5.4 Coalesced leaf spot symptoms of bacterial blight (*Pseudomonas cannabina* pv. *alisalensis*) of broccoli raab. (Courtesy of I. Rubio.)

production of **toxins** (Figure 5.3b). "**Shot holes**" are perforations in the leaf when leaf spot tissue splits or dries and falls out (Figure 5.3c). Lesions or infected areas can also cause discolored sections on leaf **petioles** and fruit **pedicles** (Figure 5.5). Twig and branch **die-back** and **cankers** (sunken and discolored areas in the wood and breaks and fissures in woody tissue) are corresponding symptoms that occur on the woody parts of trees and shrubs (Figure 5.6). Flowers can also develop spots, lesions, and blights from infections. Bacterial disease symptoms are very important as they cause significant economic losses when spots, lesions, and scabs form on fruit (Figure 5.7), and some bacterial pathogens cause seed decay. Other common symptoms include plant **wilting** (Figure 5.8) and development of mushy and **soft rots** (Figure 5.9).

In addition, some unusual symptoms can also be caused by bacterial pathogens. For example, the crown **gall** pathogen (*Agrobacterium tumefaciens*) causes galls to form on woody hosts (Figure 5.10). Research on the basic biology of this pathogen led to new fields of research including **genetic engineering** of plants. The aster yellows pathogen "*Candidatus* Phytoplasma asteris," one of the fastidious prokaryotes mentioned earlier, causes a series of virus-like symptoms such as leaf and stem deformities, foliage chlorosis, stunting, **phyllody** (growth of leaf structures on flower or fruit parts), and strange twisting of plant leaves and stems (Figure 5.11); in fact, for many years aster yellows was considered to be a viral disease. Pierce's disease of grapes causes a complex of symptoms including yellowing and reddening of foliage, eventual drying of leaves, shriveling of grape clusters, and stunting of shoots. Bacterial diseases of roots are less common. Corky root disease of lettuce is an example in which root tissues become green to brown in color and lose the ability to

FIGURE 5.5 Petiole lesion symptoms of bacterial spot (*Xanthomonas campestris* pv. *vesicatoria*) of tomato. (Courtesy of S.T. Koike.)

FIGURE 5.6 Canker symptoms of shallow bark canker (*Brenneria* [=*Erwinia*] *nigrifluens*) of walnut. (Courtesy of S.T. Koike.)

FIGURE 5.7 Fruit scab symptoms of bacterial spot (*Xanthomonas campestris* pv. *vesicatoria*) of pepper. (Courtesy of S.T. Koike.)

FIGURE 5.8 Symptoms of bacterial wilt (*Ralstonia solanacearum*) of tomato. (Courtesy of A.I. Huerta.)

effectively transport water and nutrients in the plant (Figure 5.12).

It is risky to attempt plant disease diagnosis solely on the basis of symptoms and host affected. However, when it comes to identifying plant diseases caused by bacteria, there are some key symptoms and signs that may indicate bacteria are the causal agents. First, with few exceptions, bacterial plant diseases primarily affect the aboveground portion of plants, including leaves, stems, flowers, and fruits. Second, the invading bacteria initially cause these affected leaf tissues to have a translucent, water-soaked, or greasy appearance (Figure 5.3d). The leaf, stem, or

FIGURE 5.9 Symptoms of bacterial diseases. (a) Mushy rot symptoms of soft rot (*Pectobacterium carotovorum*) of celery. (b) Soft rot symptoms of bacterial ring rot (*Clavibacter michiganensis* subsp. *sepedonicus*) of potato. ([a] Courtesy of S.T. Koike. [b] 1049 APS PRESS. With permission.)

FIGURE 5.10 Gall development from crown gall disease (*Agrobacterium tumefaciens*) on willow. (Courtesy of S.T. Koike.)

flower symptom may later take on a different look as disease progresses; however, this initial water-soaked symptom is very characteristic of bacterial diseases. Third, the shape of bacterial infections of plants can be very diagnostic. Once inside plant host tissue, pathogenic bacteria are fairly passive and cannot actively cross or penetrate through certain tissues, major veins in a leaf or other internal boundaries. Therefore, bacterial disease symptoms on dicot plants may often have an angular or rectangular appearance with the affected area having straight-sided boundaries (Figure 5.13). For monocot plants, because the major leaf veins are arranged in parallel lines that run the length of the leaf, bacterial infections will often appear as elongated, narrow lesions with straight-sided boundaries

FIGURE 5.11 Deformed celery foliage caused by the aster yellows phytoplasma ("*Candidatus* Phytoplasma asteris"). (Courtesy of S.T. Koike.)

(Figure 5.14). Keep in mind that exceptions do occur. For example, downy mildew diseases, caused by fungus-like organisms, often have similar straight-sided leaf lesions. Bacterial infections on tomato and pepper fruit will not have the angular appearance because the fruits do not have veins or other internal structures that create boundaries for the invading bacteria.

Another clue that the disease in question is caused by bacteria is the presence of a sign. In plant pathology, a "sign" is the actual visible presence of the pathogen itself (Chapter 1). Of course, we cannot see one solitary, single-celled, microscopic bacterium. However, under certain circumstances, these bacteria will mass together into incredible numbers inside plant tissues, ooze to the surface of the plant host, and be visible as cream- to yellow-colored exudates (drying to an orange- or honey-colored crystalline deposit) that collect outside the plant (Figure 5.15a and b). With the use of a compound microscope (Chapter 3), we can see a second type of bacterial sign. If we cut out a small piece of infected tissue and observe it under the microscope, we will see millions of bacteria oozing out from the cut surfaces called "bacterial streaming;" such bacterial rivers are another visible sign that indicates bacteria may be the causal agents of the disease. In some cases, the bacteria can be seen with the naked eye oozing from the cut surface (Figure 5.16) or streaming into liquid directly from cut stems (Figure 5.17; Procedure 5.4). It is good to remember that bacterial streaming alone does not prove that the problem is of bacterial origin. Plant tissues that are very rotted

FIGURE 5.12 Root lesion symptoms of corky root disease (*Sphingomonas suberifaciens*) of lettuce (healthy root on left). (Courtesy of S.T. Koike.)

FIGURE 5.13 Vein delimited, angular leaf spot symptoms of bacterial blight (*Xanthomonas campestris* pv. *syngonii*) of syngonium. (Courtesy of S.T. Koike.)

FIGURE 5.14 Vein delimited, linear leaf lesions of bacterial blight (*Pseudomonas syringae* pv. *porri*) of leek. (Courtesy of S.T. Koike.)

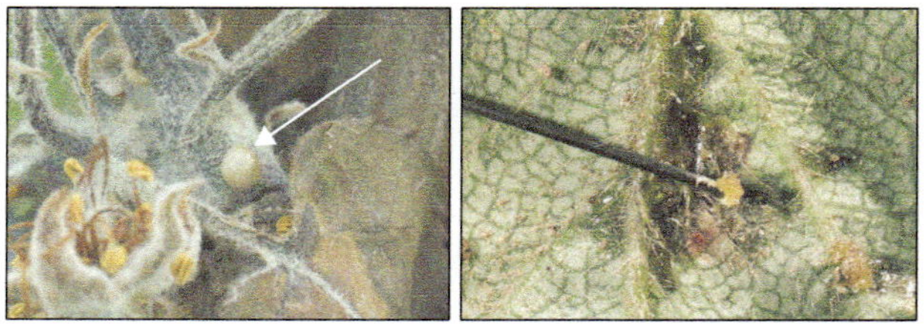

(a) (b)

FIGURE 5.15 Signs of bacterial pathogens. (a) Bacterial ooze (arrow) of fire blight (*Erwinia amylovora*) forming on surface of apple leaf. (b) Dried bacterial ooze scraped from angular leaf spot (*Xanthomonas fragariae*) of strawberry. (Courtesy of S.T. Koike.)

FIGURE 5.16 Bacterial ooze of *Ralstonia solanacearum* from an infected tomato stem. (Courtesy of A.I. Huerta.)

FIGURE 5.17 Bacterial streaming from bacterial wilt of tomato. (Courtesy of A.B. Zacaroni.)

and old will be colonized by dozens of bacterial species that are not pathogenic, but are only breaking down and feeding on the decayed plant tissues. In this case, bacterial streaming develops from secondary decay bacteria that were not involved in causing the disease.

ESTABLISHMENT OF THE ETIOLOGY OF NOVEL PLANT DISEASES

Although bacteria have been investigated as plant pathogens for over 100 years, novel diseases continue to appear. If the disease is not identified or readily diagnosed from among previous reports, researchers use Koch's postulates (Chapter 1, Figure 1.13) to determine if the recovered pathogen is the cause (**etiology**) of the disease. Sample collection sheets and photographs are used to record the symptoms of the disease (Postulate 1) as the primary data in a diagnosis. It is important to some research to photograph the lesion or symptom being used for bacterial isolations to refer to later in the process. Organisms are isolated from symptomatic tissue using standard methods (Procedure 5.1) by spreading macerated tissue that has been surface **disinfested** onto a general bacterial growth medium. After incubation one of two results are generally seen: either the bacteria that grow are uniform in color, shape, and size (Figure 5.18a) after approximately 3 days; or there is a mixture of colonies of different shapes, textures, colors, and sizes (Figure 5.18b). Occasionally, a semiselective medium is used to prevent the growth of unwanted fungal and bacterial saprophytes if researchers have some information about the disease prior to sampling. However, even using a semiselective medium, mixtures of colonies may grow from symptomatic tissue. Researchers have the most confidence moving forward

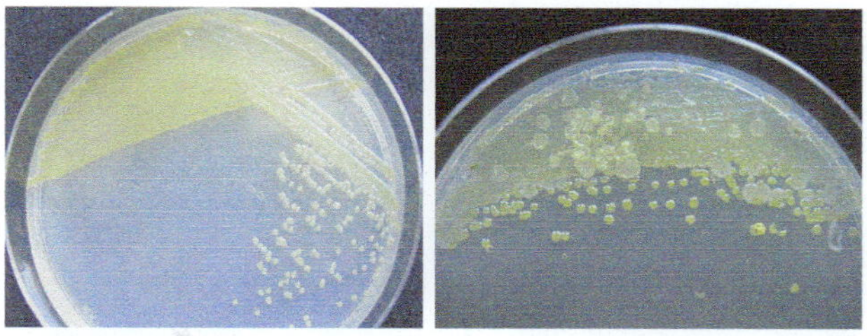

FIGURE 5.18 Isolations from surface-disinfested symptomatic leaf spot parsley tissue spread onto a semiselective medium. (a) All colonies appear uniform in color, shape, and size. (b) A mixture of colonies. (Courtesy of T.M. Jardini.)

with additional research (pathogenicity tests and bacterial identification) if the **isolates** grow from the tissue as **pure cultures** and are even more confident if similar pure cultures are obtained from multiple plants with the same disease symptoms.

Koch's second postulate (Figure 1.13) requires the suspect pathogen to be identified. DNA fragment banding patterns (often called **fingerprints**) generated by subjecting DNA amplified in **rep-PCR** reactions to **gel electrophoresis** can be used to distinguish bacteria (Figure 5.19a). These unique fingerprints serve to identify the bacteria before and after pathogenicity tests.

It would be ideal to test all isolates from all outbreaks for pathogenicity and to identify each and every isolate; however, student researcher time (contrary to their advisors' beliefs), space, and money are limited, thus the number of isolates that can be identified are limited. If all isolates from a single outbreak or multiple outbreaks have identical rep-PCR fingerprints (Figure 5.19), then two representative isolates from each outbreak (or a minimum of five) are evaluated further.

Isolated bacteria must be tested for pathogenicity to confirm that the isolated bacteria caused the disease (Figure 1.13). Methods for testing pathogenicity vary depending on the type of infection and disease. For foliar pathogens, the strongest evidence of pathogenicity comes from tests in which inoculum is sprayed onto the same species and cultivar of plant from which the organism was isolated. Inoculum is sprayed, rather than injected or rubbed, to mimic natural infection processes and to prevent false positives that can occur when mechanical wounding accompanies inoculation. However, some plants have strong protective barriers like a thick cuticle or waxy layer that make spray inoculation less practical. After inoculation, plants are subjected to high humidity to keep the stomata of the leaves open, which assists bacterial entry into the plant. Generally after 2 days, the plants are removed from the mist chamber and watered as needed. Symptom development is documented and described. In order for Koch's postulates to be fulfilled, symptoms

(a) (b)

FIGURE 5.19 (a) DNA fragment banding patterns of a DNA size standard (lane 1), known pathogens of the Apiaceae (lanes 2–3), and isolates from different fennel plants (lanes 5–15). (b) Fennel leaf streak symptoms from which one isolate was obtained. ([a] Courtesy of S.J. Mauzey. [b] Courtesy of S.T. Koike.)

must be similar to those seen in the original outbreak (Figure 1.13).

Final demonstration of the etiology of the disease requires that the bacteria isolated from symptomatic artificially inoculated plants be identical to those from the original isolate that was used to inoculate the plants. Bacterial isolation for this step is identical to the first step in this process (Procedure 5.1) in that the symptomatic tissue to be sampled is photographed, tissue is surface disinfested and macerated, and then the macerated tissue is spread onto a bacterial growth medium. Rep-PCR is used to compare the DNA fragment banding patterns of the original isolates to those of the **reisolates** from the artificially inoculated plants. Showing that the reisolate rep-PCR fingerprints are identical to those of the original strains is sufficient for the final step in the demonstration of Koch's postulates.

IDENTIFICATION OF PLANT PATHOGENIC BACTERIA

Only after pathogenicity has been demonstrated do we invest in classifying and naming the bacterial plant pathogen. The tests needed for identification are different for each **taxon**. In general, tests used to classify species are those needed to distinguish among the species within a given genus. Thus, the **primary literature** on the taxonomy of the organism will be essential in deciding what tests to use. Although researchers need to choose among tests capable of differentiating genera and species, pathovars are always distinguished by host range testing. DNA sequence-based methods, especially MLST, have accelerated the identification process in that sequence analysis can rapidly allocate isolates to different taxa and the analysis can be done without having the strains needed for comparison in the laboratory. Comparison with the **type** or **pathotype strains** is essential for identification for publication. The type and pathotype strains are the taxonomic reference strains. Researchers who do not use these taxonomic reference strains—who, for example, compare isolates to a colleague's isolate of a particular pathogen—risk misidentifying the pathogen, because the colleague's isolate could itself be misidentified.

Matching up your isolates with previously described organisms is relatively easy if the MLST sequences of (or data from) the type or pathotype strains are already available. All *P. syringae* pathotypes and many type strains and pathotypes of plant pathogenic *Xanthomonas* spp. are available in the PAMDB; soon even more species, pathovars, and genera will be available. Also, rep-PCR patterns of these strains will be very similar and usually identical to those of the type or pathotype strains. The number and type of tests conducted for identification will be dependent on the goal of the identification. If the goal is identification of novel pathogens for publication of a new taxon, full taxonomic analysis is required. Greater speed and less precision is needed if the goal is to identify an outbreak in a commercial agricultural field, for the purpose of applying appropriate disease control tactics before heavy yield losses occur.

DISEASE DEVELOPMENT AND EPIDEMIOLOGY

Disease development relies on the perfect storm of suitable environmental conditions, susceptible host, and the presence of a sufficient population of virulent cells (Figure 1.9; Disease Triangle). The environmental conditions that enable bacterial diseases to occur typically are those that favor the growth and survival of the pathogen causing the disease. Weather greatly influences the susceptibility of the plant to the pathogen. The patterns and frequency of disease are likely to be altered by global climate change.

Information about the biology, ecology, and habits of bacteria is important to understand disease development. For example, knowing where the **primary inoculum** comes from (i.e., where pathogens reside when they are not causing infections in a host) is important to understand disease development and control. Bacteria are ephemeral and in general do not produce survival structures that allow them to survive drastic changes in environmental conditions. Even then, bacteria may survive in soil, irrigation water, contaminated seed plants, seeds, on tools or workers' clothes, inside insect vectors, on overwintering crops, and on weed hosts; these sources can serve as the primary inoculum for bacterial diseases. Plant pathogenic bacteria have even been found in snow pack and cloud vapor. Identifying the primary inoculum source and reducing those populations of bacteria are important strategies for disease control. Preventing or eliminating the buildup of inoculum for the next cycle of disease is another important disease control strategy.

MANAGEMENT OF PLANT DISEASES CAUSED BY BACTERIA

Managing plant diseases caused by bacteria follows the traditional principles as set forth in integrated pest management (IPM) systems (Chapter 28). IPM systems rely on the use of multiple strategies for controlling the disease as well as minimizing impacts on the environment and considering long-term, sustainable practices. IPM options for controlling bacterial diseases include the following: using resistant cultivars; eliminating primary inoculum sources and otherwise excluding the bacteria; practicing good sanitation measures; farming while using appropriate cultural practices; applying bactericides; and managing vectors that spread bacterial pathogens. These practices can be easily remembered using the mnemonic REPEAT (resistance, exclusion, protection, eradication, avoidance, and therapy).

First, the use of a cultivar that is resistant or tolerant to the bacterial pathogen is a first-choice measure for growers. If a lettuce cultivar is resistant to bacterial leaf spot, or if a cucurbit cultivar is resistant to angular leaf spot, then use of these plants will eliminate the need to implement other measures. Second, exclude the bacterial pathogen from plant materials. Primary inoculum of bacterial pathogens often enters the production system as a result of being a contaminant on propagative plant material; therefore, the use of pathogen-free materials is critical. Use seeds that have been tested and deemed pathogen-free (certified seed) or seeds that have been treated to reduce bacterial contamination below the economic threshold. Use cuttings, transplants, crown divisions, and other vegetative materials that are free from the pathogen and that are not already diseased.

When growing transplants, seedlings, or a crop to maturity, practice good sanitation measures to prevent contamination from bacteria. Examples of such sanitation measures include the following: washing and sanitizing transplant trays or pots used to grow seedlings; sanitizing pruning shears and other tools used to prune and cut stems and branches; wearing and sanitizing gloves when handling, pruning, or pinching plants; and removing plants that exhibit disease symptoms (this is especially helpful in a propagation or greenhouse crop setting). Next, implement appropriate cultural practices when growing plants. The most important cultural practice for managing bacterial diseases is choosing the appropriate irrigation method. Because bacteria mostly spread plant-to-plant via splashing water, try to minimize or eliminate the use of overhead sprinkler irrigation; watering plants by furrow or drip irrigation can significantly reduce the incidence and severity of bacterial pathogens. Driving equipment into fields and having crews work in fields only if foliage is dry can also reduce plant-to-plant spread of pathogens. Another important cultural practice is to implement crop rotations so that susceptible crops are not grown back-to-back. In some cases, the pathogen that was present on the first crop can survive in the soil on crop residues for short periods of time and infect the next planting if that second crop is susceptible. Plant pathogenic bacteria usually do not survive for an extended period of time in soil, but such organisms can persist in crop residues and on diseased volunteer crop plant and weed hosts; therefore, eliminate these possible sources of bacterial inoculum.

Apply and use bactericides as available and appropriate. In some cases, copper- or antibiotic-based pesticides can provide some control if used preventively. However, for many bacterial diseases, such applications are of marginal use and do not provide sufficient control. In particular, bacteria become resistant to antibiotics rapidly due to their fast **generation time**. Finally, manage insect vectors, preferably by applying IPM strategies. Most of the fastidious bacteria, for example, the causal agents of aster yellows and Pierce's diseases, are vectored by insects and only infect plants when the vector feeds on the plant.

SUMMARY

Phytobacteriology is a fascinating scientific field that combines practical field agriculture, horticulture, and ecology; investigative problem solving; modern microbiology techniques; and the latest in high-tech molecular biology. This field is dynamic, having changed much over the past decades, and continues to change and remake itself as new scientific discoveries are made. Phytobacteriology is also an important field. The study of diseases caused by plant pathogenic bacteria assists growers and producers throughout the world who are producing the food and horticultural crops that we all need. Discoveries in phytobacteriology help us produce high-quality, high-yielding crops. Such findings also help make food safer, since phytobacteriology is closely aligned with food safety and food microbiology fields of study.

LABORATORY EXERCISES

Experiment 1. Potassium Hydroxide String Test for Estimating Gram Stain Character

Gram-positive bacteria are separated from Gram-negative bacteria by cell wall structure and the ability to retain differential dyes. Many researchers no longer conduct Gram stain tests and have switched to a test measuring lysis upon treatment with potassium hydroxide (KOH). Gram-negative cells will rapidly lyse when treated with KOH and release viscous string-like DNA that is easily visualized (Figure 5.1). The KOH test is easier than the Gram stain technique and corresponds to the results of Gram staining. The majority of the bacterial plant pathogens are Gram-negative, including those from these important genera: *Acidovorax*, *Agrobacterium*, *Erwinia*, *Dickeya*, *Pantoea*, *Pectobacterium*, *Pseudomonas* (see Case Study 5.1), *Ralstonia* (see Case Study 5.2), *Xanthomonas*, and *Xylella*. Important Gram-positive genera include *Clavibacter* (see Case Study 5.3), *Clostridium*, *Corynebacterium*, *Curtobacterium*, and *Rathayibacter*. The purpose of this exercise is to determine the Gram-stain character of a purified bacterial culture using KOH.

Materials

The following materials will be needed for each student or team of students:

- Controls including one Gram-positive and one Gram-negative bacterial strain
- Purified test cultures
- Glass Petri dish or microscope slide
- Inoculation loop
- KOH 3% (w/v)

Follow the protocols outlined in Procedure 5.1 to complete this exercise.

Anticipated Results

If there is no string formation within 60 s (KOH negative; Figure 5.1b), the cells are likely to be Gram-positive. If the suspension becomes viscous and form strings, the isolate is Gram-negative.

Questions

- Based on your results, to what genera of plant pathogenic bacteria are your isolates likely to belong?
- Why are known Gram-positive and Gram-negative bacteria used as controls?
- What would your interpretation be if all other data indicated that the sample was Gram-negative but the KOH indicated it was Gram-positive?
- What would you change if both the controls reacted the same way?

Experiment 2. Bacteriocin Typing

Bacteriocins are toxins produced by bacteria that inhibit closely related bacteria but not the strains producing the toxin (producer strains). This procedure can be used to determine the subspecies of *Clavibacter michiganensis* (see Case Study 5.1) if appropriate producer and indicator strains are used as controls.

Materials

Each student or team of students will require the following materials:

- Producer strains, the strains producing bacteriocins
- Control strains, these strains are also called the indicator strains because their sensitivity to

Procedure 5.1

Potassium Hydroxide String Test

Step	Instructions and Comments
1	Prepare fresh cultures of the purified test isolates and controls. Cultures grown on nutrient agar or another general bacteriological medium for 18–26 h are usually fine.
2	Scrape a single colony of bacteria off the Petri dish with an inoculation loop.
3	Place a drop of KOH (3%) on a microscope slide and mix the bacteria with the KOH for 60 s.
4	Slowly lift the inoculation loop from the slide and the bacterial suspension.
5	Observe and document whether or not a viscous string is formed between the suspension and the inoculation loop (Figure 5.1a). A dark background helps with the observation.

CASE STUDY 5.1

PSEUDOMONAS SYRINGAE PV. *MACULICOLA* AND *PSEUDOMONAS CANNABINA* PV. *ALISALENSIS*

Taxonomy of the Pathogen

- McCulloch first described *Bacterium maculicolum* as the pathogen causing pepper spot of cauliflower in 1911 (Figure 5.20). It was reclassified as *Pseudomonas maculicola* by Stevens in 1913. In 1980, it was consolidated with other fluorescent pseudomonads into *P. syringae* and designated *P. syringae* pv. *maculicola*. It has been described and reported on a variety of Brassicaceae plants.
- Historically, disease outbreaks on crucifers caused by organisms related to *P. syringae* were automatically attributed to *P. syringae* pv. *maculicola*. However, a second fluorescent pseudomonad with a broader host range (including monocots), *P. cannabina* pv. *alisalensis* (formerly known as *P. syringae* pv. *alisalensis*), was distinguished from *P. syringae* pv. *maculicola* at the turn of this century (Koike et al., 1998; Cintas et al., 2002; Bull et al., 2010).
- Some pathogens previously reported in the literature to be *P. syringae* pv. *maculicola* have been shown to be *P. cannabina* pv. *alisalensis*. These include a strain used as a part of a model system for molecular plant microbe interactions (strain B70, also named ICMP4326, ES4326, 4326, PmaM4, M4, CFBP1637) and strains reported from around the world.

Impact of the Pathogen

- *Pseudomonas syringae* pv. *maculicola* has a broad host and geographic range. Symptoms of the pathogen consist of oily dark spots on cotyledons, leaves, stems, peduncles, and seed pods. This pathogen reduces plant quality and causes production and economic losses.
- Bacterial blight caused by *P. cannabina* pv. *alisalensis* is an economically important disease of radish, broccoli raab, and arugula, resulting in 80%–100% loss in some cases, and it can significantly influence the quality of seed crops (Figure 5.21). Disease outbreaks have been reported in North America, Europe, and Australia. Diseases on crucifers reported as *P. syringae* pv. *maculicola* in the literature before 2002 may have been caused by *P. cannabina* pv. *alisalensis*. Early symptoms of the pathogen consist of water-soaked flecks on leaves. Over time, these flecks expand and become surrounded by bright yellow halos (Figure 5.3b), which eventually coalesce to form large necrotic areas referred to as blight (Figures 5.4 and 5.21a).

Disease Management

- Similar techniques and strategies are used to control *P. syringae* pv. *maculicola* and *P. cannabina* pv. *alisalensis*.
- Copper sprays have been used for the control of these two pathogens.
- Crop rotations away from crucifers have been used to manage these diseases.
- Additional recommendations include the use of clean seed and limiting overhead irrigation.
- Sanitation measures are helpful, including disinfestation of all tools before use and removal of infected plants and debris.

Recent Research

- In 2002, a bacteriophage, PBSPCA1 (Figure 5.22), capable of lysing *P. cannabina* pv. *alisalensis*, but not *P. syringae* pv. *maculicola*, was used to develop a bioluminescent reporter bacteriophage for the specific and rapid detection of *P. cannabina* pv. *alisalensis* from environmental samples.
- Additional *P. syringae* pathovars have been shown to be more closely related to *P. cannabina*, and the species is being emended to include *P. syringae* pv. *coriandricola* and *P. syringae* pv. *philadelphi*.
- The genomes of strains of *P. cannabina* pv. *alisalensis* and *P. cannabina* pv. *cannabina* have been sequenced, and the comparisons provide insights into host specificity in this species.

FIGURE 5.20 Pepper spot of crucifers (*Pseudomonas syringae* pv. *maculicola*) on cabbage. (Courtesy of S.J. Mauzey.)

(a) (b)

FIGURE 5.21 Bacterial blight of crucifers (*Pseudomonas cannabina* pv. *alisalensis*) on broccoli raab. Severe symptoms (a) on an individual plant and (b) in a field. (Courtesy of N.A. Cintas and S.J. Mauzey.)

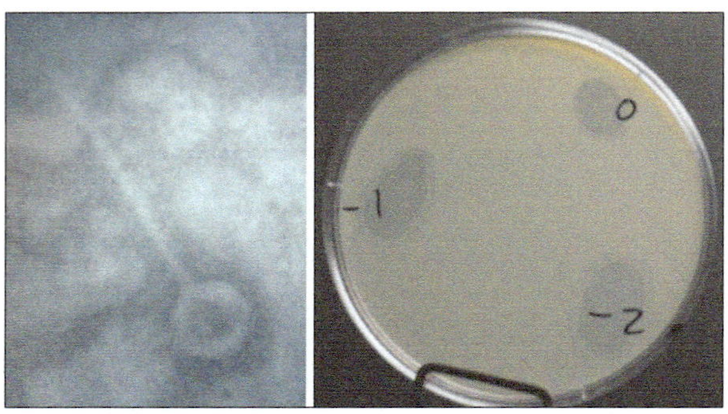

(a) (b)

FIGURE 5.22 Bacteriophage. (a) Electron micrograph of bacteriophage PBSPCA1. (b) Lysis of *Pseudomonas cannabina* pv. *alisalensis* by bacteriophage PBSPCA1. ([a] Courtesy of I. Rubio and Hsing-Yeh Liu. [b] Courtesy of C. Bull.)

bacteriocins indicates to which subspecies they belong

- Test isolates
- General growth agar media in sterile glass Petri dishes only
- Inoculation loop or pipette
- 2.5 mL of melted 0.7% agar (soft agar)
- Chemical exhaust hood
- Chloroform (refer to the Material Safety Data Sheet for chloroform when working with this chemical)
- Incubator

Follow the instructions listed in Procedure 5.2 to complete this experiment.

Anticipated Results

If using this procedure to determine the subspecies of *C. michiganensis*, then use the following producer strains:

- *C. michiganensis* subsp. *michiganensis 1379*
- *C. michiganensis* subsp. *nebraskensis CN74-1*
- *C. michiganensis* subsp. *insidiosus M1B*
- *Rathayibacter iranicus 2253*
- *Curtobacterium flaccumfaciens subsp. flaccumfaciens CV3 and CV6*

Use the following chart to determine the identity of the indicator strain.

	Producer Strains					
Indicator	**CN74-1**	**1379**	**M1B**	**CV6**	**2253**	**CV3**
C. michiganensis subsp. *sepedonicus*	+	+	+	+	+	–
C. michiganensis subsp. *insidiosus*	+	+	–	–	+	–
C. michiganensis subsp. *michiganensis*	V	–	+	V	+	–
C. michiganensis subsp. *nebraskensis*	–	+	+	–	+	–

Questions

- What is the purpose of using the producer and indicator strains in a bacteriocins experiment?
- If you get a negative result in the bacteriocin experiment, does that mean that the producer strain does not produce bacteriocins? Why?

CASE STUDY 5.2

Ralstonia solanacearum

Taxonomy of the Pathogen

- The pathogen was first described as *Bacillus solanacearum* in 1896 by E.F. Smith and soon renamed *Bacterium solanacearum* (Smith 1896) Chester 1897. It was subsequently renamed *Pseudomonas solanacearum* (Smith 1896) Smith 1914 based on the limited taxonomic techniques available at the time.
- This pathogen has had many additional synonyms, in part because the species is very heterogeneous. *Ralstonia solanacearum* is considered a **species complex**, which is composed of thousands of strains.
- A series of modern taxonomic studies led to the transfer of the organisms to *Burkholderia solanacearum* (Smith 1896) Yabuuchi et al. 1992. After further study, Yabuuchi et al. (1995) proposed a new genus for these strains leading to the most recent name *Ralstonia solanacearum*.
- The species is subdivided into four genetically distinct phylotypes based on **Ribosomal DNA ITS** region sequences.
- The four genetic groups identified correspond to the strains' geographic origins:
  - **Phylotype I** corresponds to strains of Asian origin, often thriving in tropical environments.
  - **Phylotypes IIA and IIB** correspond to warm temperate and temperate strains of the Americas.
  - **Phylotype III** strains are from Africa.
  - **Phylotype IV** strains have all been isolated in Indonesia.
- Prior to the phylotype classification scheme, *R. solanacearum* strains were divided on the basis of **race** (corresponding to host range, which is equivalent to the term pathovar used for other plant pathogenic bacteria) and biovar (corresponding to the ability to produce acid from several disaccharides and sugar alcohols).
- Under the current classification system, some of the phylotypes are further divided into **sequevars** according to differences in alleles for the endoglucanase gene.

Range of Diseases

Four major diseases are caused by different phylotypes of the pathogen.

- Brown rot of potato results in staining of the vascular ring starting at the stolon end of potato tubers. Severe infection may lead to complete rot and collapse of the tuber. Aboveground symptoms include wilting of one or a few leaves at the stem apex; as disease progresses the entire plant wilts and dies.
- Southern bacterial wilt of tomato (Figure 5.8) manifests as a sudden wilt of tomato foliage followed by stem collapse. As the disease progresses, the entire plant collapses and dies. Additional symptoms include browning of the stem, leaf chlorosis, and stem rot.
- In Moko disease of banana, infections caused by the soilborne *R. solanacearum* strains are first observed on older leaves, which exhibit chlorosis, wilt, and ultimately leaf collapse. Disease symptoms spread to young leaves, which develop pale green or whitish panels before becoming necrotic. Fruit development is arrested, and fingers may ripen prematurely and split. Internally, the fruit is discolored and eventually rots.
- For the insect-transmitted strain of moko disease, symptoms are first perceived as blackened and shriveled flower buds and peduncles. The bacteria travel into the fruit, leading to premature fruit ripening and rot. With time, bacteria enter the vascular tissue causing vascular browning and plant death.
- Granville wilt of tobacco is expressed on young plants as wilting of one or two leaves that usually recover as temperatures cool overnight. As infection advances in larger, more developed plants, leaves display one-sided wilting. Initial symptoms are often apparent on only one side of the leaf mid-vein. Over time, leaf veins turn brown and the leaf dies. Infected leaves become light green to yellow and occasionally appear scaly. Vascular browning appears on longitudinal sections of tobacco stalks, and the underground symptoms include blackening and decay of roots.

(Continued)

Disease Cycle of Bacterial Wilt

- In the presence of a susceptible host, the soilborne bacteria can move toward root exudates and enter their host via lateral root emergence sites and injured roots.
- The pathogen multiplies to high cell densities (often 10^{10} CFU/g of tissue) in the xylem vessels.
- The cells produce copious amounts of **extracellular polysaccharide**, a primary virulence factor, and the host defense response to infection leads to the blocking of water transport in the xylem vessels and to the predominant wilting symptom, from which the disease gets its name.

Impact of Bacterial Wilt

- Phylotype IIB strains causing bacterial wilt on tomato and brown rot on potato (historically known as race 3 (biovar 2)) are highly virulent strains of *R. solanacearum* that are capable of adapting to cooler climates in temperate zones. Therefore, these strains pose a threat to national agriculture in the United States and Europe and have been placed on the select agent list making them restricted pathogens for these regions.
- Bacterial wilt results in economic losses on tomato, potato, tobacco, pepper, eggplant, groundnut, and *Musa* spp. and also affects over 200 other plant species in over 50 families. Its wide geographic distribution and soilborne nature make it one of the most destructive plant pathogens in the world.
- Bacterial wilt has become a reemerging problem with 50% yield losses on the eastern shore of Virginia (the sixth largest fresh market tomato producer in the USA).

Disease Management

- Bacterial wilt is a difficult disease to control due to its soilborne nature, wide geographic distribution, and host range. It can survive in/on plant debris, soil, water, and on solanaceous weeds and other hosts.
- Prior to 2005, the soil-applied fumigant methyl bromide was an effective means of controlling the disease by reducing the initial primary inoculum in agricultural fields; however, this ozone depleting fumigant has been phased out, and no effective chemical replacement has been found.
- Exclusion, including the planting of certified material, is the best way growers can protect their crops. Phytosanitary measures are essential because once the pathogen is established in a field, it is very difficult to eradicate.
- Conventional breeding against bacterial wilt, the most efficient control tactic against yield loss, has been conducted for economically important crops like tomato, potato, eggplant, tobacco, pepper, and peanut. However, generating effective resistance in some of these crops has been slowed due to interacting factors such as pathogen variability, availability of resistant sources, environmental factors, and desired agronomic traits.

Recent Research

- Historical research of *R. solanacearum* focused on the nature of the pathogen and pathogenesis, epidemiology, disease management, diversity and detection of the pathogen, host plant response, breeding and deployment of wilt-resistant crops, and pathogen genetics (Buddenhagen and Kelman, 1964).
- With the advent and ease of sequencing DNA of the various strains from each of the four phylotypes of the *R. solanacearum* species complex, researchers have focused on the intimate interaction between host and pathogen at the molecular level.
  - The vast amount of data from the sequenced strains has led researchers to conduct comparative genome analyses to identify and study key metabolic pathways that contribute to strain fitness, pathogenicity, and virulence.
  - Comparative transcriptomic analyses are underway to identify genes in the host and the pathogen that account for latent infections and host resistance.
- Grafting tomato scions on *R. solanacearum*-resistant eggplant and tomato rootstocks planted in highly infested fields decreases bacterial wilt incidence.

CASE STUDY 5.3

Clavibacter michiganensis

Taxonomy of the Pathogen

- *Clavibacter michiganensis*, originally described in 1910 by E.F. Smith, was isolated from infected tomatoes grown in Grand Rapids, MI. He named the pathogen *Bacterium michiganense*. The pathogen has been renamed, the most recent name being *C. michiganensis*.
- The genus *Clavibacter* was described in 1984 and consisted of five species, formerly from the genus *Corynebacterium*, distinguishable from other *Corynebacterium* species by the presence of diaminobutyrate in the bacterial cell wall. The genus *Clavibacter* was later amended to describe only bacteria that have unsaturated menaquinones between 9 and 12 units in length in addition to the presence of diaminobutyrate, thus excluding all species except *C. michiganensis*.
- *C. michiganensis* has five subspecies that differ in colony morphology and pigmentation, biochemical test reactions, salt tolerance, and production of bacteriocins. They also differ in abilities to infect various host plants.
    - *C. michiganensis* subsp. *michiganensis* (Smith 1910) Davis et al., 1984 comb. nov., causes bacterial wilt and canker of tomato.
    - *C. michiganensis* subsp. *sepedonicus* (Spieckermann and Kotthoff, 1914) Davis et al., 1984 comb. nov., causes ring rot of potato (Figure 5.9b).
    - *C. michiganensis* subsp. *insidious* (McCulloch 1925) Davis et al., 1984 comb. nov., causes wilting and stunting in alfalfa.
    - *C. michiganensis* subsp. *nebraskensis* (Vidaver and Mandel 1974) Davis et al., 1984 comb. nov., causes wilt and blight of maize.
    - *C. michiganensis* subsp. *tessellarius* (Carlson and Vidaver 1982) Davis et al., 1984 comb. nov., causes leaf freckles and leaf spots in wheat.

Impact of the Pathogen

- All of the *C. michiganenesis* subspecies are vascular pathogens that enter host plants through wounds or from contaminated seed. After infection, bacteria colonize the xylem. Blockage of the xylem caused by bacterial growth results in the characteristic wilt and stunting symptoms.
- The two most economically important subspecies are *C. michiganensis* subsp. *michiganensis* and *C. michiganensis* subsp. *sepedonicus*.
    - *C. michiganensis* subsp. *michiganensis* has been known to cause up to 70% yield loss in both field and greenhouse grown tomatoes.
    - *C. michiganensis* subsp. *sepedonicus* has been known to cause up to 50% yield loss in potato, the most widely grown vegetable crop in the United States.
- The occurrence of *C. michiganensis* outbreaks is rare but can have devastating consequences.

Disease Management

- Both *C. michiganensis* subsp. *michiganensis* and *C. michiganensis* subsp. *sepedonicus* are quarantine pathogens on the European Union Plant Health Legislation list.
- The use of certified clean seed for tomato and certified clean seed pieces for potato is essential.
- Resistant commercial varieties do not currently exist.
- Sanitation practices include disinfestation of all tools and equipment after contact with the pathogen and removal of infected plants as symptoms develop.
- Crop rotation may be effective for the control of *C. michiganensis* subsp. *michiganensis* if a 2-year break in planting of a susceptible host is maintained.

(Continued)

Recent Research

- Historical research on *C. michiganensis* focused on virulence factors such as plasmids, bacteriophages, and bacteriocins.
- Current research is focusing on host pathogen interactions at the molecular level (Eichenlaub and Gartemann, 2011).
  - Both *C. michiganensis* subsp. *michiganensis* and *C. michiganensis* subsp. *sepedonicus* have fully sequenced genomes.
  - Completion of the *C. michiganensis* subsp. *michiganensis* genome identified a pathogenicity island that controls virulence of the pathogen.
  - Microarray data are currently being generated to identify genes in the *C. michiganensis* genomes that may be involved in pathogenicity.

Procedure 5.2

Bacteriocin Typing

Step	Instructions and Comments
1	Grow producer strains to log phase in liquid media. Producer strains produce bacteriocins to be tested.
2	Place ~5 μL of the producer strain suspension on nutrient agar or other general growth medium using an inoculation loop or pipet. Place five isolates on one plate separated from each other as far as possible.
3	Incubate for 4 days at 20°C or at the optimum temperature for the growth of the producers or production of the bacteriocin.
4	Grow indicator strains to log phase in liquid media. At the same time, grow your test isolates, as listed in the materials section.
5	Invert glass Petri dish of producer strains over 2 mL of chloroform for 2 h. Chloroform will melt plastic so be sure to use glass Petri dishes.
6	Adjust the indicator strains to a concentration of 10^8 CFU/mL.
7	Add 100 μL of the indicator strain to 2.5 mL of melted 0.7% agar. Do this step only when you are ready to proceed immediately with the next step.
8	Pour the inoculated melted agar over the producer plate.
9	Incubate for 2–3 days at 30°C or appropriate temperature for the growth of the indicator and test strains.
10	Read results. Cloudy zones or complete growth indicate that the bacteriocin was unable to inhibit the growth of the indicator strain (negative result; Figure 5.23a). Clear zones indicate the ability of bacteriocin to inhibit the indicator strain (positive result; Figure 5.23b).

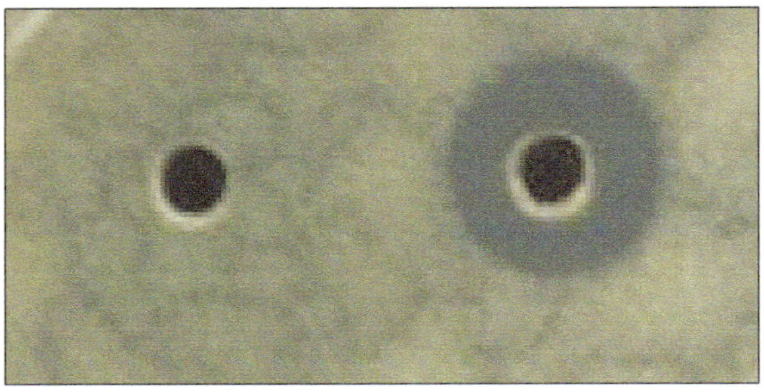

(a) (b)

FIGURE 5.23 Zone of inhibition caused by bacteriocin producing *Ralstonia solanacearum* Phylotype II sequevar 7 strain K60 on (a) bacteriocin susceptible target *R. solanacearum* Phylotype I sequevar 18 GMI1000 (lawn of growth) and (b) a negative control of *Pectobacterium carotovorum* subsp. *carotovorum*. (Courtesy of A.I. Huerta.)

- What would happen in this experiment if the producers were not treated with chloroform?
- If you did not have *C. michiganensis* strains in your laboratory, what strains could you use to demonstrate bacteriocin production?

Experiment 3. Bacteriophage Isolation

Bacteriophages are ubiquitous in the environment and can often be isolated from environments in which their host is present. It is therefore relatively easy to isolate bacteriophage from puddles left from irrigation runoff, from plants, or from soil directly below plants infected with the pathogen. Bacteriophages can be useful in biological control and detection of bacterial plant pathogens.

Materials

The following materials will be needed for each student or team of students:

- Soil or irrigation puddle water from fields in which the pathogen of interest is present
- Overnight culture of the host pathogen in nutrient broth (NB)
- 10-mL tubes containing soft agar (0.7% w/v agar)
- Nutrient agar plates or another appropriate growth medium
- Centrifuge
- Chloroform (refer to the Material Safety Data Sheet for chloroform when working with this chemical)
- Chemical fume hood

Follow the protocols listed in Procedure 5.3 to complete this experiment.

Anticipated Results

- This procedure may be used to isolate bacteriophage for a variety of bacteria. Expected results for this exercise are plaques forming in soft agar overlays containing samples and bacteria (Figure 5.22). Serial dilutions of the phage sample should result in a serial reduction of the number of plaques present on the plates. If isolation of a phage is successful, additional steps will be needed to generate high-titer stocks to be used in subsequent experiments.

Procedure 5.3
Phage Isolation

Step	Instructions and Comments
1	Collect soil or liquid from fields or locations where the pathogen is known to be present.
2	Prepare an overnight culture of the host bacteria. It is best if the culture is in log phase, but an overnight culture will be fine.
3	Enrich bacteriophage by mixing and incubating the host bacteria with the environmental sample. • Combine 5 mL of host bacterial suspension with 10 mg of soil in a 50-mL flask of NB. Alternatively, add 50 mL of field water to 5 mL of NB culture of the host and incubate overnight with gentle agitation. The addition of $CaCO_3$ (1 g) to each flask may help the enrichment process.
4	Allow the soil to settle for 10 min and then pour (or decant) the liquid into a centrifuge bottle. Centrifuge this liquid at 1500 × g for 20 min to remove the remaining bacteria and debris. Save the resultant supernatant.
5	In a fume hood, add 1 mL of chloroform to the supernatant and mix gently for 10 min.
6	Purify single plaques as described below or store. The sample can be stored over chloroform indefinitely if the container is resistant to chloroform. Alternatively, the upper phase can be transferred to a sterile container and the chloroform disposed of appropriately.
7	Prepare an overnight NB culture of the host bacteria.
8	Add 100 μL of the host bacterial culture and serial dilute 100 μL aliquots in 1:10 dilutions of the phage sample to melted 10 mL of soft agar (0.7% agar) tube. Make sure that the soft agar tube is not so hot that it kills the bacteria nor so cool that it solidifies before use.
9	Pour a soft agar overlay containing the bacteria and phage over the top of a nutrient agar plate. Allow the soft agar overlay to solidify.
10	Incubate at 27°C or the appropriate temperature overnight.
11	Observe spots of no growth ("plaques") that occur on the plates, record the number of plaques on each plate, and estimate the phage titer (Figure 5.22b).
12	Remove a single plaque from the plate using a sterile spatula or cork borer.
13	Macerate the agar sample containing the single phage plaque in 500 μL of NB.
14	Add the macerated phage plaque and 100 μL of host bacterial culture to 10 mL of melted soft agar (0.7% agar) in a tube, mix, and pour over nutrient agar as before.
15	Repeat Steps 10 through 14.
16	Use the final purified plaque containing agar plug to generate a high titer phage stock for use in further experiments.

Questions

- Which is the ideal place to collect liquid or soil samples for bacteriophage on the field?
- Why are there serial isolations of individual plaques in this procedure?
- What should be done if either no plaques are seen or if it appears that phage has lysed all the bacteria on the plate?

Experiment 4. Bacterial Streaming

Although individual cells of bacteria cannot be seen with the unaided eye, liquid with large concentrations of bacteria appear turbid. Likewise, a turbid stream of bacteria is often released into liquid if severely symptomatic plant tissue is sliced open and placed in liquid. This is one diagnostic character that can indicate that a particular disease is likely to have been caused by bacteria and is a precursor to bacterial isolation in some cases.

Materials

The following materials will be needed for each student or team of students:

- Plant material infected with bacterial pathogens. For leaf tissue, recommended examples are bacterial leaf diseases of lettuce, cilantro, parsley, celery, pepper, tomato, cucumber, and strawberry. For stem tissue, recommended examples are bacterial wilt of tomato, *Musa* species, eucalyptus, and cucurbits.
- Scalpel
- Microscope slides
- Water
- Coverslips
- Microscope with phase contrast

Follow the protocols in Procedure 5.4A and B to complete this experiment.

Anticipated Results

If the disease is caused by bacteria, visible streaming (cloudy strings, mist, or material) from the cut edge of leaf lesions will appear. For infected stems, a stream of turbid bacteria should move from the cut stem toward the bottom of the beaker due to gravity (Figure 5.17). Potential reasons that streaming will not be seen from tissue infected with bacteria include not observing soon

Procedure 5.4a

Bacterial Streaming from Leaf Tissue

Step	Instructions and Comments
1	Select a leaf spot that is suspected of being caused by bacteria. If possible, select spots that are not completely dried out and crispy; streaming may be more readily seen in tissues that are still fresh.
2	Cut a small (3–6 mm in diameter) piece from the margin of the leaf spot. A cut edge of the tissue should include the transition zone between symptomatic and nonsymptomatic tissue.
3	Place a drop of clean water on the microscope slide.
4	Place the piece of leaf tissue in the water and place the coverslip over it.
5	Look at the sample under a microscope to observe bacterial streaming along the cut edge of the leaf. Streaming should be visible within 2–5 min.

Procedure 5.4b

Bacterial Streaming from Stem Tissue

Step	Instructions and Comments
1	Select a stem from a clearly infected plant.
2	Fill a beaker/glass with clean water.
3	Cut a piece of the stem in pieces 5 cm above the soil or substrate. If the stem is already cut, make a fresh cut at one end of the stem.
4	Place the stem vertically in water with the freshly cut end immersed in water. Bind the stem to the beaker to keep it in place.
5	Look for bacterial streaming out of the stem (Figure 5.17). Streaming should be visible within 5–10 min, but can be seen directly after immersion. Altering the lighting conditions can aid in the visualization.

enough or for a long enough period, not evaluating the correct organ, or low bacterial populations in that particular lesion.

Questions

- Why do we need to observe streaming from tissue of the transition zone of a lesion?
- Does a lack of observable bacterial streaming prove that bacteria do not cause the disease?
- Do endophytic bacteria stream? If so, how would you know if the streaming is from an endophytic organism or a pathogenic organism?
- Some plants stream substances are similar to bacterial streaming. How can these be differentiated?

REFERENCES

Buddenhagen, W., and A. Kelman. 1964. Biological and physiological aspects of bacterial wilt caused by *Pseudomonas solanacearum*. *Annu. Rev. Phytopathol.* 2:203–230.

Bull, C. T., C. Manceau, J. Lydon, H. Kong, B. A.Vinatzer, and M. Fischer-Le Saux. 2010. *Pseudomonas cannabina* pv. *cannabina* pv. nov., and *Pseudomonas cannabina* pv. *alisalensis* (Cintas, Koike, and Bull 2000) comb. nov., are members of the emended species *Pseudomonas cannabina* (ex Šutič & Dowson 1959) Gardan, Shafik, Belouin, Brosch, Grimont & Grimont. 1999. *Syst. Appl. Microbiol.* 33:105–115.

Cintas, N. A., S. T. Koike, and C. T. Bull. 2002. A new pathovar, *Pseudomonas syringae* pv. *alisalensis* pv. nov., proposed for the causal agent of bacterial blight of broccoli and broccoli raab. *Plant Dis.* 86:992-998.

Davis, M. J., A. G. Gillaspie, A. K. Vidaver, and R. W. Harris. 1984. *Clavibacter:* A new genus containing some phytopathogenic coryneform bacteria, including *Clavibacter xyli* subsp. *xyli* sp. nov., subsp. nov. and *Clavibacter xyli* subsp. *cynodontis* subsp. nov., pathogens that cause Ratoon stunting disease of sugarcane and Bermudagrass stunting disease. *Int. J. Syst. Bacteriol.* 34:107–117.

Eichenlaub, R., and K. -H. Gartemann. 2011. The *Clavibacter michiganensis* subspecies: Molecular investigation of gram-positive bacterial plant pathogens. *Annu. Rev. Phytopathol.* 49:445–464.

Koike, S. T., D. M. Henderson, H. R. Azad, and D. A. Cooksey. 1998. Bacterial blight of broccoli rabb: A new disease caused by a pathovar of *Pseudomonas syringae. Plant Dis.* 82:727-731.

McCulloch, L. 1911. A spot disease of cauliflower. *USDA Bur. Plant Ind. Bull.* 225:1–15.

Smith, E. F. 1920. *An Introduction to Bacterial Diseases of Plants*. Philadelphia, PA: W. B. Saunders Co.

Yabuuchi E., Y. Kosako, I. Yano, H. Hotta, and Y. Nishiuchi. 1995. Transfer of two *Burkholderia* and an *Alcaligenes* species to *Ralstonia* gen. nov.: Proposal of *Ralstonia pickettii* (Ralston, Palleroni and Doudoroff 1973) comb. nov., *Ralstonia solanacearum* (Smith 1896) comb. nov. and *Ralstonia eutropha* (Davis 1969) comb. nov. *Microbiol. Immunol.* 39:897–904.

6 Plant-Parasitic Nematodes

Ernest C. Bernard and James P. Noe

Concept Box

- Nematodes are members of the animal kingdom. They are the only group of animals studied in plant pathology.
- There are numerous species of plant parasitic nematodes and many are important pests in most plant-production systems. Almost every plant is attacked by some species of nematode, and several have very wide host ranges.
- Most plant-parasitic nematodes are obligate parasites, meaning that they can only feed on living plant cells.
- Without a living host, many plant-parasitic nematodes will die of starvation. As a result, they almost never kill their hosts.
- Nematodes that are plant-parasitic always have a hardened spear-like stylet for feeding on living plant cells. However, not all stylet-bearing nematodes are plant parasites.
- Root-knot nematodes (*Meloidogyne* species) are the most common and widespread plant-parasitic nematodes. They cause diagnostic galls or "knots" on roots with masses of nematode eggs on the outside of the galls.
- Most nematode species do not produce any obvious diagnostic symptoms on plants. Because of difficulties in diagnosis, crop losses due to plant-parasitic nematodes often go undetected.
- Plant damage resulting from nematodes is dependent on the number of nematodes or eggs present while planting. Soil and root assays can provide information on nematode populations prior to planting that may be used to predict crop loss and formulate management strategies.

Nematodes are a phylum of animals (Nematoda) comprised of unsegmented, pseudocoelomate, worm-like (vermiform) organisms that are bilaterally symmetrical, but with a superficial radial symmetry on the head end. A unique characteristic is the presence of a pair of **amphids** on or near the head (Figure 6.1). Amphids provide sensory input to the nematode and help it orient to food sources. Nematodes are often called roundworms, but they are not related to the true worms, Annelida. The word "nematode" is derived from Greek words meaning "thread-like," which is a good descriptor for their general appearance. The body of a nematode can be thought of as a tube consisting of a tough, flexible cuticle that contains the organs. Nematodes possess most of the organ systems of higher animals, including digestive, reproductive, excretory, and nervous systems, but lack circulatory and respiratory systems; gas exchange occurs passively through the cuticle. Passive gas exchange limits the practical diameter of nematodes; hence, the very largest mammal-parasitic species are about the width of a pencil. The digestive system consists of an anterior stoma (mouth), muscular esophagus (pharynx), a set of salivary (esophageal) glands, a long, simple intestine, rectum, and anus. The reproductive system consists of one or two gonads lying alongside the intestine and opening to a vulva in females and a cloaca in males. Nematodes also possess several types of muscles that enable movement, feeding, and reproduction.

Nematodes hatch from eggs and grow by molting the old cuticle. Nearly all species have four juvenile stages followed by the adult stage. The adults are larger than **juveniles**, and in some groups, females may be swollen, primarily for increased egg production. Most nematodes are observed easily under a dissecting microscope at 40–60× magnification. The detailed observations needed to identify nematode species, however, must usually be made at much higher magnifications (600–1000×). The egg-to-egg life cycle of plant-parasitic nematodes takes a few weeks to more than a year, depending on nematode

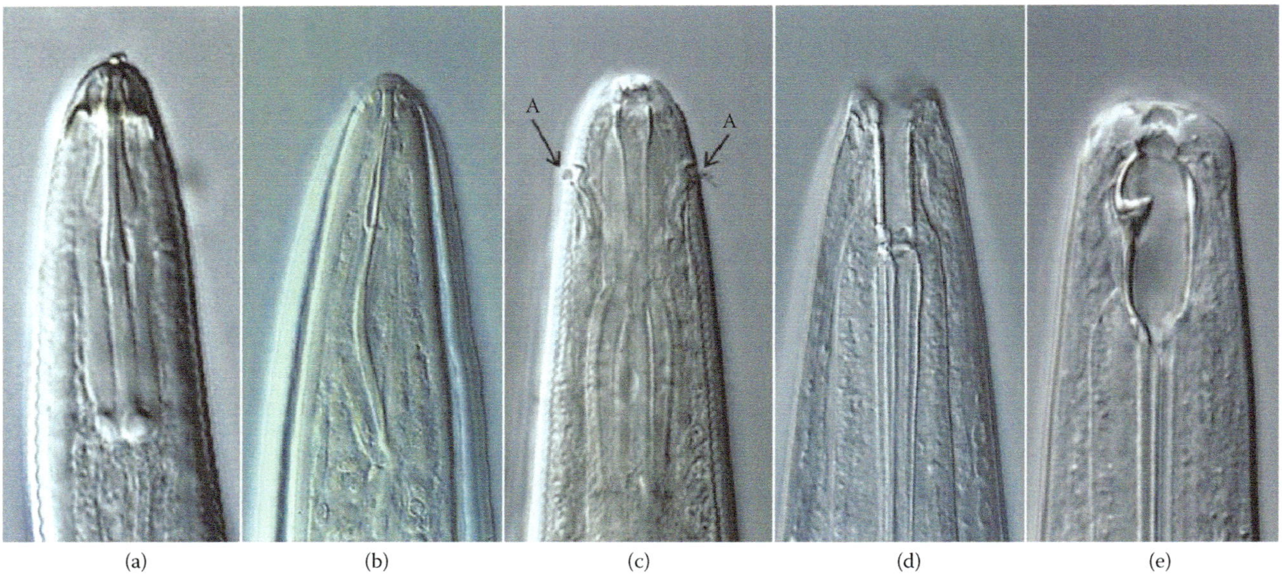

FIGURE 6.1 Head ends of various nematodes. (a) A spiral nematode (plant parasite). (b) Stubby-root nematode (plant parasite). (c) Plectid nematode (bacterial feeder). (d) Rhabditid nematode (bacterial feeder). (e) Mononchid nematode (predator). Amphids (A) are visible on the plectid nematode. ([a, b, e] Courtesy of E.C. Bernard. [c, d] Courtesy of Haley S. Smith. With permission.)

species, plant host status, and environmental conditions, especially soil temperature. Some bacteria-feeding nematodes can complete their life cycle in as little as 4 days.

All nematodes feed on living organisms. No species are known that survive on dead organic matter alone. Nematologists divide the thousands of species into several **trophic groups**, differing in their preferred food, ecological niche, and taxonomic relationships: **microbivores** (feeding on fungi and bacteria), predators (feeding on live prey, including other nematodes), **omnivores** (having several food sources or switching from one to another during maturation), plant parasites, aquatic and marine nematodes, invertebrate parasites, and vertebrate parasites. The first four groups are important in terrestrial soil ecosystems.

Microbivorous, **omnivorous**, and **predacious species**, which are collectively called **free-living nematodes**, are often found in soil samples in large numbers along with plant-parasitic nematodes (Figure 6.1). These nematodes are distinguished from plant-parasitic nematodes by their lack of a stylet or by the absence of knobs or flanges at the base of the stylet. A few species can feed and survive on both plant tissue and fungi. When observing fresh specimens collected from soil, free-living nematodes are usually moving more rapidly than plant-parasitic nematodes, which tend to lie on the bottom of the container and move slowly and gracefully in a sine-wave manner. Free-living nematodes feed primarily on microorganisms associated with the decomposition of organic matter. These nematodes recycle nutrients in soil and are an important component of the ecosystem. Types and numbers of free-living nematodes in soil samples are often used by ecologists as indicators of the health of an ecosystem. Greater numbers of free-living nematodes, and more importantly, more diversity in the number of species present, usually indicate better soil health.

With few exceptions, plant-parasitic nematodes are all **obligate parasites**, meaning that they can feed and survive only on living plant tissue. They are usually found in the soil and in plant roots, but some groups attack aboveground parts of the plant. Most species are microscopic with lengths of 0.3–4 mm and diameters of 15–35 μm, within the range of large fungal hyphae.

All plant-parasitic nematodes have a hardened spear-like feeding structure (**stylet**) in the anterior portion of their head region (Figure 6.1). In most forms, this stylet is hollow, and it looks and operates much like a hypodermic needle, allowing the removal of nutrients from within plant cells. The stylet is also used to penetrate plant tissues directly, allowing the nematode to burrow through the tissue and move toward preferred feeding sites. The successful penetration of the cell wall triggers a process called **extracorporeal digestion**. Digestive enzymes produced in one of the salivary glands are injected through the stylet into the cell. These enzymes quickly digest and liquefy the cell contents. The muscular part of the esophagus, the metacorpus or median bulb, then begins to pump rhythmically, creating a partial vacuum and pulling the digested material down through the stylet and toward the intestine. The intestine functions primarily as a storage and absorptive organ, not as a digestive organ. Plant cells fed on by nematodes are damaged, but frequently are not killed. Esophageal gland secretions sometimes cause extreme structural and physiological modifications of the targeted cells (see root-knot and cyst nematode sections).

Within the phylum Nematoda, plant-parasitism has evolved independently at least three times in three separate classes. The class Tylenchida contains the vast majority of plant-parasitic species and is characterized by a stomatostylet, the highly modified mouth apparatus that is used to puncture cell walls. Stomatostylets are hollow to allow the passage of fluids, and they typically have three knobs at the base to which the stylet protractor muscles are attached (the other ends of these muscles are attached to the head capsule). When these protractor muscles contract, they pull the stylet forward to thrust against the plant cell wall. Repeated thrusts result in puncturing and feeding. Plant parasites in the class Dorylaimida (dagger and needle nematodes) have hollow odontostylets, which function the same way, but have a markedly different structure. The odontostylet is formed in a cell in the esophagus and migrates into place in the stoma (mouth) during molting. The flanged odontophore is attached to the bottom of the odontostylet, with the protractor muscles attached to the flanges. Finally, the class Triplonchida contains the stubby-root nematodes, which have a thin, arched, solid stylet (Figure 6.1).

Regardless of stylet type, plant-feeding nematodes have evolved three basic approaches to obtaining nutrition. The least sophisticated are the **migratory ectoparasites**. These nematodes stay in soil, move along roots and graze on epidermal or cortical cells, only rarely partially entering roots. Some species are capable of causing noticeable plant damage, but most have little or no effect except for those involved in disease complexes. Some ectoparasites, most notably the ring (family Criconematidae) and pin (Paratylenchidae) nematodes, are sometimes called sedentary ectoparasites because they may settle for hours or days at a feeding site before moving.

A second approach to nutrition is that of the **migratory endoparasites**, which enter roots and then move through and feed on cells of the cortex. These nematodes may live their entire lives within a root, but if the root becomes crowded with nematodes or begins to die, these nematodes will leave it and search for another suitable host root to penetrate. The lesion nematodes (*Pratylenchus* spp.) and the burrowing nematodes (*Radopholus* spp.) of the family Pratylenchidae are among the most important migratory endoparasites.

The third and the most sophisticated approach is that of the sedentary endoparasites. This approach has evolved independently several times in the Tylenchida, but in all cases the nematode invades a root and establishes a specialized feeding site, but does not kill the affected cells. These specialized cells are a form of **transfer cell**. In nematodes, transfer cells facilitate the movement of photosynthate into the cell, resulting in more food for the nematode. This phenomenon is explained more thoroughly in the root-knot, cyst, and reniform nematode sections later in text. The adult female swells to many times the juvenile size and produces many more eggs than migratory species. Once these nematodes begin to enlarge in the root, they are no longer able to move about and their survival depends on the fate of the root.

Most plant-parasitic nematodes spend large portions of their life cycles in the soil environment. Soil texture, structure, and chemical composition are of primary importance in determining the number and types of plant-parasitic nematodes found in a given geographical area on a suitable host plant. On a given host, such as cotton, root-knot nematodes (*Meloidogyne incognita*) are usually a problem in sandier soils with relatively large spaces between the soil particles and good drainage. In contrast, reniform nematodes (*Rotylenchulus reniformis*) are found more often on cotton grown in clay soils with finer textures, relatively small spaces between the soil particles, and high bulk densities. Specific soil relationships have been demonstrated for many nematode species. This type of information is valuable to crop managers and advisors when determining the potential risk for a specific nematode problem.

Within the soil, plant-parasitic nematodes can move only a short distance, whereas long-distance dissemination is usually by the movement of soil, water, or plant-propagative parts. Movement of soil across fields, within regions, countries, and even globally is a major concern for the dispersal of plant-parasitic nematodes. Increased emphasis on biosecurity and prevention of importation of new pests has led to numerous quarantines on the movement of plants or any other product that may contain agricultural soil.

Plant-parasitic nematodes attack most economically important plants in agriculture, horticulture, ornamentals, and turf. Nematode problems are usually more prevalent in warmer, humid climates, but some species attack plants even in the coldest and most arid climates. The prevalence in warmer climates is largely because nematodes cannot regulate their internal body temperatures. Rates of development and maturation are dependent on soil temperatures. Nematodes are usually more active and reproduce more rapidly at warmer temperatures. Also, host plants are usually available longer and are growing more rapidly in warmer climates, which is a critical factor to these obligate parasites.

Aboveground symptoms attributed to plant-parasitic nematodes include generally vague and nonspecific growth problems related to root impairment and reduction in nitrogen uptake, such as stunting, yellowing, and wilting. Root symptoms may include galls, lesions, stunting, stubby appearance (stubby root), excessive branching, and a generalized darkening or rotting of the root tissues (observed with very high numbers of nematodes). Of this list, root galls and stubby root are generally diagnostic; the other symptoms indicate that some problem is present but that further investigation is needed to discover the cause. Crop losses due to nematode attack typically have a range of 10%–60% of potential yield depending on species, host status, and environmental conditions. Losses in more developed countries range from 5% to 10% each year, whereas losses in less developed countries may be 30%–60%, particularly in tropical and subtropical regions. Similarly, some crops experience much less nematode damage than other crops, even without nematode management. Many plant-parasitic nematodes enhance crop losses by forming **disease complexes** with other soilborne pathogens such as fungi and bacteria (Case Study 6.1 and Chapter 16), making it difficult to decide exactly how much damage the nematodes are causing.

Most kinds of plant-parasitic nematodes do not produce obvious or easily distinguishable symptoms on their host plants. Since most nematodes are microscopic and feed below ground, nematode problems often go undiagnosed. A key indication of a nematode problem is the occurrence of one or more of the typical root symptoms of nematode damage in an irregular or patchy distribution within a growing area. Patches of chlorotic (yellowed) plants within a field are usually a symptom of nitrogen deficiency that can be caused by nematode damage to roots. Plant-parasitic nematodes are almost never evenly distributed across a field, and the damage they cause will reflect that uneven distribution. An assay of soil samples by technicians trained in **nematology** is usually the only way to confirm a nematode diagnosis. Proper collection of soil samples is essential for reliable and accurate diagnosis of nematode problems. Plant-parasitic nematodes are found typically in the root zones of suitable host plants with most of the population occupying the top 20 cm of soil. Samples must be collected from within this root zone for effective diagnosis. Soil samples for nematode diagnosis are also more reflective of the targeted growing area if numerous small samples are collected systematically across the entire area and then composited (bulked) for analysis.

ROOT-KNOT NEMATODES

Root-knot nematodes (*Meloidogyne* species) are a widespread and diverse group of sedentary endoparasitic nematodes. Although more than 100 species have been

CASE STUDY 6.1

Slow Decline of Citrus—A World of Trouble

- Slow decline of citrus, caused by the citrus nematode, *Tylenchulus semipenetrans*, is the most significant nematode problem on citrus and is found throughout the world, wherever citrus is grown.
- In the United States, the number of groves infested varies from state to state but ranges from 50% to 90% of commercial groves.
- Resistant citrus root stocks are available, but the citrus nematode occurs in numerous physiological host races that can attack resistant rootstocks.
- The citrus nematode has a life cycle and feeding habit very similar to the reniform nematode, *R. reniformis*, although the two species are not closely related.
- The citrus nematode causes a very slow, generalized decline in the health and vigor of citrus.
    - The citrus nematode interacts strongly with water, nutrient, and temperature stress and with root rot caused by the fungus *Phytophthora parasitica*.
    - It is very difficult to assess the damage caused by the citrus nematode, because of interacting factors, and because the damage observed in the present growing season may be due to nematode attack in the previous season or even several years earlier.
- Although most citrus-producing areas are infested with the citrus nematode, not all groves within these areas are infested.
    - Sanitation, preventing the movement of soil and plants from one grove to another is essential to maintaining healthy production areas.
    - Planting stock from nurseries must be certified as disease-free before use.
- Control of the citrus nematode relies primarily on the use of chemical nematicides.
    - Nematicides that can be used after citrus trees are planted are usually systemic in the plant and water-soluble.
    - Use of these post-plant nematicides has been problematic in terms of human health risks and contamination of groundwater.
- Groves infested with the citrus nematode should not be replanted in citrus, because this nematode can live in the soil for several years without a host.

described, four root-knot species (*Meloidogyne arenaria*, *Meloidogyne hapla*, *Meloidogyne incognita*, and *Meloidogyne javanica*) cause most of the damage reported on agricultural crops. Of these four, *M. incognita* is most prevalent globally, perhaps because it has a very wide host range among commonly grown crops. Root-knot nematodes are found worldwide but are more common in warmer climates and in sandier soils. Plants infected with these nematodes may be stunted and chlorotic and show moderate wilting during the hottest part of the day. Galls or "knots" formed on host roots due to nematode infection are diagnostic for the presence of *Meloidogyne* spp. (Figure 6.2). This unique symptom, along with the wide host range and near-global distribution of the major species, probably accounts for the perception that *Meloidogyne* spp. are the most common plant-parasitic nematodes.

Root-knot nematode females lay eggs into a gelatinous matrix, and the first-stage **juveniles** undergo a molt while still in the egg. Vermiform second-stage juveniles (15 μm diameter and 400 μm length) hatch from the eggs by using their stylets to puncture and break through the tough egg shell. The second-stage juveniles are the only infective stage of root-knot nematodes. After penetrating suitable host roots near the root tips, juveniles migrate to the developing vascular cylinder and begin feeding on several cells near the endodermis. Secretions from the esophageal glands of the nematode are injected through the stylet into nearby cells. Primary xylem and phloem cells recently divided from the root meristem are especially affected. These cells enlarge to many times the size of their unaffected neighbors (**hypertrophy**); the nucleus divides without subsequent cell division, and the daughter nuclei then repeatedly divide synchronously so that dozens may be in the same cell. These hypertrophied, multinucleate cells serve as feeding sites for the rest of the nematode life cycle. The enlarged cells induced by root-knot nematodes are called **giant cells** (Figure 6.3). Giant cells are a form of transfer cell. These specialized feeding cells have extensive cell wall ingrowths consisting of protuberances and branching cell wall projections, which increase the surface area of the cellular membrane. This increased surface area allows the feeding cells to transfer large amounts of nutrients into the cell to feed the nematode. The nutrient demands of giant cells due to the high metabolic activity of the many nuclei within each cell leads to a greater share of nutrients going to support giant cells and nematodes and less to the rest of the plant, eventually resulting in symptoms of poor growth. Each successful juvenile produces two to seven giant cells. A gall quickly begins to develop around the feeding in the second-stage juvenile, often within a day, as a result not only of giant cell enlargement but also of increased

(a) (b) (c)

FIGURE 6.2 Root systems infected with root-knot nematodes. (a) Tomato root system infected with the southern root-knot nematode. (b) Canary creeper and (c) onion roots infected with the clover root-knot nematode, note different gall shapes and sizes. (Courtesy of E.C. Bernard.)

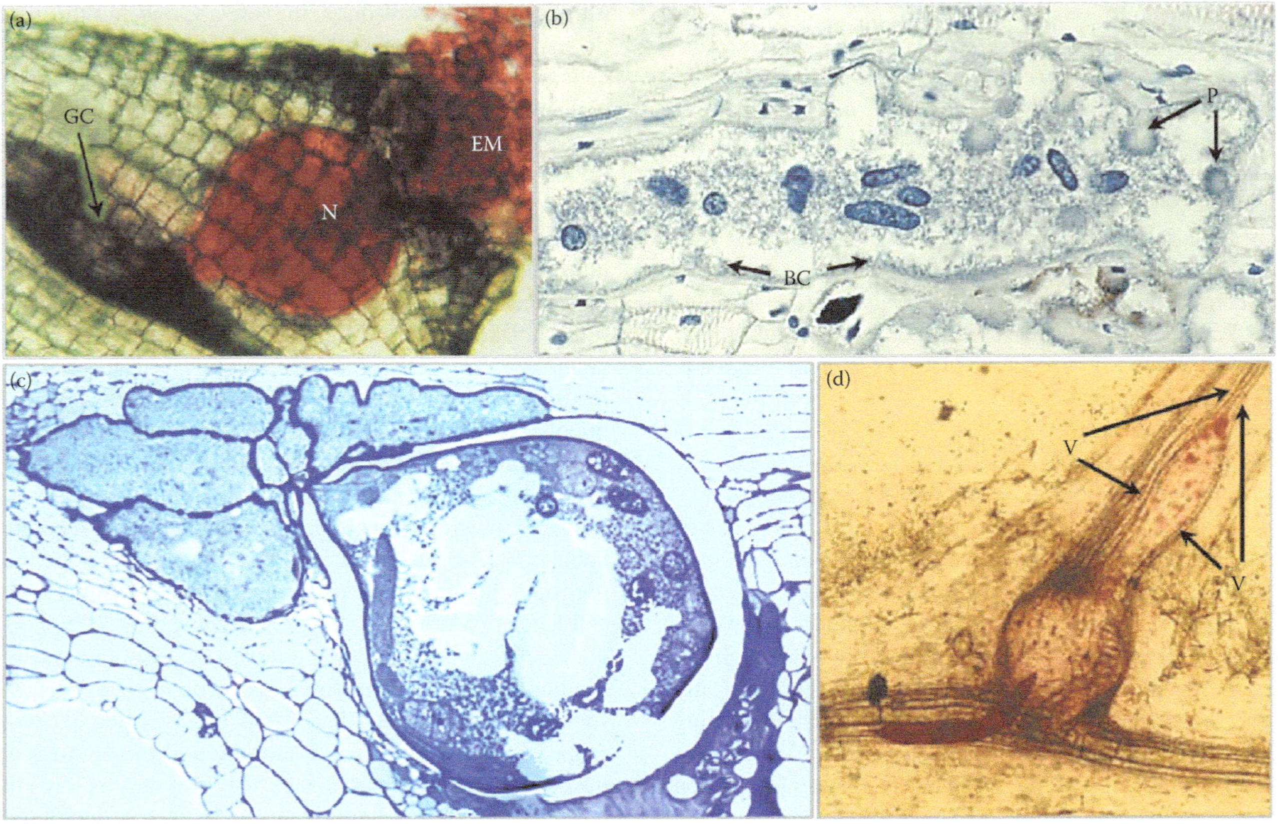

FIGURE 6.3 Anatomy of a root-knot nematode gall. (a) Stained nematode in cleared tomato root showing giant cell complex (GC), nematode (N), and egg mass (EM). Giant cells are specialized feeding cells in the vascular tissues of the plant, induced by nematode secretions. (b) Thin section of a giant cell with dense cytoplasm and multiple nuclei. The inner side of the cell wall is covered by protuberances (P) and innumerable branching cell wall ingrowths (BC) over which the cell membrane is stretched and expanded, allowing for greater movement of nutrients into the giant cell. (c) Thin section through gall showing nematode with head at the base of the giant cell complex it produced. Note dense cytoplasm and numerous nuclei. (d) Stained juvenile root-knot nematode ("sausage stage") and its giant cells, in which some nuclei (stained red) can be seen. Note that the giant cells are enclosed by vascular strands (V), thus ensuring a steady supply of nutrients to the giant cells and nematode. ([a] Courtesy of R.S. Hussey. With permission. [b, d] Courtesy of E.C. Bernard. [c] From Wyss, U., *Nemapix* Vol. 2, J.D. Eisenback and U. Zunke [Eds.], 1999. With permission.)

division of the surrounding cortical cells (**hyperplasia**). Once a feeding site is established, the juvenile begins to enlarge (called the "sausage stage" due to its appearance) and loses the ability to move.

Once juveniles begin feeding, they undergo a series of three additional molts. The third- and fourth-stage juveniles are short-lived stages and do not differ much in size from the second stage. Root-knot nematodes exhibit **sexual dimorphism** at maturity. Females (Figure 6.4) become greatly enlarged and spherical or pyriform (flask-shaped) with a body diameter of about 400 μm. Females have a slender anterior ("neck") region of varying length, depending on host root thickness and degree of crowding within a root. Males are completely different. Swollen fourth-stage males revert to a vermiform shape when they molt and are quite long (1.4 mm) and slender (30 μm of width). In many of the important species, males are a curiosity, as the females reproduce parthenogenetically (without mating); however, many of the species that live in natural habitats do require mating for successful egg production. The posterior end of the adult female usually protrudes from the surface of the root gall, where an egg mass containing 300–500 eggs (exceptionally up to 2000) is deposited into a gelatinous matrix (Figure 6.4). The life cycle typically requires 21–50 days, depending on root-knot species, plant host, and environment. Root-knot nematodes survive intercrop and plant-dormancy periods primarily as eggs in the soil. Survival rates are enhanced by protection in egg matrix and host plant debris, in addition to the large number of offspring that ensures at least a few will reach new roots.

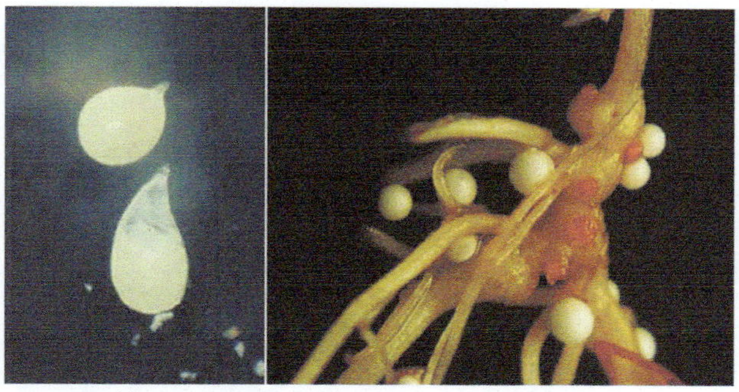

(a) (b)

FIGURE 6.4 (a) Mature root-knot nematode females excised from galled roots. (b) Root-knot nematode egg masses (stained red) protruding from galls. White Styrofoam beads are approximately 0.6 mm in diameter. (Courtesy of E.C. Bernard.)

CYST NEMATODES

Cyst nematodes (*Heterodera*, *Globodera*, *Cactodera*, *Punctodera*, and *Vittatidera* species) are sedentary endoparasites that infect vegetables, small grains, soybean, corn, and legumes. *Heterodera glycines*, the soybean cyst nematode, is one of the most studied plant-parasitic nematodes; it occurs in most soybean production areas and from year-to-year is generally the most important pathogen of soybeans in the United States. Soybean plants infected with *H. glycines* often are stunted, exhibit foliar chlorosis, and have necrotic roots and reduced root **nodulation**. Both seedlings and older plants may be killed in fields with high numbers of soybean cyst nematode at the beginning of the growing season. Other significant cyst nematodes are the potato cyst nematodes (*Globodera. rostochiensis* and *Globodera pallida*), of great economic importance in Europe, and the sugar beet cyst nematode (*Heterodera schachtii*). In general, cyst nematodes are more diverse and common in cooler regions of the temperate zones and at higher elevations in the tropics.

Although cyst nematodes are not closely related to root-knot nematodes, they have evolved similar life cycles (converged) because of similar requirements for being sedentary endoparasites. As with nearly all plant-parasitic nematodes, the life cycle consists of egg, four juvenile stages (second-stage hatching), and adult. In at least some species, egg hatch is stimulated by specific host root exudates. Second-stage juveniles may penetrate fine feeder roots anywhere, but tend to favor the region behind the root tip. Cyst juveniles penetrate roots by moving through cells, causing more destruction than do root-knot juveniles, which move between cells. These juveniles reach vascular tissues, begin feeding, and become swollen and sedentary. Cyst nematode feeding results in the production of very large specialized feeding cell called a **syncytium** (Figure 6.5). A syncytium differs from the giant cells formed by root-knot nematodes in that syncytia are formed by dissolution of the cell walls of adjacent cells (up to 250 cells) to create a large composite feeding cell. In contrast, root-knot nematodes induce individual cells to enlarge without division.

Cyst nematodes are sexually dimorphic as adults, with fourth-stage males reverting to a slender vermiform shape as adults, whereas females remain greatly enlarged. The bodies of adult cyst females eventually erupt through the root surface and can be seen easily with the naked eye or with the aid of a hand lens. Adult males are necessary for reproduction in most cyst species. Females often produce a small gelatinous egg mass and deposit eggs in it, as in root-knot nematodes, but most eggs (200–600) remain inside the female's body. As the female ages the outer cuticle changes from white to yellow and becomes brown at death (the cyst) (Figure 6.5). Cysts protect the eggs from unfavorable environmental conditions, and eggs within the cysts may survive for more than 7 years. The cysts are an excellent means of dispersal, readily spread by soil movement, wind, and water.

LESION NEMATODES

Lesion nematodes (*Pratylenchus* species) attack many crops including vegetables, row crops, legumes, grasses, and ornamentals (Figure 6.6). Important species include *Pratylenchus brachyurus*, *Pratylenchus penetrans*, *Pratylenchus scribneri*, and *Pratylenchus vulnus*. These nematodes can be found in host roots in large numbers, up to 3000 per gram of root, and cause large, spreading necrotic lesions on fibrous or coarse roots. Lesions may cover the entire root system, and root pruning may occur in heavily infested fields. Lesion nematodes are among the relatively few nematode species that cause economic damage on woody ornamentals and trees. Lesion nematodes have also been associated with **disease complexes** involving other plant pathogens. *P. penetrans* and the fungus *Verticillium dahliae* together produce a syndrome called potato early dying, which severely suppresses potato yields in many parts of the world.

The life cycle of lesion nematodes is typical of many plant-parasitic nematodes. Adult females lay eggs singly in root tissue or in the soil. The first-stage juvenile molts to the second stage within the egg. The second-stage juvenile hatches from the egg and then molts three more times to become an adult. The presence of host roots stimulates egg hatch. All juvenile stages outside the egg and adults can infect host roots. Lesion nematodes are migratory endoparasites; they enter roots and migrate

FIGURE 6.5 Cyst nematode feeding and reproduction. (a) Cyst nematode on host plant showing size of syncytium (modified feeding cells). (b) Part of syncytium showing partially dissolved plant cell walls (DW), numerous nuclei (N), dense cytoplasm, and adjacent vascular tissue (V). (c) Soybean cyst nematode females with small egg masses (EM). (d) Transformation of living white female to egg-filled brown cyst. (e) Crushed cyst showing eggs inside. ([a, e] From Zunke, U., *Nemapix* Vols. 1 and 2, J.D. Eisenback and U. Zunke [Eds.], 1999. With permission. [b, c, d] Courtesy of E.C. Bernard.)

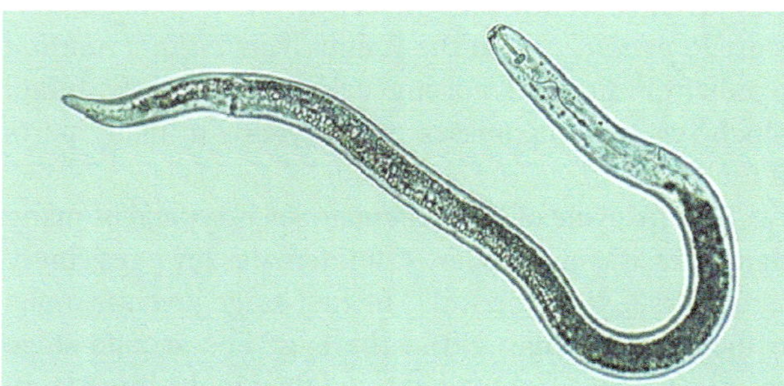

FIGURE 6.6 A female lesion nematode. Note the stout, well-developed stylet of this migratory endoparasite. (From Vrain, T., *Nemapix* Vol. 2, J.D. Eisenback and U. Zunke [Eds.], 1999. With permission.)

through the cortex, often killing the cells on which they feed. Invasion of roots may occur at root tips, the root hair region, and in young lateral root junctions. As these nematodes migrate through the cortex of host roots, using their stout, well-developed stylets to destroy cells in their path, they briefly feed on nearby cells and then continue moving. Death of cells is caused by nematode movement and feeding activities, resulting in elongated, spreading lesions just below the root surface. Lesion nematodes, as well as other migratory endoparasites, may leave and reenter roots several times. With large numbers of lesion nematodes, the lesions may completely encircle the roots causing death of the distal portion of the root segment.

RING NEMATODES

Ring nematodes (Figure 6.7) are commonly found on perennial hosts, such as turf, woody ornamentals, and trees, but also are abundant and diverse in natural environments. There are about 450 species of ring nematodes with most of the economically important species in the genus *Mesocriconema*. Ring nematodes have prominent body **annulations**, or grooves, around the outside of their cuticles that appear to be segments, or rings around the body. This character makes them easy to identify under a dissecting microscope. Most nematodes have these annulations, but usually they are very fine. Ring nematodes are migratory ectoparasites with relatively stout bodies (30–60 μm diameter; 400–700 μm length), giving them a cigar-like shape. Their wide shape and slow rate of movement prevents their reliable detection in assays that depend on movement of nematodes for collection of the specimens. With improved assay methods, ring nematodes have been found to be quite common and can occur at very high densities in the soil on some hosts.

Life cycle completion for ring nematodes takes 25–35 days with second-stage juveniles through adult stages feeding on host roots. Most populations do not produce males. Each ring nematode female may lay only three to five eggs per day in the soil, but on their typically perennial hosts, these nematodes have the capacity to increase to extremely high population densities. Ring nematodes have long, stout stylets (50–120 μm length) and feed ectoparasitically from the outside of plant roots on cells near the root surface. The nematode stops moving and may feed for an extended period of time on a single feeding cell. In some cases, the attacked plant cells take on some characteristics of transfer cells similar to those made by root-knot nematodes, and so ring nematodes are sometimes called sedentary ectoparasites. There is very little cell death at ring nematode feeding sites, and it usually takes high numbers of nematodes to visibly damage a host plant. Nevertheless, ring nematodes have been implicated in a number of disease complexes with pathogenic bacteria and fungi. *Mesocriconema xenoplax* is a common and widely distributed ring nematode with a large host range, including most plant species of *Prunus* (peach, plum, apricot, cherry, and almond), as well as walnut and apple. This species predisposes peach trees to bacterial canker, caused by the aboveground pathogenic bacterium *Pseudomonas syringae*, and is a critical component of the peach tree short life disease complex, interacting with *P. syringae* and several fungal plant pathogens. The related sheath nematodes (*Hemicycliophora*) have long, slender stylets and can cause terminal galls on lemon and other host plants.

RENIFORM NEMATODES

The reniform nematode, *R. reniformis* (Figure 6.8), may have the largest and most diverse host range of any single, economically important, plant-parasitic nematode species. In terms of global economics, it is well worth discussing this single species at the same level of importance as the previously described major groups of plant-parasitic nematodes. The reniform nematode is found throughout the world in tropical and subtropical areas and has more than 140 economically important hosts spread across 30 plant families. It causes severe damage on most of its hosts. Where reniform nematodes

(a) (b)

FIGURE 6.7 (a) A ring nematode. The annules are prominent and give a ringed appearance to the body. Note the strong, heavy stylet. (b) Anterior end of a different ring nematode. (Courtesy of K.J. Whitlock. With permission.)

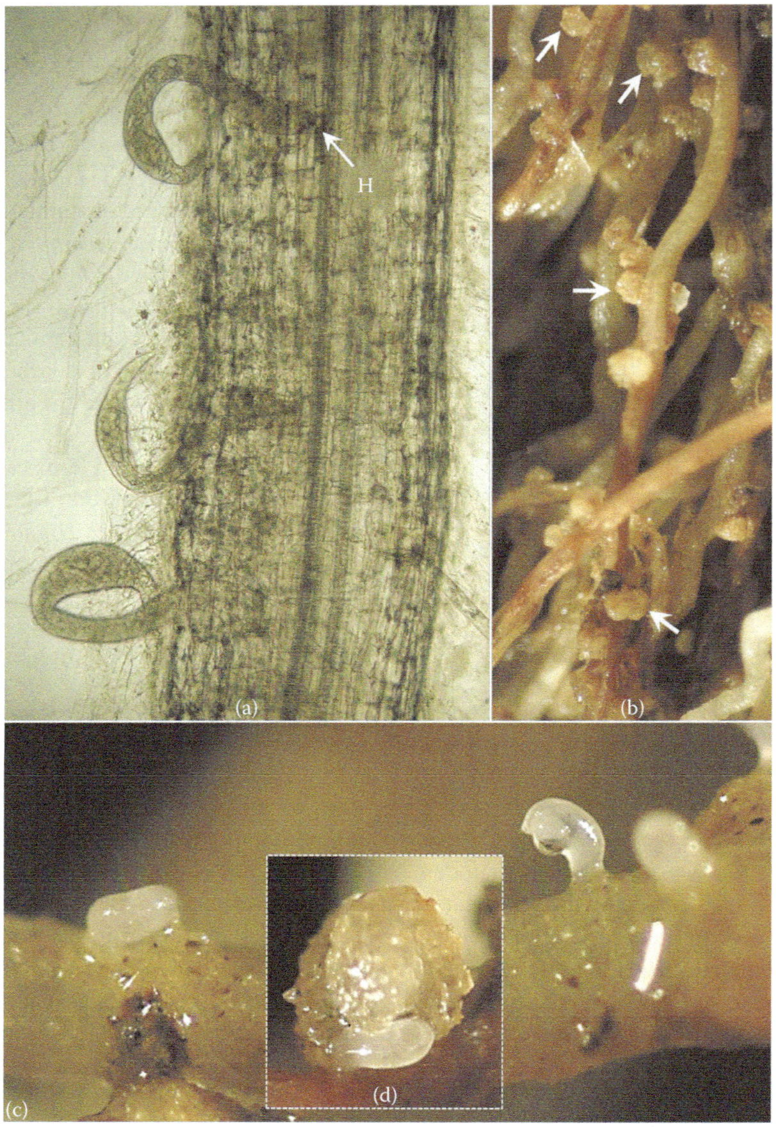

FIGURE 6.8 (a) Young female reniform nematodes embedded in host root. Note head end (H) near the vascular tissue. (b) Numerous reniform egg masses (arrows) on cotton roots. (c) Kidney-shaped females of reniform nematode after removal of egg masses. (d) Inset: relative size of female and egg mass. (Courtesy of E.C. Bernard.)

occur on any of these susceptible crops, growers have little choice except to treat their production areas with chemical nematicides. Crops on which reniform nematodes are a major limiting factor in production include cotton in the southern United States and pineapple in Hawaii. Other hosts include banana, cassava, citrus, coffee, kale, lettuce, mango, papaya, soybean, sweet potato, and tea. Aboveground symptoms of reniform nematodes include the typical root-impairment symptoms of stunting, nutrient deficiencies, and wilting. Below ground, the roots are stunted and coarse and may be discolored and decayed with high population densities of nematodes. Reniform nematodes also increase the severity of symptoms from fungal wilt diseases in cotton, caused by either *Verticillium albo-atrum* or *Fusarium oxysporum* f. sp. *vasinfectum*; infection by reniform nematodes has also been reported to cause Fusarium wilt-resistant cultivars of cotton to become susceptible.

The life cycle of reniform nematodes takes 24–30 days to complete. The first molt occurs within the egg. Second-stage juveniles hatch and undergo three molts in the soil without feeding. Young adult females are the only infective stage of reniform nematodes. Males never feed, but the species reproduces sexually. The young females retain previous layers of cuticle after molting, and this sheath of old cuticles is thought to aid the nematode in surviving in soil without feeding for up to 2 years. Vermiform young adult females (15 µm diameter; 500 μm length) penetrate host roots with only the anterior portion of their bodies, leaving the posterior portion in the soil outside the root. The nematode begins to feed on a single cell of the root endodermis, which is then transformed into a large syncytial feeding cell, incorporating up to 200 plant cells by causing the adjacent cell walls to dissolve. After the feeding site is formed, the posterior part of the female becomes swollen (100 μm diameter) and reniform (kidney-shaped), and up to 200 eggs are laid by each female into a sticky, gelatinous mass extruded on the surface of the root.

STING NEMATODES

Sting nematodes (*Belonolaimus* species) can have a devastating effect on the growth and yields of vegetables, row crops, ornamentals, and turf. The most important species of sting nematode is *Belonolaimus longicaudatus*. These long (up to 2.5 mm) nematodes also have long stylets (up to 150 μm) (Figure 6.9). They often become important pests in very sandy soils (sand content >80%). Plants parasitized by these nematodes have reduced growth and appear to be nutrient-deficient. Seedlings in heavily infested areas are usually severely stunted and

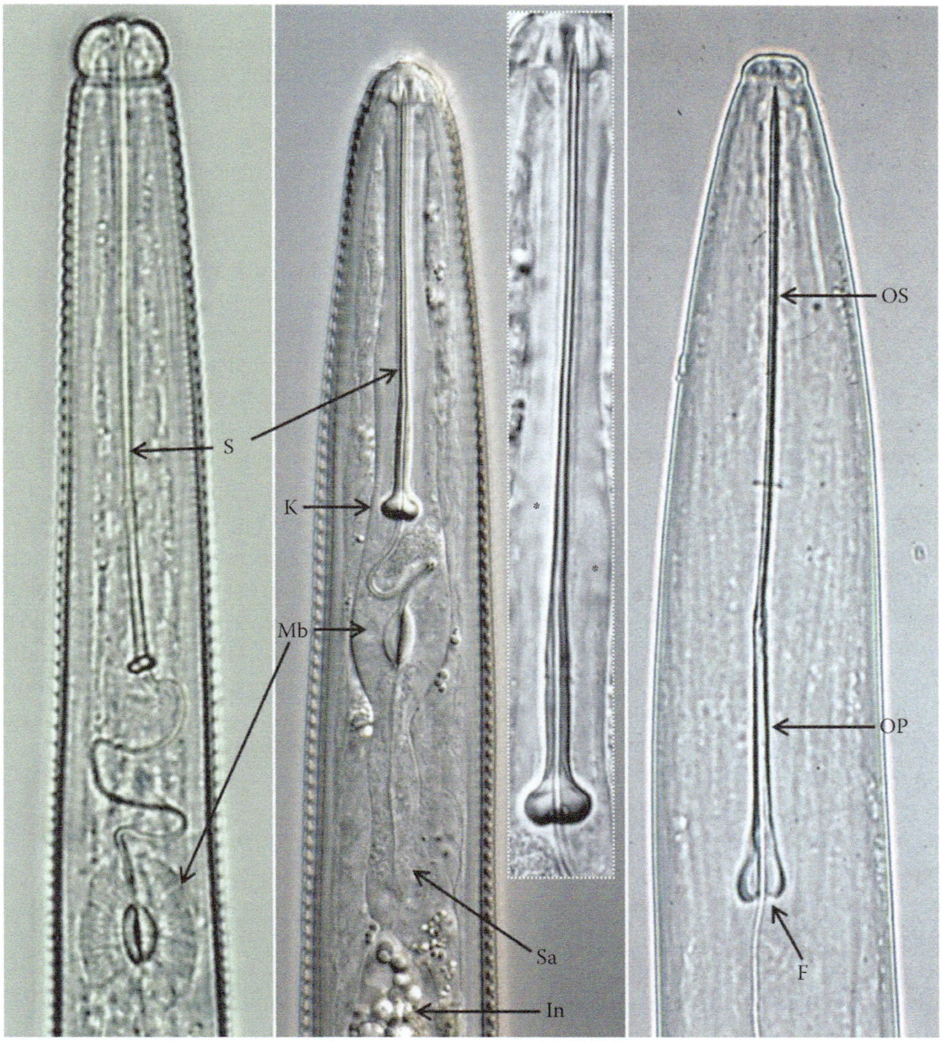

FIGURE 6.9 Plant parasitic nematodes with long stylets. (a) Anterior end of a sting nematode (Tylenchida). The very long stylet is used to penetrate deeply into root tips. (b, c) Anterior end of a sheath nematode. Root cells are pierced by repeated thrusts of the stomatostylet (S) due to contraction of the stylet protractor muscles (asterisks in inset) attached to the head end and the stylet knobs (K). After enzymes produced in the esophageal or salivary glands (Sa) have been secreted into the cell, liquefied contents are withdrawn by rhythmic contractions of the metacorpus, or median bulb (Mb), which sends the food into the intestine (In). (c) Inset shows the chiseled tip of the stylet and the lumen, or space that runs the length of the stylet, resembling a hypodermic needle. (d) Anterior end of a dagger nematode (Dorylaimida). The stylet of dagger nematodes, called an odontostylet (OS), is attached to a stylet extension, the odontophore (OP) is equipped with flanges at the end for attachment of the protractor muscles. ([a] Courtesy of J.D. Eisenback. With permission. [b, c] Courtesy of E.C. Bernard. [d] Courtesy of G. Windham. With permission).

have a much reduced root system, which can lead to plant death. Belowground symptoms consist of roots with coarse, stubby branches with dead apical meristem tissue and necrotic lesions on the root surface. These nematodes are such severe pathogens in the sandy coastal plain area of the southeastern United States that the presence of one sting nematode in a soil assay sample may signal the need for management with **nematicides**.

The life cycle of *B. longicaudatus* is similar to that of other ectoparasitic nematodes. Reproduction is sexual with males comprising about 40% of the population. The necessity for mating in this nematode requires these long nematodes to find each other, which is most efficient in sandy soils with large, connected pores; thus sting nematodes are rarely reported from heavier soils. Adult females deposit eggs in soil and the first-stage juvenile molts once before emerging from the egg. Sting nematodes feed from the root surface and do not enter root tissue. They favor root tips, and their feeding can kill the apical meristem, leading to many short, blind branches; this distinctive symptom is called stubby root.

DAGGER NEMATODES

Dagger nematodes (*Xiphinema* species) are ectoparasites and are commonly found associated with fruit or nut trees and in vineyards. Agriculturally important species include *Xiphinema americanum*, *Xiphinema californicum*, *Xiphinema index*, and *Xiphinema rivesi*. Root systems of plants parasitized by dagger nematodes may be stunted and discolored and have a limited number of feeder roots. In 1958, *X. index* was documented as the first soil nematode to **vector** a plant virus (grapevine fanleaf virus). Nematodes of the *X. americanum* group transmit the following North American nepoviruses: cherry rasp leaf virus, tobacco ringspot virus, tomato ringspot virus, and peach rosette mosaic virus. There are very few reports of direct crop damage from dagger nematodes, and most of their economic importance derives from their roles as plant virus vectors.

Dagger nematodes are in a different taxonomic class (Dorylaimida) within the nematode phylum and are only distantly related to tylenchid and triplonchid plant-parasitic nematodes. Dagger nematodes are slender, up to 4-mm long, and have a long, slender **odontostylet** attached to a flanged **odontophore**. These flanges serve the same function as stylet knobs in the stomatostylet. This combination may be 130-μm long (Figure 6.9). Females lay eggs singly in the soil near host plants and the nematodes hatch as first-stage juveniles. These nematodes have three or four juvenile stages depending on the species group, and juvenile stages are easily separated by the lengths of their developing stylets. Male dagger nematodes are very rare in most species. *Xiphinema* spp. can live up to 3 years under favorable environmental conditions.

STUBBY-ROOT NEMATODES

Stubby-root nematodes, *Paratrichodorus* and *Trichodorus* species, are plant-parasitic nematodes in the order Triplonchida. Their economic importance is greatest on corn, vegetable crops, and turf in the coastal plain region of the southeastern United States. The most important species is *Paratrichodorus minor*. These nematodes are most widely distributed in sandy soils but have also been found in organic-based soils. The characteristic symptom of damage by stubby-root nematodes is a stunted, stubby-root system, very similar to the symptoms produced by the unrelated sting nematodes. Aboveground symptoms are similar to plants deprived of a root system. Plants are stunted and chlorotic, wilt easily and have little ability to withstand drought. *Paratrichodorus* species are vectors of the following tobraviruses: tobacco rattle virus, pea early browning virus and pepper ringspot virus.

Stubby-root nematodes are migratory ectoparasites and feed almost exclusively at the root tips. As these nematodes feed in the root apical meristem area, root elongation and growth are stopped, leading to the characteristic stubby-root symptoms. Feeding by stubby-root nematodes is different from other plant-parasitic nematodes in that food is not ingested through a hollow tube in the stylet. These nematodes have a dorsally curved stylet referred to as an **onchiostylet** (Figure 6.1), with a simple groove on one side. This stylet is used to scrape and puncture root cell walls, then a minute tube is secreted through which the nematode feeds. The life cycle of this nematode consists of an egg stage, four juvenile stages, and the adult stage and can be completed in 16–17 days. Soil populations of stubby-root nematodes can increase rapidly and then decline with equal abruptness. They are also vertically migratory with the seasons. These nematodes have commonly been found at deep soil levels, which may create sampling problems and limit detection of these parasites.

ADDITIONAL PLANT-PARASITIC NEMATODES

The awl nematode, *Dolichodorus heterocephalus*, is an ectoparasite of vegetables and is similar to the sting nematode in general appearance. This nematode is generally found in fields with high soil moisture and causes symptoms on roots similar to the sting nematode. The lance nematode (*Hoplolaimus columbus*) is a serious pest of cotton and soybean in the coastal plain regions of Georgia, South Carolina, and North Carolina in the United States. *Hoplolaimus galeatus* feeding on shortleaf and loblolly pine may create points of entry for the fungus-like oomycete *Phytophthora cinnamomi* by damaging the protective ectomycorrhizal mantle on the finer roots. This damage followed by *P. cinnamomi* root infection results in littleleaf disease of those pine species. Lance nematodes have a thick, powerful stylet with tulip-shaped knobs, and feeding causes significant wounds on roots. These nematodes are migratory endoparasites in the root cortex.

Parasitism of aboveground plant tissues has evolved independently several times. In the family Anguinidae, the wheat gall nematode, *Anguina tritici*, invades florets of rye and wheat and can survive for long periods in **cockles** formed from the aborted seed. This nematode is easy to manage and has been all but eliminated from developed countries by maintaining supplies of certified nematode-free seed for growers. There are still active quarantine measures to limit its spread and reintroduction. The stem and bulb nematode, *Ditylenchus dipsaci*, is an important pest of alfalfa, onion, strawberry, and nursery crops such as narcissus and phlox. This nematode has a fairly low temperature optimum and is common in cool, moist temperate regions of the world.

Plant parasites in the family Aphelenchoididae also specialize in feeding in aboveground plant tissues. Species of *Aphelenchoides*, the bud and leaf nematodes, attack the foliar parts of ornamentals and floral crops, strawberry, rice, and many other crops. These nematodes require a moist, humid environment to move up the outside of shoots and leaves, entering the plant through stomata. Bud and leaf nematodes often cause problems in greenhouse ornamental and floral production areas. *Bursaphelenchus* species, including *Bursaphelenchus xylophilus*, the pine wilt nematode, typically have insect vectors and colonize vascular tissue in the stems (see Case Study 6.2). The pine wilt nematode feeds on the vascular tissue, reducing nutrient flow and distribution within the tree, and also attacks the vascular parenchyma cells that support the functions of the resin canals, the destruction of which reduces the movement of metabolites into the resin system and causes the tree to poison itself.

NEMATODE MANAGEMENT

Plant-parasitic nematodes typically cause noticeable crop damage only at high population densities. Management is directed at keeping nematode numbers below the damaging levels instead of trying to **eradicate** them from soil. Host plants are often able to compensate for the damage caused by moderately high numbers of nematodes (**tolerance**), especially where plants are grown under optimum environmental conditions. Although nematodes such as root-knot nematodes may establish hundreds of galls on a root system, very large numbers of these nematodes are usually required to cause significant crop losses. This density-dependent damage relationship for plant-parasitic nematodes is the foundation of nematode advisory programs. In these programs, data from soil assays are used to predict the nematode hazard to anticipated crops and to recommend control practices if necessary. In order to accurately advise farmers of nematode hazards, a relationship between nematode numbers in the soil and crop performance must first be established. This information is best obtained from infested fields, but controlled studies in greenhouse pots are also useful.

Nematode management approaches in agricultural fields consist primarily of the application of chemical pesticides (nematicides) to soil or the use of resistant cultivars. Nematicides are expensive, often dangerous to apply, and are cause for serious environmental concerns. Nematicides do not eliminate all the nematodes from a field, but are intended to reduce numbers sufficiently (typically >95%) for profitable crop production. Fields treated with nematicides may even have higher numbers of nematodes at the end of the growing season compared to fields not treated with chemicals. This rebound effect may be because of a much larger root system that allows for increased reproduction later in the growing season. Nematicides include both fumigant and nonfumigant compounds. The fumigant types are applied by liquid injection into the soil, where they volatilize to the gaseous phase in the soil. These compounds are typically very toxic to nontarget organisms in the soil, as well as to humans and other vertebrates overly exposed to them. These compounds must usually be applied well in advance of planting because they may damage the crop if not dispersed from the soil. Most nonfumigant nematicides work by attacking the nematode nervous system. These compounds usually become systemic within the plant and can be applied during or after planting time, but are extremely toxic to animals and people, whose nervous systems are affected equally as severely. In greenhouses and limited acreage high-value crops, soil sterilization with heat or application of a broad-spectrum biocidal fumigant is commonly practiced.

For most field crops, resistance is the only other practical means for controlling plant-parasitic nematodes. Resistant crops are economical in that little or no additional costs are assessed by the grower and no special equipment is needed to plant or harvest resistant plants. From the standpoint of impact on the environment, resistant cultivars are considered the best choice for nematode management, since their use does not involve the application of toxic materials to the soil that leach into groundwater or volatilize and contribute to atmospheric ozone depletion. Cultivars resistant to sedentary endoparasitic nematodes, such as root-knot and cyst nematodes, are available for some but not all crops. Use of resistant plants is complicated by the existence of physiological host races with extensive genetic variation within root-knot

CASE STUDY 6.2

Pine Wilt Nematode—A Triple Threat

- Pine wilt nematode, *Bursaphelenchus xylophilis*, causes a rapid wilt and death in as little as 3 weeks on susceptible pine trees.
  - The symptoms are caused by nematodes feeding in resin canals in the tree, in association with the blue stain fungus, *Ophiostoma piceae*, and possibly other microorganisms.
  - Pine wilt nematodes reproduce rapidly, from egg to adult in 4–5 days, and spread rapidly throughout the tree, inhibiting the functioning of the xylem through their feeding activities.
- The pine wilt nematode is not a typical plant parasite, in that it can also feed on fungi, like many members of its nematode family. Also, its feeding activities cause rapid death of the affected cells.
- The pine wilt nematode is believed to be native to the United States. American pine species are extremely tolerant to the nematode and typically show no symptoms.
- Pine wilt nematodes are carried from tree to tree by pine sawyer beetles (Cerambycidae, *Monochamus* spp.) in a very specific biological association.
  - The pine sawyer beetles are attracted to dead or declining trees to lay their eggs.
  - In the presence of the pine sawyer pupae, pine wilt nematodes form a fourth-stage juvenile specialized for dispersal that migrates into the insect tracheoles.
  - The newly emerged pine sawyer adults bore out of the tree and fly to healthy trees to feed, thus transporting the pine wilt nematodes to their next host.
- The considerable economic damage this nematode has caused comes from the following three sources:
  - The nematode has been introduced from America to Asia, where it has devastated natural stands of native Asian pine species. Up to 50% of the trees have died in infested areas.
  - Pine wood nematode is also killing imported exotic pines in the United States, particularly valuable landscape specimens in the Midwest and Northeast.
  - The threat posed by this nematode has led to a complete ban on the import of untreated pine products from the United States and Canada into the European Union and China.
- Researchers have developed a heat treatment (56°C for 30 min) that will free the wood products from pine wilt nematodes, but this treatment adds additional costs for the producers.

and cyst species. The use of a resistant cultivar is based on the prevention of reproduction by the dominant nematode race in a field. The existence of these physiological races means that some individuals in a field planted to a resistant cultivar will be able to reproduce successfully on the resistant cultivar, thus overcoming host resistance and possibly building up to damaging levels in the future. A continual battle must be waged to derive new sources of resistance to newly discovered nematode races. Very little progress has been made in identifying cultivars resistant to ectoparasitic plant-parasitic nematodes.

Other methods used to control plant-parasitic nematodes include **sanitation**, **cultural practices** (Chapter 25), **biological control** (Chapter 27), quarantines, and the use of organic soil amendments (Chapter 29). Sanitation methods prevent the spread of nematodes from infested fields or plant materials to uninfested fields by cleaning or sanitizing tools, implements, farm and nursery equipment, and any other source of nematode-infested soil and plants. Cultural control practices include tillage practices to destroy plant roots at the end of the growing season to reduce intercrop nematode survival rates and crop rotation with nonhosts. Crop rotation in particular can be quite effective when the nematode of concern has a small host range. For instance, corn–soybean rotations effectively reduce soybean cyst nematode as corn is a nonhost and hatched infective juveniles will starve. Two or more years of corn reduce soybean cyst nematode-infective juveniles enough to plant a high-yielding susceptible cultivar. Inclusion of a cyst-resistant soybean cultivar into this rotation further enhances its effects. Rotation crops available to growers, however, are severely limited by the wide host ranges of some plant-parasitic nematodes and by production economics that preclude the use of a good rotation crop with no economic market.

Biological control methods that use bacterial or fungal pathogens to control nematodes are still under development, but may be used in combination with other control methods in the future to enhance nematode management. Difficulties have been encountered with deployment of biological control agents due to failure of the biological agents to establish successfully in the soil environment. Incorporation of organic amendments into nematode-infested soil stimulates population increases of soil organisms that attack plant-parasitic nematodes. Decomposition products from the organic amendments may further reduce nematode population densities through direct toxic effects. Safer, more effective options for the control of nematodes are badly needed. Nematode problems continue to worsen globally as more agricultural production is concentrated on smaller areas of arable land, and urbanization of agricultural production areas reduces the availability of toxic nematicides.

LABORATORY EXERCISE

Plant-parasitic nematodes are found in the root zones of most of the plants grown in urban and rural areas. A diverse mixture of plant-parasitic and free-living nematodes (nonparasitic) are present in most soil samples. These nematodes can be readily isolated from both soil and root samples for viewing and identification. In this laboratory experiment, students will isolate nematodes from soil samples and observe the diverse nematode populations present in several settings.

Experiment 1. Isolation of Plant-Parasitic and Free-Living Nematodes from Soil

The objectives of this experiment are to isolate plant-parasitic nematodes from soil and to demonstrate the diversity of nematode populations in row crop fields, ornamental plantings, woodlands, and other plant communities. Nematodes can be extracted from soil in several ways. The two most common are the Baermann funnel method and the sugar flotation-centrifugation method, which is described here. The Baermann funnel method is simpler and is described in Hussey and Bernard (1975).

Materials (Sugar Flotation-Centrifugation Method)

This method was described in its present form in 1964 (Jenkins, 1964) and is widely used in nematology labs around the world. The protocol consists of the following two main steps: (1) concentrating the nematodes at the bottom of a centrifuge tube along with a small amount of soil and (2) suspending them in a sugar solution that has the same density as nematode bodies, while the denser soil particles end up at the bottom of the tube.

Each student or group of students will require the following items to complete the experiment:

- *Pictorial Key to Genera of Plant-Parasitic Nematodes* (Mai and Mullin, 1996).
- Soil sampling tubes or trowels for taking soil samples, 1-gallon plastic bags, plastic buckets, and ice chest.
- Metal sieves for separating the nematodes from soil: two sieves are required, 250-μm openings (60 mesh) and 38-μm openings (400 mesh), each with a 20 cm diameter and 5 cm depth. A 325-mesh sieve can be used in place of the 400. A small (8 cm diameter) 400-mesh sieve is very useful in the last step of the procedure but not absolutely necessary.
- Centrifuge equipped with a hanging bucket rotor and 50-mL centrifuge tubes. Tubes do not have to be sterile.
- Large lab spatula or slotted spoon, small lab spatula (e.g., a "spoonula").
- Vortex mixer (optional).
- 1- or 2-L plastic beakers, 500-mL wash bottles, small funnels, small beakers (50-mL to 150-mL sizes).
- Sucrose solution (454 g grocery-store sugar dissolved in 1 L of water).
- Dissecting microscopes with substage light source.
- Top-loading balance.
- Small Petri dishes or Syracuse watch glasses.

Sampling and Preextraction Procedures

Nearly any area with plant cover can be sampled, but in general, the longer it has been undisturbed the more diverse the nematode community will be. On the other hand, samples from recently planted or changed areas or from organic matter, such as mulch or sawdust piles, will contain pioneer species that are the first to colonize new habitats. Locate a row crop or vegetable field that has been in continuous production for 5–10 growing seasons that can be sampled during the laboratory period. Also, find some 10- to 20-year-old foundation plantings (shrubs) that can be sampled on the same day. Home lawns may have surprisingly large numbers of nematodes if they are several years old. Make the sucrose solution mentioned previously in the materials section and refrigerate (4°C) until needed.

Follow the protocol outlined in Procedure 6.1 to complete this experiment.

Anticipated Results

Students should be able to distinguish between free-living nematodes (nonparasitic types) and plant-parasitic nematodes. Free-living nematodes do not have stylets and are usually more active than plant-parasitic nematodes. Predatory nematodes, which are also free-living, have rasp-like teeth or a stylet without knobs and may be present in samples. In contrast, plant-parasitic nematodes

Procedure 6.1

Isolation of Parasitic and Free-Living Nematodes from Soil

Step	Instructions and Comments
1	Collect 20 soil cores (15- to 20-cm deep) in the root zone of an annual crop from a selected field, place in a plastic bucket, and mix thoroughly. Soil (500–1000 cm^3) should be sealed in a plastic bag, labeled, and placed in an insulated ice chest. Soil samples around shrubs, forest trees, turf, or other plant site can be collected and handled in the same manner. If a trowel is used instead of a soil sampling tube, collect small quantities of soil to a 15- to 20-cm depth and handle as above.
2	Add 100 (heavier, clayey soils) to 200 cm^3 of soil (sandier soils) to 600 mL of water in a 1-L beaker and mix vigorously for 30 s with a large spatula or spoon. Allow soil to settle for 30 s, then pour the suspension slowly through a 60-mesh sieve nested on top of a 325- or 400-mesh sieve. Remove the 60-mesh sieve. Use a wash bottle to concentrate the nematodes and soil in one sector of the finer sieve. Use a wash bottle to backwash nematodes and soil through a funnel into a centrifuge tube. Leave at least 1 cm of space in the tube. Soil should fill no more than the bottom fourth of the tube. Use two tubes if necessary.
3	Make sure tubes are balanced using a top-loading balance—the weight of the tubes and contents should be within 1 g of each other. Weight may be adjusted using a squeeze bottle of water. Centrifuge at 420 × *g* for 4 min (the brake may be used on most bench top-type centrifuges because they decelerate the rotor slowly, but for more powerful centrifuges do not use the brake). Gently pour water from tubes. Nematodes are in the soil pellet in the bottom of the tubes. Refill tubes with sucrose solution and mix with a vortex mixer or gently with a small spatula to resuspend the nematodes. Centrifuge for 15 s at 420 × *g*. Decant sucrose solution/nematode suspension onto the 400-mesh sieve (pour very slowly); a small 8 cm diameter sieve works better than the larger one. Gently rinse the screen with water to remove sugar solution and backwash nematodes into a small beaker. Beaker should contain no more than 25 mL of water. CAUTION: Follow manufacturer's directions on centrifuge use. Do not open the lid while the centrifuge is spinning. The centrifuge MUST be properly balanced before operation. In a six-bucket centrifuge, tubes must be arranged symmetrically and balanced. An unbalanced centrifuge may leap from the lab bench. Do not try to steady or turn off a vibrating or moving centrifuge, serious injury may result. Do not try to catch one that may be falling off a bench, its torque is stronger than you. Move away from it and pull the plug. Until everyone is experienced with its use, station a person at the electrical outlet to disconnect the plug in case of problems.
4	Pour samples into Petri dishes or Syracuse watch glasses and place on a dissecting scope. Nematodes are best observed using a light source from underneath. Use a pictorial key to identify nematode genera.
5	Disposal: Waste sugar and nematode samples can normally be poured down the sink drain. If the drain has a soil trap, sugar solution should be poured into a separate container and disposed in a separate sink. Sugar added to a soil trap will encourage growth of bacteria that produce noxious sulfides.

have stylets that have knobs or flanges on the posterior end and are usually quite sluggish (see Figures 6.1, 6.6, 6.7, and 6.9). If students identify the different trophic groups, differences will be seen among the samples. For instance, because of the build-up of organic matter, samples from long-established shrubs should contain more free-living nematodes than samples from agricultural fields. Some plant-parasitic nematodes should be isolated from any samples from under plants. Some of the larger plant-parasitic nematodes such as dagger, ring, or lance may be found in the samples, and students should be able to identify them using the pictorial key.

Questions

- What characteristics and/or behaviors distinguish free-living nematodes (nonparasitic types) from plant-parasitic nematodes?
- Are any predatory nematodes present in the samples?
- Which sample has more free-living nematodes? Which sample has more plant-parasitic nematodes?
- Can any of the plant-parasitic nematodes be identified using the pictorial key?

Acknowledgment

The authors and editors gratefully acknowledge the contributions of Dr. Gary Windham to this chapter.

SUGGESTED READINGS AND ONLINE RESOURCES

Baldwin, J. G., S. A. Nadler, and B. J. Adams. 2004. Evolution of plant parasitism among nematodes. *Annu. Rev. Phytopathol.* 42:83–105.

Barker, K. R., G. A. Pederson, and G. L. Windham, eds. 1998. *Plant-nematode Interactions. Monograph 36.* Madison, WI: American Society of Agronomy/Crop Science Society of America.

Coyne, D. L., J. M. Nicol, and B. Claudius-Cole. 2007. *Practical Plant Nematology: A Laboratory and Field Guide.* Cotonou, Benin: SP-IPM Secretariat, International Institute of Tropical Agriculture (IITA). http://www.uark.edu/ua/onta/info/2010%20Nematodes%20Manual%20ENGLISH.pdf.

Dropkin, V. H. 1989. *Introduction to Plant Nematology.* New, York: John Wiley.

Evans, K., D. L. Trudgill, and J. M. Webster, eds. 1993. *Plant Parasitic Nematodes in Temperate Agriculture.* Wallingford, UK: CAB International.

Ferris, H. 1999–2014. Nemaplex: The nematode-plant expert information system. A virtual encyclopedia on soil and plant nematodes. http://plpnemweb.ucdavis.edu/nemaplex/

Gaugler, R., and A. L. Bilgrami. 2004. *Nematode Behavior.* Wallingford, UK: CAB International.

Hussey, R. S., and E. C. Bernard. 1975. Soil-inhabiting nematodes. *Amer. Biol. Teacher* 37:224–226, 233.

Jenkins, W. R. 1964. A rapid centrifugal-flotation technique for separating nematodes from soil. *Plant Dis. Rep.* 48:692.

Jones, M. G. K. 1981. Host cell responses to endoparasitic nematode attack: Structure and function of giant cells and syncytia. *Ann. Appl. Biol.* 97:353–372.

Luc, M., R. A. Sikora, and J. Bridge, eds. 1990. *Plant Parasitic Nematodes in Subtropical and Tropical Agriculture.* Wallingford, UK: CAB International.

Maggenti, A. 1981. *General Nematology.* New York, NY: Springer-Verlag.

Mai, W. F., and P. G. Mullin. 1996. *Plant-parasitic Nematodes. A Pictorial Key to Genera.* 5th ed. Ithaca, NY: Cornell University Press.

Perry, R., and M. Moens. 2013. *Plant Nematology.* Wallingford, UK: CAB International.

Southey, J. F., ed. 1986. *Laboratory Methods for Work with Plant and Soil Nematodes.* London: Her Majesty's Stationery Office.

Starr, J. L., R. Cook, and J. Bridge, eds. 2002. *Plant Resistance to Parasitic Nematodes.* Wallingford, UK: CAB International.

Wilson, M. J., and T. Kakouli-Duarte. 2009. *Nematodes as Environmental Indicators.* Wallingford, UK: CAB International.

Zuckerman, B. M., W. F. Mai, and M. B. Harrison. 1985. *Plant Nematology Laboratory Manual.* Amherst, MA: University of Massachusetts, Agricultural Experiment Station.

7 An Overview of Plant Pathogenic Fungi and Fungus-Like Organisms

Ann Brooks Gould

Concept Box

- Fungi are eukaryotic, heterotrophic, and absorptive organisms with cell walls.
- Plant pathogenic organisms that are collectively grouped as fungi have been placed into the following three different kingdoms: Stramenopila (Oomycota), Protozoa (Plasmodiophoromycota), and Fungi (Chytridiomycota, Zygomycota, Ascomycota, and Basidiomycota).
- Chitin is a primary component of cell walls of true fungi and is not found in the cell walls of fungus-like organisms in Oomycota. Species in Oomycota contain cellulose in their cell walls; cellulose is not found in the cell walls of true fungi.
- Most plant pathogenic fungi can live as saprophytes or parasites (facultative saprophytes). However, some plant pathogenic fungi require living plant cells for their nutrition (biotrophs).
- The sexual state of a fungus is known as the teleomorph and the asexual state is known as the anamorph.
- The anamorphic state of most fungi is the conidium (Ascomycota and Basidiomycota) and nonmotile sporangiospores (Zygomycota). In Plasmodiophoromycota (kingdom Protozoa), Chytridiomycota (kingdom Fungi), and most species of Oomycota (kingdom Stramenopila), the anamorphic state is the motile sporangiospore, referred to as a zoospore.
- Examples of teleomorphic spores include oospores (Oomycota), zygospores (Zygomycota), ascospores (Ascomycota), and basidiospores (Basidiomycota).
- Most fungi are dispersed as spores through air currents, water, and animals (primarily insects). Fungi may also be spread in or on infected plant parts, movement of soil, and on agricultural equipment.

Of the biotic plant disease agents known to humankind, fungi are the most prevalent, and the study of these organisms has some very rich history. Fungi were once considered to be plants and thus were in the domain of botanists. Fungi are actually microorganisms, however, and differ from plants such that they are now placed in their own kingdom (Table 7.1). With the development of techniques needed for studying microorganisms, the discipline of mycology (Greek: *Mycos* = fungus + *-logy* = study) developed, and now the fungi encompass organisms from many different groups. Indeed, not all fungi that cause plant disease are classified as true fungi now.

Although the formal, systematic study of fungi is merely 250 years old, fungi have played important roles in the history of humankind for thousands of years. In their most important role, fungi are agents of decay, breaking down complex organic compounds into simpler ones for their own use or by other organisms. Fungi are important in the production of food (wine, leavened bread, and cheese) and can also be a source of food. Fungi may be poisonous, may produce hallucinogens, and have played a role in religious rites in many cultures. Fungi can attack wood products, leather goods, fabrics, petroleum products, and foods at any stage of their production, processing, or storage. Some produce toxins (called **mycotoxins**), are a source of antibiotics, cyclosporin (an immunosuppressant), vaccines, hormones, and enzymes, and can also cause diseases in

TABLE 7.1
Some Differentiating Characteristics of Plants, Animals, and Fungi

Characteristic	Animals	Plants	Fungi
Mode of obtaining food	Heterotrophy	Autotrophy	Heterotrophy
Chloroplasts	No	Yes	No
Chief components of cell wall	No cell wall	Cellulose and lignin	Chitin, glucans, or cellulose (in fungus-like organisms)
Chief sterols in cell membrane	Cholesterol	Phytosterols (ß-sistosterol, campesterol, stigmasterol)	Ergosterol
Food storage	Glycogen	Starch	Glycogen
Chromosome number in thallus	Diploid	Haploid, diploid, or polyploid	Haploid, diploid, or dikaryotic
Alternation of generations	No	Yes	Yes
Specialized vascular tissues for transport	Yes	Yes	No

both animals and plants. Those fungi found consistently in association with a particular plant disease are called fungal **pathogens**.

Fungi can destroy crops, and the economic consequences of this have been enormous throughout human history. Fungi reduce yield, destroy crops in the field and in storage and produce toxins poisonous to humans and animals. Blights, blasts, mildews, rusts, and smuts of grains are mentioned in the Bible. The Greek philosopher Theophrastus (370–286 BC) recorded his speculative, but not experimental, studies of grain rusts and other plant diseases. Indeed, wheat rusts have been important from ancient times until the present wherever wheat is grown. The Roman religious ceremony, the Robigalia, appealed to the rust gods to protect grain crops from disease.

When societies depend on a single crop for a major degree of sustenance, plant diseases can have a devastating impact. The potato was the major food of Irish peasants in the 1800s; the disease late blight of potato destroyed the potato crop in 1845–1846, resulting in massive starvation and emigration for years to come. The discipline of plant pathology was born over the scientific and political controversy caused by this disease. Until that time, fungi (or fungus-like organisms as is the case for the late blight pathogen) seen in plants were thought to be the result of plant disease, rather than the cause of plant disease. The famous German botanist Anton deBary worked with the late blight pathogen and proved experimentally that this was not the case. The causal agent of light blight of potato is now known as *Phytophthora infestans*. Other fungal and fungus-like diseases that played major roles in human history include chestnut blight, coffee rust, downy mildew of grape, stem rust of wheat, and white pine blister rust.

So, what are these fascinating organisms, and what role do they play in plant disease?

CHARACTERISTICS OF FUNGI AND FUNGUS-LIKE ORGANISMS

A **fungus** (plural: fungi) is a **eukaryotic**, **heterotrophic**, absorptive organism that develops a microscopic, diffuse, branched, tubular thread called a **hypha** (plural: hyphae). A group of hyphae is collectively known as **mycelium** (plural: mycelia), which is often visible to the unaided eye. Mycelium makes up the vegetative (nonreproductive) body or thallus of the fungus.

Fungi have definite cell walls and no complex vascular system and, except for a few groups, are not motile. The thallus of some fungi is single-celled (as in the yeasts). Many fungi can be grown in pure culture, facilitating their study, and they reproduce sexually or asexually by means of spores. As a group, fungi encompass a way of life shared by organisms of different evolutionary backgrounds.

The hyphae of most fungi are microscopic and differ in diameter among species (3–4 μm to ≥30 μm wide). **Septa** (crosswalls) may or may not be present and usually contain small pores to maintain continuity with other cells. Those hyphae without crosswalls are called aseptate or **coenocytic** (Chapters 1 and 8). Hyphae may branch to spread over a growing surface, but branching usually occurs only when nutrients near the tip become scarce. In culture, fungi form **colonies**, which can be discrete or diffuse, circular collections of hyphae or spores, or both, that arise from one cell or one grouping of cells.

Hyphal cells are bound by a cell envelope called the plasma membrane or **plasmalemma**. The plasmalemma of fungi differs from plants and animals in that it contains the sterol ergosterol, not cholesterol as in animals, or a phytosterol as in plants. Outside the plasmalemma is the **glycocalyx**, which is manifested as a slimy sheath (as in slime molds, kingdom Protista) or as a firm cell wall (as in most other fungi and fungus-like organisms). This cell wall is composed chiefly of polysaccharides. In the fungal phyla Ascomycota and Basidiomycota, **chitin** (ß-[1→4] linkages of *n*-acetylglucosamine) and **glucans** (long chains of glucosyl residues) are major cell wall components. Zygomycota contain chitin-chitosan and polyglucuronic acid; and fungus-like organisms in the Oomycota contain **cellulose** (ß-[1→4] linkages of glucose) and glucans. Chitin, glucans, and cellulose form strong fibers called microfibrils, which, when embedded, in a matrix of glycoprotein and polysaccharide, lend support to the hyphal wall. Cell walls also contain proteins and, in some fungi, dark pigments called **melanins**.

Fungal cells may contain one or many nuclei. A hyphal cell with genetically identical haploid nuclei is **monokaryotic**; cells with two genetically different but compatible haploid nuclei are **dikaryotic** (a characteristic of fungi in the Basidiomycota). Compared with animals and plants, the nuclei of fungi are small with fewer chromosomes or number of DNA base pairs. **Plasmids** (extrachromosomal pieces of DNA that are capable of independent replication) are found in fungi; the plasmid found in *Saccharomyces cerevisiae*, the common yeast fungus used to make wine and leavened bread, has been intensively studied. **Vacuoles** in hyphal cells act as storage vessels for water, nutrients, or wastes or may contain enzymes such as nucleases, phosphatases, proteases, and trehalase. Cells accumulate carbon reserve materials in the form of lipids, **glycogen**, or low molecular weight carbohydrates such as **trehalose**. **Mitochondria** (energy-producing organelles) in fungi vary in size, form, and number.

Fungal cells from different hyphal strands may often fuse in a process called **anastomosis**. Anastomosis, which is very common for some fungi in the Ascomycota and Basidiomycota, results in the formation of a three-dimensional network of hyphae and permits the organization of some specialized structures such as **rhizomorphs**, **sclerotia**, and fruiting structures, also known as **sporocarps**.

HOW FUNGI AND FUNGUS-LIKE ORGANISMS INFECT PLANTS

Nutrition

Fungi have the advantage over organisms in the plant kingdom in that they do not have chlorophyll and are not dependent on light to manufacture food. Thus, fungi can grow in the dark and in any direction as long as there is an external food source and water.

Fungal hyphae elongate by apical growth (from the tip). Unlike animals, fungi do not ingest their food and then digest it; instead, they obtain their food through an absorptive mechanism. As the fungus grows through its food, hyphae secrete digestive **exoenzymes** into the external environment (Chapter 32). Nutrients are carried back through the fungal wall and stored in the cell as glycogen. Free water must be present to carry the nutrients back into fungal cells.

Fungi can grow rapidly over a surface stratum and may penetrate it or may produce an aerial mycelium. Fungi have the potential to utilize almost any carbon source as a food. This is restricted only by what exoenzymes the fungus produces and releases into the environment. Some of these enzymes are listed in Table 7.2.

Fungal growth may continue as long as the appropriate nutrients and environmental conditions are present. Growth ceases when the nutrient supply is exhausted or the environment is no longer favorable for development. Nutrients required by fungi include a source of carbon in the form of sugars, polysaccharides, lipids, amino acids, and proteins; nitrogen in the form of nitrate, ammonia, amino acids, polypeptides, and proteins; sulfur, phosphorus, magnesium, and potassium in the form of salts; and trace elements such as iron, copper, calcium, manganese, zinc, and molybdenum.

TYPES OF PATHOGENS: HOW DO FUNGI AND FUNGUS-LIKE ORGANISMS OBTAIN THEIR FOOD?

Most fungi are **saprophytes** in that they use nonliving organic materials as a source of food. These organisms are important scavengers and decay organisms and along with bacteria recycle carbon, nitrogen, and essential mineral nutrients. Most plant pathogenic fungi are versatile organisms that can live as saprophytes or parasites. These fungi are called **facultative saprophytes**. They attack their hosts, grow and reproduce as parasites, and then act as saprophytes, along with the normal nonpathogenic soil microflora, between growing seasons. This makes them very difficult to control.

Many plant pathogenic fungi are classified as **necrotrophs**. These fungi are usually saprophytes and survive well as sclerotia, spores, or mycelia in nonliving host material. Given the opportunity, however, these fungi become parasitic, kill, and then feed upon dead plant tissues. Necrotrophs produce **secondary metabolites** that are toxic to susceptible host cells. Fungal enzymes degrade tissues killed by the toxins, and the cell constituents are used as food. Disease symptoms caused by necrotrophs are manifested as small to very large patches of dead, blackened, or sunken tissue. A classic example of a necrotroph is the fungal pathogen *Monilinia fructicola*, which causes brown rot of peaches.

Biotrophic pathogens (sometimes called obligate parasites) grow or reproduce only on or within a suitable living host. The biotrophic relationship is highly host specific. Biotrophic pathogens, especially fungi that attack foliar plant parts, derive all their nutrients from the host. They invade host tissues, but do not kill them (or may kill

TABLE 7.2
Exoenzymes Produced by Some Fungi

Exoenzyme	Substrate	Utility
Cutinase	Cutin: A long-chain polymer of C_{16} and C_{18} hydroxy fatty acids, which, with waxes, forms the plant cuticle	Facilitates direct penetration of host cuticle by pathogenic fungi
Pectinase (pectin methyl esterase, polygalacturonase)	Pectin: Chains of galacturonan molecules (α-[1→4]-D-galacturonic acid) and other sugars; main components of the middle lamella and primary cell wall	Facilitates penetration and spread of pathogen in host; causes tissue maceration and cell death
Cellulase (cellulase C_1, C_2, C_x, and ß-glucosidase)	Cellulose: Insoluble, linear polymer of ß-(1→4) linkages of glucose; skeletal component of plant cell wall	Facilitates spread of pathogen in host by softening and disintegrating cell walls
Hemicellulase (e.g., xylanase, arabinase)	Hemicellulose: Mixture of amorphous polysaccharides such as xyloglucan and arabinoglucan that vary with plant species and tissues; a major component of primary cell wall	Role of these enzymes in pathogenesis is unclear
Ligninase	Lignin: Complex, high-molecular-weight polymer (made of phenylpropanoid subunits); major component of secondary cell wall and middle lamella of xylem tissue	Few organisms (mostly basidiomycetes) degrade lignin in nature; brown rot fungi degrade lignin, but cannot use it as food; white rot fungi can do both
Lipolytic enzymes (lipase, phospholipase)	Fats and oils (fatty acid molecules): Major component of plant cell membranes; also stored for energy in cells and seeds and found as wax lipids on epidermal cells	Fatty acid molecules used as a source of food by pathogen
Amylase	Starch: Main storage polysaccharide in plants; enzymes hydrolyze starch to glucose	Glucose readily used by pathogen as a source of food
Proteinase or protease	Protein: Major components of enzymes, cell walls, and cell membrane; enzymes hydrolyze protein to smaller peptide fractions and amino acids	Disruption of cell membrane and enzymatic activity affects host cell function; role of these enzymes in pathogenesis is unclear

them gradually), thus ensuring a steady supply of nutrients. These fungi may produce special penetration and absorption structures called **haustoria**. Haustoria penetrate the host cell wall, but not the plasmalemma, essentially remaining outside the host cell while nutrients are transferred across the host plasmalemma into the fungus. Classic examples of pathogenic biotrophs include the powdery mildew fungi (Chapter 12), downy mildew oomycetes (Chapter 8), and the smuts and rusts (Chapter 14).

Some biotrophs are part of a mutually beneficial symbiosis with other organisms. These are **mutualists** and include **mycorrhizal fungi** (fungi that grow in association with plant roots) and **endophytes**. Another group of organisms, the **hemibiotrophs**, functions as both biotrophs and necrotrophs during their life cycle. An example of a hemibiotroph is the soybean anthracnose pathogen, *Colletotrichum lindemuthianum*. This fungus first lives as a biotroph by growing between the plasma membrane and the cell wall of living cells. The fungus then suddenly switches to a necrotrophic phase and kills all the cells it has colonized.

DISEASE SYMPTOMS CAUSED BY FUNGI AND FUNGUS-LIKE ORGANISMS

Details of the infection process are described in Chapters 1 and 24. The visual manifestation of this process (or **symptoms**) varies depending on the infected plant part, the type of host, and the environment. Symptoms caused by fungal pathogens can be similar to those caused by other biotic and abiotic disease agents. Generally described, these symptoms are the following:

- **Necrosis**—Cell death; affected tissue may be brown or blackened, sunken, dry or slimy; in leaves, often preceded by yellowing (chlorosis, or a breakdown in chlorophyll).
- **Permanent wilting**—Blockage or destruction of vascular tissues by fungal growth, toxins, or host defense responses; affected tissue wilts and dies, and associated leaves may scorch and prematurely abscise.
- Abnormal growth—**Hypertrophy** (excessive cell enlargement), **hyperplasia** (excessive cell division), **etiolation** (excessive elongation); affected tissue appears gall- or club-like, misshapen, or curled.
- Leaf and fruit **abscission**—Premature defoliation and fruit drop.
- Replacement of host tissue—Plant reproductive structures replaced by fungal hyphae and spores.
- **Mildew**—Surfaces of aerial plant parts (leaves, fruit, stems, and flowers) are covered with a white to gray mycelium and spores.

Common manifestations of these symptoms are listed in Table 7.3.

TABLE 7.3
Symptoms of Plant Diseases Caused by Fungi

Symptom	Manifestation	Description	Example
Necrosis	Leaf spot	Discrete lesions of dead cells on leaf tissue between or on leaf veins, often with a light-colored center and a distinct dark-colored border, and sometimes accompanied by a yellow halo; fruiting structures of fungus often evident in dead tissue	Common leaf spot of strawberry (*Mycosphaerella fragariae*)
	Leaf blotch	Larger, more diffuse regions of deaf or discolored leaf tissue	Horse chestnut leaf blotch (*Guignardia aesculi*)
	Needle cast	Needles of conifers develop spots, turn brown at the tips, die, and prematurely fall (cast) to the ground	Rhabdocline needle cast of Douglas-fir (*Rhabdocline pseudotsugae*)
	Anthracnose	Sunken lesions on leaves, stems, and fruit; on leaves, regions of dead tissue often follow leaf veins and/or margins	Anthracnose (or cane spot) of brambles (*Elsinoë veneta*)
	Scab	Discrete lesions, sunken or raised, on leaves, fruit, and tubers	Apple scab (*Venturia inaequalis*)
	Dieback	Necrosis of twigs that begins at the tip and progresses toward the twig base	Sudden oak death, Ramorum leaf blight (*Phytophthora ramorum*)
	Canker	Discrete, often elliptical lesions on branches and stems that destroy vascular tissue; can appear sunken, raised, or cracked; causes wilt, dieback, and death of the branch; fruiting structures of fungus often evident in dead tissue	Cytospora canker of spruce (*Cytospora kunzei* or *Leucostoma kunzei*)
	Root and crown rot	Necrosis of feeder roots to death of the entire root system; often extends into the crown and can girdle the base of the stem; causes wilt, dieback, and death of the canopy	Rhododendron wilt (or Phytophthora root rot) (*Phytophthora cinnamomi*; *P. parasitica*)
	Cutting rot	Cuttings in propagation beds are affected by a blackened rot that begins at the cut end and travels up the stem, rapidly killing the cuttings	Blackleg of geranium (*Pythium* species)
	Damping-off	Seeds and seedlings are killed before (preemergent damping-off) or after (postemergent damping-off) they emerge from the ground	Damping-off caused by species of *Pythium*, *Rhizoctonia*, and *Fusarium*
	Soft rot	Fleshy plant organs such as bulbs, corms, rhizomes, tubers, and fruit are macerated and become water-soaked and soft; tissues may eventually lose moisture, harden, and shrivel into a mummy; common postharvest problem	Rhizopus soft rot of papaya (*Rhizopus stolonifer* syn. *R. nigricans*)
	Dry rot	Dry, crumbly decay of fleshy plant organs	Fusarium dry rot of potato (*Fusarium sambucinum*)
Permanent wilting	Vascular wilt	Vascular tissue is attacked by fungi, resulting in discolored, nonfunctioning vessels and wilt of canopy	Dutch elm disease (*O. ulmi*)
Abnormal growth	Leaf curl	Leaves become discolored, distorted, and curled, often at the leaf edge	Peach leaf curl (*Taphrina deformans*)
	Club root	Swollen, spindle- or club-shaped roots	Clubroot of crucifers (*Plasmodiophora brassicae*)
	Galls	Enlarged growths, round or spindle-shaped, on leaves, stems, roots, or flowers	Cedar-apple rust (*Gymnosporangium juniperi-virginianae*)
	Warts	Wart-like outgrowths on stems and tubers	Potato wart disease (*Synchytrium endobioticum*)
	Witches' broom	Profuse branching of twigs that resembles a spindly broom	Witches' broom of cacao (*Crinipellis perniciosa*)
	Etiolation	Excessive shoot elongation and chlorosis, induced in poor light or by growth hormones	Bakanae of rice (foolish seedling disease) (*Gibberella fujikuroi*)
Leaf and fruit abscission		Infection of petioles, leaves, and fruit cause tissues to drop prematurely	Anthracnose of shade trees (*Apiognomonia* species)
Replacement of host tissue		Plant reproductive structures are replaced by fungal hyphae or spores, or both	Common smut of corn (*Ustilago maydis*)
Mildew		Surfaces of aerial plant parts (leaves, fruit, stems, and flowers) are covered with white to gray mycelium and spores	Powdery mildew of rose (*Podosphaera pannosa*)

FUNGAL REPRODUCTION

The reproductive unit of most fungi is the **spore**, which is a small, microscopic unit consisting of one or more cells, that provides a dispersal or survival function for the fungus. Spores are produced in asexual or sexual processes.

Asexual Reproduction

Asexual reproduction is the result of mitosis, and progeny are genetically identical to the parent. Some of the true fungi (Ascomycota and Basidiomycota) produce nonmotile, asexual spores called **conidia** (Figure 7.1). Conidia are produced at the tip or side of a stalk or supporting structure known as a **conidiophore**. Conidiophores can be arranged singly or in fruiting structures. A **synnema** consists of a group of conidiophores that are fused together to form a stalk. An **acervulus** is a flat, saucer-shaped bed of short conidiophores that grow side by side within host tissue and beneath the epidermis or cuticle. A **pycnidium** is a globose or flask-shaped structure lined on the inside with conidiophores. A pycnidium has an opening called an **ostiole** through which spores are released.

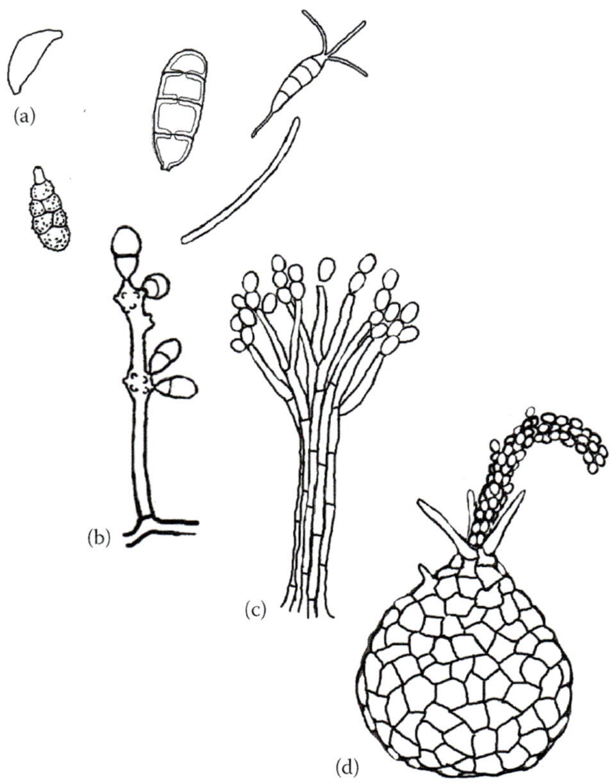

FIGURE 7.1 Asexual reproduction in mitosporic fungi. (a) Single conidia varying in shape, color (hyaline or brown), and number of cells. (b–d) Asexual fruiting structures: (b) conidia produced on a distinct conidiophore, (c) synnema, and (d) pycnidium. (Courtesy of N. Shishkoff).

Another type of asexual spore is the **chlamydospore**. Chlamydospores are thick-walled conidia that form when hyphal cells round up and separate. These spores function as resting spores and are found in many groups of fungi. Some members of the Oomycota (Chapter 8) produce a motile spore called a **zoospore** in a sac-like structure called a **sporangium**. The stalk that supports the sporangium is called a **sporangiophore**. Zygomycota (Chapter 10), best known for *Rhizopus stolonifer*, the fungus that causes bread mold and soft rot of fruit and vegetables, produces asexual spores called **sporangiospores**, also borne in sporangia.

Sexual Reproduction

Genetic recombination is the result of sexual reproduction, and offspring are genetically different from either parent. The heart of the sexual reproductive process is fertilization. In fertilization, sex organs called **gametangia** produce special sex cells (gametes or gamete nuclei) that fuse to form a zygote. The fertilization process occurs in the following two steps: (1) **plasmogamy**, when the two nuclei, one from each parent, join together in the same cell; and (2) **karyogamy**, when these nuclei fuse together to form a zygote.

In most fungi, the vegetative body or thallus has one set of chromosomes (haploid). The diploid (having two sets of chromosomes) zygote that results from fertilization must therefore undergo meiosis (a chromosome reduction division) before it can develop into a haploid, multicellular organism. In contrast, most plants and animals, and some fungus-like species in the kingdom Stramenopila, possess a diploid thallus. Meiosis, therefore, occurs in the gametangia as the gametes are produced.

Homothallic fungi produce both male and female gametangia on a single mycelium that are capable of reproducing sexually (i.e., they are self-fertile). This is analogous to a monoecious plant, where both male and female flowers are found on the same plant. Homothallic gametangia may be obviously differentiated into male and female structures (**antheridia** and **oogonia**, respectively) or may be morphologically indistinguishable (sexually undifferentiated). Homothallic species occur in all phyla of fungi.

Conversely, the sexes (or **mating types**) in **heterothallic** fungi are separated in two different individuals, which are self-sterile. This is analogous to dioecious plants, where male and female flowers are produced on different plants. The different mating types in heterothallic fungi are usually differentiated as plus (+) or minus (–), or by using letters (e.g., *A* and *A'*). Heterothallic fungi are not as common as homothallic ones.

Examples of sexual spores include **oospores**, **zygospores**, **ascospores**, and **basidiospores**. The name of a particular spore type usually reflects the parent structure that produced it. For example, ascospores are produced in a parent structure called an ascus, and basidiospores are produced from a parent structure called a basidium.

FUNGI AND THE ENVIRONMENT

Fungi are ubiquitous and grow in many different habitats and environmental conditions. Of these, a source of moisture is most critical for growth and reproduction, and most fungi grow best in a damp environment. As stated previously, moisture is needed to move nutrients into hyphal cells. Indeed, some thin-walled species require a continuous flow of water to prevent desiccation. For foliar pathogenic fungi, free moisture and high relative humidity are important for germination and penetration of leaf tissue. For soil fungi, free moisture in the soil is important for dispersal as well.

Some fungi can adapt to very low moisture availability by regulating the concentration of solutes (or osmotic potential) in cells. A cellular osmotic potential higher than that of the environment will cause water to enter cells. When the environment becomes drier, concentration of solutes outside the cell becomes higher than that within the cells and water leaves the cells. The net result is that the cell desiccates and the plasmalemma shrinks. Active release or uptake of solutes in hyphal cells helps certain fungi maintain adequate hydration under fluctuating conditions. Other fungi produce resting structures with impermeable walls that withstand drying. These structures germinate when conditions become favorable for growth.

Since oxygen is required for respiration (generation of energy), most fungi do not grow well when submerged. Fungi that require oxygen for respiration are called **aerobic** or obligately oxidative. Those that do not need oxygen for respiration are **anaerobic** or fermentative. Fungi that can derive energy by oxidation or fermentation are **facultative fermenters**. Most plant pathogenic fungi are aerobic or facultative fermenters.

Fungi must adapt to, or tolerate, considerable temperature fluctuations that occur daily or seasonally. Most fungi are **mesophilic** and grow well between 10°C and 40°C with the optimum temperature for most species between 25°C and 30°C. Some species are **thermophilic** (grow at 40°C or higher) or **psychrophilic** (grow at less than 10°C). Fungi function best at a range of pH 4–7 and some produce melanins in cell walls to protect against damage from sunlight.

SURVIVAL AND DISPERSAL

With the exceptions of biotrophs, most plant pathogenic fungi spend part of their life cycle as parasites and the remainder as saprophytes in the soil or on plant debris. When a fungus encounters adverse environmental conditions, drains its environment of nutrients, or kills its host, growth usually ceases. Inactive hyphae are subject to desiccation or attack by insects or other microbes, so pathogenic fungi must survive in a reduced metabolic or dormant state in plant debris or soil until conditions improve, or disperse to find other hosts.

Fungi produce spores both to survive adverse environmental conditions (e.g., chlamydospores) and for dispersal (e.g., zoospores). Indeed, the different reproductive strategies exhibited by many fungi have different ecological advantages. Generally speaking, fungi reproduce asexually when food sources are abundant and dispersal and spread is of prime importance, and they reproduce sexually at the end of the season when food is limited or in cases where dispersal is not as critical. For example, the biotrophic powdery mildew pathogens produce abundant asexual spores (conidia) during the growing season when environmental conditions are favorable and there is plentiful host material. These spores are easily dispersed in air and serve to spread disease rapidly over the summer months. At the end of the growing season, environmental and host factors trigger sexual reproduction of the fungus. The ascocarp produced (called a **cleistothecium** or **perithecium**; see Chapter 12) overwinters in plant parts or debris releasing ascospores (spores that result from meiosis) to initiate new infections during the following spring.

Spore dispersal can be critical for fungal survival; spores released quickly from fruiting structures have a better chance of finding new hosts and initiating new infections. Spores can be passively or actively liberated from the parent mycelium. Passive spore release mechanisms include rain splash, mechanical disturbance (wind, animal activity, and cultivating equipment), and electrostatic repulsion between a spore and its **sporophore** (supporting stalk). Spores can also be actively released; ascospores are forcibly discharged from the ascus in many members of the Ascomycota (e.g., *Claviceps purpurea*), and sudden changes in cell shape may launch spores into the air (e.g., the aeciospore stage of the rust fungus *Puccinia*).

Most spores, once liberated, are passively dispersed (or **vectored**) through air or soil over short or long distances. The most common modes of dispersal for plant pathogenic fungi include air currents, water (via rain splash and dispersal through flowing water), and animals (notably insects). Insects have an intimate role in the life cycle and spread of fungi that cause many major diseases. For example, beetles that vector the Dutch elm disease pathogen *Ophiostoma ulmi* pick up fungal spores on their bodies and inoculate new hosts during feeding (see Case Study 7.1). Some plant pathogenic fungi such as *Tilletia caries* (stinking smut of wheat) are dispersed on seed, which conveniently places them with their host

CASE STUDY 7.1

Dutch Elm Disease

- Dutch elm disease is an example of a vascular wilt disease. Pathogens that cause vascular wilts live in the xylem and disrupt the flow of water and nutrients.
- **Significance**: In two twentieth-century epidemics, Dutch elm disease destroyed over 40 million American elm trees planted throughout the United States as street tree monocultures.
- **Pathogens**: *Ophiostoma ulmi, O. novo-ulmi.*
- **Hosts**: American elm (*Ulmus americana*) is most susceptible; winged elm, slippery elm, rock elm, and cedar elm vary from susceptible to tolerant; Siberian and Chinese elms are the most resistant.
- **Symptoms and signs**:
  - In summer, leaves flag (or wilt) on individual branches, followed by branch death shortly thereafter.
  - Affected xylem is discolored in a characteristic streaking pattern just beneath the bark.
- Transmission:
  - The native elm bark beetle (*Hylurgopinus rufipes*) and the smaller European bark beetle (*Scolytus multistriatus*) serve as vectors. Female insects lay eggs in dead or dying elm wood. Eggs hatch into larvae that feed in wood to form characteristic tunnel patterns called galleries. If the fungus is present, conidia stick to the adult beetles as they emerge in spring to feed on healthy twigs. Spores deposited in xylem during feeding germinate and form a mycelium that penetrates xylem pit openings. Conidia are produced, and they disperse through the xylem sap.
  - The pathogens also spread from infected to healthy trees through root grafts.
- **Management**: Dutch elm disease is best managed by combining cultural and chemical controls.
  - Sanitation measures include prompt removal of dead or dying elms as well as diseased limbs in mildly affected trees. Wood must be debarked or burned prior to beetle emergence the following spring.
  - Insecticide is used to control the beetle vector, and valuable trees may be injected on a preventive basis with fungicide.
  - Trenching midway between diseased and healthy trees is necessary to prevent root graft transmission.
  - From breeding programs, hybrids of American elm and elms of Asian origin, which are resistant to Dutch elm disease, have been released. These trees are disease resistant yet possess the desired horticultural qualities of the American elm. In addition, a number of American elms with a moderate level of tolerance to Dutch elm disease have been recently introduced to the horticultural trade.

when the time is appropriate for infection to commence. Soil pathogens that produce motile zoospores (some Oomycota and plasmodial slime molds) can more actively move through soil. Soil zoospores, attracted to roots by root exudates (**chemotaxis**), encyst and penetrate the root cortex to initiate new infections.

Some fungi, such as the basidiomycete *Armillaria mellea*, disperse through potentially inhospitable environments by producing rhizomorphs, which are root-like structures composed of thick strands of somatic hyphae. Rhizomorphs have an active meristem that grows through soil from diseased plants to healthy roots, thus facilitating dispersal of the fungus to new substrates. In cross section, a rhizomorph has a dark, outer rind of thick-walled, nonliving cells and an inner cortex of thin-walled cells and active cytoplasm.

More commonly, however, fungi produce resting structures called sclerotia. Sclerotia are spherical structures that are 1 mm to 1 cm in diameter with a thick-walled rind and a central core of thin-walled cells that have abundant lipid and glycogen reserves. Sclerotia may remain viable in adverse conditions for months or years. Many plant pathogenic fungi in the Ascomycota, Basidiomycota, and mitosporic fungi produce sclerotia, especially those that infect herbaceous plants as a means of surviving between crops. Sclerotia may germinate to form sexual structures as in *C. purpurea* or asexual conidia, as in *Botrytis cinerea*.

CLASSIFICATION OF PLANT PATHOGENIC FUNGI

The classification of organisms (taxonomy) including fungi is based on criteria that changes as more information becomes known about these organisms. Taxonomy of fungi has been classically based on sexual and asexual spore morphology and how they are formed, followed by secondary considerations such as hyphal and colony (the fungus in culture) characteristics. Molecular techniques that examine and compare the genetics of organisms are increasing in importance to taxonomists.

As previously mentioned, the "fungi" as a group used to be classified with the plants in the kingdom Planta, but plant pathogenic species are now placed in three different kingdoms (Table 7.4). The endoparasitic slime molds belong to the kingdom Protozoa. The water mold fungi belong to the phylum Oomycota within the kingdom Stramenopila. These two groups are more closely related to protozoans (endoparasitic slime molds) or brown algae (water molds) than to true fungi and are often referred to in the literature as fungus-like, lower fungi, or pseudofungi and are discussed further in Chapters 8 and 9. The other fungal groups (e.g., Chytridiomycota, Zygomycota, Ascomycota, and Basidiomycota) are members of the kingdom Fungi (or Mycota). These fungi are also often referred to as the true fungi or higher fungi. Interestingly, the Chytridiomycota, once placed in the kingdom Protozoa, may be regarded as ancestors of the other true fungi.

Fungi are further divided into subcategories. The most basic classification of a fungal organism is the **genus** and **species** (or specific epithet). The generic name and specific epithet are constructed in Latin according to internationally recognized rules. Fungal names are also associated with an authority, which is the name of the person who discovered or named the species. For example, the species *C. purpurea* (Fr.:Fr.) Tul. was described by Elias Fries and modified by Tulanse.

In the classification scheme, related genera are grouped into families, families into orders, orders into classes, classes into phyla, and phyla into kingdoms. Each grouping has a standard suffix (underlined). For example, the Dutch elm disease pathogen, *O. ulmi*, is classified as follows:

Kingdom **Fungi** (or **Mycota**)
Phylum **Asco<u>mycota</u>**
Class **Asco<u>mycetes</u>**
Order **Ophiostomat<u>ales</u>**
Family **Ophiostomat<u>aceae</u>**
Genus ***Ophiostoma***
Species (and authority)
Ophiostoma ulmi
(Buisman) Nannf.

Fungi are probably best identified by the phylum to which they belong. They are placed in phyla based on the sexual phase of the life cycle. Brief descriptions of the different phyla of plant pathogenic fungal or fungus-like organisms, the kingdoms to which they belong, and examples of each are as follows.

KINGDOM PROTOZOA

Plasmodiophoromycota (Chapter 9) is one of four groups of slime molds including the endoparasitic slime molds that produce zoospores with two flagella and are obligate parasites. Sexual reproduction in this group includes fusion of zoospores to form a zygote. Some of these fungi vector plant viruses (Chapter 4).

Plasmodiophora brassicae (clubroot of crucifers) has a multinucleate, amoeboid thallus (plasmodium) that lacks cell wall. The plasmodium invades root cells, inducing hypertrophy and hyperplasia (thus increasing its food supply). The "clubs" that result interfere with normal root function (absorption and translocation of water and nutrients). Plants affected by this disease are wilted and stunted. *P. rassicae* survives in soil between suitable hosts as resting spores (product of meiosis) for many years (Chapter 9).

KINGDOM STRAMENOPILA

Many members of the **Oomycota** (Chapter 8) produce motile, biflagellate zoospores that have a tinsel flagellum and a whiplash flagellum. The thallus is diploid and their cell walls are coenocytic and contain cellulose and glucans. Oomycetes reproduce asexually by producing motile zoospores in a zoosporangium (Figure 7.2); a few species produce nonmotile sporangiospores. Chlamydospores occur in some species. Sexual reproduction results in an oospore (zygote) produced from contact between male (antheridia) and female (oogonia) gametangia (gametangial contact). Most members of the Oomycota attack roots, but some species of *Phytophthora*, the downy mildews, and white rusts also attack aerial plant parts. The Oomycota are often called water molds because most species have a spore stage that swims (biflagellate zoospores) and thus require free water for dispersal.

The ubiquitous soil water mold *Pythium* causes seed rots, seedling damping-off, and root rots of all types of plants and is especially troublesome in greenhouse plant production and in turfgrasses. *Pythium* species are also responsible for many soft rots of fleshy vegetable fruit and other organs that are in contact with soil particles in the field, in storage, in transit, and at market.

Phytophthora species cause a variety of diseases on many hosts ranging from seedlings and annual plants to fully developed fruit and forest trees. They cause root rots; damping-off; rots of lower stems, tubers, and corms; bud or fruit rots; and blights of foliage, twigs, and fruit. Some species of *Phytophthora* are host specific, whereas others have a broad host range. The genus is best known for its root and crown rots (most *Phytophthora* species), late blight of potato and tomato (*P. infestans*), Rhododendron wilt (*P. cinnamomi*), and sudden oak death (*Phytophthora ramorum*).

TABLE 7.4
Characteristics of the Different Phyla of Plant Pathogenic Fungi

	Kingdom					
	Protozoa	Stramenopila	Fungi (Mycota)			
	Plasmodiophoromycota	Oomycota	Chytridiomycota	Zygomycota	Ascomycota	Basidiomycota
Habitat	Aquatic	Aquatic	Aquatic/terrestrial	Terrestrial	Terrestrial	Terrestrial
Form of thallus	Plasmodium (multinucleate mass of protoplasm without a cell wall)	Coenocytic hyphae	Globose or ovoid thallus (lacks a true mycelium)	Well-developed coenocytic hyphae; rhizoids, stolons	Well-developed septate hyphae; some are single-celled (yeasts)	Well-developed septate hyphae
Chromosome number in thallus	Haploid (cruciform nuclear division)	Diploid	Diploid	Haploid	Haploid	Dikaryotic, haploid
Chief component of cell wall	Cell wall of thallus lacking	Cellulose/glucans	Chitin	Chitin-chitosan	Chitin/glucans	Chitin/glucans
Motile stage	Zoospores with two anterior, unequal, whiplash flagella	Zoospores with two flagella: one anterior tinsel and one posterior whiplash	Zoospores with one posterior whiplash flagellum	None	None	None
Pathogenic relationship	Obligate parasites	Facultative or obligate parasites	Obligate parasite	Facultative parasites; symbionts	Facultative or obligate parasites; symbionts	Facultative or obligate parasites; symbionts
Asexual reproduction	Secondary zoospores in zoosporangia	Zoospores in sporangia; chlamydospores	Holocarpic (entire thallus matures to form thick-walled resting spores)	Sporangiospores in sporangia	Conidia on conidiophores produced on single hyphae or in fruiting bodies; budding (yeasts)	Budding, fragmentation, arthrospores, oidia, or formation of conidia
Sexual reproduction	Resting spores with chitin in cell walls, the result of fusion of zoospores to form a zygote	Oospores, the result of fusion between male antheridia and female oogonia	Not confirmed	Fusion of gametangia to produce a zygospore	Formation of ascospores within an ascus	External formation of four basidiospores on a basidium; fusion of spermatia and receptive hyphae (rusts); fusion of compatible mycelia (smuts)
Major plant diseases (or groups of diseases)	Clubroot of crucifers, powdery scab of potato	Root, stem, crown, seed, and root rots, white rust, downy mildew	Potato wart, crown rot of alfalfa	Soft rot, fruit rot, bread mold, postharvest disease	Powdery mildew, root, foot, crown, and corm rot, vascular wilt, canker, wood decay, anthracnose and other foliar diseases, needle cast, apple scab, postharvest disease	Root rot, heart rot, white and brown wood decay, fruit rot, smuts, rusts, snow mold of turfgrass, fairy ring

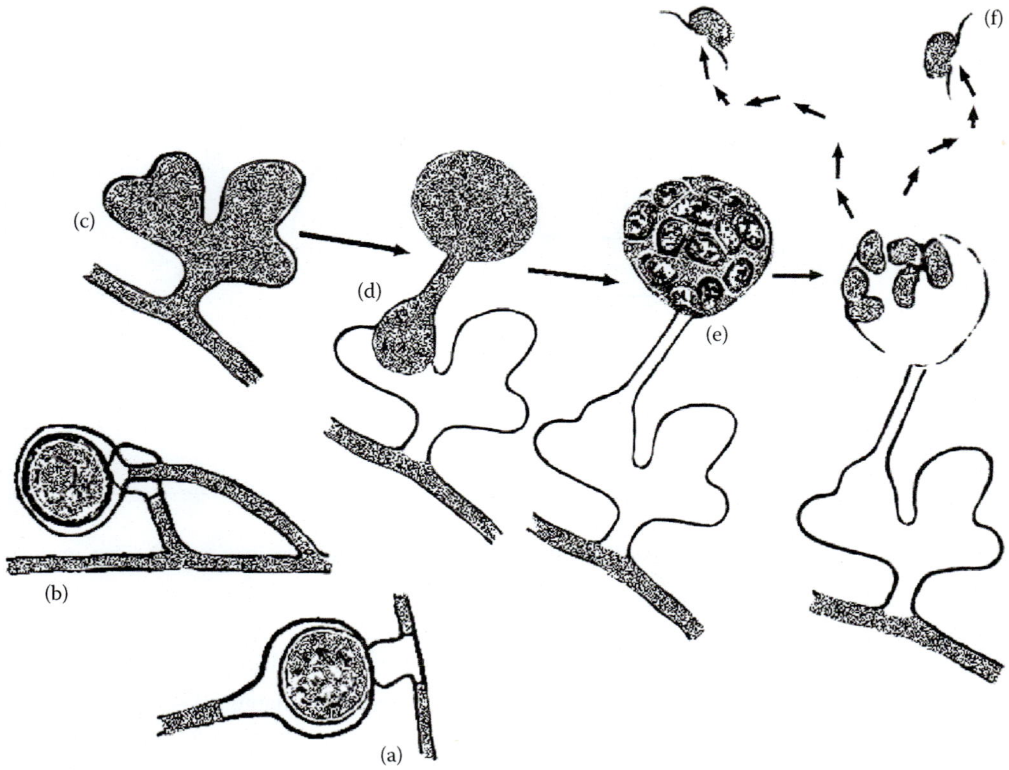

FIGURE 7.2 Reproduction in *Pythium aphanidermatum* (Oomycota), a root, stem, seed, and fruit rot pathogen with a very wide host range. (a and b) Sexual stage: oogonium (female gamete) with an (a) intercalary or (b) terminal antheridium (male gamete). (c–f) Asexual stage: (c) inflated, lobate sporangium, (d) sporangium with vesicle, (e) sporangium with vesicle containing zoospores, and (f) biflagellate zoospores. (Courtesy of N. Shishkoff).

The downy mildews (e.g., *Plasmopara* spp.) are obligate parasites that occur on many cultivated crops. They primarily cause foliage blights that attack and spread rapidly in young and tender green leaf, twig, and fruit tissues. These fungi produce sporangia (asexually) on distinctively dichotomously branched sporangiophores, which emerge through stomata on the lower surfaces of leaves, forming a visible mat with necrotic lesions on leaf surfaces. The downy mildews require a film of water on the plant and high relative humidity during cool or warm, but not hot, weather. Downy mildew of grape (*Plasmopara viticola*) destroyed the grape and wine industry in Europe after the disease was imported from the United States in 1875. It led to the discovery of one of the first fungicides, **Bordeaux mixture** (copper sulfate and lime; Chapter 26). Downy mildew fungi are also troublesome for the ornamentals industry; recently, downy mildew of impatiens (*Plasmopara obducens*) has threatened a staple of the bedding plant industry, *Impatiens walleriana*.

KINGDOM FUNGI

Chytridiomycota

Chytridiomycota (Chapter 9), the chytrids, inhabit water or soil. They are coenocytic and lack a true mycelium. The thallus is irregular, globose, or ovoid; diploid; and its cell walls contain chitin and glucans. They have motile cells (zoospores) that possess a single, posterior, whiplash flagellum. The mature vegetative body transforms into thick-walled resting spores or a sporangium.

Olpidium brassicae is an **endobiotic** parasite of the roots of cabbage and other monocots and dicots. The entire thallus converts to an asexual reproductive structure (**holocarpic**); the sexual cycle in this organism, however, has not been confirmed. *Olpidium* is a known vector of plant viruses such as *tobacco necrosis virus* and *lettuce big vein virus*.

Zygomycota

The **Zygomycota** are true fungi with a well-developed, coenocytic, haploid mycelium with chitin-chitosan in their cell walls. Zygomycetes produce nonmotile spores called sporangiospores. Zygospores (sexual spores) are produced by fusion of two morphologically similar gametangia of opposite mating type (heterothallic) (Figure 7.3). These fungi are terrestrial and are saprophytes or weak pathogens, causing soft rots or molds. Members of this group also include parasites of insects (Entomophthoralean fungi), nematodes, mushrooms, and humans, as well as dung fungi and a few ectomycorrhizal fungi (see discussion on mutualists).

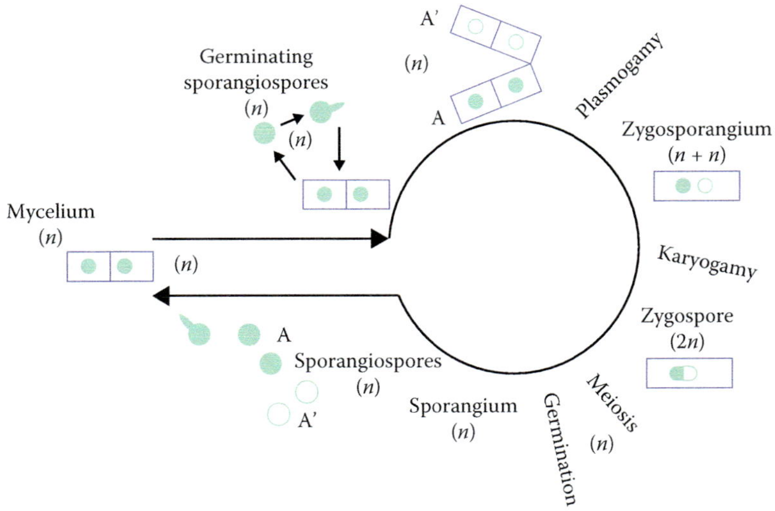

FIGURE 7.3 Generalized life cycle of the Zygomycota. This life cycle is for heterothallic members of the Zygomycota. (Courtesy of A.B. Gould.)

Rhizopus stolonifer is a ubiquitous saprophyte, notably recognized as a common bread mold, which also causes a soft rot of many fleshy fruit, vegetables, flowers, bulbs, corms, and seeds. Black, spherical sporangia that produce asexual sporangiospores (also known as mitospores) appear on the swollen tip (**columella**) of a long, aerial sporangiophore. The fungus produces **rhizoids** (short branches of thallus that resemble roots) within its food substrate and hyphal **stolons** that skip over the surface of the substratum. *Rhizopus* does not penetrate host cells; it produces cellulases and pectinases that degrade cell walls, causing infected areas to become soft and water soaked and absorbs the nutrients released by the enzymes. The fungus emerges through the broken epidermal tissue and asexually reproduces, forming a fluffy covering of fungal growth. When the food supply diminishes, gametangia combine to form a zygosporangium with a single zygospore inside.

Like *Rhizopus*, species of *Mucor* are common soil inhabitants and are often associated with decaying fruits and vegetables. They differ from *Rhizopus* fungi, however, in that they lack rhizoids. *Mucor amphibiorum* causes a dermal and systemic infection of toads, frogs, and platypus, and other species in this genus are human parasites and are implicated in dermal and respiratory allergies. Post-harvest rots of apple, peach, and pear are caused by *M. piriformis*. This species is a soil inhabitant that contaminates fruit that falls to the orchard floor. The fungus penetrates at the calyx or through wounds and within 2 months in cold storage causes complete decay of the fruit. Fungal mycelia can often be seen emerging in tufts from the surface of the affected fruit. The disease is managed by destroying fallen fruit and using sanitation practices where contaminated soil is not allowed to contact harvested fruit.

Ascomycota

The **Ascomycota** (Chapters 11–13) are true fungi with a well-developed, septate mycelium containing chitin in their cell walls. The sexual structure of the ascomycetes is an **ascus**, which is sac-shaped and contains the products of meiosis, called **ascospores**, which normally number eight per ascus. Asci are either **unitunicate** (with a single wall) or **bitunicate** (with a double wall).

Ascospores are forcibly liberated from asci, which are produced in fruiting structures called **ascocarps**. Ascocarps produced by various ascomycetes include the following:

- **Cleistothecia**—Completely closed ascocarps lined with one or more asci.
- **Perithecia**—Flask-shaped ascocarps with an ostiole at the tip (Chapter 13).
- **Apothecia**—Open, cup-shaped ascocarps where asci are exposed (Chapter 13).
- **Ascostroma** or **pseudothecia**—Asci are produced in a cavity or **locule** buried within a matrix (stroma) of fungal mycelium (Chapter 13).

Asci not produced in an ascocarp are called **naked asci** (Chapter 11).

Asexual reproduction in Ascomycota includes fission, budding, fragmentation, or formation of conidia or chlamydospores. Conidia are produced on conidiophores. The species of Ascomycota are a diverse group of fungi that can use many different substrates and occupy a variety of niches. Some members of this phylum form ectomycorrhiza.

One of the most infamous diseases caused by an already introduced ascomycete is ergot of rye and wheat.

The pathogen *C. purpurea* occurs worldwide. Grain in the seed head is replaced with fungal sclerotia (called **ergots**) that are poisonous to humans and animals. When ingested, fungal alkaloids restrict blood vessels and affect the central nervous system, causing gangrene in the extremities, hallucination, and miscarriage. The resulting disorder in humans, known as ergotism or St. Anthony's fire, was common in the Middle Ages, especially in peasants who subsisted on bread made with contaminated rye flour.

Claviceps purpurea is homothallic. In the spring, sclerotia germinate on fallen seed heads to form perithecia at the periphery. Each perithecium produces many asci, which forcibly discharge eight ascospores that are wind-disseminated to infect the ovaries of young flowers. Within a week, droplets of conidia exude in a sticky liquid (honeydew) from the young florets of the infected heads. Insects are drawn to the honeydew, and they carry the conidia on their body parts to new flowers. Conidia are also spread by splashing rain. The ergots mature at the same time as seed and are harvested or fall to the ground where they overwinter. Grain containing as little as 0.3% of sclerotia by weight can cause ergotism.

Basidiomycota

Basidiomycota (Chapters 14 and 15) are true fungi that often have large, macroscopic fruiting bodies. They have a well-developed, septate mycelium with chitin in their cell wall. The mycelium of basidiomycetes can be dikaryotic, each cell containing two haploid nuclei. The sexual structure is called a **basidium** (or club), on which four haploid basidiospores (the result of karyogamy and meiosis) are produced on stalks called sterigmata. Basidia are produced on **basidiocarps**. Basidiomycetes reproduce asexually by budding and fragmentation or by formation of conidia, **arthrospores** (formed when hyphal fragments break into unicellular sections), or **oidia** (produced by short branches, or oidiophores, that cut off oidia in succession from the tip). Like ascomycetes, basidiomycetes utilize a variety of substrates and occupy many different niches, and some form ectomycorrhiza. Some of the most important causes for plant diseases, the rusts and smuts (Chapter 14), are basidiomycetes.

Some basidiomycetes are best known for the production of the classic mushroom, which is a basidiocarp. A mushroom basidiocarp consists of a **pileus** (cap) lined with **gills** and a **stipe** (stalk). A **volva** may remain at the base as a remnant of the universal veil that once enveloped the entire developing mushroom (button). The **annulus** is the ring around the stipe as a remnant of a partial or inner veil that once enveloped only the gills or pores on the underside of the basidiocarp. Basidia line the outer surface of gills.

Armillaria mellea affects hundreds of species of trees, shrubs, fruits, and vegetables worldwide. The pathogen is also an important decay fungus in forest ecosystems. *Armillaria* forms rhizomorphs and **mycelial fans** (fan-shaped hyphae) under the bark near the crown of infected trees. Rhizomorphs consist of cord-like "shoestring" threads of mycelium 1–3 mm in diameter. They have a compact outer layer (or rind) of black mycelium and a core of white mycelium and form a branched network that develops over wood and in the surrounding soil. Rhizomorphs initiate new infections when in contact with fresh substrate. Plants affected by the fungus show symptoms similar to those caused by other root diseases (reduced growth, small, yellow leaves, dieback, and gradual decline and death). *Armillaria* produces honey-colored mushrooms at the base of dead or dying trees in early autumn.

Mitosporic Fungi

Some fungi have been classified in an artificial group (not a true phylum) called the Mitosporic Fungi (Chapter 13). These fungi have no known sexual state in their life cycle, and until recently, had not been classified in one of the other groups. Scientists are now uncovering the **phylogenetic** relationships that these organisms have with sexually reproducing fungi through studies of gene sequences.

Mitosporic fungi usually reproduce asexually (mitotically) via conidia and are considered to be anamorphic partners of associated teleomorphs in the Ascomycota or Basidiomycota, whose identity may or may not be known. They have hyphae that possess the characteristics of the teleomorph. Some mitosporic fungi have no known reproductive process and have traditionally been called the sterile fungi or **Mycelia Sterilia**, although some of these fungi do produce sclerotia. However, molecular systematics methods are very useful for placing mitosporic fungi with their associated teleomorphs, and as new information is reported, the need for artificial names such as Mycelia Sterilia will be eliminated.

Mitosporic fungi occupy the same habitats as their teleomorph relatives, thus they are largely terrestrial. Most are saprophytes, some are symbionts of lichens, others are grass endophytes or mycorrhizal fungi, and many are plant pathogens. Historically, these fungi were divided into three informal classes: **Hyphomycetes** (spores are produced on separate conidiophores), **Agonomycetes** (Mycelia Sterilia), and **Coelomycetes** (spores are produced in fruiting structures called **conidiomata**). Classification within these groups was based on morphology, such as conidiophore, conidia, and hyphal characteristics. As our knowledge of phylogenetic relationships among the fungi increases, the use of classification schemes based solely on morphology will decline.

MUTUALISTS

Certain biotrophs, such as mycorrhizal fungi and endophytes, live all or most of their life cycles in close association with their plant host. In many cases, this symbiosis is a mutualistic one because both the fungus and the plant host benefit from the association.

TURFGRASS ENDOPHYTES

Simply stated, endophytes are organisms that grow within plants. When searching the literature, one finds that the term is often used to describe a specific relationship between certain fungi and turfgrass hosts. Unlike turfgrass pathogens such as *Magnaporthe poae*, however, endophytic fungi do not harm the host; indeed, they appear to enhance stress tolerance and resistance to feeding by certain insects.

Endophytic fungi associated with turfgrass are ascomycetes in the family Clavicipitaceae and include species of *Epichloë* (ana. *Neotyphodium* [=*Acremonium*]) and *Balansia*. Although these fungi colonize all aboveground plant parts, they are found most readily within the leaf sheath of the turfgrass host and do not colonize the roots. Turfgrass hosts include perennial ryegrass and several fescues (e.g., tall fescue, hard fescue, chewings fescue, and creeping red fescue).

The ability to reproduce sexually varies among the different types of endophytic fungi. In the heterothallic endophyte *Epichloë*, a stroma, composed of fungal mycelium embedded together with the flowers and seeds of the grass itself, forms on the grass stem just below the leaf blade. The stroma is situated in such a way that nutrients flow from living host tissues into the fungus. Clusters of spores called **spermatia** are produced on the surface of the stroma. Flies (*Phorbia* species) visit the stroma and eat the spermatia, which remain undigested in the intestinal tracts of the flies. When a female fly then visits another stroma of opposite mating type to lay eggs, she defecates and deposits spermatia all over the stroma. Fertilization follows, and then perithecia form and ascospores are released to infect nearby florets. Interestingly, the fly larvae that hatch from the eggs then eat the stroma. Some endophytes such as *Neotyphodium* do not produce stroma and cannot sexually reproduce. These fungi grow into the seed at flowering time and are thus seed transmitted. Other endophytes exhibit both sexual and asexual reproductive characteristics.

Benefits conferred to the host by this endophytic association include enhanced nutrition, drought tolerance, increased hardiness, and resistance to disease. Notably, surface feeding by insect pests is deterred. Toxic alkaloids such as peramine are produced in infected hosts, and these compounds deter feeding by insects such as chinch bug, sod webworm, and billbug. *Neotyphodium* endophytes are used in turfgrass breeding programs to enhance these beneficial characteristics in turf intended for sports or landscape (not pasture) use.

Unfortunately, the endophyte association in pastures grasses can have disastrous impact on grazing livestock. Host plants containing *Neotyphodium* endophytes produce ergot alkaloids that reduce blood flow to extremities, causing tails and hooves to rot and fall off. In other associations, ingested lolitrems cause animals to spasm uncontrollably in a syndrome known as ryegrass staggers. Indeed, sleepy grass (*Achnatherum robustrum*) infected by a *Neotyphodium* endophyte produces lysergic acid, a relative of lysergic acid diethylamide (LSD); horses that ingest relatively small amounts of such compounds go to sleep for several days.

MYCORRHIZAL FUNGI

A mycorrhiza (or "fungus-root," coined by A.B. Frank in 1885) is a type of mutualism where both partners benefit in terms of evolution and fitness (ability to survive and reproduce). This is primarily a nutritional relationship where the fungal partner increases the efficiency of nutrient uptake (notably phosphorus) by the host and in turn receives carbon made by the host during photosynthesis. The mycorrhizal association is intimate and diverse, and the fungi involved vary in taxonomy, physiology, and ecology. Mycorrhizal fungi infect more than 90% of vascular plants worldwide, including many important crop plants, and thus the association is economically, ecologically, and agriculturally significant.

Based on the morphology of the association, mycorrhizae are classified into the following two major groups: ectomycorrhizae (outside of root), and endomycorrhizae (within root).

Ectomycorrhizae

In this group, a fungal **mantle** of septate hyphae forms over the entire root surface, replacing root hairs and root cap. Hyphae proliferate between root cortex cells, forming an intercellular network, or **Hartig net**, and it is within this net that nutrients are exchanged between partners. The mycelium can extend and retrieve nutrients up to 4 m from the surface of roots. In a typical ectomycorrhizal relationship, the fungus does not penetrate cells of the cortex or stele (vascular bundle) and is very long lasting (up to 3 years).

The ectomycorrhizal association comprises about 20% of all mycorrhizal associations. Ectomycorrhizal fungi are mostly members of Basidiomycota and Ascomycota and include tooth fungi, chanterelles, mushrooms, puffballs, truffles, and club fungi. Indeed, many fungal sporocarps seen beneath the tree canopy on the forest floor are ectomycorrhizal fungi. The taxonomy of this group is based on the identification of the fungal sporocarp.

Approximately 5000 fungal species form ectomycorrhizal associations with more than 2000 hosts, including plants in the Betulaceae (birch), Fagaceae (beech and oak), Pinaceae (pine), and also with some Pteridophytes (ferns and horsetail) and eucalyptus. Note that with certain forest trees, fungal hyphae may penetrate cortical cells close to the stele resulting in a form of mycorrhiza (called ectendomycorrhiza) that is intermediate to ectomycorrhiza (with mantle and Hartig net) and arbuscular mycorrhiza (AM).

Endomycorrhizae

Endomycorrhizae, in contrast, do not change the gross morphology of the root, and hyphae proliferate both between and within the cells of the root cortex but do not penetrate the stele. There are three different types of endomycorrhizae: arbuscular, orchidaceous, and ericaceous.

Arbuscular Endomycorrhizae

The AM relationship, also known as glomeralean or, formerly, vesicular-arbuscular mycorrhiza (VAM), is worldwide in distribution and occurs in many different habitats. AM fungi are obligate symbionts that have a very broad host range; a relatively small number of species (approximately 150) in the Glomeromycota (one of seven phyla in the kingdom Fungi) form endomycorrhizae with more than 70% of angiosperms and gymnosperms.

In the infection process, chlamydospores resting in soil germinate in the vicinity of plant roots. They penetrate the epidermis, and an aseptate, irregular mycelium grows between the cells of the root cortex. Curious hyphal coils called **pelotons** may form between and within the cells of the outer cortex. **Arbuscules** (special, dichotomously branched haustoria) then develop intracellularly and serve as the site of nutrient and carbon exchange. These structures last for only 4–15 days before hyphal content is withdrawn and the arbuscules disintegrate. **Vesicles** may be produced intra- or intercellularly. These ovate to spherical structures contain storage lipids and may also serve as propagules. The fungus reproduces asexually via chlamydospores that form internally or on external hyphae that may extend as far as 1 cm from the surface of the root. As the sexual state of these fungi is unknown, the taxonomy of this group has been traditionally based on characters such as chlamydospore morphology and content. Newer molecular methods are helping to redefine phylogenetic relationships among these species of fungi.

Although there are relatively few glomeralean species, many host plants can accept more than one mycorrhizal endophyte. The ability of an endophyte to infect a certain host often depends on environmental conditions rather than host specificity. For example, a strain of a single species indigenous to a certain area may successfully compete with a nonindigenous strain of the same species introduced into the soil, suggesting that certain strains may adapt to local environmental conditions.

Orchidaceous

This group includes those fungi that form associations with members of the orchid family. In nature, fungal hyphae penetrate the protocorms of orchids during the saprophytic stage, enabling the seedlings to continue development. Hyphal pelotons that form intracellularly greatly increase the surface area needed for the movement of carbon and minerals to the orchid plant. Orchidaceous mycorrhizal fungi are primarily basidiomycetes, most of which form imperfect stages in the genus *Rhizoctonia*.

Ericaceous

In this group, fungi in Ascomycota and Basidiomycota form associations with plants in Ericales (e.g., rhododendron, blueberry, heather, Pacific madrone, bearberry, and Indian pipe). In most associations, penetration of cortical cells by the fungus occurs, but arbuscules do not form. A mantle or sheath may form on the root surface and in some cases a Hartig net may be present. An interesting example of an ericaceous mycorrhizal relationship is the one between several *Boletus* species (Basidiomycota) and *Monotropa*, or Indian pipe. This plant is a white, unifloral (produces a single flower), achlorophyllous (without chlorophyll), parasitic plant (Chapter 17) that receives all of its carbon and nutrients from trees via a connecting mycorrhizal fungus.

Significance of the Mycorrhizal Association

Mycorrhizal fungi, in essence, act as root hairs. Fungal mycelia explore large volumes of soil and retrieve and translocate nutrients that may be otherwise unavailable to the host. The success of the mycorrhizal relationship is probably due to its long evolutionary history, its economy (fungal mycelium is less "expensive" for a host to maintain than an extensive system of root hairs), and its efficiency (fungi produce phosphatase enzymes that readily solubilize phosphorus from soil particles).

Hosts associated with a mycorrhizal fungus often exhibit an increase in growth, which is attributed to an increase in phosphorus nutrition of the plant. External mycelium effectively increase the phosphorus depletion zone around each root, and the phosphorus absorbed by the fungal mycelium is translocated to the site of nutrient exchange (such as arbuscules or the Hartig net) and is released to the host. Fertilizer high in phosphorus and nitrogen reduces mycorrhizal infectivity and sporulation in host plants.

The mycorrhizal relationship benefits plants in other ways, as well. Ectomycorrhizae, for example, play a role in protecting plants from plant pathogenic organisms; the mantle acts as a physical barrier. Some ericoid mycorrhizae produce antimicrobial compounds that inhibit other competing organisms. In a broader sense, mycorrhizae serve to interconnect individuals within a plant community, thus mature plants can "nurture" seedlings in a community buffering effect.

The fungi benefit from the association in that a high percentage of photosynthate manufactured by the host is translocated to the fungal symbiont. Host sugars (glucose and sucrose) obtained by the fungus are used to produce new hyphae or are converted to glycogen and stored in vesicles or other structures.

SUGGESTED READINGS

Agrios, G. N. 2005. *Plant Pathology*, 5th ed. Burlington, MA: Elsevier Academic Press.

Blackwell, M., R. Vilgalys, T. Y. James, and J. W. Taylor. 2012. Fungi. Tree of Life Web Project. http://tolweb .org/Fungi.

Braselton, J. 2001. Plasmodiophoromycota. Pages 81–91 in: *The Mycota VII. Part A: Systematics and Evolution*, D. J. McLaughlin, E. G. McLaughlin, and P. A. Lemke (eds). Berlin: Springer-Verlag.

Barr, D. J. S. 2001. Chytridiomycota. Pages 93–112 in: *The Mycota VII. Part A: Systematics and Evolution.* D. J. McLaughlin, E. G. McLaughlin, and P. A. Lemke (eds). Berlin: Springer-Verlag.

Carlile, M. J., S. C. Watkinson, and G. W. Gooday. 2001. *The Fungi*, 2nd ed., San Diego, CA: Academic Press.

Deacon, J. 2006. *Fungal Biology*, 4th ed. Malden, MA: Blackwell Publishing.

Hawksworth, D. L., B. C. Sutton, and G. C. Ainsworth. (eds). 1983. *Ainsworth & Bisby's Dictionary of the Fungi*. Kew Surrey: Commonwealth Mycological Institute.

Kyde, M. M., and A. B. Gould. 2000. Mycorrhizal endosymbiosis. Pages 161–198 in: *Microbial Endophytes*, C. W. Bacon and J. F. White (eds). New York, NY: Marcel Dekker.

Schumann, G. L. 1991. *Plant Diseases: Their Biology and Social Impact*. St. Paul, MN: APS Press.

Schumann, G. L., and C. J. D'Arcy. 2009. *Essential Plant Pathology*, 2nd ed. St. Paul, MN: APS Press.

Sinclair, W. A., and H. H. Lyon. 2005. *Diseases of Trees and Shrubs*, 2nd ed. Ithaca, NY: Cornell University Press.

Ulloa, M., and R. T. Hanlin. 2012. *Illustrated Dictionary of Mycology,* 2nd ed. St. Paul, MN: APS Press.

Volk, T. J. 2001. Fungi. Pages 141–163 in: *Encyclopedia of Biodiversity*, Vol. 3. New York, NY Academic Press.

White, J. F., P. V. Reddy, and C. W. Bacon. 2000. Biotrophic endophytes of grasses: A systemic appraisal. Pages 49–62 in: *Microbial Endophytes*, C. W. Bacon and J. F. White. (eds). New York, NY: Marcel Dekker.

8 Oomycota

The Fungus-Like Organisms

Robert N. Trigiano, Otmar Spring, Alan S. Windham, Richard E. Baird, Steven N. Jeffers, and Kurt H. Lamour

Concept Box

- Oomycota are heterotrophic fungus-like organisms of the kingdom Stramenopila with diploid motile spores.
- The group is characterized by the production of oospores, thick-walled resting spores resulting from sexual reproduction and capable of surviving adverse environmental conditions.
- Sexual reproduction is usually by gametangial contact between oogonia and antheridia where meiosis occurs; asexual reproduction is typically by motile, biflagellate sporangiospores (zoospores), or sporangia, which may release zoospores or germinate via a germ tube.
- Hyphae are typically coenocytic (solid septa may be present at the base of reproductive structures or in old and damaged hyphae), are relatively large (6–10 μm diameter), contain cellulose in the wall, and have diploid ($2N$) nuclei.
- Species in this group ecologically range from saprophytes to obligate parasites. Most of the more important plant pathogens are classified in Peronosporales and Albuginales.

The phylum Oomycota (class Oomycetes) encompasses an incredibly diverse group of organisms that were for many years considered true fungi or at the very least were always included in the study of fungi. Most mycologists and other biologists recognized that the oomycetes, although being achlorophyllous and having mycelium-like colonies, were somewhat different and did not fit the conventional and prevailing concept of true fungi. The oomycetes have been categorized into different kingdoms and other groups in the past, and some confusion of their taxonomic status still exists today. For the purposes of this chapter, we will follow Alexopoulos et al. (1996) and place them in the kingdom Stramenopila (Ribeiro, 2013). In addition, by convention and for the purposes of this chapter, we will continue to use terms such as hyphae and mycelium, which are typically reserved for true fungi.

The oomycetes include organisms ranging along a continuum from saprophytic to obligately parasitic. Some are strictly aquatic, whereas others are terrestrial; in essence, they occupy widely divergent habitats. Whatever their mode of nutrition or ecological niche, all species share some morphological and physiological characteristics. The following is not an exhaustive list, but these primary features readily distinguish the oomycetes from all other plant pathogens, including the true fungi.

- The life cycles of oomycetes can be best described as **diplontic**; that is, for the vast majority of their life history, the ploidy level of the nuclei is diploid ($2n$). Haploid nuclei are found only in the gametangia where meiosis occurs. Compare this to the true fungi that have typical **haplontic** life cycles where nuclei are haploid (n) or may exist as a **dikaryon** ($n + n$), another functional variation of haploid. The zygote is the only diploid structure in most true fungi.
- Hyphae are generally **coenocytic**, but solid (no pores) septa or crosswalls do occur at the base of reproductive structures and in older portions of the mycelium (Figure 8.1). Hyphal walls contain some cellulose, which is not typically found in true fungi and is more characteristic of plants. Most true fungi have some amount of chitin in their walls.

- Sexual reproduction occurs by **gametangial contact** between **oogonia** and **antheridia**. The resulting thick-walled **oospore**, which is diploid, is the resting or overwintering spore for many species (Figure 8.2).

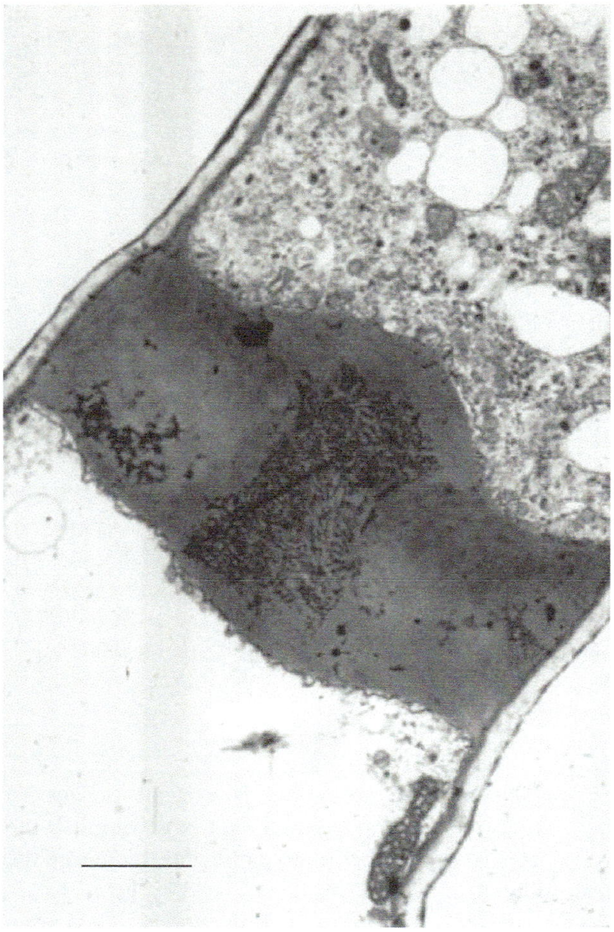

FIGURE 8.1 Transmission electron micrograph of a septum in old intercellular hyphae of *Peronospora tabacina*. Bar = 1 μm. (From Trigiano et al., *Tob. Sci.*, 29, 116–121, 1985. With permission.)

- Asexual reproduction by the vast majority of species is by **zoospores** produced in sporangia. Typically, many more-or-less kidney-shaped zoospores, each with one tinsel flagellum directed forward and one whiplash flagellum facing behind, are cleaved from the cytoplasm in the **sporangium** or in a vesicle. Some members of the downy mildews do not produce zoospores; instead, they germinate directly to form hyphae from sporangia. Note that many authors consider the "sporangia" of the downy mildews, a group of obligate plant parasites in the Oomycetes (Peronosporaceae), to be "conidia." However, for the purposes of this chapter, we will refer to these structures as sporangia.
- Several biochemical synthetic pathways and storage compounds in the oomycetes are very different from those present in true fungi.

Just as there is confusion as to the placement of members of the phylum Oomycota in a kingdom, there is also discussion on how the phylum should be divided into classes and orders. Most mycologists and plant pathologists agree that there should be about 10 recognized orders in the phylum. Most of the important plant pathogens fall into two orders: the Peronosporales and the Albuginales (Figure 8.3), and most fall into the families Peronosporaceae, Pythiaceae, and Albuginaceae (Figure 8.3 and Table 8.1). Notorious and damaging organisms such as the downy mildews, the 100+ species of *Phytophthora*, and the white blister rusts are included in these families. There are also several species of *Aphanomyces* included in the Saprolegniales (Saprolegniaceae) that can cause root rots of field crops such as alfalfa, peas, and sugar beets. Although this family is primarily composed of saprophytic organisms, some species cause diseases of fish. These organisms will not be discussed in this chapter.

The species in Peronosporales are very diverse and include soil-inhabiting facultative parasites, such

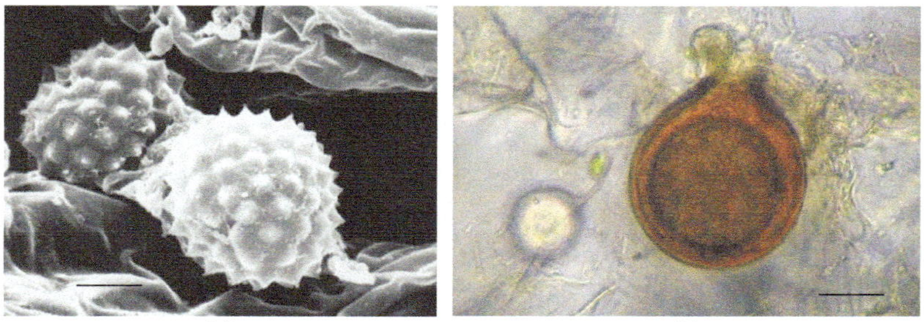

(a) (b)

FIGURE 8.2 Oospores of two downy mildew species. (a) Scanning electron micrograph of two oospores of *Peronospora tabacina* in tobacco tissue. Notice the ornamented spore wall. Bar = 10 μm. Oospores of *P. tabacina* are rarely observed. (b) An oospore of *Plasmopara obducens* in impatiens stem tissue. Notice the thick oospore wall. Bar = 10 μm. ([a] Courtesy of R.N. Trigiano. [b] Courtesy of A.S. Windham.)

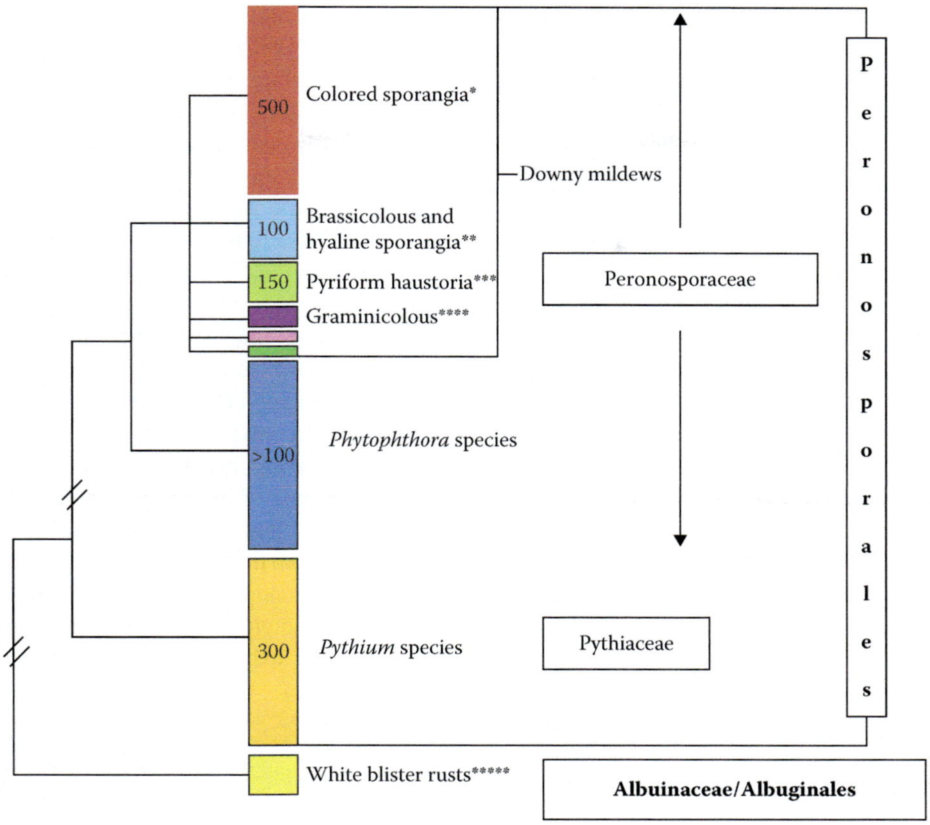

FIGURE 8.3 Phylogenetic relationships within Peronosporales compared with Albuginales. Downy mildew genera include the following: **Peronospora* and *Pseudoperonospora*; ***Hyaloperonospora* and *Perofascia*; ****Basidiophora*, *Benua*, *Bremia*, *Novotelnova*, *Paraperonospora*, *Plasmopara*, *Plasmoverna*, and *Protobremia*; *****Eraphthora*, *Graminivora*, *Peronosclerospora*, *Poakatesthia*, *Sclerophthora*, *Sclerospora*, and *Viennotia*. White blister rusts (Albuginales) genera include the following: *Albugo*, *Pustula*, and *Wilsoniana*. Numbers in the boxes indicate approximate number of species. (Adapted from Thines, M., *Phytophthora*: *A Global Perspective*, CAB International, Oxfordshire, 2013. With permission.)

TABLE 8.1

Two Classification Schemes for Plant Pathogenic Genera of Saprolegniales, Peronosporales, and Albuginales

Classification System (sensu Alexopoulos et al., 1996)	Classification System (sensu Thines, 2013; Thines and Spring, 2005)
Order: Saprolegniales	Order: Saprolegniales
Family: Saprolegniaceae	Family: Saprolegniaceae
Genus: *Aphanomyces*	Genus: *Aphanomyces*
Subclass: Peronosoporomycetidae	Subclass: Peronosoporomycetidae
Order: Peronosporales	Order: Peronosporales
Family: Pythiaceae	Family: Pythiaceae
Genera: *Pythium* and *Phytophthora*	Genus: *Pythium*
Family: Peronosporaceae	Family: Peronosporaceae
Genera: Downy mildews (e.g., *Peronospora*, *Plasmopara*, *Bremia*, etc. (about 10)	Genera: Downy mildews (e.g., *Peronospora*, *Plasmopara*, *Bremia*, etc. (about 20), *Phytophthora*
Family: Albuginaceae	Subclass: Albuginomycetidae
Genus: *Albugo*	Order: Albuginales
	Family: Albuginaceae
	Genera: *Albugo*, *Pustula*, *Wilsononia*

as *Pythium* and some *Phytophthora* species that cause seedling diseases, wilts and root and seed rots; other *Phytophthora* species are facultative saprophytes that cause aerial (stem and leaf) blights, and root rots; and the obligate parasitic downy mildews. The white (blister) rusts, classified in the genera *Albugo*, *Pustula*, and *Wilsoniana* of the Albuginales (Albuginaceae; Table 8.1), are also obligate parasites that typically attack only the aerial portions of plants.

The Peronosporales can be further subdivided into the following three major families based primarily on characteristics of sporangia and sporangiophores, ecological niche, biotroph versus necrotroph, and more recently, by molecular data: Pythiaceae, Peronosporaceae, and Salisapiliaceae (Table 8.1). The dozen or so species in the Salisapiliaceae are primary inhabitants of saline or brackish habitats, and these are not discussed further here. Albuginales is composed of a single family, the Albuginaceae. Species in Pythiaceae have sporangia that are borne variously, but not in chains, on somatic hyphae or on sporangiophores of indeterminate growth (grows continuously). This family includes plant pathogenic species in the genus *Pythium*. About 20 genera classified in the Peronosporaceae form sporangia on well-defined, branched sporangiophores of determinate growth (Hawksworth et al., 1995). This family includes the downy mildew organisms (some common genera: *Peronospora*, *Plasmopara*, *Bremia*, and *Pseudoperonospora*) and many species of *Phytophthora*. For many years, classification of *Phytophthora* species into a family was uncertain, and until recently, many considered the genus to belong in the Pythiaceae. However, with advanced molecular techniques, it became clear that the genus *Phytophthora* should be classified in the Peronosporaceae (Thines, 2013). Albuginaceae contains the following three genera: *Albugo*, *Pustula*, and *Wilsoniana*, which produce chains of sporangia on club-shaped sporangiophores of indeterminate growth below the epidermis of the host plant. The species of Albuginaceae are known as the white rusts or the white blister rusts. For a more extensive treatment of the taxonomy of the Peronosporales and Albuginales (Figure 8.3), refer to Thines (2013) and Beakes and Sekimota (2009).

PYTHIACEAE

The family encompasses 10 genera (Hawksworth et al., 1995) and includes plant pathogenic *Pythium* species. Many soilborne species of *Pythium* are not plant pathogens.

Pythium

According to Thines (2013), there are about 300 species of *Pythium* and more are being described each year. Many species cause seed and root rots and pre- and post-emergence damping-off of many greenhouse and field crops. In preemergence damping-off, the germinating seedling is attacked, colonized, and killed before breaking through to the soil surface. Often, seedlings will emerge from soil and immediately begin to wilt (postemergence damping-off). In this scenario, the young feeder roots are destroyed and cortical tissues of the primary root are invaded by the pathogen. Dark, water-soaked lesions develop on the stem at the soil line. Extracellular hydrolytic enzymes, various pectinases, cellulases, and hemicellulases produced by the pathogen degrade cell walls and cause plant tissues to lose their structural integrity. When the seedlings can no longer support themselves, they collapse and fall onto the soil surface (damping-off; Figure 8.4). Often and especially in high humidity environments such as greenhouses, white, watery-appearing mycelia can be seen growing on the rotting remains of the shoots. *Pythium* root diseases can also severely damage established plants in the greenhouse (Figure 8.5) and field. Plant roots often lack the small feeder roots, and secondary roots are discolored (brown or black). The first symptom noticed on shoots is wilting and some will possibly exhibit mineral deficiencies. Older plants

FIGURE 8.4 Damping-off of Boston ivy caused by *Pythium* species. Note the dead roots and stems near the soil line. (Courtesy of A.S. Windham.)

FIGURE 8.5 Root rot of chrysanthemum caused by *Pythium* species. Note the discoloration of roots. Healthy roots should be white. (Courtesy of A.S. Windham.)

infected with *Pythium* can also exhibit necrotic, watery-appearing lesions on the stem (Figure 8.6). *Pythium* blight has become a serious problem of turfgrass, especially on high-value sports fields and golf courses (Figure 8.7).

The life history of *Pythium debaryanum* (Figure 8.8) is generally representative of most *Pythium* and other oomycete species in that meiosis occurs in gametangia, diploid oospores are the resting or overwintering spore, and biflagellated zoospores are produced. *Pythium* species survive in soil as somatic hyphae or oospores. Zoospores from sporangia and/or oospores are attracted to roots where they encyst and germinate to infect tissue. Somatic hyphae may also directly infect roots. Hyphae grow intercellularly within host tissue and do not produce **haustoria**. The asexual part of the life cycle is completed with the formation of sporangia. Sporangia may be indistinguishable from somatic hyphae or be quite distinct depending on the species. The contents of sporangia migrate into **vesicles** where zoospores are differentiated. Vesicles burst and zoospores are liberated to the surrounding environment. Sexual reproduction is initiated with the differentiation of gametangia (antheridia and oogonia) from somatic hyphae. There are a few **heterothallic** *Pythium* species, which require different mating types, but most are **homothallic** where both gametangia involved in the sexual process may arise from the same hyphae or they may be formed on different hyphae of the same individual that are located nearby each other. One or more antheridia contact a single oogonium and meiosis occurs, reducing the ploidy of the nuclei in the gametangia to haploid. A nucleus or nuclei from an antheridium passes into the oosphere (plasmogamy) and eventually fuses with a nucleus or nuclei in the oosphere (karyogamy) to restore the diploid condition. The oosphere develops a smooth thick wall, which is capable of germinating after a resting period. Oospores may germinate directly via a germ tube or by zoospores depending on the prevailing environmental (primarily temperature) conditions to initiate infection. Some of the more important *Pythium* species include *P. aphanidermatum, P. debaryanum, and P. ultimum.*

FIGURE 8.6 Black leg of geranium caused by *Pythium ultimum*. Note the necrotic stem lesions. (Courtesy of A.S. Windham.)

FIGURE 8.7 Pythium blight of turfgrass (tall fescue). (a) Pythium blight of a tall fescue plot. Notice the irregular brown patches. (b) Mycelium of *Pythium* sp. growing on tall fescue. (Courtesy of A.S. Windham.)

Unfortunately, host resistance to *Pythium* species is not available for field and greenhouse crops. Control of these diseases is primarily through cultural practices and fungicides, although there are some biocontrol strategies for some oomycetes. Good soil drainage and management of soil moisture are important cultural practices used to control diseases caused by *Pythium* species in both field and greenhouse situations. Sound nutritional management of the crop, especially avoiding excess nitrogen fertilization, can help to prevent diseases by these organisms. Crop rotation using nonhosts may also be an effective management tactic. In nursery crops, adoption of composted hardwood bark as a container medium instead of peat has helped manage root rot diseases. Seedbeds and greenhouse benches can be treated with captan or other contact fungicides to prevent damping-off. Systemic fungicides such as metalaxyl (specific for the oomycetes) can be used to either prevent or control root rots.

PERONOSPORACEAE

PHYTOPHTHORA

There are approximately 100 species of *Phytophthora* (Thines, 2013), but with the advent of molecular techniques in recent years, this number is rapidly growing. *Phytophthora* species cause many of the same types of diseases as *Pythium* species on the same plants (Table 8.2) and thus may be easily confused with one another when only considering symptomology. In addition, they can cause a number of other diseases, including

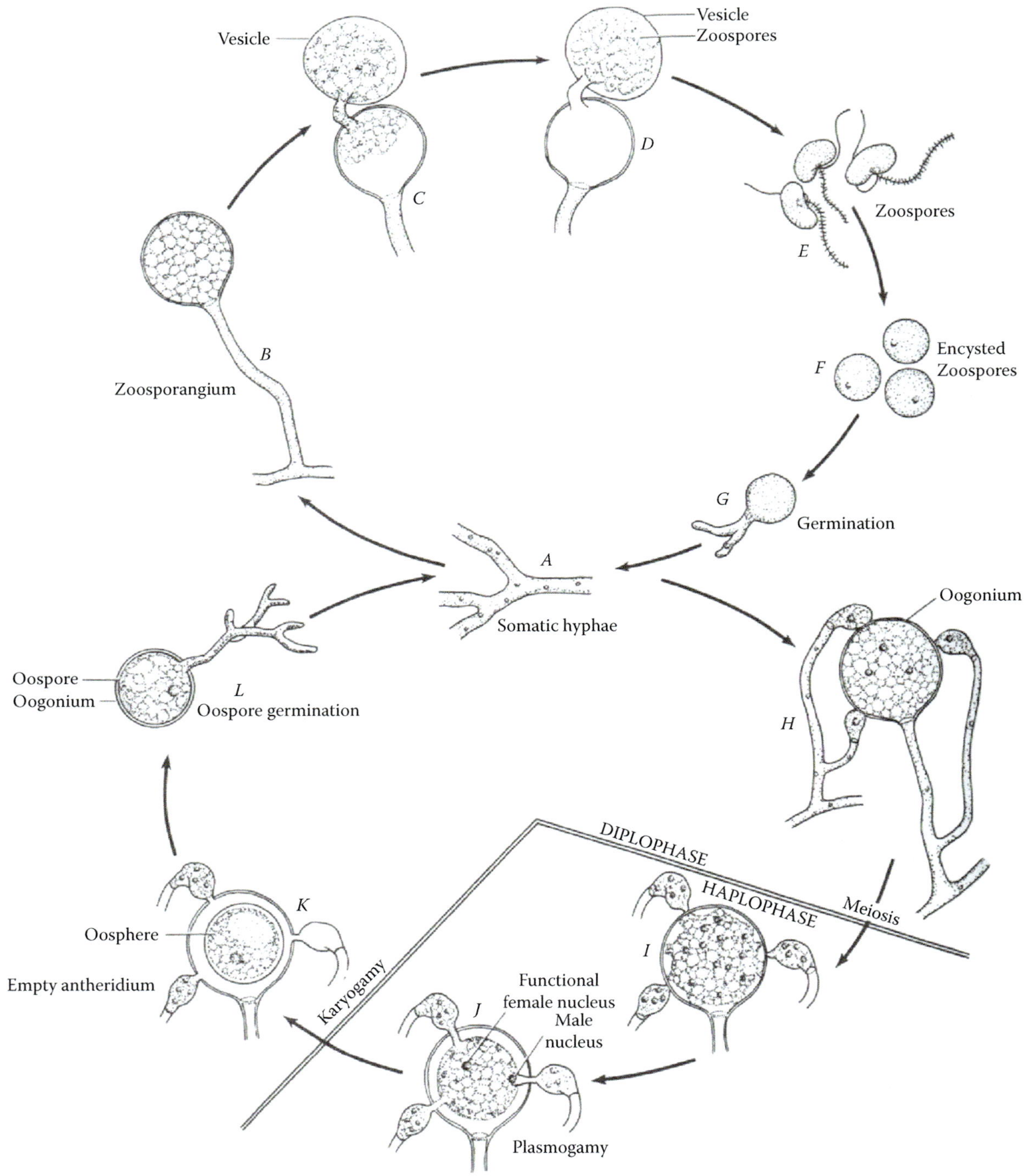

FIGURE 8.8 Life history of *Pythium debaryanum*. (Drawing by R.W. Scheetz. Reprinted from Alexopoulos, C. J., Mims, C. W., and Blackwell, M., *Introductory Mycology*, John Wiley & Sons, NY, 1996. With permission.)

foliar blights of field, nursery, forest, and greenhouse crops (Figure 8.9).

The life history of *Phytophthora* species differs in detail compared with that exhibited by *Pythium* species. The most notable difference between the two genera is that zoospores are delimited and functional within the sporangium for species of *Phytophthora*, whereas in *Pythium* species, zoospores are formed from the cytoplasm of sporangia that has migrated into a vesicle. *Phytophthora* species also have indeterminate sporangia, but many species have very differentiated sporangiophores and/or sporangia (Figure 8.10). For example, *Phytophthora infestans* produces sympodially branched sporangiophores, which have swollen nodes, and produces lemon-shaped, papillate sporangia. Some *Phytophthora* species produce haustoria, unlike *Pythium* species.

TABLE 8.2
Some Common Diseases Caused by *Phytophthora* Species

Phytophthora Species	Disease
P. fragariae	Red stele of strawberry
P. megasperma	Root and stem rot of soybean
P. parasitica	Black shank of tobacco
P. citrophthora	Foot rot of citrus
P. cinnamomi	Root rots of forest and nursery crops and avocado
P. capsici	Cucurbits (squash, pumpkins), tomato, peppers
P. ramorum	Sudden oak death (SOD)
P. infestans	Late blight of potato and tomato
P. palmivora	Black pod of cacao

FIGURE 8.9 Foliar blight of impatiens caused by *Phytophthora* species. (Courtesy of A.S. Windham.)

Phytophthora infestans: late blight of potato (and tomato). *P. infestans* is the most well-known member of the genus. The Irish famine of 1845 and 1846 caused by *P. infestans* was largely responsible for the deaths of many who were overly dependent on the potato crop and for the subsequent mass emigration to the United States. Along with a few other species, it is also the most extensively studied species because of its economic impact on agriculture and history. This organism is heterothallic and requires two mating types for sexual reproduction by gametangial contact. Sexual reproduction is not necessary for the survival of the organism as the mycelium may survive in infected tubers. The disease can be initiated when infected tubers are planted or when volunteer potatoes sprout in the spring. Under favorable environmental conditions (cool and moist), the pathogen grows into the shoot, and after sufficient time, a matter of days, sporangia on sporangiophores will emerge through stomata on the bottom surface of leaves. Sporangia are dispersed by air currents and rain to other plants and deposited on the soil surface. Typically less than 10 biflagellated zoospores form and are liberated from each sporangium. Zoospores that land on leaves encyst and then germinate via germ tubes. Germ tubes form **appressoria** that penetrate the leaf surface either directly or indirectly through stomata. Sporangia that are in soil also produce zoospores, which may infect the existing or developing tubers. Under favorable conditions, many generations or cycles of sporangia are produced to infect plants. Late blight is a **polycyclic** disease and may quickly devastate entire fields. As the pathogen grows and produces sporangia, host tissue is destroyed, which creates lesions that coalesce. Often the entire shoot portion of the plant is destroyed, and yield is dramatically reduced. Infected tubers are often small with sunken lesions, stained purple or brown, and may have an offensive odor due to secondary bacterial and fungal invaders.

Control of late blight of potatoes is accomplished using several tactics. Seed potatoes should be free of the

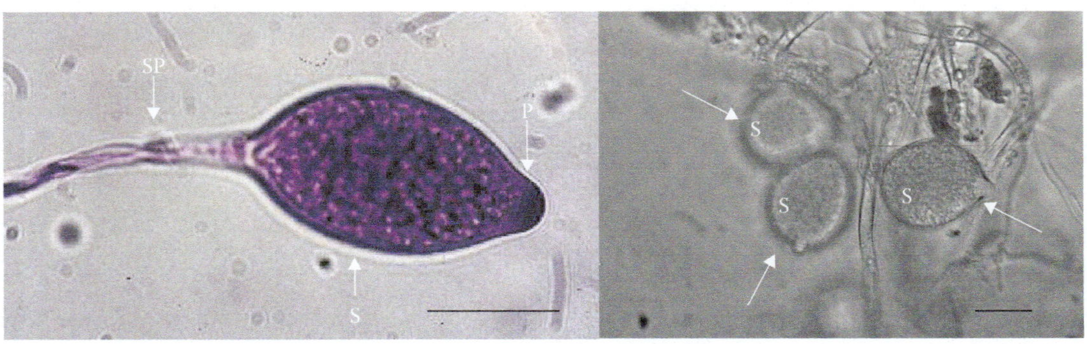

(a) (b)

FIGURE 8.10 Sporangiophores and sporangia of a *Phytophthora* species. (a) A stained sporangiophore (SP) and oblong-shaped sporangium (S) with prominent papillae (P) of *P. capsici*. Bar = 20 µm. (b) Sporangia (S) of *P. nicotianae*. Arrows point to papillae on the sporangia. The zoospores have not differentiated within the sporangia. Bar = 20 µm. ([a] Courtesy of K. Lamour. [b] Courtesy of A.S. Windham.)

pathogen; volunteer plants should be rouged and infected debris from any previous potato crop destroyed. Disease-resistant varieties should be used when possible, although resistance may be broken by environmental conditions favorable to pathogen development, or when new strains of the pathogen develop. Contact fungicides can be applied at prescribed intervals, according to disease development models based on temperature and moisture. Metalaxyl, a systemic fungicide specific for the oomycetes, has been used extensively to protect crops. However, an increasing number of reports indicate resistance to the fungicide.

Control of other diseases caused by *Phytophthora* species is similar to control tactics used for *Pythium* species. Well-drained soils and careful management of soil moisture will help minimize damage by these pathogens to many crops. For greenhouse crops, growing media and containers should be decontaminated to ensure that pathogen propagules are killed. Composted hardwood bark has been used to control disease in container nursery crops. Unlike diseases caused by *Pythium*, there are a number of crops species (fruits, etc.) in which disease-resistant varieties are available. Systemic fungicides such as metalaxyl and fosethyl-Al are commonly used to protect various crops from infection. The regulation and plant losses caused by *Phytophthora ramorum* in woody ornamental nurseries in the United States have prompted an integrated disease management program by progressive growers (see Case Study 8.1). This systems approach examines critical control points such as plant procurement, propagation, irrigation, and media storage to minimize the introduction, spread, and growth of *P. ramorum* (Figure 8.11). *Phytophthora capsici* is another pathogen that causes widespread damage to many crop species, especially cucurbits. Provided favorable environmental conditions, the pathogen can devastate crops quickly. The disease is very difficult to control (Figure 8.12, Case Study 8.2).

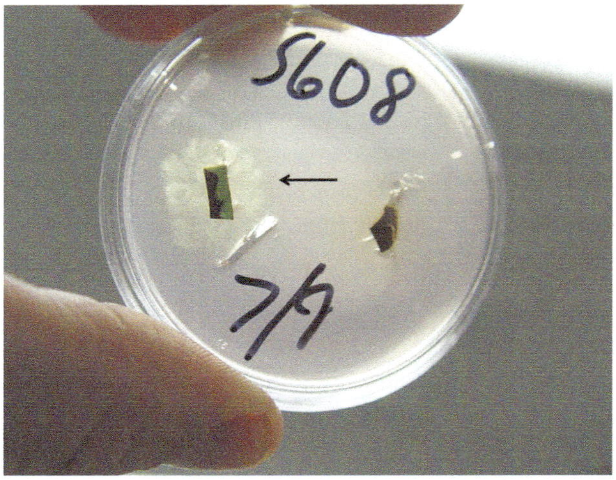

FIGURE 8.11 *Phytophthora ramorum* mycelium (arrow), the cause of sudden oak death, expanding from a piece of Rhododendron leaf onto selective agar medium. (Courtesy of K. Lamour.)

CASE STUDY 8.1

Sudden Oak Death—Neither Unique to Oaks Nor Sudden

- Sudden oak death (SOD) is caused by the oomycete *P. ramorum*, which is a species that attacks a wide variety of woody hosts and is under intense regulation by the European and the U.S. government agencies.
- *P. ramorum* was isolated from bleeding stem cankers on tanoak and coast live oak in California in the mid-1990s, which led to the common name "sudden oak death."
- The time from infection to tree death can be from months to years.
- Areas on the west coast of the United States that are known to be affected by the disease are currently under quarantine and plant production facilities and are routinely inspected and tested for the presence of *P. ramorum*.
- *P. ramorum* was primarily responsible for the "Systems Approach" movement for procuring and producing healthy woody ornamental in nurseries.
- The pathogen is primarily isolated from aboveground plant parts and has been isolated on numerous occasions from plants shipped to many other areas of the United States (Figure 8.11).
- It is spread via rain splash, infested soil, and infected plant material.
- *P.* ramorum is an obligately outcrossing species.
    - Makes abundant thick-walled asexual chlamydospores.
    - Epidemics in the United States likely comprise one to two clonal lineages of *P. ramorum*.
    - Does not appear to be completing the sexual stage in natural epidemic populations.
- The Joint Genome Institute, U.S. Department of Energy, completed a whole-genome sequence in 2004.
    - Contains a large gene family of rapidly evolving "effector" proteins that play a key role in overcoming plant defenses.

Peronosporaceae—The Downy Mildews

The **downy mildews** are obligate, biotrophic parasites of aerial portions of flowering plants and belong to Peronosporaceae. There are almost 20 genera in the family (Figure 8.3). Some genera can be distinguished by the highly differentiated branching patterns of sporangiophores that always exhibit determinate growth (Figure 8.13), host range restriction (only occurring on *Brassica* species [*Hyaloperonospora* and *Perofascia*]) with hyaline sporangia, production of colored sporangia (*Peronospora* and *Pseudoperonospora*), shape of haustoria (e.g., *Basidiophora*, *Bremia*, *Plasmopara*, and others), only occurring on grasses (*Graminivora*, *Peronosclerospora*, and others), and all genera are differentiated by molecular methodologies. Thines (2013) provides a complete listing of genera and classification. Sporangiophores emerge through stomata predominately on the lower leaf surface, and they develop completely before any sporangia are formed. Formation of sporangia on individual sporangiophores is synchronous (Figure 8.14), and mature sporangia are disseminated by air currents. The sporangia of many species germinate by producing biflagellated zoospores, characteristic of Oomycota. However, sporangia of *Pseudoperonospora* and *Peronospora* and some species of *Bremia* germinate via a germ tube and for this reason (among others), the sporangia have been termed conidia by some authors. Regardless of the mode of sporangial germination, these propagules serve to infect either healthy plants or healthy tissue remaining on an infected plant. They penetrate the leaf surfaces directly either through the epidermal wall or through stomata. Intercellular hyphae ramify throughout the mesophyll and establish haustoria within the host cells (Figure 8.15a and b). The infection may be localized in a leaf or become systemic in the plant. When sufficient energy has become available to the organisms and when environmental conditions are satisfactory, usually cool temperatures with high humidity and darkness, sporangiophores and sporangia are produced and initiate the disease cycle over again. Many crops of sporangia can be produced in a growing season, and thus diseases caused

FIGURE 8.12 Vegetable blight caused by *Phytophthora capsici*. Note the white areas on the fruit, which are millions of sporangiophores and sporangia produced by the pathogen. (Courtesy of K. Lamour.)

CASE STUDY 8.2

Vegetable Blight Caused by *Phytophthora capsici*

The oomycete plant pathogen *Phytophthora capsici* is an introduced species that causes serious crop loss to tomato, pepper, eggplant, snap, pumpkins, lima beans, and all cucurbits at locations worldwide.

- *P. capsici* attacks all parts of a plant including root, stem, foliage, and fruit.
- *P. capsici* produces millions of deciduous (caducous) lemon-shaped spores (sporangia) on the surface of the infected tissue (Figure 8.12).
- Disease is favored by warm and wet conditions. A single rainstorm can lead to the rapid and devastating spread within a field.
- *P. capsici* is one of the few outcrossing (heterothallic) species of *Phytophthora* to complete the sexual stage and produce thick-walled sexual spores (oospores) in nature.
- Thick-walled oospores remain dormant for years and germinate to produce genetically diverse strains, which are the products of meiosis.
- Once *P. capsici* has been introduced into a field, it is very difficult to eradicate.
- A recent genome project resulted in a high-quality reference genome sequence, thousands of new molecular markers, and a dense genetic linkage map.
- Disease is very difficult to control, and vegetable plant breeders are working to develop resistant vegetable varieties.

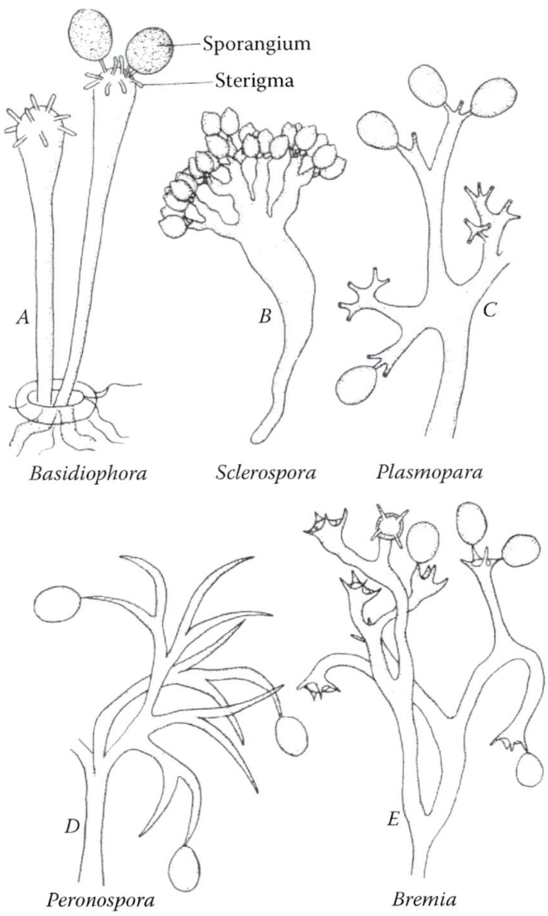

FIGURE 8.13 Sporangiophores characteristic of five downy mildew genera of the Peronosporaceae. (Reprinted from Alexopoulos, C. J., Mims, C. W., and Blackwell, M., *Introductory Mycology*, John Wiley & Sons, NY, 1996. With permission.)

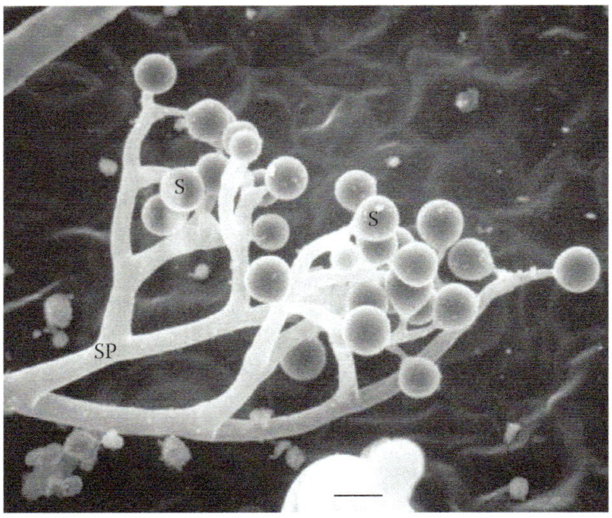

FIGURE 8.14 Scanning electron micrograph of synchronous development of sporangia (S) on a sporangiophore (SP) of *Peronospora tabacina*. Bar = 15 μm. (From Trigiano et al., *Tob. Sci.*, 29, 116–121, 1985. With permission.)

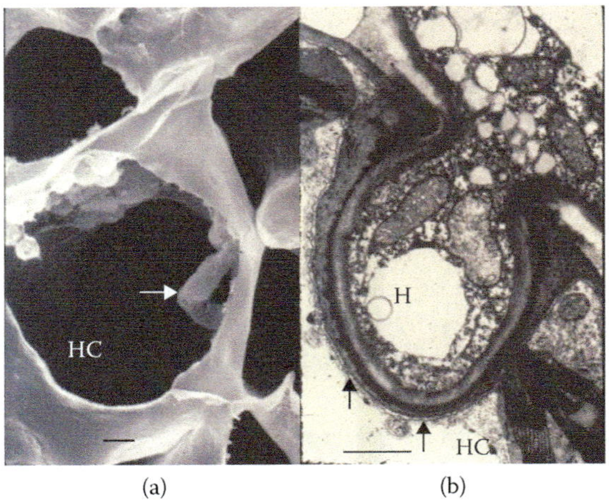

FIGURE 8.15 Haustoria of *Peronospora tabacina* in tobacco. (a) Scanning electron micrograph of a finger-like haustorium (arrow) in a host cell (HC). Bar = 2 μm. (b) Transmission electron micrograph of a developing (immature) haustorium (H) in a host cell (HC). Notice the plasmalemma (cell membrane is intact, see arrow). Bar = 1 μm. (From Trigiano et al., *Can. J. Bot.*, 61, 3444–3453, 1983. With permission.)

FIGURE 8.16 Leaf blight and necrosis caused by tobacco blue mold *Peronospora tabacina*. (Courtesy of O. Spring.)

by these organisms can be considered polycyclic. Sexual reproduction is by gametangial contact. Ornamented or nonornamented oospores are usually produced within the host tissue and serve as overwintering structures (Figure 8.2a and b). Oospores usually germinate directly via germ tubes, but sometimes indirectly by forming a germ sporangium or vesicle that produces zoospores.

Members of the downy mildews infect and cause diseases in many plants, but in general, the susceptibility is restricted to a single host species or genus, as expressed in the specific epithet of the scientific name of the pathogen (Table 8.3). Probably the most well-known disease in this group is downy mildew of grapes caused by *Plasmopara viticola* (Figure 8.17). Briefly, a root aphid

TABLE 8.3

Some Common Diseases Caused by Downy Mildews

Species	Crop
Bremia lactucae	Lettuce
Peronospora tabacina	Tobacco (blue mold)
Peronospora destructor	Onion
Peronospora antirrhini	Snapdragon
Peronosclerospora sorghi	Sorghum
Plasmopara viticola	Grape
Plasmopara halstedii	Sunflower
Plasmopara obducens	Impatiens
Pseudoperonospora cubensis	Many cucurbit species
Pseudoperonospora humili	Hops
Sclerophthora macrospora	Cereal and grass species

FIGURE 8.17 Downy mildew of grape (*Vitis vinifera*) caused by *Plasmopara viticola*. Sporangiophores and sporangia on immature berries. (Courtesy of O. Spring.)

native to North America was introduced to France in the mid-1860s with devastating consequences for the wine industry. In an attempt to control the aphid, the wine producers imported American rootstock, which had good tolerance to the insects. Unbeknown to them, they also imported *P. viticola* with the plants. The French grapes were extremely susceptible to the disease, which threatened the continued existence of the industry. In 1882, Alexis Millardet noticed some vines that were sprayed with copper sulfate ($CuSO_4$) did not have the disease. The owner of the vines had applied a mixture of copper sulfate and lime to persuade passersby not to eat his crop. Millardet, being a professor at a university, seized upon his observation and developed the first fungicide, Bordeaux mixture. Thus, the French wine industry had solutions to the aphid and downy mildew problems.

In the recent years, downy mildews have become major diseases of herbaceous ornamentals such as *Rudbeckia* and *Impatiens*. A *Plasmopara* species and *Plasmopara obducens* are pathogens of *Rudbeckia* and *Impatiens walleriana*, respectively, in greenhouses, nurseries, and landscapes (Figure 8.18). *P. obducens* has become so widespread in the Eastern United States that sales of *I. walleriana*, one of the two most popular bedding plants, have dropped by 75% or more. The movement of plant material internationally is likely responsible for the introduction of these fungi into the United States.

FIGURE 8.18 Downy mildew of garden impatiens caused by *Plasmopara obducens*. (a) A planting of impatiens defoliated by downy mildew. (b) Sporangiophores and sporangia of *P. obducens* (white) on the bottom surface of an impatiens leaf. (Courtesy of A.S. Windham.)

Varieties of some crops resistant to downy mildews are available, but contact fungicides are still used to control diseases. More recently, systemic fungicides, including metalaxyl and fosetyl-Al, which may or may not be combined with contact fungicides, have been used successfully to control downy mildew diseases. However, in the last several years, some of the species, which cause downy mildew diseases (e.g., *Pseudoperonospora cubensis*—downy mildew of cucurbits [see Case Study 8.2] and *Peronospora tabacina*—blue mold of tobacco [see Case Study 8.3 and Figure 8.18]) have developed resistance to acylalanine class fungicides.

Albuginales/Albuginaceae—The White (Blister) Rusts

The **white rusts** are obligate plant parasites on flowering plants and are represented by the genera *Albugo*, *Pustula*, and *Wilsoniana*. Genera are differentiated by host range with *Albugo* typically occurring on Brassicaceae, *Pustula* on Asteridae, and *Wilsoniana* on Caryophyllales. *Albugo candida* is the most common and economically relevant species and affects numerous crops of *Brassica*, *Raphanus*, and *Sinapis*. Another very frequently observed species is *Albugo ipomoeae-panduratae* and is found growing in sweet potatoes and, more frequently, in wild morning glory (Figure 8.19). Species are differentiated by

CASE STUDY 8.3

Peronospora tabacina—Blue Mold of Tobacco

- Blue mold of tobacco, the downy mildews. It is only a pathogen of *Nicotiana* (tobacco) species (Main, 2005).
- All downy mildews are obligate plant parasites (cannot live and reproduce without a living host) and as such form an intimate relationship with their host. These relationships are characterized by the formation of haustoria (Figure 8.15), which transport nutrients from the host to the intercellular hyphae.
- *Peronospora tabacina* seldom reproduces sexually by oospores (Figure 8.2a), but is highly reproductive asexually by sporangia (sporangiospores) borne on sporangiophores (Figure 8.14). Sporangia are the primary mode of dissemination for this pathogen.
- The disease may affect seedlings as well as older plants and derives its name from blue to grayish-blue mass of sporangia and sporangiophores, which emerge from stomata on the undersurface of leaves with chlorotic (yellow) lesions. These lesions will eventually turn brown or necrotic (Figure 8.16). Leaves or portions of leaves may be "cupped." The pathogen usually produces discrete lesions on leaves, but may also be found as systemic infections, which may be common in some regions.
- *Peronospora tabacina* probably does not overwinter in the United States and Canada in production fields. Annual outbreaks and/or epidemics of the disease are initiated by sporangia arriving on wind currents from tropical zones south of the 30th parallel.
- Rapid disease development and spread are favored by cool, wet, and overcast conditions, whereas hot and dry weather and UV irradiation in sunlight adversely affect disease development and spread.
- Some general strategies for controlling blue mold are as follows:
  - Local management of tobacco crops
  - Use more than one control method
  - Make the environment less favorable for the pathogen and for the spread of the disease
  - Use protective fungicides (see websites for details)
  - Manage beds, greenhouses, and fields to favor the crop
  - Also, see Ivors et al. (2006).

(a) (b)

FIGURE 8.19 *Albugo ipomoeae-panduratae* infecting morning glory. (a) Upper surface of leaf showing chlorotic lesions. (b) White crusty pustules containing sporangia on the bottom surface of a leaf. (Courtesy of A.S. Windham.)

host-specificity and oospore ornamentation. *Pustula* and *Wilsoniana* are differentiated from *Albugo* by having an equatorial, ring-type wall thickening in zoosporangia. *Pustula* species are typically found in the members of Asteraceae, and *Pustula helianthicola* causes a very serious disease in sunflower (Figure 8.20). *Wilsoniana potulacae* is an important pathogen of *Portulaca oleraceae* (Potulacaceae, Caryophyllales), whereas *W. amaranthi* and *W. bliti* are frequently found in *Amaranthus* species.

The life history of *Albugo* species is similar to that exhibited by many other oomycetes except in the production of sporangiophores and sporangia. The club-shaped sporangiophores are contained within the host tissue, and each sporangiophore produces several sporangia in the intercellular space under the epidermis. While the apical sporangium is globose, the subsequent sporangia in the chain are box- or cube-like. In contrast to former assumptions that the pressure from growing sporangia finally ruptures the epidermis, enzymatic digestion is the more important factor in liberating the sporangia and forming a white crust. Old pustules on morning glory may appear

FIGURE 8.20 Infection of sunflower (*Helianthusannuus*) by *Pustula helianthicola*. (a) Typical lesions caused by *P. helianthicola* on leaves of sunflower (upper leaf surface). (b) Scanning electron microscopy of a ruptured pustule of *P. helianthicola* (lower leaf surface). Bar = 100 µm. (c) Scanning electron micrograph of sporangiophores (SP) and sporangia (S) within the pustule. Bar = 20 µm. (Courtesy of O. Spring.)

orange to pink. Sporangia are disseminated by wind and rain and germinate via zoospores. Zoospores encyst, form germ tubes, and initiate infections in suitable host plants. Oospores are formed by gametangial contact and serve as resting spores.

The plant pathogens classified in Oomycota are an interesting and distinct group of organisms. They cause a number of economically important diseases and have influenced the history of humans. There is no doubt that with additional research we will understand their taxonomic position better and learn more about the diseases they cause and how to control them.

LABORATORY EXERCISES

The oomycetes occur worldwide in diverse ecosystems including estuaries, lakes, oceans, rivers, and streams. Taxa within the group, however, can also occur in agricultural fields and in forest habitats growing as saprophytes or parasites of green plants. Because of the diversity of oomycetes and because they are found in many different environments, several representative experiments have been included in this chapter to enable students to learn more about the methods to study this unique group.

Many fascinating and intriguing experiments can be conducted with the members of the oomycetes, including those that are not plant pathogens. However, we will be limiting consideration to some of the more common species that cause plant diseases such as those in the genera: *Pythium*, *Phytophthora*, and *Albugo*. Soil-inhabiting species of *Pythium* and *Phytophthora* are easy to grow and manipulate in axenic cultures, whereas species of *Peronospora* and *Albugo* are obligate parasites that require a living host to complete their life cycles and cannot be grown in culture. Obligate pathogens are difficult to work with in most experimental systems. Although these obligate pathogens are very challenging to maintain, some educational and research exercises can be completed with relative ease. As has been emphasized throughout this book, secure the proper permits to obtain and transport pathogens and then destroy all experimental materials by autoclaving or by other means when the experiments are completed.

The following experiments are designed to provide hands-on experiences for students working with *Pythium* species that cause root rots and damping-off of peas and beans, isolation of *Phytophthora* species from diseased plant tissues and directly from soil, sporangia, and oospore formation by *Pythium* and *Phytophthora* in culture, and microscopic observation of *Peronospora* and *Albugo* species in host materials.

Experiment 1. Root Rot of Bean and Peas Caused by *Pythium* Species

Several *Pythium* species cause root and seed rots of various crops. These diseases can devastate both field-grown and greenhouse crops. This experiment is designed to demonstrate the symptoms and signs of the diseases.

Materials

Each student or team of students will require the following items:

- Culture of *P. ultimum* or *P. aphanidermatum*; cultures may be obtained from ATCC or colleagues with the proper permits
- Untreated (no fungicides) seeds of any cultivar(s) of common edible beans (*Phaseolus vulgaris*) and peas (*Pisum sativum*)
- Six plastic pots of 10 cm diameter, pot labels, and permanent marker

- Sand for plastic pots and pencil or large glass rod for making holes in sand; autoclave sand prior to use
- Two plastic flats with pasteurized Promix or other soilless medium and paper towels
- Laboratory blender, scissors, and long transfer forceps
- 250-mL flask containing 125 mL of sterile cornmeal (CM) broth (Difco, Detroit, MI) and two 1000-mL beakers
- Four Petri dishes of 10 cm diameter containing CM agar (add 15 g agar to the above formulation)
- Compound microscope and microscope slides

Follow the protocol listed in Procedure 8.1 to complete this experiment.

Anticipated Results

Depending on the temperature at which the inoculated plants are grown and the amount of inoculum applied, symptoms of damping-off should be evident between 3 and 7 days with both bean and pea plants. The inoculated plants should appear wilted at first, and then necrotic and water-soaked lesions will occur on the stem at the soil line. Infected plants will not be able to maintain stature and will fall over. Plants whose roots are cut will typically display symptoms 1 or 2 days earlier than plants with uncut roots. Roots of infected plants should appear dark and very soft compared with white and firm roots from uninoculated plants. Seeds treated with *Pythium* will fail to germinate and rot, or germinate poorly with the seedling succumbing to the disease very quickly. All plants that are not treated with *Pythium*, including those

Procedure 8.1

Root and Seed Rots Caused by Pythium Species

Step	Instructions and Comments
1	Plant seeds in flats are filled with pasteurized soilless medium 7–10 days before the laboratory. Grow in a cool greenhouse or laboratory. Each team of students will require at least 12 germinated seeds of each bean and pea. Space seeds so that the roots of individual plants will not grow together and seedlings can be easily separated.
2	Inoculate the sterile CM broth in 250-mL flasks with several plugs from the margin of 5-day-old colonies of either *P. ultimum* or *P. aphanidermatum* growing on CM agar dishes 1 week before the laboratory. Incubate the liquid cultures at 18°C–22°C either on a slow (30 rpm) shaker or on a shelf in an incubator or laboratory. Prepare an equal number of uninoculated flasks (medium without *Pythium* species).
3	Autoclave the sand the day before the experiment. This step is essential for the success of the exercise.
4	Gently remove the seedlings from the flats and wash the particles of soilless medium from the roots using tap water. Store plants with roots wrapped in moist paper towels on the laboratory bench.
5	Swirl the liquid culture to remove hyphal growth from the glass and empty the contents of the flasks into the blender. Add 375 mL of sterile distilled water and homogenize the mixture with short bursts (high speed) of the blender. Pour the suspension into a 1000-mL beaker.
6	Dip the roots of three bean plants into the *Pythium* species suspension. Make three large holes in the sand in each plastic pot using a pencil or glass rod. Be sure that sand is moist. Very gently plant the beans with as little damage to the roots as possible. Label each pot: *Pythium*-intact roots. Repeat the procedure with pea seedlings and label.
7	Trim about 25% of the root length from another group of three bean plants. Dip the remaining roots in the suspension and plant as in Step 6. Label the pot: *Pythium* cut roots. Repeat the procedure with pea seedlings and label.
8	Plant three bean seeds about 2 cm (0.5–1.0 inch) deep in a pot. Pour about half of the remaining *Pythium* species homogenate onto the surface of the sand in the pot. Label the pot: *Pythium* bean seeds. Repeat the procedure with pea seeds and label.
9	Repeat Steps 6–8 using the contents of an uninoculated flask mixed with 375 mL of sterile water for the dip and drench treatments. Label pots according to the treatment.
10	Set the plants on a laboratory bench near a window and observe for symptom development. Water with distilled water and do not allow the sand to dry out.

with cut roots, should grow normally unless a contaminating pathogenic *Pythium* species is present in the potting medium. Uninoculated seeds should germinate and grow normally.

Questions

- What are the controls in this experiment and why are they necessary?
- How would you complete Koch's postulates for this experiment?

Experiment 2. Isolation of *Phytophthora* Species From Plant Tissues and Soil

Many members of Pythiaceae (*Pythium* species) and Peronosporaceae (*Phytophthora* species) are widely distributed in soils. They may survive for long periods without a host. This exercise is designed to provide experience isolating *Phytophthora* species from diseased plants and infested soil using a selective medium.

Materials

Each student or group of students will need the following items:

- Roots, stems, or leaves from diseased plants (rhododendron, azalea, soybean, tomato, pepper, or tobacco work well)
- Soil sampling tool (2.5 cm diameter)
- Ice chest
- Paper towels and plastic bags
- Scalpel with #10 blade (caution: very sharp)
- Twenty, plastic Petri dishes of 10 cm diameter containing PARPH medium (Jeffers and Martin, 1986; Ferguson and Jeffers, 1999; Table 8.4)
- Incubator set at 20°C without light
- Sieves with 4- and 2-mm openings
- Several aliquots of 100 mL of 0.3% water agar (3 g agar in 1 L of water) in 250-mL beakers
- Magnetic stirrer and stir bars
- One-mL wide-bore pipette and bulb or pump
- Top loading balance and plastic weigh boats
- Compound microscope and glass slides and coverslips

Follow the protocol outlined in Procedure 8.2 to complete this exercise.

Anticipated Results

Colonies of *Phytophthora* species should develop on PARPH from both soil and diseased tissues between 24 and 72 h. Colonies of *Pythium* species may also develop if hymexazol is omitted from the isolation medium (PARP), or if there are hymexazol-tolerant *Pythium* species present in the samples. Soil from sources without infected plants may or may not contain *Phytophthora* and *Pythium* species. *Phytophthora* colonies should not develop on PARPH from healthy plant tissues.

TABLE 8.4
Growth/Isolation Media for *Phytophthora* Species

V8 Agar (V8A) Growth and Sporulation Medium

V8 juice[a], 200 mL
$CaCO_3$, 2 g
Distilled water, 800 mL
Agar, 15 g

PARP(H) Isolation Medium

Delvocid (50% pimaricin): P, 10 mg
Sodium ampicillin: A, 250 mg
Rifamycin-SV (sodium salt): R, 10 mg
75% PCNB (Terraclor): P, 67 mg
Hymexazol[b]: H, 50 mg
Clarified V8 juice[c], 50 mL
Agar[c], 15 g
Distilled water, 950 mL

Note: All antimicrobial amendments should be added after the base agar medium has been autoclaved and cooled to 50°C–55°C (autoclaved vessel can be held in hand without discomfort).

[a] Clarified V8A (cV8A) can be made by stirring 200 mL V8 juice with 2 g $CaCO_3$ for 15 min and then centrifuging for 10 min at 4000×*g*. Supernatant may be stored frozen at −20°C and use 100 mL with 900 mL water and 15 g agar to make cV8A.

[b] Inclusion is optional: hymexazol inhibits *Pythium* species.

[c] Corn meal agar (CMA) may be substituted for clarified V8 juice and agar.

Questions

- What morphological characteristics would you use to recognize and identify *Pythium* and *Phytophthora* species?
- What are the characteristics used to distinguish *Pythium* and *Phytophthora* species from each other?
- Can a plant be infected with *Phytophthora* and *Pythium* species at the same time? Can soil be infested with both genera?

Experiment 3. Production of Sporangia and Oospores by *Phytophthora* and *Pythium* Species

Many species of *Phytophthora* and *Pythium* will form sporangia and oospores in culture if proper environmental conditions are provided. These simple experiments are designed to allow students to observe asexual and sexual reproduction in these genera.

Procedure 8.2

Isolation of *Phytophthora* Species from Infected Plant Tissues

Step	Instructions and Comments
1	Place roots and stem segments from diseased plants into plastic bags, and keep moist using damp towels. Healthy plants should also be sampled. Transport the samples to the laboratory in a cool ice chest. Samples should be kept in the dark.
2	In the laboratory, gently wash the tissues under running tap water for 5–10 min and blot excess moisture with a paper towel.
3	Cut samples into 1-cm segments and place four segments on each of five PARPH Petri dishes. Each piece should be pushed into the agar so that the tissues are surrounded by agar. Incubate the cultures at 20°C in the dark.
4	Observe colony growth on PARPH agar dishes after 48–72 h. Continue to examine the dishes for up to 1 week. Transfer small pieces of mycelium from individual whitish-colored *Phytophthora* colonies to fresh dishes containing PARPH medium and incubate at 20°C in the dark.
	Isolation of *Phytophthora* Species from Soil
1	Collect soil cores up to 20 cm deep using a sampling tool approximately 2 cm in diameter. Collect 10 core samples within 20 cm of target symptomatic plants and 10 samples of "noninfested" soil from around healthy-appearing plants of the same type.
2	Place each set of 10 soil cores into separate and individually labeled plastic bags to make one infested and one noninfested composite soil sample. Return the composite soil samples to the laboratory in a cool ice chest. All samples should be stored in a dark, cool place until the assay begins.
3	Prepare each composite soil for isolation by breaking up the clods. Remove rocks and plant debris by first using a coarse (4 mm) screen followed by a smaller screen (2 mm). Thoroughly mix the soil and return to plastic bags.
4	Add 50 mL of the infested soil to 100 mL of 0.3% water agar contained in a 400-mL beaker with a stir bar. Place on a magnetic stirrer at high speed for about 3 min. Pipette 1-mL aliquots of the suspension onto PARP(H) dishes. Evenly spread the suspension across the surface of the agar. Use up to five dishes per composite soil sample. Repeat for the "noninfested" soil.
5	Incubate the dishes at 20°C in the dark for 48–72 h. Do not enclose dishes in plastic bags or boxes. Wash soil from dishes under running tap water. Examine dishes with a dissecting microscope (30× to 50×) and count colonies. Subculture from these colonies to fresh PARP(H) medium if desired. Subcultured colonies may be saved for later identification. Compare the number of colonies from different locations and soil types.

Materials

Each team of students will require the following materials:

- Agar (CM) cultures of several homothallic species of *Pythium* and *Phytophthora* (e.g., *Phytophthora cactorum* or *Phytophthora citricola*).
- Freshly gathered grass (tall fescue, blue grass, etc.) clippings autoclaved for 20 min on two successive days.
- Sterile distilled water (to enhance sexual reproduction in *Pythium* cultures, a drop of chloroform containing cholesterol [1 mg in 10 mL chloroform] may be added to 8 mL of distilled water. Allow the chloroform to evaporate under a transfer hood before adding grass or inoculating with *Pythium*).
- Sharp scissors to cut grass.
- Fine-tipped forceps.
- Several 60-mm plastic Petri dishes.
- V8A and clarified V8A (Table 8.4).
- Centrifuge and Whatman #1 filter paper.
- Incubator or laboratory bench equipped with fluorescent lights.
- Sterile plastic drinking straws.
- Soil.
- Nonsterile soil extract solution (NS-SES)—Stir 15 g of soil in 1 L of distilled water for at least 4 h and allow the suspension to settle overnight. Decant the water and centrifuge for 10 min at 4000 × *g* followed by filtration through Whatman # 1 filter paper (Jeffers and Aldwinckle, 1987). Store the filtered soil extract in the refrigerator.
- Dissecting and compound microscopes.

- Microscope slides and coverslips.
- Lactoglycerol solution (lactic acid : glycerol 1:1 v:v) with 0.1% acid fuchsin.
- Brightly colored nail polish or Vaseline.

Follow the protocol in Procedure 8.3 to complete this experiment.

Anticipated Results

Sporangia should form in all *Pythium* and *Phytophthora* cultures in as little as 24–48 h for many *Pythium* and *Phytophthora* species in NS-SES. Gametangia and oospores usually take 2–4 weeks to form in the cultures.

Questions

- Why where homothallic species of *Phytophthora* and *Pythium* used?
- Are all species of *Phytophthora* and *Pythium* homothallic?
- What is the taxonomic significance of the origin of antheridia?
- What purpose do oospores serve in the life cycle of these organisms?
- Assign a ploidy level (haploid or diploid) to each of the structures seen in the cultures.

Experiment 4. Observing Morning Glory Leaf Tissue Infected with *Albugo ipomoeae-panduratae*

White rust of morning glory is very common. Infected morning glory leaves have chlorotic halos on the upper surface and white "pustules" on the lower surfaces of the leaves (Figure 8.19). White pustules may appear pink and very crusty in the fall.

Materials

Each student will require the following items:

- Morning glory leaves infected with *A. ipomoeae-panduratae* (Figure 8.19)
- Compound microscope, glass microscope slides, and coverslips
- Water in a dropper bottle
- Razor blades (caution: very sharp)
- 0.1% calcofluor in water
- Epifluorescense microscope equipped with 395–420 nm excitation filter and 470 nm absorption filter

Follow the protocol in Procedure 8.4 to complete this experiment.

Procedure 8.3

Production of Sporangia and Oospores by *Pythium* and *Phytophthora* Species

Step	Instructions and Comments
1	Grow cultures of *Pythium* species on CM agar in 60-mm plastic Petri dishes at 20°C in darkness for 3 days.
2	Add 8 mL of sterile distilled water to a number of empty Petri dishes of 60 mm diameter. Separate and add 5–10 autoclaved blades of grass to the water. Remove several plugs of mycelium from the margin of the *Pythium* species colony with a sterile plastic straw and cut into quarters with a sharp scalpel. Transfer four quarters to the Petri dish with the blades of grass. Make sure that the agar pieces are in contact with the grass. Incubate at 20°C in the light.
3	Cultures may be observed weekly, with either a 40× dissecting scope, or individual pieces of grass from the culture may be mounted in lactoglycerol (with or without acid fuschin) on microscope slides and viewed with a compound microscope. The slide may be made semipermanent by painting the clean, dry edges of the coverslip with nail polish. Draw and label all structures. To stimulate zoospore release from sporangia of *Pythium* species, chill (2°C–4°C) the grass cultures for a few hours and then allow them to warm to room temperature. Observe sporangia, vesicle formation, and zoospores with the aid of a dissecting microscope.
4	Transfer *Phytophthora* species cultures to V8A and incubate in the dark at 20°C–25°C for 48–72 h. Note that the colonies should be at least 2 cm in diameter. Aseptically remove agar plugs (2 mm) near the margin of the colony with a sterile drinking straw and place five of them into a sterile, empty Petri dish of 60 mm diameter.
5	Cover the agar plugs with 7–10 mL of NS-SES. Place dishes under continuous fluorescent lights at 20°C–25°C. Observe plugs with a dissecting microscope after 12–24 h for sporangia or if not present, after an additional 24 h. Draw and label all structures.
6	To initiate oospore formation in the cultures of *Phytophthora* and *Pythium* species, transfer the organisms to 60-mm Petri dishes containing cV8A and incubate at 20°C–25°C in the dark. Microscopically examine the cultures weekly for up to 6 weeks. Draw and label all structures. This procedure works well for homothallic species of both *Pythium* and *Phytophthora*. Oospores are usually present after 2 weeks and are mature by 4 weeks.

Procedure 8.4

Morphological Features of *Albugo ipomoeae-panduratae* in Infected Morning Glory Plants

Step	Instructions and Comments
1	Observe pustules on morning glory leaves using a dissecting microscope (Figure 8.19).
2	With a razor blade, cut cross sections (as thin strips as possible) from morning glory leaves infected with *Albugo ipomoeae-panduratae* and mount in water on a microscope slide. Try to include a portion of a pustule in the section. Draw sporangia and pustule morphology.
3	Cut other thin sections of leaf tissue and pustules and mount in calcofluor solution. View these sections using epifluorescense, and draw all pathogen and host structures. Caution: do not look directly at the UV light source.

Anticipated Results

Sectioned mounts of pustules should reveal many square- or rectangular-shaped sporangia. Knob-like haustoria may also be seen in host cells. Compare with a prepared slide if these structures cannot be observed in the fresh sections. If the tissue is observed with calcofluor, hyphae, sporangia, and haustoria should "glow" white. Plant cell walls will also "glow."

Questions

- Why is it relatively difficult to observe sporangiophores of *Albugo* compared with *Peronospora*?
- Why are *Albugo* sporangia box-shaped?
- Why does calcofluor stain *Albugo* structures? Would it stain structures of most oomycetes?

REFERENCES

Alexopoulos, C. J., C. W. Mims, and M. Blackwell. 1996. *Introductory Mycology*. New York, NY: John Wiley & Sons, Inc.

Beakes, G. W., and S. Sekimot. 2009. The evolutionary phylogeny of the oomycetes—Insights gained from studies of holocarpic parasites of algae and invertebrates. Pages 1–24, In: *Oomycete Genetics and Genomics: Diversity, Interactions and Research Tools,* K. Lamour and S. Kamoun (eds.). Hoboken, NJ: Wiley-Blackwell.

Ferguson, A. J., and S. N. Jeffers. 1999. Detecting multiple species of *Phytophthora* in container mixes from ornamental crop nurseries. *Plant Dis*. 83:1129–1136.

Hawksworth, D. L., P. M. Kirk, B. C. Sutton, and D. N. Pegler. 1995. *Ainsworth & Bisby's Dictionary of the Fungi*, 8th ed. Wallingford, UK: CAB International.

Ivors, K. L., K. Seebold, C. S. Johnson, and A. Mila. 2006. Blue mold control plan. http://www.ces.ncsu.edu/depts/pp/bluemold/control_2006.php [accessed July 1, 2016].

Jeffers, S. N. and H. S. Aldwinckle. 1987. Enhancing detection of *Phytophthora cactorum* in naturally infested soil. *Phytopathology* 77:1475–1482.

Jeffers, S. N., and S. B. Martin. 1986. Comparison of two media selective for *Phytophthora* and *Pythium* species. *Plant Dis*. 70:1038–1043.

Main, C. E. 2005. The blue mold disease of tobacco. http://www.ces.ncsu.edu/depts/pp/bluemold/thedisease.php.

Ribeiro, O.K. 2013. A historical perspective of *Phytophthora*. Pages 1-10, In: *Phytophthora A Global Perspective,* K. Lamour (ed.). Wallingford, UK: CAB International.

Thines, M. 2013. Taxonomy and phylogeny of *Phytophthora* and related oomycetes. Pages 11–18, In: *Phytophthora A Global Perspective*, K. Lamour (ed.). Wallingford, UK: CAB International.

Thines, M. and O. Spring. 2005. A revision of *Albugo* (Chromista, Peronosporomycetes). *Mycotaxon* 92:443–458.

Trigiano, R. N., C. G. Van Dyke, and H. W. Spurr, Jr. 1983. Haustorial development of *Peronospora tabacina* infecting *Nicotiana tabacum*. *Can. J. Bot*. 61:3444–3453.

Trigiano, R. N., C. G. Van Dyke, H. W. Spurr, Jr., and C. E. Main. 1985. Ultrastructure of sporangiophore and sporangium ontogeny of *Peronospora tabacina*. *Tob. Sci*. 29:116–121.

SUGGESTED READINGS

Agrios, G. N. 1997. *Plant Pathology*, 4th ed. New York, NY: Acad. Press.

Griesbach, J. A., J. L. Parke, G. A. Chastagner, N. J. Grunwald, and J. Aguirre. 2012. *Safe Procurement and Production Manual: A Systems Approach for the Production of Healthy Nursery Stock. Oregon Association of Nurseries*. http://www.oan.org/?page=861.

Hock, H. S. 1974. Preparation of fungal hyphae grown on agar-coated microscope slides for electron microscopy. *Stain Technol*. 49:318–320.

Martin, F. W. 1959. Staining and observing pollen tubes in the style by means of fluorescence. *Stain Technol*. 34:125–128.

9 Non-Oomycota Zoosporic Plant Pathogens

Sharon E. Mozley-Standridge, David Porter, and Marc A. Cubeta

Concept Box

- Several different unrelated lineages of plant pathogenic fungi and fungus-like organisms that reproduce by the production of zoospores are represented in this chapter.
- *Labyrinthula terrestris, Plasmodiophora brassicae,* and *Synchytrium macrosporum* have been chosen here as diverse examples of lesser-studied simple eukaryotic plant pathogens.
- The genus *Labyrinthula* in the kingdom Stramenopila shares a common ancestry with the Oomycota (Chapter 8). *Plasmodiophora* appears to be in the Cercozoa, distantly related to Foraminifera, and *Synchytrium* is a chytrid, in the kingdom Fungi.
- Knowledge and understanding of the phylogeny and taxonomy of genetically diverse assemblages of plant pathogenic microorganisms can contribute to improved diagnosis and management of plant disease.

The scientific discipline of plant pathology is focused largely on the study of microorganisms that cause disease (i.e., pathogens) on economically important species of vascular plants, but plant pathologists must be familiar with a wide variety of pathogens to ensure accurate identification of the causal agent for deploying economical and environmentally sound, disease management strategies. In addition, plant pathologists must also be able to recognize a diverse range of plant-associated organisms typically studied in related disciplines outside of plant pathology (e.g., biology, botany, mycology, microbiology, and zoology). Because of the need to be broadly trained, we present here the background information and some simple hands-on experiments for three unusual (and often overlooked) plant pathogens, *Plasmodiophora brassicae* Wor., *Labyrinthula terrestris* (DM Bigelow, MW Olsen, and Gilb), and *Synchytrium macrosporum* Karling, which all students of plant pathology should be familiar.

Plasmodiophora brassicae, *L. terrestris*, and *S. macrosporum* are often referred to as "zoosporic fungi." Zoosporic fungi are found in several phylogenetically unrelated groups of microorganisms that produce motile, flagellated **spores** (**zoospores**), usually due to **asexual reproduction** during some stage of their life (Fuller and Jaworski, 1987).

Although the phrase "zoosporic fungi" is descriptive, it does not have evolutionary or phylogenetic significance. Zoosporic fungi are found in two different kingdoms of eukaryotic organisms: the kingdom Fungi and the recently named kingdom Stramenopila (or Straminipila) (Dick, 2001). One of the most basal branches of the kingdom Fungi is the **phylum** Chytridiomycota, all of which are zoosporic fungi. The more familiar (but not zoosporic) Ascomycota, Basidiomycota, and Glomeromycota are also in the kingdom Fungi. The phylum Zygomycota (Chapter 10) is currently being revised, and students should be aware that several clades (Mucormycotina, Kickxellomycotina, Zoopagomycotina, and Entomophthoromycotina) are being used in place of the phylum until relationships can be fully resolved (Hibbett et al., 2007). The zoosporic fungi in the kingdom Stramenopila include hyphochytrids, labyrinthulids, and oomycetes. This kingdom represents an extremely diverse group of organisms, which also includes photosynthetic organisms such as brown algae, chrysophyte algae, and diatoms. Organisms classified in the phylum Chytridiomycota produce zoospores that usually have a single, posteriorly directed, and smooth **flagellum**, whereas organisms classified in the kingdom Stramenopila (stramenopiles) are characterized by the production of zoospores that usually have two flagella (biflagellate), but always with tripartite tubular hairs on the anterior flagellum (Dick, 2001; Figure 9.1). A flagellum that possesses two rows of tripartite tubular hairs is also referred to as a **heterokont** or tinsel flagellum

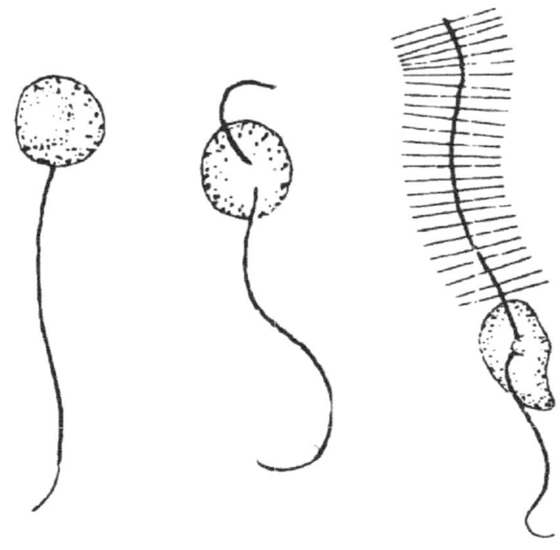

FIGURE 9.1 Zoospores of (a) *Synchytrium macrosporum*, (b) *Plasmodiophora brassicae*, and (c) *Labyrinthula zosterae*. The tripartite tubular hairs on the posterior flagellum of *L. zosterae* are not visible with a light microscope. (Illustration prepared by Lynnette J. Gray, redrawn from zoospore diagrams in Dick, M.W., *Straminipilous Fungi*, Springer, the Netherlands, 2001. With permission.)

(Kirk et al., 2001). While the flagella exhibited by zoospores in each group can be seen by light microscopy, details such as the tripartite hairs and diagnostic internal features must be viewed using an electron microscope, which can unfortunately limit proper identification.

The plasmodiophorids are often considered zoosporic fungi because of their osmotrophic mode of nutrition and production of zoospores, but are distantly related to both the stramenopiles and chytrid fungi (Dick, 2001). Recent molecular evidence suggests that the plasmodiophorids are in Cercozoa, a diverse group of protists that includes Foraminifera (Archibald and Keeling, 2004). As a group, the plasmodiophorids are obligate endoparasites (**biotrophs**) of algae, fungi, protists, and plants. Some plasmodiophorids are important pathogens of agricultural crops and include *P. brassicae* (clubroot of crucifers), *Polymxya graminis* (root diseases of cereals), and *Spongospora subterranea* (powdery scab of potato). The latter two pathogens can also transmit viruses that cause plant diseases (Chapter 4).

Labyrinthulids produce a network of fine hyaline filaments through which the characteristic spindle-shaped cells move (called an ectoplasmic net), thus prompting some researchers to classify them with the slime molds. However, morphological, ultrastructural, and recent molecular data place the labyrinthulids into the kingdom Stramenopila as a sister group to the oomycetes (Chapter 8), clearly separating them from both the plasmodiophorids and true slime molds (Porter, 1990; Baldauf and Doolittle, 1997; Honda et al., 1999; Leander and Porter, 2001; Dick, 2001; Blanton 1990; Spiegel et al., 2004; Schaap et al., 2006). The labyrinthulids include organisms that are important decomposers and parasites of algae and plants in coastal marine habitats. *Labyrinthula zosterae* Porter and Muehlstein is a pathogen of eelgrass (*Zostera marina* L), an ecologically important seagrass that forms vast subtidal meadows in estuarine communities in temperate regions of the world. Between 1934 and 1935, most populations of eelgrass present in the North Atlantic were killed due to a pandemic "wasting disease" caused by *L. zosterae*. Atlantic eelgrass beds have partially recovered, but are not as extensive as before the pandemic. *L. zosterae* is still associated with localized dieback of eelgrass. Another pathogenic species of labyrinthulid, *L. terrestris*, is responsible for rapid blight disease of turfgrass (Bigelow et al., 2005). The unexpected occurrence of a marine organism as the cause of an emerging disease of terrestrial grass may be the result of changes in our cultural practices for turfgrass management, such as the increased use of saline or reclaimed water for irrigation. Certain other labyrinthulids are parasites of marine invertebrates (Porter, 1986).

Chytrids are members of the phylum Chytridiomycota, the most basal **clade** within the kingdom Fungi (James et al., 2006). Because of their early evolutionary divergence, most chytrids exhibit certain characteristics that are not shared by fungi in the phyla Ascomycota and Basidiomycota, such as determinate growth and production of zoospores via asexual reproduction. Chytrid sporangia are microscopic in size and much smaller than the fruiting bodies of most Ascomycota and Basidiomycota. Although a large number of chytrid species are parasitic on a wide variety of other organisms including fungi, only a few plant hosts are economically important. Chytrids often appear simple in form and structure on initial examination, but their morphological characteristics can exhibit considerable variation in shape and size. Chytrids are of great ecological importance as decomposers of a wide variety of biological substrates. For example, a large number of chytrids degrade cellulose in the leaves and stems of plants, whereas others can degrade chitin, keratin, and sporopollenin, the latter is a biopolymer associated with pollen grains that is highly resistant to biological degradation (Sparrow, 1960). Chytrids present within the rumen of certain herbivores (e.g., cows and sheep) are also associated with the breakdown of cellulosic substrates and provide much needed energy and nutrients for the growth and development of animals.

Similar to other true fungi, chytrids have chitin in their cell walls, flattened plate-like mitochondrial cristae, and exhibit an absorptive mode of nutrition. However, unlike all other members of the kingdom Fungi, chytrids reproduce by the formation of zoospores and lack true mycelium. Chytrid zoospores differ from stramenopile zoospores by having only one posteriorly directed, smooth flagellum (Figure 9.1). Because chytrids are microscopic and usually determinant in their growth, they are often overlooked in the environment but can be readily found and observed using baiting techniques and a dissecting microscope. Although chytrids are sometimes referred to as aquatic fungi or water molds, chytrids are present and can be found nearly anywhere that water is available, including soil. A large number of chytrids are parasites of algae, fungi, plants, and even frogs. Many chytrids can also exist as **saprotrophs**.

The genus *Synchytrium* has over 200 species that are parasites of a wide range of algae and plants in fresh water and terrestrial habitats. *Synchytrium endobioticum* is the best-known species and is the causal agent of potato wart disease (see Case Study 9.1). The fungus can also attack tomato and noncultivated species of *Solanum*. Since the initial discovery of potato wart in Hungary in 1896, this disease has been identified in most of the continents. Because of the serious nature of potato wart, *S. endobioticum* was placed on the list of quarantined organisms established in 1912 by the United States Department of Agriculture (USDA) under the Plant Protection Act. Canada and other countries have also established quarantine laws to prevent movement of this fungus on potatoes and in soil. A recent outbreak of potato wart in a single field in Prince Edward Island (PEI) in Canada in 2000 prompted the USDA to impose a quarantine that banned importation of seed potatoes from PEI into the United States for 1 year. Potato wart was also discovered in two additional fields on PEI in 2002. The **resting spores** of *S. endobioticum* can survive for 20–30 years in soil (Agrios, 2005).

Other chytrid pathogens of economically important species of agricultural plants include *Olpidium* spp. (infects roots of many plants and serves as a vector for at least six viruses that cause plant diseases, see Chapter 4), *Physoderma alfalfae* (causal agent of crown wart of alfalfa) and *Phyllachora maydis* (causal agent of brown spot of corn).

PLASMODIOPHORA BRASSICAE

Plasmodiophora brassicae is an important pathogen of cultivated agricultural crops that belong to the mustard family (Brassicaceae) (Figure 9.2). Members of this plant family are often referred to as crucifers (because of their cross-shaped flowers) or cole (which is German for stem) crops and includes the following: broccoli, Brussels sprouts, cabbage, Chinese cabbage, canola, cauliflower, collards, kale, kohlrabi, mustard, radish, rape, rutabaga, and turnip. Several additional genera of cultivated and weed species of plants in the genera *Alyssum, Amoracia, Brassica, Camelina, Capsella, Erysimum, Iberis, Lepidium, Lobularia, Lunaria, Matthiola, Nasturtium, Raphanus, Rorippa, Sinapis, Sisymbrium*, and *Thlaspi* are also hosts for *P. brassicae* (Farr et al., 1995). *Arabidopsis thaliana*, a mustard species known as mouse-ear cress, widely employed as a model system in the genetic research of plant development and plant–microbe interactions, is also a host for *P. brassicae* (Siemens et al., 2002).

Plasmodiophora brassicae is a biotrophic (obligate) parasite that causes a devastating disease of crucifers

CASE STUDY 9.1

Wart Disease of Potato

- Potato wart is caused by the chytrid fungus *Synchytrium endobioticum* and is distributed worldwide. Wart disease is of economic significance to the potato industry on an international scale.
- *Synchytrium endobioticum* is related genetically to *S. macrosporum,* but has a narrower host range and infects plants in the family Solanaceae, such as potato and tomato.
- Potato wart disease is challenging to manage with traditional approaches such as breeding plants for resistance and crop rotation.
- The fungus can survive for many years in soil in the absence of a plant host by producing resting spores. These spores germinate to produce swimming spores (zoospores) that penetrate host cells directly to initiate disease.
- *Synchytrium endobioticum* is a biotroph and obtains nutrients inside of plant cells that result in abnormal cell enlargement (hypertrophy) and division (hyperplasia) and formation of galls (warts) on roots, stems, and tubers of potato.
- The fungus is dispersed mainly by the activities of humans by moving potatoes, animal manure, soil, and equipment that harbor the organism.
- Potato wart is managed primarily by plant inspection, legislation to establish quarantine laws, and eradication to prevent potential spread of the fungus.
- Plant and soil bioassays coupled with DNA-based molecular methods are providing useful tools for detecting and identifying *S. endobioticum.*

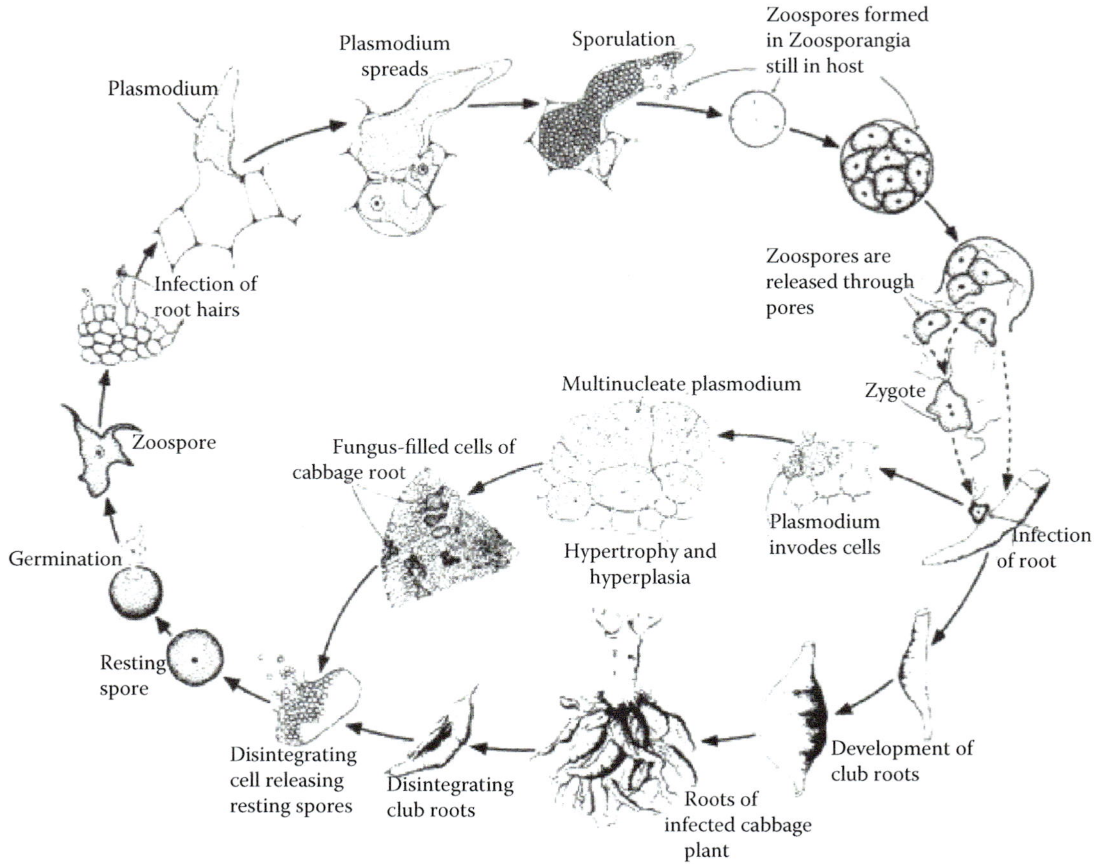

FIGURE 9.2 Life cycle of *Plasmodiophora brassicae* (From Agrios, G.N., *Plant Pathology,* Academic Press, New York. With permission.)

known as clubroot. The disease occurs throughout the world in commercial crucifer production fields, but is also a serious problem in home gardens. Clubroot has been known since the thirteenth century in Western Europe and was first studied in detail by Woronin in Russia in the late 1870s. Woronin originally described *P. brassicae* as a slime mold. Despite the tremendous amount of research on clubroot, it still remains as one of the most serious diseases of crucifers and is largely responsible for the disappearance of commercial cabbage production in many regions throughout the world.

Because *P. brassicae* produces resting spores that can persist in soil for many years, traditional approaches such as crop rotation are of limited value in managing clubroot. Some success has been achieved in breeding plants for resistance to *P. brassicae*. However, field populations of *P. brassicae* are genetically diverse and composed of many races of the pathogen. Races are represented by genetically distinct individuals and typically identified by inoculating a series of well-defined species of crucifers and observing them for disease symptoms (Williams, 1966; Cubeta et al., 1998). Although some resistant varieties of crucifers have been developed and are commercially available, there are no varieties that are resistant to all known races of *P. brassicae*. One approach for managing clubroot used for centuries involves the modification of the soil environment by adding lime (either calcium carbonate or calcium oxide) to increase soil pH to at least 7.2. This approach provides an economical and effective means of reducing the damaging effects of clubroot, and it is hypothesized that the increased soil pH interferes with zoospore movement and the initial infection process on roots. Unfortunately, a soil pH of 7.2 or higher is often not a favorable growth environment for many cultivated agricultural crops.

Life History of *Plasmodiophora brassicae*

Plasmodiophora brassicae can survive for at least 10 years in soil by forming resting spores (Agrios, 2005; Figure 9.2). During the periods of cool, wet weather when the soil becomes saturated with water, resting spores (also referred to as cysts) germinate to produce usually one primary zoospore with two hairless flagella (Figure 9.1). These zoospores swim to the root and penetrate hairs of young roots or enter the plant through wounds in secondary roots (Williams, 1966). Once inside the root, *P. brassicae* produces an amoebae-like structure called a **plasmodium** (plural: **plasmodia**), which passes through the cells and becomes established in them. In plant cells, the nucleus of each plasmodium divides and becomes transformed into a multinucleate structure called a zoosporangium (plural: zoosporangia) that contains four to eight secondary zoospores. Secondary zoospores are discharged through exit pores in the plant cell wall. They usually fuse with each other to cause more infection of roots and form additional plasmodia. These diploid plasmodia divide by meiosis and produce clusters (**sorus**, plural: sori) of resting spores with a single haploid nucleus.

As plasmodia continue to grow and develop, they ingest proteins and sugars in the plant cells as a source of nutrients while stimulating them to divide (**hyperplasia**) and enlarge (**hypertrophy**). This abnormal plant growth results in the production of small, spindle-shaped swellings on roots that later develop into larger-sized galls or clubs (Figure 9.3). Root galls interfere with nutrient and water movement in the plant, and the initial symptoms on infected plants often appear as yellowing and wilting of the lower leaves, particularly on warm, sunny days. Severely infected plants are often stunted and smaller than healthy plants. Eventually the galls become a food source for other soil-dwelling microorganisms that initiate their decay and release of resting spores into the soil.

SYNCHYTRIUM MACROSPORUM

Synchytrium macrosporum is a biotrophic pathogen of more than 1400 different species of plants representing 185 families and 933 genera ranging from liverworts (hepatophytes) to flowering plants (angiosperms), especially those in the Asteraceae, Brassicaceae, Cucurbitaceae, Fabiaceae, and Solanaceae (Karling, 1964). *Synchytrium macrosporum* has the largest and widest host range of any biotrophic fungus in the kingdom Fungi (Karling, 1964).

FIGURE 9.3 Healthy (left) and infected (right) Chinese cabbage roots. (Courtesy of Lisa A. Castlebury.)

Synchytrium macrosporum is a weak pathogen that primarily attacks young seedlings. Although most plants survive early infection and grow to maturity, some seedling death may occur in rare cases of severe infection, particularly if environmental conditions are favorable. As with *P. brassicae*, galls caused by the hypertrophy and hyperplasia of infected host epidermal cells are the most recognizable symptom produced by *S. macrosporum* on plants. In general, galls form on leaves and stems of developing plants and range in size from 0.35 to 1.3 mm. However, galls can also form on the roots and underground fruits of certain legumes. The characteristic galls are composed of a central infected cell with a single resting spore and a sheath of surrounding cells of increased size. Occasionally, some portion of the host cell cytoplasm is retained around the resting spore.

LIFE HISTORY OF *SYNCHYTRIUM MACROSPORUM*

Synchytrium macrosporum survives primarily as resting spores in soil and infected plant debris (Figure 9.4). Depending on the host and geographic location, dormant resting spores usually germinate in the presence of moisture in late winter or early spring. When resting spores germinate, they function as prosori (singular: prosorus), or container for cellular contents that will later become sori. During germination, contents of the resting spore and prosorus exit the thick-walled casing through an opening or exit pore in the wall that is eventually filled by a plug of dark pigmented material. The cytoplasm, surrounded by a plasma membrane, undergoes a number of mitotic divisions before partitioning into numerous sporangia. The number of sporangia within a single sorus can range from 120 to 800, while the diameter of an individual **sporangium** ranges from 18 to 60 μm. Sporangia within a sorus can remain dormant for 1–2 months, and the infected plant tissue needs to dry completely and be rehydrated to induce resting spores to germinate (Karling, 1960). These conditions may simulate events that occur in nature where infected material senesce, dry out and then become rehydrated with water from dew or rain.

The cytoplasm within each sporangium cleaves up into individual zoospores, each with its own nucleus. In order for zoospores to be released, the sorus has to open up (dehisces) and release individual sporangia from the soral membrane. The sporangia, in turn, release zoospores by the splitting of the sporangial inner membrane. Zoospores swim to new hosts through the thin film of water present either on the surface of a plant or in soil to infect young seedlings. Zoospores can also be dispersed in water splashed from one plant to another. The zoospores are egg-shaped (ovoid) to slightly elongate (3–3.8 × 4–4.5 μm, with a single yellowish-orange refractive lipid globule) and possess a single, posteriorly directed, hairless flagellum (12–14 μm in length). The flagellum can be seen extending out from the

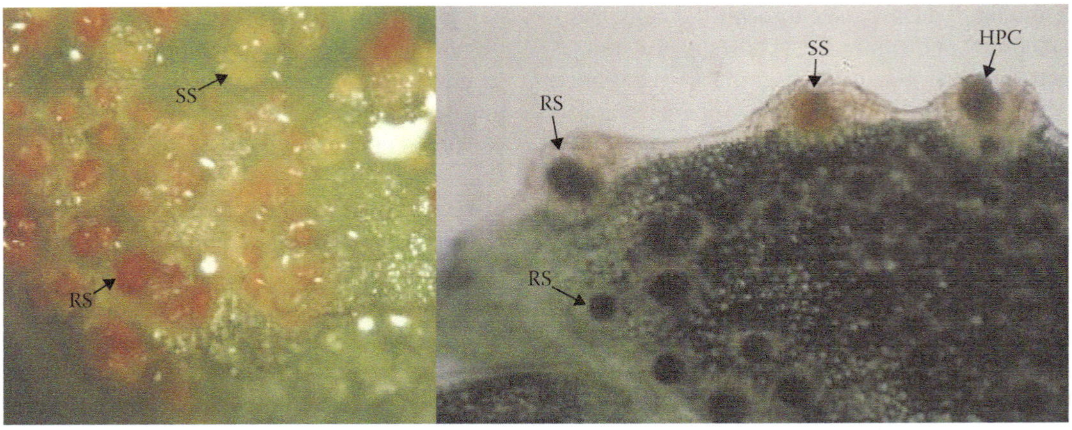

(a) (b)

FIGURE 9.4 Galls with sporangia, sori, and resting spores of *Synchytrium macrosporum* on infected chickweed leaves. (a) Galls filled with either dark orange resting spores (RS) or yellow sori containing sporangia (SS); (b) Microscopic view of galls showing hypertrophied plant cells (HPC) containing resting spores (RS) and sori with sporangia (SS). (Courtesy of M.J. Powell and P. Letcher.)

body of the zoospore, but electron microscopy is required to see internal zoospore details necessary for identification. When zoospores are released from the sporangium in the spring, they alternate between swimming and moving in an amoeboid fashion for as long as 24 h before settling down on the surface of a host. Once settled, encystment occurs and the flagellum is either retracted into the zoospore, later referred to as a zoospore cyst, or cleaved-off completely, and a membrane is produced on the outside of the zoospore. A narrow germ tube develops from the encysted zoospore and penetrates the host cell wall. Cytoplasm flows into the host cell from the zoospore cyst, and both the zoospore cyst and germ tube disintegrate. The cytoplasm can assume a variety of different shapes from round to amoeboid once inside the host and moves within the cell positioning itself near the host nucleus. The cytoplasm, now called an initial cell or uninucleate thallus, increases in size and develops a thick wall as it matures. After the thick-walled initial cell goes into a state of dormancy, it becomes a resting spore.

Resting spores can be either spherical or ovoid in shape. They range in diameter from 80 to 270 μm. The color of the resting spore wall varies from dark amber to reddish-brown and is usually 4–6 μm thick. The walls of resting spores and sporangia are yellow-orange colored, as are the characteristic galls that form because of infection. Although Karling (1960) observed the production of zygotes from the fusion of two zoospores and their nuclei, no one has substantiated this observation or determined the role that sexual reproduction plays in resting spore formation.

LABYRINTHULA TERRESTRIS

The devastating eelgrass wasting disease of 1934 and 1935 brought attention to the obscure protist *Labyrinthula*, which at the time was known only from a few nineteenth-century German publications. Muehlstein et al. (1988) demonstrated that *L. zosterae* was the causal agent responsible for the necrotic lesions on eelgrass and dieback disease symptoms observed in seagrass meadows by satisfying Koch's postulates. Recently, the identification of a species of *Labyrinthula* (*L. terrestris*) that causes rapid blight of turfgrass has stimulated interest in this unusual protist (Figure 9.5, see Case Study 9.2). Other species are not pathogens of grasses and exist primarily as saprotrophs. In contrast to *P. brassicae* and *S. macrosporum*, most species of *Labyrinthula* can be easily grown on nutrient medium in culture, and none of them appear to be biotrophs. In culture, labyrinthulids are most easily grown with yeast or bacterium coinoculated on the nutrient medium as a food organism (Porter, 1990; Figure 9.6).

LABYRINTHULA SPECIES

Species of *Labyrinthula* are known primarily from their trophic (or feeding) stage. This stage, which can be observed on nutrient medium in a Petri dish, is composed of a colony of cells within a network of interconnecting filaments (called the ectoplasmic network; Figure 9.7). The cells are spindle shaped and move with a gliding motion within the ectoplasmic network with speeds as rapid as 150 μm/min. The network plays an important role in the biology of the *Labyrinthula* colony by the following: (1) providing a structure through which the cells move, (2) aiding in the attachment of the colony to the substrate, (3) housing digestive enzymes necessary for feeding by the cells, and (4) serving as a conduit for transmitting signals within the colony to coordinate communal movement (Porter, 1990).

In *L. terrestris*, the trophic colony is the only stage of the life history that has been observed. In other closely related species, such as *L. vitellina*, a sexual life cycle

FIGURE 9.5 (a) Healthy uninfected *Poa annua* (turfgrass) and (b) *Labyrinthula terrestris*-infected turfgrass on the right showing characteristic symptoms of rapid blight. (Courtesy of Julia Kerrigan.)

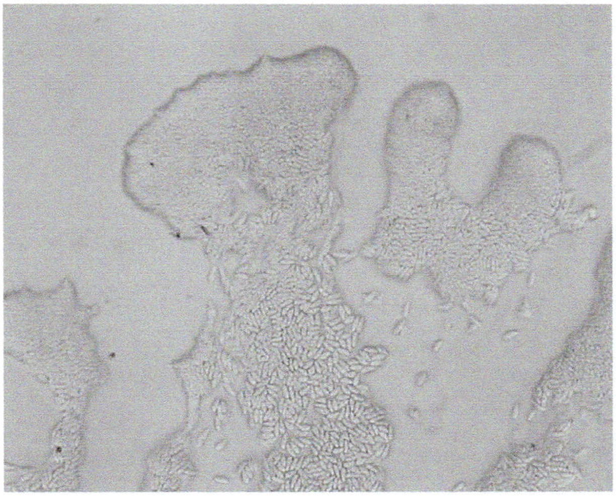

FIGURE 9.6 *Labyrinthula terrestris* from Arizona Turf Grass (*Poa annua*) growing as a pure culture on SSA. (Courtesy of D. Porter.)

CASE STUDY 9.2

Rapid Blight of Cool Season Turfgrasses—An Unexpected Pathogen

- Rapid blight is a disease of several cool season turfgrasses and a major concern to golf course managers using low-quality irrigation water in the southeast and western United States and United Kingdom.
- The symptoms include a patchy water-soaked appearance and browning followed by chlorosis and death of roughstalk bluegrass (*Poa trivialis*), annual bluegrass (*Poa annua*), perennial ryegrass (*Lolium perenne*), colonial bentgrass (*Agrostis tenuis*), and creeping bentgrass (*Agrostis stolonifera*) on golf courses during autumn and spring.
- Rapid blight is an emerging disease that was first noted in the late 1990s. The disease can quickly cause severe damage on high amenity turf and is costly and challenging to manage.
- An endobiotic chytrid fungus was originally thought to cause rapid blight, but in 2005 *Labyrinthula terrestris*, a marine net slime organism, was identified as the causal agent of this disease (Figure 9.11).
- Other species of *Labyrinthula* grow in estuarine and marine habitats where some are known to cause diseases of seagrasses. Prior to the description of *L. terrestris* on turfgrasses, none were known to be pathogens of terrestrial plants. The key link was that all affected turf areas were being irrigated with high salinity irrigation water (>2.0 dS/m).
- An obvious question: how has a pathogen, whose known relatives are inhabitants of marine habitats, become a serious and emerging disease problem on turfgrasses? How did it become established in land-locked states like Arizona, Nevada, and Utah?
- The answer is not known, but turfgrass management practices may be important in selecting for this emergent pathogen.
- Rapid blight is very challenging to manage with fungicides, and only a few are known to be effective. Fungicide use that did not suppress *L. terrestris* could have possibly reduced microbial competition in the turfgrass ecosystem.
- Golf courses are increasingly using effluent water for irrigation, which often has higher salinity.
- Frequent and close mowing allows for the entry of the pathogen into grass leaves.
- Rapid blight is an excellent reason for plant pathologists to expect the unexpected and to be broadly trained (see infected turfgrass in Figure 9.4).

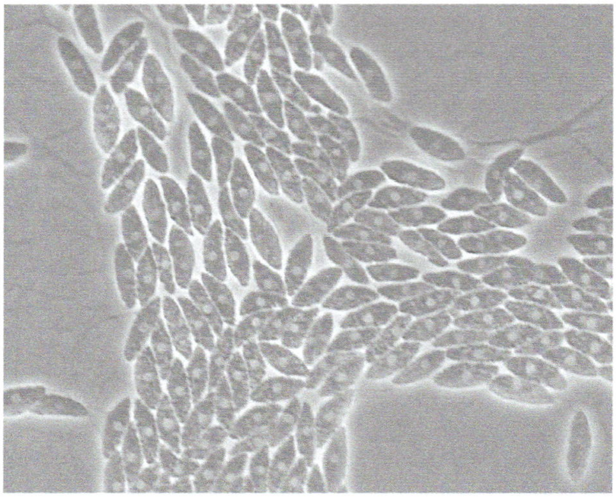

FIGURE 9.7 *Labyrinthula terrestris* on SSA. (Courtesy of D. Porter.)

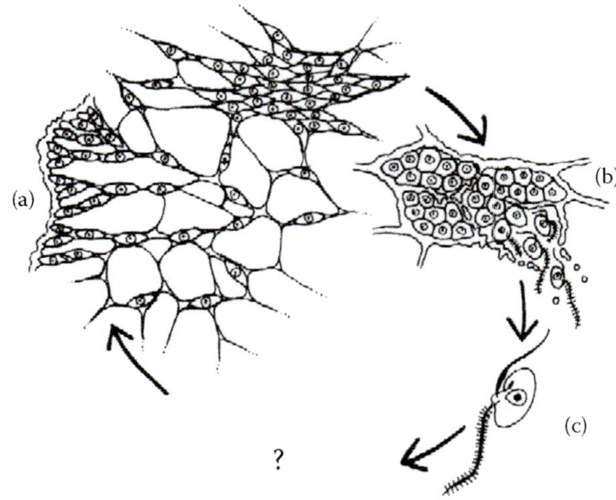

FIGURE 9.8 Life cycle of *Labyrinthula vitellina* (the sexual phase of the life cycle of *L. zosterae* has not been observed). (a) Colony of trophic cells within an ectoplasmic network; (b) sorus within which meiosis occurs and haploid zoospores (meiospores) are released; and (c) biflagellate zoospore (which possibly develops into new spindle-shaped trophic cells, although the developmental process is not known). (From Porter, D., *Handbook of Protoctista*, Jones and Bartlett, Boston, 1990. With permission.)

was observed and characterized by biflagellate, heterokont zoospores produced by meiosis (Porter, 1990). However, the fusion of gametes (syngamy) has never been observed. The life cycle of *L. vitellina* is presented in Figure 9.8 (Porter, 1990).

LABORATORY EXERCISES

Plant diseases have been observed and recorded by humans for more than 2000 years (Agrios, 2005). Many plant diseases were initially described based on the observation of visible signs (vegetative reproductive structures of the pathogen) and symptoms (reactions of the plant to infection) on fruits, leaves, roots, and stems. Since then, scientists in the discipline of plant pathology have continued to investigate the causal role that microorganisms play in plant disease and how their biology, ecology, and genomes influence pathogenesis (i.e., the disease-causing ability of an organism). The majority of scientific studies conducted by plant pathologists have focused primarily on plant species of economic importance to agriculture and fostered the development of experimental methods to examine plant pathogens and their associated diseases. In general, most agricultural crops are subject to many diseases caused by a wide array of plant pathogens. Some of these pathogens have a narrow host range (host specialists) and can infect on a single species or a variety of plant, whereas other pathogens have the ability to infect a wider range of plants (host generalists) often in genetically different families. The incidence and severity of disease can also vary depending on the environmental conditions and genetic composition of the pathogen and plant.

In the following laboratory exercises, the biology of *P. brassicae*, a well-studied pathogen of crucifers in agricultural production systems (Sherf and MacNab, 1986), *S. macrosporum*, an important pathogen of plants in natural ecosystems, and *L. terrestris*, a pathogen of high maintenance turf grasses, will be examined. The first two of these organisms produce motile spores (zoospores) that are an important component of their disease ecology (Agrios, 2005; Fuller and Jaworski, 1987; Karling, 1960). *Labyrinthula terrestris* is not known to produce zoospores, but appears to be transmitted by contact with injured plant leaves. Differences in the life cycle, feeding behavior, and symptom expression of each organism on specific host plants will provide the basis for determining differences in host susceptibility, and how these organisms cause plant disease. In each laboratory exercise, students will examine the infected plant material to familiarize themselves with the signs of each organism and the symptoms of the diseased plants. Various extraction and artificial inoculation methods (depending on the organism) will be employed to monitor and record disease development. Because *L. zosterae* can be readily isolated in pure culture on nutrient medium, the students will initially isolate this organism from infected grasses and then reisolate the organism from plants they have artificially inoculated in order to fulfill Koch's postulates and offer "proof of pathogenicity" (Chapter 1). The biotrophic feeding nature of *P. brassicae* and *S. macrosporum* that prevents their culturing on a nutrient medium also provides a unique opportunity for students to understand how Koch's postulates are modified to examine the disease-causing activities of these biotrophic plant pathogens.

Experiment 1. Susceptibility of Crucifers to *Plasmodiophora brassicae*

The selection of species or varieties of crucifers with reduced susceptibility to *P. brassicae* can often be used to manage clubroot disease. However, because of the inherent genetic diversity that exists in field populations of *P. brassicae*, no variety of plant is likely to be resistant to all genetic individuals of *P. brassicae*. In this laboratory experiment, a modification of the procedures developed by Williams (1966) and Castlebury et al. (1994) for the isolation of resting spores and plant inoculation will be employed to examine the susceptibility of different crucifers to infection by *P. brassicae*. Each crucifer host will be critically examined for the incidence and severity of disease symptoms and compared with noninoculated plants (control) (Figures 9.9 and 9.10). Students will also have an opportunity to examine the characteristic microscopic structures of *P. brassicae* in infected plant tissue (Figure 9.11).

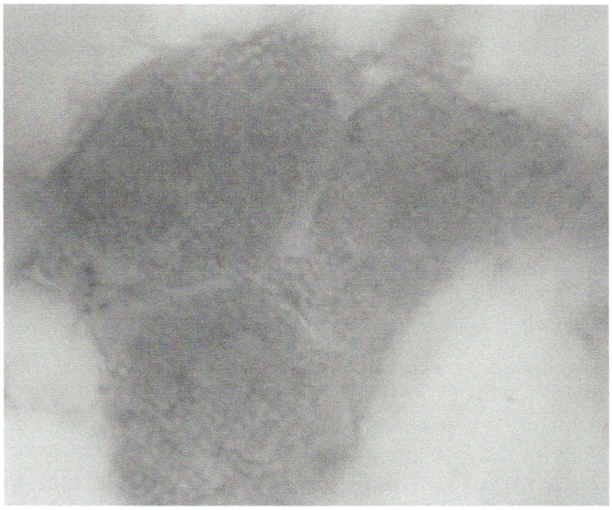

FIGURE 9.11 Resting spores of *Plasmodiophora brassicae* in root cells of Chinese cabbage. (Courtesy of Lisa A. Castlebury).

Materials

Each student or team of students will require the following laboratory items:

- Beakers (50 and 100 mL)
- Blender

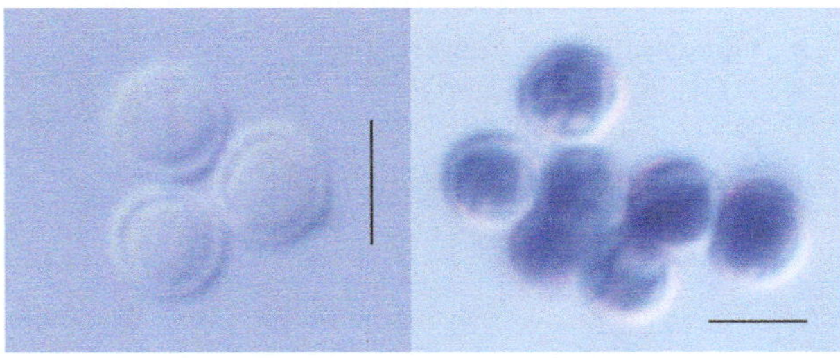

(a) (b)

FIGURE 9.9 (a) Unstained and (b) stained resting spores of *Plasmodiophora brassicae*. Bar represents 4 µm. (Courtesy of Lisa A. Castlebury.)

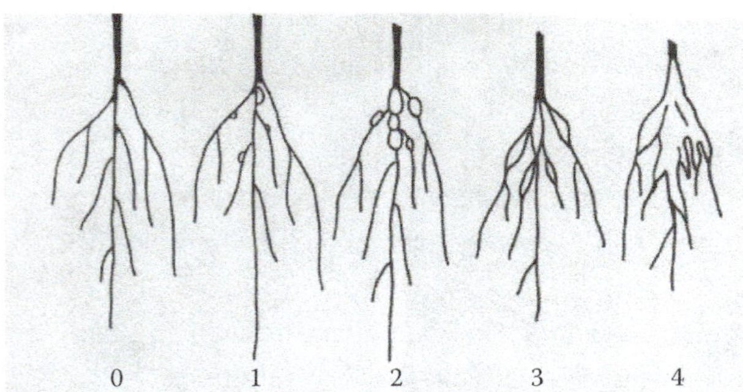

FIGURE 9.10 Clubroot disease rating scale (0–4) for crucifer seedlings. (Modified by L.J. Gray, from Williams, P.H., *Phytopathology*, 56, 624–626, 1966. With permission.)

- Centrifuge (tabletop or swinging bucket)
- Centrifuge tubes
- Cheesecloth
- Compound light microscope
- Dissecting microscope
- Distilled water
- Erlenmeyer flasks or Pyrex round media storage bottles with screw caps (500 and 1000 mL)
- Forceps
- Funnel
- Glass slide and coverslips
- Graduated cylinder (1000 mL)
- Hemacytometer
- Infected plant material—100–500 g fresh weight of infected roots from a local county agricultural extension agent or plant pathologist or student collection of infected roots from crucifers grown in a home garden or a commercial field with a history of clubroot disease (see Figure 9.3 for an infected root). Infected plant material should be placed in a paper bag and transported to the laboratory in a cooler for processing, and any infected roots not used for the laboratory exercise can be stored for 2–3 years in a nondefrosting freezer at −20°C.
- Mortar and pestle
- Paper cups (6–8 cm deep)
- Pasteur pipettes with rubber bulb
- Plastic Petri plates (9 cm)
- Plastic stakes and trays
- Potting soil (peat moss and vermiculite, 1:1, v/v)
- Seeds—obtain small packages from local garden center, Asian market, or seed company

and try to obtain as many crucifers as possible. Have each student or group of students select one or two species of plants for the inoculation experiments.
- Staining solution (0.005% cotton blue in 50% acetic acid)
- Stir plate and stir bars
- Whatman #1 filter paper (9 cm)
- Wooden Petri dish holder

Follow the instructions in Procedures 9.1 through 9.3 to complete the experiment on *P. brassicae*.

Anticipated Results

Crucifers will vary in their susceptibility to infection by *P. brassicae*, and galls of various sizes will be produced on infected roots. Galls should not be observed on roots of crucifer plants in the control treatment. Resting spores should be readily observed inside of root cells with the microscope after sectioning and staining.

Questions

- Why is it important to include a noninoculated control and replicates of each crucifer crop in the experiment?
- What was the response of each crucifer species and variety to infection? Were there differences in disease incidence and severity?
- What are some differences in morphology of infected and healthy roots based on macroscopic and microscopic examination?

Experiment 2. Characterization and Comparison of Plant Infection by the Gall-Inducing Chytrid Fungus *Synchytrium macrosporum*

The plant pathogen *S. macrosporum* has the widest host range of any known biotrophic fungus (Karling, 1960). The organism can infect more than 1400 different species of plants, most of which grow in natural rather than agricultural ecosystems. Plants infected with *S. macrosporum* can also exhibit considerable variation in symptom expression. In this laboratory exercise, students will examine diseased chickweed plants for resting spores, sporangia, and sori of *S. macrosporum* and will observe the unique swimming pattern of zoospores associated with this organism. Subsequent experiments will be conducted to compare and contrast the disease symptoms of asparagus, bean, corn, and turnip artificially inoculated with *S. macrosporum*. The inability to culture *S. macrosporum* on nutrient medium also requires the use of a modified method for isolation of resting spores and inoculation of plants.

Materials

Each student or team of students will require the following laboratory items:

- Commercial peat-based soil mix—well drained!
- Commercial soluble fertilizer
- Compound light microscope
- Dissecting microscope
- Dissecting needles
- Glass slides and coverslips

Procedure 9.1

Germination of Mustard Family Seeds

Step	Instructions and Comments
1	Prepare five plates each of the following seeds: broccoli (*Brassica oleracea* L. var. *italica Plenk*), cabbage (*B. oleracea* var. *capitata* L.), canola (*Brassica napus* L.), cauliflower (*B. oleracea* var. *botrytis* L.), Chinese cabbage (*Brassica pekinensis* [Lour.] Rupr.), collard (*B. oleracea* L. var. *acephala* DC), kale (*B. oleracea* L. var. *acephala* DC), mustard (*Brassica nigra* L.), radish (*Raphanus sativus* L.), rutabaga (*B. napus* L.), and turnip (*B. rapa* L. [=*B. campestris* L.]).
2	Place a piece of Whatman #1 filter paper into a plastic Petri plate of 9 cm diameter and moisten with distilled water. Arrange seeds (30–50 per plate) on filter paper 1- to 2-cm apart and gently press each seed into the filter paper with a pair of forceps. Offset seeds in each row to allow roots to grow straight and not become entangled.
3	Incubate seeds at 20°C–25°C with 12 h of supplemental lighting from two cool watt fluorescent bulbs placed 10 cm above plates (250 μmol s^{-1} m^{-2}). After seeds have germinated (usually 24–48 h), carefully place Petri plates on wooden holder at a 45° angle (Figure 9.12).
4	Check plates daily to ensure that the filter paper remains moist. If the filter paper begins to dry out, remove lid and add 1–3 mL of distilled water at the base of the plate.

Procedure 9.2

Isolation and Quantification of *Plasmodiophora brassicae* Resting Spores

Step	Instructions and Comments
1	To release resting spores of *P. brassicae* from roots, place 100 g of diseased roots (either fresh or frozen) in 400 mL of sterile distilled water, and macerate in a blender for 2 min at high speed.
2	Place a glass funnel with five layers of cheesecloth at the top of a 1000-mL Erlenmeyer flask and collect filtrate from blended solution. If cheesecloth becomes clogged with plant material and does not filter properly, remove the plant material and liquid from the cheesecloth, and repeat blending and filtering procedure.
3	Remove plant debris from cheesecloth and place it in a mortar with 5–10 mL of distilled water. Grind debris with a pestle for 1–2 min, filter the solution as described earlier, and combine with previously collected filtrate.
4	Place equal volumes of filtrate into centrifuge tubes (10- or 50-mL tubes depending on the size of the rotor) and adjust their weight by adding appropriate amounts of distilled water to each tube. Once centrifuge tubes have a similar weight and are balanced, place them in the rotor and centrifuge for 10–15 min at 2000 × *g* at room temperature.
5	Gently remove centrifuge tubes and place in a rack. Carefully remove the top, gray-colored fraction with a Pasteur pipette (this fraction contains the resting spores and will often appear above a whitish-colored layer in the middle of the tube) and place into another centrifuge tube. Repeat the centrifuge process as needed, particularly if the filtrate is cloudy or contaminated with excessive plant material. Place the collected filtrate containing the resting spores in a glass beaker with a stir bar for quantification.
6	Gently mix the collected filtrate using a stir plate to distribute the resting spores evenly. Pipette a small drop of the collected filtrate on a clean glass slide, add a coverslip, and examine with a microscope at 400× to 1000× magnification. Resting spores are round and approximately 4 µm in diameter (Figure 9.7). For better resolution of resting spores, add one drop of staining solution (0.005% cotton blue in 50% acetic acid) to the slide preparation prior to examination (Figure 9.7). If resting spores are observed, determine their concentration in the collected filtrate with the following procedure. Spores without cytoplasm are not viable and should not be counted.
7	Measure the volume of the collected filtrate in a graduated cylinder, pour the collected filtrate into a glass beaker, and gently swirl the mixture to distribute the resting spores evenly. Place one drop of spore solution at the edge of a coverslip on a hemacytometer counting chamber with a ruler and allow solution to be drawn into area between coverslip and glass slide. Let the slide sit for 2–3 min to allow spores to settle on the slide.
8	Examine the center of the slide and locate the area of the ruler with 25 cells (5 rows and 5 columns), consisting of 16 smaller squares each. Count the number of resting spores in each corner and center squares (5 squares total). Calculate mean number of spores per square, apply correction factor for the area (2.5 × 105) and determine the concentration of the resting spores per mL of the collected filtrate. For example: upper left square = 74; upper right square = 56; lower left square = 45; lower right square = 60; and center square = 80; (74 + 56 + 45 + 60 + 80)/5 = 63; 63 × (2.5 × 105) = 1.575 × 107 resting spores per mL of the collected filtrate.
9	After determining the concentration of the resting spores, adjust spore concentration to 1×10^7 spores per mL and pour equal amount of stock solution into 50-mL beakers. The number of beakers required is determined by the total number of plant species and the varieties to be inoculated. Freshly extracted resting spores are preferred for inoculation. However, the resting spore solutions may be stored for 3–5 days at 4°C or 3–4 months at –20°C prior to use.

- Kimwipes®
- Pasteur pipettes with rubber bulb
- 50-mL beakers
- 10-cm plastic Petri dishes
- Plastic bags with twist ties
- Plastic pots (10 or 15 cm diameter)
- Plastic or wooden stakes
- Seeds (asparagus, bean, corn, and turnip)

Procedure 9.3

Inoculation of Seedlings and Collection of Data

Step	Instructions and Comments
1	For each crucifer examined, fill 20 paper cups (6- to 8-cm deep) with a potting mixture consisting of 1 part peat moss and 1 part vermiculite. Poke several small holes at the bottom of the cup to allow for adequate drainage and water potting mixture until moist. Place one set of 10 cups into a plastic tray and repeat this process for a second set of cups (two replicates).
2	Remove 20 seedlings from Petri dishes and dip roots into quantified resting spore solution of *P. brassicae* for 10 s. Create a 5-cm deep hole with a pencil and transplant one seedling into each individual paper cup. In a separate set of 20 cups, also include seedlings dipped in sterile distilled water as a control. Incubate seedlings at 18°C–28°C for 6–8 weeks. During the first week of incubation, keep potting mixture saturated with water. Thereafter, water seedlings as needed and fertilize after 3 weeks (1 g of 20-20-20 fertilizer in 3.78 L of water).
3	After incubation, remove the inoculated seedlings and gently wash soil from roots in running tap water. Examine roots of each seedling for galls. Determine the percentage of infected seedlings based on gall symptoms (disease incidence) and severity of infection with the 0–4 rating scale (Figure 9.9). Follow the example presented in Table 9.1. Repeat the aforementioned procedure for each of two replicates of each plant species examined and then calculate a separate average for disease incidence and severity. Also, examine seedlings in each control treatment for the incidence and severity of clubroot disease.
4	Cut a thin section of healthy and diseased roots with a sharp scalpel or razor blade. Place the section in distilled water on a glass slide and observe cells for the presence of resting spores (Figure 9.10).
5	Cut the infected roots from the remaining seedlings, place in a plastic freezer bag, and store at –20°C for future laboratory exercises.
6	At the completion of the experiment, place all paper cups, plant materials, and soil in an autoclave for 1 h at 121°C.

FIGURE 9.12 Wooden apparatus for holding Petri dishes at a 45° angle. (Courtesy of Brian R. Cody.)

- Tape (double-sided)
- Toothpicks (wooden)
- Tween 80 (0.1% solution)
- 9-cm Whatman #1 filter paper

Follow the instructions in Procedure 9.4 to observe *S. macrosporum* resting spores, galls, sori, and zoospores. This will prepare you to identify *S. macrosporum*-infected chickweed and zoospores, both of which are key to completing this experiment. After Procedure 9.4, you need to complete Procedure 9.5 in order to obtain the necessary *S. macrosporum* inoculum (zoospores) required for Procedure 9.6. In addition, before starting Procedure 9.6, you need to prepare the host plants for inoculation by allowing the seeds to germinate. Use the following list of germination times: bean (*Phaseolus vulgaris*) 6–10 days; corn (*Zea mays*) 5–7 days; turnip (*Brassica rapa*) 7–10 days, and asparagus (*Asparagus officinalis*) 14–21 days. Asparagus seeds should be soaked in warm water for 48 h prior to planting. Be sure to replenish warm water when it is cooled down at least two or three times during the 48-h soaking period. Asparagus seeds germinate slowly and should be planted 2–3 weeks before planting bean, corn, and turnip seeds. This will ensure that all species of plants can be inoculated with *S. macrosporum* at the appropriate growth stage. Also, note that when growing the host plants, the peat soil mix should be kept moist but not saturated with water. Host plants can be grown in a greenhouse or in the laboratory at room temperature with supplemental fluorescent lighting photoperiod and 0.3 m from the light source. When plants are 21-days-old, fertilize once a week with a 1/2 recommended rate of a complete

TABLE 9.1
Example Data Sheet to Calculate Disease Incidence and Severity

Crucifer tested	Number of healthy plants with rating = 0	Number of diseased plants with rating = 1	Number of diseased plants with rating = 2	Number of diseased plants with rating = 3	Number of diseased plants with rating = 4
Cabbage	2	1	3	2	2

Notes: Disease incidence = (number of diseased plants/total number of plants) × 100 = (1 + 3 + 2 + 2)/(2+1 + 3 + 2 + 2) × 100 = 80%.

Disease severity = (number of diseased plants in each rating category × correction factor associated with each disease rating category)/total number of plants = [(0 × 2) + (1 × 1) + (3 × 2) + (2 × 3) + (2 × 4)]/(2 + 1 + 3 + 2 + 2) = 2.1.

Procedure 9.4
Squash Mount Procedure

Step	Instructions and Comments
1	Collect several chickweed (*Stellaria media*) plants infected with *S. macrosporum*. Diseased plants will have galls with dark amber to reddish-brown resting spores or bright orange-yellow sori from germinated resting spores on aboveground stems (Figure 9.10a and b). Infected chickweed plants can be collected from February to March in the Southeastern United States (e.g., Alabama and Georgia). This may change into a later date for areas farther north depending on when chickweed seeds germinate.
2	Place a piece of infected chickweed tissue in a clean Petri dish, secure with double-sided tape and then affix the Petri dish on the stage of a dissecting microscope with double-sided tape.
3	Add six separated drops of sterile double-distilled water on an inverted Petri dish lid and place the lid next to the dissecting scope. Flame sterilize the tips of two dissecting needles, cool for 10 s, and carefully remove orange sori and resting spores from tissue.
4	Rinse sori and resting spores by serially running them through the six drops of sterile water before placing on a clean glass slide in a small drop of sterile water.
5	Carefully place a clean coverslip over the drop of water and gently tap it with the eraser end of a pencil to break open the sori and release the individual sporangia. Examine the prepared slide with a compound light microscope.
6	Place slide in a Petri dish with a moist Kimwipe® folded into the bottom. Take a small wooden toothpick, break in half, and use the two halves to hold the slide above the moistened Kimwipe®.
7	Examine the slide after 1 h and again after every 2 h. Sporangia from mature sori are usually ready to release zoospores after 1–2 h. Moisten the Kimwipe® periodically by adding a small amount of water to ensure that the slide does not dry completely.

soluble fertilizer. Plant six seeds of each host plant in a 10- or 15-cm diameter pot and thin to three plants after seedlings have emerged from the soil. Because *S. macrosporum* is not available commercially, cultures for use in the laboratory exercise will be prepared from infected chickweed plants collected from the field. Infected plant tissue can be dried and stored at room temperature in a low-humidity environment to provide a source of viable resting spores for future experiments. Once you have the required inoculum (zoospores), and you have prepared the host plants, then carry out Procedure 9.6 to complete Experiment 2. The inoculation procedure is a modification of the method developed by Karling (1960).

Anticipated Results

After completing Procedure 9.4, students should observe the multicellular sheath produced from the host epidermal cells surrounding both sporangial and resting spore galls. The students should also observe zoospore discharge from individual sporangia, be able to distinguish different parts of the life cycle, and observe the "jerky" swimming pattern

Procedure 9.5

Preparation of Inoculum of *Synchytrium macrosporum*

Step	Instructions and Comments
1	Soak fresh or dried leaves of chickweed infected with *S. macrosporum* in sterile distilled water for 5–7 days to soften tissues around prosori with resting spores of the fungus.
2	With dissecting needles and a dissecting microscope, tease prosori from tissue and rinse gently in distilled water.
3	Transfer prosori to a Petri plate lined with two layers of moistened, Whatman #1 filter paper and incubate in the laboratory for 2–3 weeks or until prosori germinate and sori mature.
4	Transfer 25–50 mature sori individually with a fine dissecting needle into a drop of distilled water on a clean glass slide.
5	Cover the suspension of sori with a coverslip and press gently with a pencil eraser to break open the sori to release zoosporangia.
6	Zoospores are released from zoosporangia within 1–2 h and can be collected by gently washing the slide with distilled water into a 10-mL beaker 2–2.5 h after zoospores have formed.

Procedure 9.6

Inoculation of plants with *Synchytrium macrosporum*

Step	Instructions and Comments
1	Swab the emerging leaves of each seedling with Tween 80 (0.1% v/v) and rinse with sterile distilled water. This procedure provides a wet surface for zoospores to swim on the leaf. The Tween solution should be applied to leaves of similar age 1–2 weeks prior to conducting the experiment to determine whether it is toxic to each species of plant. If phytotoxicity is observed, dilute the Tween solution with water to a concentration that does not damage the leaves.
2	Dilute the zoospore mount from Steps 5 and 6 of the Procedure 12.5 with 5 mL of sterile distilled water and place a drop of the zoospore mount onto the treated emerging leaf. Inoculate nine plants (three plants in three separate pots). Also, include another pot of three seedlings on which a few drops of sterile distilled water are placed on emerging leaves (control treatment).
3	Place a bell jar or plastic container over the seedling to maintain high humidity. If seedlings are too large to cover, then place a wet pad of absorbent cotton around the inoculated leaf. For corn and asparagus, several pots can be covered with a plastic dome or large plastic bag with a twist tie.
4	Each seedling should be inoculated once a day for six consecutive days.
5	Once the inoculations are completed, remove the bell jar or plastic container, bag, or dome and maintain plants in a greenhouse or laboratory for 3–4 weeks after inoculation. Galls should be apparent 2 weeks after infection and will mature in 2–4 months.

of the zoospores, which is typical for chytrid fungi. After carrying out Procedure 9.6, students should see a difference in infection levels between the different host plants: *S. macrosporum* should not infect asparagus, but will usually cause moderate to severe infection of bean, corn, and turnip. The type, size, and structure of galls produced by *S. macrosporum* will vary on chickweed, corn, bean, and turnip.

Questions/Activities

- Draw an infected chickweed plant and label the *S. macrosporum* galls. Try to determine the size, shape, and color of the galls. Why might it be important to note such features of the infection? What kind of damage do the galls cause to the plant?

- Draw resting spores, sori, and sporangia observed with a light compound microscope. Try to determine the color, shape, and size of each structure. Do you see any unique features for each? Why might it be important to note such features?
- What kind of swimming pattern do the zoospores have and how are they released from the sporangia? Can you see how many flagella each zoospore has? Why might this be important to know?
- Using a life cycle diagram for *S. macrosporum*, label the structures observed (Figure 9.4). Were you able to observe a complete life cycle? Why is it important when diagnosing disease to know the complete life cycle of a pathogen?
- What differences can you note in the appearance of galls on asparagus, bean, corn, and turnip? Are these galls similar in appearance to the galls observed on chickweed?
- Do you think these differences are attributable to the plant, the fungus, or both? Explain your answer.
- Complete the data sheet in Figure 9.13 Answer yes or no in the "Infected" column. Put "YES" in the appropriate box corresponding to each species of plant if galls, resting spores, and/or prosori are present on the plant tissue and "NO" if no symptoms (i.e., galls) and/or signs of the fungus are evident. In the "Degree of Infection" column, indicate the severity of disease with the following scale; sparse = 2–10 galls per plant, moderate = 10–20 galls per plant, heavy = 20–100 galls per plant, and severe ≥100 galls per plant. In the "Type of Gall" column, indicate whether galls were single-celled (e.g., simple) or multi-celled (e.g., composite). If the degree of infection was severe, then place the term "Confluent" in the type of gall column. Confluent galls occur when several epidermal cells adjacent to one another become infected and the individual sheaths of enlarged host cells that form around the developing gall merge together (Karling, 1960).
- Summarize the results of your observations in Figure 9.13 What can you conclude from these observations?
- Karling advocated the use of host range for the identification of species of *Synchytrium*. What modern techniques could be used to aid in the identification of a fungus rather than conducting a host range study? Do you think host range studies are still important? Explain your answer.

Experiment 3. Observations of Eelgrass Wasting Disease

Labyrinthula zosterae is an important pathogen of seagrass in estuarine environments. The organism can be grown in pure culture by using yeast cells to supplement their nutrition. In this laboratory exercise, students will observe eelgrass (*Zostera marina*) wasting disease and isolate *L. zosterae* from diseased eelgrass exhibiting typical symptoms. To demonstrate that *L. zosterae* is the causal agent of eelgrass wasting disease, students will follow Koch's postulates to establish "proof of pathogenicity."

Materials

Each student or team of students will require the following laboratory items:

- Agar (Difco)
- Air pump and tubing
- Alcohol lamp
- Antibiotics (ampicillin, penicillin G, and streptomycin sulfate)
- Artificial or natural seawater—20–30 L for the SSA+ medium if being used

For Procedure 9.8

- Buchner funnel with rubber cork
- 25-L Carboy
- Clorox (sodium hypochlorite, NaOCl)
- Compound light microscope
- Culture of yeast (any nonfilamentous, nonmucoid yeast—preferably one cultured from nonsurface disinfested and decaying eelgrass leaves)

Species	Infected	Degree of Infection	Type of Gall
Asparagus officinalis			
Brassica rapa			
Phaseolus vulgaris			
Zea mays			

FIGURE 9.13 Datasheet for Recording Susceptibility of Asparagus, Bean, Corn, and Turnip to *Synchytrium macrosporum*.

- Dissecting microscope
- Dissecting needles and scissors
- Eelgrass leaves (*Z. marina*)—collect fresh leaves from drift near seagrass meadows at lowest tides (easiest to wade or swim to). Make sure to check with the local authorities before collecting as most seagrass beds are protected by law. Can be found along both the North American Atlantic coast (Labrador to North Carolina) and the North American Pacific Coast (Alaska to Baja California).
- 2-L Erlenmeyer flasks with cotton stoppers and glass tube for aeration
- Ethanol (80% EtOH)
- Filters (≤0.4 μm pore diameter)
- 2-L sidearm flasks
- Forceps
- Germanium dioxide (GeO_2)
- 5-mL glass pipettes
- Glass slides and coverslips
- Paper towels (sterile)
- 4.5 and 10-cm-diameter plastic Petri dishes
- 50-mL plastic screw top tubes
- 3.78-L plastic Ziploc bags
- Scalpel
- Serum seawater plus medium (SSA+)—can be made according to Procedure 9.8 or a prepackaged mix can be obtained commercially with the correct chemistry and made from distilled water and Instant Ocean.
- Sterile distilled water
- Tygon tubing
- Vacuum pump with rubber tubing
- 9-cm Whatman #1 filter paper

Follow the instructions in Procedures 9.7 through 9.11 to complete the experiment on *L. zosterae*.

Anticipated Results

Wasting disease should be readily observed on eelgrass, and *L. zosterae* should be easily cultured from infected eelgrass leaves on SSA+ isolation medium. Eelgrass plants inoculated with *L. zosterae* will produce typical wasting disease symptoms of leaves, followed by isolation and observation of the microorganism from diseased leaves on SSA+ isolation medium. No symptoms should be observed, and *L. zosterae* should not be isolated from eelgrass plants in the control treatment.

Questions

- Why is it necessary to clip a noninoculated, disinfested leaf piece to a green eelgrass plant as an experimental control?

Procedure 9.7

Preparation and Filtration of Seawater from Natural Sources

Step	Instructions and Comments
1	Place a Buchner funnel containing two pieces of Whatman #1 filter paper into a 2-L side-arm flask connected to a vacuum source with flexible rubber tubing.
2	Once the vacuum has been established, slowly pour the seawater into the Buchner funnel and proceed until the flask is full.
3	Filtered seawater should be stored at 4°C–10°C in clean plastic carboys and covered with black plastic. For the production of sterile seawater, use a filtration system (0.4 μm pore diameter) or autoclave for 20 min at 121°C.

Procedure 9.8

Preparation of Serum Seawater Plus (SSA+) Isolation Medium

Step	Instructions and Comments
1	Add 12 g of agar and 3 mg of germanium dioxide (a diatom inhibitor) to 1 L of filtered seawater (Procedure 9.7). Autoclave the medium for 20 min at 121°C.
2	After the medium has cooled to 50°C, add 4–10 mL of sterile horse serum (1% v/v, BBL or Gibco) and 250 mg each of the antibiotics ampicillin, penicillin G, and streptomycin sulfate. Gently swirl to mix the medium and pour into 10-cm plastic Petri dishes. Caution: All of the SSA+ medium should poured immediately into Petri dishes because reheating (remelting) of this medium after it has solidified will cause the horse serum protein to coagulate and render it useless as an isolation medium.

Procedure 9.9

Collecting Healthy and Diseased Samples of *Zostera marina*

Step	Instructions and Comments
1	One week before beginning the isolation procedure, streak several plates of SSA+ medium with a culture of yeast (see the list of materials provided previously). The yeast will serve as a food source for *Labyrinthula zosterae*. Monthly transfer on SSA without antibiotics can maintain monoxenic cultures of *L. zosterae* and yeast.
2	Collect approximately 20 healthy plants (without disease symptoms manifested by the appearance of blackened or dead necrotic areas present on leaves) with at least three inner (youngest) green leaves from the eelgrass bed.
3	While collecting, also include a portion of the rhizome and roots with each collected shoot. Place shoots in a 3.78-L plastic Ziploc® bag with a paper towel dampened with seawater.
4	While transporting, keep the plants cool, but not directly on ice.
5	After collecting healthy plants, carefully examine plants in the eelgrass beds for wasting disease symptoms. Plants infected with *L. zosterae* will have black streaks and patches of necrotic tissue (lesions) on older leaves (Figure 9.7).
6	Collect approximately 30 leaves with disease symptoms by selecting leaves with some areas of healthy green tissue adjacent to the necrotic lesions. Place the diseased leaves into a new 3.78-L plastic Ziploc® bag, and keep it moist and cool until ready to isolate the pathogen as described earlier.
7	If you are unable to collect healthy and diseased eelgrass plants, request them from a colleague in a coastal area. Eelgrass plants can be shipped overnight in an insulated container and successfully used for this laboratory experiment.

Procedure 9.10

Isolation of *Labyrinthula terrestris* from Eelgrass (*Zostera marina*)

Step	Instructions and Comments
1	Dip forceps and scalpel into a 50-mL tube containing 70% ethanol and flame to disinfest.
2	Cut eelgrass leaves into small pieces (5–10 mm^2) using the disinfested forceps and scalpel. Use leaf pieces from the edge of the blackened, necrotic lesions where *L. zosterae* is likely to be most active.
3	Place cut leaf pieces to a sterile 4.5-cm Petri dish, add 0.5% sodium hypochlorite to cover leaf pieces and gently swirl them for 1 min. Aseptically transfer each disinfested leaf piece to new 4.5-cm Petri dish, add sterile distilled water and gently swirl for 2 min.
4	Repeat the rinsing process with sterile, filtered seawater. Transfer leaf pieces to a sterile paper towel or filter paper to remove excess water and place four to five disinfested pieces of eelgrass leaf tissue on a Petri dish of 10 cm diameter containing SSA+ medium.
5	Observe each SSA+ plate daily for the growth of *L. zosterae* from each piece of eelgrass tissue (Figure 9.11).
6	Continue to observe plates for at least 1 week or until they become overgrown by bacteria and/or fungi, making observation and isolation of *Labyrinthula* difficult.
7	When an appropriate *Labyrinthula* colony is located and ready for transfer, first streak a clean dish of SSA+ with a small amount of yeast from an actively growing culture.
8	Then transfer a portion of the actively growing *Labyrinthula* colony to the yeast streak on the new Petri dish of SSA+.

- Low salinity has been reported to inhibit wasting disease. How could you test this hypothesis with the inoculation apparatus set up for this laboratory experiment?
- How could you determine whether necrotic lesions on leaves of other seagrasses are caused by *L. zosterae* or other plant pathogenic microorganism(s)?

Procedure 9.11

Demonstrating *Labyrinthula zosterae* as the Causal Agent of Eelgrass Wasting Disease Using Koch's Postulates

Step	Instructions and Comments
1	Fill 2-L Erlenmeyer flasks to the neck with artificial or natural sterilized seawater. Place a single healthy eelgrass shoot in each flask and weigh down with a short piece of rubber tubing slipped over their rhizome. Plug the flasks with cotton through which a glass tube extends to near the bottom of the flask (a sterile 5-mL pipette is a good substitute). Attach an air pump or airline with a cotton plug filter to the glass tube and adjust the rate to deliver about one bubble per second.
2	Place several 1-cm pieces of green eelgrass leaf in distilled water and autoclave for 20 min at 121°C. Place some of these sterilized leaf pieces in cultures of *Labyrinthula* isolated by using Procedure 12.10. Place them on the agar surface adjacent to the spreading colonies of *Labyrinthula* but not on the top of the yeast cells. Allow the *Labyrinthula* cells to grow into the leaf piece for 24–48 h.
3	Cut 0.5-cm pieces of thin Tygon® tubing and slit the tube wall along one radius to create a small clip to attach the inoculated leaf pieces to healthy eelgrass plants. Clips should be sterilized before use.
4	Remove a green shoot of eelgrass from a flask and place on a sterile paper towel. Pick up a piece of inoculated carrier leaf using sterile forceps and clip it to a green leaf of the eelgrass shoot with a Tygon® clip. Replace the inoculated shoot into the flask.
5	As an experimental control, clip a sterilized piece of leaf tissue to a green leaf of an eelgrass shoot in a separate flask.
6	Repeat the experimental and control inoculations for at least five flasks each.
7	Place the flasks on a lighted bench or greenhouse where the plants will receive at least 20% full sunlight.
8	Observe plants for wasting disease symptoms 1–7 days after inoculation (Figure 9.7).
9	Complete Koch's postulates by reisolating *L. zosterae* on SSA^+ medium from diseased leaves with necrotic lesions in the flasks.

ACKNOWLEDGMENTS

- Paul H. Williams, Emeritus Professor, University of Wisconsin, for his many useful suggestions and clubroot resistance screening protocol that served as a template for developing Laboratory Experiment 9.1
- Timothy James (University of Michigan) and Lisa Castlebury (USDA Systematic Botany and Mycology Laboratory) for presubmission review and L.C. for providing photographs of resting spores and diseased cabbage
- Jim Kerns and Gerald Lee Miller for presubmission review of the information presented on *L. terrestris*
- Celeste A. Leander for providing comments and suggestion for Laboratory Experiment 9.3
- Andy Tull (University of Georgia) for advice on growing plants for the laboratory exercises
- Lynnette J. Gray for illustrations
- Bryan R. Cody for digital images
- Pamela E. Puryear (North Carolina State University) for assistance in literature searches

REFERENCES

Agrios, G. N. 2005. *Plant Pathology*, 5th ed. New York, NY: Academic Press.

Archibald, J. M., and P. M. Keeling. 2004. Actin and ubiquitin protein sequences support a Cercozoan/Foraminiferan ancestry for the plasmodiophorid plant pathogens. *J. Eukaryot. Microbiol.* 51:113–118.

Bigelow, M., D. Olsen, and R. L. Gilbertson. 2005. *Labyrinthula terrestris* sp. nov., a new pathogen of turf grass. *Mycologia.* 97:165–190.

Blanton, R. L. 1990. Phylum Acrasea. Pages 75–87 in: *Handbook of Protoctista.* L. Margulis, J. O. Corliss, M. Melkonian and D. J. Chapman (eds.), Boston, MA: Jones and Bartlett.

Castlebury, L. A., J. V. Maddox, and D. A. Glawe. 1994. A technique for the extraction and purification of viable *Plasmodiophora brassicae* resting spores from host root tissue. *Mycologia.* 86:458–460.

Cubeta, M. A., B. R. Cody, and P. H. Williams. 1998. First report of *Plasmodiophora brassicae* on cabbage in Eastern North Carolina. *Plant Dis.* 81:129.

Dick, M. W. 2001. *Straminipilous Fungi.* Dordrecht, the Netherlands: Springer Science + Business Media.

Farr, D. F., G. F. Bills, G. P. Chamuris, and A. Y. Rossman. 1995. *Fungi on Plants and Plant Products in the United States.* St. Paul, MN: APS Press.

Fuller, M. S., and A. Jaworski, A. 1987. *Zoosporic Fungi in Teaching and Research.* Athens, GA: Southeastern Publishing Corp.

Hibbett, D. S., M. Binder, J. F. Bischoff, M. Blackwell, P. F. Cannon, O. E. Erikkson, S. Huhndorf, T. James, P. M. Kirk, P. Lücking, H. Thorsten Lumbsch, F. Lutzoni, P. B. Matheny, D. J. McLaughlin, M. J. Powell, S. Redhead, C. L. Schoch, J. W. Spatafora, J. A. Stalpers, R. Vilgalys, M. C. Amie, A. Aptroot, R. Bauer, D. Begerow, G. L. Benny, L. A. Castlebury, P. W. Crous, Y. Dai, W. Gams, D. M. Geiser, G. W. Griffith, C. Gueidan, D. L. Hawksworth, G. Hestmark, K. Hosaka, R. A. Humber, K. D Hyde, J. E. Ironside, U. Kõljalg, C. P. Kurtzman, K.-H. Larsson, R. Lichtwardt, J. Longcore, J. Mi⊠dlikowska, A. Miller, J.-M. Moncalvo, S. Mozley-Standridge, F. Oberwinkler, E. Parmasto, V. Schüßler, J. Sugiyama, R. G. Thorn, L. Tibell, W. A. Untereiner, C. Walker, Z. Wang, A. Weir, M. Weiss, M. M. White, K. Winka, Y.-J. Yao, and N. Zhang. 2007. A higher-level phylogenetic classification of the fungi. *Mycol. Res.* 111:509–547.

Honda, D., T. Yokochi, T. Nakahara, S. Raghukumar, A. Nakagiri, K. Schaumann, and T. Higashihara. 1999. Molecular phylogeny of labyrinthulids and thraustochytrids based on the sequencing of 18S ribosomal gene. *J. Eukaryot. Microbiol.* 46:637–647.

Karling, J. S. 1960. Inoculation experiments with *Synchytrium macrosporum. Sydowia.* 14:138–169.

Karling, J. S. 1964. *Synchytrium.* New York, NY: Academic Press.

Kirk, P. M., P. F. Cannon, J. C. David, and J. A. Stalpers (eds). 2001. *Ainsworth & Bisby's Dictionary of the Fungi.* Cambridge, UK: Cambridge University Press.

Leander, C. A., and D. Porter. 2001. The Labyrinthulomycota is comprised of three distinct lineages. *Mycologia.* 93:459–464.

Muehlstein, L. K., D. Porter, and F. T. Short. 1988. *Labyrinthula* sp., a marine slime mold, produces the symptoms of wasting disease in eelgrass, *Zostera marina. Marine Biol.* 99:465–472.

Porter, D. 1986. Mycoses of marine organisms: an overview of pathogenic fungi. Pages 141–151 in: *Biology of Marine Fungi*, S. T. Moss (ed). Cambridge: Cambridge University Press.

Porter, D. 1990. Phylum Labyrinthulomycota. Pages 388–398 in: *Handbook of Protoctista*, L. Margulis, J. O. Corliss, M. Melkonian, and D. J. Chapman (eds). Boston, MA: Jones and Bartlett.

Sherf, A. F., and A. A. MacNab. 1986. *Vegetable Diseases and their Control.* New York, NY: John Wiley & Sons.

Schaap, P., T. Winckler, M. Nelson, E. Alvarez-Curto, B. Elgie, H. Hagiwara, J. Cavendar, A. Milano-Curto, D. E. Rozen, T. Dingermann, R. Mutzel, and S. L. Balduf. 2006. Molecular phylogeny and evolution of social amoebas. *Science.* 314:661–663.

Siemens, J., M. Nagel, J. Ludwig-Mueller, and M. D. Sacristan. 2002. The interaction of *Plasmodiophora brassicae* and *Arabidopsis thaliana*: Parameters for disease quantification and screening of mutant lines. *J. Phytopath.* 150:592–605.

Sparrow, F. K. 1960. *Aquatic Phycomycetes.* Ann Arbor, MI: University of Michigan Press.

Spiegel, F. W., S. L. Stephenson, H. W. Keller, D. Moore, and J. C. Cavendar. 2004. Mycetozoans. Pages 547–576 in: *Biodiversity of Fungi, Inventory and Monitoring Methods.* G. M. Mueller, G. F. Bills, and M. S. Foster (eds). Amsterdam, the Netherlands: Elsevier.

Williams, P. H. 1966. A system for the determination of races of *Plasmodiophora brassicae* that infect cabbage and rutabaga. *Phytopathology.* 56:624–626.

10 Plant Pathogenic Zygomycetes

Kathie T. Hodge

Concept Box

- Relatively few zygomycetes are plant pathogens. Those that are, belong to the Mucormycotina, an early diverging subphylum of Fungi.
- Fungi in this group grow as a haploid **coenocytic** mycelium and produce haploid asexual spores in stalked **sporangia**. They also produce sexual spores called **zygospores** at the site of fusion of **gametangia** of opposite mating types.
- The genus *Rhizopus* includes key pathogenic species.
- Most are **necrotrophic** pathogens that cause diseases called wet or soft rots. They infect fruit and other plant parts, causing significant postharvest losses.
- They are typically soilborne and can persist in soil.
- Infected plants and fruits develop a coarsely hairy appearance from the abundant, stalked, asexual sporangia.

The fungi called zygomycetes are really a number of disparate groups with ancient origins that predate the ascomycetes and basidiomycetes. For a long time, we have been uncertain about the evolutionary relationships of these groups that arise near the base of the fungal tree of life. However, this murky picture is of little concern to plant pathologists, as all of the significant plant pathogenic zygomycetes belong to a single lineage, the fungal subphylum Mucoromycotina, order Mucorales.

Zygomycetes that do not belong to the Mucoromycotina fall into a few other major lineages. Their various lifestyles include parasitizing other soil microfauna like amoebae (e.g., order Zoopagales), growing as parasites in the guts of insects (e.g., orders Harpellales and Asellariales), colonizing animal dung (e.g., order Kickxellales), living in soil (e.g., *Mortierella*), and forging mycorrhizal partnerships with mosses (*Endogone* spp.). Arbuscular mycorrhizal fungi, which are all obligate symbionts of plant roots, were once classified among zygomycetes, but now boast their own distinct phylum, Glomeromycota. It is likely that other "zygomycete" groups will also be recognized as new phyla as we learn more about them.

The order Mucorales is a well-defined group of fungi with **coenocytic** hyphae. That is, the hyphae generally lack septa, although septa may be produced occasionally to delineate reproductive organs or wall off damaged areas or older mycelium. The nuclei within the hyphae are haploid. Mucoralean fungi reproduce asexually by producing **sporangia** containing one or often many sporangiospores. The globose sporangia are held aloft on long stalks called **sporangiophores**, which can be tall enough to be seen with the naked eye—a mass of them lends a colony a distinctly hairy appearance.

These fungi are typically **heterothallic**; each haploid strain is one of two alternate mating types. The sexual cycle involves mating between strains of compatible mating types via special inflated cells called **gametangia**. The gametangia fuse at their tips, merging their cytoplasm, and for a brief time they are **dikaryotic**, then diploid as the nuclei go through **karyogamy**. At the point of fusion, a globose cell or sporangium develops, in which a single, dark **zygospore** is formed (Figure 10.1). Inside the zygospore, meiosis occurs at some point prior to the germination of the zygospore via sporangium production. Zygospores are typically thick-walled and may serve as resistant structures during winter or drought. In mucoralean fungi, the alarmingly hairy patches of asexual sporangia on the host surface represent the dominant reproductive state, and the sexual stage is of minor importance in the disease cycle. However, genetic recombination during zygospore formation is likely key to the evolution of novel traits such as fungicide resistance or cold temperature tolerance.

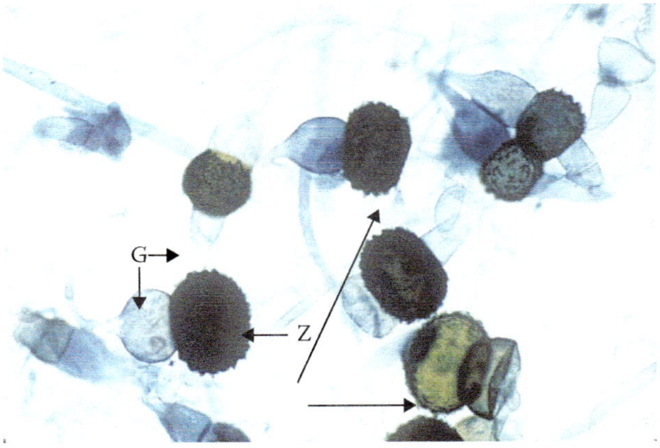

FIGURE 10.1 Dark, sexual zygospores (Z) of *Rhizopus* sp., formed at the point of fusion between two gametangia (G) of compatible mating types. (Courtesy of Curtis Clark; [CC-BY-SA-3.0 (http://creativecommons.org/licenses/by-sa/3.0)], via Wikimedia Commons; http://commons.wikimedia.org/wiki/File:Rhizopus_zygospores.jpg.)

ZYGOMYCETES AS PLANT PATHOGENS

Most members of Mucorales are saprobic, growing on dead plant parts, dung, or in soil. A few have plant pathogenic abilities. All economically significant plant pathogenic zygomycetes belong to the order Mucorales, in the fungal subphylum Mucoromycotina. These fungi have often been called "pin molds" because of their long-stalked fruiting bodies with rounded heads full of asexual spores.

The genus *Rhizopus* includes the most significant plant pathogens; a few others are minor players. *Rhizopus* species are amazing fungi. They include opportunistic human pathogens that need to be diagnosed and treated quickly because of their horrifying rate of growth through human flesh. One species is used to prepare tempeh, an Indonesian food made from cooked soybeans fermented by *Rhizopus oligosporus*. The common bread mold *Rhizopus stolonifer* can fill a Petri dish and burst out under the lid in just 4 or 5 days, and it often eats my strawberries in the refrigerator before I can.

Pathogenesis

Rhizopus and kin are **necrotrophic** pathogens that do not directly invade living plant cells. Some are important postharvest pathogens, particularly of fruits. They produce and excrete a small arsenal of enzymes that break down the **pectins** that make up the middle lamella of a plant's cell wall. This process causes plant cells to fall apart, and then the unrestrained turgor of the cell makes them **lyse**—this process gives the zygomycete rots a wet appearance. Only one species, *Rhizopus microsporus*, is known to produce a toxin that is a pathogenicity factor—its intriguing story is described in Case Study 10.1.

Invasion of a host is often through wounds, but under wet conditions, the fungi may be able to invade directly through the host cuticle. In some cases, fungi that infect the flowering heads may become seedborne and thus are vertically transmitted. Under permissive conditions, the fungus emerging from the seed after germination may cause damping-off of the seedling. When conditions are not permissive, the fungus may nevertheless be introduced to the field where the seeds are planted and persist in the soil.

The ability of *Rhizopus* species to degrade cell walls makes them useful in flax retting, in which flax fibers are separated from other stem materials by microbial fermentation. Some species see industrial use as sources for pectin-degrading enzymes. These fungi are not particularly adept at degrading cellulose, but they do swiftly degrade many plant sugars and polysaccharides, and under ideal conditions, they grow extremely rapidly.

Important Zygomycetes

Relatively few zygomycetes are plant pathogens. Most belong to the genus *Rhizopus*, named for its characteristic "root feet" (Figure 10.2). The genera *Choanephora, Gilbertella, Blakeslea*, and *Mucor* include a few other plant pathogens. The most significant of this small group is *R. stolonifer.* It can grow very swiftly when conditions are right, and it sporulates profusely on the plant surface. It can be impressively hairy (Figure 10.3).

Rhizopus and Mucor Rots

Wet rots of many different fruits, vegetables, and ornamentals are caused by a number of zygomycetes. Among them, *R. stolonifer* is the main cause, but other species of *Rhizopus* and *Mucor* can cause similar symptoms. These fungi are ubiquitous in soil and may sporulate profusely on decayed fruits on the soil surface.

Wounds may predispose fruits to infection, but unwounded plants can also be susceptible when conditions are wet. Fruit flies and other insects may spread spores among ripe fruits, and birds can distribute spores, particularly on confectionary sunflowers, which birds love even more than people do.

Rhizopus rots affect over 300 kinds of fruit, flowers, and vegetables. In strawberries, it is often called "strawberry leak," because the affected fruits bleed copious amounts of red juice as they rapidly decay (Figure 10.3). Other victims include jackfruit, cherries, flowers, seeds (sunflower and amaranth), and vegetables like summer squash and sweet potato. On ornamentals such as poinsettia, Rhizopus rot is infrequent, but can be devastating when conditions are right (Figure 10.4). Wet conditions with little airflow promote infection, and the disease may develop rapidly when ornamentals are packed in plastic sleeves for shipment.

CASE STUDY 10.1

Rhizopus microsporus and Its Symbiont-Derived Toxins

- Rhizopus seedling blight of rice, caused by the fungus *R. microsporus*, is an important disease of rice seedlings in Asia.
- This fungus, unique among zygomycetes, produces a toxin that acts as a pathogenicity factor, **rhizoxin** (Figure 10.5).
- Rhizoxin inhibits cell division, causing characteristic swelling of the rice root and facilitating invasion by the fungus.
- Surprised researchers found that the toxin is not produced by the fungus itself but by bacteria (*Burkholderia rhizoxinica*) that live in the lumen of hyphae as endosymbionts. Without its bacterial symbiont, the fungus cannot infect rice.
  - The bacteria live within fungal cells (Figure 10.6).
  - They produce rhizoxin, a toxin that inhibits mitosis in the plant host by binding beta-tubulin and disrupting mitotic spindle formation.
  - The bacterial toxin contributes to the necrosis of rice cells, allowing the fungus to invade.
  - The fungus is resistant to rhizoxin, permitting it to harbor the bacteria without itself being damaged.
  - The bacteria constrain host reproduction—a fungus that loses its endobacteria cannot produce sporangia.
- Researchers are looking for ways to control the disease by controlling the bacterium itself.
- Because it inhibits mitosis and cell division, rhizoxin is a chemical of some promise in the treatment of human cancers.
- Note that some other strains of *R. microsporus* produce **rhizonins**, potent cyclopeptide toxins that damage mammalian livers. These toxins do not seem to be important in infecting the plant host, but they may contribute to food spoilage. They are produced by a different endosymbiotic bacterium, *Burkholderia endofungorum*.

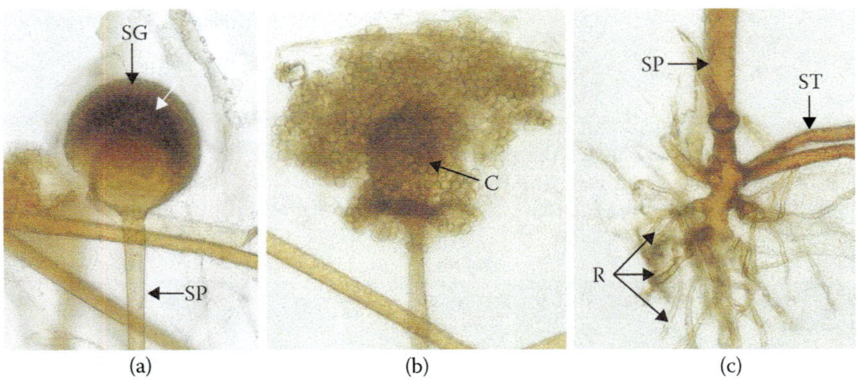

FIGURE 10.2 *Rhizopus stolonifer.* (a) A spherical sporangium (SG) is held aloft on a stout stalk (sporangiophore: SP) as the spores (white arrow) develop inside. (b) Its fragile outer wall breaks soon, releasing a mass of spores and revealing the lollipop-like columella inside, which is obscured by spores. (c) A cluster of root-like rhizoids (R) anchor the sporangium to the substrate. The fungus often produces arching side-shoots or stolons (ST), on which new sporangia are formed. (Courtesy of K. Hodge.)

Good field sanitation may diminish the amount of *Rhizopus* inoculum. Because the disease is most often a postharvest problem, fungicides and temperature control can also be important in disease control. Temperature control is especially key in controlling postharvest rots of stone fruits and pears caused by *Mucor piriformis*, which can grow at lower temperatures than most of its kin (see Case Study 10.2). Water used to hydrocool fruits after harvest can spread spores if not kept clean and chlorinated. A fungicide with good activity against zygomycetes may be needed to protect ripe fruit such as cherries on their way to market. Dichloran, one such fungicide, is often added to agar media that is used to culture a range of food spoilage fungi. It preferentially inhibits the growth of most *Rhizopus* and *Mucor* strains, which can grow so quickly that they grow right over other fungi that develop more slowly.

Most Rhizopus and Mucor rots proceed in a similar manner: host cells die because of enzymatic damage to cell walls. However, *R. microsporus* relies on a different mechanism of pathogenesis. It causes rice seedling blight by using toxins to kill cells, which it can then digest. It was a great surprise to researchers to find that these toxins, **rhizoxins**, are not produced by the fungus itself, but by a bacterial endosymbiont.

Rhizopus stolonifer is a versatile and powerful fungus. It is often found in homes, growing in damp places, and is common in compost and soils worldwide. It and several related species can be opportunistic human pathogens, causing a disease called **zygomycosis**. People and animals with healthy immune systems are very rarely affected, but patients with poor immunity or extensive injuries occasionally acquire Rhizopus or Mucor infections that advance rapidly through tissues—they can be quickly fatal. Do not worry; your chances of acquiring an infection from a fuzzy strawberry are vanishingly small.

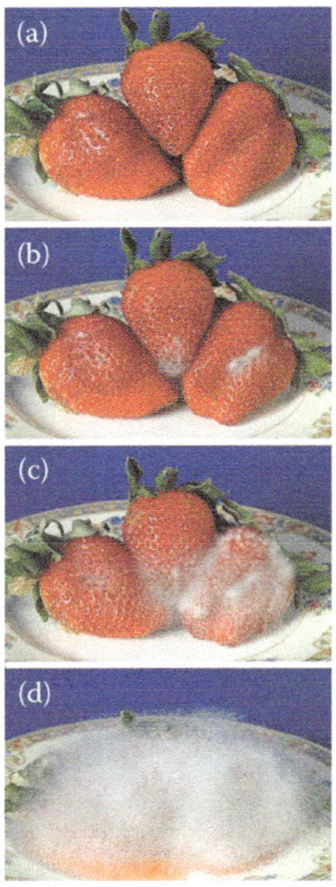

FIGURE 10.3 Time sequence of Rhizopus rot of strawberry, also called "strawberry leak." Few fungi can match the speed of *Rhizopus stolonifer* once it starts. (a) Time zero, (b) 36 h, (c) 44 h, and (d) 58 h elapsed. (Courtesy of Kent Loeffler.)

Choanephora

Choanephora is a genus of only two species, both of which are plant pathogens of relatively minor importance. It differs from *Rhizopus* and *Mucor* in producing a second type of asexual sporangium called a **sporangiole**. A sporangiole is small, lacks a columella, and contains a small number of spores. *Choanephora infundibulifera* causes Choanephora leaf blight of soybean and may cause a pod rot of other legumes.

Choanephora cucurbitarum is best known for causing a rot of summer squashes (Figure 10.7) and other cucurbits. It can attack a variety of other crops, including amaranth, various brassicas, peppers, and ornamentals including dahlia, hibiscus, petunia, and zinnia. It is often acquired in the field by water splash or wind dispersal of the spores, but can also become seedborne when it affects flowers. In the seed, the fungus may kill the embryo, preventing germination, or it may bide its time and later cause damping-off of a seedling. Seed infection may introduce the fungus into soil in locations where Choanephora diseases were previously absent.

Gilbertella

Gilbertella persicaria causes rot in fruit such as peaches, nectarines, pears, and tomatoes. It resembles *Rhizopus*, but lacks rhizoids, and is further distinguished by the unusual method by which its sporangia open—they split

(a) (b)

FIGURE 10.4 (a) Rhizopus rot of poinsettia. (b) Wetness due to packing or poor ventilation of ornamentals favors the development of Rhizopus stem rots. (Courtesy of Margery Daughtrey.)

CASE STUDY 10.2

MUCOR PIRIFORMIS: SHALL I EAT A PEACH?

- Mucor wet rot can be an important postharvest disease of peaches and other stone fruit in cold storage, particularly in western North America.
- Zygomycetes really like peaches: Fruit rots caused by *M. piriformis* greatly resemble rots caused by *Rhizopus* and *Gilbertella*.
- Unlike most fungi, *M. piriformis* can grow in cold storage at temperatures as low as 3°C.
- When temperatures in cold storage are too high, *M. piriformis* can quickly spread through whole cases of fruit.
- It may also be a problem during long-distance shipping when temperature is not well controlled.
- Good sanitation, postharvest fungicide application, and careful temperature control are key in managing this disease.

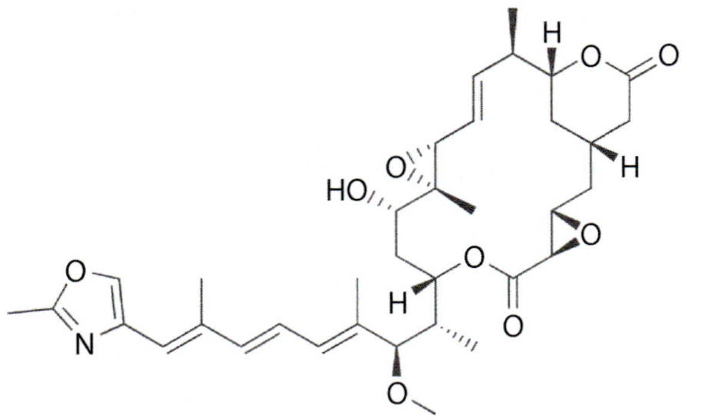

FIGURE 10.5 Rhizoxin, a toxin produced by a bacterium, *Burkholderia rhizoxinica*, lives within the cells of *Rhizopus microsporus*, enabling it to infect rice. From https://commons.wikimedia.org/wiki/File:Rhizoxin.png [accessed Jul 2, 2016].)

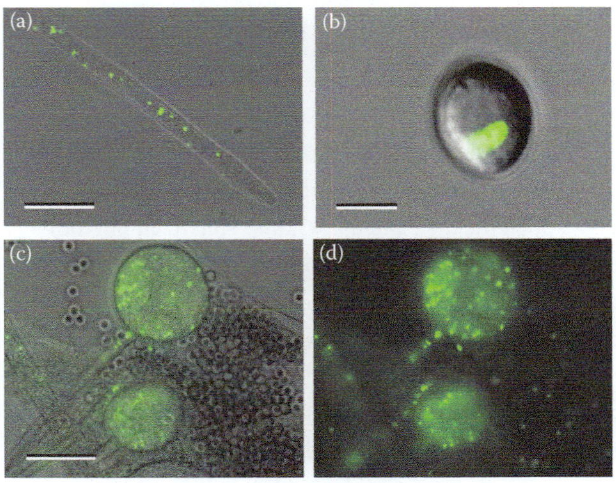

FIGURE 10.6 Confocal laser scanning micrographs of fungi after microinjection with green fluorescent protein (GFP)-labeled bacteria. (a) Hypha showing bacteria migrating to the tip (scale bar represents 25 μm). (b) A single vegetative spore containing a rod-shaped GFP-labeled endobacterium (scale bar represents 2 μm). (c and d) Sporangiophores, sporangia, and spores formed after restitution of the symbiosis (white light and fluorescence mode, scale bar represents 30 μm). (From Partida-Martinez, L.P., et al., *Curr. Biol.*, 17, 773–777, 2007. With permission.)

FIGURE 10.7 Choanephora rot of summer squash caused by *Choanephora cucurbitarum*. (Courtesy of Michelle Grabowski.)

in half longitudinally and release spores that are crowned with slender appendages. In peach orchards, Gilbertella, Mucor, and Rhizopus rots are often found together and strongly resemble each other on fruit: gray and hairy. As the three fungi differ in their sensitivity to fungicides, it may be important to know which is prevalent in a given orchard. Wounding facilitates infection. The pathogens persist in soil, so soil contact must be avoided. After harvest, hydrocooling the fruit in chlorinated water and application of postharvest fungicides can help suppress fruit rot in vulnerable growing regions. Chlorine can kill spores, but will not halt an infection that is already in progress.

Other Zygomycete Pathogens

Blakeslea trispora has been reported as a plant pathogen in India and southeastern Asia. It is more often found as a soil-associated saprobe. Its greatest claim to fame, however, is in industrial production of beta-carotene and lycopene as food additives and nutritional supplements. The fungus synthesizes a precursor to these compounds, trisporic acid, as a part of its sexual cycle.

LABORATORY EXERCISES

Experiment 10.1. Observation of *Rhizopus stolonifer* Asexual Structures

Identification of plant pathogenic zygomycetes requires knowledge of their key asexual characteristics. The purpose of this exercise is to learn to mount and observe zygomycete structures using a compound microscope. Because their structures tend to be larger than most other plant pathogenic fungi, observation using a dissecting scope can help students develop a sense of the characteristic colonies and mycelia.

Cultures in Petri dishes are appropriate for this exercise, but infected fruit may also be used. Strawberries can be easily infected with *R. stolonifer* by using a scalpel or pin tool to insert spores or a fragment of an agar culture into a wound. Place the inoculated strawberry in a moist chamber, and after 3–5 days at room temperature, asexual sporangia can be observed on the fruit surface.

Materials

The following materials are needed for each student or pair of students:

- Petri dish cultures of one or more species of *Rhizopus*, *Mucor*, or *Phycomyces*
- Potato dextrose agar (PDA) medium
- Pencil and paper
- Compound and dissecting microscopes
- Glass slides and coverslips
- Dropper bottle containing a mounting medium such as water or lactic acid cotton blue
- Tissues or bibulous paper for blotting excess mounting medium
- Two pin tools and/or scalpels with #11 blade
- Alcohol lamp or Bunsen burner for sterilizing tools and a striker or match

Follow the instructions outlined in Procedure 10.1 to complete the experiment.

Anticipated Results

Rhizopus and kin grow rapidly in culture on PDA and other media and readily produce asexual sporangia. Sexual structures are not expected to develop because these fungi are usually heterothallic. Students should be careful to distinguish the outer sporangial wall from the columella: the outer wall often breaks down, releasing its spores, and the smaller columella may be mistaken for a young sporangium (see Figure 10.2).

Questions

- Describe the color and texture of the culture, and note its age. Could you recognize a mucoralean plant pathogen using only a dissecting microscope?
- Using the compound microscope, describe and draw the asexual reproductive structures of the fungus: coenocytic hyphae, sporangia,

Procedure 10.1

Observation of Asexual Structures of Zygomycetous Fungi

Step	Instructions and Comments
1	Observe the culture by placing the Petri dish on the stage of the dissecting microscope. Begin at low power, and increase magnification. Notice the pin-shaped asexual sporangia and other features of the culture.
2	While leaving the culture on the stage of the dissecting scope, open the dish and use a flame-sterilized pin tool or scalpel to remove a small amount of material including the sporangia. Avoid scraping up much agar, and in general aim for a small wisp of fungus. Transfer the material to a drop of mounting medium (water) on a glass slide. If it forms a clump, use two pin tools to tease it apart a little. Mounting too much material is a common cause of frustration—you want to be able to see through your sample and observe individual structures. Once the material is spread, gently apply a cover slip, tapping it down with a pencil eraser or the handle of the pin tool. Do not use your finger—fingerprints look rather interesting under a microscope, but they will obscure your view. If there is excess mount fluid, blot it away at the edge of the cover slip with bibulous paper or a tissue.
3	Observe your slide under the compound microscope, beginning with a low power lens such as 4× or 10× objective. First, move the stage around to look at what you have captured and get a sense of how the fungus is organized. Next, increase the magnification to see more detail: the 40× objective is probably sufficient. Look for the characteristic structures of Mucoralean fungi: coenocytic hyphae, the sporangium, the sporangiophore topped with an inflated columella, and the sporangiospores. If you are examining a *Rhizopus* species, look also for stolons and rhizoids. Once you have seen these structures, prepare a drawing, including a scale bar to indicate size.

sporangiophores, columellae, and in the case of *Rhizopus* spp., rhizoids and stolons.

- Are the spores of the fungus smooth or striated?
- Observe the attachment of the sporangium wall to the columella. This feature is often important in the identification of mucoralean fungi. In some, the columella is entirely inside the sporangium; in others, the columella bulges out the bottom. The latter is called an **apophysate** sporangium. Does your fungus have apophysate sporangia?

Experiment 10.2. Observation of Mating in *Phycomyces*

Heterothallic mating and formation of sexually recombinant zygospores are key features of the life cycle of mucoralean fungi. Although this stage is not essential to the disease cycle, it may allow a pathogenic fungus to respond to strong selection pressures such as fungicide application, cold temperature storage, or host resistance.

Although *Phycomyces* species are saprobes and not plant pathogens, they make a tractable and rather spectacular system for observing zygosporogenesis in Petri dish culture. Alternatively, this experiment could be done with two compatible strains of a plant pathogenic *Rhizopus* species. Compatible *Phycomyces* strains will mate readily on many different media that contain at least some thiamine. We suggest cornmeal agar (CMA) because it is relatively weak and therefore suppresses aerial growth that might conceal the mating structures.

Materials

The following materials are needed for each small group of students:

- Petri dish cultures of *Phycomyces blakesleeanus* or *Phycomyces nitens*: one of each mating type
- CMA
- Parafilm®
- Pencil and paper
- Permanent marker
- Compound and dissecting microscopes
- Glass slides and coverslips
- Dropper bottle containing a mounting medium such as water or lactic acid cotton blue
- Tissues or bibulous paper for blotting excess mounting medium
- Two pin tools and/or scalpels with #11 blade
- Alcohol lamp or Bunsen burner for sterilizing tools and a striker or match

Follow the instructions outlined in Procedure 10.2 to complete the experiment.

Anticipated Results

Phycomyces species are heterothallic and readily undergo sexual reproduction in the lab when compatible strains are paired on CMA or other media (must contain a trace of thiamine). Students will be able to observe the stages of mating and zygospore formation at the interface between the colonies of compatible mating types (see Figure 10.1).

Procedure 10.2

Observation of Mating and Sexual Structures of ***Phycomyces***

Step	Instructions and Comments
1	Approximately 2 weeks before the laboratory, obtain or revive two cultures of *Phycomyces* species: one of each mating type (+ and –). They should be growing vigorously before they are coinoculated on a mating plate. At this stage, the choice of culture medium is unimportant: PDA or a similar rich medium works well. The fungus will likely produce its impressively long sporangiophores in culture.
2	Prepare the mating plate by labeling the Petri dish containing CMA: On the agar side of the dish, use a marker to write the fungus name, the date, and time near the edge. Mark "+" and "–" at the points on the underside of the agar plate where you will inoculate the plate with strains of the two mating types. Aim for the strains to be 3–4 cm apart near the center of the dish.
3	Working under sterile conditions in a transfer hood, use a flame-sterilized scalpel to cut a small cube of the "+" culture from near the growing edge of the colony. Place it on the agar surface of the mating plate over the "+" mark. Do the same for the "–" culture.
4	Use a dissecting microscope to monitor the plate over the next week, paying close attention to the zone of contact between the two strains.
5	Observe and draw the gametangia, the developing zygospore, and the spiny, branching appendages.

Expect mating to begin about 4 days after inoculating the mating plate, and zygospores begin to form 4–7 days after that. Ideally, students should observe their mating plates several times during the process, but in general, each plate will contain a number of stages at the same time in the week after the two strains grow together.

Questions

- Because of the twining of basal hyphae at the agar surface, you will not be able to be certain which parent gametangium is "+" and which is "–." Is there any difference between the two gametangia in each mating?
- Each parent mycelium is haploid. Consider the nuclear events going on during mating. Where do meiosis and karyogamy occur?
- Although we will not observe it, after dormancy is broken, the zygospore will germinate by forming a small sporangium. What is the nuclear state and mating type(s) of the sporangiospores inside?

SUGGESTED READINGS

Battaglia, E., I. Benoit, J. van den Brink, A. Wiebenga, P. M. Coutinho, B. Henrissat, and R. P. de Vries. 2011. Carbohydrate-active enzymes from the zygomycete fungus *Rhizopus oryzae*: A highly specialized approach to carbohydrate degradation depicted at genome level. *BMC Genomics* 12:38.

Benny, G. L., and R. K. Benjamin. Zygomycetes.org (Accessed August 2012).

Kirk, P. M. 1984. A monograph of the Choanephoraceae. *Mycological Papers* 152:1–61.

Lackner, G., and C. Hertweck. 2011. Impact of endofungal bacteria on infection biology, food safety, and drug development. *PLoS Pathogens* 7(6):e1002096.

Liu. X., H. Huang, and R. Zheng. 2007. Molecular phylogenetic relationships within Rhizopus based on combined analyses of ITS rDNA and pyrG gene sequences. *Sydowia* 59:235–253.

Ma, L.-J., A. S. Ibrahim, C. Skory, M. G. Grabherr, G. Burger, M. Butler, M. Elias, A. Idnurm, B. F. Lang, T. Sone, A. Abe, S. E. Calvo, L. M. Corrochano, R. Engels, J. Fu, W. Hansberg, J. -M. Kim, C. D. Kodira, M. J. Koehrsen, B. Liu, D. Miranda-Saavedra, S. O'Leary, L. Ortiz-Castellanos, R. Poulter, J. Rodriguez-Romero, J. Ruiz-Herrera, Y.-Q. Shen, Q. Zeng, J. Galagan, B. W. Birren, C. A. Cuomo, and B. L. Wickes. 2009. Genomic analysis of the basal lineage fungus *Rhizopus oryzae* reveals a whole-genome duplication. *PLoS Genetics* 5(7):e1000549.

Michailides, T. J., and R. A. Spotts. 1990. Postharvest diseases of pome and stone fruits caused by *Mucor piriformis* in the Pacific Northwest and California. *Plant Diseases* 74:537–543.

Partida-Martinez, L. P., C. F. de Looss, K. Ishida, M. Ishida, M. Roth, K. Buder, and C. Hertweck. 2007. Rhizonin, the first mycotoxin isolated from the Zygomycota, is not a fungal metabolite but is produced by bacterial endosymbionts. *Applied and Environmental Microbiology* 73:793–797.

Partida-Martinez, L. P., M. Shamci, K. -O. Greulich, and C. Hertweck, 2007. Endosymbiont-dependent host reproduction maintains bacterial-fungal mutualism. *Current Biology* 17:773–777.Shtienberg, D. 1997. Rhizopus head rot of confectionery sunflower: Effects on yield quantity and quality and implications for disease management. *Phytopathology* 87:1226–1232.

11 Taphrinomycete and Saccharomycete Pathogens

Margery L. Daughtrey, Kathie T. Hodge, and Nina Shishkoff

Concept Box

- *Taphrina* and *Protomyces* are two genera of Taphrinomycetes, an early diverging clade of Ascomycota.
- Both *Taphrina* and *Protomyces* have a saprobic yeast stage and a parasitic mycelial stage during which asci are formed.
- The asci of *Taphrina* and *Protomyces* are naked (not contained in an ascocarp).
- Typical symptoms of diseases caused by *Taphrina* or *Protomyces* are galls, leaf curls, and leaf spots.
- Peach leaf curl is a common disease caused by *T. deformans*.
- *Eremothecium* is unusual in that it is a plant pathogen within the Saccharomycetes (ascomycetous yeasts).

Taphrinomycete and saccharomycete **pathogens** cause a number of obscure diseases as well as some commonly recognized problems such as peach leaf curl and oak leaf blister. Many of the diseases discussed in this chapter affect weeds or native plants rather than cultivated species, which tend to receive the most attention from plant pathologists. The plant–pathogen interactions in these groups are unique and quite fascinating.

Phylogenetic studies based on DNA sequences reveal three major lineages (subphyla) within Ascomycota: Taphrinomycotina, Saccharomycotina, and Pezizomycotina (Hibbett et al., 2007; Kurtzman, 2011). Taphrinomycotina (roughly approximate to the previous name Archiascomycetes) appear to have diverged before Saccharomycotina (ascomycetous yeasts) and Pezizomycotina (filamentous ascomycetes).

TAPHRINOMYCETE PATHOGENS

There are four classes in the subphylum Taphrinomycotina: Neolectomycetes, Pneumocystidomycetes, Schizosaccharomycetes, and Taphrinomycetes; each of these is composed of a single order (Hibbett et al., 2007). Although they are known to associate with trees, it is not known whether Neolectomycetes are plant pathogens. Pneumocystidomycetes are yeast-like animal pathogens; and Schizosaccharomycetes are fission yeasts. The plant-parasitic Taphrinomycetes undergo sexual reproduction that is **ascogenous**, but lack **ascogenous hyphae** and **ascocarps**. The single order in this class, Taphrinales, includes two families, Protomycetaceae and Taphrinaceae, both of which contain exclusively plant pathogens. Members of these families produce both yeast-like and mycelial states; they lack defined fruiting bodies; and **asexual reproduction** occurs by budding.

Protomycetaceae

Protomycetaceae includes six genera (*Burenia*, *Protomyces*, *Protomycopsis*, *Saitoella*, *Taphridium*, and *Volkartia*) (Table 11.1). The genus *Protomyces* causes galls on leaves, stems, flowers, and fruit of plants in Apiaceae (dill family) and Asteraceae (composite family). *Protomyces macrosporus*, for example, causes a leaf gall of wild carrot (*Daucus carota*); it has been reported to occur in 26 plant genera, including 14 in Asteraceae. *Protomyces gravidus* causes a stem gall on giant ragweed (*Ambrosia trifida*) in the United States (Figure 11.1) and affects tickseed (*Bidens*). Of the other members of the family Protomycetaceae, *Burenia* and *Taphridium* occur only in Apiaceae, whereas *Protomycopsis* and *Volkartia* occur only in Asteraceae. These three genera cause color changes or galls and spots on their hosts.

TABLE 11.1
Classification of *Taphrina, Burenia, Protomyces, Protomycopsis, Saitoella, Taphridium*, and *Volkartia*

Kingdom	Fungi	Fungi
Phylum	Ascomycota	Ascomycota
Subphylum	Taphrinomycotina	Taphrinomycotina
Class	Taphrinomycetes	Taphrinomycetes
Order	Taphrinales	Taphrinales
Family	Taphrinaceae	Protomycetaceae
Genus	*Taphrina*	*Burenia, Protomyces, Protomycopsis, Saitoella, Taphridium, Volkartia*

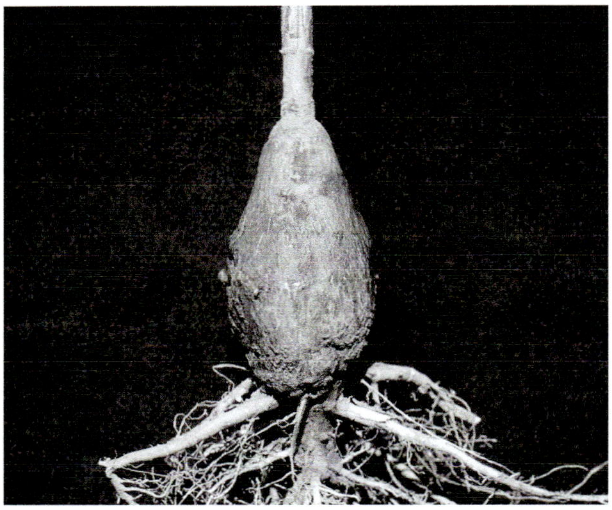

FIGURE 11.1 Diseases caused by *Protomyces* and *Taphrina*. Stem gall of giant ragweed (*Ambrosia trifida*) caused by *Protomyces gravidus*. (From Holcomb, G.E., *Plant Dis.*, 79(8), 1995, cover. With permission.)

Protomyces produces an intercellular, probably **diploid**, mycelium and thick-walled **intercalary** ascogenous cells that form in the host tissue. These cells are sometimes mistaken for the spores of **smuts**, such as the **white smut**, *Entyloma* (Preece and Hick, 2001). The ascogenous cells of *Protomyces* overwinter and in the spring either produce a **synascus** themselves or form a vesicle that serves as a synascus. A synascus is a compound ascus. Inside the multinucleate synascus, **meiosis** apparently occurs for each nucleus, and then the four spore mother cells thus formed from each nucleus continue to divide mitotically to form many ascospores. Hundreds of these spores are forcibly released en masse, and they continue budding after release. Fusion of two compatible ascospores forms a (probably diploid) cell that grows into a hypha that can infect leaves once again.

Taphrinaceae

The genus *Taphrina* was created in 1832 by Fries; it is the only genus of the Taphrinaceae (Table 11.1). Species concepts in *Taphrina* have changed over the years—it originally included some 95 species (Mix, 1949), but lately 28 species have been defined on the basis of molecular analysis (Fonseca and Rodrigues, 2011); additional species will likely be accepted after further research. *Taphrina* species are **parasites** of members of many plant families. Those on ferns generally differ from species that attack **angiosperms** in having thin, **clavate** asci. The fern hosts of *Taphrina* show some very elaborate symptoms: *Taphrina cornu-cervi* causes antler-shaped galls on the fern *Polystichum aristatum* Presl. and *T. laurencia* causes highly branched bushy protuberances on the fronds of *Pteris quadriaurita* in Sri Lanka. The fern-parasitizing *Taphrina* species are in need of further research.

Although *Taphrina* spp. affect many herbaceous and woody plants, the symptoms that they cause are often overlooked. The woody plant hosts that occur in the United States and Canada are shown in Table 11.2. Some of the most frequently encountered Taphrina diseases are those affecting the genus *Prunus*—a number of species occur on plums, apricots, and cherries. *Taphrina pruni* causes deformed fruits called "plum pockets" or "bladder plums" on plums and related species. The **witches' brooms** caused by *T. wiesneri* on Japanese flowering cherries caused some concern in Washington, DC, in the 1920s, but these perennial infections were successfully eradicated by pruning (Sinclair et al., 1987).

Taphrina species often cause small, yellow leaf spots that may or may not be thickened or blistered into concave or convex areas; these often turn brown with age (Figure 11.2). Spots on maples (*Acer* species) caused by a number of different *Taphrina* species may be brown to black. Asci generally appear as a white bloom on one or both leaf surfaces within the areas colonized by *Taphrina*. Twig deformation, galls, witches' brooms, and distorted **inflorescences** of fruit are also possible. Alder cones infected by *T. robinsoniana* show antler-shaped outgrowths (Figure 11.3).

Within the host, *Taphrina* forms a dikaryotic mycelium that is intercellular and **subcuticular** or that stays within the epidermal wall. Eventually the hyphae form a mass beneath the cuticle and rounded ascogenous cells

TABLE 11.2
***Taphrina* Species and Their Woody Plant Hosts Reported in the United States and Canada**

Plant Host	Pathogen Species
Acer spp.	*Taphrina aceris*, UT
	T. carveri, c, s USA, ONT
	T. darkeri, nw USA, w Can
	T. dearnessii, c, e NA
	T. letifera, e, nc USA, NS
	T. sacchari, c, e USA, QUE
Aesculus californicus	*T. aesculi*, CA, TX
Alnus spp.	*T. alni* (catkin hypertrophy), AK, GA, c Can, Eur, Japan
	T. japonica, nw NA, Japan
	T. occidentalis (catkin hypertrophy), nw NA, QUE
	T. robinsoniana (catkin hypertrophy), c, e NA
	T. tosquinetii (catkin hypertrophy), NH, NS, Eur
Amelanchier alnifolia	*T. amelanchieris* (witches' broom), CA
Amelanchier sp.	*T. japonica*, CA
Betula spp.	*T. americana* (witches' broom), NA
	T. bacteriosperma, n hemis
	T. boycei, NV
	T. carnea, ne USA, e, w Can, Eur
	T. flava, ne USA, e Can
	T. nana, WY, e, w Can, Eur
	T. robinsoniana, NA
Carpinus caroliniana	*T. australis*, e USA, ONT
Castanopsis spp.	*T. castanopsidis*, CA, OR
Corylus	*T. coryli*, USA, Japan
Malus	*T. bullata*, WA, Eur
Ostrya virginiana	*T. virginica*, c, e US, e Can
Populus spp.	*T. johansonii* (catkin hypertrophy), USA, e Can, Eur, Japan
	T. populi-salicis, nw NA
	T. populina, n hemis
	T. rhizophora NY, WI, Eur
Potentilla spp.	*T. potentillae*, USA, Eur
Prunus spp.	*T. armeniacae*, USA
	T. communis, (plum fruit hypertrophy), NA
	T. confusa (chokecherry fruit hypertrophy, witches' broom), USA, e Can
	T. farlowii (cherry leaf curl, fruit and shoot hypertrophy), c, e USA
	T. deformans, cosmopolitan
	T. flavorubra (cherry and plum fruit and shoot hypertrophy)
	T. flectans (cherry witches' broom), w USA, BC
	T. jenkinsoniana, NV
	T. mirabilis, e and s USA
	T. pruni (plum pockets), n hemis
	T. pruni-subcordatae, (plum witches' broom), w USA
	T. thomasii (witches' broom on holly-leaf cherry), CA
	T. wiesneri (cherry witches' broom) cosmopolitan
Pyrus sp.	*T. bullata*, WA, BC, Eur
Quercus spp.	*T. caerulescens*, NA, Eur, n Africa
Rhus spp.	*T. purpurescens*, c, e USA, Eur

(*continued*)

TABLE 11.2 (*CONTINUED*)
Taphrina **Species and Their Woody Plant Hosts Reported in the United States and Canada**

Plant Host	Pathogen Species
Salix laevigata	*T. populi-salicis*, CA
Sorbopyrus auricularia	*T. bullata*, BC
Ulmus americana	*T. ulmi*, c, e USA, QUE, Eur

[a] Diseases caused by the fungi listed for each host genus are leaf blisters or curls unless otherwise noted. Geographic distributions are identified with postal abbreviations of states and provinces or more broadly as follows: c, central; e, east; n, north; s, south; w, west; Can, Canada; Eur, Europe; hemis, hemisphere; NA, North America. A number of fungi in this table have not yet been confirmed as distinct species by molecular analysis.
Source: Courtesy of W.A. Sinclair, Cornell University.

FIGURE 11.2 *Taphrina purpurescens* leaf blister of dwarf sumac (*Rhus copallina*). Affected leaves display red-brown blisters. (Courtesy M.L. Daughtrey.)

FIGURE 11.3 Catkin hypertrophy on hazel alder, *Alnus serrulata*, caused by *Taphrina robinsoniana*. (Courtesy M.L. Daughtrey.)

develop, in which karyogamy takes place. Each of these cells is then divided by a septum into two: the lower cell becomes the stalk and the upper will become the ascus. Meiosis occurs in the upper cell, then a second mitotic division, yielding eight **haploid** nuclei that are then packaged into eight ascospores. There are no ascocarps, purely naked asci (Figure 11.4). Ascospores in many species bud within the ascus, packing the ascus with **blastospores**. This budding may continue after spores are forcibly discharged from the ascus. Much of the *Taphrina* life cycle is spent in the **saprotrophic** yeast state, during which the fungus is haploid and uninucleate. For some of these haploid yeasts, the dikaryotic state and the host plant it parasitizes are not known (Fonseca and Rodrigues, 2011). *Taphrina* species generally overwinter as haploid blastospores in bud scales or on bark, although in a few cases overwintering is accomplished by means of perennial **dikaryotic** mycelium within the host. For many *Taphrina* species, the details and location of mating and dikaryotization are not known.

Cultures made from ascospores or blastospores of *Taphrina* grow slowly on artificial media as pale pink yeast

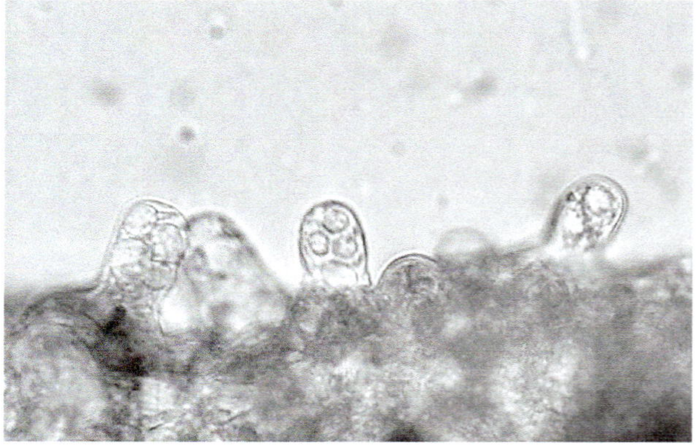

FIGURE 11.4 Asci of *Taphrina purpurescens*. The asci are "naked," that is, they are produced directly from the host tissues and not inside an ascocarp. Several ascospores can be seen in the ascus at the center. (Courtesy M.L. Daughtrey.)

colonies. Without the knowledge of the parasitic phase of the organism, the fungus in culture might be mistaken for one of the pink basidiomycetous yeasts that are common in plant tissue. Molecular methods are important for the identification of such cultures. They have also been employed for the detection of *T. deformans* in washings from symptomless peach buds (Tavares et al., 2004). PCR fingerprints have been used to distinguish a number of species of *Taphrina* occurring on alder (*Alnus* spp.) that would otherwise be inseparable based on yeast phase characteristics (Bacigalova et al., 2003).

Some important diseases are caused by *Taphrina*. Most diseases caused by *Taphrina* attract the attention of only the most curious observer. Two stand out as more obvious and damaging to cultivated plants: oak leaf blister caused by *T. caerulescens* (see Case Study 11.1 and Figure 11.5) and peach leaf curl caused by *T. deformans* (see Case Study 11.2 and Figure 11.6).

Taphrina deformans causes the best-known disease in this group, peach leaf curl. This disease is notorious because it causes striking symptoms and significant damage on economically important hosts. Peaches, nectarines, and almonds are affected (possibly by different strains); the disease occurs all over the world where peaches are grown, but has a more limited geographic occurrence on its other hosts. Tissues of peach leaves that are infected when they are still undifferentiated later show yellow, pink or reddish areas that are buckled and curled into weird shapes; entire leaves may be distorted (Figure 11.6). Cynthia Westcott (1953), a noted ornamental pathologist, described peach leaves as sometimes looking "as if a gathering string had been run along the midrib and pulled tight." The leaf distortion results from growth-regulating chemicals secreted by the fungus: **cytokinins** and **auxins** are produced by some *Taphrina* species. Leaf drop may lead to poor fruit quality and weaken trees. The young peach fruits may be distorted or show reddish warty spots lacking the usual fuzz; infected twigs are swollen. The blistered portions of leaves or fruit develop a powdery gray coating of naked asci on the upper surface. After the ascospores are released, the diseased leaves sometimes turn brown, wither, and drop.

Ascospores that land on the surface of a peach tree will bud to produce a saprotrophic, **epiphytic** yeast phase. During the summer, fall, and winter, the fungus lives thus invisibly and harmoniously on the peach tree. This haploid inoculum can persist on the host plant for several years. In the spring, expanding leaf buds are subject to fungal invasion when plant surfaces are wet

CASE STUDY 11.1

Taphrina caerulescens: Oak Leaf Blister

- Oak leaf blister affects about 50 species of white and red oaks, but is particularly significant on red oaks in the southern United States.
- Haploid ascospores overwinter on bud scales.
- Infection occurs during wet periods in springtime.
- Affected leaves display yellow to brown blisters on the leaves, leaf curling, and leaf drop (Figure 11.5).
- A layer of naked asci forms between the leaf cuticle and epidermis and then swells to break the cuticle before forcibly discharging ascospores.
- Control measures are not usually necessary for oak leaf blister because the disease does not usually result in leaf drop or reduced vigor.

FIGURE 11.5 Leaf blister of red oak (*Quercus* sp.) caused by *Taphrina caerulescens*. (Courtesy M.L. Daughtrey.)

FIGURE 11.6 Peach leaf curl caused by *Taphrina deformans*. (Courtesy M.L. Daughtrey.)

for a minimum of 12.5 h (Rossi et al., 2006). Cool temperatures (lower than 16°C) during a wet period allow infection, and temperatures below 19°C are required for disease development. Disease susceptibility is negatively correlated with the rate of host shoot development. The fusion of two yeast cells is thought to form the mycelium that infects the plant. Asci later form beneath the cuticle of the host and then burst through to release wind-dispersed, haploid ascospores (Rossi et al., 2007).

Although the pathogen stays in constant contact with its host, infection by the parasitic phase of *T. deformans* occurs only on immature leaves, during a series of infection periods in early spring (Rossi et al., 2007). Thus, by the time the symptoms are noticed, weeks later, it is generally too late to achieve any disease control in that same year. Environmental factors such as the timing of spring rains and the severity of the preceding winter will influence how extensive disease symptoms are from one year to the next. Because the fungus overwinters on the surface of the host, this is one of the few diseases that can be controlled with spray treatments at the close of the growing season. Dormant sprays in late fall or treatments before bud swell in the spring can curb infection. Lime–sulfur and **Bordeaux mixture** have traditionally been used in this fashion. Other effective fungicides include chlorothalonil, fixed coppers, ferbam, or ziram. Chemical control is used for only a few other diseases caused by *Taphrina* species, primarily those affecting plums and cherries.

SACCHAROMYCETE PATHOGENS

Most ascomycetous yeasts (Saccharomycetes) are not plant pathogens, but the genus *Eremothecium* (syn. *Ashbya*, *Holleya*, *Nematospora*) is an exception (Kurtzman and Sugiyama, 2001) (Table 11.3). *Eremothecium* species are filamentous fungi that also produces yeast cells within the host (Batra, 1973). **Asci** that are formed directly from the **mycelium** and not enclosed in any fruiting body contain needlelike **ascospores**.

CASE STUDY 11.2

Taphrina deformans: Peach Leaf Curl

- Developing leaves are infected, resulting in blistered, puckered, and discolored areas (Figure 11.6).
- Premature leaf drop can hurt tree vigor; rarely, young fruits may be infected and drop.
- Years with mild winters followed by cool, rainy springs foster this disease and increase the chance of negative effects on peach yield.
- If it is unusually cool while peach leaves are emerging from buds, damage from *T. deformans* is increased, because leaves mature more slowly; leaves are no longer vulnerable once they reach a certain stage of maturity.
- Often in home gardens, the disease is not noticed until it is too late in the season to take any action that will reduce the problem for that year.
- Pesticide applications made after the onset of symptoms are wasted because mature leaf tissue is not susceptible.
- The unusual life cycle of the pathogen makes it vulnerable to disease management actions taken during the tree's dormant season.
- A single preventive spray of fungicide during dormancy eliminates the overwintering stage of *T. deformans* and thus controls the disease. Trees can be sprayed either after 90% of the leaves have fallen in the autumn or before bud swell in the spring.

TABLE 11.3
Classification of *Eremothecium*

Kingdom	Fungi
Phylum	Ascomycota
Subphylum	Saccharomycotina
Class	Saccharomycetes
Order	Saccharomycetales
Family	Dipodascaceae
Genus	*Eremothecium*

Sucking insects, especially the true bugs, often vector pathogens in this genus. *Eremothecium coryli* fruit rot associated with stink bug feeding was responsible for losses of more than 30% in field tomatoes in California (Miyao et al., 2000). In addition to tomatoes, *E. coryli* infects cotton, citrus, hazelnuts, and soybeans, and *Eremothecium sinecaudii*, a related species, infects the seeds of mustards. The three species *Eremothecium cymbalariae*, *Eremothecium ashbyi*, and *Eremothecium gossypii* all infect *Hibiscus* species and coffee and cause surface lesions on citrus fruits and cotton bolls. *E. ashbyi* is also used for the industrial production of riboflavin (vitamin B_2) via fermentation process.

CONCLUSION

Although most of the curious swellings, discolorations, and spots due to fungal pathogens in Taphrinomycetes and Saccharomycetes remain little studied and largely ignored, the interaction of *T. deformans* with peach trees is considered a major disease because it can cause significant economic loss. Societal value placed on the host plant often determines whether a disease is of minor or major import—it is not just a matter of how many plants are killed or disfigured, but which plant species are affected determines whether control efforts and research dollars will be focused on the problem.

REFERENCES

Bacigalova, K., K. Lopandic, M. G. Rodrigues, A. Fonseca, M. Herzberg, W. Pinsker, and H. Prillinger. 2003. Phenotypic and genotypic identification and phylogenetic characterisation of *Taphrina* fungi on alder. *Mycol. Prog.* 2:179–196.

Batra, L. B. 1973. Nematosporaceae (Hemiascomycetidae): Taxonomy, pathogenicity, distribution, and vector relations. *U.S. Dept. Agric., Agric. Res. Ser. Tech. Bull.* 1469:1–71.

Fonseca, A., and M. G. Rodrigues. 2011. *Taphrina* Fries (1832). In: C. P. Kurtzman, J. W. Fell, and T. Boekhout (eds). *The Yeasts: A Taxonomic Study: Descriptions of Teleomorphic Ascomycetous Genera and Species*, vol. 2, pp. 823–858. Amsterdam, the Netherlands: Elsevier (online).

Hibbett, D. S., M. Binder, J. F. Bischoff, M. Blackwell, P. F. Cannon, O. E. Eriksson, S. Huhndorf, T. James, P. M. Kirk, R. Lücking, H. T. Lumbsch, F. Lutzoni, P. B. Matheny, D. J. McLaughlin, M. J. Powell, S. Redhead, C. L. Schoch, J. W. Spatafora, J. A. Stalpers, R. Vilgalys, M. C. Aime, A. Aptroot, R. Bauer, D. Begerow, G. L. Benny, L. A. Castlebury, P. W. Crous, Y. C. Dai, W. Gams, D. M. Geiser, G. W. Griffith, C. Gueidan, D. L. Hawksworth, G. Hestmark, K. Hosaka, R. A. Humber, K. D. Hyde, J. E. Ironside, U. Kõljalq, C. P. Kurtzman, K. H. Larsson, R. Lichtwardt, J. Longcore, J. Miadlikowska, A. Miller, J. M. Moncalvo, S. Mozley-Standridge, F. Oberwinkler, E. Parmasto, V. Reeb, J. D. Rogers, C. Roux, L. Ryvarden, J. P. Sampaio, A. Schüssler, J. Sugiyama, R. G. Thorn, L. Tibell, W. A. Untereiner, C. Walker, Z. Wang, A. Weir, M. Weiss, M. M. White, K. Winka, Y. J. Yao, and N. Zhang. 2007. A higher-level phylogenetic classification of the fungi. *Mycol. Res.* 111:509–547.

Kurtzman, C. P. 2011. Discussion of teleomorphic and anamorphic Ascomycetous yeasts and yeast-like taxa. In: C. P. Kurtzman, J. W. Fell, and T. Boekhout (eds). *The Yeasts: A Taxonomic Study: Classification of the Ascomycetous Taxa*, vol. 2, pp. 293–307. Amsterdam, the Netherlands: Elsevier (online).

Kurtzman, C. P., and J. Sugiyama. 2001. Ascomycetous yeasts and yeastlike taxa. In: D. J. McLaughlin, E. G. McLaughlin, and P. A. Lemke (eds). *The Mycota: Systematics and Evolution*, pp. 179–200. Berlin, Germany: Springer-Verlag.

Mix, A. J. 1949. A monograph of the genus *Taphrina*. *Univ. Kan. Sci. Bull.* 33:1–167.

Miyao, G. M., R. M. Davis, and H. J. Phaff. 2000. Outbreak of *Eremothecium coryli* fruit rot of tomato in California. *Plant Dis.* 84:594.

Preece, T. E., and A. J. Hick. 2001. An introduction to the Protomycetales: *Burenia inundata* on *Apium nodiflorum* and *Protomyces macrosporus* on *Anthriscus sylvestris*. *The Mycologist* 15:119–125.

Rossi, V., M. Bolognesi, and S. Giosuè. 2007. Seasonal dynamics of *Taphrina deformans* inoculum in peach orchards. *Phytopathology* 97:352–358.

Rossi, V., M. Bolognesi, L. Languasco, and S. Giosuè. 2006. Influence of environmental conditions on infection of peach shoots by *Taphrina deformans*. *Phytopathology* 96:155–163.Sinclair, W. A., H. H. Lyon, and W. A. Johnson. 1987. *Diseases of Trees and Shrubs*. Ithaca, NY: Cornell University Press.

Tavares, S., J. Inacio, A. Fonseca, and C. Oliveira. 2004. Direct detection of *Taphrina deformans* on peach trees using molecular methods. *Eur. J. Plant Pathol.* 110:973–982.

Westcott, C. 1953. *Garden Enemies*. New York, NY: D. Van Nostrand Co., Inc.

12 The Powdery Mildews

Margery L. Daughtrey, Kathie T. Hodge, and Nina Shishkoff

CONCEPT BOX

- Powdery mildews are obligate parasites that show interesting morphological adaptations to herbaceous versus woody hosts.
- Powdery mildew fungi have specialized feeding cells called haustoria that absorb nutrients from their hosts.
- Molecular genetic studies and scanning electron microscope studies have led to a recent reversal in taxonomic thought and a new paradigm: the anamorphs reflect phylogeny much better than the teleomorphs.
- Many powdery mildew fungi have recently been renamed to reflect the newly apparent relationships.
- Powdery mildews injure many ornamental crops and garden plants because of the highly conspicuous colonies that cause aesthetic injury; vegetable, field, and fruit crops suffer from yield and quality reduction.

The **powdery mildew** fungi (order Erysiphales with the single family Erysiphaceae) in the phylum Ascomycota cause easy-to-recognize diseases. The fungus grows across the surface of the host in conspicuous colonies, creating whitish circular patches that sometimes coalesce until the entire leaf surface is white. The colonies may form on either the upper or the lower surface of leaves, as well as on stems, flower parts, and fruits. Because energy from **photosynthesis** is diverted into growth of the pathogen, infected plants may be stunted and produce fewer or smaller leaves, fruits, or grain. The impacts of powdery mildews on their hosts may be mainly esthetic or may reduce yield or quality. Many floral and nursery crops (Figure 12.1) as well as fruit, vegetable and field crops are affected (Figure 12.2).

Powdery mildews are spread over great distances by wind and can be moved around the world on plants with inconspicuous or latent infections. Within the powdery mildew family, Erysiphaceae, some members have very narrow host ranges, whereas others have broad host ranges and affect plants in multiple plant families. Powdery mildews differ from the vast majority of other fungal pathogens in that, with few exceptions, the mycelium grows superficially over host tissues. Only specialized feeding cells, called haustoria (Chapter 7), invaginate the host epidermis (Figure 12.3). Powdery mildew life cycles (Figure 12.4) are entirely biotrophic. No species has been grown in axenic culture apart from its host for any significant duration, and none can grow on dead plant material.

Conidiophores develop directly from the mycelium on the host surface throughout the growing season, producing infective conidia one at a time (Figure 12.5), or in short chains. In tropical climates or greenhouses, this may be the only spore stage in the life cycle. Conidia are windborne or splash-dispersed and serve as secondary inocula during the growing season. Crops grown in greenhouses may suffer repeated cycles of infection

FIGURE 12.1 Powdery mildew of poinsettia caused by *Erysiphe poinsettiae*. White colonies on the red bracts of this Christmas favorite make the plants unsaleable. The sexual state of this pathogen is unknown, but can be deduced from the *Pseudoidium*-type conidiophore. (Courtesy of M.L. Daughtrey.)

The hosts listed are only suggestions; as long as ascocarps are present, any mildew from any plant can be used for a laboratory exercise. However, these hosts are commonly infected over much of the United States, so they are particularly dependable sources of study material.

Dogwood: Host to *Erysiphe* (sect. *Microsphaera*) *pulchra* (ascocarps with multiple asci and dichotomously branched appendage tips; conidia borne singly) and to *Phyllactinia corni* (ascocarps with multiple asci and spine-like appendages with bulbous bases; conidia in chains, the first formed with a different shape from the rest)

Grapevine: Host to *Erysiphe* (sect. *Uncinula*) *necator* (ascocarps with multiple asci and hooked ends to the appendages; conidia borne singly)

Grasses: Host to *Blumeria graminis* (ascocarps with multiple asci and threadlike appendages; conidia in chains borne on conidiophores with bulbous bases)

Lilac: Host to Erysiphe (sect. *Microsphaera*) *syringae* (ascocarps with multiple asci and dichotomously branched appendages; conidia borne singly)

Maple: Host to *Takamatsuella circinata* (ascocarps with multiple asci and hooked ends to the appendages; no known conidial state)

Monarda (bee-balm) and other members of the mint family: Host to *Neoerysiphe galeopsidis* (ascocarps with multiple asci and threadlike appendages; conidia in chains and bearing ridgelike striations)

Oak: Commonly host to three powdery mildews. On the west coast, live oaks are hosts to *Cystotheca lanestris* (ascocarp with a single ascus, threadlike appendages; conidia in chains, crystalline inclusions [fibrosin bodies] visible in cytoplasm). On the east coast, oaks are host to *Erysiphe* (sect. Microsphaera) *extensa* (ascocarps with multiple asci and dichotomously branched appendages; conidia borne singly) and to *Phyllactinia angulata* (ascocarps with multiple asci and spinelike appendages with bulbous bases; conidia in chains, the first-formed with a different shape from the rest)

Phlox: Host to *Golovinomyces magnicellulatus* (ascocarps with multiple asci and threadlike appendages; conidia in chains)

Plane tree (*Platanus*): Host to *Erysiphe* (sect. *Microsphaera*) *platani* (ascocarps with multiple asci and dichotomously branched appendages; conidia borne singly)

Squash, pumpkin: Host to *Podosphaera* (sect. *Sphaerotheca*) *xanthii* (ascocarp with a single ascus and threadlike appendages; conidia in chains, crystalline inclusions [fibrosin bodies] visible in cytoplasm).

Rose: Host to *Podosphaera* (sect. *Sphaerotheca*) *pannosa* (ascocarp with a single ascus and threadlike appendages; conidia in chains, crystalline inclusions [fibrosin bodies] visible in the cytoplasm)

Willow (and aspen): Host to *Erysiphe* (sect. *Uncinula*) *adunca* (ascocarps with multiple asci and with hooked ends to the appendages; conidia borne singly)

Zinnia: Host to Golovinomyces *spadiceus* (ascocarps with multiple asci and threadlike appendages; conidia in chains)

FIGURE 12.2 A List of Common Host Plants of Powdery Mildews

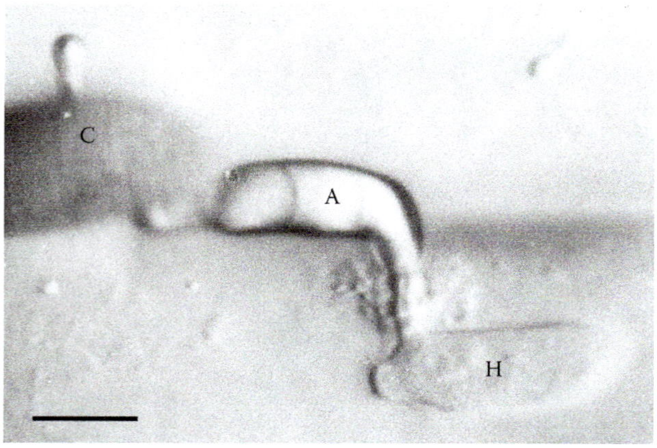

FIGURE 12.3 Penetration structures of *Blumeria graminis* f. sp. *hordei*, powdery mildew of barley. A conidium (C, upper left) that is slightly out of focus on the leaf surface has germinated to produce a primary (top) and a secondary (lower) germ tube. The secondary germ tube has produced a large appressorium (A, center) and penetrated the epidermal cell wall. A haustorium (H, bottom right) has been formed inside the epidermal cell. This basic unit of infection may be replicated in hundreds of epidermal cells underlying a single mildew colony. (Courtesy of J.R. Aist, Cornell University.)

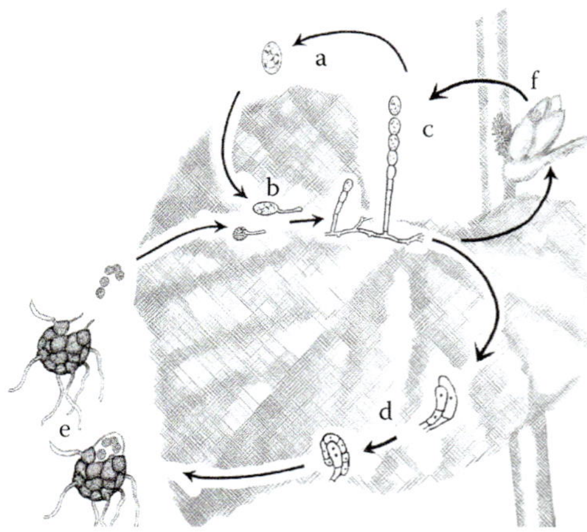

FIGURE 12.4 (a)–(f) Life cycle of a powdery mildew. (Courtesy of N. Shishkoff.)

via conidia year-round. Spore release typically follows a diurnal pattern, with the greatest number of conidia released around midday, in response to decreased relative humidity. Outdoors, rain may serve as a trigger for the release of conidia. Powdery mildews are unusual among fungi in that their conidia do not require large amounts

FIGURE 12.5 Conidiophores and conidia of *Erysiphe poinsettiae*, cause of powdery mildew of poinsettia. (Courtesy of Gail Celio, University of Georgia.)

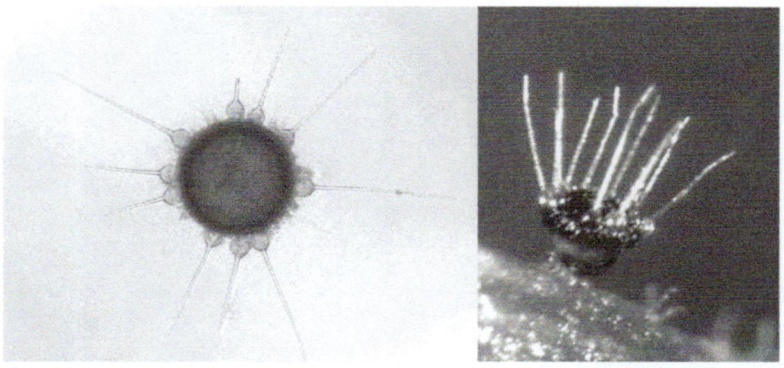

(a) (b)

FIGURE 12.6 The ascocarp (chasmothecium) of *Phyllactinia angulata*, the fungus causing one of the powdery mildew diseases of oak (a) as it appears when first formed, with the bulbous-based appendages lying flat against the oak leaf surface (b) after liberation. The appendages have successfully flexed to lift the ascocarp from the leaf undersurface. After a brief tumble, the top of the liberated fruiting body has adhered by its sticky cushion to the nearby plant material. It will overwinter in this upside-down position. In spring, the ascocarp will split open around its equator, revealing asci in its "lid" in the perfect position to discharge infective spores into the air. (Courtesy of M.L. Daughtrey.)

of free water to germinate. Significant disease spread can thus occur even during dry weather.

Most species of the Erysiphales are **heterothallic**; two compatible individuals must mate before asci can be formed. Compatibility is genetically determined by the mating type locus. Two different **alleles** confer two possible mating types. Some homothallic species are known, but the genetic basis of homothallism has not been determined.

After mating, **ascocarps** are produced superficially on the host, typically toward the end of summer (Figure 12.6). Ascocarps can be an important overwintering stage in temperate climates as they provide the primary inoculum at the beginning of each growing season for a number of powdery mildews. In arid climates, they aid in survival during hot, dry periods. In direct contrast to conidia, free water is required to stimulate ejection of the ascospores and allow their germination. In some species, which attack perennial hosts, such as *Podosphaera* (sect. *Sphaerotheca*) *pannosa* on rose, the fungus can survive as mycelium inside the infected buds. In powdery mildews of tropical climates, the ascocarps are of little importance and have apparently been lost altogether in some species.

The appendages on the ascocarps function in dispersal by helping ascocarps to adhere to plant surfaces and **trichomes**. For powdery mildews on woody plants, the elaborate appendages allow the ascocarp to cling to rough bark surfaces. In *Phyllactinia*, which causes the powdery mildew of alders and numerous other hardwoods, the moisture-sensitive appendages play a unique role in dispersal (Figure 12.6). On maturity, the appendages press down on the substrate, breaking the ascocarp away and releasing it into the wind. Sticky secretions produced from a second set of appendages on the top of the ascocarp promote its adherence when it lands. The ascocarp passes the winter upside down. In spring, the ascocarp splits around the circumference and flips open completely, like a hinged jewel box, so that the formerly upside-down asci in the "lid" now point upward to discharge their infective spores. In other powdery mildews, the appendages are less complex, and the ascocarps typically open from the pressure of asci against the upper surface.

HOST RELATIONSHIPS

Powdery mildews engage in fascinating interactions with their host. Unlike other plant pathogens, a single individual penetrates its host at many different sites and does not proliferate within the plant tissues. A few powdery mildews are atypical in that they do penetrate the stomatal space and form a limited **hemiendophytic** mycelium. The complex organs of infection formed by powdery mildews are among the best-studied host–pathogen interfaces in plant pathology (Figure 12.3).

THE INFECTION PROCESS

Powdery mildew conidia deposited on a hydrophobic leaf surface excrete an adhesive matrix within minutes of contact (Nicholson and Kunoh, 1995). This matrix is believed to mediate host recognition. In a compatible interaction, conidia rapidly germinate by producing one or more germ tubes that follow the contours of the host surface (Figure 12.7). At the apex of the germ tube, an **appressorium** (Figures 12.3, 12.7, and 12.8) is produced at the site where penetration will occur. In different species, the appressoria can be undifferentiated, simple, forked, or lobed, and these morphological variations can

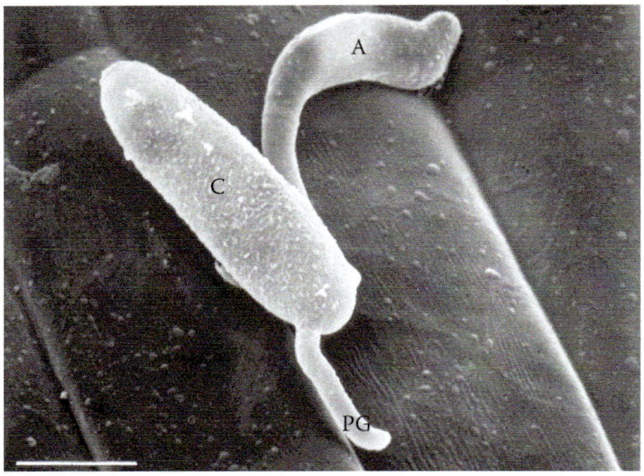

FIGURE 12.7 Scanning electron micrograph of a conidium (C) of *Blumeria graminis* f. sp. *hordei* (the cause of powdery mildew of barley) on a leaf surface. A primary (PG) germ tube and a secondary germ tube are visible; the secondary germ tube (top, right) has started to differentiate into an appressorium (A). In this species, the primary germ tube serves a sensing function, and the secondary germ tube leads to penetration of the host. (Courtesy of J.R. Aist, Cornell University.)

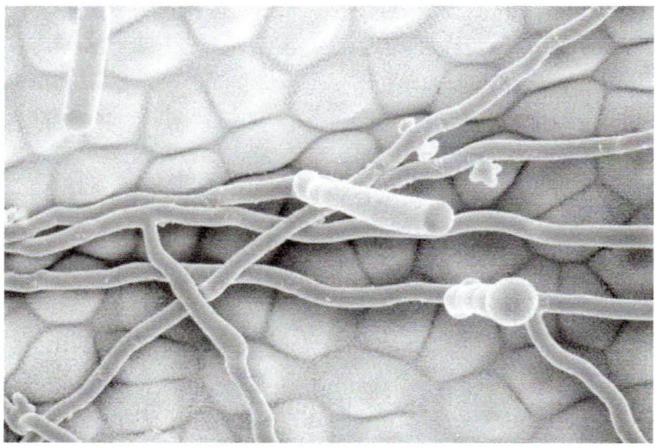

FIGURE 12.8 Scanning electron micrograph of a powdery mildew colony on a leaf surface. The small, lobed appressoria mark points of penetration into the leaf. Several conidiophores can be seen projecting upwards off the leaf surface. (Courtesy of Gail Celio, University of Georgia.)

be useful in identification. The fungus also continues to grow across the host's surface, forming the distinctive superficial mycelium.

The difficult traverse of the plant cell wall is the next step in infection. At this stage, a hypha with narrow diameter, known as a **penetration peg**, grows from the bottom surface of the appressorium. Through a combination of enzymatic and mechanical actions, it penetrates the cell wall of an epidermal cell. Having gained entry, the fungus produces a **haustorium** (Figure 12.3), a special feeding cell that assimilates nutrients from the plant. Haustoria can be simple, lobed, or digitate. The haustorium is not in direct contact with the host cytoplasm; rather, it is surrounded by a gel-like extrahaustorial matrix and the cell membrane of its host cell. The fungus induces changes in the plant membrane surrounding the haustorium that result in leakage of nutrients that are then taken up by the fungus. The invaded cell remains alive and functioning for some time and receives nutrients from surrounding cells that are then leaked to the fungus. This basic pathogenic interaction is repeated in many different cells, and the superficial mycelium spreads across the leaf surface and occasionally develops appressoria, each leading to penetration and parasitism of a single cell. Although the host is seldom killed outright, the continual drain of nutrients often depresses growth and sometimes causes puckering, and chlorotic, or necrotic spotting of the leaf surface.

HOST RESPONSES

Host plants resist powdery mildew attack through passive and active defense systems. Leaf surface features, such as the thick waxes that protect the undersurfaces of some grass leaves, may inhibit germination. The cuticle and cell wall present formidable barriers to penetration because their thickness and durability affect the penetration ability of fungal germ tubes. Many plants actively respond to penetration attempts by forming **papillae** (singular: papilla), tiny, thick deposits of callose and other materials that are deposited on the inner surface of the cell wall. Papillae are a physical barrier to invasion and serve as foci for the release of antimicrobial compounds. The **hypersensitive response** may be induced in incompatible interactions, resulting in the death of an invaded host cell and starvation of the pathogen. All of these mechanisms are genetically controlled and provide important sources of genetic resistance for plant breeders.

TAXONOMY

Powdery mildews are classified in the class Leotiomycetes (with primitive cup fungi), order Erysiphales, and family Erysiphaceae. For years, the taxonomy of the powdery mildews relied largely on the morphology of the ascocarps and their appendages because they showed obvious differences (Braun, 1987; 1995). Recent advances in molecular phylogeny have shown the morphology of the **anamorph** to be a better indicator of relationships (Saenz and Taylor, 1999; Takamatsu et al., 1999). Characteristics of the conidial chains and the ornamentation of the conidial surface are particularly useful (Cook et al., 1997). Now it is thought that differences in the ascocarps merely reflect adaptation to particular hosts. Braun and Cook (2012) have summarized the changes in the classification of powdery mildews over time and proposed a revised classification with 6 tribes and 16 genera.

Conidia and Conidiophores

Conidia (meristem arthrospores) are produced on unbranched conidiophores, singly or in chains. In the past, the name *Oidium* had been used to refer to any conidial state of a powdery mildew because conidia of many of the genera looked the same. The electron microscope work of Cook et al. (1997), however, showed that most powdery mildew genera could be identified by unique characteristics of their conidia, so it is now usually possible to identify each genus from its conidia. The tribes that are delimited by conidial characteristics are described in Figures 12.9 and 12.10

Ascocarp (Chasmothecium)

The ascocarps of powdery mildews are initially fully closed, and they have often been referred to as **cleistothecia**. They differ from other cleistothecial fungi, however, in that the ascocarps do eventually rupture and the ascospores are forcibly discharged. Developmental features also suggest that the ascocarps are more similar to **perithecia** than cleistothecia. Therefore, the term "**chasmothecium**" was created for powdery mildew ascocarps.

Mycologists once assumed that primitive powdery mildews were *Erysiphe*-like, with undifferentiated appendages on the ascocarps (Figure 12.9). New genetic evidence suggests that the primitive powdery mildews lived on woody hosts and possessed curled appendages (Takamatsu et al., 2000). Species on herbaceous hosts typically lack elaborate appendages. This suggests that appendages may be "costly" to maintain, valuable in anchoring ascomata to persistent plant parts, but lost quickly when the plant host has no persistent parts on which to anchor. Related species may thus differ in ascocarp appendage morphology while retaining the same conidial morphology.

Identification of Powdery Mildews

Changes in taxonomy (Braun and Cook, 2012) as well as revisions to the International Code of Botanical Nomenclature are sure to be confusing for a time because many familiar names have been changed in order to reflect our new understanding of relationships among the powdery mildew fungi. In this system, the family Erysiphaceae is divided into 6 tribes and 17 current genera. A simple overview of modern powdery mildew genera is presented in Figure 12.9. It is easy enough to recognize that a disease is caused by a powdery mildew fungus by noting the white powdery colonies on infected plant parts. Morphological differences separating the species can be subtle, however, so the best route to diagnose powdery mildew diseases is through the use of a thorough host index, such as Farr et al. (1989) or Farr and Rossman (online, updated index), together with Braun and Cook (2012).

IMPORTANT DISEASES CAUSED BY POWDERY MILDEWS

Powdery Mildew of Grape

Powdery mildew of grape is caused by *Erysiphe* (sect. *Uncinula*) *necator*. The disease can reduce yield and fruit quality, as well as stunt vines and decrease their winter hardiness. *Vitis vinifera* and hybrids such as Chardonnay and Cabernet Sauvignon are more susceptible to powdery mildew than are varieties derived from the American species. Wine quality can be affected when as few as 3% of berries are infected. Leaves, green shoots, and fruits are susceptible to infection. White mildew colonies appear on upper and lower surfaces of leaves and may coalesce. Immature leaves may be distorted and stunted, shoot infections appear as dark patches, and infected berries can display white colonies (Figure 12.11), abnormal shape, or rust-colored spots and may split open, rendering them vulnerable to fruit rot by *Botrytis cinerea*. Infected leaves may be cupped-up or scorched and may fall prematurely.

Erysiphe necator overwinters as ascocarps in crevices on the surface of the bark of vines or as mycelium in buds (Gadoury and Pearson, 1988). Ascospore release is triggered by even minuscule amounts of rain in the spring when the temperature is warm enough. The ascospores are spread by wind to the green tissues of the grapevine from budbreak through bloom. The short time required from infection until the production of new inoculum gives this powdery mildew powerful epidemic potential. Because little free moisture is needed for conidial germination, powdery mildew can be a serious problem even in years when black rot and downy mildew are hampered by dry conditions. Relative humidity of 40%–100% is required for the formation of microdroplets of water used by conidia for germination and infection. Optimum temperature for disease development is from 20°C to 25°C, but infection is possible over a wide temperature range of 15°C–32°C.

All cultural practices that promote the drying of plant surfaces will reduce powdery mildew. Full sun exposure is desirable; alignment of rows with the direction of the prevailing wind is recommended. Vines should be pruned and trained to reduce shading and allow air circulation within the canopy. Excessive nitrogen should be avoided. Irrigation should be done with trickle systems rather than over-the-vine impact sprinklers. Control of infections during the immediate pre-bloom through early post-bloom period is stressed for grapes because this is when the fruits are most susceptible to the disease. The more susceptible *V. vinifera* and hybrid cultivars require continued suppression of powdery mildew on the foliage

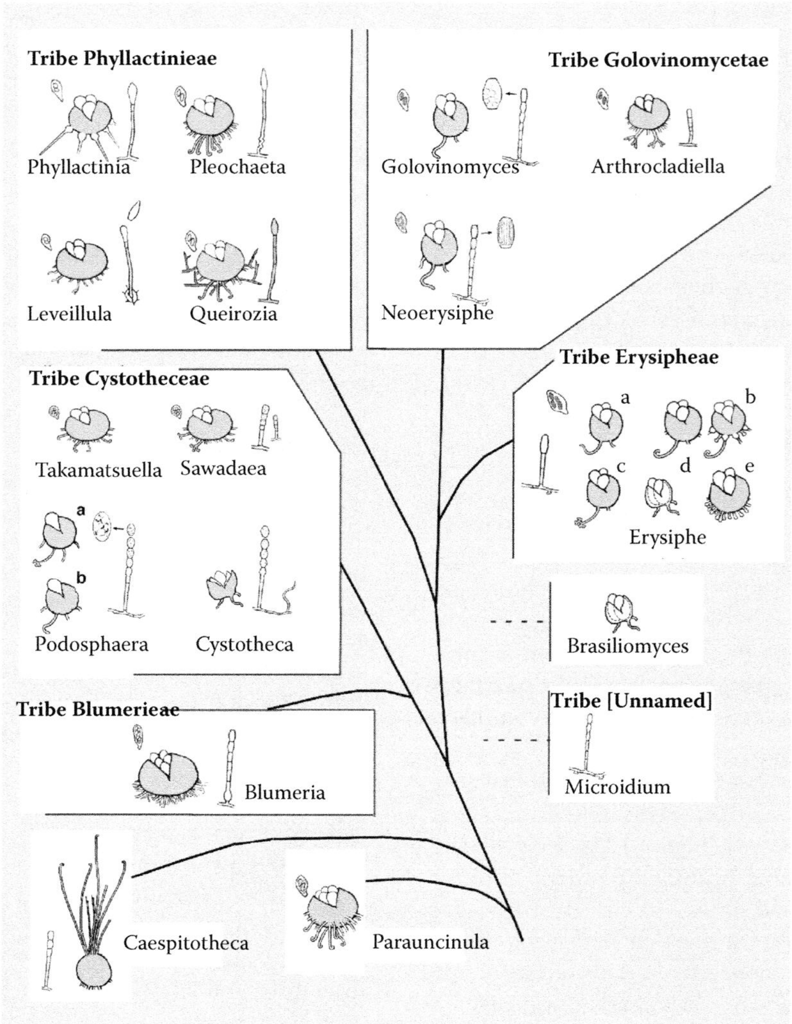

FIGURE 12.9 Classification of powdery mildews. Braun and Cook (2012) divided powdery mildews into 17 genera, each characterized by diagnostic vegetative characteristics of the conidiophores and conidia (which they had categorized using the then-existing anamorph generic names) as well as ascocarp morphology. The family *Erysiphaceae* is divided into six tribes, with two genera (*Caespitotheca* and *Parauncinula*) basal to the clade and not placed in tribes. Tribe *Erysipheae*: *Erysiphe* is shown with pseudoidium-type conidiophores (i.e., with conidia borne singly, not in chains, which defined the the anamorph genus *Pseudoidium* in Braun and Cook [2012]). The current genus *Erysiphe* embraces several former genera now considered as morphological sections: (a) *Erysiphe* sect. *Erysiphe*, with hair-like appendages; (b) *Erysiphe* sect. *Uncinula*, containing the former genera *Uncinula* and *Bulbouncinula*, with curled ends to the appendages; (c) *Erysiphe* sect. *Microsphaera*, with dichotomously branched appendage tips; (d) *Erysiphe* sect. *Californiomyces*, with reduced ascomata; and (e) *Erysiphe* sect. *Typhulochaeta*, with club-shaped appendages. *Brasilomyces* (below) is phylogenetically similar to *Erysiphe*, but its exact relationship to other genera is still unclear. Tribe *Golovinomycetae*: *Golovinomyces* has two-spore asci and euoidium-type conidiophores (with conidia in chains). *Neoerysiphe* has asci that only mature after overwintering and striatoidium-type conidiophores (with striate conidia). *Arthrocladiella* has dichotomously branched appendages and graciloidium-type, small conidiophores. Tribe *Phyllactinieae*: *Phyllactinia* has bulbous-based appendages and ovulariopsis-type conidiophores (conidiophores long, conidia large, clavate). *Pleochaeta* has curled appendages and ovulariopsis-type conidiophores. *Leveillula* is unusual in having internal mycelium and conidiophores that grow out of stomatal pores, has undifferentiated appendages and oidiopsis-type conidiophores (emerging from plant stomata, conidia dimorphic, the first-formed conidium lanceolate, the rest ellipsoidal or cylindrical). *Queirozia* has pigmented conidiophores and conidia, and the mycelium contains pigmented aerial hyphae. Tribe *Cystotheceae*: *Podosphaera* has fibroidium-type conidiophores (conidia borne in chains and with cytoplasmic inclusions). This genus now consists of (a) *Podosphaera* sect. *Podosphaera* (with branched appendages), and (b) *Podosphaera* sect. *Sphaerotheca* (with hair-like appendages). *Sawadaea* has curled appendages (that can be uni-, bi-, or trifurcate) and octagoidium-type dimorphic conidiophores. *Cystotheca* has an ascomatal outer wall that easily splits in two, and setoidium-type conidiophores (scattered among thick-walled aerial hyphae). Tribe *Blumerieae*: *Blumeria*, sole genus in this tribe, is found on grasses; it has oidium-type conidiophores (bulbous base to conidiophores; conidia in chains). Tribe [currently unnamed]: created to accommodate the genus *Microidium*, with no known ascocarp stage.

1. The first-formed conidium in a chain has a pointed tip, whereas subsequent conidia are blunt at both ends .... Tribe Phyllactinieae (genera Phyllactinia, Pleochaeta, and Leveillula)
1. All conidia formed have a similar shape ....Go to 2
2. Conidia formed singly and are ready to break off, although they may adhere to each other in a pseudo-chain; there is a clear demarcation between the conidiogenous cell and the mature conidium .... Tribe Erysipheae (genus Erysiphe)
2. Conidia formed in chains and mature gradually, while tightly connected to each other. Only the most mature conidia at the tip are ready to break off; no clear demarcation between conidiogenous cell and the first conidium in the chain .... Go to 3
3. Conidia have crystalline inclusions in the cytoplasm called fibrosin bodies that are highly refractive, particularly with phase microscopy ....Tribe Cystotheceae (genera Podosphaera, Sawadaea, Takamatsuella and Cystotheca)
3. Conidia may have vacuoles or small particles in the cytoplasm, but no crystalline inclusions ....Go to 4
4. Conidiophores with bulbous base; only found on grasses ....Tribe Blumerieae (genus Blumeria)
4. Conidiophores cylindrical; found on hosts other than grasses ....Go to 5
5. Conidiophores stout, with a width of 7 to 15 μm ....Tribe Golovinomyceteae (genera Golovinomyces, Neoerysiphe, and Arthrocladiella)
5. Conidiophores narrow, with a width of 4 to 7 μm ....Tribe [unnamed] (genus Microidium)

FIGURE 12.10 A Key to Powdery Mildew Tribes

FIGURE 12.11 Powdery mildew of grape, caused by *Erysiphe* (sect. *Uncinula*) *necator*. The powdery white bloom on the fruit makes these grapes unfit for consumption or processing. (Courtesy of Wayne Wilcox, Cornell University.)

until at least **veraison**, in order to maintain a functional leaf canopy to ripen the crop and avoid premature defoliation. Controlling later-season infection also helps to reduce the overwintering **inoculum** and may improve winter hardiness in colder regions.

Elemental sulfur has long been used internationally for powdery mildew suppression, although it is phytotoxic to some native American grapes (e.g., Concord) and hybrid cultivars. Modern fungicides such as the sterol inhibitors (both the demethylation inhibitor [DMI] and, more recently, morpholine groups) and strobilurins are also widely used in disease management programs (Chapter 26). Certain nontraditional products (e.g., oils, potassium bicarbonate and monopotassium phosphate salts, and dilute hydrogen peroxide) also provide suppression and are used to variable extents where they are labeled for this purpose. It appears that much of the activities of such materials are eradicative, presumably due to the susceptibility of the exposed fungal colony to topical treatments of these substances.

Powdery Mildew of Cereals

Powdery mildew diseases affect **monocots** as well as **dicots**. The pathogen *Blumeria graminis* has several host-specialized **formae speciales** (f. sp.) that infect different cereal crops. An example is *B. graminis* f. sp. *tritici*, a pathogen of wheat (*Triticum aestivum*) (Bockus et al., 2010).

White colonies of powdery mildew are most common on the upper surface of older wheat leaves. The haustoria of *Blumeria* are unusual because they have long finger-like projections. The conidiophores generate long chains of conidia that can germinate over a wide temperature range (1°C–30°C), with an optimum temperature of 15°C–22°C and an optimum relative humidity of 85%–100%. A complete disease cycle takes place in 7–10 days, so epidemics can develop quickly. Wheat plants are most susceptible when they are heavily fertilized and rapidly growing.

As wheat plants mature, ascocarps begin to appear on the leaf surface as long as both mating types are present. The ascocarps are important, because sexual recombination allows new races of powdery mildew to be produced. These new races may be able to grow on wheat cultivars that were not susceptible to the original powdery mildew population. Ascocarps on wheat stubble also function as overwintering structures, and in mild climates, the mycelium itself can survive. Infections by ascospores require rain and may occur in midsummer or fall. Powdery mildew of wheat prospers in both humid and semi-arid climates. Yield is decreased by powdery mildew, because photosynthesis is reduced, whereas transpiration and respiration are increased. Wheat yields may be reduced by as much as 40%.

In spite of the strong impact of this disease, it is not always cost-effective to use fungicides to control powdery mildew on wheat. Systemic fungicides are used for powdery mildew control in Europe, primarily. Host-plant resistance is used worldwide for powdery mildew management. Plant breeders work to supply wheat lines that are not susceptible to the powdery mildew **races** in a given geographic area; new lines are needed frequently to keep up with adaptations of the pathogen. Rotating crops and eliminating "volunteer" wheat plants and crop debris aid in disease management.

Powdery Mildew of Rose

Powdery mildew diseases are especially important on ornamental plants, on which any visible infection may lower the aesthetic appeal. Powdery mildew of rose (*Rosa* × *hybrida*) affects roses in gardens as well as in field or greenhouse production. It is the most common disease of roses grown in greenhouses for cut flowers and on potted miniature roses. Young foliage and pedicels are especially susceptible and may be completely covered with mildew (Figure 12.12). Young shoots and flower petals may be infected. The rose powdery mildew is *Podosphaera* (sect. *Sphaerotheca*) *pannosa*. The host range of this fungus includes primarily plants in the genera *Rosa* and *Prunus*, but also a few plants in other families, such as forsythia and smoke-tree (*Cotinus*).

FIGURE 12.12 Powdery mildew of rose. The flower buds and pedicels in the foreground are whitened by the sporulating mycelium of the fungus. (Courtesy of M.L. Daughtrey.)

Infections may develop quickly, and conidia may start to germinate within just a few hours of landing on the host (Horst, 1983). Chains of conidia are produced only 3–7 days after the initial infection. Disease development is optimum at 22°C. The fungus overwinters in buds unless the climate is so mild that conidia are continually produced year-round, as in greenhouses. Control of rose powdery mildew in greenhouses is achieved by air circulation with fans plus heating and ventilation to reduce humidity, coupled with the use of fungicides. Strobilurins, DMIs, and morpholine fungicides are commonly utilized for disease control in greenhouses. Products featuring potassium bicarbonate, botanical extracts, and biocontrols, including the fungus *Pseudozyma flocculosa*, have recently been developed. Gardeners wishing to circumvent the problem of powdery mildew on outdoor roses can seek disease-resistant cultivars and species.

Powdery Mildew on Cucurbits

Numerous vegetable crops are susceptible to powdery mildew, but cucurbits are arguably the group most severely affected. The powdery mildew on cucurbit crops (*Cucumis* species, including squash, pumpkin, and cucumber) is *Podosphaera* (sect. *Sphaerotheca*) *xanthii* (referred to *Sphaerotheca fuliginea* or *S. fusca* in earlier literature). Infections on both upper and lower leaf surfaces occur, and these reduce yield by lowering plant vigor and increasing the number of sun-scorched fruits (Figure 12.13). Resistant cultivars of cucumbers and melons are available, but most

FIGURE 12.13 Powdery mildew on squash begins as discrete white colonies. (Courtesy of M.L. Daughtrey.)

squash varieties are susceptible. Susceptible melons may have significantly poorer fruit quality due to low sugar content. Even the flavor of winter squash can be harmed because the fruits of mildew-infected plants have fewer stored soluble solids that affect taste. The color and handle quality of pumpkins can be ruined by powdery mildew, making them less desirable as "Jack-o'-lanterns." The disease develops on cucurbits during warm summer weather (a mean temperature of 68°F–80°F is most favorable). Conidia are thought to be airborne over long distances to initiate infections, with diseased plants in southern states releasing clouds of conidia in early spring that are carried north with prevailing winds.

The fungicides used for disease management in cucurbits are often systemic materials with specific, single-site, modes of action that make them vulnerable to the development of resistance in the pathogen (McGrath, 2001). The fungicide benomyl is no longer effective against cucurbit powdery mildew because of the development of resistance. Although it is unclear if sexual recombination occurs to a significant extent in U.S. populations of *P. xanthii*, development of resistance to new fungicides can be very rapid. Mutations in the DNA of conidia are uncommon, but because there are so many conidia, there are many chances for mutations to occur. Because powdery mildews can carry out so many secondary cycles in one season, an advantageous mutation can spread rapidly. Growers of any crop with high susceptibility to a powdery mildew are encouraged to rotate among chemical classes with different modes of action to delay the development of resistance. The use of materials with multisite modes of action, such as chlorothalonil, copper, sulfur, horticultural oil, and potassium bicarbonate, is also a good strategy to avoid resistance.

An **integrated pest management** (IPM) program for cucurbit powdery mildew can be very effective (McGrath and Staniszewska, 1996; McGrath, 2001; Chapter 28). Elements of such a program include using resistant varieties when available, scouting for the first colonies beginning at the time of fruit initiation, using air-assist sprayers to maximize spray coverage on difficult-to-cover lower leaves, and using fungicides according to a resistance management strategy. For greenhouse-grown cucumbers, silicon amendment of the nutrient solution has reduced the disease. A number of organisms, including the fungi *Pseudozyma flocculosa*, *Sporothrix rugulosa*, *Tilletiopsis* spp., *Ampelomyces quisqualis*, and *Verticillium lecanii*, have been used for biocontrol (Chapter 27) of cucumber powdery mildew in greenhouses (Dik et al., 1998). Mechanisms of antagonism used by these fungi include **hyperparasitism** or **antibiosis**. As powdery mildews may thrive at lower relative humidity than the antagonistic fungi that might be deployed against them, environmental conditions in a given situation may determine whether biocontrol is effective.

Powdery Mildew on Flowering Dogwood

Many woody plant species are subject to one or more powdery mildews, but these diseases rarely have a significant economic impact. Lilac foliage, for example, frequently shows conspicuous powdery mildew in late summer (Figure 12.14), but the shrubs are attractive again during bloom the following spring. Powdery mildew has been noted on dogwoods (*Cornus* species) since the 1800s, but a serious powdery mildew disease on *Cornus florida*, the flowering dogwood, became apparent in the eastern United States only in the mid-1990s (Figure 12.15). *Erysiphe* (sect. *Microsphaera*) *pulchra* (Cooke and Peck) is the powdery mildew that

FIGURE 12.14 Powdery mildew of lilac is one of the most common diseases in the home landscape. (Courtesy of Robert Kent.)

FIGURE 12.15 Dogwood powdery mildew causes curling, discoloration, and deformation of leaves at the tips of branches; the fungus may entirely coat the leaves. The eastern flowering dogwood (*Cornus florida*) is quite susceptible, but most cultivars of the Kousa dogwood (*Cornus kousa*) are quite resistant. (Courtesy of M.L. Daughtrey.)

now affects flowering dogwood. Aesthetic injury is caused on landscape trees by whitening and twisting of the leaves at the tips of branches in mid-summer, while economic injury is seen during nursery production when young trees with powdery mildew are severely stunted. In many cases, infection of landscape trees is subtle, such that the thin coating of mycelium on the leaves is often overlooked. Symptoms of reddening or leaf scorch are often seen on infected leaves in dry summers, and it is thought that the mildew may contribute to the drought stress on the host tree.

Ascocarps on leaf debris are the only documented overwintering mechanism for this powdery mildew. The abundance of ascocarps varies from year to year, perhaps due to variation in weather conditions during the fall when they are maturing. The non-native Kousa dogwood, *Cornus kousa*, is not appreciably affected by powdery mildew. Several selections of powdery mildew-resistant cultivars of *C. florida* have been developed. Fungicides such as propiconazole and chlorothalonil are used for disease suppression during nursery crop production.

In some ways, powdery mildews seem to be mild diseases, because the fungi only rarely kill the plants that they parasitize. Powdery mildew fungi and their hosts come closely together in a carefully balanced parasitic relationship that results in nutrient flow to the powdery mildew without extensive death of plant host cells. Perennial plants may be infected by powdery mildew summer after summer, and still leaf out vigorously every spring. Annual plants usually bear fruit and set seed in spite of powdery mildew. Yet to the horticultural producers aiming to have high yield, high-quality crops, or to gardeners wanting attractive and vigorous plants, powdery mildews are formidable diseases. Curbing powdery mildew epidemics on economic hosts is essential to maximize the growth potential of the plants and maintain the marketability of the products harvested from them.

LABORATORY EXERCISES

Experiment 1. Powdery Mildew Morphology

This lab will familiarize the student with the variations in sexual and asexual spore structures in some common powdery mildew fungi. Students should work alone if possible.

Materials

Students will require the following materials to complete this exercise:

- Dissecting microscopes and compound microscopes (the latter with at least 40× magnification; 100× oil immersion is necessary if Step 5 is attempted).
- Microscope slides, coverslips, dissecting tools, water, and 70% ethanol in dropper bottles (and lacto-cotton blue stain in dropper bottles and immersion oil if Step 5 is attempted. The recipe for lacto-cotton blue mounting medium is 250 mL of glycerol, 100 mL of 85% lactic acid, 50 mL of water, and 3 mL of cotton blue stock [1% cotton blue crystals in lactic acid]).
- Fresh or dried plant materials with powdery mildew colonies on it, including ascocarps. Ideally, there should be a variety of fresh materials for Steps 1 and 5, and then a number of fresh or dried specimens labeled as unknowns ("Unknown #1," "Unknown #2," etc.) for Steps 2–4.

The instructor can easily collect fresh plant materials with powdery mildew colonies on it, outdoors in spring and summer, or from a greenhouse all year long. To collect materials that include ascocarps, collect host plants at the end of summer or in early fall for immediate use or to dry and store this material for use all year long. Figure 12.2 helps the instructor find ascocarps of diverse morphological types.

Follow the instructions outlined in Procedure 12.1 to complete the exercise.

Anticipated Results

In Step 1, the students should gain an appreciation for the superficial nature of the colony and (if fresh materials of *Erysiphe* and *Golovinomyces* [or *Podosphaera*] are present) be able to identify whether conidia are borne singly or in chains. Colonies of *Erysiphe* grown in humid conditions may have conidia in pseudochains (conidia simply sticking together after formation) rather than in the true chains of *Golovinomyces* or *Podosphaera* (conidia remaining attached to each other while gradually maturing). In Step 2, if *Podosphaera* is present, fibrosin bodies should be easily visible (especially with phase microscopy). If *Phyllactinia* or *Leveillula* is present, the student should be able to find the two types of conidia. In Step 3, students should be able to key out specimens to tribe based on conidiophore characteristics. In Step 4, students should be able to confirm their identification of tribe based on the morphological features of ascocarps (appendage type and number of asci per ascoma). In Step 5, students should be able to detach and stain the epidermal peel, see appressoria, and determine if they are simple or lobed. They should also be able to see a haustorium and appreciate how little of the body of the fungus is within the plant host.

Procedure 12.1
Powdery Mildew Morphology

Step	Instructions and Comments
1	**Asexual stages**: Examine plant material provided for this step (fresh if possible) under a dissecting microscope to identify superficial mycelium, **conidiophores**, and conidia.
2	**Conidia**: Pick one of the unknown mildews provided and remove a small piece of plant material with a powdery mildew colony on it. Touch the material to a water droplet on a microscope slide to detach conidia into the droplet. A coverslip can be placed on top of the droplet to examine the conidia. Mounting in water alone can cause conidia to aggregate around air bubbles in the preparation. While in general this is bad microscope technique, in the case of powdery mildews, air bubbles often allow the texture of the conidial surface, any bumps or ridges (such as the striations on the outer wall of conidia of *Neoerysiphe*), to be more apparent along the air: conidium interface. Air bubbles can be removed by adding a drop of 70% ethanol to the edge of the coverslip and waiting for the alcohol to wash through the water. This eliminates the bubbles and liberates the spores into the suspension. The conidia can now be examined for the presence of more than one shape of conidium (dimorphic conidia is a feature of the tribe Phyllactinieae) and for internal features such as crystal inclusions (fibrosin bodies in the conidia of members of the tribe Cystotheceae).
3	**Conidiophores**: Using the same unknown mildew used in Step 2, take forceps and scalpel and cut off a small piece of leaf tissue that includes conidiophores of the powdery mildew. Mount in water or lacto-cotton blue and examine at 40× magnification. Are conidia borne singly or in chains? With the observations from Steps 2–3, it is now possible to use the key in Figure 12.10 to identify the tribe to which the unknown powdery mildew belongs.
4	**Ascocarps**: Using the same unknown mildew, examine the sexual stage with the compound microscope. Remove ascocarps from the plant material using a dissecting needle or forceps and place in a drop of water on a microscope slide. Gently place a coverslip on top of the droplet and examine the ascocarps for presence and type of appendages (70% ethanol may again be needed to drive out air bubbles, particularly if dried material is used). Press the coverslip gently to crush the ascocarps, and determine the number of **asci** per ascocarp. The number of ascospores per ascus will be apparent in mature specimens. Figure 12.9 can be used to determine if the ascocarp type is consistent with the tribe identified using the key in Figure 12.10.
5	If fresh material is available, make an epidermal peel and examine the powdery mildew colony for superficial **appressoria** and **haustoria** within the epidermal cells. Select the host with the thickest, most succulent leaf, and use a scalpel to carve a square into the upper or lower surface of a leaf where powdery mildew is present. Center the square so that it includes the very margin of the colony, where there is superficial mycelium but not yet conidiophores. Using the finest forceps available, grab a corner of the cut epidermis and peel it off, depositing it in a small droplet of water and using dissecting needles to smooth it out with the intact epidermis side up. A drop of lacto-cotton blue can be placed over it for a few minutes, until the fungal mycelium turns blue, but the plant tissue remains mostly unstained. The cotton blue can be removed by placing the tip of a paper towel to the edge of the droplet and wicking it up. More water is added and a coverslip can be placed over the epidermal peel. Using an oil immersion objective, examine the superficial mycelium and appressoria, and by carefully focusing down from the appressoria, observe haustoria (which may have become stained by the cotton blue).

Questions

- Why do you think the conidia are borne on the top of conidiophores instead of forming directly from the superficial mycelium on the epidermis of the plant?
- What features of the ascocarps make them ideal as overwintering structures?
- Why does it make sense that powdery mildews rarely kill their hosts, but root rot fungi often do? Figure 12.10 is a key that roughly follows phylogenetically significant morphological

characteristics, in order to give you a sense of the relationships among powdery mildews as well as to help you identify them. However, if you were trying to help a grower with a vexing powdery mildew problem, you might want a fast and easy key. How would you design one?

Experiment 2. Powdery Mildew Control

Powdery mildews are difficult to study because they are obligate biotrophs, but it is possible to study them in the laboratory—even to compare their susceptibility to fungicidal agents. Students may work individually or in small teams to complete this exercise.

Materials

The following materials are needed for this exercise:

- Two-week-old pumpkin or summer squash seedlings with large cotyledons (optimum age for use may differ slightly by season, region, or cultivar, therefore several batches of seeds should be sown a few days apart to ensure that plants are ready on the day of the laboratory exercise). "Seneca Prolific" is a summer squash cultivar that is very susceptible to powdery mildew.
- A cucurbit (or detached leaves from a cucurbit) infected with *Podosphaera xanthii*. Such infected plant materials may be readily available in spring from retail garden centers accidentally selling infected transplants and are more likely available in summer from home gardens or farm fields of pumpkin or squash. Detached leaves covered with powdery mildew can be stored with moist toweling in a cooler or refrigerator overnight for use the next day. Infected materials should be stored far away from the test seedlings, and anyone handling the infected plants should not go near the test seedlings.
- For safety, hazardous chemicals should be avoided. A commercially available refined mineral oil, such as JMS Stylet Oil, may be used if compatible with college safety protocols. Otherwise, a homemade preparation of 0.1% emulsified oil could be prepared with peanut oil, corn oil, canola oil, grape seed oil, safflower oil, or sunflower oil (5 mL of oil + 4.5 mL of distilled water + 0.5 mL of Tween 80 shaken vigorously and diluted to a 0.1% oil emulsion). The test substance should be applied by the instructor according to label instructions (if a commercial product is used) and with appropriate safety equipment, including safety glasses, lab coats, and nitrile gloves. Two spray bottles should be available: one for the oil emulsion and one for the water control.
- Cork borers (approximately 1-cm-diameter, dissecting tools (including forceps and dissecting needles), laboratory marking pens, and jars of 70% ethanol. Alcohol lamps or gas burners should be available for burning ethanol off tools after they have been sterilized, but if this is not possible, tools can be air-dried before use.
- Plates of 1% water agar with the bottom plate surface divided into approximate halves, one-half for the oil-treated discs, one for the water-treated controls.
- Dissecting microscopes.
- Computer programs as available and appropriate: a spreadsheet program and a statistical analysis program.

Follow the instructions outlined in Procedure 12.2 to complete this exercise.

Anticipated Results

Students should observe a reduction in the percent colonization of oil-treated cotyledon discs. It is also possible that a reduction in incidence will be evident. They may also notice a difference in the "quality" of infection. This lab presents, in miniature, what many professional plant pathologists do for a living: evaluate the effect of substances on fungal growth. The instructor has great latitude in making this a simple lab or one that incorporates a large number of plant pathological concepts. If a commercial oil emulsion product is used, students should be taught the concepts of pesticide safety and proper understanding of pesticide labels. Students should be taught the basics of experimental design appropriate to their level of knowledge; they should at least understand the use of control treatments and replications. They should also be able to understand the difference between an *in vitro* assay such as this one and a field trial. Finally, they should be introduced to whatever level of statistics is appropriate to determine if the treatment has had an effect. The instructor could leave them to devise their own spreadsheets or tell them how to organize the data (e.g., the percent coverage data could be averaged over the four discs and the agar plates used as replicates for a mixed analysis of variance [ANOVA], with data normalized using an arcsine square root transformation). The instructor could let students analyze their data as individual experiments of three replicates or take the entire classroom's data and reanalyze them (possibly including the students as a variable to let them recognize that students may differ in the way they assess percent colonization), and discuss whether these analytical choices change the results.

Procedure 12.2
Powdery Mildew Control

Step	Instructions and Comments
1	Divide cucurbit seedlings into control and treatment groups. Following pesticide label instructions and wearing appropriate safety equipment, the instructor should treat plants in a chemical fume hood (or outdoors): one set of plants should be sprayed with the test solution and the other with water. Plants can be handled after the applied test solutions have dried. (This step can be performed by the instructor a day before the laboratory exercise or the morning of the exercise.)
2	Students (wearing gloves) should cut discs from treated and control cotyledons with cork borers (sterilized with 70% alcohol and flamed before use and between treatments). Begin with the control treatment to avoid accidentally contaminating the cotyledons with oil on the cork borer or forceps. Cut 6–8 discs per cotyledon, for a total of 12 discs per treatment, and a few extra control discs to practice inoculation (Step 3). The control discs can be placed in groups of four, face-up, in each of three water agar plates; they should be placed onto one-half of the plate and set firmly so that the bottom surface of each disc is in good contact with the agar and will not move or dry out. Place four discs from oil-treated plants on the other half of the plate, making sure the treatments are clearly labeled.
3	Examine an infected leaf under the dissecting microscope and identify the conidiophores. Are conidia borne singly or in chains? Practice removing spores from a single chain using the tip of a dissecting needle, then placing them gently on the top surface of a cotyledon disc. When this inoculation technique has been mastered, inoculate each disc in the center with a few conidia, trying to do this as uniformly as possible, with the set of three agar plates.
4	Wait for a week or so for the colonies to grow. Plates should be set under grow lights (about 1 foot away from the lights). When control discs have colonies that cover approximately 80% of the disc, students should rate the amount of coverage for each disc and tabulate the data.

Questions

- What is the purpose of the control treatment?
- Why did the plates need to be incubated under lights?
- Why did you inoculate three sets of plates instead of one?
- What are some of the drawbacks of such simple, *in vitro* assays compared with field trials? What are some advantages?
- What are some of the difficulties in determining percent coverage of leaf tissue? Is percent coverage the best way to measure disease severity? What other ways could you assess damage from powdery mildew? Are there other diseases where other ways of assessing damage would be more useful?

REFERENCES

Bockus, W. W., R. L. Bowden, R. M. Hunger, W. L. Morrill, T. D. Murray, and R. W. Smiley. 2010. *Compendium of Wheat Diseases and Pests*, 3rd edition. St. Paul, MN: American Phytopathological Society.

Braun, U. 1987. *A Monograph of the Erysiphales (Powdery Mildews)*. Berlin-Stuttgart, Germany: J. Cramer.

Braun, U. 1995. *The Powdery Mildews (Erysiphales) of Europe*. Jena, Germany: G. Fischer Verlag.

Braun, U., and R. T. A. Cook. 2012. *Taxonomic Manual of the Erysiphales (Powdery Mildews)*. Utrecht, the Netherlands: CBS-KNAW Fungal Biodiversity Centre.

Cook, R. T. A., A. J. Inman, and C. Billings. 1997. Identification and classification of powdery mildew anamorphs using light and scanning electron microscopy and host range data. *Mycol. Res.* 101:975–1002.

Dik, A. J., M. A. Verhaar, and R. R. Belanger. 1998. Comparison of three biocontrol agents against cucumber powdery mildew (*Sphaerotheca fuliginea*) in semi-commercial scale glasshouse trials. *Eur. J. Plant Path.* 104:413–423.

Farr, D. F., G. F. Bills, G. P. Chamuris, and A.Y. Rossman. 1989. *Fungi on Plants and Plant Products in the United States*. St. Paul, MN: American Phytopathological Society.

Farr, D. F., and A. Y. Rossman. 2016. Fungal Databases, Systematic Mycology and Microbiology Laboratory, ARS, USDA. http://nt.ars-grin.gov/fungaldatabases/ [accessed Jul 3, 2016].

Gadoury, D. M., and R. C. Pearson. 1988. Initiation, development, dispersal and survival of cleistothecia of *Uncinula necator* in New York vineyards. *Phytopathology* 78:1413–1421.

Horst, R. K. 1983. *Compendium of Rose Diseases*. St. Paul, MN: American Phytopathological Society.

McGrath, M. T. 2001. Fungicide resistance in cucurbit powdery mildew: Experiences and challenges. *Plant Dis*. 85:236–245.

McGrath, M. T., and H. Staniszewska. 1996. Management of powdery mildew in summer squash with host resistance, disease threshold-based fungicide programs, or an integrated program. *Plant Dis*. 80:1044–1052.

Nicholson, R. L., and H. Kunoh. 1995. Early interactions, adhesion, and establishment of the infection court by *Erysiphe graminis*. *Can. J. Bot*. 73(SUPPL): S609–S615.

Saenz, G. S., and J. W. Taylor. 1999. Phylogeny of the Erysiphales (powdery mildews) inferred from internal transcribed spacer (ITS) ribosomal DNA sequences. *Can. J. Bot*. 77:150–169.

Takamatsu, S., T. Hirata, Y. Sato, and Y. Nomura. 1999. Phylogenetic relationship of Microsphaera and Erysiphe sect. Erysiphe (powdery mildews) inferred from the rDNA ITS sequences. *Mycoscience* 40:259–268.

Takamatsu, S., T. Hirata, and Y. Sato. 2000. A parasitic transition from trees to herbs occurred at least two times in tribe Cystotheceae (Erysiphaceae): Evidence from nuclear ribosomal DNA. *Mycol. Research* 104:1304–1311.

13 Plant Pathogenic Species of the Ascomycota

Pezizomycotina

Ning Zhang, Richard E. Baird, and Robert N. Trigiano

Concept Box

- The single, most diagnostic characteristic of species classified in Ascomycota is an ascus, the saclike structure containing ascospores that resulted from the sexual process.
- Pezizomycotina contains the filamentous (mycelium) fungi that constitute the majority of the species in Ascomycota.
- The ploidy of these fungi is primarily haploid with very short-lived dikaryotic and diploid phases.
- In this subphylum, persistent asci and ascospores develop within a fruiting body that may be cleistothecium, perithecium, pseudothecium, or apothecium.
- Fungi in this group may reproduce asexually only, sexually only, or both asexually and sexually.
- Prior classification systems provided a name for the sexually reproductive phase (ascus: teleomorph) and one or more names for the asexually reproductive phase (conidium: anamorph). This has led to some confusion and two names for the same organism do not conform to rules of systematic nomenclature.
- "One Fungus = One Name," a major change in fungal nomenclature, which has not been adopted yet, will affect the scientific names of many plant pathogenic fungi that have both sexual and asexual states.

Ascomycota is the largest phylum in the kingdom Fungi and contains over 65,000 species, which constitutes 66% of all known fungal species (Kirk et al., 2008). The unifying diagnostic characteristic of species in Ascomycota is the **ascus**, a sac-like structure containing ascospores that resulted from sexual reproduction and meiosis (Figures 13.1 and 13.2). Asci and sometimes paraphyses (sterile hyphae) typically originate from the **hymenium,** which is termed the fertile layer in the fruiting body. Based on molecular phylogenetic analyses and morphology, Ascomycota is divided into the following three subphyla (Figure 13.3): Taphrinomycotina ("Archiascomycetes"), Saccharomycotina (Chapter 11), and the Pezizomycotina ("Ascomycotina"). Most members of the first two subphyla are yeasts or yeast-like organisms. This chapter focuses on Pezizomycotina (excluding the powdery mildews: Chapter 12), which are the filamentous ascomycetes that constitute the majority of Ascomycota, and their asexual reproductive modes that have been treated previously as "Deuteromycetes" or "mitosporic fungi." We no longer use the two former classification schemes for conidial forms of Ascomycetes and, instead, treat them as asexual forms of species in Ascomycota. This change of classification systems will be explored in more detail in subsequent paragraphs.

Members of Ascomycota are ubiquitous and cosmopolitan. They play important roles in all ecosystems virtually as saprotrophs in decomposition and nutrient cycling, as endophytes, pathogens of plants and animals, and as mycoparasites attacking other fungi. Most plant pathogens in Ascomycota are found in the classes Sordariomycetes, Leotiomycetes, and Dothideomycetes. These include some of the best-known plant pathogens such as *Cryphonectria parasitica* (the causal agent of chestnut blight), *Pyricularia* (*Magnaporthe*) *oryzae* (the cause of rice blast), and various species of *Fusarium*, which cause many different wilt and root rot diseases) (Alexopoulos et al., 1996).

About 42% of known Ascomycota species are lichenized (Kirk et al., 2008). The lichenized ascomycetes have an obligate symbiotic relationship with photosynthetic symbionts, which is either a unicellular green alga or a cyanobacterium (blue-green alga). The ascomycete member is never found without its photosynthetic partner under natural conditions, whereas the photosynthetic partner sometimes can be free-living in nature (Curry and Baird, 2007).

Endophytes refer to fungi that live inside the leaves and/or stems of apparently healthy plants (Alexopoulos et al., 1996). Numerous endophytic fungal survey studies have demonstrated that species in Ascomycota are the most frequently isolated fungal phylum from terrestrial plant foliage. The best-studied endophytes belong to Hypocreales of the Sordariomycetes (e.g., *Balansia* and *Epichloë*). The infected host plants often benefit from increased drought resistance, reduced feeding by insects, and limited pathogen infections (Alexopoulos et al., 1996). There are also many ascomycetes associated with plant roots, such as *Phialocephala fortinii–Acephala applanata* species complex, known as the dark-septate endophytes, but the functions of these fungi are still not well understood.

Some members of Ascomycota (e.g., *Aspergillus fumigatus*, *Sporothrix schenkii,* and *Fusarium solani* species complex) are associated with opportunistic infections of humans and other animals. As symbionts of arthropods, ascomycetes comprise a diverse assemblage of species that range from antagonistic to mutualistic. For example, species of *Ceratocystis* and *Ambrosiella* are associated with bark beetles, which disperse fungal spores to other individuals of the host. *Cordyceps* species directly parasitize a broad range of arthropods. As an alternative to chemical pesticides, species of *Trichoderma* have been developed into commercial products for plant disease management (Chapter 27).

Saprobic ascomycetes function in the decomposition and nutrient cycling of plant litter including wood, herbaceous stems, and dung. For example, *Chaetomium* species, an important cellulolytic organism, is responsible for the destruction of paper and fabrics. Ascomycota also contains species known as producers of some of the most important fungal secondary metabolites. These include trichothecene mycotoxins formed by many members of Sordariomycetes and ergot and other alkaloids produced by the grass endophytes *Claviceps* and *Epichloë*.

REPRODUCTION AND LIFE CYCLE OF ASCOMYCOTA

Three reproductive categories are recognized in the Ascomycota (Curry and Baird, 2007). First, some ascomycetes only reproduce sexually, for example, *Anisogramma anomala*, the eastern filbert blight fungus. Second, for some ascomycetes, such as *Fusarium oxysporum*, only asexual reproduction is known, whereas in the third category, some species produce both ascospores and mitotic or asexual spores. The life cycle of a typical plant pathogenic ascomycetous fungus, *Pyricularia* (*Magnaporthe*) *oryzae*, is shown in Figure 13.1. The sexual state of a fungal species is called **teleomorph**, and **anamorph** refers to the asexual state.

For the sexually reproducing species, some are **homothallic**; that is, they reproduce sexually by itself without the aid of another individual. Those that are self-sterile and require another strain to reproduce sexually are called

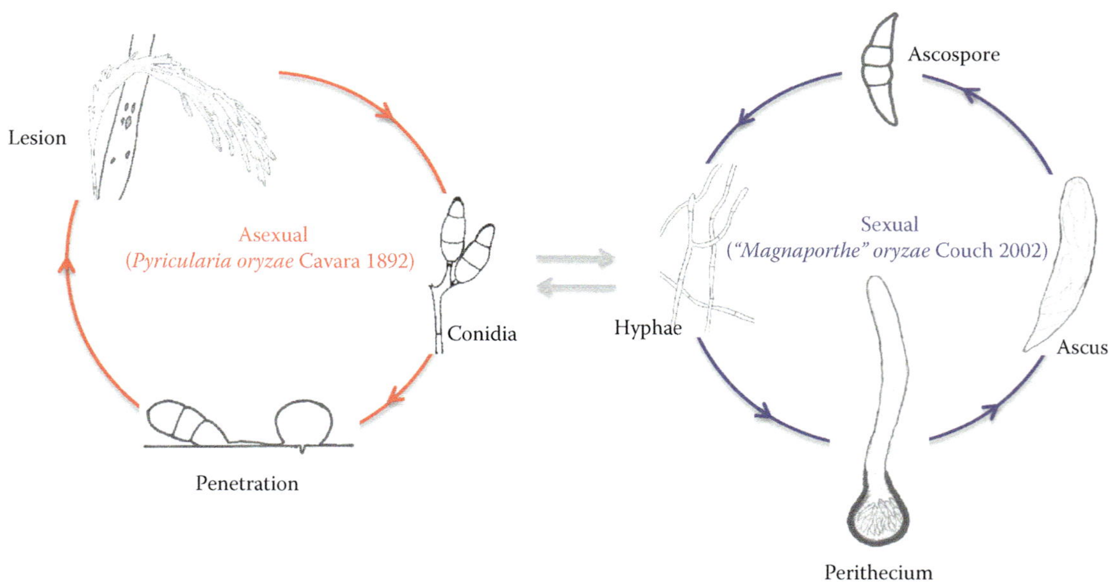

FIGURE 13.1 Life cycle of the rice blast fungus. (Courtesy of N. Zhang.)

heterothallic. The mating system in Ascomycota is controlled by the *MAT* locus, which specifies one of two alternative mating types, for example, *MAT1-1* and *MAT1-2*. Sexual reproduction in heterothallic species occurs only between individuals of different mating types. Asci usually arise from ascogenous hyphae through the mediation of a crozier, a structure that functions to ensure that only two compatible nuclei are sequestered in the developing ascus (Alexopoulos et al., 1996).

ASEXUAL FORMS OF ASCOMYCOTA

Ascomycota is an anamorph-rich phylum with great diversity and complexity of asexual forms. Many lichens reproduce asexually by vegetative propagules called soredia, a structure composed of algal cells surrounded by fungal hyphae. For the non-lichen-forming ascomycetes, asexual reproduction may be accomplished by fragmentation of mycelium, formation of chlamydospores or **conidia** (Figure 13.4). In earlier classification systems, Ascomycete species (and some basidiomycetes: Chapter 15) that had no known sexually produced spore were placed in "Deuteromycetes" or mitosporic fungi, which refers to conidia or spores formed by the asexual process of mitosis. Although solely asexually reproducing species do not form the diagnostic asci and ascospores of Ascomycota, molecular phylogenetic analyses and ultrastructural features of the hyphae and conidia support their placement in Ascomycota. Now that the asexually reproducing species are integrated into the classification scheme for other ascomycetes, "Deuteromycetes" are discontinued as a formal taxonomic term. However, although the older classification system is not phylogenetically correct, it is still very useful for the identification of pathogens and other fungi in Ascomycota. Therefore, a consideration of various asexually reproducing forms represented in Ascomycota is useful.

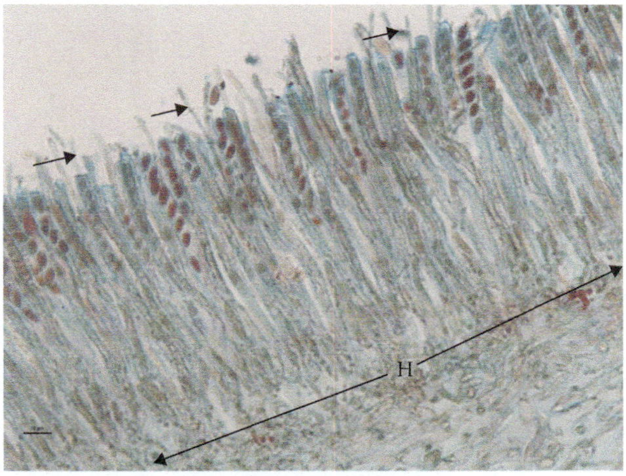

FIGURE 13.2 Asci and ascospores of *Sclerotinia* sp. Bar (lower left) = 10 μm. Note the sterile, hairlike hyphae (paraphyses: arrows) between and intermixed with asci. Not all species in Ascomycota have paraphyses. The hymenium (H) is the layer from which asci and paraphyses originate. (Courtesy of Triarch, Inc., Wisconsin [Slide 3-42B].)

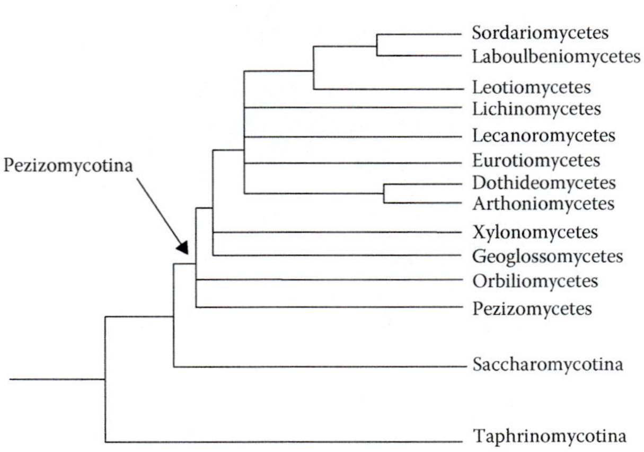

FIGURE 13.3 Phylogeny of major groups of Ascomycota. (Courtesy of N. Zhang.)

Many ascomycete fungi reproduce asexually by means of conidia—nonmotile, mitotic spores formed at the tip or side of sporogenous (conidiogenous) cells. A **conidiogenous cell** refers to the hyphal cell from which a conidium is formed directly (Figure 13.4). A **conidiophore** is a simple or branched, specialized hypha that bears one or more conidiogenous cells. Conidiophores may be formed singly or grouped together to form specialized fruiting bodies (**conidiomata**), such as pycnidia, acervuli, sporodochia, or synnemata (Alexopoulos et al., 1996). Many dichotomous keys to genera of fungi as well as descriptions of species and genera refer to these structures. Asexually reproducing fungi that produce conidia in pycnidia or acervuli were classified as Coelomycetes, whereas in contrast, the Hyphomycetes produced conidia on hyphae that are not enclosed in a structure composed of fungal or host tissue. Lastly, there were a few species in this phylum that, in addition to producing conidia, produce resting or overwintering reproductive structures called sclerotia.

A **pycnidium** (Figure 13.5) is a spherical or inversely pear-shaped hollow conidiomata internally lined with conidiophores and composed entirely of fungal tissues.

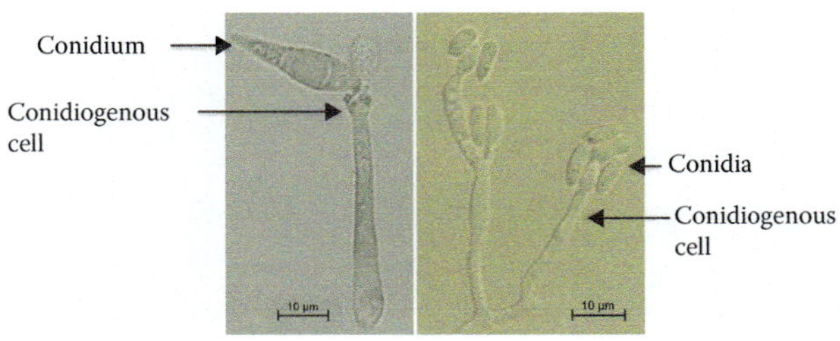

FIGURE 13.4 Examples of conidia and conidiogenous cells. (Courtesy of Jing Luo and N. Zhang.)

Pycnidia may be formed, either embedded in host tissue with only the opening (ostiole) showing, or on/in a fungal "mat" on the surface of the plant organ. Some genera of fungi produce sterile, hairlike structures (setae) within the pycnidium or at the opening of the body. Pycnidia bear a superficial resemblance to perithecia (an ascocarp type discussed later) and can only be identified correctly by the presence of conidia and the absence of asci and ascospores. Generally, mature conidia are released from an opening (ostiole) at the top. *Phoma*, *Phyllosticta*, and *Septoria* are common examples of genera that produce pycnidia.

An **acervulus** is a flat- or saucer-shaped cushion of conidiophores embedded just below the epidermal or cuticle layer of plant tissue (Figure 13.6). This structure can look much like the pycnidium. The difference is the "cover" that is composed of host plant epidermal tissue. Acervuli may or may not contain setae, which can be important for the identification of genera and species. Conidia are released from acervuli after the host epidermis rupture. *Colletotrichum*, *Pestalotia*, and *Discula* are older names for genera that produce acervuli.

A **sporodochium** is a small, dense structure composed of hyphae arranged as a tissue, stroma, and bears conidiophores and conidia either in or on the surface (Figure 13.7). There is no specific arrangement of individual conidiophores on this structure and sporodochia are commonly produced by *Fusarium*, *Myrothecium*, and *Epicoccum* species.

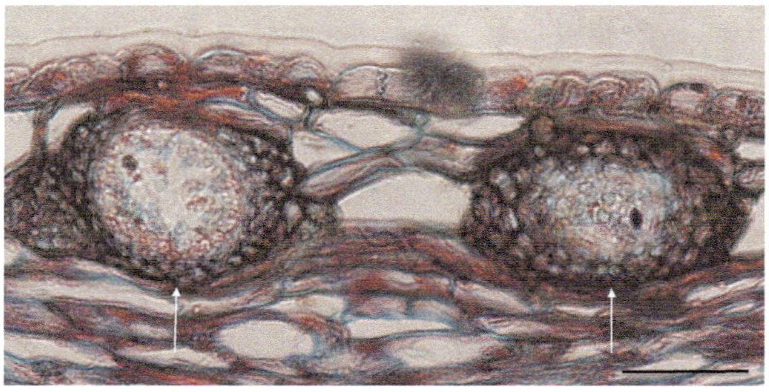

FIGURE 13.5 Pycnidia (arrows) of *Phyllosticta* sp. Bar = 50 μm. (Courtesy of Triarch, Inc., Wisconsin [Slide 3-42B].)

FIGURE 13.6 Acervuli (arrows) of *Colletotrichum* sp. Bar = 50 μm. (Courtesy of Triarch, Inc., Wisconsin [Lab 815].)

A **synnematum** is a collection of conidiophores that are fused at their bases and along the shaft (Figure 13.8a and b). It resembles bundles of wheat stems and produces conidia along the length of the conidiophores and/or only at the apex. Common genera that produce synnemata are *Graphium*, *Stilbella*, and *Dematophora*.

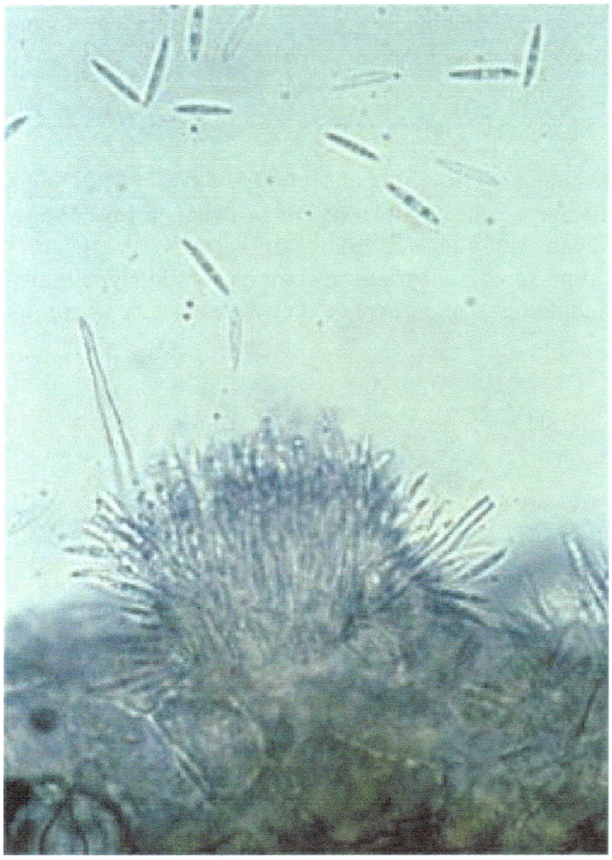

FIGURE 13.7 Sporodochium of *Volutella pachysandrae*. (Copyright 2015 American Phytopathological Society. [http://www.apsnet.org/edcenter/illglossary/Article%20Images/Forms/DispForm.aspx?ID=712].)

(a) (b)

FIGURE 13.8 Light (a) and scanning electron (b) micrographs of synnemata of *Ophiostoma (Graphium) ulmi*. (Copyright 2015 American Phytopathological Society. [http://www.apsnet.org/edcenter/illglossary/Article%20Images/Forms/DispForm.aspx?ID=738].)

A **sclerotium** is a mass of hyphae that is composed of very compact, hard tissues and may be differentiated into distinctive layers (Figure 13.9). Sclerotia are produced in various sizes and shapes and are useful in some classification schemes. Sclerotia are resistant to adverse environmental conditions and can permit the fungus to survive for extended periods without a suitable host. Some common genera that produce sclerotia are *Sclerotium*, *Claviceps*, and *Verticillium* (microsclerotia).

A **chlamydospore** is typically a large, thick-walled nonornamented survival structure rich in lipid reserves and is produced either at the tip or intercalary (in the middle) of a hyphae (Figure 13.10). Many times chlamydospores are darkly pigmented. Some genera that form chlamydospores are *Fusarium*, *Thielaviopsis*, and *Aspergillus*, as well as many other phyla of fungi and fungus-like organisms.

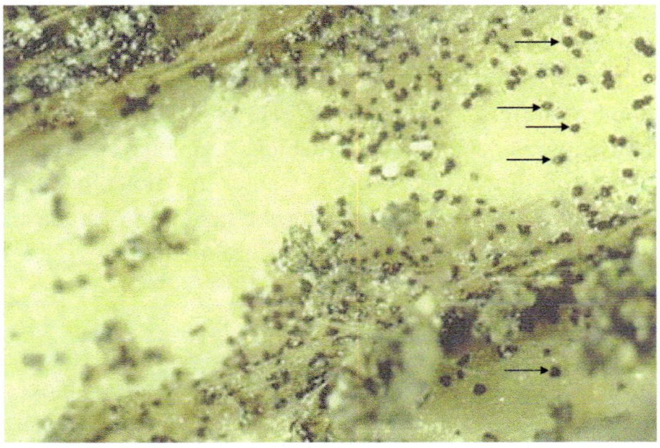

FIGURE 13.9 Numerous black microsclerotia (sclerotia) (arrows) of *Verticillium dahliae* on an infected potato stem. (https://www.cals.ncsu.edu/course/pp728/Verticillium/Vertifin.htm.)

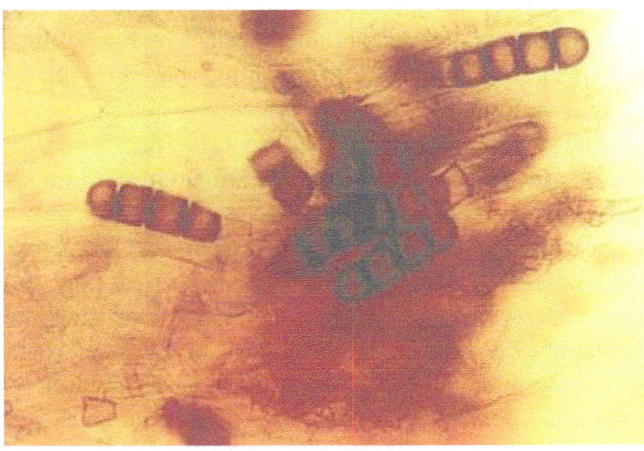

FIGURE 13.10 Chlamydospores of *Thielaviopsis basicola*. (Article Images: chlamydospore; www.apsnet.org; 400 × 267 search by image.)

TAXONOMY

Traditionally, classification of ascomycetes was based on morphological and ontogenetic characters. For example, the position of ascocarps in relation to substrates (Figure 13.11) was used in the family delimitation of Diaporthales (an order in the Sordariomycetes), but this classification was not supported by more recent molecular phylogenetic analyses. The shape, size, color, and texture of ascospores (Figure 13.11) has been used in generic-level classification of ascomycetes, but many molecular phylogenetic studies suggest that these generic delimitations do not always reflect evolutionary history. For instance, in Magnaporthaceae, genera based on ascospore shape are not monophyletic (a single lineage), likely due to convergent evolution. In some ascomycetes, asexual states seem to be more informative in defining monophyletic taxa because the sexual states tend to be more conserved.

Subcellular or ultrastructural and biochemical characters shed light on fungal evolution too. As part of the Assembling the Fungal Tree of Life project, a searchable database for selected fungal taxa is available at http://aftol.umn.edu, which includes descriptions and illustrations of characters such as nuclear division, septum/pore cap, sterol data, and hyphal tip organization.

Current classification of species included in Ascomycota is based primarily on molecular sequence data combined with morphological, ontogenic, and other characters. The major lineages of known ascomycetes, which include Taphrinomycotina, Saccharomycotina, and Pezizomycotina, are illustrated in Figure 13.3. In this chapter, three classes (Sordariomycetes, Leotiomycetes, and Dothideomycetes) in Pezizomycotina that include important plant pathogens will be introduced. More information on Ascomycota systematics can be found from the Deep Hypha *Mycologia* special November/December issue in 2006 and Lumbsch and Huhndorf (2010).

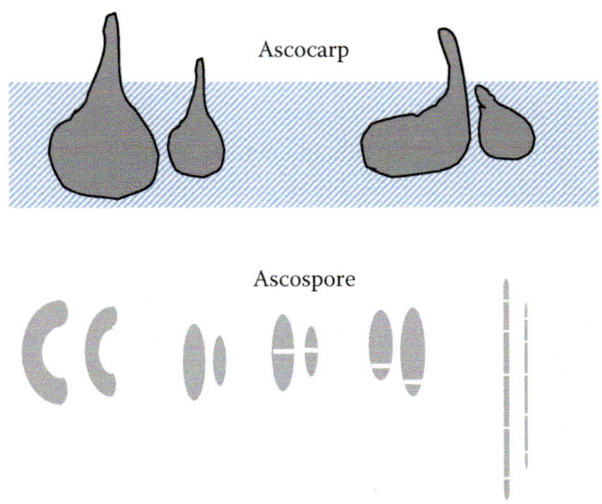

FIGURE 13.11 Orientation of ascocarps (upper) and morphologies of ascospores (lower). (Courtesy of N. Zhang.)

ONE FUNGUS = ONE NAME, A MAJOR CHANGE IN FUNGAL NOMENCLATURE

Many ascomycetes have both sexual and asexual states (pleomorphic). Connections between teleomorphs and anamorphs of the same species were frequently unknown because often the two do not co-occur temporally or physically. For convenience, the International Code of Botanical Nomenclature previously made exceptions for fungi and allowed unique scientific names (Latin binomials) for both the teleomorph and anamorph with the teleomorph name taking priority. These practices for fungi are now discontinued because modern molecular tools such as polymerase chain reaction (PCR) and DNA sequencing enable us to integrate or link the asexual and sexual phases of an organism into a single species, and thus unify the name. According to the International Code of Nomenclature for algae, fungi, and plants (Melbourne Code, available online at http://www.iapt-taxon.org/nomen/main.php), the dual naming system for fungi is replaced with one scientific name for each species based on priority or most commonly occurring form. All legitimate fungal names published prior to January 1, 2013, compete equally for priority, and the sole correct name is now the earliest legitimate name, regardless of the life history state of the type.

It was estimated that about 10,000–12,000 species names of pleomorphic (more than one form, i.e., sexually and asexually reproducing phases) fungi, including many plant pathogenic species, might be affected by the "One Fungus = One Name" new nomenclature (Hawksworth, 2012). The change will result in the use of some scientific names that are unfamiliar to plant pathologists and other user communities. Although the transition is difficult, the purpose of the new nomenclature is for the stability of nomenclature and systematics in the long term. Moving to one scientific name for each species of fungus aligns the fungi with the other groups of organisms governed by codes of nomenclature including the International Code of Zoological Nomenclature and the International Code of Nomenclature of Bacteria. No other group of organisms was allowed to have more than one scientific name (Zhang et al., 2013).

During this transitional period, which may take several years or longer, plant pathologists, mycologists, geneticists, and the broad user communities are working together to determine which names to use for pleomorphic fungi. For example, the International Sub-commission on the Taxonomy of *Colletotrichum* determined in 2012 that the asexual genus name *Colletotrichum* would be used for the large group of important fungi that cause anthracnose and black spot diseases, rather than the sexual name *Glomerella*. This decision was based, not only on priority, but because *Colletotrichum* is a more commonly used name in the applied sciences. The International Commission on *Penicillium* and *Aspergillus* has also decided to use the asexual generic names *Penicillium* and *Aspergillus* because both predate the names of genera associated with either of the sexual states.

The name to use for the rice blast fungus has not been determined as of this writing. The asexual genus name *Pyricularia* has been used for the rice blast fungus since 1892. Based on the morphological similarity, the sexual state of the rice blast fungus was believed to belong to *Magnaporthe* and was named *M. oryzae* in 2002. However, recent phylogenetic analyses demonstrated that the rice blast fungus does not belong to *Magnaporthe*. As the oldest and legitimate generic name for the rice blast fungus, *Pyricularia* should be used. However, because this species is a widely used model system and has large impacts, researchers from all over the world are still debating on whether to use *P. oryzae* or conserving *M. oryzae* for the rice blast fungus (Zhang et al., 2013).

As scientific names of plant pathogenic fungi are integrated, accurate scientific names will be placed on websites such as *Index Fungorum* (http://www.indexfungorum.org/Names/Names.asp), MycoBank (http://www.indexfungorum.org/), and the USDA-ARS SMML Fungal Nomenclature (http://nt.ars-grin.gov/fungaldatabases/nomen/nomenclature.cfm), which emphasizes plant-associated fungi. The National Center for Biotechnology Information (NCBI) also manages a taxonomy database, and it will be updated as decisions are made.

SORDARIOMYCETES

Sordariomycetes is a large class in Ascomycota with over 10,000 described species. It contains most nonlichenized ascomycetes-producing **perithecia** (flask-shaped ascocarps; Figures 13.12 and 13.13) or less frequently **cleistothecia** (completely closed ascocarps; Figure 13.12), **inoperculate** (lacks a hinged cap), and **unitunicate asci** (ascus wall with a single functional layer) (Alexopoulos et al., 1996). The term "Pyrenomycetes" was used formerly to unite fungi with perithecial ascocarps and unitunicate asci, but its use was discontinued because many perithecial species do not belong to this group based on phylogenetic analyses.

Unique ultrastructure characters of the Sordariomycetes include the following: (1) immature ascogenous hypha/ascus, endoplasmic reticulum associated with toroid occlusion (donut-like with central pore), and (2) for mature ascogenous hypha/ascus, there is a subspherical pore cap membrane, which is a simple membrane enclosing the cytoplasm. Leotiomycetes and Sordariomycetes appear to have the same spindle pole body form and nuclear envelope organization during nuclear division, that is, an unlayered disc with an intact nuclear envelope and an internal microtubule-organizing center (Celio et al., 2006).

Sordariomycetes share some features with Dothideomycetes and Leotiomycetes, but are unique in having

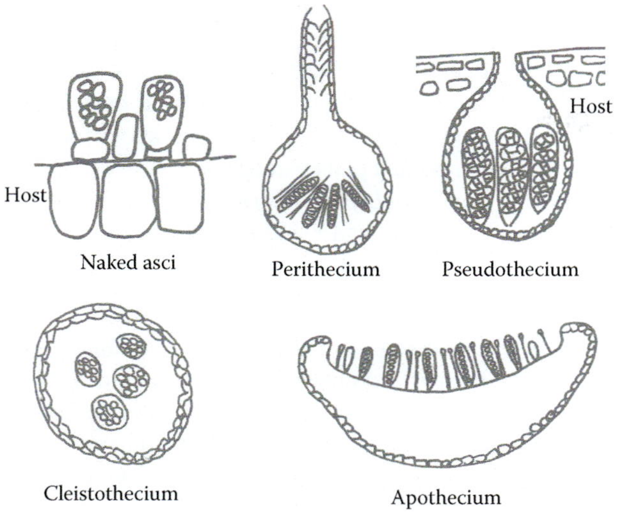

FIGURE 13.12 Five ways in which fungi in Ascomycota bear their asci. (Courtesy of N. Zhang.)

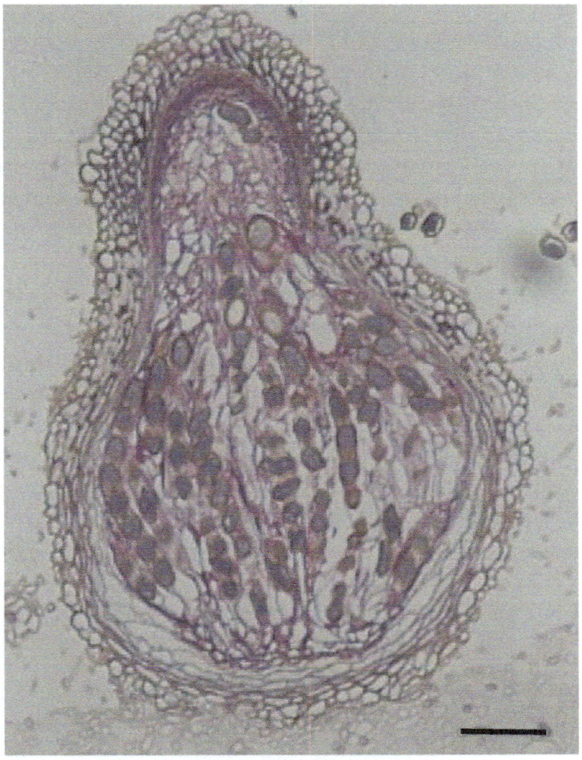

FIGURE 13.13 Longitudinal section of a perithecium of *Sordaria* sp. Bar = 50 μm. (From Carolina Biological Supply, Burlington, North Carolina. With permission.)

a true perithecium and inoperculate, unitunicate asci. Many fungi in Sordariomycetes are plant pathogens, including *Fusarium* spp. (wilt diseases of various crops), *Colletotrichum* spp. (anthracnose of many plants), *Claviceps purpurea* (ergot of rye), *Discula destructiva* (dogwood anthracnose), and *Gaeumannomyces graminis* var. *tritici* (take-all disease of cereals).

Rice Blast Fungus

The rice blast fungus, *Pyricularia* (*Magnaporthe*) *oryzae*, recently segregated from *Magnaporthe grisea*, is one of the most devastating threats to food security worldwide. This fungus, which belongs in the order Magnaporthales of Sordariomycetes, is better known for its asexual state, *P. oryzae*, which has been studied extensively as a model system for aerial infection and host–pathogen interactions. *Pyricularia oryzae* infects host plants in a manner typical of other foliar pathogens. Conidia are dispersed during the growing season. They adhere tightly to the leaf surfaces by spore tip mucilage. The fungus enters the host leaves using a sophisticated method that increases turgor pressure in the **appressorium** (a hyphal, pressing organ) to force an infection peg through the leaf surface into the plant (Figure 13.1).

Numerous studies have been conducted on appressorium development in the rice blast fungus. However, the root-infecting capacity of this pathogen was not discovered until 2004. A number of genes, such as the *MAP* kinase and *CPKA* gene, are essential for the formation of functional appressoria for leaf penetration (Xu et al., 1997).

The rice blast pathogen can cause lesions on all parts of the susceptible plant, such as leaf, panicle, pedicel, and seed, but the most commonly observed symptom is the diamond-shaped lesion with a gray center with a brown border on leaves. Pear-shaped conidia are produced on the lesion and released to begin a new infection cycle. The sexual state of the fungus has not been observed in nature, but can be produced in the laboratory when pairing isolates of different mating types (*MAT1-1* and *MAT1-2*). Controls for rice blast include crop rotation, breeding of resistant varieties, and fungicide sprays.

Chestnut Blight Fungus

Cryphonectria parasitica (Diaporthales, Sordariomycetes) is the causal agent of chestnut blight, which was likely introduced to North America from Asia with the host material. The disease was first reported in the early 1900s, and in a few decades, it destroyed most American chestnut trees (*Castanea dentata*) in eastern North America. Ascospores or conidia can serve as inoculum, which enter through wounds often caused by insects, and hyphae grow into the vascular cambium. Cankers form around the stem killing the vascular cambium, which girdles the tree and prevents the formation of new vascular tissues. In the 1950s, the pathogen was found in Europe, but with less virulence. Further investigations led to the discovery of a hypovirulence factor, a double-stranded RNA virus in the fungal cytoplasm that reduced the virulence of the fungal pathogen to the host plant. Use of hypovirulent viruses as biocontrol agents did not succeed at the population level in North American because in nature the viruses do not spread easily. Other efforts to restore American chestnut are being made to create resistant cultivars through breeding and genetic engineering.

LEOTIOMYCETES

Leotiomycetes refers to the "inoperculate discomycetes", a group of nonlichenized ascomycetes characterized by the production of **apothecia** (Figure 13.12, cup- or disc-shaped ascocarps) and unitunicate, inoperculate asci with an apical pore from which ascospores are released. Some Leotiomycetes cause plant diseases. For example, *Rhabdocline pseudotsugae* and *Rhabdocline wierii* are the causal agents of needle blights of Douglas fir. *Sclerotinia sclerotiorum*, the cause of white mold, can infect various vegetables. *Rhytisma acerinum* causes tar spot of maple. One representative plant pathogen in Leotiomycetes is described in detail here. Powdery mildews are omitted in this chapter and discussed in Chapter 12.

Brown Rot Pathogen

Monilinia fructicola is the cause of brown rot of peach and other stone fruits. Young peaches are usually resistant to this pathogen but become increasingly susceptible as they approach maturity. Fungal invasion occurs through trichome sockets, insect punctures, and other wounds. Mycelium spreads through the host tissue rapidly and destroys plant cell walls by secreting pectinases and other extracellular hydrolytic enzymes (Chapter 32). Mycelium reemerges on the surface of the fruit to produce conidia that are spread by wind, water, and insects to other fruit. Some fruit falls on the ground where they quickly disintegrate under the action of saprotrophic fungi. Heavily infected fruit remaining on the tree loses water and becomes a shriveled mass of plant tissue and mycelium—a mummy. The mummy may persist on the tree through the winter or it may fall on the ground and the fungus may survive in the soil for up to 3 years. In either case, the pathogen overwinters in the mummy. The following spring, mummies on trees produce conidia via mitosis, whereas mummies on the ground produce ascospores via meiosis, and a new infection cycle is initiated. Control of the brown rot pathogen is primarily by fungicide sprays on a time schedule depending on weather and orchard location. Cultural practices, such as sanitation of mummies and pruning and removal of infected materials, can substantially reduce disease levels. Insect control, especially for those pests that directly damage fruit, can also reduce disease incidence (Curry and Baird, 2007).

DOTHIDEOMYCETES

Formerly known as "Loculoascomycetes," Dothideomycetes is a large class of Ascomycota, characterized by the production of asci within locules in a preformed stroma and **bitunicate asci** with fissitunicate dehiscence (jack-in-the-box). Bitunicate asci consist of a thin inextensible outer layer and a thick extensible inner layer. Ascospores are usually released by the extension of the inner ascus wall and the rupture of the outer wall. Numerous fungi in this class are important plant pathogens, such as species of *Alternaria*, *Bipolaris*, *Leptosphaerulina*, *Mycosphaerella*, *Phyllosticta*, *Venturia*, and *Zymoseptoria*.

Apple Scab Pathogen

Venturia inaequalis attacks various plants, including apple, crab apple, firethorn, and hawthorn, and is considered one of the most important pathogens of apple. Both leaves and fruit of the apple tree can be infected. The fungus begins its life cycle in the spring by ejecting ascospores from pseudothecia (Figure 13.12), which were produced in the diseased leaf tissue lying on the ground since the winter. Ascospores, which consist of two cells of unequal size (hence the name "inaequalis"), are dispersed by air. Those that land on suitable host surfaces produce short germ tubes with appressoria by which they penetrate the host cuticle. Mycelium grows in host tissue for a few days, after which, hyphae reemerge at the surface to form acervuli and conidia (Figure 13.14). Conidia are produced during the spring and summer months. Similar to ascospores, conidia also infect via appressoria and penetration of the host cuticle. The asexual state was named *Spilocaea pomi*, but now because of one fungus one name, both sexual and asexual states are named *V. inaequalis*. Lesions on the leaves begin as velvety, olive-green lesions and then get darker and "scabby" in appearance, with distinct margins. When the cells in the leaf die, the mycelium forms pseudothecia and ascospores to overwinter. This fungus is heterothallic and requires two different mating types for sexual reproduction. Control of the apple scab pathogen is primarily with fungicides, removal of infected or dead plant tissues, and use of resistant varieties.

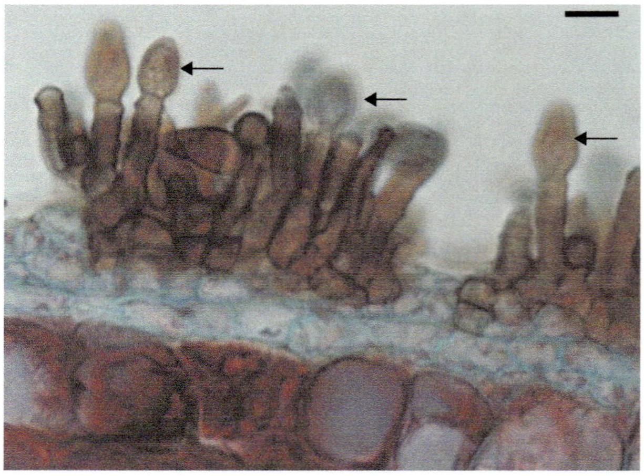

FIGURE 13.14 *Venturia inaequalis* conidia (arrows). Bar = 10 μm. (Courtesy of Triarch, Inc., Wisconsin [slide 3-48B].)

CULTURE AND MAINTENANCE

Saprotrophic and necrotrophic species of Ascomycota can usually be cultured on synthetic media. Sporulation (production of ascospores or conidia) is sometimes enhanced by adding natural substrates to synthetic agar, such as sterilized host plant tissue. For example, carnation leaf can enhance the production of conidia for some *Fusarium* species. However, the biotrophic ascomycetes, such as the powdery mildews, require live host tissue to grow

Formulations of a number of general and selective media for phytopathogenic fungi were provided by Singleton et al. (1992). Preservation of fungal cultures in a metabolically inactive state in liquid nitrogen should be used to minimize mutation for genetic or pathogenicity studies. Many laboratories also store fungal cultures in glycerol solution at –80°C. Other methods include lyophilization and storage in sterile soil or sterile distilled water and on agar slants (Singleton et al., 1992).

GENOMES OF ASCOMYCOTA

Genomics is the study of the genomes of organisms. Many ascomycetes are model systems for studying reproduction, gene gain and loss, horizontal gene transfer, as well as gene structure and regulation. For example, *Neurospora crassa*, the red bread mold, is a classic model organism for genetic studies. The genome of *N. crassa* was completely sequenced in 2003; it contains 43 megabases and approximately 10,000 genes.

Genomics is a fast evolving field with new branches emerging quickly. Functional genomics is the characterization of functions of genes, RNA transcripts, and protein products. Structural genomics refers to the dissection of the architectural features of genes and chromosomes. Comparative genomics is the study of the relationship of genome structure and function across different biological species or strains. Metagenomics is the study of genetic material recovered directly from environmental samples, bypassing the need for isolation and lab cultivation of individual species. Phylogenomics is the analysis of genome data for evolutionary reconstruction.

According to the Genome Online Database (www.genomesonline.org), as of November 4, 2013, there have been 1569 complete and ongoing genome projects for Ascomycota, which constitute 24% of the 6578 genome projects for all Eukarya. Ascomycetes being sequenced for genomes are mostly economically important, and many of them are plant pathogens. For instance, the genome project for *Pyricularia* (*Magnaporthe*) *oryzae*, the rice blast fungus, was completed by the International Rice Blast Genome Consortium and the BROAD Institute (http://www.broadinstitute.org). With high-throughput next-generation sequencing technologies, an exponential growth of the genomic data for Ascomycota in the next decades is expected.

LABORATORY EXERCISES

The "mitosporic" (Deuteromycetes) fungi (asexual phases of ascomycetes) are a heterogeneous and artificial (does not reflect phylogeny) group erected on the production of asexual spores or conidia that are formed on conidiophores, which develop either free on mycelium or are enclosed in structures called conidiomata (fruiting structures). Laboratory exercises for the mitosporic fungi are designed to teach students how to recognize the different asexual reproductive structures and the role of environmental conditions or nutrition on the shapes and forms of the conidiomata. Note that some members of this group do not produce conidia and form sclerotia or no structures. However, these will not be considered in these laboratory exercises.

Also included in this section is an experiment using *Sordaria fimicola*. This ascomycete readily produces perithecia and ascospores in culture. There are also mutants that express different ascospore colors. Because of the mutants, this fungus is ideally suited for studying crossing-over events during meiosis and can be used to genetically map the color trait.

Experiment 1. Identification of Asexually Reproducing Ascomycetes by Observing Reproduction Structures

When attempting to identify a fungus, it is important to note morphological features of the fungus under the dissection microscope. Fungi placed on microscope slides often fragment or spores do not remain attached to conidiophores, making identification difficult. Morphological features to observe using the dissection microscope include the following: conidia borne singularly, in chains, clusters, or clumps; conidia borne in slime droplets; conidia borne at the apex or laterally on conidiophores; and presence of pycnidia, sporodochia, synnemata, or acervuli. The purpose of this experiment is to demonstrate variations in conidiomata that are used in the identification of the asexually reproducing ascomycetes in host material and on artificial media. This exercise will train students how to recognize the different asexual structures involved in spore production that are important for identifying these fungi. Fungi representative of each conidiomata type can be compared during this exercise (Table 13.1). Depending on availability, many other fungi can be substituted.

Materials

Each student or team of students will require the following materials:

- Cultures of at least two species of fungi showing each of the four types of asexual development in Table 13.1
- Dissection and light microscopes

TABLE 13.1
Genera of Fungi Used to Demonstrate Types of Conidiophores and Conidiomata

Naked Hyphae	Sporodochia	Pycnidia	Synnemata	Acervuli
Alternaria	*Epicoccum*	*Diplodia*	*Briosia*	*Colletotrichum*
Bipolaris	*Fusarium*	*Macrophoma*	*Graphium*	*Discula*
Cladosporium	*Tubercularia*	*Phomopsis*	*Harpographium*	*Pestalotia*
Penicillium	*Volutella*	*Septoria*	*Trichurus*	*Melanconium*

- Dissecting needles or probes
- Slides, coverslips, distilled water for mounting hyphae and spores
- A stain such as aqueous analine blue (0.05% or 5 mg/L)
- Eye dropper with bottle
- Cellophane tape
- Single-edge razor blade or scalpel with #11 blade
- Pencil eraser
- Alcohol lamp for surface-sterilizing instruments

Follow the protocol listed in Procedure 13.1 to complete this exercise.

Anticipated Results

Students should be able to differentiate the various ways in which conidia are produced, either naked as individual conidiophores or clustered together into synnemata, pycnidia, acervuli, and sporodochia.

Questions

- Describe and draw the following structures: individual conidiophores, single or multiple branched, spore or conidia attachments; sporodochia, conidiophore arrangements, locations on host tissues (living materials).
- How do pycnidia differ from sporodochia or acervuli?
- Where are the conidiophores located in a pycnidium?
- Does the pycnidium have a pore (ostiole) in the neck (when present)? What is the purpose of an ostiole?
- Are the pycnidia embedded in or on the surface of the host tissue?
- What is a synnema?
- How do the conidiophores in synnemata differ from the conidiophores in sporodochia?
- Where are the spores borne on synnemata and how are they attached?

Experiment 2. Effects of Light and Temperature on Growth and Sporulation of Fungi

The purposes of this experiment are to show how temperatures affect the growth of fungi and to compare sporulation and growth potential of fungi when placed under different lighting regimens. Through direct observation, students will learn that growth conditions will affect spore production, cultural characters, and growth rates of fungi.

Common fungi that can be used in this exercise include *Alternaria* species (single conidiophores), *Epicoccum nigrum* or *F. solani* (sporodochia types), *Colletotrichum graminicola* or *Pestalotia* spp. (acervuli types), and *Phoma* or *Phomopsis* (pycnidia types). Many other fungi may be substituted in this experiment. Experiments typically require up to 14 days to complete, but will obviously vary depending upon the fungi included in the study.

The fungi should be grown on potato dextrose agar (PDA; Difco) for several days, perhaps a week, at room temperature to provide sufficient inoculum. The number of conditions (treatments) may vary depending upon the available space and number of replicate plates (V-8 medium) used. We suggest that the class be divided into four groups, each with their own species of fungi.

Materials

Each team of students will need the following items to complete the experiment:

- Four isolates of common asexually reproducing fungi (one for each group). Inoculate each fungus onto five Petri dishes of 10-cm-diameter containing PDA (Difco, Lansing, MI)
- A minimum of 21 Petri dishes of 10-cm-diameter containing V-8 agar medium (Diener, 1955: 200 mL of V-8 medium (Campbell Soup Co.), 3 g of $CaCO_3$, 15 g of Difco agar, and 800 mL of water)
- Aluminum foil
- Parafilm®
- Sterile cork borer, dissection needle, or spatula for subculturing fungi

Procedure 13.1
Observation of Asexual Structures of Ascomycete

Step	Instructions and Comments
1	Observe the fungal culture with a stereo dissection microscope at low power. Note the shape and form of the fruiting structures. Fruiting structures can be observed more easily on the hyphae nearest to the margin of active growth in culture. Conidia in mass may obscure fruiting structures. The mycelium (hyphae) becomes much thicker and often obscures the asexual structures near the center of the dishes.
2	Viewing asexual fruiting structures of mitosporic fungi in plant material is often essential to identify the pathogen. Transition areas in the zone between healthy and necrotic plant tissue are usually good areas to look for structures. If fruiting structures are not evident in this area, move the tissue piece so that areas with greater tissue damage are viewed. If fruiting structures are not observed within the lesions, place the plant tissue in a moist chamber overnight. To obtain fruiting bodies, place a moistened tissue paper at the bottom of a plastic food storage container (a plastic bag will also work), place the plant tissue onto tissues and close the box. Store at room temperature and check in 12–24 h for sporulation. Adding a drop of water directly to the lesion may also stimulate sporulation.
3	After viewing diseased plant material with a dissection microscope, a closer examination with a compound or light microscope is necessary to identify the fungus. Slice thin sections (1 mm) of a leaf lesion with a single-edge razor blade or scalpel and mount on a microscope slide with a drop of analine blue or water. Look for conidia and/or fruiting bodies on the plant tissue and floating in the mounting medium.
4	To observe sporulation of a fungal colony growing in a Petri dish, place a drop of water or analine blue to the area of interest, and add a coverslip directly onto the agar medium. Fruiting structures can be observed with minimal damage to them. For closer observation, mount fruiting structures on a slide in a drop of distilled water and a coverslip. If fruiting bodies cannot be seen on the agar plate, simply remove a tiny portion of the mycelium (≤1 cm) and place it into the water on the slide. Flame-sterilize the dissection needles and allow cooling before removing mycelium from the Petri dish. Sometimes the coverslip does not appear directly to the microscope slide when a mass of mycelium and agar are mounted. If this happens, use a pencil eraser to gently press on the coverslip and flatten the preparation. First, use the lowest power (generally a 4× or 10× objective lens) on a specific area containing the fruiting structures and proceed to a high, dry objective lens (40×). As with the stereo microscope, observe the hyphae nearest to the growing margin and move inward until light can no longer penetrate through the mycelium. Observe and draw the structures and spores in detail.
5	A third method of viewing asexually reproducing fungi is with cellophane tape. This method is particularly useful for obtaining fruiting structures and spores from fresh material. Place a piece (1–2 cm) of cellophane tape with the adhesive side downward onto the plant material. Press slightly and remove from the tissue. Take the piece of cellophane, place the side that was in contact with plant tissue downward onto a microscope slide containing water. Observe spores and/or fruiting structures with the 10× and 40× objective lenses and record your observations.

- Alcohol burner, matches, and container of 95% ethanol for cooling subculturing tools
- Incubators set for different temperatures and lighting conditions
- Millimeter ruler for measuring growth of colonies
- Light source for observing cultures

Follow the protocol outlined in Procedure 13.2 to complete the experiment.

Anticipated Results

Students should observe that changes in temperatures affect growth and sporulation of fungi. However, different fungal species used in the experiment may not respond with the same growth and sporulation patterns as the other species when compared under similar growth conditions.

Questions

- How did growth vary between species? Hint: graph the data and calculate a growth rate.
- Did the fungi exhibit the same growth rate throughout the experiment? If not, what may have caused the changes in growth rates?
- When was sporulation first observed for each fungus? How did it vary per light regime?
- How did the effects of different light regimes vary between species (consult with other groups)?
- Compare different ways in which you might be able to collect data on sporulation (e.g., number of conidiophores per unit area and spore counts).

Procedure 13.2

Temperature and Light Influence Growth and Sporulation of Fungi

Step	Instructions and Comments
1	Use a flame-sterilized cork borer to cut agar pieces from cultures growing on PDA medium. If possible, remove plugs from actively growing margins. Mycelium closest to the original inoculation plugs can differ physiologically from mycelium nearest to the margin resulting in different growth rates.
2	Each group should transfer a single plug (mycelium side down) of their fungus onto the center of a Petri dish containing V-8 agar. Inoculate a minimum of 21 dishes and wrap with Parafilm®. Label dishes with the name of the fungus and number them from 1 to 21.
3	Each group should place at least three of the inoculated Petri dishes in incubators set at either 10°C, 20°C, 30°C, or 40°C. Record the number of Petri dishes at each temperature. If insufficient incubator space is available, we suggest that the 20°C and either 10°C or 30°C treatments be used. Cultures should be exposed to light for a minimum of at least 8 h per day; otherwise, sporulation could be adversely affected.
4	Each group should place at least three of the inoculated Petri dishes in incubators set for the following light treatments: continuous artificial light, alternating light and darkness (12 h each), and continuous darkness. The continuous dark treatment may also be achieved by wrapping the cultures in aluminum foil. All incubators should be maintained at the same temperature (e.g., 20°C–25°C).
5	Record growth data every 2 days for 14 days (maximum) or until the mycelium reaches the edge of the Petri dishes for both temperature and light treatments. Using a millimeter ruler, measure the radial growth from inoculation plug to the perimeter of the colony. The growing edge of the mycelium can be viewed easily by placing the dishes toward a light source. Repeat in three locations for each dish and find the average to the nearest millimeter for each dish. Now average the radial growth for all three replicates for each fungus at each temperature. Plot these data on a graph with the growth in millimeter in *y*-axis and days in culture in *x*-axis. Calculate the standard deviation (SD) for each point using the following formula: $$\mathrm{SD} = \sqrt{\frac{\sum (x-\bar{x})^2}{(n-1)}} = \sqrt{\frac{X^2}{(n-1)}}$$ where x = sample measurement, $\bar{x}$ = calculated mean of data, $X = x - \bar{x}$, and n = number of samples used to calculate the mean.
6	Sporulation should also be recorded when growth is measured. A rating scale for sporulation can be developed such as none = 1, slight = 2, moderate = 3, and heavy = 4. Find the average rating for each fungus at each treatment. Plot sporulation on the *y*-axis and days on the *x*-axis.
7	Prepare a final report from the data. Include graphs showing the growth and sporulation of species under various temperatures and lighting conditions.

Experiment 3. *Sordaria fimicola* Perithecia, Asci, and Ascospore Formation and Calculation of the Distance of the Locus (Gene) for Ascospore Color from the Centromere

S. fimicola is a homothallic, saprophytic ascomycete fungus typically found inhabiting dung. It is easily cultured on artificial media and has been used to study genetics and inheritance. *Sordaria* is an ideal organism for studying the typical physical characteristics of members of the Sordariaceae (perithecia with persistent asci and forcibly discharged ascospores) as well as demonstrating some genetic concepts. It does not produce conidia and therefore will not contaminate the work area. The wildtype color of ascospores is black and mutants have tan- or gray-colored ascospores. Although, *Sordaria* is homothallic, if the wildtype and mutant type are cocultured, sexual hybridization and recombination will occur between them. The ascospores in individual asci will be of mixed colors in a 1:1 ratio; mating between individuals of the same type will result in all ascospores of identical color.

Materials

Each student or team of students should have the following items:

- Cultures of black wildtype (156290) mutant gray-type (156292), and mutant tan-type (156294) ascospores. Cultures may be purchased from Carolina Biological Supply, Burlington, NC (1.800.334 .5551): http://www.carolina.com/fungi/sordaria-fimicola-fungi-cultures/FAM_156290.pr?catId=10567&mCat=10476&sCat=&ssCat=&question=#family-details

Procedure 13.3

Sordaria fimicola **Perithecia, Asci, Ascospore Formation and Calculation of the Distance of the Locus (Gene) for Ascospore Color from the Centromere**

Step	Instructions and Comments
1	Pour 20 mL of sterile nutrient agar into each of 6 Petri dishes of 10-cm-diameter and allow agar to cool and solidify.
2	Using a sterile #5 cork borer, cut a number of inoculum plugs from the edges of growing colonies of both wildtype and mutant cultures—do not use the same cork borer for each culture. Inoculate two dishes for each combination of wildtype and mutant as shown in Figures 13.15 and 13.16a, seal with Parafilm®, and incubate at room temperature. Observe growth and formation of perithecia (black structures: Figure 13.16b) using a dissecting microscope.
3	With a sterile scalpel, remove a few perithecia from the zones indicated in Figure 13.16a and place in a drop of water on a microscope slide. Cover with a glass coverslip and gently apply pressure with a flat pencil eraser to spread the contents of the perithecia (Figure 13.17).
4	Observe the mount using a compound microscope. Draw the asci (including color) and ascospore arrangements for each of the three inoculated cultures (Figure 13.17). Examine at least 100 asci and count the variations (crossing over and no crossing over events) of ascospore arrangements (Figure 13.17). Note that if the ascospores are all colorless to green-gray, reincubate and reexamine the cultures in a few days.
5	Calculate the percentage of crossing over events: (# of crossing over/100) × 100. Now calculate the distance (in map units) between the centromere and the gene for color: % crossing-over events/2. The higher the number of crossing-over events, the more distance the gene locus for ascospore color is from the centromere. Compare your results to other students or groups and calculate a class mean map unit from the centromere. For a very good explanation of this concept, see Pearson LabBench Activity; Analysis of Results II for *Sordaria*: http://www.phschool.com/science/biology_place/labbench/lab3/analysis2.html.
6	Incubate the unexamined cultures for several additional days and observe the inside lid of the top of the Petri dishes every other day. At some time, discharged ascospores should be apparent on the inner surface of the Petri dish lid.

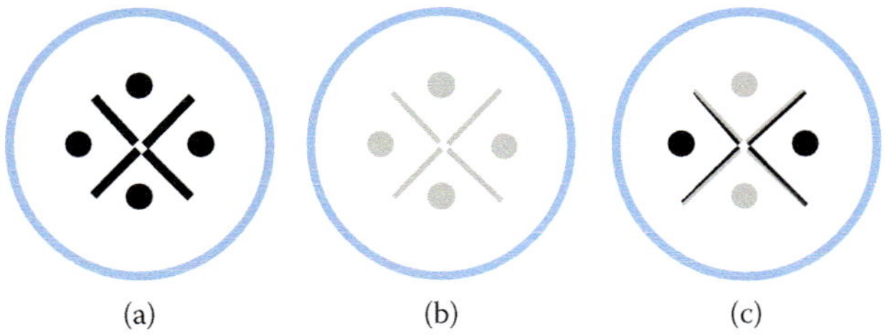

FIGURE 13.15 Culture pairings (circles) of black (wildtype) and tan (mutant) isolates of *Sordaria fimicola*. (a) Wildtype with wildtype. (b) Mutant with mutant. (c) Wildtype with mutant; ascospores of both colors in an ascus (hybridization) are expected, and crossing-over events should be evident. Perithecia sample should be collected along the colored lines between circles; the color of the line indicates expected color of ascospores. (Courtesy of R.N. Trigiano.)

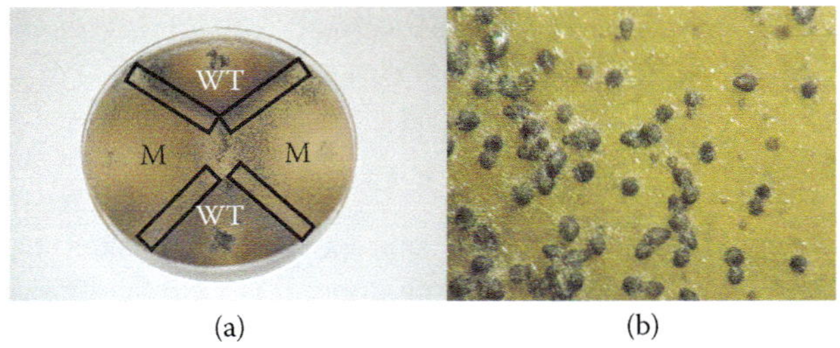

FIGURE 13.16 *Sordaria fimicola*. (a) Paired mating between wildtype (WT) and color mutant (M) of *S. fimicola* (see Figure 13.15c). Boxes indicate areas of potential mating between WT and M and where perithecia should be harvested for examination. (b) Perithecia formed on the surface of the agar between the two color types. (Courtesy of Mary M. Dee.)

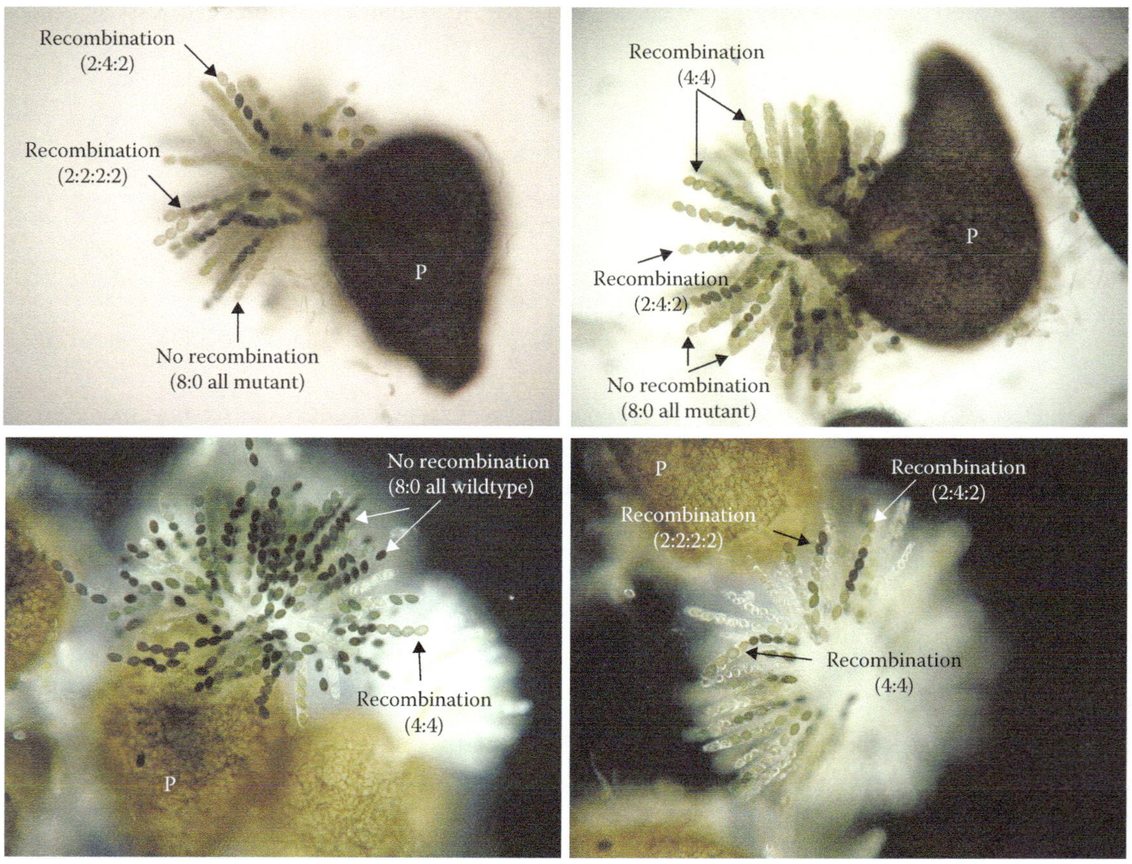

FIGURE 13.17 Examples of whole mounts of *Sordaria fimicola* squashes showing perithecia (P) with asci containing ascospores exhibiting recombination and crossing-over events. Numbers in parentheses equal number of either mutant or wildtype color spores. (Courtesy of Mary M. Dee.)

Grow cultures on nutrient agar in Petri dishes of 10-cm-diameter for one week. These cultures will be used for inoculating student cultures.

- Nutrient agar (nutrient broth [Difco, Detroit, MI] 8 g and 20 g/L agar): 20 mL in plastic Petri dishes of 10-cm-diameter
- Two sterile #5 cork borers
- Dissecting and compound microscopes
- Slides and coverslips
- Dropper bottle with water
- Pencil with new, flat eraser
- Computer with Internet capabilities
- Colored (black and tan) pencils

Follow the protocol listed in Procedure 13.3 to complete this experiment.

Anticipated Results

Perithecia should form within 7–14 days after inoculation depending on the incubation temperature (Figure 13.14). When mating wildtype with wildtype cultures, only black ascospores should be present in asci. In the case of mating mutant (gray or tan) with mutant, only gray or tan ascospores should be present. Mating a wildtype (black) with a mutant (gray or tan) should result in some asci containing ascospores of the same color, and others with 1:1 ratio of mutant to wildtype colored ascospores (Figure 13.11b). Ascospores of different colors in the same ascus (four black then four tan) indicate "hybridization" between the two color types. Alternating patterns of colors of spores indicate crossing-over "recombination" events (Figures 13.11b and 13.12a and b). The frequency of crossing-over events is related to the distance of the locus (gene) controlling color of the ascospore from the centromere of the chromosome.

Questions

- Why is the inner surface of the lid of the Petri dish black?
- *S. fimicola* is often found growing on dung (coprophilus), what is the advantage in forcibly discharging ascospores?
- How many patterns of different colors of ascospores are present? Why are there always two spores of the same color next to each other?
- What would happen if all three variants (wildtype [black], and gray and tan mutants) were grown on the same Petri dish? Would you expect all colors to be in an individual ascus?
- How might the frequency of crossing-over events help you map the color trait?
- Why is sexual reproduction and recombination important?

ACKNOWLEDGMENT

The work was supported in part by the National Science Foundation (grant number DEB 1145174) to N. Zhang.

REFERENCES

Alexopoulos, C. J., C. W. Mims, and M. Blackwell. 1996. *Introductory Mycology*. New York, NY: John Wiley & Sons, Inc.

Celio, G. J., M. Padamsee, B. T. M. Dentinger, R. Bauer, and D. J. McLaughlin. 2006. Assembling the fungal tree of life: Constructing the structural and biochemical database. *Mycologia* 98:850–859.

Curry, K. J., and R. E. Baird. 2007. Ascomycota: The filamentous fungi forming perithecia, apothecia, and ascostromata. In: R. N. Trigiano, M. T. Windham, and A. S. Windham (eds). *Plant Pathology Concepts and Laboratory Exercises* (2nd edition, pp. 127–132). Boca Raton, FL: CRC Press.

Diener, U. L. 1955. Sporulation in pure culture by *Stemphylium solani*. *Phytopathology* 45:141–145.

Hawksworth, D. L. 2012. Managing and coping with names of pleomorphic fungi in a period of transition. *IMA Fungus* 3:15–24.

Kirk, P. M., P. F. Cannon, D. W. Minter, and J. A. Stalpers, eds. 2008. *Ainsworth and Bisby's Dictionary of the Fungi*. Wallingford, UK: CAB International.

Lumbsch, H. T., and S. M. Huhndorf. 2010. Life and earth sciences no. 1 Fieldiana. *Myconet* 14:1–64.

Singleton, L. L., J. D. Mihail, and C. M. Rush, eds. 1992. *Methods for Research on Soilborne Phytopathogenic Fungi*. St. Paul, MN: APS Press.

Xu, J. R., M. Urban, J. A. Sweigard, and J. E. Hamer. 1997. The *CPKA* gene of *Magnaporthe grisea* is essential for appressorial penetration. *Molecular Plant-Microbe Interactions* 10:187–194.

Zhang, N., A. Y. Rossman, K. Seifert, J. W. Bennett, G. Cai, L. Cai, B. Hillman, K. D. Hyde, J. Luo, D. Manamgoda, W. Meyer, T. Molnar, C. Schoch, M. Tadych, J. F. White, Jr. 2013. *Impacts of the International Code of Nomenclature for Algae, Fungi, and Plants (Melbourne Code) on the Scientific Names of Plant Pathogenic Fungi*. Online. APS net Feature. St. Paul, MN: American Phytopathological Society.

14 Rust and Smut Diseases

Lori M. Carris and Larry J. Littlefield

Concept Box

- Rust and smut diseases are caused by teliospore-forming, biotrophic fungi in the phylum Basidiomycota.
- Most rust and smut pathogens are highly host specific, often able to infect a limited number of closely related hosts.
- Rust fungi attack a wider range of host plants than smut fungi and cause economically important diseases on grasses and herbaceous and woody dicot hosts.
- Nearly two-thirds of the hosts of smut fungi are grasses and sedges, and smut diseases are of greatest importance on cereal crops.
- Infection of leaves and stems by rust and smut pathogens can indirectly affect yield by reducing the photosynthetic ability of the plant; replacement of cereal grains by smut spores directly impacts yield.
- Rust fungi have more complex life cycles than smut fungi, with up to five distinct spore stages that may require infection of only one host (autoecious rusts) or two unrelated hosts (heteroecious rusts). The rust life cycle generally includes urediniospores, a repeating spore stage that can contribute to massive disease epidemics.
- The lack of a repeating stage in smut fungi limits disease potential to levels of inoculum present at the onset of the growing season.
- Both rust and smut are controlled with resistant host cultivars and/or fungicides. Cultural practices, including elimination of alternate hosts, are also used to control rust pathogens.

The pathogens responsible for rust and smut diseases belong to two of the three groups (subphyla) of Basidiomycota (Figure 14.1). The majority of species in these subphyla, Pucciniomycotina (rusts) and Ustilaginomycotina (smuts), are plant parasites. In nature, rust and smut fungi are **obligate biotrophs**, requiring living plant hosts for nutrition and to complete their life cycles. Rust and smut fungi produce **basidiospores** and **basidia** similar to other Basidiomycota, but their basidia emerge from thick-walled **teliospores** (Figures 14.2 and 14.3) rather than being formed on or in fruiting bodies such as mushrooms and puffballs. In rust and smut fungi, the parasitic phase of the life cycle is filamentous (mycelial) and **dikaryotic**, with paired nuclei of compatible mating types (sometimes designated "$n + n$"). Dikaryotic teliospores are formed from this stage, and nuclear fusion (karyogamy) occurs in the teliospore, resulting in a brief diploid phase prior to germination to produce a basidium. Meiosis occurs in the basidium, and haploid basidiospores are formed. Some smut fungi also have a haploid saprotrophic phase in which they grow in a yeastlike manner and can be cultured on artificial media. Rust and smut fungi cause important plant diseases, but rust fungi, in general, are more important economically worldwide than are smuts.

Not all members of Pucciniomycotina are rust fungi, nor are all members of Ustilaginomycotina smut fungi. For example, *Microbotryum* is a group of plant parasites called "anther smuts" because they produce smutlike teliospores in the anthers of infected hosts, but this group of pathogens is more closely related to an evolutionary sense to the rust fungi in Pucciniomycotina than to smut fungi. The subphylum also includes morphologically and ecologically diverse groups of yeastlike fungi including *Sporidiobolus* and *Rhodosporidium*, called "red yeasts" because of their distinctive pink to red colonies formed on agar medium; parasites of fungi and ferns; and an unusual group of scale insect parasites in the genus *Septobasidium*

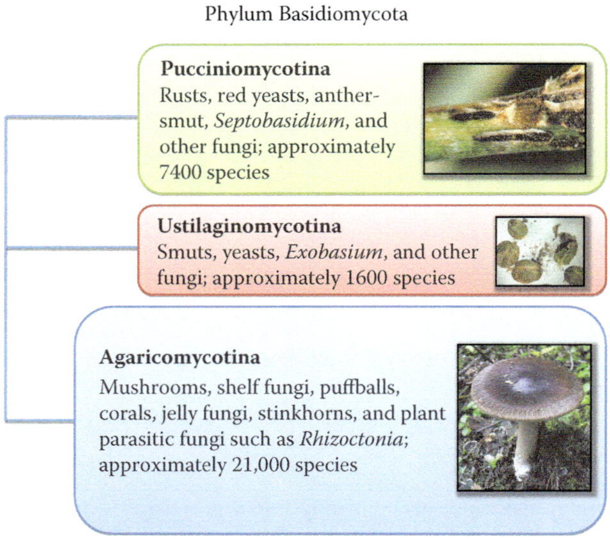

FIGURE 14.1 Rust and smut fungi belong in subphyla Pucciniomycotina and Ustilaginomycotina, respectively, in phylum Basidiomycota. Examples of different types of fungi and the approximate number of species in each subphylum are provided. Inset photos are Pucciniomycotina—telial stage of asparagus rust, *Puccinia asparagi*; Ustilaginomycotina—sori of dwarf bunt of wheat, *Tilletia contraversa*; and Agaricomycotina—*Amanita pachycolea*. (Courtesy of L. Carris.)

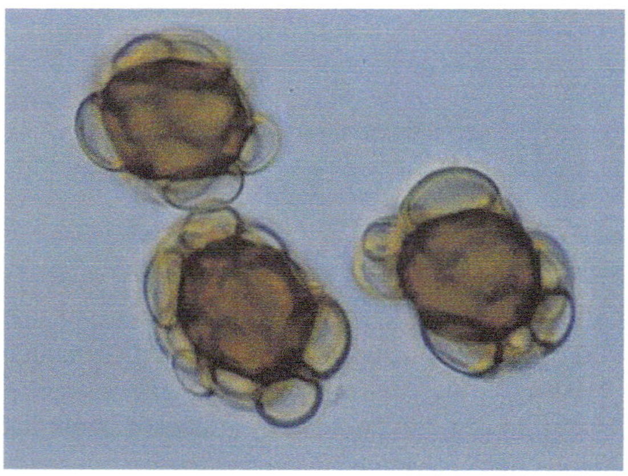

FIGURE 14.2 Spore balls of *Urocystis poae*. Dark brown teliospores are surrounded by pale brown sterile cells. (Courtesy of L. Carris.)

(Aime et al., 2006). However, the vast majority of the species in Pucciniomycotina are rust fungi placed in the order Pucciniales. The genus *Exobasidium*, a group of plant parasites on hosts including *Rhododendron* and *Vaccinium* that form basidia directly from infected host tissues, is within Ustilaginomycotina. Some members of Ustilaginomycotina are animal parasites, for example, several different yeasts in the genus *Malassezia* are associated with disorders such as dandruff and dermatitis in humans and other warm-blooded animals (Begerow et al., 2006).

FIGURE 14.3 Two-celled teliospores of *Puccinia* species on *Elymus* species. (Courtesy of Lisa Castlebury.)

This chapter focuses on the plant pathogens traditionally called rust and smut fungi. Only a few important examples illustrating particular concepts of pathogen biology, host–pathogen relationships, and disease management are presented, but suggested readings are provided for greater breadth and depth of information.

RUST DISEASES

Rust fungi have plagued agriculture for millennia and are responsible for some of our most devastating plant diseases. The ancient Romans celebrated an annual spring festival called the Robigalia in an attempt to appease the rust god "Robigus" and spare their wheat from devastation by the pathogen we now know as stem rust, *Puccinia graminis*. Western Europe was largely spared from stem rust epidemics during the early era of wheat cultivation, until the common barberry (*Berberis vulgaris*) was introduced, probably from Asia. The French legislature in the 1600s passed laws aimed at eradicating barberry, nearly two centuries before the German botanist Anton DeBary provided scientific proof connecting barberry and wheat stem rust. Similar legislation was passed by the American colonies in the 1700s. A different type of rust fungus, *Hemileia vastatrix*, destroyed British coffee plantations on the island of Ceylon (now Sri Lanka) in the late 1800s. Plantation owners quickly turned to the production of tea, and many credit the contemporary British preference for tea as the beverage of choice to the nineteenth-century destruction of the coffee plantations by this fungus. In the twentieth century, white pine blister rust, *Cronartium ribicola*, caused massive destruction to the white pine forests of North America. Development of rust-resistant white pines may one day facilitate the return of the white pine as an important timber species.

Rusts continue to be major pathogens of a wide range of cultivated plants. A race of stem rust of wheat, called Ug99, was discovered in Uganda in 1998, which now poses a major threat to global food security because 90% of the world's wheat cultivars are susceptible to this race (Singh et al., 2011). One of the most severe outbreaks of coffee rust occurred in Central America in 2012, providing a reminder that despite widespread use of fungicides, this pathogen remains a serious threat to one of our most popular beverages.

Symptoms and signs of cereal and other herbaceous rusts are commonly seen as shriveled seed and erumpent lesions (pustules) on leaves and stems. Many rust fungi have highly complex life cycles containing up to five different spore stages; some of these rusts require two unrelated, alternate hosts for completion of their life cycle. The five sequential stages of the complete rust life cycle, beginning and ending with teliospores, are shown in Figure 14.4. The overwintering stage (or oversummering stage, depending on the rust and the climate in which it exists) is that of teliospores borne in telia. Teliospores germinate to form one or more basidia and basidiospores. In most rust fungi, basidiospores are forcibly discharged via a mechanism that is identical to that found in most mushrooms. The forcible discharge enables basidiospores to become airborne and initiate infection if they land on a susceptible host. Basidiospore-initiated infections result in the production of small, one-celled, **spermatia** (also called **pycniospores**) inside flask-shaped **spermagonia** (also called pycnia) that develop embedded in host tissue. Spermatia are formed in a sweet, sticky exudate that attracts insects, which transmit spermatia from one spermagonium to another. Spermatia fuse with **flexuous hyphae** emerging from another spermagonium of the opposite, compatible mating type, thereby accomplishing fertilization and establishing the dikaryotic nuclear condition. In rusts that lack a spermagonial stage, dikaryotization occurs through hyphal fusion of compatible strains within the host tissue. Dikaryotic hyphae grow through the host tissue, eventually forming **aeciospores in aecia**; that stage is followed by the production of **urediniospores** in **uredinia** and, eventually, by **teliospores** in **telia**, to complete the cycle. The uredinial stage is sometimes called the "**repeating stage**," because urediniospores infect the same host on which they are formed (Figure 14.5). Nuclear fusion occurs in the teliospore, meiosis occurs during teliospore germination, and haploid **basidiospores** are formed. By convention, the different stages of rust fungi are often designated by the following Roman numerals: spermagonia/spermatia (0), aecia/aeciospore (I), uredinia/urediniospore (II), telia/teliospore (III), and basidia/basidiospore (IV). The life cycle, showing the nuclear condition of different hyphal and spore stages, is summarized in Figure 14.4. Rusts that complete their entire life cycle on two unrelated host species are termed **heteroecious** rusts, and those that require only one host species are termed **autoecious** rusts. In the restricted sense, the "**primary host**" of a heteroecious rust is the plant on which the uredinial and/or telial stages are formed, and the spermagonial and aecial stages are formed on the "**alternate host**." Hence, in *P. graminis*, wheat, or a related species, is the primary host (Figure 14.5) and common barberry is the alternate host (Figures 14.6 and 14.7). Similar cycles requiring two unrelated hosts occur in other hetereocious cereal rusts: for example, wheat leaf rust (*Puccinia recondita* f. sp. *tritici*) and oat crown rust (*Puccinia coronata*). The life histories of these cereal rusts have been well known for more than a century, but that of the economically important stripe rust of wheat (*Puccinia striiformis*) presented a scientific mystery. The stripe rust pathogen produces teliospores that germinate to form basidia and basidiospores, as in the other important cereal rust fungi, but there was no known alternate host infected by basidiospores and hence these spores were not thought to have a role in the life cycle. In 2010, this puzzle was solved when two species of barberry were shown to be alternate hosts in the stripe rust life cycle (Jin et al., 2010).

Autoecious rusts such as those on asparagus, bean, flax, or rose complete their entire life cycle on just one host. Rusts with aecial, uredinial, and telial stages in their life cycle, whether they are autoecious or heteroecious, are called **macrocyclic** rusts. **Demicyclic** and **microcyclic** rusts, also called **short cycle rusts**, lack one or more spore stages in their life cycle. Demicyclic rusts lack the uredinial stage, and microcyclic rusts lack both the aecial and uredinial stage. The spermagonial stage may be absent in both macrocyclic and short cycle rusts. Because they lack the uredinial (repeating) stage of the life cycle, short cycle rusts do not have the potential to increase in the dramatic manner of macrocyclic rusts. For example, cedar-apple rust, *Gymnosporangium juniperi-virginianae*, a demicyclic rust, does not have "explosions" of disease, but significant damage can occur in apple orchards resulting from massive primary infection by huge numbers of basidiospores originating from cedar (botanically, *Juniperus*), the alternate host (Figure 14.4c). In Figures 14.8 and 14.9, a juniper heavily infected with *Gymnosporangium betheli* is shown. The alternate host for *G. betheli* is black hawthorn (Figure 14.9). Hollyhock rust (*Puccinia malvacearum*) produces teliospores that overwinter and germinate to form basidiospores capable of initiating new infections resulting in telial formation the following year, exemplifying a microcyclic rust. Other abbreviated forms of rusts lack the sexual stage completely, or it is yet to be discovered, in some cases because it does not occur in the predicted stage of the life

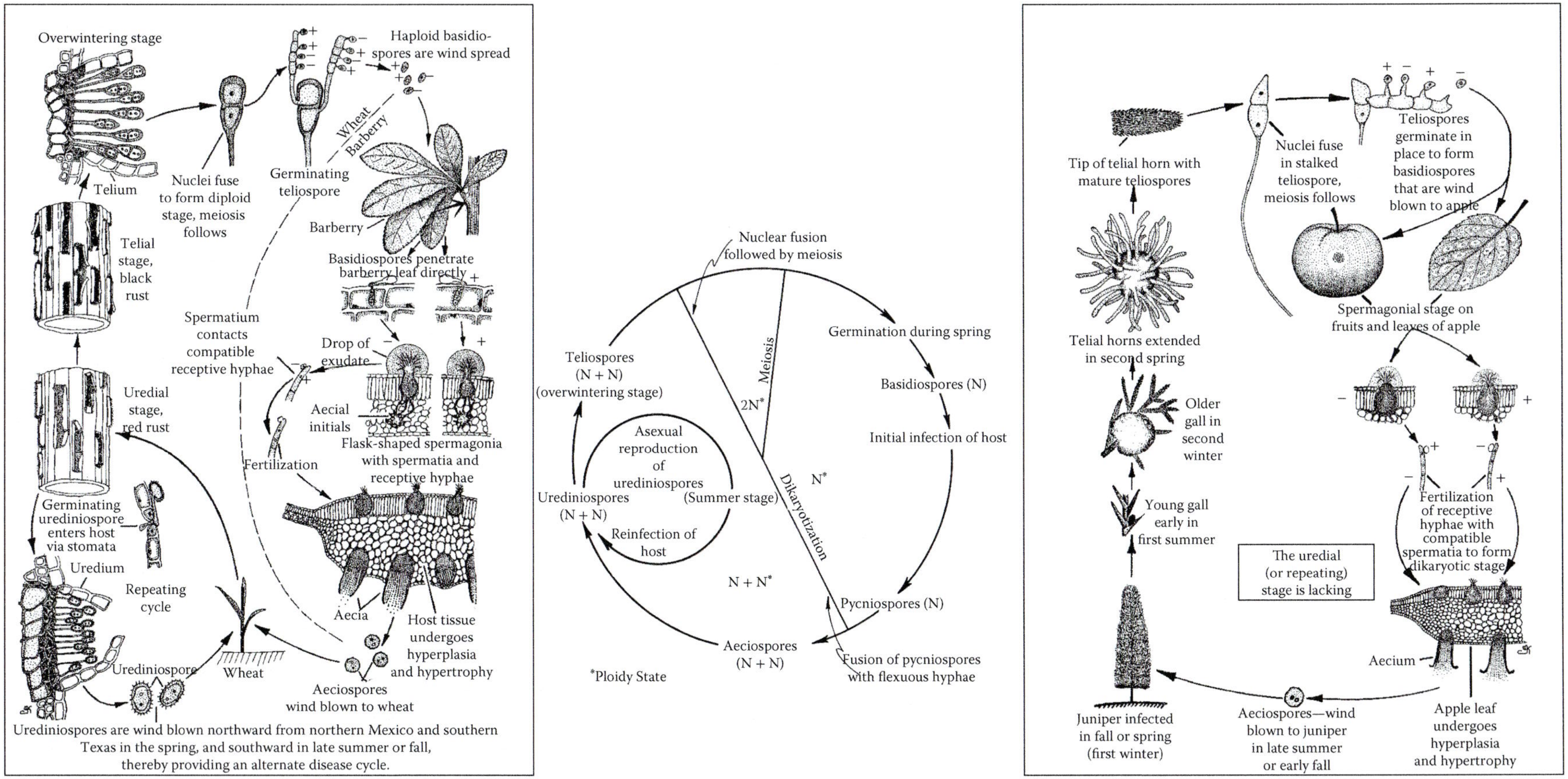

(a) (b) (c)

FIGURE 14.4 (a) Disease cycle of wheat stem rust (*Puccinia graminis* f. sp. *tritici*). (From Kenaga et al., *Plant Disease Syllabus*, Baltimore Publication, Lafayette, 1971. With permission.) (b) Generalized life cycle of rust fungi showing the sexual and asexual reproductive stages and ploidy number of those stages. (From Littlefield, L.J., *Biology of the Plant Rusts: An Introduction*. Iowa State University Press, Ames, 1981. With permission.) (c) Disease cycle of cedar-apple rust (*Gymnosporangium juniper-virginianae*). (From Kenaga et al., *Plant Disease Syllabus*, Baltimore Publication, Lafayette, 1971. With permission.)

FIGURE 14.5 Uredinial stage of stem rust, *Puccinia graminis* on wheat. (Courtesy of Xianming Chen.)

FIGURE 14.6 Aecial stage of *Puccinia graminis* on barberry leaves (*Berberis* sp.). (Courtesy of Xianming Chen.)

cycle. For example, coffee rust (see Case Study 14.1) has long been thought to survive successfully in the uredinial stage; teliospores are rarely formed and are not thought to be important in the life cycle. However, a recent study showed that sexual reproduction does occur in the coffee rust life cycle, but via meiotic nuclear division during germination of urediniospores in a process termed "cryptosexuality" (Carvalho et al., 2011).

In many rusts, including those on alfalfa, asparagus, bean, cereals, flax, sunflower, and many other herbaceous angiosperm hosts, the most damaging effect on yield results from the uredinial stage irrespective of the rust being autoecious or heteroecious. The subsequent telial stage on the same host species is essential for the completion of the life cycle, but typically, it develops too late in the growing season to cause a serious reduction in yield. The uredinial stage is not only the major cause of severe yield reduction in herbaceous hosts; it is also the spore stage most commonly and widely spread by wind, potentially resulting in epidemics spanning

FIGURE 14.7 Close-up of aecial stage on the underside of barberry leaf. Each cup-shaped aecium is filled with chains of aeciospores that will become aerially dispersed. Aeciospores can only infect wheat. (Courtesy of Xianming Chen.)

FIGURE 14.8 Rocky Mountain juniper (*Juniperus scopulorum*) heavily infected with the rust pathogen *Gymnosporangium betheli*. The orange structures are the telial stage of the fungus. (Courtesy of R. Alan Black.)

FIGURE 14.9 Gelatinous orange telial stage and aecial stage of *Gymnosporangium betheli* on Rocky Mountain juniper, *Juniperus scopulorum*, and black hawthorn, *Crataegus douglasii*, respectively. (Courtesy of R. Alan Black.)

hundreds of miles during a growing season. The annual south-to-north dispersal of both wheat leaf and stem rust urediniospores occurs along the well-documented "Puccinia Path." Both the pathogens can overwinter as uredinial stage infections on winter wheat in southern Texas and Mexico; urediniospores produced there in early spring are blown northward and have been trapped at elevations up to 4300 m (approximately 14,000 ft). Progressive infections of susceptible winter wheat occur from northern Texas through South Dakota and, eventually, into susceptible spring wheat in Minnesota, North Dakota, and central Canada by midsummer. Similarly, the rapid spread of coffee rust in Asia and Africa resulted largely from windborne urediniospores, as in much of South America, following the introduction of rust into the western hemisphere, possibly carried there on plants from Africa in 1970. Another more recent example of the long-distance dispersal potential of urediniospores is Asian soybean rust (*Phakopsora pachyrhizi*), a pathogen that made its way from Asia to Hawaii by 1994, to South America in 2002, and then to the continental United States in 2005, with the latter jump likely occurring via urediniospores carried by the winds of a tropical storm.

In contrast to rusts of most agronomic crops, some conifer rusts cause the greatest damage in the spermagonial and aecial stages of the life cycle. Trunk and branch cankers, and galls, resulting from infections 2 or more years old, initiated by basidiospores from an alternate host, can eventually girdle infected regions, resulting in death of all tissue beyond the canker. Several species of pine rusts cause millions of dollars loss annually in both northern and southern forest of the United States, as well as in fir, larch, pine, spruce, and other timber species worldwide. In the white pine blister rust (*C. ribicola*) life cycle, pine, the spermagonial and aecial host, is commonly referred to as the primary host because it is more economically important than *Ribes*, the host that supports the uredinial and telial stages. In contrast, an important example of a uredinial stage rust in commercial trees is leaf rust of poplars (*Melampsora larici-populina*), which causes severe defoliation, reduced pulpwood production, and is a major factor limiting the use of poplars for bioenergy programs.

Rust pathogens are highly host specific, with most species being limited to one or only a few closely related hosts. Within some rust species, there are also "specialized forms" (Latin plural, *formae speciales*; abbreviated "f. sp.") that are limited to certain host species, for example, *P. graminis* f. sp. *tritici* on wheat and barley, *P. graminis* f. sp. *avenae* on oats, *P. graminis* f. sp. *secalis* on rye, and several others. There are also "physiologic races" within the species of several rusts, for example, *Melampsora lini* on flax and in many *formae speciales* of rusts (e.g., *P. graminis* f. sp. *tritici*) that can attack only certain cultivars of host species, as described by Stakman in 1914. The identity and, sometimes, the actual genotypes of the physiologic races are determined by inoculating defined sets of "differential varieties" or "host differentials," each containing one or more known race-specific resistance genes, and approximately 10–14 days later, observing the various degrees or types of immune, resistant, or susceptible responses of those varieties. The distribution of reaction types across the set of differential varieties tested determines the identity of the rust races so tested. Ug99, the race of wheat stem rust previously mentioned as a major global threat to wheat, possesses novel virulence to resistance genes from rye and wheat relatives, as well as virulence to most of the other known resistance genes from wheat. The reason for this lethal combination of virulence is that most of the wheat cultivars grown around the world are susceptible to Ug99.

Rust fungi obtain nutrition through specialized branches called **haustoria** that grow from intercellular hyphae into living host cells. Although haustoria develop inside host cells, they are surrounded by an extrahaustorial membrane derived from the host plasma membrane and do not have direct contact with the host cytoplasm. Nutrients absorbed by haustoria are transported through the intercellular hyphae, supporting further expansion of the

CASE STUDY 14.1

Coffee Rust

- Coffee rust, caused by *Hemileia vastatrix* and *H. coffeicola*, is the most economically important disease of arabica and robusta coffee (*Coffea arabica* and *Coffea canephora*, respectively) and is present in all major coffee-producing regions of the world.
- Coffee is the most important agricultural commodity in trade value and is second only to crude oil value as a global export commodity.
- Coffee was already a popular beverage in the nineteenth-century Europe. Coffee rust was first reported in the former British colony of Ceylon (now Sri Lanka) in 1869 and, within a few years, had spread to all of the coffee plantations on the island. No longer able to produce coffee, plantation owners switched to tea production, and tea consumption remains popular today in Britain.
- Coffee rust infects leaves and causes premature defoliation that weakens the plant and reduces the size and number of berries in the current year of infection. The next year's crop may also be reduced, and severe infection can kill plants.
- Coffee rust reduces yields as much as 80%, but average annual losses due to coffee rust are estimated at 15%.
- Only the wind-dispersed uredinial stage is important in the life cycle, and since no alternate hosts have been identified, coffee rust is autoecious.
- Urediniospores require free water on leaves in order to germinate.
- Each lesion formed on leaves by the rust fungus consists of many sori; a single lesion can produce up to 400,000 spores over a 5-month period.
- Movement of coffee rust from one continent to another is via infested plant material.
- The use of fungicides combined with management practices including proper pruning, plant spacing, and weed control has been successful in the control of coffee rust. Inconsistent use and reduced effectiveness of fungicides have been blamed for major outbreaks of coffee rust.
- Genes from wild coffee species are being used to breed coffee cultivars that are resistant to coffee rust.

fungus colony and the production of reproductive spores. Nutrients diverted from the host into the pathogen contribute to reduced yield or quality of the plant's economic product. Haustoria are also now known to be involved in the secretion of small proteins called **effectors** that play an important role in the suppression of host defenses.

Resistance to rusts is often expressed as "hypersensitivity" or the "hypersensitive reaction," a localized reaction in plants in response to haustorium formation in which the invaded host cell and a few surrounding cells are killed, thus depriving the obligate parasite of nutrition. Other mechanisms, for example, accumulation of toxic compounds (phytoalexins) can contribute to the death of a pathogen in incompatible interactions with some types of rust fungi such as crown rust in oats. Other types of resistance include "tolerance," where the host is capable of producing an acceptable yield in spite of rust infection, and "slow rusting" or partial resistance, when rust development in host tissues is significantly retarded rather than halted (killed), as in hypersensitivity. These latter types of resistance are thought to result from more widely based, multigenic responses than hypersensitivity, which often results from the action of a single resistance gene in the host. Rapid development and spread of new physiologic races of rusts results by the planting of resistant cultivars over large regions of crop production. The genetic control of resistance and susceptibility of many hosts, and the pathogenicity of many rust pathogens, respectively, are governed by the "gene-for-gene" relationship, first demonstrated by Flor in the 1950s with flax rust.

A wide array of management techniques for rust diseases is dictated by the many variations in life cycles of rust fungi, the disease cycles of the different diseases, the diversity of host plants and differences in their cultural methods, the variably important significance of wind dispersal of spores, different economic impacts of rust diseases on different hosts, and differences in breeding techniques for herbaceous hosts versus timber species. Examples of some more commonly used methods are discussed in the following text.

Genetic Resistance

Wherever genetically resistant cultivars are available, they are generally the most effective and least expensive means of managing rust diseases. Many cultivars of rust-resistant agronomic, orchard, and vegetable crops are available to commercial growers and homeowners. If they are not available, when new races of rust arise, different means of disease management must be utilized, including the following.

Foliar Fungicidal Sprays

Both protective and eradicant sprays are widely recommended for control of many rust diseases of herbaceous and orchard crops and ornamentals; many are quite effective. Unfortunately, the widespread and continued use of such fungicides sometimes provided selection pressure for the evolution of new, fungicide-resistant, or fungicide-tolerant strains of rust pathogens. Judicious use of foliar sprays is thus essential.

Cultural Practices

Practices such as planting susceptible conifers in low-risk environments, based on environmental evaluations, is recommended for some conifer rusts and is suggested even for some cereal rusts. Sanitation procedures, for example, inspecting and destroying infected seedlings, and pruning out infected branches well below cankers or galls, another type of cultural practice, are recommended for some pine rusts. Other examples include providing adequate greenhouse ventilation to reduce high humidity favorable for geranium rust; similarly, for producers of organic asparagus, planting crop rows parallel to the prevailing wind is recommended to facilitate air movement and to reduce humidity in asparagus fields for the management of rust disease. Removing or plowing under crop debris, including overwintering teliospores, is recommended for some vegetable diseases. However, such efforts may be futile if other growers in the region fail to do the same, thus providing the region an ample supply of windborne inoculum, similar to the near impossibility of controlling corn smut by sanitation.

Elimination of Alternate Hosts

This has long been a recommended practice for many heteroecious rusts, dating back to the seventeenth-century French legislation and similar laws in the American colonies noted previously. Influenced by a severe rust epidemic in 1916 and crucial demands for North American wheat during World War I, massive barberry eradication campaigns were implemented in 18 northern wheat-producing states in the United States and three adjacent Canadian provinces. That campaign lasted for some 50 years, although less in some states and provinces than in others, and resulted in over 100 million barberry bushes being destroyed. The practice greatly reduced localized infections of wheat rust early in the growing season, which definitely increased yields in the participating states and provinces and reduced the potential for the development of new races by genetic recombination on the barberry host. However, barberry eradication did little to reduce the northward advance of urediniospores and the consequent wheat stem rust infections along the Puccinia path, as the latter was more a result of meteorological than biological factor. Removal of junipers near apple orchards was formerly recommended as a control for cedar-apple rust, but the impracticality of that, given the widespread distribution of junipers and the windborne distribution of basidiospores from that host, was later realized. Similarly, in the United States, only partial success was achieved by eradicating wild currant bushes in the western states for the control of white pine blister rust, although the practice was more successful in the eastern states.

SMUT DISEASES

The term "smut" comes from the dark masses of teliospores formed in host organs that give the infected plant the appearance of being soiled or burnt. The life cycles of most smut fungi are similar. Smut fungi, similar to rust fungi, produce one teliospore stage in each growing season. Teliospores in most smut fungi are one-celled, with darkly pigmented, usually ornamented, and thickened walls, and represent the overwintering or perennating stage in the life cycle (Figure 14.10). Teliospores can remain viable for many years in smut fungi such as the dwarf bunt pathogen of wheat (*Tilletia contraversa*), and this longevity has presented challenges to the development of effective control measures. Teliospores of the most common smut fungi form a dark, powdery mass inside the structures formed in or on plant hosts called **sori** (singular: **sorus**). A sorus is covered with a persistent or ephemeral peridium composed of host plant and/or fungal tissues. In addition to teliospores, smut sori may

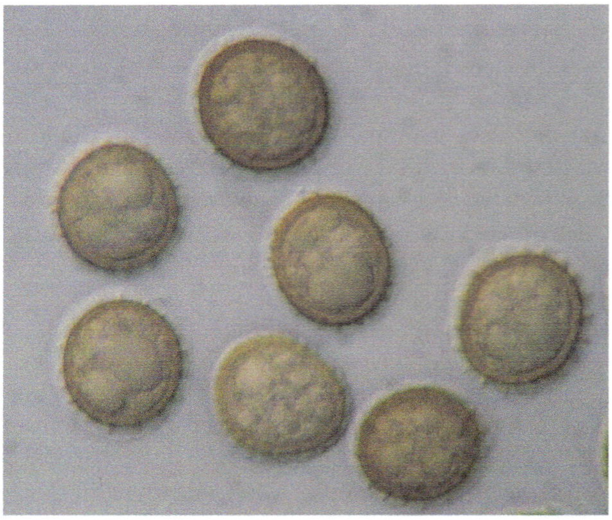

FIGURE 14.10 Teliospores of *Tilletia caries*. (Courtesy of L. Carris.)

also include a central columnlike structure composed of host tissue called a columella and/or sterile cells. Sterile cells are hyaline, sporelike structures interspersed with teliospores in some genera of smut fungi including *Tilletia* and *Urocystis*. Unlike teliospores, sterile cells do not germinate, and their function is not known. Teliospores are dispersed primarily by wind, but dispersal may also involve animals (including humans) or water. Teliospore germination requirements for moisture, temperature, pH, and, in some species, light vary among different species. During teliospore germination, a basidium (also called a **promycelium**) emerges through a crack in the teliospore wall. Basidiospores (also called primary **sporidia**) are formed on the basidium in most types of smut fungi. Basidiospores can fuse with an adjacent basidiospore of a compatible mating type, germinate directly to produce hyphae, or produce a secondary sporidium. Haploid spores, called secondary sporidia, grow saprotrophically on soil and plant surfaces. In smuts such as common corn smut, *Ustilago maydis*, secondary sporidia proliferate in a yeastlike manner. Conjugation occurs between compatible primary or secondary sporidia, and the resulting dikaryotic hyphae infect the host plant. Smut fungi such as *Ustilago tritici*, the fungus causing loose smut of wheat, lack a basidiospore or sporidial stage, and the dikaryon is formed between compatible germ tubes emerging from individual cells of the basidium. Within the host, dikaryotic smut hyphae advance in an intracellular manner by invaginating the host plasma membrane, or via intercellular hyphae and intracellular haustoria depending upon the smut. Unlike rust fungi, smut fungi lack a repeating asexual stage that results in secondary infection during the growing season.

Major resistance genes have been identified in wheat and barley hosts of smut pathogens. Fifteen different bunt resistance genes (*Bt*) are known from different wheat cultivars, with corresponding avirulence genes (*avr*) in wheat bunt pathogens *Tilletia caries*, *T. controversa*, or *T. laevis*. Races of wheat bunt are identified by inoculating strains of the pathogen on differential wheat cultivars, with each cultivar containing a different set of resistance genes. Prior to the widespread use of fungicide seed treatments, bunt-resistant wheat cultivars were rendered ineffective by the rapid emergence of new races of wheat bunt. Six resistance genes have been identified in barley cultivars with the corresponding avirulence genes in *Ustilago hordei*. Interestingly, in the well-studied corn smut pathosystem, no major resistance genes have been identified. Complete annotation of the *U. maydis* genome has revealed fascinating insights into how this biotrophic pathogen infects and grows inside its host plant. Unlike a necrotrophic pathogen, *U. maydis* has relatively few genes coding for plant cell wall-degrading enzymes but has a number of novel secreted effector proteins that may be involved in restructuring the host plant metabolic pathways to produce nutrients needed for growth and reproduction of the fungus.

Smut fungi are categorized by where teliospores are formed in the host plant—inflorescence, leaf, stem, or root. Inflorescence smuts are the most common type. In some inflorescence smuts, the sorus retains the shape of the infected host organ. The common bunt pathogens of wheat, *T. caries* and *T. laevis*, are examples of this type of inflorescence smut where the wheat seed is replaced by a dark, powdery mass of teliospores (Figure 14.11). Other types of inflorescence smuts destroy the developing florets and transform the inflorescence into a mass of teliospores, with only the rachis left intact, as in loose smut of wheat caused by *U. tritici*. Leaf smuts infecting grasses and other monocots form elongated, subepidermal sori that look like thin, black stripes on the infected leaves; this type of smut is also called "stripe smut." As the teliospores develop in leaf smuts, the host epidermis ruptures to reveal masses of teliospores. Onion smut (*Urocystis magica*) and flag smut of wheat (*Urocystis agropyri*) are examples of this type of leaf smut. In other types of leaf smuts such as those caused by species of *Entyloma*, the teliospores remain embedded in the host leaves, and the sori may be evident as pale lesions on the infected leaves. Stem

FIGURE 14.11 Comparison of bunted and healthy wheat seeds. Wheat seed infected by the dwarf bunt pathogen, *Tilletia contraversa* (top left) are hard and rounded compared with those infected by the common bunt pathogen, *T caries* (top right), which retain the shape of the uninfected wheat seeds (bottom). (Courtesy of L. Carris.)

smuts form sori that are typically confined to the stem of the host plant, usually grasses. Root smuts include species of *Entorrhiza* that form sori in swollen parts of roots of sedges (Cyperaceae) and rushes (Juncaceae). An additional group of smuts are those that form large, tumorlike growths (galls) as in common corn smut (*U. maydis*) (Figure 14.12). Sori of gall-forming smuts are usually not confined to a single organ of the infected host plant.

Smut fungi are also categorized by the location of the infection in the host plant—seedling, embryo, shoot, and local infection. These types of infection are described in detail, with examples.

Seedling Infection

The wheat pathogens causing common bunt, *T. caries* and *T. laevis*, dwarf bunt, *T. contraversa,* and flag smut, *U. agropyri*, are examples of seedling-infecting smuts. Teliospores of common and dwarf bunt may remain viable in the soil for a decade or more under optimal conditions, and those of flag smut can remain viable for at least 4 years. Teliospore germination results in the formation of an aseptate, hyphallike basidium and long, thin, basidiospores that conjugate with other basidiospores of compatible mating type shortly after formation. The conjugated basidiospores, called **"H-bodies"** because of their distinctive shape, germinate to produce secondary sporidia or infective dikaryotic hyphae. The infective hyphae of the common and dwarf bunt fungi penetrate the seedling, grow intercellularly to the meristem of the plant, and are carried upward as the wheat plant develops. When the wheat inflorescence begins to develop, the hyphae proliferate within the developing ovaries, converting every kernel in an inflorescence into a sorus containing millions of dark, powdery teliospores enclosed with a persistent host pericarp. These sori, also called "bunt balls" (Figure 14.13), are broken open during harvest, releasing windborne teliospores that can contaminate uninfected wheat seed.

FIGURE 14.12 Ears of corn infected by corn smut, *Ustilago maydis*, ready for preparation in a restaurant. Corn smut is a delicacy in Mexico, where it is called "*Cuitlacoche*." The gray, gall-like structures are the sori of the smut fungi that have developed within individual ovaries of the corn inflorescence. When sori mature, they will be filled with masses of dark, powdery teliospores. (Courtesy of Tsutomi Arie.)

Common bunt is effectively controlled by the combined use of resistant wheat cultivars and fungicide seed treatments and is now rare in conventionally grown wheat in most developed countries. It remains a major limitation to wheat production in low-input agriculture and has reemerged as an important disease in organic wheat production in northern and western Europe. In flag smut, the fungus grows to the meristem of the wheat seedling and in the spring grows systemically within the plant and forms elongated gray or black sori in the developing leaves. The term "flag smut" comes from the formation of sori in the last leaf (the "flag" leaf) that is formed prior to the emergence of the inflorescence. The flag leaf is particularly important as it makes up nearly three-quarters of the leaf area on the wheat plant that contributes to grain fill in the developing seed head. Infected plants are often stunted and twisted in appearance, and the seed head may fail to develop or emerge. Flag smut is effectively controlled through the use of systemic fungicide seed treatments.

Embryo Infection

Loose smut of wheat (*U. tritici*) and barley (*Ustilago nuda*) are examples of embryo-infecting smuts. Embryo-infecting smuts survive as dormant mycelium in the seed until seed germination, and the fungus then grows to the developing inflorescence. The fungus grows inter- and intracellularly into all the developing inflorescence

FIGURE 14.13 Bunted wheat inflorescence. Each of the wheat kernels is replaced by a sorus of *Tilletia caries*. (Courtesy of Janet Matanguihan.)

tissues except the rachis and awns. Usually the entire inflorescence is destroyed and converted to a sorus, and by the time the inflorescence emerges from the leaf sheath, the head has been converted into a mass of dark, powdery teliospores. The teliospores are dispersed by wind and rain to developing flowers on uninfected plants. The teliospores germinate, and infective dikaryotic hyphae are formed through fusion of adjacent cells of the basidium, without formation of primary or secondary sporidia. After infection, the fungus grows into the developing ovaries, and the resulting seeds show no external sign of the dormant smut mycelium contained within. Because the mycelium is inside the wheat seed, conventional fungicide seed treatments are not effective, but hot water or heat treatment of infected seed and use of clean seed and systemic fungicides are effective control measures.

Shoot Infection

Shoot-infecting smuts infect the host plant through host shoots or buds. An example of this type of smut is sugarcane smut, Ustilago scitaminea, which infects buds of sugarcane and transforms the host floral stem into a "smut whip"—a thin, whip-like sorus up to a meter in length that emerges from the apex of the plant. Each whip is covered with a delicate, silvery membrane that ruptures easily to expose the powdery mass of teliospores. A whip can release up to ten billion teliospores in a single day. Teliospores germinate to produce a four-celled basidium that forms basidiospores that either fuse to form the infective dikaryon, or bud to form sporidia. The dikaryon may also form by conjugation of basidial cells. Control of sugarcane smut is primarily through the use of resistant varieties and cultural practices including the use of healthy planting material and removal of infected shoots.

Local Infection

In local-infecting smuts, fungal hyphae and sori are confined to the site of infection. Common corn smut, *U. maydis*, is the best-known example of a local-infecting smut. Any meristematic tissue on an infected corn plant can be transformed into gall-like sori up to several centimeters in diameter. Sori form in tassels, leaves, brace roots, and stems, but are most common on the corn ear, where they often form in clusters. Initially, the sori are firm and covered in a white periderm, but as the teliospores mature, the periderm often ruptures to reveal the dark powdery mass of spores. Teliospores are windborne and contaminate the soil, where they overwinter. In the spring, the teliospores germinate to produce a four-celled basidium and basidiospores that proliferate in a yeast-like manner. The sporidia are wind- and water-dispersed and, after fusing with a compatible strain to form the infective dikaryon, can infect any aboveground meristematic tissues, even in the seedling stage. During infection, the fungal hyphae grow both intra- and intercellularly, and massive proliferation of the hyphae and production of teliospores occurs 5 or 6 days following infection to form the distinctive sori. Resistant cultivars are available for field (dent) corn, and some cultivars of sweet corn are tolerant, but none are resistant to corn smut. Fungicide seed treatments are not effective in controlling corn smut.

The immature gall-like sori of corn smut have long been considered a delicacy in Mexico, where they are known as "*Cuitlacoche*" or the alternate spelling "*Huitlachoche*." Cuitlacoche is described as having an earthy, smoky, mushroom taste with a hint of corn and is used in quesadillas, tamales, soups, and other dishes (Figure 14.14). Cuitlacoche is sold fresh and in cans, and is developing a following in the United States.

Corn smut is also an important model organism for studying virulence and dimorphism—the ability of a fungus to switch between a mycelial and yeastlike growth form. In 2006, *U. maydis* was the first smut fungal genome to be completely sequenced.

Karnal bunt of wheat (*Tilletia indica*) and rice kernel smut (*T. horrida*) are also local-infecting smuts, but these fungi only infect through developing ovaries, and often only a few or one seed in the inflorescence is infected. This type of smut is also referred to as a partial bunt. Teliospores germinate to produce basidia and a large number of basidiospores (up to 150) that do not conjugate but germinate to form hyphae or secondary sporidia. The sporidial stage proliferates on plant leaves and on the surface of soil or water (for the rice kernel smut); sporidia become airborne through a forcible discharge mechanism. In wheat, infection by the Karnal bunt pathogen occurs from the time that the florets begin to emerge from the boot stage, up to early development of the wheat kernel called the soft dough stage. Because the fungus infects the developing floret, seed treatments are not effective, and control efforts for Karnal bunt and rice kernel smut focus on the development of resistant wheat and rice varieties, respectively. See Case Study 14.2.

FIGURE 14.14 Corn smut in tortilla as served in restaurant in Mexico. (Courtesy of Tsutomi Arie.)

CASE STUDY 14.2

Karnal Bunt Of Wheat—A Minor Pathogen With A Major Impact

- Karnal bunt of wheat, caused by the smut fungus *Tilletia indica*, poses a substantial threat to the ability of U.S. growers to export wheat.
- Karnal bunt infects species of wheat and triticale, a wheat × rye hybrid.
- The Karnal bunt pathogen requires cool, wet weather during the heading stage of the host plant for infection to occur.
- Karnal bunt is a local-infecting smut that infects through the developing floret; typically, only one or a few kernels per head are affected.
- Direct yield losses are generally minor, but the presence of 3% or more infected kernels results in flour with an objectionable color, taste, and smell.
- Over 70 countries, including the United States, have implemented quarantines to prevent the introduction and spread of the Karnal bunt pathogen.
- Karnal bunt is spread primarily via infected seed.
- Control of Karnal bunt using cultural or chemical methods is generally not effective when environmental conditions are favorable for infection.
  - Teliospores can survive in the soil for 5 years or more.
  - Fungicide seed treatments are not effective.
  - Chemical control of Karnal bunt using fungicides applied to the foliage is cost-prohibitive.
- Resistance to Karnal bunt has been identified in wheat and wheat relatives, and wheat varieties with high levels of resistance have been developed.

LOOSE AND COVERED SMUTS

The terms "covered smut" and "loose smut" refer to whether the **peridium** that encloses the sorus is persistent or ephemeral, respectively. The peridium may be of host origin or of host and fungal origin, depending upon the smut. In a covered smut, exemplified by the common bunt pathogens *T. caries* and *T. laevis*, the peridium remains intact. In a loose smut, such as *U. tritici*, the peridium is thin and delicate and ruptures easily to expose the powdery mass of teliospores.

Economic Impact of Smut Fungi

Smuts are most important on cereal crops, for example, barley, corn, oats, sorghum, and wheat. Other cultivated plants that are infected by smut fungi include forage grasses, onions, and sugarcane. Most of the 1600 species of smut fungi parasitize wild host plants, infecting more than 4000 plants belonging to 75 different families. Over two-thirds of smut fungi parasitize grasses (Poaceae; 57% of known hosts) and sedges (Cyperaceae; 12% of known hosts). Most smut fungi have restricted host ranges, infecting only a few, and usually closely related plant species. Exceptions include *T. controversa*, with a host range reported to encompass 19 genera of grasses, and *Ustilago striiformis*, reported to infect grass hosts belonging to 44 different genera.

The most common smut diseases in North America are corn smut, stinking smut, or common bunt of wheat, and loose smut of wheat, barley, and oats. The economic impact of smut infection is most evident when the smut fungus replaces all of the individual kernels of cereals with millions of dark, powdery teliospores, or when it causes individual kernels, as in corn smut, to become grossly enlarged and distorted. With the development of highly effective chemical seed treatments for small grains, and smut-resistant corn varieties, smuts have considerably less economic impact than 60 years ago. However, several smut pathogens of wheat—dwarf bunt, Karnal bunt, and flag smut—have been targets of international quarantines and have negatively impacted the export of wheat in the United States and other wheat-exporting countries. For example, grains and other plant materials cannot be exported if teliospores of the quarantined pathogen are detected in the shipment. Karnal bunt of wheat, considered a minor pathogen of wheat based on yield impact, was restricted to northwestern India and other parts of Asia until 1972 when it was detected in Mexico. The United States and at least 30 other countries quickly established quarantines on import of wheat from countries where Karnal bunt was known to occur. In 1996, Karnal bunt was discovered in the southwestern United States, and the U.S. Department of Agriculture declared an "extraordinary emergency." A national survey for Karnal bunt was initiated involving every wheat-growing county in the country, preserving the ability of U.S. growers to export wheat. As of 2011, 37 states were still participating in this survey.

LABORATORY EXERCISES

Smut and rust fungi are not easy organisms to adapt to laboratory exercises in courses offered at institutions that do not have ongoing research programs and ready access to these pathogens. Availability of smut and rust inoculum will depend upon the types of crops that are grown in the region. These fungi, as for other plant pathogens, cannot be shipped across state lines without a permit from the U.S. Department of Agriculture, Animal and Plant Health Inspection Service (APHIS). However, smut and rust fungi are often locally abundant on different types of noncultivated plants, and their conspicuous symptoms make many of these pathogenic fungi relatively easy to find in the field. Collection, observation, and identification of smut and rust fungi make an effective laboratory exercise. The exercises provided below depend on the local availability of inoculum, but use host plants that should be readily available as seed.

Experiment 1. Bean Rust

Bean rust, caused by *Uromyces appendiculatus*, is a common disease of dry and snap beans, lima beans, and scarlet runner beans. Pustules containing the reddish-brown spores develop on bean leaves and pods. The total time required to complete the following experiment is approximately 3–5 weeks. The objectives of this experiment are as follows:

- To demonstrate an inoculation method for a rust fungus
- To observe the development of rust pustules on leaves
- To observe urediniospore and teliospore morphology under the microscope

Materials

The following materials are needed per student or group of two or three students:

- Two plastic pots of 10-cm-diameter
- Sufficient commercial potting mix for two pots
- Edible dry pinto beans available in grocery stores provide inexpensive, susceptible seeds for this experiment—three seeds per pot
- 4–5 mL dry talc
- Microspatula
- Tween-20 solution and household water-mist applicator bottle if mist inoculation method is used
- Bean rust fungus, *U. appendiculatus*

Follow the instructions provided in Procedure 14.1 to complete this experiment.

Questions

- In bean rust, what spore stage would have been produced if you had used aeciospores rather than urediniospores as inoculum?
- Describe the morphological features of urediniospores produced because of the inoculation, including pigmentation, spore ornamentation, and number and distribution of germ pores.
- What effect did incubation temperature have on the development of rust? What appeared to be the optimum temperature for the development of rust?

Experiment 2. Corn Smut

Corn smut is one of the most widely distributed of the smut fungi, and fresh material can usually be obtained from home gardeners or plant disease diagnostic clinics. Corn smut teliospores will remain viable for years if stored in a –20°C freezer. Golden Cross Bantam Hybrid is a classic hybrid sweet corn that is highly susceptible to corn smut and is available from commercial sources. Corn seedlings are inoculated when they are 1–3 weeks old, and galls (sori) begin to form within 2 weeks of inoculation. This experiment requires inoculation of two *Ustilago maydis* strains of compatible mating types. Single sporidial lines can be produced from plates of germinating teliospores and used in different combinations to demonstrate compatibility of strains. Cultures of the yeastlike sporidial stage are maintained on potato dextrose agar (PDA) slants with transfers every 1–2 weeks. For long-term storage, sporidial lines can be suspended in sterile 75% glycerol in cryopreservation tubes and stored at –80°C. The total length of time required to complete the experiment is 6–8 weeks. If corn is planted in advance and emerging seedlings are provided, the experiment can be completed in 3–4 weeks. The objectives of this experiment are as follows:

- To demonstrate an inoculation method for corn smut
- To observe the development of corn smut galls on leaves
- To observe the germination of corn smut teliospores
- To observe how corn smut grows as a yeastlike saprotroph on nutrient agar

Procedure 14.1

Production of Bean Rust in the Teaching Laboratory

Step	Instructions and Comments
1	Grow pinto bean seedlings in potting mix in pots of 10-cm-diameter until seedlings are 2–3 weeks old. Thin to two plants per pot before inoculation. Soak bean seeds in wet paper towels 24 h prior to planting will speed up germination. For best disease development, inoculate plants when leaves are approximately 50%–70% expanded. Seedlings can be placed in the refrigerator to reduce the rate of leaf expansion without hurting the plants.
2	Prepare inoculum by mixing 4–5 mL dry talc with 1 mL or less of urediniospores scraped from infected bean leaves. Place inoculum into a test tube or screw top vial covered with —four to five layers of cheesecloth secured with a rubber band. Inoculate plants with a mixture of *Uromyces appendiculatus* urediniospores and dry talc by shaking the test tube above the leaves, allowing the urediniospores and talc to settle on the leaves.
3	Alternatively, mix dry spores in 20–30 mL of water with two drops per L Tween-20. Use atomizer to moist-inoculate plants.
4	Very lightly mist the inoculated plants if inoculated with dry spores and talc, being careful not to wash the inoculum off the leaves. Place pots and a water-soaked paper towel into a large plastic bag, seal the bag, and incubate at room temperature for 18–24 h. Partially open the bags to allow the plants to dry gradually over the next 3–4 h, and then transfer pots to a greenhouse or growth chamber.
5	By maintaining inoculated plants at different temperatures ranging from 15°C to 20°C, the effect of temperature on rust development can be observed over the next 7–14 days. Spores can be examined microscopically by mounting in water on microscope slides.

Procedure 14.2

Production of Corn Smut in the Laboratory

Step	Instructions and Comments
1	Grow corn seedlings in pots of 10-cm-diameter containing commercial potting mix (three to five plants per pot) until three to four leaves have emerged. Depending upon conditions, seedlings should emerge from soil 3–4 weeks after planting.
2	One day before inoculation, start sporidial lines in potato dextrose broth in a 50-mL flask and grow for approximately 18 h at 28°C on a rotary shaker at 200 rpm. After 18 h, spin down cells on a centrifuge and pour off the broth. Resuspend in water and combine the compatible strains.
3	Inject approximately 0.1 mL of the sporidial suspension into the middle of the whorl of corn seedlings by inserting the needle into the "v" formed by the cotyledon. Corn smut galls should begin to form within 14 days and teliospores after an additional 10 days.
4	To observe teliospore germination, sprinkle teliospores onto PDA plates and incubate at room temperature. Teliospores should begin to germinate in 2–3 days. Germinating teliospores, basidia, and sporidia should be mounted in a drop of water on a microscope slide with a coverslip and observed under the microscope.

Materials

The following materials are needed by students to complete this exercise:

- Seeds of Golden Cross Bantam Hybrid sweet corn—10 seeds per lab group
- Plastic pots of 10-cm-diameter—two pots per lab group
- Sufficient commercial potting mix for pots
- Compatible *U. maydis* sporidial lines
- Sterile syringe and 0.1-mL hypodermic needle
- PDA plates
- Microscope slides and coverslips

Follow the protocol listed in Procedure 14.2 to complete this experiment.

Anticipated Results

Galls should begin to form on inoculated corn plants within 2 weeks. Developing galls consist of pale, firm masses of fungal hyphae covered by a peridium. As teliospores develop, they are evident as dark streaks developing within the galls. Eventually the peridium ruptures to reveal the mass of dark, powdery teliospores. Teliospores will germinate on PDA, producing basidia and budding sporidia. Teliospore germination and sporidial formation are readily observed under the microscope.

Questions

- What is the source of inoculum for corn smut under field conditions?
- What result would you expect if a single sporidial line was used for inoculation? Why?
- What attributes of corn smut have contributed to it for becoming an important model organism?

REFERENCES

Aime, M. C., P. B. Matheny, D. A. Henk, E. M. Frieders, R. H. Nilsson, M. Piepenbring, D. J. McLaughlin, L. J. Szabo, D. Begerow, J. P. Sampaio, R. Bauer, M. Weiss, F. Oberwinkler, and D. Hibbett. 2006. An overview of the higher level classification of Pucciniomycotina based on combined analyses of nuclear large and small subunit rDNA sequences. *Mycologia* 98:896–905.

Begerow, D., M. Stoll, and R. Bauer. 2006. A phylogenetic hypothesis of Ustilaginomycotina based on multiple gene analysis and morphological data. *Mycologia* 98:906–916.

Jin, Y., L. J. Szabo, and M. Carson. 2010. Century-old mystery of *Puccinia striiformis* life history solved with the identification of *Berberis* as an alternate host. *Phytopathology* 100:432–435.

Kenaga, C. B., E. B. Williams, and R. J. Green. 1971. *Plant Disease Syllabus*. Lafayette, IN: Baltimore Publication.

Littlefield, L. J. 1981. *Biology of the Plant Rusts: An Introduction*. Ames, IA: Iowa State University Press.

Singh, R. P., D. P. Hodson, J. Huerta-Espino, Y. Jin, S. Bhavani, P. Njau, S. Herrera-Foessel, P. K. Singh, S. Singh, and V. Govindan. 2011. The emergence of Ug99 races of the stem rust fungus is a threat to world wheat production. *Ann. Rev. Phytopath.* 49:465–481.

SUGGESTED READINGS

Bockus, W. W., R. L. Bowden, R. M. Hunger, W. L. Morrill, T. D. Murray, and R. W. Smiley. 2010. *Compendium of Wheat Diseases and Pests*, 3rd ed. St. Paul, MN: APS Press.

Brefort, T., G. Doehlemann, A. Mendoza-Mendoza, S. Reissmann, A. Djamei, and R. Kahmann. 2009. *Ustilago maydis* as a pathogen. *Ann. Rev. Phytopath.* 47:423–445.

Carris, L. M. 2001. Smut fungi. Pages 919–921 in: *Encyclopedia of Plant Pathology*. O. C. Maloy and T. D. Murray (eds). New York, NY: John Wiley & Sons, Inc.

Carris, L. M., L. A. Castlebury, and B. J. Goates. 2006. Nonsystemic bunt fungi—*Tilletia indica* and *T. horrida*: A review of history, systematics and biology. *Ann. Rev. Phytopath.* 44:113–133.

Dodds, P. N., M. Rafiqi, P. H. P. Gan, A. R. Hardham, D. A. Jones, and J. G. Ellis. 2009. Effectors of biotrophic fungi and oomycetes: Pathogenicity factors and triggers of host resistance. *New Phytol.* 183:993–1000.

Fischer, G. W., and C. S. Holton. 1957. *Biology and Control of the Smut Fungi*. New York, NY: The Ronald Press Company.

Kushalappa, A. C., and A. B. Eskes. 1989. Advances in coffee rust research. *Ann. Rev. Phytopath.* 27:503–531.

Maloy, O. 2001. White pine blister rust. Online. *Plant Health Progress*. doi:10.1094/PHP-2001-0924-01-HM.

Money, N. P. 2007. *The Triumph of the Fungi. A Rotten History*. New York, NY: Oxford University Press.

Mueller, O., R. Kahmann, G. Aguilar, B. Trejo-Aguilar, A. Wu, and R. P de Vries. 2008. Review: The secretome of the maize pathogen *Ustilago maydis*. *Fungal Genet. Biol.* 45:S63–S70.

Pataky, J. K., and K. M. Snetselaar. 2006. Common smut of corn. *The Plant Health Instructor*. doi:10.1094/PHI-I-20006-0927-01.

Webster, J., and R. W. S. Weber. 2007. *Introduction to Fungi*. 3rd ed. New York, NY: Cambridge University Press.

Wilcoxson, R. D., and E.E. Saari (eds). 1996. *Bunt and Smut Diseases of Wheat: Concepts and Methods of Disease Management*. Mexico City: CIMMYT.

15 Basidiomycota

Diverse Complex of Saprophytic, Parasitic, and Symbiotic Fungi

Richard E. Baird, C. Elizabeth Stokes, and Alan S. Windham

Concept Box

- Fungi in this group produce haploid **basidiospores** on **basidia**. Basidia line the surface of gills (typical mushroom), tubes (boletes), pores and teeth (shelf and conks), and internally (puffballs).
- Some fleshy basidiomycetes cause heart, butt, and root rots of trees.
- The hyphae of many basidiomycetes have **clamp connections**.
- Hyphal cells are separated by a very complex **dolipore septum**.
- Many fleshy basidiomycetes are associated with **wood decay**; some of them form **ectomycorrhizae** with plant partners; relatively few species that form mushrooms are **parasitic**.

Approximately 30% of all known fungi are classified into the phylum Basidiomycota (basidiomycetes; Alexopoulos et al., 1996). A large percentage of this total belongs to the group known as the fleshy basidiomycetous fungi that produce soft to flexible fruiting bodies that decay or decompose rapidly. Commonly called mushrooms (Figure 15.1), fleshy basidiomycetes include gilled and pored fungi, the puffballs, bird's nest fungi, jelly fungi, and the stinkhorns (Christensen, 1950). Species that form structures associated with wood decay are generally shelf- to conk-like (lacking a stalk; Figures 15.2 through 15.4) or resupinate (appressed to the plant tissue; Figure 15.5).

In nature, basidiomycetes exist as somatic mycelium (hyphae grown in a mass or layer) on wood, litter, and in soil. When environmental conditions are favorable, they form fruiting bodies of various sizes and shapes. Fleshy basidiomycetes occur worldwide in diverse habitats in the cold, temperate, and tropical regions. Individual species, however, are restricted in distribution based on habitats or flora constituents.

Although the majority of basidiomycetous fungi are saprophytic, certain mushroom-forming species are necrotrophic and still others, the rusts and smuts (Chapter 14), are obligate biotrophic plant pathogens. Many species form mycorrhizal associations with woody plants and forest trees while some species of fleshy basidiomycetes are associated with aquatic or wetland ecosystems such as bogs (Alexopoulos et al., 1996). In addition to the types of basidiomycetous fungi mentioned earlier, the genera *Ceratobasidium* and *Thanatephorus*, the teleomorphs of *Rhizoctonia* species (Case Study 15.1), produce basidia on extremely small mycelial mats on soil or plant surfaces. A discussion of *Ceratobasidium* and *Thanatephorus*/*Rhizoctonia* is included later in this chapter due to the economic importance of these genera to agriculture.

MORPHOLOGY

While species of fleshy basidiomycetes range from saprophytic to parasitic, the morphological structures of the fruiting bodies within this group are classified by several important characteristics. One of the most important is the presence of **basidia** (Figure 15.6), of which several forms or types occur. The basidium (singular) is the site where karyogamy and meiosis occur. The resulting nuclei that form following meiosis migrate through

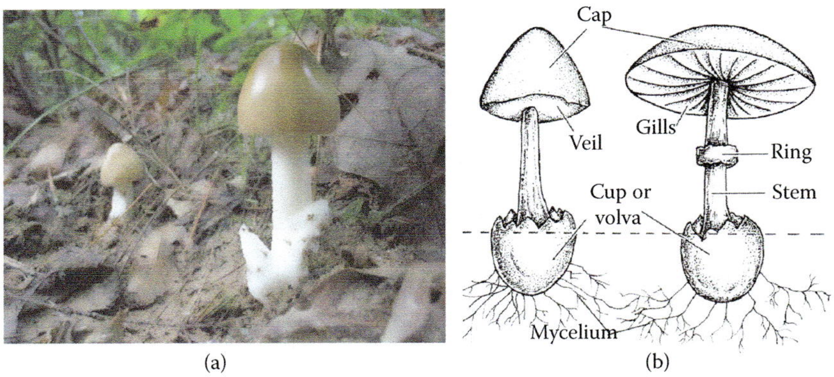

FIGURE 15.1 Fleshy mushroom. (a) *Amanita* species. (Courtesy of R. Baird.) (b) Diagrammatic representation of a mushroom. (Courtesy of Joe McGowen, Mississippi State University.)

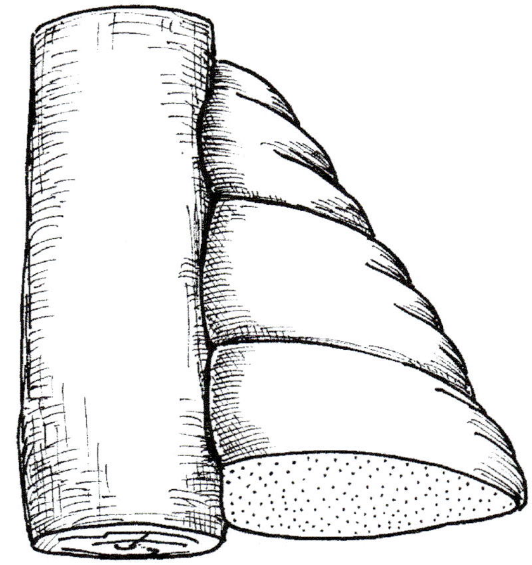

FIGURE 15.2 Woody mushrooms called conks are shelflike structures where basidiospores are formed. Conks grow on living or dead woody plants or trees. (Courtesy of Joe McGowen, Mississippi State University.)

FIGURE 15.4 Conk of *Fomes fomentarius*. (Courtesy of A. Windham.)

FIGURE 15.3 *Ganoderma* species, a conk fungus, causes butt rot of numerous tree species. (Courtesy of A. Windham.)

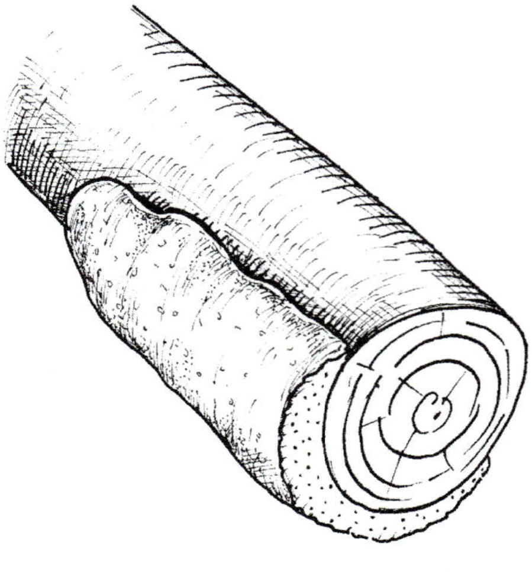

FIGURE 15.5 Resupinate basidiocarp growing on a dead branch. (Courtesy of Joe McGowen, Mississippi State University.)

CASE STUDY 15.1

Rhizoctonia solani

- *Rhizoctonia solani* is an important soilborne pathogen that can cause severe economic loss to agricultural crops. The fungus causes damping-off of seedlings for crops such as cotton, soybeans, and vegetables. The pathogen is also important in ornamental and forest nurseries.
- The pathogen is particularly damaging when growth conditions of seedlings are poor. If the germinating seed is not destroyed below ground, diseased seedlings have sunken dark brown cankers at the ground surface. Often they will appear to be windblown and are broken at the site of the canker (damping-off).
- Different **anastomosis groups (AG)** of *R. solani* occur and attack a wide range of hosts or are host specific.
- The fungus does not produce asexual spores, but basidiospores are produced on exposed basidia formed directly on soil or plant debris. *R. solani* form sclerotia as a means for surviving in soil and plant debris. The *R. solani* AG that attacks rice has adapted to survive in water by forming a thick outer layer around the sclerotia.
- During the subsequent growing season, sclerotia germinate and can reinfect the host crop planted in the field. Because this fungus attacks many hosts, rotation to economically important nonhost crops is not a very effective means of control.
- Control is primarily by use of fungicide seed treatments and when appropriate, in-furrow or drenches with fungicides. Growing conditions that increase plant health reduce postemergence damage. Biological controls using bacterial species (*Bacillus* spp.) and other fungi (*Trichoderma* spp.) have also been successful as seed treatments.

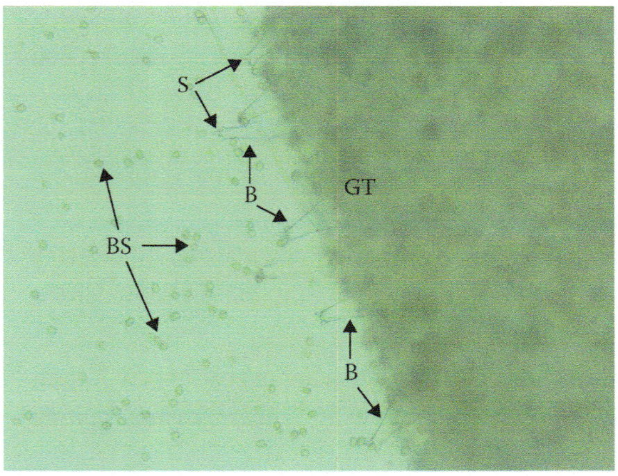

FIGURE 15.6 Basidia (B) and numerous basidiospores (BS) of *Pleurotis* species. Note the four sterigma (S) on which the basidiospores are attached. GT, gill tissue or trama. (Courtesy of R. Baird.)

sterigmata, small stalklike outgrowths, on which spores are formed externally. The **basidiospores** (haploid or N) are formed from the products of meiosis and are also called meiospores.

Hyphae of the basidiomycetes appear to be similar to other phyla such as Ascomycota except that the hyphae contain a specialized septum called the **dolipore septum**. The dolipore septum (Figure 15.7) is unique to most members of Basidiomycota and restricts the movement of nuclei and other organelles between the hyphal cells. The exception is the rust fungi (Chapter 14); these basidiomycetes have cross walls similar to the species of Ascomycota. Another important characteristic that distinguishes the basidiomycetous fungi from other fungal phyla is the presence of **clamp connections** (Figure 15.8) on the hyphae where septa have formed. This structure occurs in less than 50% of all species of basidiomycetes. Clamp connections help ensure the dikaryotic (N + N) condition of new cells during compartmentalization (septa formation) of the mycelium. This enables the basidiomycete fungi to reproduce sexually.

Although most fleshy basidiomycetes are associated primarily with the decay of organic materials, other species form unions with plants and forest trees in what is called a mycorrhizal or symbiotic association. However, several mushroom-forming genera can be pathogenic on economically important crops such as corn, vegetables, grasses, and forage crops. Members included in the genus *Marasmius, Xeromphalina*, and *Armillaria* (Case Study 15.2) have also been reported to be pathogenic on agricultural and horticultural crops.

BASIDIOMYCETE TAXONOMY

Basidiomycota is a diverse phylum of fungi. Based on molecular studies, mycologists have subdivided the phylum into subphyla, which include Pucciniomycotina,

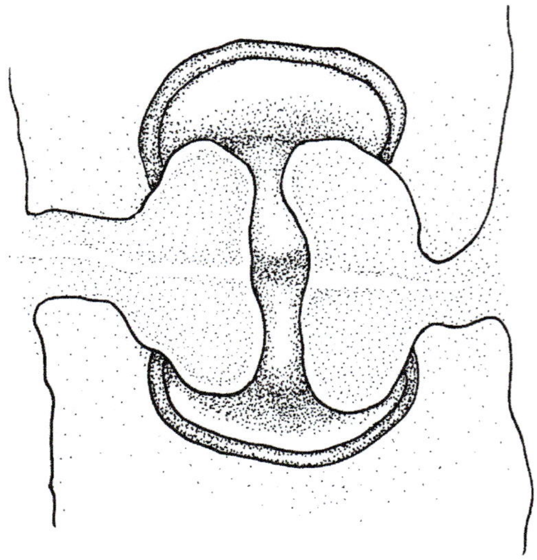

FIGURE 15.7 Hyphal cells separated by a dolipore septum that occurs only in certain genera and species of basidiomycetes. The structure of the septum restricts the movement of nuclei between the hyphal cells and maintains the nuclear condition. (Courtesy of Joe McGowen, Mississippi State University.)

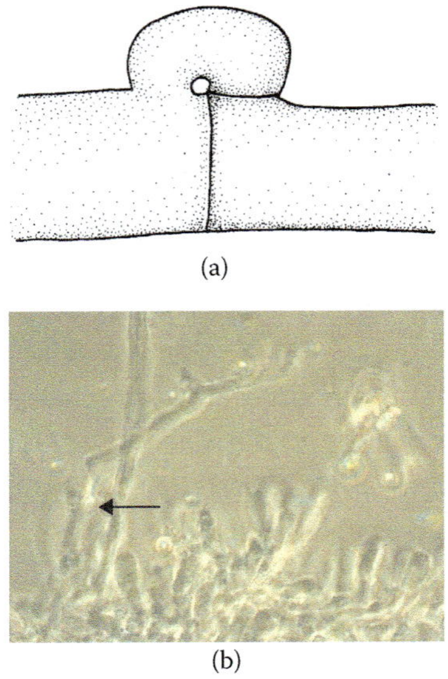

FIGURE 15.8 (a) Hyphae with clamp connections that occur only in some basidiomycetes. (Courtesy of Joe McGowen, Mississippi State University.) (b) Arrow indicates a clamp connection. (Courtesy of R. Baird.)

CASE STUDY 15.2

Honey Mushroom Causes Armillaria Root Disease

- Honey mushrooms or *Armillaria* spp. causes root disease throughout temperate and tropical regions attacking conifers and declining hardwood tree species such as oaks. These fungi belong to the order Agaricales and form basidiocarps that typically have caps, gills, and stems.
- This mushroom-forming species is found in clusters, honey or yellow-brown in color with small scales on the caps, and white gills and spore prints. A complex of different species of *Armillaria* can be found throughout the United States.
- The honey mushrooms, which are common in forests of the United States, are also known as the "shoestring fungus." They become established on the roots of trees and attack sapwood colonizing the tissue under the bark moving along the boles as black rhizomorphic strands or "shoestrings."
- Long-range basidiospore dispersal is by wind. Short-range infections occur by mycelial or rhizomorphic contact with roots. After establishment, the fungus can survive saprophytically up to 50 years in stumps waiting for the next generation of trees to infect.
- Control of Armillaria root disease is by stump removal following tree harvest. Foresters can plant resistant tree species if management practices allow for those species. Increased plant vigor and planting practices that increase root growth reduce damage by the pathogen.

Ustilaginomycotina (both discussed in Chapter 14), and Agaricomycotina. Agaricomycotina includes mushroom-forming fungi (Agaricomycetes), jelly fungi (Dacrymycetes and Tremellomycetes), and yeasts. This chapter focuses on the class Agaricomycetes. These fungi produce aggregates or clusters of basidia on a layer of specialized hyphae called a **hymenium**. All are fleshy basidiomycetes. Agaricomycetes includes gill and tube mushrooms, puffballs, bird's nest fungi, earthstars, stinkhorns, polypores, chanterelles, tooth and coral fungi, shelf and bracket fungi. The latter group contains many wood decay fungi. Economically important asexual fungi, such as *Rhizoctonia* spp., are linked to sexual states in Basidiomycota, order Ceratobasidiales. The asexual genus *Rhizoctonia* is linked to two sexual or basidia-forming genera, that is, *Ceratobasidium*, which has two nuclei per cell, and *Thanatephorus*, which has more than two nuclei per cell.

Agaricales or gilled fungi belong to an order that is referred to as toadstools or mushrooms, which are the fruiting bodies. Although there is an extreme variation in morphology, mushrooms generally consist of a stalk (stipe) and a cap (pileus). The cap and sometimes the stalk support gills (lamellae) or pores where the basidia and basidiospores are produced. When a stalk is absent, the fruiting bodies are called woody conks, resupinate or bracket fungi. The gills or pores have a hymenial layer where basidia and basidiospores are formed (Figure 15.1).

HABITAT

Members of Agaricales are found on a broad range of plants that are commonly associated with the decay of dead plants and woody tissues (lignicolous). Others types are more specific or occur in select ecosystems such as the coprophilous (dung inhabiting) fungi or the hallucinogenic mushroom species. *Psilocybe cubensis* is a very common coprophilus species found in pastures in the southern United States and is often collected for its hallucinogenic properties.

Certain mushroom-forming species are extremely host specific. Examples of this specific host–fungi relationship is the saprophytic mushroom-forming species *Strobilurus esculentus*, which forms mushrooms only on fallen spruce cones, or *Auriscalpium* species (tooth-forming fungus) observed only on cones of pine and firs. The mycorrhizal-associated mushroom species *Suillus pictus* and *S. americanus* are found primarily where white pine trees grow (Figure 15.9).

FIGURE 15.9 *Suillus americanus*—ectomycorrhizal with white pine. (Courtesy of A. Windham.)

MYCORRHIZAE

Almost all vegetation and trees require a beneficial symbiotic relationship with certain genera and species of fungi. In a mycorrhizal relationship (Chapter 10), fungal hyphae envelope the thin rootlets of trees, but do not harm the roots. The symbiotic relationship occurs when carbohydrates or sugars from photosynthesis are passed from the plant to the fungal associate, whereas the fungus acts as a secondary root system providing water and nutrient uptake to the plant or the tree host (Figure 15.10). The transport or movement of the materials to either partner in the symbiotic association involves complex chemical pathways not discussed here. Generally two types of mycorrhizal fungal associations are considered—endomycorrhizae and ectomycorrhizae. Hyphae of endomycorrhizal fungi grow into the cortical cells of the root and penetrate the root cell walls, but do not penetrate the root cell membrane. They also do not form a thick layer of hyphae around the root hairs. Endomycorrhizal fungi are taxonomically diverse and are found in several different fungal phyla.

Alexopoulos et al. (1996) estimated that there were over 2000 species of ectomycorrhizal fungi, and the majority of these were fleshy basidiomycete mushroom-forming species, such as *Russula* species

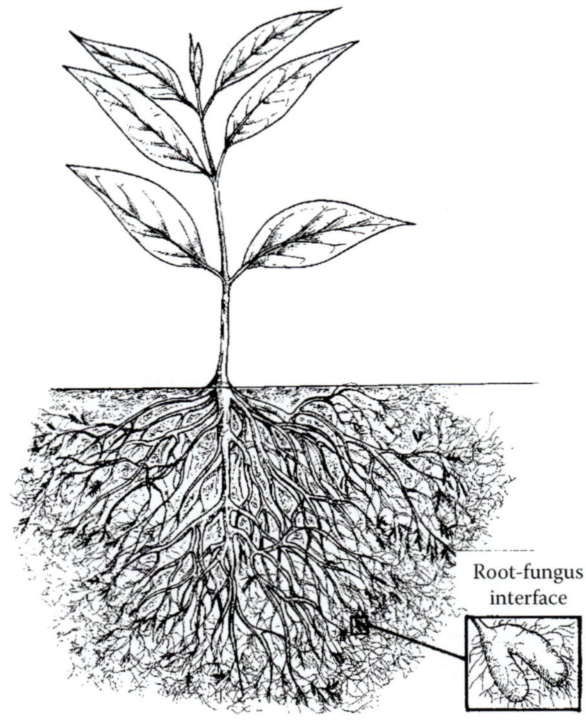

FIGURE 15.10 Ectomycorrhizal associations between a fungus and a tree. Fungus acts as a "secondary root system" to provide nutrients and water for the host tree species. (Courtesy of Joe McGowen, Mississippi State University.)

FIGURE 15.11 Ectomycorrhizal mushroom forming *Russula* species. (Courtesy of A. Windham.)

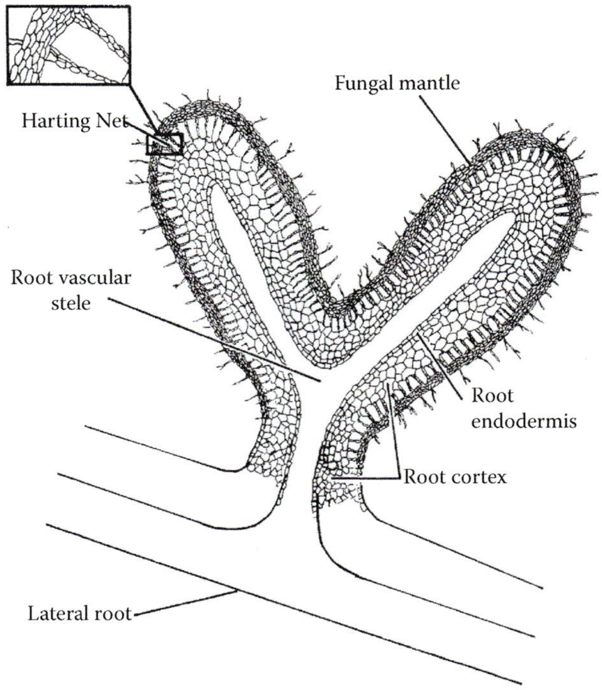

FIGURE 15.12 "Hartig Net" (box) from an ectomycorrhizal association. Hyphae from the fungus form an outer layer around the root and grow between the cortical cells. (Courtesy of Joe McGowen, Mississippi State University.)

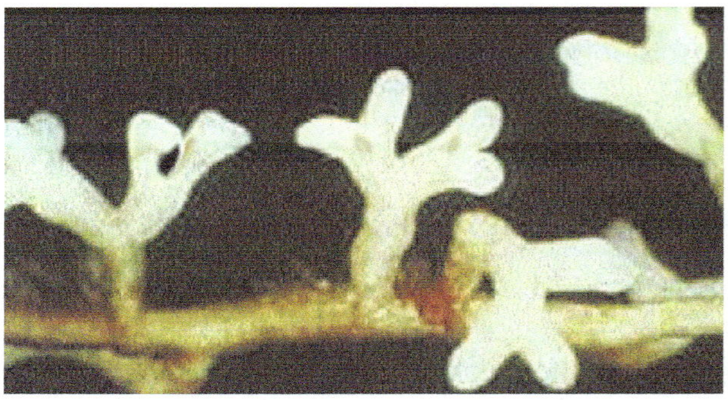

FIGURE 15.13 Feeder roots with "Hartig Net" of ectomycorrhizal fungus. (Courtesy of A. Windham.)

(Figure 15.11). The hyphae of these fungi attach to the feeder rootlets of plants forming a fungal mantle or thick layer of mycelium around the surface of the roots. In a true ectomycorrhizal association, the fungus initially forms a mantle (Figures 15.10, 15.12, and 15.13) around the roots before growing between the outer cortical cell layers to form the "Hartig Net" (Wilcox, 1982). The color of the mantle varies depending upon the fungal species. This network of fungi is the main distinguishing feature that defines ectomycorrhizae, since a number of fungi might be able to form what could be interpreted as a mantle, but do not have intercellular growth. The root hairs in an ectomycorrhizal association become forked or multiforked or are irregularly shaped. Other fungi that form ectomycorrhizal associations are members of the phylum Ascomycota, including the cup fungi and truffles, but the number of species and genera are fewer in number than that of the phylum Basidiomycota.

Ectomycorrhizal fungi are associated with many plant and tree species. Important tree partners include members of Pinaceae (pines, true firs, hemlock, spruce, and others), Fagaceae (oaks, beach, and others), Betulaceae (birch, alders, and others), Salicaceae (poplars, willows, and others), and Tiliaceae (basswood). On a given tree root, many ectomycorrhizal-forming species can occur. A specific tree species can partner with many different fungal species at one time. For example, Douglas fir in the northwestern United States has more than 200 ectomycorrhizal associates. In turn, a single fungus can associate with a broad range of plants or be specific to select tree species. An example is the mushroom-forming species *Suillus americanus* that only occurs under white pine. In contrast, *Amanita vaginata* occurs under many species of pines and hardwoods.

PATHOGENIC BASIDIOMYCETES

Pathogenic genera and species of fleshy basidiomycetous fungi can infect agricultural crops and forest tree species. For agricultural crops, a complex of fungi called the "sterile white basidiomycetes," containing genera that produce white mycelium with clamp connections in culture, have been identified as mushroom-forming species including *Marasmius* and *Xeromphalina*. Cultures of these fungi have been reported to infect roots and lower stems of corn, sorghum, millet, and vegetable crops such as snap beans.

Mushroom, conk, or bracket-forming basidiomycetous fungi are pathogenic on forest trees. A classic example of pathogenicity by a mushroom-forming species is *Armillaria mellea* or *A. tabescens* (shoestring fungus-honey mushrooms). The fungus becomes established on trees or woody plants by forming rhizomorphs, black shoestringlike structures that transport nutrition and water, extending out from the mycelial mat. Rhizomorphs are a clustering of hyphae into strands forming a complex mycelial structure (Figure 15.14). *A. mellea* is one of the most common, best known, and most damaging fungi that cause diseases of forest, shade, and ornamental trees and shrubs. It flourishes on an extraordinary broad host range and under diverse environmental conditions. *A. mellea* becomes established when host plants are under stress (e.g., nutrients and water), whereas *A. tabescens* can be highly pathogenic. The latter fungus invades bark and cambium tissues of the roots and the root collar of coniferous tree species. The extent and speed of invasion depends on the host species and the degree of stress to the plants. Plant stress increases by plant competition and by mechanical injury resulting in declining root systems. After the host dies, the fungus continues to decay the host tissue and may survive in the soil for decades.

Another fleshy basidiomycete species pathogenic to forest trees is *Heterobasidion araucariae* (formerly named *H. annosum*, which causes annosum root rot). This basidiomycetous fungus occurs in coniferous forest ecosystems and can invade tree roots when under stress from environmental conditions such as drought or when management practices are minimal. The fungus occurs throughout the United States and Canada but causes severe economic damage in the southeastern United States. Young seedlings can be killed by *H. araucariae*, but in older trees the heartwood is invaded and chronic infections can continue for years. When roots of different trees graft, the fungus can invade uninfected trees. The conks or shelflike woody mushrooms can develop on infected living trees or on dead hosts. Conks are generally found slightly buried in the humus or can be seen just above the ground surface on the butt of the tree (Figure 15.3). Recognition of the basidiocarps or conks of *H. araucariae* is useful for diagnosis.

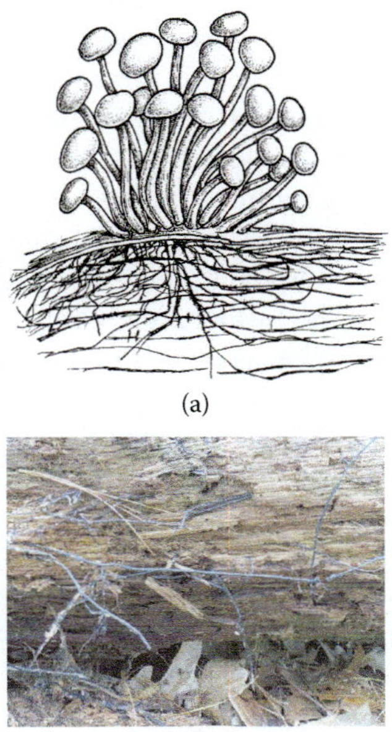

(a)

(b)

FIGURE 15.14 Rhizomorphs are shoestring-like structures deriving nutrients and water for fungal growth and sporulation. ([a] Courtesy of Joe McGowen, Mississippi State University. [b] Courtesy of R. Baird.)

WOOD ROTS CAUSED BY FLESHY BASIDIOMYCETES

The majority of fleshy basidiomycetes are saprophytic and often have the ability to decay wood. Fungi in this group derive nutrients from dead plants or woody tissues by enzymatically and chemically degrading cellulose, hemicellulose, and/or lignin (Fergus, 1966). Some of these fungi occur on living trees, but they generally decay only the dead tissues of the heartwood. Two groups of fleshy basidiomycetous fungi are most important in deriving nutrition in their role as wood rotters. The groups include the resupinate fungi (form flat basidiomata on host; Figure 15.5) placed into the Thelephoraceae or Corticiaceae (Alexopoulos et al., 1996), and the remainder were traditionally placed into the group known as the Polypores. Within these groups, the fungi are referred to as white rot or brown rot decay species.

White Rots

White rot fungi can generally utilize all components of the wood including lignin, cellulose, and hemicellulose. The wood degraded by these fungi is light in color due to the loss of lignin (brown color). Since lignin gives strength to the wood, the tissue becomes spongy, stringy, or mottled. Genera in the Agaricales, such as *Agaricus*, *Coprinus*, and *Pluteus*, are common plant and wood decay fungi. Many genera in Aphyllophorales such as polypores (pored fungi), chanterelles, tooth and coral fungi, and the corticoid-forming fungal species also decay wood. An example of a white rot fungus is *Oxyporus populinus* that attacks numerous hardwood tree species and causes trunk and limb rots. *Trametes versicolor* attacks the sapwood of hardwood trees such as sweet gum, *Liquidambar styraciflua*.

Brown Rots

Brown rot fungi, also known as dry rot, generally degrade cellulose and hemicellulose and either lack or have limited ability to utilize lignin. Lignin is more or less brown and remains after other wood polymers have been decayed, so wood decomposed by these fungi appears brown, cubicle, and crumbly. There is a defined margin between healthy and decayed wood. This form of rot is more common in coniferous trees than hardwood tree species. *Poria placenta* and *Phaeolus schweinitzii* are often associated with brown rot of timber and lumber.

FAIRY RINGS

Fairy rings (Figure 15.15) are fleshy mushroom-forming basidiomycetes that grow in circular zones. Although there are many species that form fairy rings, some of the more common genera include *Coprinus*, *Chlorophyllum*, *Lepiota*, and *Marasmius*. The rings are readily observed in lawns or golf courses, but can be found in forests and other ecosystems. The rings occur from the expansion or outward circular growth of the mycelium that initially originated from one central point. The mushrooms, however, almost always form near the outer edge of active growth zones. Some fairy rings are hundreds of years old and extend over very long distances. A ring can be disrupted due to physical barriers or loss of the nutrient base, forming arcs or partial zones. In golf course greens or home lawns, fairy rings may be considered unsightly or may affect the growth of the grass causing obvious discolored patches and dead tissues within the zones. The mycelium from the fungus can act as a physical barrier preventing adequate moisture and reduced vigor of the invaded lawn. Control or prevention of rings includes reducing thatch and maintaining adequate moisture to the thatch layers. Labeled fungicides are effective for controlling these fungi.

FIGURE 15.15 Fairy ring. (a) *Chlorophyllum molybdites* mushrooms forming a fairy ring on a grassy lawn. (b) *C. molybdites* mushrooms. (Courtesy of A. Windham.)

OTHER BASIDIOMYCETES (NONFLESHY)

An unusual plant pathogenic basidiomycete genus, *Exobasidium* (Figure 15.16), forms basidia directly from dikaryotic mycelium. A layer of hyphae forms between the epidermal cells of the plant tissue causing tissue distortion including swelling, blistering, and galling. Infections can occur on host species including azalea, rhododendron, camellia, blueberry, ledum, and leucothoe, and other ornamental and forest plants. Hyphae of *Exobasidium* grow intercellularly and invade cells by the formation of haustoria. Controls include removal or sanitation of galls in landscape situations or small greenhouses. Fungicides can be used, but success depends on applications prior to bud break and follow-up sprays. Resistant host plants are generally not available.

The genus *Rhizoctonia*, the asexual stage of a Basidiomycete, is one of the most economically important groups of pathogenic species in the world. Billions of dollars are lost annually from seed decay, root rots, seedling disease, and aerial or foliar blights. Although *Rhizoctonia* is known primarily for disease of agronomic, vegetable, and ornamental crops, several species were reported to form mycorrhizal associations with plants (e.g., orchids). *Rhizoctonia* species do not produce conidia in culture and can be difficult to identify. Isolates of *Rhizoctonia* species may vary greatly in number and size of sclerotia (survival structures), hyphal growth (appressed or aerial) and color in culture, and presence of a dolipore septum. Without conidia, these cultural characters alone make it difficult to identify *Rhizoctonia* species. *Rhizoctonia* spp. consists of numerous mating types forming a pathogen or disease complex that attacks one to many host plants depending upon which type is present. The mating type forms a complex that can be very confusing since the mating types can be divided based on their formation of different sexual structures. Species of *Rhizoctonia* are distinguished by color of the hyphae, numbers of nuclei per cell, and teleomorphic forming structures.

Identification is further complicated by different mating types, called anastomosis groups (AG) (Carling, 1996). The multinucleate isolates of *R. solani* vary in shape, color, hyphal thickness, and the number and size of sclerotia (survival structures) in culture. These types will only mate by hyphal fusion with compatible AGs. During complete anastomosis, the hyphal cell walls of the two compatible types fuse, lysis occurs, and the nuclei are exchanged between cells. Isolates of *Rhizoctonia* that do not mate are called vegetatively incompatible.

The hyphae of the different groups have different numbers of nuclei per cell as stated previously. The binucleate (two nuclei per cell) group is known as the *Ceratobasidium* anastomosis group (CAG) and is named after the teleomorphic state for this group of *Rhizoctonia* species. Approximately 20 different mating types have been identified in this group. The second group, which includes *R. solani*, has three or more nuclei per cell (multinucleate) and typically has large diameter, "right angle" hyphal branching (Figure 15.16), and isolates are divided among 13 AGs. However, as nature always provides more variation than what is thought or believed, it was later determined that numbers of nuclei per cell are not always indicative of a specific teleomorph grouping. Isolates of binucleate *Rhizoctonia* spp. that form *Ceratobasidium* teleomorphs have been reported, and new groupings of combined AG and CAG are now accepted by many researchers.

The number of nuclei per cell and anastomosis pairing studies have been the most useful and effective methods for the identification of *Rhizoctonia* species. Recently developed molecular techniques are more accurate in the identification of CAG and AG forms of *Rhizoctonia*. The anastomosis pairing groups have been correlated with groups defined by the molecular techniques (Vilgalys and Cubeta, 1994; Gonzalez et al., 2001).

The teleomorph for the AG forms of *Rhizoctonia* has been identified as *Thanatephorus cucumeris*. At least two other teleomorphic genera have been identified for *Rhizoctonia*, but will not be mentioned here. The teleomorphs of the CAG and AG types of *Rhizoctonia* species are generally nondescript. A hymenium layer of hyphae may form on the soil surface or occasionally in culture with the basidia and basidiospores being produced directly from that area (Sneh et al., 1991). Sexual structures are either rare or extremely difficult to find in nature.

Because of the importance of *Rhizoctonia* spp. mating types, a review of several important AGs are included later in text. Root and hypocotyl rot pathogens of agronomic, horticultural, and vegetable crops are known to cause major economic losses by *R. solani* AG 4 on cotton, snapbeans, soybeans, and sweet corn; however,

FIGURE 15.16 Exobasidium leaf gall of azalea. (Courtesy of A. Windham.)

many environmental and management factors that cause stress to the plants are associated with the extent of losses. Other AG groups, such as AG 2-2, can also be found on soybeans, but probably occur when associated with a prior crop. For example, AG 2-2 causes crown root and brace root rot on corn and significant losses when corn and sugar beets are rotated. The prior crop, such as corn, can enable the pathogenic AG 2-2 population to increase by the end of season; a subsequent crop of sugar beets can then lead to major losses. *R. solani* AG or CAG can cause disease specific to certain parts of plants. In the case of soybeans, AG roots and hypocotyl rots are primarily caused by AG 4 or AG 2-2 and foliar blights by AG 11 or the further subdivided mating type AG 2-3. Furthermore, AG 2-2 causes large patch disease resulting from infected leaf sheaths of zoysia grass showing the wide plant host range. Many ornamental crops, annual or perennial or woody, are affected by AG and CAG groups. Azalea and *Rhododendron* spp. are susceptible to foliar or aerial blight caused by different CAG groups such as CAG 3 and 7, but select AGs such as AG 4 were more common on stems and roots. For vegetable crops such as French bean, several pea species, and tomatoes, CAG 3-5 were found to cause root and hypocotyl disease of seedlings. In onion, the same three CAGs cause seed decay, hypocotyl, and stem disease resulting in significant losses.

Target spot of tobacco, also called Rhizoctonia leaf spot, is caused by basidiospore infections from *Thanatephorus cucumeris*. The fungus overwinters in soil or plant debris as sclerotia. The sclerotia germinate and hyphae grow saprophytically in the soil. Under favorable environmental conditions, basidiospores are formed on tiny resupinate mycelial tissues that form a hymenial surface bearing basidia and basidiospores. The spores are windborne and disseminated to host tobacco plant foliage. The spores germinate, and appressoria form from the growing hyphae, allowing direct entry of the fungus and subsequent colonization of the tissue. Once established in the host, target-shaped lesions develop from these initial infections. New basidiospores are produced within the lesions causing secondary infections and ultimately resulting in defoliation of the tobacco plant. New sclerotia form and overwinter in the dead host tissue or soil, completing the disease cycle.

The fleshy basidiomycetes include a very large and diverse group of fungi. Species within the group occur in different ecosystems including agricultural fields, forests, and ornamental landscapes. These fungi survive and reproduce as either saprophytes or parasites and are involved in mycorrhizal associations. Fruiting bodies of basidiomycetes can vary in shape and size, ranging from mushroomlike to those that form woody conks on dead wood, or species can be very tiny and wispy with only a hymenial layer present in soil or on plant tissues. The key morphological structures used by taxonomists to group basidiomycetes together include the basidium, dolipore septum, and, when present, clamp connections. The basidium is where meiosis and subsequent basidiospore formation occur.

LABORATORY EXERCISES

Rhizoctonia species are soilborne pathogenic fungi that are distributed widely in agricultural field soils and synthetic growth media used in greenhouses. They cause a number of diseases including damping-off of many seedlings in the field and cuttings of ornamental plants in the greenhouse, aerial and web blights, and seed rots. Depending on the species of *Rhizoctonia*, hyphal cells can be multinucleate or binucleate, and sclerotia may or may not be produced as survival and/or overwintering structures. The sexual stage (produces basidia and basidiospores) is inconspicuous and seldom observed, but known to initiate a foliar disease of tobacco called "target spot."

This section lists two experiments. The first is designed to familiarize students with the general morphological characteristics of *Rhizoctonia* species, which have multinucleate hyphal cells and the binucleate Rhizoctonia-like fungi (BNRLF). The second experiment is geared to provide students with experience in visualizing and counting nuclei, which is necessary for correctly identifying these fungi.

Experiment 1. Identification of *Rhizoctonia* Species Using Hyphal Characteristics

Rhizoctonia species do not form conidia, but are identified based on cultural characters such as mycelial shapes and color, sclerotia formation, and hyphal morphology. The hyphae are brown, elongated, inflated, and branched at right angles. This exercise will familiarize students with general cultural characteristics of this important genus.

Materials

The following materials are needed for each student or group of students:

- Several Petri dishes (10-cm-diameter) of *R. solani* growing on potato dextrose agar (PDA)
- Several Petri dishes (10-cm-diameter) of a BNRLF growing on PDA
- Dissection needle or microspatula for transferring mycelium from the Petri dish to slides
- Dissection and light microscopes, slides, coverslips, and distilled water for mounting the hyphae

Follow the protocol outlined in Procedure 15.1 to complete these observations.

Anticipated Results

Morphological differences between the two groups of *Rhizoctonia* include thickness and color of the hyphae (mycelium), presence of sclerotia, and patterns of sclerotial occurrence in the Petri dishes. Hyphae are relatively wide, branched at right angles, and are constricted at septa (Figure 15.17).

Questions

- What do various *R. solani* isolates look like in culture?
- Discuss how sclerotia vary in shape and size and the patterns they formed in the Petri dishes.
- What distinguished this group of fungi from others?

Procedure 15.1

Identification of *Rhizoctonia solani*

Step	Instructions or Comments
1	Compare cultures of *R. solani* and BNRLF for the following characteristics (see Step 2): • *R. solani* hyphae are generally darker than hyphae of BNRLF. • Cells of BNRLF hyphae are elongated and often slightly inflated. • *R. solani* hyphae branch at right angles and are constricted at the septa (see Figure 15.15). • Sclerotia of *R. solani* are light golden-brown at first, becoming dark brown to black with age, and are variable in size in cultures. • Asexual fruiting bodies are absent in BNRLF cultures.
2	Cut a small plug (5 mm^2) of agar containing fungal mycelium and place on microscope slide in a drop of water. Cover with a plastic coverslip and flatten.
3	Observe the preparation with a compound microscope at 100× and 400×.

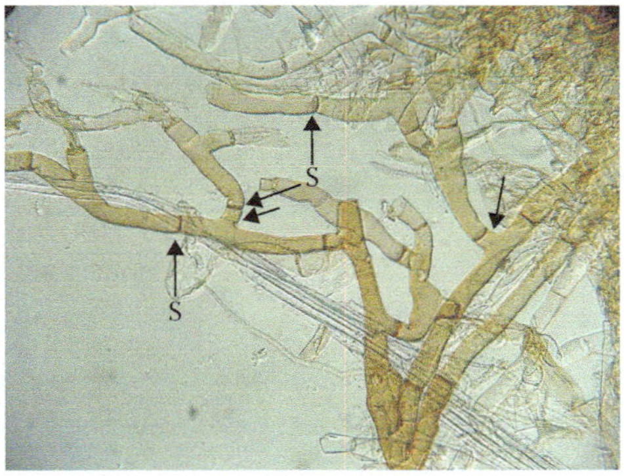

FIGURE 15.17 *Rhizoctonia solani.* This species of *Rhizoctonia* is characterized by large diameter and "right angle" branching hyphae (arrows). Hyphae are typically constricted at septa (S). (Courtesy of A. Windham.)

Experiment 2. Determine the Number of Nuclei in Hyphal Cells of *Rhizoctonia* Species

The genus *Rhizoctonia* is divided into groups based on the number of nuclei (multinucleate or binucleate) per hyphal cell. The multinucleate isolates are *R. solani* and are further divided on the basis of **AGs**. The binucleate isolates are either *R. cerealis* or BNRLF and belong to the **CAG**. Nuclear stains can be used to differentiate the groups.

Materials

Each student or team of students will need the following materials:

- Petri dishes (10-cm-diameter) containing four to six different AG and CAG of *Rhizoctonia* species growing on PDA
- Petri dishes (10-cm-diameter) containing 5 mL of water agar
- Dissection needle or spatula for transferring mycelium to slides
- Dissection and light microscopes
- Slides and coverslips
- 0.5% aniline blue in lactophenol or 0.05% Trypan blue in lactophenol in dropper bottles
- Alcohol lamps

Follow the protocol outlined in Procedure 15.2 to complete the experiment.

Anticipated Results

The nuclear condition of the hyphal cells should be easily determined for different isolates. From these observations, CAG or AG isolates can then be determined for the unknowns.

Questions

- How many nuclei did you observe in hyphal cells of *R. solani*?
- How many nuclei did you observe in hyphal cells of CAG types?

Procedure 15.2

Determining Number of Nuclei in Cells of *Rhizoctonia* Species

Step	Instructions and Comments
1	Inoculate isolates of *Rhizoctonia* species onto thin layers of water agar in Petri dishes of 10-cm-diameter and incubate at room temperature for 48–72 h. Label dishes with the species name.
2	After the fungi have grown at least 5 cm from the inoculation plug, place a drop of 0.5% analine blue in lactophenol or 0.05% Trypan blue in lactophenol onto the hyphae at the edge of the colony. Place a coverslip over the stained area, lightly press the coverslip down and onto the mycelium. Observe the hyphal cells at 400× using a compound microscope. Be careful not to let the objective lens touch the agar. Count the number of nuclei in each hyphal cell for each isolate.
3	Another method for nuclear staining is to remove a section (1 cm^2) of mycelium from the thin-layer water agar plates and place onto a microscope slide. Then add dye as specified in Step 2. Place a coverslip over the material, heat slightly with a small alcohol burner until the agar begins to melt. Place the slide onto the microscope stage, view at 400×, and count nuclei as in Step 2.

REFERENCES

Alexopoulos, C. J., C. W. Mims, and M. Blackwell. 1996. *Introductory Mycology.* 4th edition. New York, NY: John Wiley & Sons, Inc.

Carling, D. E. 1996. Grouping in *Rhizoctonia solani* by hyphal anastomosis reaction. Pages 37–47 in: *Rhizoctonia Species: Taxonomy, Molecular Biology, Ecology, Pathology and Disease Control.* B. Sneh, S. Neate, and S. M. Dijst (eds). Dordrecht, the Netherlands: Kluwer Academic Pub.

Christensen, C. M. 1950. *Keys to the Common Fleshy Fungi.* 4th edition. Minneapolis, MN: Burgess.

Fergus, C. L. 1966. *Illustrated Genera of Wood Decay Fungi.* Minneapolis, MN: Burgess.

Gonzalez, D., D. E. Carling, S. Kuninaga, R. Vilgalys, and M. A. Cubeta. 2001. Ribosomal DNA systematics of *Ceratobasidium* and *Thanatephorus* with *Rhizoctonia* anamorphs. *Mycologia* 93:1138–1150.

Sneh, B., L. Burpee, and A. Ogoshi. 1991. *Identification of Rhizoctonia species.* St. Paul, MN: APS Press.

Vilgalys, R., and M. A. Cubeta. 1994. Molecular systematic and population biology of *Rhizoctonia. Annu. Rev. Phytopathol.* 32:132–155.

Wilcox, H. E. 1982. Morphology and development of ecto- and ectendomycorrhizae. Pages 103–113 in: *Methods and Principles of Mycorrhizal Research.* N. C. Schenck (ed). St. Paul, MN: APS Press.

SUGGESTED READINGS

Berres, M. E., L. J. Szabo, and D. J. McLaughlin. 1995. Phylogenetic relationships in auriculariaceaous basidiomycetes based on 25S ribosomal DNA sequences. *Mycologia* 87:821–840.

McLaughlin, D. J., M. E. Berres, and L. J. Szaba. 1995. Molecules and morphology in basidiomycetes phylogeny. *Can. J. Bot.* 73:684–692.

Moore, R. T. 1987. The genera of *Rhizoctonia*-like fungi: *Asorhizoctonia, Ceratorhiza* gen. nov., *Epulorhiza* gen. nov., *Moniliopsis* and *Rhizoctonia. Mycotaxon* 29:91–99.

Ogoshi, A. 1975. Grouping of *Rhizoctonia solani* Kühn and their perfect stages. *Rev. Plant. Protect. Res.* 8:93–103.

Parmeter, J. R., Jr., and H. S. Whitney. 1970. Taxonomy and nomenclature of the imperfect state. Pages 7–19 in: *Biology and Pathology of Rhizoctonia solani.* J. R. Parmeter Jr. (ed). Berkeley, CA: University of California Press.

Swann, E. C., and J. W. Taylor. 1993. Higher taxa of basidiomycetes: An 18S rRNA gene perspective. *Mycologia* 85:923–936.

Weber, N. S., and A. H. Smith. 1988. *A Field Guide to Southern Mushrooms.* Ann Arbor, MI: University of Michigan Press.

16 Soilborne Plant Pathogens

Bonnie H. Ownley and D. Michael Benson

Concept Box

- Soilborne plant pathogens include fungi, bacteria, nematodes, viruses, and parasitic plants that are capable of surviving in soil for extended periods of time in the absence of a host plant.
- Soilborne plant pathogens survive in soil either passively by the production of dormant propagules or actively by the saprophytic colonization of host debris.
- Management of soilborne pathogens is aimed at either reducing the residual amount of inoculum available to infect the next crop or preventing germination of pathogen propagules in the host infection court.
- Environmental factors in soil, including soil pH, macro- and micro-elements, water, aeration, temperature, and organic matter affect the ability of soilborne plant pathogens to survive and cause disease.
- Disease management strategies to control soilborne plant pathogens include plant resistance, cultural practices, biological controls, conventional fungicides, fumigation, and compost amendments.

Soilborne plant pathogens are a diverse group of bacteria, fungi, nematodes, parasitic higher plants, and viruses that survive in soil between infection cycles of their hosts. Many soilborne fungal pathogens are either **hemibiotrophs** that invade host cells but keep the cells alive for a period prior to killing the cell or **necrotrophs** that kill the host cells prior to invading the cells. A few soilborne fungal pathogens, such as the club root pathogen, *Plasmodiophora brassicae*, are **obligate biotrophs** that infect only living host cells.

Fungal pathogens may be **soil inhabitants**, which have a saprophytic phase that allows for growth and colonization of crop debris in soil or **soil invaders**, which survive in soil, but do not grow saprophytically on crop debris prior to host infection. Soil invaders produce spores or survival structures that may persist in soil in only one season or for many years depending on the pathogen and survival structure formed. Soil invaders only increase in population by spores or survival structures produced on or in their host.

Life in the soil for soilborne plant pathogens is regulated by environmental factors such as temperature, moisture, oxygen, and pH, but foremost, by the availability of nutrients for spore germination and growth toward the plant host. It was recognized in the mid-twentieth century that many fungal pathogens survive in soil in a dormant state as spores, chlamydospores, sclerotia, or other survival structures. This phenomenon, termed **soil fungistasis**, was demonstrated to occur in natural soils in all climates, from temperate to tropical to desert to arctic environments. The effect was greatest in the upper few inches of soil where the number and diversity of microorganisms was greatest; the effect diminishes with soil depth. Soil fungistasis is overcome when a source of nutrients, primarily carbon and nitrogen, becomes available to the fungal pathogen. Fungal spores that are "nutrient-dependent" and "nutrient-independent" are both subject to soil fungistasis. Nutrient-independent spores, once removed from the fungistatic conditions in soil and placed in a drop of water on a microscope slide, are able to germinate without **exogenous** nutrients. Since these spores do not germinate in soil, but do germinate in a water droplet on a microscope slide, inhibitory substances must also be present in soils that prevent spore germination until enough exogenous nutrients are available to overcome soil fungistasis.

Due to spatial distribution of fungal propagules in the three-dimensional soil environment, most propagules never encounter the nutrients needed to overcome soil fungistasis and eventually die, without ever infecting a host plant. The propagules that will successfully infect the host lie either in the **rhizosphere**, the zone under the influence of the root, or on the **rhizoplane**, the root surface (Figure 16.1). What is unique about the rhizosphere/rhizoplane and root infection? As the root system of a plant develops, **root exudates** in the form of simple sugars (carbon), amino acids (nitrogen), and many other types of compounds found in the plant diffuse into soil from gaps between cortical cells in the **zone of root elongation** just behind the **root cap** (Figure 16.1). Compared with nonrhizosphere soil, that is, bulk soil, the number of bacteria is 20–40 times greater in the rhizosphere, whereas the number of fungi is several folds greater. The large population of microorganisms in the rhizosphere is supported by root exudates. Competition for these nutrients by soil microorganisms, primarily bacteria and fungi, is intense, and thus, the rhizosphere effect is limited to only 1–2 mm from the root in most cases. For propagules of soilborne plant pathogens in the rhizosphere or on the rhizoplane, these nutrients provide the energy needed for the propagules to overcome soil fungistasis, germinate, and infect the host (Figure 16.2). This same "exudates phenomenon" occurs in the vicinity

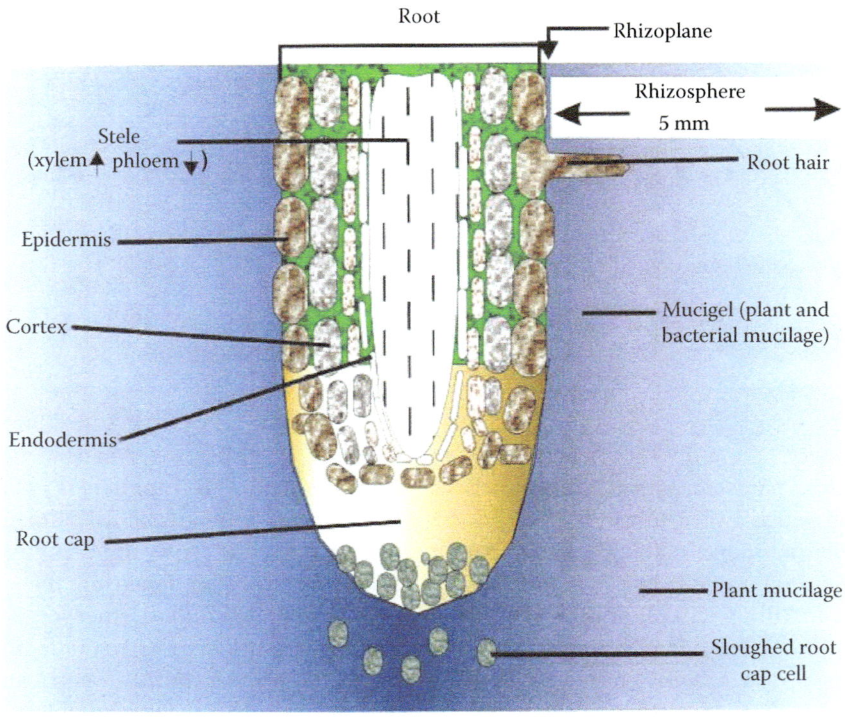

FIGURE 16.1 The rhizosphere. (From Maier, R.M., et al., *Environmental Microbiology*, Academic Press, San Diego, 2000. With permission.)

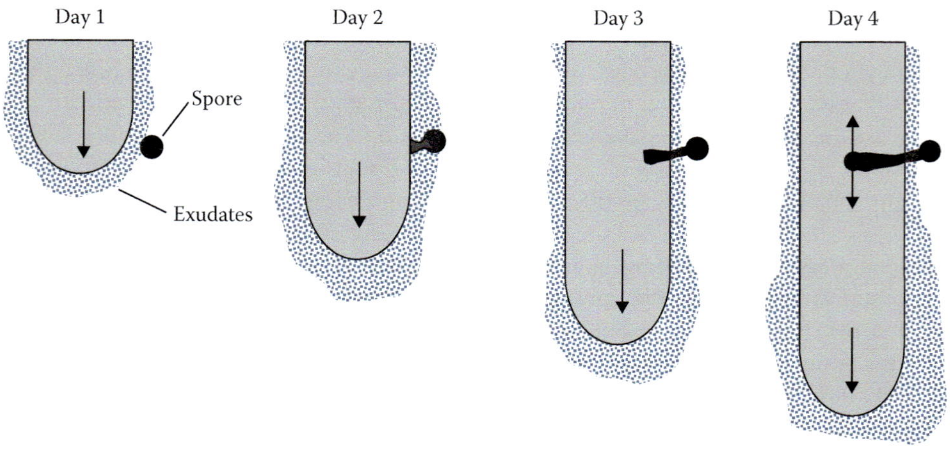

FIGURE 16.2 Exudate from a peanut root tip stimulates the germination of a chlamydospore of *Fusarium oxysporum*. The germ tube grew toward the source of the exudate and penetration occurred behind the root tip. (Redrawn from Bruehl, G.W., *Soilborne Plant Pathogens*, Macmillan Publishing Co., New York, 1986. With permission.)

of hypocotyls of seedlings and the **spermosphere** zone around germinating seeds. For these various host situations, the **infection court** for soilborne pathogens becomes the zone of root elongation, seedling hypocotyl, and/or seed coat.

The number of propagules of a pathogen in a soil sample whether in the form of spores, sclerotia, or chlamydospores is termed **inoculum density** and is expressed as propagules per gram soil. For example, in cotton field soil, the inoculum density of *Verticillium dahliae*, which causes verticillium wilt, may range from 20 to 40 microsclerotia per gram of soil. Given a favorable environment and a susceptible host, this is more than enough inoculum to cause a severe epidemic of verticillium wilt.

From the standpoint of plant disease control, it is important to know the type of propagule, that is, **inoculum**, surviving in soil for a given soilborne pathogen and the number of propagules per unit volume of soil, that is, **inoculum density**. For many soilborne fungal pathogens, inoculum density is correlated with root infection and subsequent **disease incidence**. For instance, an inoculum density of 0.8 propagules per gram of soil at 26°C resulted in 50% pre-emergence damping-off of radish by *Rhizoctonia solani* (Case Study 16.1), whereas 7 propagules per gram of soil were required to cause the same 50% damping-off at 15°C, a temperature unfavorable for *R. solani* (Benson and Baker, 1974). In both the cases, it took only one propagule to infect, but the chance of that one propagule infecting was much less at 15°C.

SURVIVAL

Mechanisms of Survival

The success of soilborne plant pathogens from crop to crop and year to year relies exclusively on the ability to survive between crops and years. Soilborne plant pathogens exhibit a variety of survival mechanisms as a group, but most individual pathogens utilize only one or two mechanisms. In general, survival of soilborne plant pathogens may be active or passive. Active forms of survival include parasitic survival on alternate hosts, commensal survival in the rhizosphere of nonhosts, and saprophytic survival. Dormant survival may be imposed by the environment (i.e., soil fungistasis) or inherent in the genetic make-up of the pathogen.

Soilborne plant pathogens like *R. solani* have a highly **competitive saprophytic ability** to survive in soil. Characteristics of these pathogens include rapid germination and growth on crop debris in soil; enzymes to decompose complex carbohydrates (Chapter 32); production of antimicrobial compounds; and tolerance of antimicrobial compounds produced by other microorganisms challenging for colonization of the crop debris. Other pathogens like *V. dahliae* colonize parenchyma cells of host stems and roots by saprophytic growth as the host is dying from

CASE STUDY 16.1

Rhizoctonia Root Rot Caused by *Rhizoctonia solani*—Problems in Poinsettia Propagation

- Rhizoctonia stem rot is the most significant problem in the propagation of poinsettia (*Euphorbia pulcherrima*) (Benson et al., 2002).
- *Rhizoctonia solani* has a wide host range and survives as sclerotia or mycelium in infested crop debris in greenhouse production facilities.
- The pathogen attacks juvenile host tissue. Lesions from infections caused by *R. solani* develop on the stem at the surface of the rooting cube. Lesions expand rapidly and girdle the stem. Infected cuttings can collapse within 5–7 days.
- Root rot develops in the rooting cube or on rooted cuttings transplanted into pots. As lesions expand from stem infections initiated during rooting, crown rot may occur. Stunting is common. Foliar symptoms of infection include chlorosis, leaf necrosis, wilting, defoliation, and plant death (Figure 16.3).
- Several fungicides are registered for use against Rhizoctonia diseases in the greenhouse. Rooting strips can be soaked in fungicide solution to protect cuttings from disease, but care should be taken for the selection of fungicide and cultivar of poinsettia because some fungicides can reduce rooting. Fungicide drenches can be used after transplanting to prevent crown and root rot.
- In greenhouse production systems, it is important to practice strict sanitation. All sources of infested crop debris should be removed. During propagation, avoid wetting the foliage of newly made turgid cuttings. Potted plants should be spaced so that ventilation is adequate to avoid moisture conditions favorable to *R. solani*.

FIGURE 16.3 Poinsettia cutting with stem lesion in leaf abscission area (left) and close up of hyphae of *Rhizoctonia* growing out into rooting cube (right). (From Benson, D.M., et al., *Plant Health Progress*, 2002. doi:10.1094/PHP-2002-0212-01-RV.)

FIGURE 16.4 Black rhizomorphs of *Armillaria mellea*. (From Worrall, J., *The Plant Health Instructor*, 2004. doi:10.1094/PHI-I2004-0706-01. With permission.)

the parasitic effects of the pathogen in xylem tissues. In this situation, *V. dahliae* has a "timing" advantage in colonizing the cotton crop debris compared with other microorganisms in soil. *Gaeumannomyces graminis*, which causes take-all of wheat and other cereals and grasses, survives as mycelium on roots and plant debris invaded during the parasitic phase of take-all disease. Survival in wheat debris is shortened in moist, well-aerated soil at moderate to warm temperatures; these conditions hasten the decomposition of crop debris in soil. *G. graminis* survives longer in dry or compacted soil, conditions that are not favorable for microbial decomposition of organic debris. Pathogens of woody tissues, such as *Armillaria mellea*, are well adapted to survive as mycelium or rhizomorphs on diseased trees or decaying roots. **Rhizomorphs** are thick, cordlike threads of mycelium (1–3 mm diameter) with a compact black outer layer of mycelium and an inner white or colorless core (Figure 16.4). Survival of pathogens on woody tissues depends on how long the substrate will last. These pathogens can survive on tissues that are high in carbon and low in nitrogen. Often, they are very tolerant of tannic acid that is naturally found in woody tissues. Tannic acid, which inhibits the growth of most fungi, stimulates rhizomorph production by *A. mellea* in culture (Cheo, 1982). Generally, successful saprophytic survival requires possession of the plant debris until a host infection court, such as new seed, seedling, hypocotyl, or root, invades the space near the colonized residue.

Survival Structures

Soilborne pathogens that depend on dormant survival produce some type of resting propagule such as a thick-walled conidium, **chlamydospore**, **microsclerotium**, **oospore**, or **sclerotium** (Table 16.1). *Cochliobolus sativus* causes common root rot in wheat and survives between crops as conidia in soil. Chlamydospores are thick-walled, asexual structures produced by pathogens like *Fusarium*, *Phytophthora*, and *Thielaviopsis* that survive in soil for several years. A number of soilborne pathogens like *Verticillium*, *Rhizoctonia*, *Macrophomina*, *Sclerotinia*, *Sclerotium*, and *Phymatotrichopsis* produce sclerotia or microsclerotia as resting propagules. Sclerotia may be rudimentary or well organized into different tissue layers. Sclerotia of most soilborne pathogens survive many years in soil in the absence of a host. Oospores produced by some species of *Phytophthora* and *Pythium* are thick-walled, sexual spores formed on or in infected host tissues (Chapter 8). Fertilization of a nucleus in the oogonium by antheridial nuclei results in the development of the oospore within the oogonium. Oospores may have an inherent dormancy so that some are available for germination over a period of several years.

Survival over Time

Whether a soilborne plant pathogen survives by active or passive mechanisms, over time the number of surviving propagules of the pathogen will decline in soil. The pattern of decline can be visualized in the form of one of three idealized curves (Figure 16.5). For pathogens that survive as conidia or chlamydospores in soil, the population of those propagules enters a logarithmic death phase wherein most of them die quickly over a relatively short period, such as over winter (Figure 16.5, Curve A). However, a small proportion of the population of propagules is very successful in survival and persists for a long period resulting in a residual population. For pathogens with multicellular propagules like sclerotia, there is an initial lag phase as individual cells in the

TABLE 16.1
Characteristics of Selected Soilborne Plant Pathogens

Name of Pathogen	Pathogen Type	Survival	Dispersal	Infection Court	Symptoms	Hosts	Parasitism
Agrobacterium tumefaciens[a]	Prokaryote	Cells in soil, plant galls	Cultivation, infected plants, irrigation water	Wounds on roots and stems	Galls, low vigor, stunting	Many	Hemibiotroph
Armillaria mellea	Basidiomycete	Hyphae and rhizomorphs in infected woody tissues	Root contact between infected and healthy trees, rhizomorph growth in soil, windborne basidiospores	Roots	Basal stem cankers, collar rot, root rot	Many woody angiosperms and gymnosperms, some nonwoody plants	Necrotroph
Bipolaris sorokiniana (teleomorph = *Cochliobolus sativus*)	Ascomycete	Thick-walled conidia	Windborne conidia	Seeds, coleoptiles, subcrown internode	Root rot, seedling blight, whiteheads	Wheat, barley, rye, other grass species	Hemibiotroph
Cylindrocladium parasiticum (teleomorph = *Calonectria ilicicola*)	Ascomycete	Microsclerotia	Cultivation, windborne plant debris	Root tips	Black root rot, leaf chlorosis, stem necrosis	Peanut, soybean, blueberry, tea, hardwoods	Necrotroph
Fusarium oxysporum (anamorphic Nectriaceae)	Species complex, Ascomycete	Chlamydospores	Cultivation, plant debris, infected transplants	Root tips	Stunting, vascular wilt	Many	Hemibiotroph
Fusarium roseum (anamorphic Nectriaceae)	Species complex, Ascomycete	Chlamydospores, mycelia	Cultivation	Basal stems	Foot and crown rot	Many	Necrotroph
Fusarium solani (anamorphic Nectriaceae)	Species complex, Ascomycete	Chlamydospores	Cultivation, plant debris	Hypocotyls	Root rot	Many	Necrotroph
Gaeumannomyces graminis	Ascomycete	Hyphae in crop debris	Cultivation, plant debris	Roots	Chlorosis, low vigor, stunting, whiteheads	Wheat, barley, other grasses	Necrotroph
Hymenula cerealis (synonym *Cephalosporium gramineum*)	Ascomycete	Hyphae in crop debris	Cultivation, crop debris	Wounds on roots and crown	Chlorotic stripes, stunting, vascular wilt, whiteheads	Wheat	Hemibiotroph
Macrophomina phaseolina (anamorphic Botryosphaeriaceae)	Ascomycete	Microsclerotia	Cultivation, plant debris	Root tips	Charcoal rot of roots and stems	Many	Necrotroph
Pectobacterium carotovorum (formerly *Erwinia carotovora* subsp. *carotovora*)	Prokaryote	Cells in soil, potato tuber debris, water	Irrigation water, aerosols, potato seed pieces	Wounds on potato tubers, enlarged lenticels on tubers	Soft rot of tubers	Potato	Necrotroph
Phymatotrichopsis omnivora (formerly *Phymatotrichum omnivorum*) (anamorphic Rhizinaceae)	Ascomycete	Sclerotia	Growth of hyphal strands in soil	Roots	Bronzing of leaves, permanent wilt	Many dicotyledonous plants (>2000)	Necrotroph

(Continued)

TABLE 16.1 (*CONTINUED*)
Characteristics of Selected Soilborne Plant Pathogens

Name of Pathogen	Pathogen Type	Survival	Dispersal	Infection Court	Symptoms	Hosts	Parasitism
Phytophthora cinnamomi	Oomycete	Chlamydospores	Cultivation, water, infected plants	Root tips	Root rot, low vigor	Many hosts (>900)	Hemibiotroph
Pseudocercosporella herpotrichoides (synonym for *Oculimacula yallundae*)	Ascomycete	Hyphae in crop debris	Rain splash, cultivation	Basal leaf sheaths	Eyespot, foot rot, lodging	Wheat	Hemibiotroph
Pythium spp.	Oomycete	Sporangia, oospores, mycelium	Cultivation, water, infected plants, seeds	Root tips	Root rot and damping-off	Many	Necrotroph
Ralstonia solanacearum	Prokaryote	Cells in soil and plant debris, contaminated water sources	Cultivation	Wounds on roots	Permanent wilt	Solanaceous crops	Hemibiotroph
Rhizoctonia solani (teleomorph = *Thanatephorus cucumeris*)	Basidiomycete	Sclerotia, mycelium in soil	Cultivation, crop debris	Seeds, stems, hypocotyls, roots	Damping-off, stem rot, root rot	Many	Necrotroph
Sclerotinia spp.	Ascomycete	Sclerotia	Cultivation, crop debris	Stem	Blight	Many	Necrotroph
Sclerotium rolfsii (teleomorph = *Athelia rolfsii*)	Basidiomycete	Sclerotia	Cultivation, crop debris	Lower stem	Blight	Many	Necrotroph
Thielaviopsis basicola (anamorphic Microascales)	Ascomycete	Chlamydospores	Cultivation, infected plants	Root tips	Root rot, chlorosis	Many	Hemibiotroph
Verticillium dahliae (anamorphic Plectosphaerellaceae)	Ascomycete	Microsclerotia	Cultivation, infected plants	Root tips	Wilt, vascular necrosis	Many	Hemibiotroph

Source: Young, J.M. et al., *Int. J. Syst. Evol. Microbiol.*, 51, 89–103.
[a] Proposed name change = *Rhizobium radiobacter*.

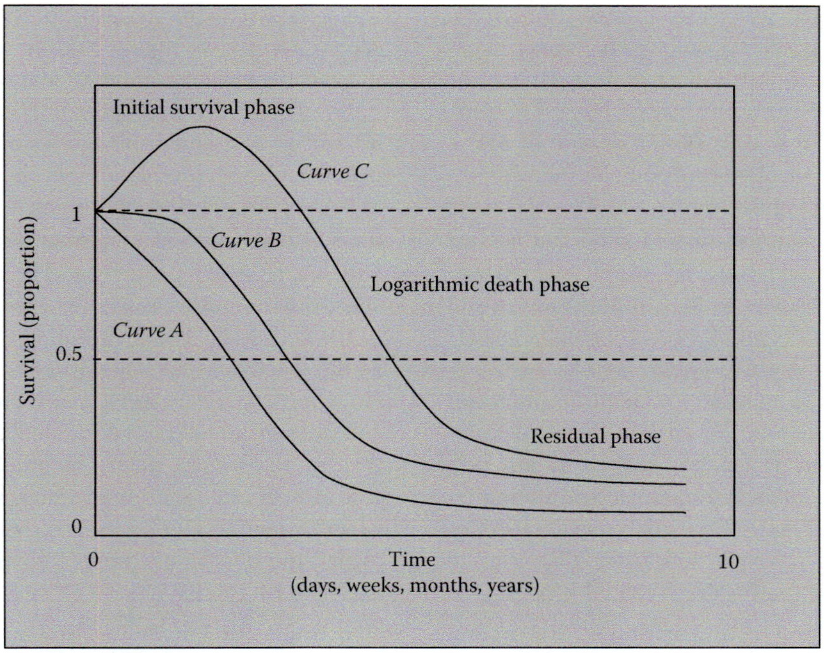

FIGURE 16.5 Idealized survival curves for soilborne plant pathogens. Curve A represents inoculum that enters the logarithmic death phase immediately after dissemination, Curve B represents multicellular inoculum, such as sclerotia, that has a lag phase before entering the logarithmic death phase, and Curve C represents saprophytic growth of inoculum after dissemination to soil. (From Campbell, C.L., and Benson, D.M., *Epidemiology and Management of Root Diseases*, Springer-Verlag, Berlin, 1994. With permission.)

sclerotium die before the population of sclerotia enters the logarithmic death phase (Figure 16.5, Curve B). For pathogens with high competitive saprophytic ability, the pathogen population may actually increase in soil as the pathogen colonizes new substrates before entering the logarithmic death phase (Figure 16.5, Curve C). In all three curves, however, the number of surviving propagules, that is, the inoculum density of the pathogen in the residual phase of survival, determines the initial amount of disease that will develop in a subsequent crop. For most soilborne plant pathogens, this residual inoculum density is usually more than adequate to initiate an epidemic. Therefore, plant disease management practices are aimed at reducing this residual inoculum density even lower or by preventing the germination of pathogen propagules in the host infection court.

ENVIRONMENT AND SOILBORNE PLANT PATHOGENS

Soil Water and Disease

The amount of water present in soil can have a profound effect on soilborne pathogens and the diseases they cause. Just as plant growth is affected by the amount of water available for root uptake, some soilborne pathogens are uniquely suited to take advantage of soil water availability and cause disease. While many soilborne pathogens cause disease across a normal range of soil moisture for good plant growth, some like the water molds, that is, *Phytophthora* and *Pythium* spp., and others like *Fusarium* spp. cause epidemics under conditions of saturated and drought conditions, respectively.

Soil Water Potential

The energy of soil water is measured in terms of **soil water potential**, which is composed of gravitational, matric, osmotic, and other forces. The amount of energy that an organism, for example, fungus or plant root cell, would expend to take up pure, free water is arbitrarily set at a water potential of 0. However, in soil there are several forces that bind water to soil particles, and the amount of energy required to overcome these forces is defined in terms of suction or pressure and expressed in bars or pascals. The unit of pressure in the International System of Units is the pascal (Pa). One bar equals approximately 0.1 mega pascal (MPa). As an analogy, imagine the "energy" or force, that is, suction in this case, that you would have to exert on a drinking straw to remove water from a glass. **Gravitational potential** is potential energy due to gravity. Gravitational water is water in excess of that which is held by soil particles. Gravitational water passes through soil to deep within the profile and may reach the water table. After gravitational water has drained through the soil profile, the soil is at field capacity with a water potential of about −0.3 bar (−0.03 MPa) (Figure 16.6). The water potential at field capacity varies with soil texture. For sandy soils, field capacity is at −0.1 bar (−0.01 MPa). In most soils, the permanent wilting point for many plants is −15 bar (−1.5 MPa) (Figure 16.6).

Matric potential is a measure of the force needed to remove adsorption and capillary water held on and between soil particles. Many biological processes in soil are affected by matric potential. For example, as matric potential decreases, the rhizosphere shrinks as water films become thinner (Figure 16.6). **Osmotic potential** is the pressure that needs to be applied to move water from a hypotonic (more water, less solutes) to a hypertonic (less water, more solutes) solution across a semipermeable membrane. It results from the concentration of ions and molecules dissolved in soil water. The osmotic potential component of total water potential is negligible in all but the most saline soils. Matric potential dominates total water potential in soils and has the greatest impact on water uptake by roots and growth and infection by soilborne pathogens. Inside the plant tissues, osmotic potential dominates and can influence pathogen growth and colonization after infection.

In soil, water potential can be described in three categories. These include **gravitational water** from 0 to −0.03 MPa water potential, which drains through the soil profile after rain or irrigation, **capillary water** from −0.03 to −3.1 MPa, which is held by forces of adsorption on soil particles and capillary forces in pores between soil particles, and **hygroscopic water** with a water potential below −3.1 MPa. Gravitational water is available for root uptake only briefly, as the profile drains while hygroscopic water is so tightly held on soil particles that it is unavailable to roots. Gravitational water can carry propagules of soilborne pathogens deeper into the soil profile, but in poorly drained soils with slow percolation, oxygen deficiency can harm plant roots and aerobic microbes.

As gravitational water drains through the soil profile, the soil pores, or space between soil particles, are filled with air as water percolates. The diameter of the pore space determines whether or not the pore can retain water as the soil dries. The distribution of soil pore sizes, that is, the space between soil particles, varies with soil type. However, the relationship between pore size diameter and water potential has been calculated. At field capacity, that is, −0.03 MPa water potential, pores with a diameter of 8.79 µm would hold water, whereas at −1.5 MPa water potential, the permanent wilting point for most plants, the pore diameter that continues to retain water is only 0.19 µm. Even at −1.5 MPa, however, the relative humidity of the soil is 98.9%.

In soil, capillary water between −0.03 and −1.5 MPa water potential is available for plant growth. Water potential gradients exist in soil as plant roots take up water during the day such that in bulk soil the water potential can be −0.5 MPa, dropping to −0.8 MPa in the rhizosphere, and −1.4 MPa at the root surface. At night, the system would re-equilibrate to about −0.5 MPa water potential.

Soil Water Effects on Disease

Phytophthora root rot caused by a number of *Phytophthora* spp. and Fusarium foot rot caused by

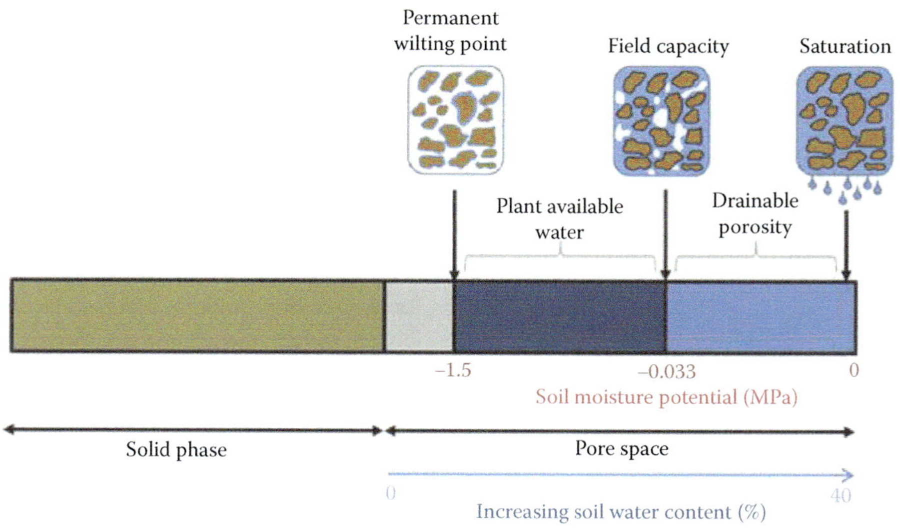

FIGURE 16.6 Soil water potential, moisture content, and air-filled pore space at saturation, field capacity, and permanent wilting point of field soil. Saturated soil drains quickly due to gravity. After drainage, water is most available to plants at field capacity (-0.033 MPa). As soil water potential decreases, soil pores become filled with air and water content decreases. Water becomes unavailable to plants at approximately -1.5 MPa (permanent wilting point). (Courtesy of B.H. Ownley.)

Fusarium culmorum illustrate two diseases where wet soils and dry soils, respectively, are important in disease development.

Phytophthora Root Rot

The biology of *Phytophthora* spp. that cause root rot and crown rot is very tightly keyed into soil water and, more importantly, changes in soil water status resulting from rainfall and irrigation. Although chlamydospores and oospores are responsible for long-term survival of *Phytophthora* spp. in soil, once soil temperature is favorable, changes in soil water can trigger sporulation resulting in the formation of sporangia. Typically, sporangia of most *Phytophthora* species are formed in soil within 12 h of when matric potential reaches between −0.01 and −0.015 MPa; remember field capacity is near −0.03 MPa water potential, indicating that sporangia form under very wet soil conditions. Sporangia can persist in soil for a few days or at most a few weeks; however, when soil reaches 0 MPa water potential (saturated conditions), sporangia are triggered to produce and release motile **zoospores**. Exudates of carbon compounds in the rhizosphere serve as a chemotactic gradient that the motile zoospores follow as they swim through water-filled pores to the rhizoplane where they encyst, germinate, and infect the root. Rather than swim in a straight line along the nutrient gradient to the root, zoospores exhibit a helical swimming pattern that requires a turning diameter of 110 μm, in the case of *Phytophthora cinnamomi*. This means that soils at matric potentials below about −0.0015 MPa water potential do not have water-filled pores that is large enough to accommodate the swimming zoospores. However, soils need to stay saturated (0 MPa water potential) only for a few hours for zoospores to find and infect plant roots. For management of Phytophthora root rots, strategies that prevent sporangium formation can be the key to control of this pathogen.

Fusarium Foot Rot

In the Pacific Northwest and other geographical areas with low annual rainfall (20–40 cm) and dryland farming practices, Fusarium foot rot of wheat caused by *F. culmorum* can limit crop production. The pathogen survives as chlamydospores in the crop debris mulch layer and infects roots 2–3 cm below the soil surface where secondary roots emerge. Under adequate moisture conditions, no further disease development occurs, but under water stress conditions (very low water potentials), the pathogen colonizes the crown roots and moves up the stem one to three internodes causing a chocolate brown discoloration of the inner tissues while the leaf sheath wrapped around the internodes remains apparently healthy. The symptoms are not apparent until the plant has formed heads, at which time bleached, empty "whiteheads" take the place of seed-filled heads.

In dryland farming, precipitation is collected in the mulch soil profile for one season before the wheat crop is planted in fall for summer harvest the following year. Fusarium foot rot was first recognized in the Pacific Northwest when semidwarf wheat cultivars were grown on close plant spacings, and high nitrogen fertilization rates were applied to promote yield. These crop production practices resulted in soil and leaf water potentials that stressed the plant and favored infection and colonization of plant tissues by *Fusarium*. As shown in Figure 16.7, wheat plants grown under high rates of nitrogen had soil water potentials in the upper 120 cm of the soil profile that were −7 to −8 bar (−0.7 to −0.8 MPa) drier than plants grown under low nitrogen regimes. Likewise, the leaf water potential (osmotic potential) was −5 to −8 bar (−0.5 to −0.8 MPa) drier in the stems of high nitrogen plants than in low nitrogen plants. This difference in soil and leaf water potential was enough to result in more infections and subsequent severe disease development with corresponding loss of yield. To solve the problem, growers are advised to fertilize at a conservative rate and to use a row spacing width of 40–45 cm. These practices will not result in plant water stress later during crop growth.

Soil Aeration

Organic matter and microorganisms are found mainly in the surface layers of soil because plant roots predominate in the upper layers. The relative abundance of oxygen is the main factor responsible for root distribution because the roots of higher plants are aerobic. In general, soils range from 40% to 60% pore space, and air content of soil varies inversely with water content. If all the soil pores are filled with water, then air is excluded. Oxygen deficiencies often develop when porosity is reduced to 10% or less for an extended time period (Bruehl, 1986).

The rate at which organic matter (such as a dead plant infected with a plant pathogen) decomposes is strongly influenced by aeration and temperature. When the soil is warm, moist, and well-aerated, organic matter decomposes quickly. Bringing plant crowns and roots to the surface through cultivation hastens decomposition, mainly by the effect of increased aeration on the activity of soil microbes that are responsible for decomposing the crop debris. Rapid decomposition of diseased plant material can greatly reduce the ability of a plant pathogen to survive as a saprophyte (Bruehl, 1986).

Soil Temperature

The metabolism, growth, and interactions of plants, plant pathogens, and other microorganisms in the soil are

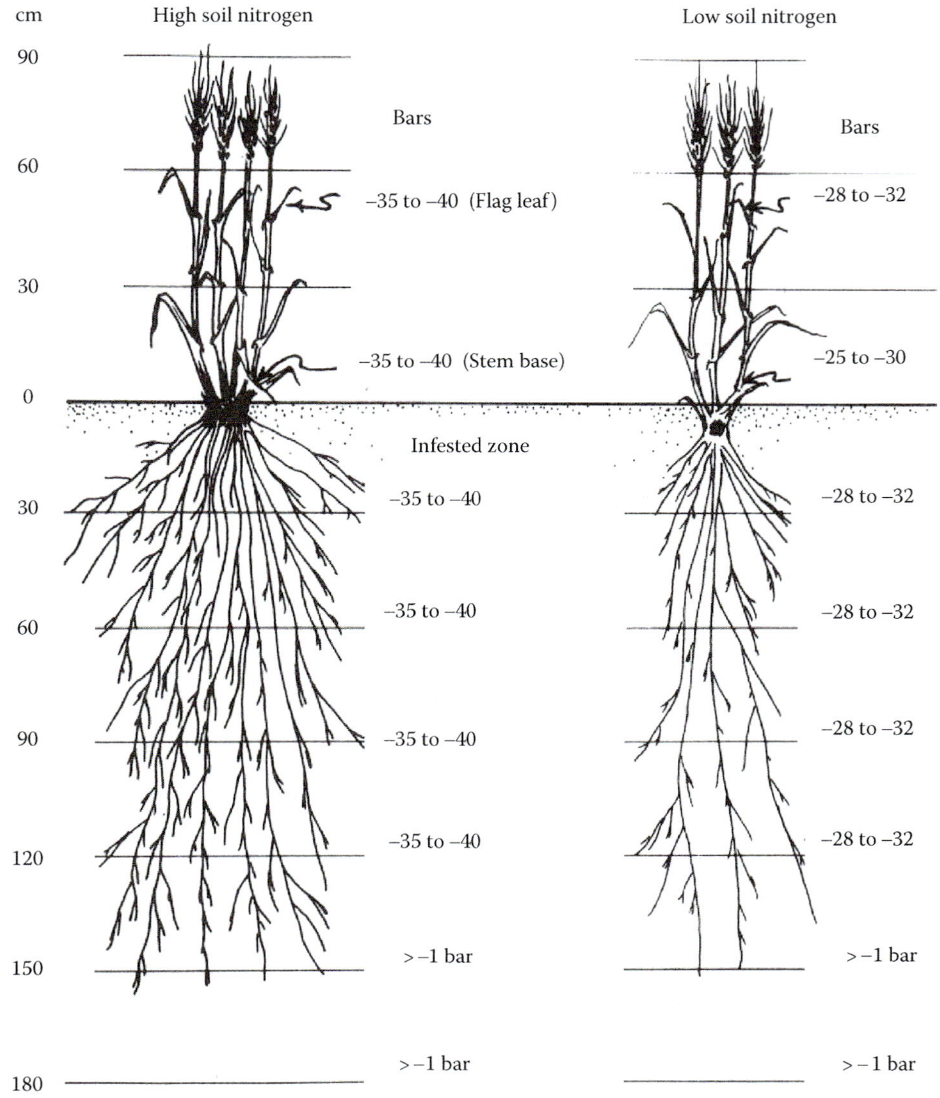

FIGURE 16.7 Typical soil and plant water potentials recorded in the field for *Triticum aestivum* "Nugaines" (winter wheat) at mid-dough stage, grown under Northwest USA dryland conditions with high (60–120 lb N/acre) vs. low (20–40 lb N/acre as residual in the profile, none applied) nitrogen fertility, in the presence of soil infestation of *Fusarium roseum* f. sp. *cerealis*. (From Cook, R.J., *Phytopathology*, 63, 451–458, 1973. With permission.)

directly affected by soil temperature. Generally, the lower temperatures that we associate with late fall, winter, and early spring in the Northern Hemisphere are below the minimum required for disease-causing activity of most pathogens. Plant diseases are not usually initiated during that time or if they are in progress, they often come to a halt. However, pathogens do differ in their preference for higher or lower temperatures, whereas the majority of plant diseases develop more rapidly when higher temperatures prevail, and others develop best with cooler temperatures.

Vascular wilt disease, caused by *V. dahliae* and *Verticillium albo-atrum*, can be severe at soil temperatures of 20°C–28°C. *V. albo-atrum* is more important in cooler, humid, northern regions, and *V. dahliae* causes significant disease further south, particularly on irrigated crops. Both the fungi are unimportant or absent in the humid tropics and semitropics (Bruehl, 1986).

V. dahliae is more widely distributed because it is adapted to warmer climates and produces abundant, long-lived microsclerotia (Figure 16.8). *V. albo-atrum* overwinters as microsclerotial-like hyphae, rather than microsclerotia. To illustrate how small differences in response to temperature can be important, consider this example. *V. albo-atrum* is actually more virulent on cotton than *V. dahliae*, but *V. albo-atrum* is found in areas with cool soil temperatures, fails to grow at 30°C, and has not been isolated from the major cotton-producing areas with warm soil temperatures in the United States.

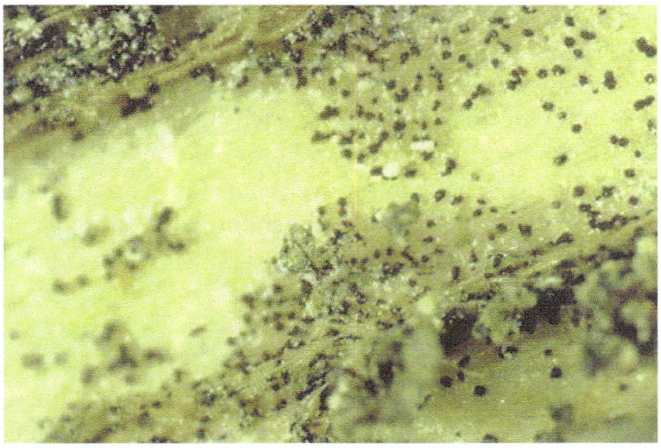

FIGURE 16.8 Microsclerotia of *Verticilliumdahliae* on an infected potato stem. (From Berlanger, I., and Powelson, M.L., *The Plant Health Instructor.* 2005, doi:10.1094/PHI-I-PHI-I-2000-0801-01. With permission.)

On the other hand, *V. dahliae* is an important pathogen of cotton in the southern and southwestern United States. Both species are pathogenic on potatoes from 12°C to 28°C, but *V. dahliae* alone is still pathogenic at 30°C (Bruehl, 1986).

In some cases, the optimum temperatures for disease development, growth of the pathogen, and growth of the plant host are quite different. For example, the optimum temperature range for the development of black root rot of tobacco caused by *Thielaviopsis basicola* is 17°C–23°C. Tobacco grows best at 28°C–29°C, and plant resistance to black root rot increases as soil temperature rises. Similarly, the pathogen grows fastest in culture at 28°C–30°C. Neither the host nor the pathogen grows well at 17°C–23°C. However, the host is more negatively affected than the pathogen, so that even a weak isolate of *T. basicola* can cause maximum disease development at 17°C–23°C (Bruehl, 1986).

Temperature also affects the geographic distribution of soilborne plant pathogens. The fungal pathogen *Phymatotrichopsis omnivora*, which causes root rot of many dicotyledonous hosts, is not found in northern climates. In contrast, temperature plays a minimal role in the geographic distribution of *Fusarium solani* f. sp. *phaseoli*, which can be found wherever bean is grown. After the bean host dies, *F. solani* f. sp. *phaseoli* survives as chlamydospores in soil until favorable environmental conditions return.

Soil pH

Soil pH influences the stability and activity of antibiotics and enzymes, the adsorption of substances and bacteria to soil colloids, the balance between total populations of bacteria and fungi, microbial diversity, and the availability of mineral nutrients to higher plants and microorganisms. Low soil pH makes iron, manganese, zinc, copper, and cobalt more available. When soil pH is less than 5.5, nitrogen, calcium, magnesium, phosphorus, potassium, sulfur, and molybdenum become less available. When soil pH is greater than 7, copper and zinc become less available. Chronic high soil pH often leads to shortages of iron, manganese, cobalt, copper, and zinc. Soils with pH values between 5.6 and 7 are best for most crop plants.

Many soilborne diseases are affected by soil pH. The effect of soil pH on clubroot disease of cabbage, caused by the zoospore-forming obligate biotroph *P. brassicae*, is a well-known example (Chapter 9). Development of clubroot is strongly favored by acid soils. Clubroot does not occur in heavily limed, moist soils. In fact, adding lime to soil to control clubroot disease has been a management practice for more than 200 years. In agriculture, the term “lime” is generic and refers to several forms of calcium and magnesium. Calcium carbonate ($CaCO_3$) is the most widely distributed naturally occurring form of lime and is found in limestone, chalk, and calcite (Campbell and Greathead, 1990).

Actively growing cabbage roots reduce rhizosphere pH, which favors infection. In a study by Dobson et al. (1983), in which equivalent amounts of various forms of lime were added to increase the soil pH, the fineness of the material used had an effect on the final pH and subsequently the percentage of infected plants. The form of lime, that is, calcium hydroxide or calcium carbonate, was not important. With the addition of the coarser lime materials, acidic microsites could develop and disease was favored.

How does “liming” reduce clubroot disease? The pH increases and the added calcium appears to be important factors. Maturation of zoosporangia and release of zoospores are delayed or prevented by higher levels of calcium. The effect of calcium is enhanced by higher soil pH. There is also an effect on the host–pathogen interaction. With lower calcium levels, infections may occur, but they are inhibited from further development. Aside from the role of calcium-rendering intercellular plant pectins more resistant to enzymatic degradation, calcium may act as a cellular messenger that affects various regulatory mechanisms in higher plants (Campbell and Greathead, 1990).

Take-all of wheat, caused by *G. graminis* var. *tritici*, is favored by alkaline soils. The severity of disease can also be greatly affected by the type of fertilizer applied. Take-all disease is less severe with NH_4-N compared with NO_3-N forms of fertilizer. When wheat plants are supplied with ammonium nitrogen, the pH of rhizosphere soil is lower than bulk soil. In contrast, when nitrate-nitrogen is applied, there is an increase in rhizosphere pH compared with bulk soil (Smiley and Cook, 1973). Apparently, when plants are given nitrate-nitrogen, OH^- and HCO_3^- are excreted from the root, which raises the rhizosphere pH. When plants are given ammonium-nitrogen, H^+ ions are excreted, resulting in a decrease in rhizosphere pH. The rhizosphere pH can deviate from bulk soil pH by 1.2 units, and this difference can have a large effect on the severity of take-all. A decrease in rhizosphere pH is also associated with increased availability of iron, manganese, phosphorous, and zinc. Increased levels of these elements have been linked to decreases in take-all severity by making the plant less susceptible to disease.

Black root rot of burley tobacco, caused by *T. basicola*, occurs in western North Carolina. However, in some fields the disease does not develop. By comparing the exchangeable Al^{+3} across fields, it was determined that fields with at least 1 milliequivalent (meq) of Al^{+3} per 100 g of soil were suppressive to black root rot. Soils with a pH below 5 have soluble Al^{+3} in soil solution that controls soil pH. Amending soil with an alkaline source of calcium, such as calcium hydroxide, raises soil pH as the soluble Al^{+3} is replaced by insoluble $Al(OH)_3$. In an experiment, a black root rot suppressive soil at pH 4.4 with 0.9 (meq) of Al^{+3} per 100 g of soil became conducive to disease when calcium hydroxide was used to raise the pH to 5.8, at the same time lowering Al^{+3} to 0.3 (meq) of Al^{+3} per 100 g of soil (Meyer and Shew, 1991). In Brazilian soils with soil pH in the range of 4.8, *P. capsici*, which causes Phytophthora crown and root rot in pepper, is not a disease problem. However, if soils are limed to raise the pH, severe disease develops. In Petri plate culture, mycelial growth and sporulation of both *T. basicola* and *P. capsici* are inhibited in the presence of aluminum. The mechanism of aluminum toxicity to fungi is not completely understood, but critical protein synthesis pathways may be inhibited.

CULTURAL PRACTICES

Successful crop production requires integration of a number of cultural practices that result in high yield and elimination or suppression of weeds, insects, and plant pathogens. For soilborne plant pathogens, cultural practices such as avoidance, crop rotation, tillage and residue management, water management, crop management, and sanitation have proven useful to limit diseases caused by these pathogens. A holistic approach to plant health involves giving ample consideration to cultural practices. See Chapter 25 for additional information on cultural control of plant diseases.

AVOIDANCE

A common form of avoidance is to delay the planting date. For example, warm-weather crops like corn and cotton are stressed when planted in cool soil early in the spring. Planting after soils have warmed avoids prolonged exposure in cold soils to damping-off pathogens like *Pythium* and *Thielaviopsis*. In the case of root-knot nematodes, a winter crop of carrots planted in the southeastern United States between November 18 and December 1 has about 90% marketable roots, whereas only 50% are marketable if planted by October 16. As long as the seeding date can be altered economically, a grower has some control over choice of planting date and its corresponding soil temperature, and subsequent potential effects on disease.

CROP ROTATION

Farmers have practiced crop rotation since ancient times. Crop rotation to a nonhost is the most effective cultural practice for crop production because soil nutrients essential for a given crop can be replaced and pathogen populations in soil may decline during the nonhost period of the rotation due to the lack of nutrients. Generally, crop rotation is more effective against soil invaders than soil inhabitants because soil invaders may not persist in soil very long in the absence of a host. The effectiveness of crop rotation is dependent on the longevity of survival structures of soilborne pathogens. For example, 3- to 4-year rotations are sufficient for the control of *Phoma lingam,* which causes blackleg of cabbage, whereas soilborne pathogens like *Rhizoctonia*, *Sclerotinia*, *Verticillium*, and *Thielaviopsis* produce survival structures that persist in soil for many years and are not likely to be controlled through crop rotation.

Crop rotation can also have preventative value. For example, in Washington State, new lands were brought under irrigation to grow dry beans. For several years, yields were very high, and growers did not practice crop rotation. Eventually, *F. solani* f. sp. *phaseoli* became a serious disease problem and yields declined. Then the growers started practicing crop rotation to nonhosts; however, by that time, populations of *F. solani* f. sp. *phaseoli* had reached 2000–3000 chlamydospores per gram of soil. Once populations were established to this extent, inoculum levels of *F. solani* f. sp. *phaseoli* capable of economic loss were maintained, even with crop rotation. In other words, once certain pathogens are increased to very high population levels, they appear to be immortal (Bruehl, 1986). If crop rotation had been followed in the beginning, this problem could have been avoided.

TILLAGE AND RESIDUE MANAGEMENT

The adoption of reduced tillage and conservation tillage has had a tremendous negative impact on the management of soilborne pathogens like *Sclerotium*, *Cephalosporium*, and *Pyrenophora tritici-repentis* that survive in crop debris. Unlike conventional tillage where the practice of moldboard plowing would bury crop debris containing these pathogens, reduced tillage provides a favorable microclimate for the pathogens in crop residues on the soil surface. Reduced tillage or no tillage results in moderation of soil temperatures and conservation of soil moisture, factors that favor survival of soilborne pathogens. In addition, crop residues serve as a food base for sporulation and/or growth of the pathogen prior to infection. Although several soilborne pathogens have become more severe under reduced tillage, the downside of conventional tillage is reduced soil moisture, increased labor and energy costs, and wind and water erosion of bare fields. In Kansas, loss of 2.5 cm of top soil resulted in 100 kg/ha/year reduction in soybean yield presumably due to loss of nutrients.

Fusarium root rot of bean, caused by *F. solani* f. sp. *phaseoli*, is particularly severe in low rainfall areas and soils with compacted hard pans that restrict rooting of the bean plant. Farmers who used a chisel plow to loosen the compacted soil prior to planting beans enjoyed an almost threefold increase in yield because the root system was able to develop to a greater depth for water uptake and compensate for the *Fusarium* infection on the shallow roots. On deep soils, the practice of deep plowing to depths of 30 cm compared with a conventional depth of 15 cm has the potential to bury inoculum beyond the host infection court. *Sclerotinia minor* causes lettuce drop disease and produces sclerotia that survive many years in soil. In lettuce fields that were deep plowed, the incidence of lettuce drop was reduced 50% in the first year. However, the extra cost of deep plowing was not recovered as disease incidence returned to pre-deep plow levels only after 2 years because sclerotia became redistributed throughout the soil profile. On the other hand, *Sclerotium rolfsii*, which causes southern blight in peanuts and other crops, is effectively controlled with conventional tillage (Case Study 16.2).

Solarization

Soil solarization is a control strategy for soilborne plant pathogens, initially developed in Israel, whereby soil prepared for planting is covered with a clear plastic tarp during the summer. Each day as the sun warms the soil under the cover, temperature in solarized soils may be 10°C warmer at a 10-cm depth than that in uncovered soil, with soil temperature reaching nearly 50°C. Soils are usually left covered for about 6 weeks. Solarization works by direct thermal inactivation of survival propagules of soilborne plant pathogens such as sclerotia, chlamydospores, conidia, and hyphae in crop debris. In general, the higher the soil temperature, the shorter the time needed to kill propagules; however, there is a temperature below which a particular propagule can survive. To reduce the number of pathogen propagules, solarized soils must reach the minimum temperature needed to kill a particular propagule and remain at that temperature long enough for the killing effect to take place. Many diseases caused by soilborne plant pathogens such as *Verticillium* spp., *Fusarium* spp., *R. solani*, and *Sclerotium* spp. have been controlled through soil solarization. Nematodes and weeds are also controlled with this method.

Solarization has several advantages including the following: a nonchemical approach to disease control with no worker exposure concerns, less total costs than fumigation, easy to implement with conventional farm equipment, may provide benefits for multiple years, effective against a range of plant pathogens, and may not kill thermal-tolerant beneficial microorganisms. The disadvantages of soil solarization include the following: the costs of field preparation are the same as fumigation; effective implementation is limited to specific geographic areas with hot, dry summers; it may reduce or eliminate

CASE STUDY 16.2

Southern Blight Caused by *Sclerotium rolfsii*—Basic Disease Management Strategies Should Not Be Overlooked

- *S. rolfsii* produces sclerotia that enable it to survive in soil.
- *S. rolfsii* occurs primarily in the warm humid southeastern United States.
- In the late 1990s, the pathogen was reported in the Midwest; more recently, it was reported in greenhouses in Colorado.
- Symptoms include wilting, yellowing, and browning of lower leaves of infected plants.
- Signs of the fungus are extensive white mycelial growth starting at the base of plant stems and extending upward; sclerotia (tan spheres the size of mustard seeds) also develop (Figure 16.9).
- Fungicides that control *S. rolfsii* include flutolanil and pentachloronitrobenzene.
- Basic management practices including immediate removal and destruction of infected plants, pasteurization of potting soil before reuse, ensuring adequate drainage, avoiding overhead irrigation, and using pathogen-free stock can prevent disease outbreaks. In field crops, tillage can reduce inoculum by burying sclerotia and reducing their survival (Pottorff, 2003).

FIGURE 16.9 Eruptive germination of two sclerotia of *Sclerotium rolfsii* during soil assay on moist paper toweling. In eruptive germination, note that from each central sclerotium two strands composed of numerous hyphae each have emerged from the sclerotium, one above and one below the sclerotium with subsequent diffuse growth. Inset shows close-up of eruptive germination. (Courtesy of Barbara Shew, North Carolina State University.)

beneficial non-thermal-tolerant microorganisms such as mycorrhizal fungi; costs of plastic disposal; and loss of crop production for 4–6 weeks.

Biofumigation

In the early 1980s, scientists discovered that residues left over from cabbage harvest could be used to control the soilborne pathogen *Fusarium oxysporum* f. sp. *conglutinans*, which causes cabbage yellows disease. When dry cabbage residues were amended to soil at 1% (w/w) and covered with a plastic tarp for 8 weeks, the incidence of cabbage yellows in a subsequent cabbage crop was nil. When soil infested with *F. oxysporum* f. sp. *conglutinans* was suspended above the moistened cabbage residues in a closed jar, the propagules in the infested soil were killed. Thus, a volatile fungicidal compound was suspected. Later, this volatile was identified as isothiocyanate. Biofumigation takes advantage of fungitoxic compounds produced during decomposition of crop residues, normally crucifer crops, by holding these volatiles in soil for a period of time with a clear plastic tarp. In addition to the toxic effect of the decomposing residue, plant pathogens are also subject to solarization when biofumigation is carried out during warm periods of the year.

FUMIGATION

Fumigants are biocides that kill cells on contact whether the cells are bacterial, fungal, nematode, or plant in origin. Fumigants include two classes of compounds, the halogenated aliphatic hydrocarbons like methyl bromide, chloropicrin, and 1-3 dichloropropene (1,3 D) and the methyl isothiocyanate liberators like metam-sodium, dazomet, and of course the cabbage residues mentioned earlier. Since fumigants are biocides, they must be applied before the crop is planted or in a way to avoid contact with crop roots. Due to cost, fumigants are generally used only on high-value crops like vegetables and some ornamental crops. A number of fumigants discovered after 1940 have since been banned in the United States due to human health or environmental concerns.

Methyl bromide has been one of the most effective fumigants ever discovered. At one time, virtually all fields in California that were used to grow strawberries were treated with methyl bromide. However, due to its high volatility, a positive attribute for diffusion through the soil profile, using methyl bromide has been correlated with the depletion of the earth's ozone layer. Accordingly, the United States joined with other countries in signing the Montreal Protocol that calls for the elimination of agricultural uses of methyl bromide in developed countries. Although critical use exemptions have been granted by the international governing body for crops where alternatives have not been found, growers will eventually have to rely solely on alternative controls to methyl bromide fumigation.

Once released in soil, fumigants diffuse throughout the soil profile based on their inherent vapor pressure, soil temperature, soil moisture, and soil structure (clay versus sand, clods versus well-tilled soil). Fumigants are less effective in cold, wet, and poorly tilled soils. The concentration of the toxicant, and exposure time for the pathogen propagule, determine when death of the propagule occurs. Propagules in the lower portion of the root zone may not be exposed to a sufficient concentration of fumigant to be killed. Fumigants like chloropicrin are most effective against fungal pathogens, whereas 1,3 D is more effective for nematode control. Methyl bromide is effective against both fungal and nematode pathogens as well as weed seeds.

PLANT RESISTANCE

Selection of disease-resistant cultivars, if available, is an important component of disease management strategies for soilborne plant pathogens. The advantages of plant resistance include reduction or elimination of applications of disease control products, compatibility with other disease management practices, and the long-lasting nature of many disease-resistant genes. For example, use of resistant cultivars is the best management tool for the control of wilt diseases caused by *Fusarium* and *Verticillium*. There are also disadvantages associated with host resistance. In some cases, disease-resistant cultivars may lack the more desirable horticultural traits or potential for high yields found in disease susceptible cultivars; and resistance to one disease does not imply resistance to another. Depending on the host–pathogen system, disease resistance may be short-lived as new strains of the pathogen develop that overcome the plant's resistance genes. Unfortunately, due to the broad host range and large genetic variability of many soilborne plant pathogens, resistant cultivars are not available for all diseases on all crops. Examples of pathogens for which there is little or no genetic resistance available include *Gaeumannomyces*, *Pythium*, *Sclerotinia*, and *Sclerotium*. For additional information on host resistance, see Chapters 20 and 23.

BIOLOGICAL AND CHEMICAL CONTROLS

Biological control products (also called biopesticides) and conventional fungicides are both available for the control of many soilborne plant pathogens. The commercial products available to growers often change as new research leads to the development of more efficacious and environmental-friendly products. Both the topics are treated thoroughly elsewhere in this text (see Chapters 26 and 27).

ORGANIC CROP PRODUCTION

In October 2002, the U.S. Department of Agriculture adopted national certification standards for organic production and processing. In organic crop production, the certification standards do not allow the use of conventional synthetic pesticides and fertilizers and plant growth regulators. Disease management systems developed for organic crops include selection of resistant plants and cultivars (produced by traditional breeding, not genetically modified), selection of planting site, exclusion, many of the cultural practices described earlier, application of biological controls, and soil amendment with composted organic materials. A more in-depth treatise of organic agriculture and plant disease is found in Chapter 29.

Site selection involves consideration of previous soilborne disease problems associated with specific fields, as well as physical features, and soil and water characteristics of the site. For example, well-drained sites reduce the risk of damping-off and root rot caused by soilborne *Pythium* and *Phytophthora* spp. Exclusion, as exemplified in Case Studies 16.1 and 16.2, involves preventing diseased plants and planting materials and other production materials contaminated with pathogens, such as soil, potting mix, pots, or trays, from entering the crop production system. Water can also be a source of pathogen contamination. Incorporation of composts into soil is a fundamental practice in organic crop production. Benefits include enhancing soil fertility, increasing organic matter, and increasing the populations and diversity of soil microflora. Depending on the nature of the composted materials, and the combination of plant host and pathogen, specific benefits to the management of soilborne diseases have also been reported.

LABORATORY EXERCISES

The laboratory exercises will provide an introduction to soilborne pathogens and some methods for detecting and quantifying the diseases they cause. Most of the experiments can be completed in one or two lab periods with minimum preparations.

GENERAL CONSIDERATIONS

The laboratory exercises can be performed by individual students, or teams of two or more, depending on resources and space. In some exercises, the fungi will be isolated directly from natural populations in soil. For other exercises, the fungi can be purchased from the American Type Culture Collection (ATCC; http://www.atcc.org/). An approved U.S. Department of Agriculture, Animal Plant Health Inspection Service, Plant Protection and Quarantine (USDA, APHIS, PPQ) application and permit are required to purchase or for interstate transport of plant pathogens (http://www.aphis.usda.gov/wps/portal/aphis/home/).

Experiment 1. Detection of *Rhizoctonia solani* in Soil

R. solani is a common soilborne fungal pathogen with a very large host range that causes pre- and post-emergence **damping-off**, root rot, and stem and crown rots in plants (Case Study 16.1). Most agricultural soils contain inoculum of *R. solani* in the form of **hyphae** associated with colonized organic matter and sclerotia. In this experiment, you will learn a method to detect and quantify *R. solani* in soil and then later use cultures of the *R. solani* that you have collected for additional experiments on inoculum density and disease.

Follow the protocol listed in Procedure 16.1 to complete this experiment.

Materials

Each student or team of students will require the following items:

- Aniline blue (0.5%) or Trypan blue (0.05%) in lactophenol solution in dropper bottles.
- Bunsen burner or alcohol lamp and ethanol (95%) for sterilizing transfer needle.
- Dissecting and light microscopes.
- Flat wooden toothpicks—birch if possible.

Procedure 16.1
Isolation of *Rhizoctonia* from Soil

Step	Instructions and Comments
1	Prior to lab day, students should collect an assortment of soil samples from various agricultural fields, cultivated gardens, and landscape beds. Collect 1 L of soil in a resealable plastic bag to a depth of 10 cm from several different locations in the sampling area (field, garden, and flower bed). Break up any clods and mix the soil sample thoroughly by rotating the bag for a few minutes. Store at room temperature away from direct sunlight. If you have access to a commercial greenhouse, sample potting mixes debris left on and under benches, along walkways, and/or other production areas.
2	If collected soil is moist, use as is. If dry, moisten the plastic bag by adding small volumes of water and mixing until the soil is moist enough to hold together when squeezed, but it easily breaks apart and does not form a mud ball.
3	Fill plastic pot with test soil. Mark the pot label with the soil source and your name, and place in pot. Incubate the test soils in pots for 1 or 2 days at 16°C, if possible. It may be necessary to cover the pot surface with plastic wrap to keep the soil from drying out during the incubation period.
4	After the incubation period, insert six toothpicks to a depth of 5 cm into the soil equidistant around the soil surface. Only the end of the toothpick should be visible so it can be retrieved after incubation in the test soil. Incubate the soil at room temperature for 2 days and cover with a moist paper towel if needed to prevent from drying.
5	Mark the bottom of the *Rhizoctonia*-selective medium dishes with the soil source and your name. Use forceps to retrieve the toothpicks after incubation in the test soil. Space out three toothpicks per dish of *Rhizoctonia*-selective medium. Place the other three toothpicks on a second agar dish. Incubate dishes at room temperature.
6	One day later, examine the agar dishes under a dissecting microscope for colonies of *R. solani* growing from the toothpicks (Figure 16.10). *Note:* It is critical that the toothpicks be examined after 1 day on the agar dishes, because the colonies will grow together after this time.
7	Place the 5-mm grid guide under the Petri dish of *Rhizoctonia*-selective medium and count the number of 5-mm squares adjacent to the toothpick that have *Rhizoctonia* colonies growing in them. Count the total number of colonies for the six toothpicks. Assume that in 24 h only one colony of *Rhizoctonia* will have grown over one 5-mm square.
8	Determine the inoculum density (propagules per gram) of *Rhizoctonia* in the soil sampled. Each toothpick has a volume of 0.075 cm^3 so the total volume of soil displaced by the six toothpicks is 0.45 cm^3. Therefore, one colony per six toothpicks is 1 propagule/0.45 cm^3 of soil. Assuming the soil has a bulk density of 1, then 1 propagule/0.45 cm^3 of soil is 1 propagule/0.45 g of soil. Convert to propagules per gram of soil by multiplying propagules per 0.45 g by 2.22. Thus, for six toothpicks the number of colonies counted multiplied by 2.22 is the number of propagules per gram of soil.
9	Compare the growth of hyphal colonies from the toothpicks under the dissecting microscope with the characteristics of *Rhizoctonia* shown in Figure 16.11. Observe right-angle branching with hyphal constriction at the branch hypha.
10	Follow Procedure 15.2 for staining nuclei of *Rhizoctonia* and observe with a light microscope (Chapter 3) to confirm the identity of the colony.
11	Use the transfer needle and aseptic technique to transfer a small block of agar containing hyphae of a suspected *Rhizoctonia* colony to the center of a Petri dish of PDA. Repeat, if the second dish of toothpicks also has a *Rhizoctonia* colony.
12	Store the pure cultures of *Rhizoctonia* for future experiments by sealing the edges with stretched-out Parafilm.

Source: Adapted from Paulitz, T.C. and Schroeder, K.L. *Plant Dis.* 89, 767–772, 2005.

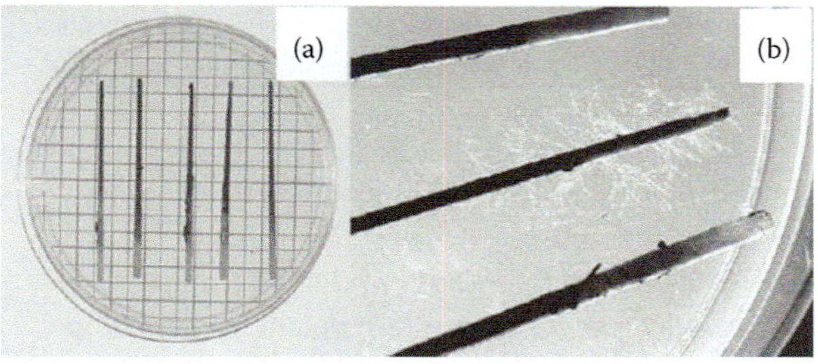

FIGURE 16.10 Toothpick assay. (a) Five toothpicks were removed from soil and placed on a plate of Rhizoctonia selective medium, with 5-mm counting grid underneath. (b) Colony of *Rhizoctoniasolani* growing out of toothpick after 24 h. (From Paulitz, T.C., and Schroeder, K.L., *Plant Dis.* 89, 767–772, 2005. With permission.)

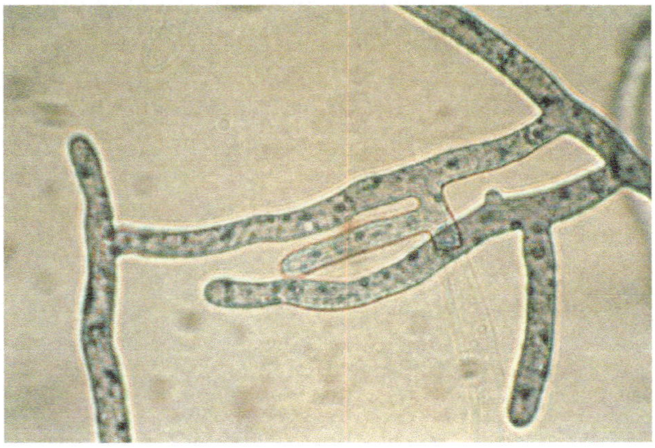

FIGURE 16.11 Right-angle branching of hyphae of *Rhizoctonia solani*. (From Fundamental FUNGI Image Collection & Teaching Resource, APS Press Slide Collections. Photomicrograph courtesy of W. E. Batson, Jr. With permission.)

- Forceps.
- Paper towels.
- Parafilm strips (1-cm-wide) for sealing agar dishes of *Rhizoctonia* pure colonies.
- Plastic pots (10-cm-diameter), about 450 cm^3 capacity.
- Plastic wrap.
- Pot labels and marking pens.
- Potato dextrose agar ([PDA]; 1 agar dish; Table 16.2).
- Resealable plastic bags (3.78-L) for soil samples.
- *Rhizoctonia* selective medium (four agar dishes; Table 16.2) per soil assayed per student or team.
- Soil samples.
- Transfer needle.
- Transparent guide (10-cm-diameter) with 5-mm squares. *Note*: The guide can be made by photocopying graph paper with 5-mm grids onto transparency sheets then cutting to fit the size of Petri dishes.

TABLE 16.2

Recipes for *Rhizoctonia* Selective Medium, Potato Dextrose Agar, 10% Carrot Agar, and 1.5% Water Agar

***Rhizoctonia* Selective Medium**
Difco Agar, granulated, 15 g
Deionized water, 1 L
Combine in 2-L flask and autoclave at 121°C for 15 min
Cool to 50°C–55°C, then add
Thiophanate-methyl fungicide, 1 mg/L (*Note:* Cleary's 3336, OHP 6672, Fungo Flo)
Chloramphenicol, 100 mg/L
Pours about 60 Petri dishes (10-cm-diameter)

Potato Dextrose Agar (PDA)
Difco Potato Dextrose Agar, 39 g
Note: PDA contains the infusion from 200 g Irish potatoes, 20 g dextrose, and 15 g agar
Deionized water, 1 L
Combine in 2-L flask and autoclave at 121°C for 15 min
Pours about 60 Petri dishes (10-cm-diameter)

10% Carrot Juice Agar
Difco Agar, granulated, 15 g
Fresh carrot juice, 100 mL
Deionized water, 900 mL
Combine in 2-L flask and autoclave at 121°C for 15 min
Pours about 40 Petri dishes (10-cm-diameter)

1.5% Water Agar
Difco Agar, granulated, 15 g
Deionized water, 1 L
Combine in 2-L flask and autoclave at 121°C for 15 min
Pours about 60 Petri dishes (10-cm-diameter)

Note: After cooling, poured dishes should be stored in sealed plastic bags in the dark until use.

Anticipated Results

Colonies of *R. solani* and possibility binucleate *Rhizoctonia* fungi will grow from the toothpicks within 24 h after culturing on the *Rhizoctonia* selective medium (Figure 16.10). Under the dissecting microscope, *Rhizoctonia* hyphae will appear slightly brownish and glistening. Under the light microscope, right-angle branching of *R. solani* hyphae will be evident (Figure 16.11). Nuclear staining will be necessary to separate *R. solani* from binucleate *Rhizoctonia*. Isolates of *R. solani* should be recovered from all soil types but not all soils tested, so a large variety of different soil samples should ensure success for the class. Inoculum densities may range from less than 1 to more than 100 propagules per gram of soil.

Questions

- Why does the toothpick method work for detecting *R. solani* in soil but not a soilborne pathogen like *P. cinnamomi*?

- Were all of the *Rhizoctonia* hyphae detected multinucleate?
- What soils had the highest inoculum density of *R. solani*?

Experiment 2. Rhizoctonia Diseases

This experiment is designed to demonstrate pre- and post-emergence damping-off as well as stem rot caused by *R. solani*. For the most part, *R. solani* attacks young plant tissues before the host has developed mature tissues that are more resistant. Germinating seeds and seedlings are the prime targets of *R. solani*. Vegetatively propagated plant material can also be severely diseased by this pathogen. In the following exercise, traditional and enzyme-linked immunosorbent assay **(ELISA)** methods will be used to recover *R. solani* from infected plant tissue. Refer to the chapter on viruses (Chapter 4) for an explanation of how ELISA detects antigens of a pathogen.

Materials

Each student or team of students will require the following items for Procedures 16.2A through C, 16.3, and 16.4:

- Balance (one per class).
- Beaker (500-mL).
- Bedding plant seeds—snapdragons, impatiens, vinca, etc.
- Bent wire, stiff.
- Blender, cookie sheet, sieve with 2-mm openings (USA Standard Testing Sieve, No. 10), aluminum foil squares.
- Clear plastic bag that is large enough to hold cuttings in a strip of rooting cubes
- Commercial peat moss potting mix.
- ELISA kit for *Rhizoctonia*. Pre-order from Neogen Corporation at http://plant.neogeneurope.com/.
- Erlenmeyer flask (125-mL) with 50 mL of sterile deionized water.
- Forceps and scalpel.
- Four plastic bags (3.78-L).
- Graph paper (1 sheet).
- Greenhouse flats (four) with drainage sufficient to accommodate 64 seeds on 2.5-cm grid, that is, about 25 × 25 cm with a depth of 2.5 cm.
- Hand spray bottle for misting.
- Inoculum of *R. solani* on PDA from toothpick isolations described in Procedure 16.1.
- Long grain white rice, 25 g; deionized water, 18 mL; 250-mL Erlenmeyer flask; aluminum foil cover for flask.
- Paper towels.
- PDA (two agar dishes, Table 16.2).
- Pot labels and marking pens.
- *Rhizoctonia* selective medium (two agar dishes, Table 16.2).
- Ruler (15-cm).

Procedure 16.2A

Pre- and Post-Emergence Damping-off of Vegetable and Bedding Plants Caused by *Rhizoctonia solani*—Preparation of *R. solani* Inoculum

Step	Instructions and Comments
1	Combine 18 mL deionized water with 25 g rice grains in 250-mL Erlenmeyer flask with aluminum foil cover. Autoclave for 40 min.
2	Next day, use stiff bent wire to "stir" up rice grains to break apart the solid mass. Autoclave again for 40 min. Do not add any more water.
3	When cool, add three to five agar disks from *Rhizoctonia* culture to the rice grains. Incubate on lab bench at room temperature.
4	Each day, "bump" round-bottomed edge on thick mouse pad or palm of your hand to keep the mycelium from binding rice grains together.
5	*Rhizoctonia* colonized rice is ready to use after 7 days of incubation.
6	Reserve 10 intact rice grains colonized by *R. solani* for use in Procedure 16.3.
7	Add the remainder of the rice grains from one flask to a blender, then pulse blend until rice grains are pulverized.
8	Screen the pulverized rice grains through the 2-mm mesh sieve onto the cookie sheet and store temporarily in folded aluminum foil square.
9	Weigh three aluminum foil packets of rice grains; one with 0.05 g of pulverized rice gains, a second packet with 0.5 g, and the third packet with 1.0 g of grains. *Note*: Once rice grains are removed from the flask, they must be used immediately, that is, within 2 h.

Procedure 16.2B

Pre- and Post-Emergence Damping-off of Vegetable and Bedding Plants
Caused by *Rhizoctonia solani*—Infesting Soilless Mix and Planting Seeds

Step	Instructions and Comments
1	Transfer 1500 cm^3 of dry soilless potting mix to a 3.78-L plastic bag and rotate the bag slowly 25 times to mix the contents. Pour mix into a greenhouse flat and mark the pot label "Trt 1" (untreated control).
2	To a second lot of 1500 cm^3 of dry potting mix in a 3.78-L bag, add 0.05 g of pulverized rice grain inoculum by sprinkling across the surface of the mix. Rotate the bag slowly 25 times and then pour the infested mix into a flat. Label "Trt 2" (0.05 g inoculum rate).
3	Repeat Step 2 for the 0.5 g rate of inoculum and label "Trt 3" (0.5 g inoculum rate).
4	Repeat Step 3 for the 1 g rate of inoculum and label "Trt 4" (1 g inoculum rate).
5	After filling each flat, smooth the soil mix surface. Water gently to wet the mix and let drain.
6	Make a small depression (5-mm deep) with a pencil or pen and make an 8 × 8 grid at 2.5-cm spacing.
7	Plant one seed per depression and resmooth the surface. Gently rewater the flat to settle the potting mix and place on a greenhouse bench, or shelf in a lab with daylight.
8	Water the flats very lightly every day. Do not create craters in the mix when watering, or seeds will be washed out. On the other hand, the mix must remain moist for good seed germination. A hand spray bottle set for misting works well.

Procedure 16.2C

Pre- and Post-Emergence Damping-off of Vegetable and Bedding Plants
Caused by *Rhizoctonia solani*—Assessing Disease

Step	Instructions and Comments
1	Most vegetable and bedding plant seeds will germinate and emerge in 7–10 days. Take stand counts when the first seedlings emerge. Record separate counts for healthy seedlings and seedlings that appear diseased, that is, damped-off. Record counts by Trt (= treatment) number.
2	Continue recording stand counts every other day for the next 2 weeks.
3	At the end of the 2-week stand count period, calculate pre- and post-emergence damping-off percentages for each treatment as disease should have reached a maximum level. For pre-emergence damping-off, subtract the total number of seedlings emerged (whether healthy or diseased) from the stand count in the untreated control. This value is the numerator to be divided by the stand count in the untreated control. Multiply the resulting fraction by 100 to obtain percentage pre-emergence damping-off. Repeat calculation for each inoculum density of *Rhizoctonia*. For post-emergence damping-off, divide the number of diseased seedlings emerged by the total number of seedlings whether diseased or healthy at that inoculum density. Multiply the fraction by 100; the resulting percentage is post-emergence damping-off. Repeat calculation for each inoculum density of *Rhizoctonia*.
4	Plot the inoculum density (in this case 0.05 g rice/1500 cm^3, 0.5 g rice/1500 cm^3, and 1.0 g rice/1500 cm^3 mix) on the x-axis and percent damping-off on the y-axis. Use separate symbols for the pre- and post-emergence damping-off data. Connect the data points and compare the positions of the resulting lines. Compare your results with that of students who used the same or different hosts.
5	Examine diseased seedlings and observe symptoms and signs of disease.
6	Transfer four to six diseased seedlings to a 125-mL Erlenmeyer flask with 50 mL of sterile deionized water. Swirl to remove potting mix adhering to the seedlings. Use forceps to transfer the seedlings to paper towels to blot excess moisture. Finally, transfer the seedlings to Petri dishes of PDA (two to three seedlings per dish).
7	Observe the culture dishes for mycelial growth characteristic of *Rhizoctonia*.

Procedure 16.3

Rhizoctonia Stem Rot of Poinsettia Caused by *Rhizoctonia solani*

Step	Instructions and Comments
1	Make five vegetative cuttings from a poinsettia stock plant. Cuttings should be 8- to 10-cm long with the cut made about 2 cm below the second node of leaves from the terminal. Remove the second node of leaves by breaking off the petiole at the node.
2	Rooting strips may be 13–20 cubes. To save space, break or cut them into five-cube sections and saturate the strip with water.
3	Place one of the cuttings in the hole provided in the strip and proceed to do the same with the rest of the cuttings.
4	With a pair of curved forceps, make a small depression about the length and depth of a rice grain in the cube at the junction line between the first and second cubes, then again between the fourth and fifth cubes.
5	Use the forceps to transfer an intact rice grain colonized by *R. solani* (see Procedure 16.2A, Preparation of *R. solani* Inoculum) to each of the two depressions. Mark a pot label with your name and date, and stick it in the cube at one end.
6	Prepare a second set of cuttings without rice grain inoculum as an untreated control. Only a few of these are needed for the class.
7	Incubating the cuttings. a. If a mist chamber dedicated to plant pathology research is available, place the rooting strip under a mist system which cycles on for 1 min/h, dawn to dusk. b. If a mist chamber is not available, moisten several layers of paper towels and place them in a plastic bag. Put the strip of cuttings in the bag and seal the bag. With the bag method, it will be necessary to place a second pot label at the opposite end of the strip of cuttings to hold the bag away from the cuttings. Incubate the bags at room temperature on a shelf with daylight but not direct sun.
8	Examine the cuttings weekly for signs and symptoms of Rhizoctonia stem rot. Record your observations.
9	With a scalpel, cut a small piece (3- to 5-mm long) of diseased tissue and place in the center of a Petri dish of PDA. After 2–3 days, check the PDA dish for growth typical of *R. solani*. Record your observations.
10	Excise four pieces of the rooting cube (3- to 5-mm long) at various points along the entire strip and place them on a dish of PDA or *Rhizoctonia* selective medium. After 2–3 days, check the cube pieces on the Petri dish for growth typical of *R. solani*. Record your observations.

Procedure 16.4

Detection of *Rhizoctonia solani* from Diseased Tissues Using ELISA

Step	Instructions and Comments
1	Collect samples of suspected *Rhizoctonia*-diseased material from the Procedures 16.1 and 16.2.
2	Follow manufacturer's directions for grinding the tissue sample and extracting the antigen in buffer.
3	Place the extract in Alert test wells per manufacturer's instructions, followed by various solutions as directed.
4	Observe color change and compare with positive and negative controls.

- Strip of polyfoam rooting cubes in plastic liner. *Note*: Most horticultural laboratories will have a supply of these for vegetative propagation research or courses.
- Vegetable seeds—tomato, cucumber, etc.
- Vegetative poinsettia plants to provide five cuttings.

Follow the methods outlined in Procedures 16.2A through C to complete the first part of this experiment.

Anticipated Results

As a variety of hosts were used, the percentage of pre- and post-emergence damping-off will vary by crop. In general, the more inoculum that was used, the more disease that should develop up to a limit, that is, 100% disease. Often the maximum is lower for host and environmental reasons. Pre-emergence damping-off symptoms may be hard to observe unless you can find the diseased seed among the mix and note its decomposing state, otherwise this disease is noted by the difference in percentages of emergence between the untreated control and the flats with *Rhizoctonia* inoculum. Post-emergence damping-off is easy to observe if you look at the flats every 2 days. Seedlings that were apparently healthy during the last observation period will be collapsed and possibly covered with mycelium of *Rhizoctonia*. Thus, stand count of healthy seedlings in flats with *Rhizoctonia* will decrease over the observation period. Transfer of damped-off seedlings to Petri dishes of PDA should result in cultures with brownish mycelium typical of *R. solani*.

Questions

- Did the percent of pre- and post-emergence damping-off increase with greater amounts of *Rhizoctonia* inoculum?
- Did the different hosts tested vary in their susceptibility to pre- and post-emergence damping-off?

Continue with this experiment by following the protocol listed in Procedure 16.3.

Anticipated Results

Assuming that the toothpick cultures isolated from the soil were indeed *R. solani*, symptoms and signs of Rhizoctonia stem rot will be very obvious in 3–21 days depending on temperature and method of incubation. Cuttings incubated in plastic bags will develop symptoms first, while those under a mist system will take 5–14 days for initial symptoms. Disease will progress from cuttings nearest to the rice grains with *Rhizoctonia* to those furthest from the inoculum. In the bag method, cuttings will be totally collapsed and mushy brown in appearance especially those nearest to the rice grains with *Rhizoctonia*. Hyphae of *R. solani* will be apparent entwining the diseased tissue and growing from and along the rooting cube. Separating the plastic liner of the rooting strip from the polyfoam rooting cubes will reveal the presence of *Rhizoctonia* hyphae on the sides and bottom of the rooting strip. If a mist chamber is used, initial symptoms of stem rot will be apparent as brownish white lesions on the stems of the cuttings at the cube surface level. Aerial hyphae growing between the strip and the stem will be observed. As the lesions expand and girdle the stem, the cutting will become necrotic and collapse. Leaf lesions that are necrotic and water soaked may develop on leaves that touch the rooting strip and make contact with inoculum on the strip. Cultures made from diseased tissue should be typical of *R. solani*, but there may be some fungal and bacterial contamination depending on the technique. The control cuttings without *Rhizoctonia*-colonized rice grains should remain healthy. Cuttings under the mist system will root in 3–4 weeks.

Questions

- Did you observe hyphae of *Rhizoctonia* growing on the cube?
- Did you recover *Rhizoctonia* on the PDA dish from pieces of the strip that were furthest from the rice grains?
- Did this experiment fulfill Koch's postulates?
- What control methods do you think a grower should adopt to prevent Rhizoctonia stem rot?

Finish this set of experiments by completing the protocol listed in Procedure 16.4.

Anticipated Results

Plant tissue infected with *Rhizoctonia* will result in a color change to blue after a few minutes.

Questions

- Did you observe a blue positive color change for your sample?
- How long did the ELISA procedure take to run?
- Were the results of the ELISA in agreement with the isolations on culture dishes for *Rhizoctonia*?

Experiment 3. Detection of *Sclerotium rolfsii* in Soil

S. rolfsii is a soil inhabitant that attacks the stems of a wide range of host plants at or near the soil surface (Case Study 16.2). This pathogen survives as sclerotia free in soil or associated with crop debris. Only a few sclerotia per kilogram of soil are needed to incite disease in most hosts. A food base is needed for plant infection in addition to a favorable microenvironment in the plant canopy. In row crops, canopy closure results in environmental conditions favorable for infection. Volatile compounds such as methanol produced from remoistened hosts materials, such as peanut hay, stimulate the germination of sclerotia and possibly provide an energy source for growth. Thus, a very simple method can be used to detect the pathogen in field soils.

Materials

Each student or team of students will require the following items:

- Aluminum baking pan (22 × 30 cm, or larger)
- Bucket

- Collection of field, landscape, and garden soils for the detection of *S. rolfsii*
- Paper towels
- Plastic bag sized for aluminum pan
- Plastic beaker (400-mL) for measuring soil and water
- Resealable plastic bags (3.78-L)
- Sieve with 425-µm openings (USA Standard Testing Sieve, No. 40)
- Tap water, 1 L
- Trowel
- Wash bottle with 1% (v/v) aqueous methanol solution
- Wooden stick (~30 cm)

Optional

- Bunsen burner or alcohol burner with ethanol (95%), forceps, transfer needle
- Parafilm or resealable plastic bag for storage of cultures on PDA
- PDA (two dishes, Table 16.2)

Follow the instructions provided in Procedure 16.5 to complete this experiment.

Anticipated Results

Rapidly spreading colonies of *S. rolfsii* with white coarse mycelium and developing sclerotia will be seen in 48 h. The mycelium will develop additional tan sclerotia over

Procedure 16.5

Assay for Detecting *Sclerotium rolfsii*

Step	Instructions and Comments
1	Visit agricultural fields, home gardens, and landscape beds where crops such as ajuga, chrysanthemum, carrots, hosta, melons, tomato, peanut, pepper, pumpkin, corn, and wheat have grown. *Note*: *S. rolfsii* is common in southern states and California, but the pathogen is not normally found in regions where the average winter temperature falls below 0°C. However, *S. rolfsii* var. *delphinii* is cold tolerant and found in Iowa and surrounding Midwest states on hosta and other herbaceous ornamentals. Collect a total of 1 L of soil from within the top 5 cm of the soil surface at several locations in a given area and store in a resealable plastic bag. Prior to assay, break up clods. If soil will not be used within the week, it should be air-dried for long-term storage.
2	Line the aluminum pan with three to four layers of dry paper towels.
3	Place 250 cm^3 of soil in the bucket and add about 1 L of tap water. Suspend the soil in the water by stirring or swirling the bucket, then decant through the sieve slowly so that the sieve does not overflow. It may be necessary to add more water and repeat to transfer all of the soil onto the sieve. Bumping the sieve or gently rubbing the underneath side of the sieve will speed up the flow of soil solution through it. Next, tilt the sieve and wash the soil to the lower side of the sieve. Transfer the soil residue to the aluminum pan by bumping the sieve upside down on one half of the pan. Use a pot label to spread out the soil residue and then thoroughly moisten the soil residue and paper towels with the methanol solution. Repeat this process with another 250 cm^3 of soil, until the entire 1-L sample has been processed.
4	Use the wash bottle with 1% methanol solution to transfer the organic debris and soil particles on the sieve to the aluminum pan, spreading the debris as evenly as possible.
5	Place the plastic bag around the aluminum pan and seal.
6	In 48 h, count the number of developing colonies of *S. rolfsii*. Colonies will have coarse white mycelium growing several centimeters from the origin (Figure 16.9).
7	Compare the results from the various soils with different crops.
	Optional: Obtaining Pure Cultures of *S. rolfsii* for Other Experiments
1	Use forceps and aseptically transfer individual sclerotia from the aluminum pan to Petri dishes of PDA.
2	In 24–48 h, aseptically transfer a small block of agar with mycelium from the edge of the developing colony to the second dish of PDA to obtain a pure culture.
3	Seal the PDA dish with Parafilm or place in resealable plastic bag and store for future use.

time. If the sample contained more than two to three sclerotia per liter of soil, the developing colonies may grow together. For *S. rolfsii*, an inoculum density of two to three sclerotia per liter is enough to cause severe epidemics in susceptible crops.

Questions

- What is the purpose of the 1% methanol solution?
- Why was a 425-µm mesh sieve used?
- Describe the change in appearance that hyphae of *S. rolfsii* go through as sclerotia develop on the mycelium.

Experiment 4. Student Pet Pathogen Project

Most soilborne plant pathogens are microscopic and hard to observe. However, *S. rolfsii* produces coarse mycelium and relatively large, gold to black-colored sclerotia that can be seen easily with the naked eye (Figure 16.12). Although a plant pathogen, *S. rolfsii* has a very strong saprophytic habit that allows it to grow on a broad range of crop residues whether of host origin or not. Over the course of a few weeks you will be "feeding" your pet culture of *S. rolfsii* various residues to determine just how wide ranging this ability to colonize substrates saprophytically really is.

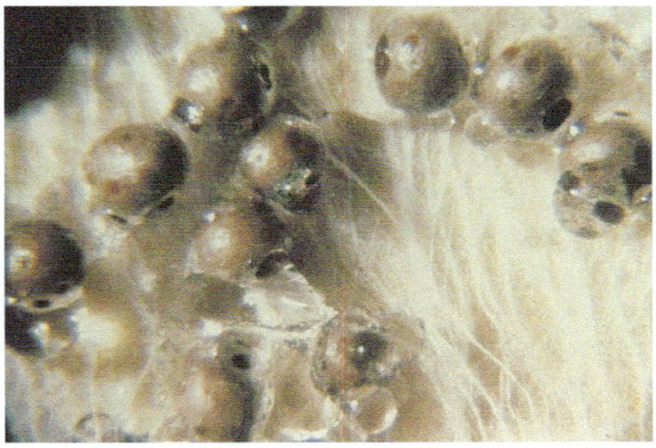

FIGURE 16.12 Brown sclerotia and white mycelia of *Sclerotium rolfsii*. (From Fundamental FUNGI Image Collection & Teaching Resource, APS Press Slide Collections. Photo courtesy of L. F. Grand. With permission.)

Materials

Each student or team of students will require the following items:

- Bunsen burner or alcohol burner with ethanol (95%), transfer needle
- Clear plastic box with lid (22 × 30 × 10 cm deep)
- Culture of *S. rolfsii* on PDA
- Paper towels
- Wash bottle with tap water
- Various sources of organic matter such as broccoli, carrot slices, chewing gum, and wood chips

Follow the protocol outlined in Procedure 16.6 to complete this experiment.

Anticipated Results

The nutrients in the PDA culture of *S. rolfsii* will serve as an initial food base as the fungus starts to grow across the paper towels. Since *S. rolfsii* has a high competitive saprophytic ability, mycelium and sclerotia will eventually colonize the residues and spread throughout the box including the sidewalls. Size and number of sclerotia on the debris will depend on the nutrient value of that debris to the pathogen.

Questions

- Describe the color changes that sclerotia go through from initiation to maturity.
- What type of residues did your pet prefer?

Experiment 5. *Thielaviopsis* Black Root Rot

T. basicola (synonym *Chalara elegans*), a soilborne fungal pathogen, causes black root rot and damping-off of a

Procedure 16.6
Keeping Your Pet *Sclerotium rolfsii* Culture Alive

Step	Instructions and Comments
1	Line the plastic observation box with several layers of paper towels and moisten.
2	Use the transfer needle to cut a 3 × 3 cm square of agar from the center of the PDA dish with *S. rolfsii* and place the agar square at one end of the observation box.
3	Place the various types of organic residues throughout the box.
4	Store the box at room temperature on the lab bench away from direct sunlight.
5	Observe the growth and development of *S. rolfsii* after 2 and 7 days and then for the next few weeks with particular attention to the number and size of sclerotia produced on the various residues and the extent of mycelia growth. It may be necessary to rewet the paper towels weekly.

wide range of host plants, including vegetables, ornamentals, and field crops such as cotton. The fungus reproduces asexually and produces two very different spore types: **aleuriospores** (also referred to as chlamydospores) and **endoconidia** (also called **phialospores** and **phialoconidia**) (Figure 16.13). Aleuriospores produced by *T. basicola* are darkly pigmented, thick-walled, and ovoid to subglobose in shape and occur in chains of five to eight cells. In contrast, endoconidia are hyaline (colorless), barrel-shaped or cylindrical, and occur in chains of three to five cells. The endoconidia are produced by large conidia-producing cells called **phialides**. In culture, copious quantities of both the spore types can be produced. However, growth on nutrient-poor medium tends to promote the production of endoconidia. Aleuriospores are the primary overwintering or survival propagules, but both the types of spores can incite disease. The following exercises are designed to familiarize students with the following: use of a **hemacytometer** to quantify fungal spore inoculum, signs and symptoms of black root rot caused by *T. basicola*, and a method to recover this pathogen from soil and diseased roots.

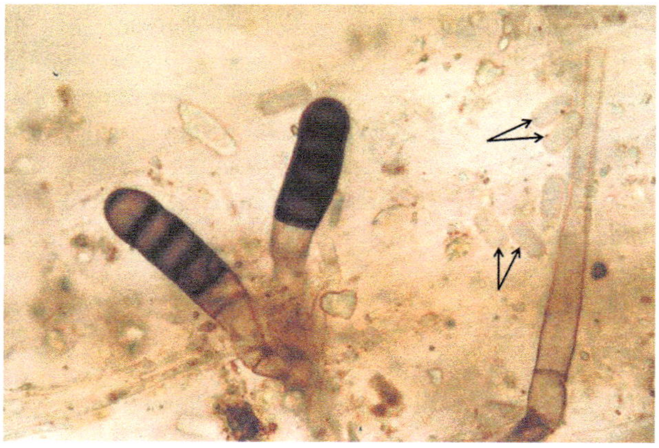

FIGURE 16.13 Darkly pigmented, globose aleuriospores and hyaline, barrel-shaped endoconidia of *Thielaviopsis basicola*. (From Fundamental FUNGI Image Collection & Teaching Resource, APS Press Slide Collections. Photomicrograph courtesy of L. F. Grand. With permission.)

Materials

Each student or team of students will require the following items for Procedures 16.7 through 16.9:

- Carrot
- Cheesecloth
- Culture of *T. basicola* on 10% carrot juice agar
- Culture tube (20-mL)
- Deionized water
- Dissecting and light microscope; microscope slides and coverslips
- Hand counter
- Hand sprayer for misting
- Hemacytometer (cleaned with ethanol) and glass coverslip
- NaOCl (1.0%)
- Pansies (4-week-old plants in cell pack)
- Plastic pots (10-cm-diameter)
- Pot labels and markers
- Potting soil
- Spreader
- Sterile filter paper disk in sterile Petri dishes
- Sterile knife
- Streptomycin sulfate (50 μg/mL)
- Transfer pipet

A hemacytometer is a glass counting chamber with etched grids that can be viewed under a microscope. Although originally designed for quantifying red blood cells, it can be used to determine how many fungal spores are in a suspension. *Note*: The procedure described applies to hemacytometers with Neubauer rulings (grid) (Becker et al., 1996). See Chapter 3 for the details on the use of microscopes.

Follow the protocol described in Procedure 16.7 to complete this part of the experiment.

Anticipated Results

The sample of spores in the suspension will be primarily endoconidia since filtration through cheesecloth will remove the mycelia and aleuriospores.

Procedure 16.7

Quantification of *Thielaviopsis basicola* Spore Inoculum Using a Hemacytometer

Step	Instructions and Comments
1	Flood 3-week-old culture of *T. basicola* growing on 10% carrot juice agar (Table 16.2) with 10 mL of deionized water.
2	Lightly scrape the culture surface with a spreader to free fungal spores. The inoculum will likely be a mixture of endoconidia and aleuriospores.
3	Remove the spore suspension with a transfer pipet; filter the suspension through four layers of cheese cloth, and then transfer the suspension to a 20-mL culture tube.

(Continued)

4	Place the cover glass over the hemacytometer and carefully transfer a small amount of the spore suspension into one of the V-shaped wells with a transfer pipet (Figure 16.14a). Allow the mirrored surface of the hemacytometer to be covered by capillary action. *Note*: It is important to avoid overflow.
5	The etched grid of the hemacytometer is divided into nine large squares, each with a surface area of 1 mm^2. With the coverslip in place, the chamber depth is 0.1 mm. This represents a total volume of 0.1 mm^3 or 10^{-4} cm^3 or 10^{-4} mL. Place the hemacytometer on the microscope stage and focus under low power. Using the hand counter, count the number of spores in the four large squares in the outer four corners (Figure 16.14b). Count only those spores that are within the boundaries of the four counting squares. Calculate the mean total number of spores in four squares for three different samples of the spore suspension. The average total number of spores in the four squares should be in the range of 60–200. *Note*: If many of the spores are clumped, dilute the suspension and repeat the process. Alternatively, too few spores will be observed if the spores settled to the bottom of the tube before the sample was withdrawn from the suspension.
6	Calculate the number of spores per milliliter of your suspension according to the following: a. Multiply the total number of spores counted in the four squares by 2500 to yield the number of spores per milliliter of suspension. b. If the spore suspension was diluted, multiply the calculation from Step 6a by the dilution factor. For example, if a 1:2 dilution factor was used, multiply by 2.

Procedure 16.8

Black Root Rot of Pansy Caused by *Thielaviopsis basicola*

Step	Instructions and Comments
1	The mixture of endoconidia and aleuriospores or the suspension of endoconidia following filtration through cheesecloth (Procedure 16.7) can be used as inoculum. Add the 10 mL spore suspension of inoculum to 90 mL deionized water.
2	Uproot a pansy plant, remove soil to expose the roots, and dip the roots into the spore suspension of *T. basicola*. Uprooting the plant will likely cause minor wounding, which will enhance disease. Repot the pansy plant. Pour the remaining spore suspension around the plant roots. *Note*: To enhance disease severity further, the pH of the growing medium should be 6 or greater, soil temperature should be maintained below 25°C, and planting medium should be kept moist during the incubation period.
3	Write your name and date on a pot label and place it in the pot.
4	Controls should be prepared that do not receive any *T. basicola* inoculum. Only a few controls are needed per class.
5	Observe plants weekly and compare with untreated control plants for foliar symptoms of disease. If different forms or rates of inoculum were used, that is, mixture of aleuriospores and endoconidia or endoconidia alone, or different rates of endoconidia as determined with a hemacytometer, compare the rate of disease progress based on the development of foliar symptoms.
6	When symptoms appear, gently remove plants from pots, wash potting soil from roots, and observe roots for signs and symptoms of black root rot.
7	Observe roots with a dissecting microscope.
8	Cut a small section of blackened root tissue and prepare a squash mount for viewing with 40×, 100×, or 400× magnification with a light microscope (assuming a 10× magnification eyepiece in addition to objectives). To prepare a squash mount, place the tissue in a drop of deionized water, cover with a glass coverslip, and press gently with a pencil eraser to flatten the preparation. Observe under the lowest power before proceeding to higher magnification.

Procedure 16.9

Detection of *Thielaviopsis basicola* in Rotted Pansy Roots and Infested Soil Using a Carrot Root Disk Assay

Step	Instructions and Comments
1	Surface-disinfest the whole carrot by dipping it in 1.0% NaOCl for 20 min, then rinse with deionized water to eliminate pathogens from the carrot surface.
2	Cut the carrot into disks (5-mm thick) with a sterile knife.
3	Place carrot disks onto filter paper disks inside sterile Petri dishes. Moisten the filter paper disks with sterile deionized water.
4	Prepare untreated carrot disk controls by incubating moistened disks as described below, but not adding soil or diseased roots.
5	Inoculate the carrot disks with roots or soil. a. Using the rotted pansy roots from Procedure 16.8, wash the roots free of potting soil in running tap water, blot dry, and place a clump of roots on the carrot disks. b. Using the infested potting soil from Procedure 16.8, or agricultural, garden, or landscape soils, spread soil sample thinly over the surface of the carrot disks. Mist the soil to moisten it, but avoid free water on the surface. Preincubate the carrot disks for 2–4 days at 22°C–24°C in the dark. Wash carrot disks free of soil and place in sterile Petri dishes with moistened filter paper.
6	Seal the Petri dishes containing carrot disks with rotted roots, or disks preincubated with soil with Parafilm and incubate for 7–10 days, at 22°C–24°C in the dark. The carrot disks should be moist, but not wet during the incubation period. *Note*: If bacterial soft rot develops, mist the carrot disks with 50 µg/mL of streptomycin sulfate.

Note: Selective culture media for *T. basicola* have been developed (Specht and Griffin, 1985); however, they contain toxic ingredients and may not be suitable for classroom laboratory exercises.

Source: Adapted from Yarwood, C.E., *Phytopathology* 38, 346–348, 1946.

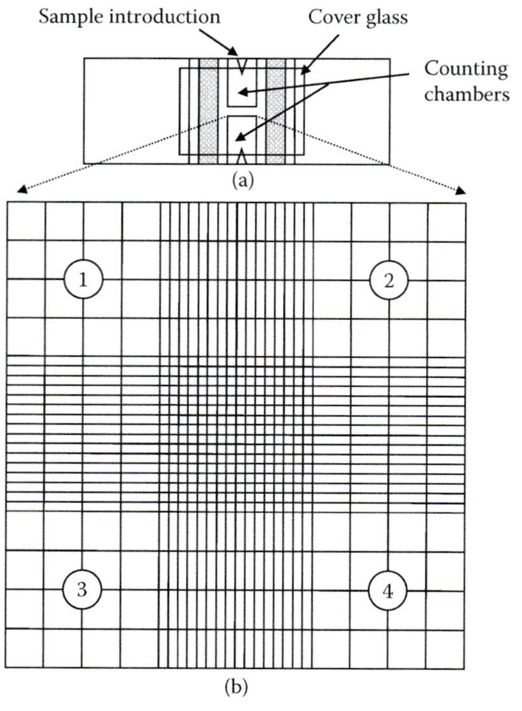

FIGURE 16.14 Hemacytometer. (a) Top view. (b) Enlarged etched grid pattern of hemacytometer. (Courtesy of B. Ownley.)

Question

- A dilution of 1:10 of a spore suspension was made prior to counting the spores in a hemacytometer. In the four outer squares, a total of 120 spores were counted. Calculate the number of spores per milliliter in the spore suspension.

Follow Procedure 16.8 to complete the experiment with black root rot of pansy.

Anticipated Results

Foliar symptoms of black root rot of pansy include stunting, wilting, and yellowing (chlorosis). Symptoms on roots will range from dark spots visible on normally white roots to large sections of blackened, water-soaked roots. With magnification, the darkly pigmented aleuriospores of *T. basicola* should be visible. If different rates of inoculum are compared, disease onset should be earlier and disease should be more severe with increasing rates of inoculum.

Questions

- What was the time span between inoculation and symptom expression?

- Describe the condition and size of diseased roots relative to the untreated control.
- Based on the condition of the roots, do you think that the fungus progressed beyond the cortex into the vascular tissue?

Follow the instructions in Procedure 16.9 to complete the rest of this experiment.

Anticipated Results

After incubation of the carrot disks, *T. basicola* colonies will appear as grayish colonies. The colonies will become black as masses of aleuriospores are produced.

Question/Discussion

- Following incubation, describe the appearance of carrot disks with roots or soil applied relative to the untreated control carrot disks.

ACKNOWLEDGMENTS

The authors gratefully acknowledge the advice of Barbara Shew and Larry Grand, Department of Plant Pathology, North Carolina State University, in the design of the experiments with *S. rolfsii*.

REFERENCES

Berlanger, I., and Powelson, M.L. 2–5. *The Plant Health Instructor*. 2005, doi:10.1094/PHI-IPHI-I-2000-0801-01.

Becker, J. M., G. A. Caldwell, and E. A. Zachgo. 1996. Determination of cell number. In: *Biotechnology—A Laboratory Course*. New York, NY: Academic Press, pp. 221–222.

Benson, D. M., and R. Baker. 1974. Epidemiology of *Rhizoctonia solani* preemergence damping-off of radish: Inoculum potential and disease potential interaction. *Phytopathology* 64:957–962.

Benson, D. M., J. L. Hall, G. W. Moorman, M. L. Daughtrey, A. R. Chase, and K. H. Lamour. 2002. The history and diseases of poinsettia, the Christmas flower. Online. *Plant Health Progress*. doi:10.1094/PHP-2002-0212-01-RV.

Bruehl, G. W. 1986. *Soilborne Plant Pathogens*. New York, NY: Macmillan Publishing Co.

Campbell, C. L., and D. M. Benson (eds). 1994. *Epidemiology and Management of Root Diseases*. Berlin, Germany: Springer-Verlag.

Campbell, R. N., and A. S. Greathead. 1990. Control of clubroot of crucifers by liming. In: A. W. Engelhard (ed). *Soilborne Plant Pathogens: Management of Diseases with Macro- and Microelements*. St. Paul, MN: APS Press, pp. 90–101.

Cheo, P. C. 1982. Effects of tannic acid on rhizomorph production by *Armillaria mellea*. *Phytopathology* 72:676–679. doi:10.1094/Phyto-72-676.

Cook, R. J. 1973. Influence of low plant and soil water potentials on diseases caused by soilborne fungi. *Phytopathology* 63:451–458.

Dobson, R. L., R. L. Gabrielson, A. S. Baker, and L. Bennet. 1983. Effects of lime particle size and distribution and fertilizer formulation on clubroot disease caused by *Plasmodiophora brassicae*. *Plant Diseases* 67:50–52.

Maier, R. M., I. L. Pepper, and C. P. Gerba (eds). 2000. *Environmental Microbiology*. San Diego, CA: Academic Press.

Meyer, J. R., and H. D. Shew. 1991. Soils suppressive to black root rot of burley tobacco, caused by *Thielaviopsis basicola*. *Phytopathology* 81:946–954.

Paulitz, T. C., and K. L. Schroeder. 2005. A new method for the quantification of *Rhizoctonia solani* and *R. oryzae* from soil. *Plant Diseases* 89:767–772.

Smiley, R. W., and R. J. Cook. 1973. Relationship between take-all of wheat and rhizosphere pH in soils fertilized with ammonium vs. nitrate-nitrogen. *Phytopathology* 63:882–890.

Specht, L. P., and G. J. Griffin. 1985. A selective medium for enumerating low populations of *Thielaviopsis basicola* in tobacco field soils. *Canadian Journal of Plant Pathology* 7:438–441.

Worrall, J. 2004. Armillaria root disease. *The Plant Health Instructor*. doi:10.1094/PHI-I2004-0706-01.

Yarwood, C. E. 1946. Isolation of *Thielaviopsis basicola* from soil by means of carrot disks. *Phytopathology* 38:346–348.

Young, J. M., L. D. Kuykendall, E. Martinez-Romero, A. Kerr, and H. Sawada. 2001. A revision of *Rhizobium* Frank 1889, with an emended description of the genus, and the inclusion of all species of *Agrobacterium* Conn 1942 and *Allorhizobium undicola* de Lajudie *et al.* 1998 as new combinations: *Rhizobium radiobacter*, *R. rhizogenes*, *R. rubi*, *R. undicola* and *R. vitis*. *International Journal of Systematic and Evolutionary Microbiology* 51:89–103.

SUGGESTED READINGS

Bockus, W. W. 1998. The impact of reduced tillage on soilborne plant pathogens. *Annual Reviews of Phytopathology* 36:485–500.

Bockus, W. W., and N. A. Tisserat. 2000. Take-all root rot. *The Plant Health Instructor*. doi:10.1094/PHI-I-2000-1020-01.

Daughtrey, M. 2003. New and re-emerging diseases in 2003. In: *Proceedings of the 19th Conference on Insect and Disease Management, Society of American Florists*, pp. 51–60. Alexandria, VA.

Duniway, J. M. 1980. Water relations of water molds. *Annual Reviews of Phytopathology* 17:431–460.

Engelhard, A. W. (ed). 1990. *Soilborne Plant Pathogens: Management of Diseases with Macro- and Microelements*. St. Paul, MN: APS Press.

Gamliel, A., and J. J. Stapleton. 1993. Characterization of antifungal volatile compounds evolved from solarized soil amended with cabbage residues. *Phytopathology* 83:899–905.

Hall, R. (ed). 1996. *Principles and Practice of Managing Soilborne Plant Pathogens*. St. Paul, MN: APS Press.

Katan, J. 1981. Solar heating (solarization) of the soil for control of soilborne pests. *Annual Reviews of Phytopathology* 19:211–236.

Katan, J. 2000. Physical and cultural methods for the management of soil-borne pathogens. *Crop Protection* 19:25–731.

Kuack, D. 2003. Plants could be dangerous to the health of your business. *Greenhouse Management & Production* 23:2.

Lockwood, J. L. 1988. Evolution of concepts associated with soilborne plant pathogens. *Annual Reviews of Phytopathology* 26:93–121.

Pottroff, L. P. 2003. Foil the soilborne diseases. *Nursery Management & Production* 19:47–50.

Ramirez-Villapuda, J., and D. E. Munnecke. 1987. Control of cabbage yellows (*Fusarium oxysporum* f. sp. *conglutinans*) by solar heating of field soils amended with dry cabbage residues. *Plant Diseases* 71:217–221.

Rodriquez-Kabana, R., M. K. Beute, and P. A. Backman. 1980. A method for estimating number of viable sclerotia of *Sclerotium rolfsii. Phytopathology* 70:917–919.

Rosskopf, E. N., D. O. Chellemi, N. Kokalis-Burelle, and G. T. Church. 2005. Alternatives to methyl bromide: A Florida perspective. Online. *Plant Health Progress*. doi:10.1094/PHP-2005-1027-01-RV. http://www.plantmanagementnetwork.org/sub/php/review/2005/methylbromide/

Shew, H. D., and J. R. Meyer. 1992. Thielaviopsis. In: L. L. Singleton, J. D. Mihail, and C. M. Rush (eds). *Methods for Research on Soilborne Phytopathogenic Fungi*. St. Paul, MN: APS Press, pp. 171–174.

Singleton, L. L., J. D. Mihail, and C. M. Rush (eds). 1992. *Methods for Research on Soilborne Phytopathogenic Fungi*. St. Paul, MN: APS Press.

Stapleton, J. J. 2000. Soil solarization in various agricultural production systems. *Crop Protection* 19:837–841.

Subbarao, K. V., S. T. Koike, and J. C. Hubbard. 1996. Effects of deep plowing on the distribution and density of *Sclerotinia minor* and lettuce drop incidence. *Plant Diseases* 80:28–33.

Weber, R. W. S., and H. T. Tribe. 2004. Moulds that should be better known: *Thielaviopsis basicola* and *T. thielavioides*, two ubiquitous moulds on carrots sold in shops. *Mycologist* 18:6–10. doi:10.1017/S0269915X04001028.

17 Parasitic Plants

Daniel L. Nickrent and Lytton J. Musselman

CONCEPT BOX

- The haustorium of dodder is a complex organ that effects the morphological and physiological connection between parasite and host.
- Unlike some other parasitic plants, *Cuscuta* seed germination does not depend on host molecular signaling.
- Dodder seedling undergoes circumnutation and responds to the presence of host plants through various cues including light and volatile chemicals. These cues help the plant orient toward and attach to the host.
- Dodder is a powerful physiological sink, outcompeting host sinks such as fruits that normally receive the lion's share of photosynthates.

Parasitic organisms represent a significant proportion of the earth's biodiversity, and flowering plants are no exception. Parasitic plants attach to their hosts by means of a **haustorium**, and this feeding mode has evolved independently in angiosperms 12 times (Table 17.1). There are no parasitic monocots, but haustorial parasites occur in the magnoliid and eudicot clades on the angiosperm phylogenetic tree (Figure 17.1). In an analogous fashion, a type of parasitism involving a mycorrhizal fungus has repeatedly evolved, and these plants are called **mycoheterotrophs**. The hallmark of green plant evolution is photosynthesis; however, this process has been lost repeatedly in all but two lineages of haustorial parasites (and many mycoheterotrophs). These plants are termed **holoparasites**, whereas those that retain photosynthesis are called **hemiparasites**. Approximately 4400 species in 276 genera of parasitic plants exist (Table 17.1; Parasitic Plant Connection website), but among these relatively few are pathogens, which cause diseases in plants of economic importance used by humans. About 11% of the genera are pathogens, and species in the following five genera inflict the most damage: *Arceuthobium*, *Cuscuta, Orobanche*, *Striga*, and *Cuscuta* (Figures 17.2 and 17.3).

Species of the genus *Cuscuta*, known by the English common name of dodder, are common throughout much of the world and are easily recognized by their twining, tangled masses of yellow or orange stems often blanketing their host plants (Figure 17.2). Some, like *Cuscuta campestris*, are serious pathogens of crops such as alfalfa (*Medicago sativa*), carrots (*Daucus carota*), lentils (*Lens esculentus*), and a diversity of other crops, weeds, and native plants. Except for onion (*Allium cepa*), monocots are not attacked. The only plants that can be confused with *Cuscuta* are species of *Cassytha* (Lauraceae), a genus of tropical parasites bearing a stunning resemblance to dodder. In addition to differences in morphology of their tiny flowers and the kind of fruit, the two genera differ in stem pubescence: dodder has glabrous stems, whereas *Cassytha* has pubescent stems. These two parasitic plants are good examples of **convergent evolution**, not only in overall morphology, but also in other aspects of their biology such as seed germination and host impact.

Parasitic plants have evolved three times on the asterid I clade (Figure 17.1). In addition to *Cuscuta* in Convolvulaceae, the order Lamiales contains Orobanchaceae, one of the most economically important parasitic plant families. Although the vast majority of Orobanchaceae are benign root parasites, two genera, *Striga* (witchweeds) and *Orobanche* (broomrapes) together cause more damage to crop plants than all other parasitic plants combined. *Striga asiatica* (Figure 17.3), a pathogen of maize (*Zea mays*), is native to Africa, but was found in the eastern United States in the 1950s. Since then, a governmental program has been successful in eradicating witchweed from the affected regions. Several species of *Orobanche* are agricultural pathogens throughout the Middle East and Europe, and some species have been accidentally introduced to North America. Like *Striga*, *Orobanche* seeds are tiny and require specific molecules exuded by host roots to stimulate germination. The final group of economically important parasitic plants that will

TABLE 17.1
The 12 Clades of Haustorial Parasites in Flowering Plants

Clade	Order	Family	Example genera	Genera	Species
1	Laurales	Lauraceae	*Cassytha*	1	16
2	Piperales	Hydnoraceae	*Hydnora, Prosopanche*	2	15
3	Santalales	19 Families	*Phoradendron, Santalum, Viscum*	167[a]	2347
4	Saxifragales	Cynomoriaceae	*Cynomorium*	1	2
5	Zygophyllales	Krameriaceae	*Krameria*	1	18
6	Malpighiales	Rafflesiaceae	*Rafflesia, Rhizanthes, Sapria*	3	32
7	Malvales	Cytinaceae	*Bdallophyton, Cytinus*	2	7
8	Cucurbitales	Apodanthaceae	*Apodanthes, Pilostyles*	2	10
9	Ericales	Mitrastemonaceae	*Mitrastemon*	1	2
10	Boraginales	Ehretiaceae	*Lennoa, Pholisma*	2	5
11	Solanales	Convolvulaceae	*Cuscuta*	1	200
12	Lamiales	Orobanchaceae	*Orobanche, Pedicularis, Striga*	93	1725
			Total parasitic plants	276	4379

[a] Only parasitic genera are counted (order includes 12 additional nonparasitic genera).

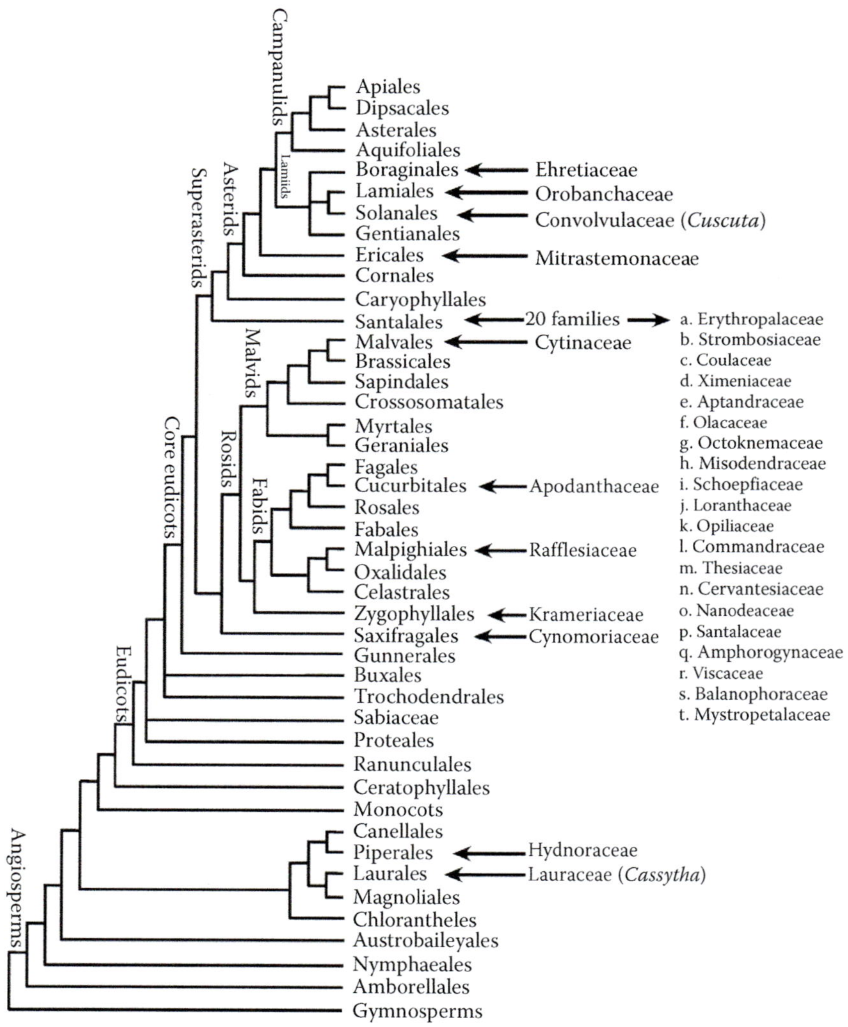

FIGURE 17.1 Phylogenetic relationships among the angiosperm orders indicating the locations of the 12 haustorial parasite clades. (Tree topology was adapted from many sources.)

FIGURE 17.2 Field of sunflowers in Turkey heavily parasitized by dodder (*Cuscuta campestris*). (Courtesy of L. Musselman.)

FIGURE 17.3 *Striga asiatica* (Asiatic witchweed). (a) *S. asiatica* emerging from soil next to its host plant, maize (*Zea mays*). (b) Although the flowers are showy, they self-pollinate prior to opening and upon fruiting, and produce thousands of tiny seeds. (Courtesy of L. Musselman.)

be considered here are the mistletoes (Mathiasen et al., 2008). Taxonomically, mistletoes can be found in five different families within the sandalwood order (Santalales), but in North America, only *Arceuthobium* (Figure 17.4) and *Phoradendron* (Figure 17.5) (both Viscaceae) occur. As parasites of conifers in the pine family in North and Central America, the dwarf mistletoes (*Arceuthobium*) cause billions of dollars of damage on commercial timber trees. One example is *A. douglasii*, which causes significant losses to Douglas fir (Figure 17.4).

LABORATORY EXERCISES

The purposes of these exercises are (1) to familiarize you with the morphology and anatomy of mature *Cuscuta* plants parasitizing a host; (2) to observe seed germination and seedling behavior; (3) to document the growth, twining, and attachment of dodder stems on different hosts; and (4) to demonstrate how this parasitic plant acts as a physiological sink for host nutrients. You (or your instructor) will first need to locate a population of dodder. During summer and fall in North America, several *Cuscuta* species are commonly seen, particularly *C. campestris* and *C. gronovii*. The first species is often seen growing on clover and alfalfa along roadsides, whereas the latter favors edges of ponds and lakes, often growing on water willow (*Justicia americana*), buttonbush (*Cephalanthus occidentalis*), and jewelweed (*Impatiens capensis*). Although commonly occurring on these hosts in nature, both the species will parasitize a wide range of other hosts when cultivated. Bear in mind that laboratory experiments taking place in the spring may not have mature *Cuscuta* plants available from nature; however, with planning, flowering and fruiting plants can be made available by cultivating them under greenhouse conditions. Fruits (capsules) are greenish when immature, but become brown and translucent when mature. Only these capsules should be used when collecting seeds for cultivation and experimentation. Fortunately, these seeds remain viable for years when kept cool and dry. If a good crop of fruits is found, it is worthwhile to harvest many, remove the seeds, and store these in glass jars in the refrigerator. The germination percentages are never 100%, and the rate drops off in successive years.

EXPERIMENT 1. MORPHOLOGY OF MATURE *CUSCUTA*

When a population of *Cuscuta* is located, cuttings of the host with attached dodder can be brought into the lab for further observation. Make sure the host cutting is about 20 cm long and has some well-developed leaves. Place the cutting immediately in water in a large plastic cup. If the plants are not to be studied the same day, the cup with the cutting can be enclosed in a plastic bag and kept at room temperature for a few days or at 4°C for up to a week. This infected host plant cutting can be used to infect other host plants in the greenhouse. If you have grown *Cuscuta* in the greenhouse, you can bring the entire potted host plant with parasite into the laboratory for observation and dissection. If the dodder has flowers and/or fruits are present, these can be dissected and studied. It is easy to grow *C. campestris*

(a) (b)

FIGURE 17.4 *Arceuthobium douglasii* (Douglas-fir dwarf mistletoe). (a) Parasitized Douglas fir (*Pseudotsuga menziesii*) showing a massive witches' broom in its lower half. The mistletoe endophyte induces the host to branch profusely. (b) Close-up of Douglas-fir dwarf mistletoe with young fruits. The mature fruit explosively dehisces the seed, thus no animal agents are required for dispersal to new host branches. (Courtesy of D. Nickrent.)

(a) (b)

FIGURE 17.5 *Phoradendron leucarpum* (leafy mistletoe). (a) Leafy mistletoe infesting a tree. (b) Close-up of *P. leucarpum*. (Courtesy of D. Nickrent.)

to flowering, but this will require an adequate supply of hosts as well as care to ensure that the parasite growth does not spread to unintended plants. Flowering usually begins 3 weeks after host attachment. The induction of flowering in *Cuscuta* depends upon environmental cues such as short days (technically, long night of 14 h) and flowering of the host upon which it is attached (Baldev, 1962; Fratianne, 1965).

Observe the dodder plant attached to the host. The mature stems are yellow to orange, indicating that this plant has very little chlorophyll and is generally referred to as a holoparasite (Figure 17.6a). However, seedlings and other parts of the plant can be greenish, and some photosynthetic activity has been detected (reviewed in Dawson et al., 1994). If desired, the chlorophyll extraction method outlined in Procedure 21.1 can be tried with dodder and compared with a typical green plant. You will find that the amount of chlorophyll present is very low and, in fact, insufficient to provide sufficient sugars to maintain the plant. For this reason, *Cuscuta* is considered an **obligate parasite** because it ultimately depends upon a host for survival.

Notice that the plant lacks expanded leaves, but does have small-scale leaves at the nodes where branching occurs. Also notice that not all the stems are the same—some are coiled and some are not. Young stems are 1 mm or less in diameter, whereas well-nourished, attached stems can be 2 mm in diameter. The distal parts of the stems that are not coiled undergo **circumnutation**, a circular movement in a counterclockwise direction. If this stem encounters a support, it will coil around it. Such a twining stem can also be called a **tendril**. In *C. campestris*, the main stems do not twine, whereas the tendrils, formed from the axils do. In *C. gronovii*, when a stem branches, either branch can twine around the host (Costea and Tardif, 2006).

When the tendril has coiled, it will form small bumps called **prehaustoria** on the inner surface of the curved stem (Figure 17.6a). Coiling and prehaustorium formation respond to mechanical stimulation and the type of light striking the plant (Lane and Kasperbauer, 1965; Orr et al., 1996; Haidar et al., 1997; Li et al., 2010). Prehaustoria have specialized elastic cells that allow them to closely adhere and attach to the host (Vaughn, 2002). At this point, cells inside the prehaustorium differentiate from haustorial initials and begin growing through the prehaustorium tissue and then into the host tissue. These are called **hyphal cells**. Find some dodder

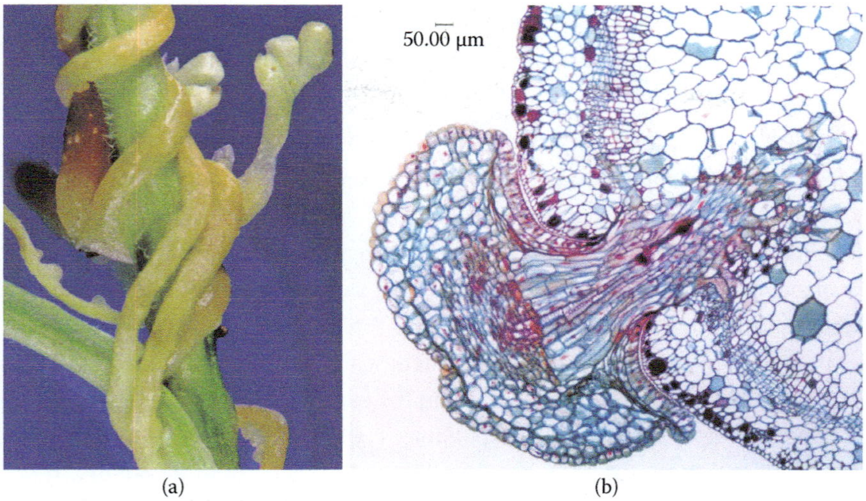

FIGURE 17.6 (a) *Cuscuta gronovii* (scaldweed) parasitizing the stem and leaf of its host, *Eupatorium rugosum*. (Courtesy of D. Nickrent.) (b) Cross section through the haustorium of *C. campestris* showing host tissue penetration. (Courtesy of M. Costea.)

stems that have just begun to attach to the host. Pull on the stems with prehaustoria and notice that you feel the resistance. Continue pulling until you have removed the parasite stem from the host. Examine the prehaustorium with the stereo microscope and you will see that that they are oval in shape and that the tissue in the center appears different. The central part represents the hyphal cells of the haustorium that have torn away from the host.

Further development of the haustorium requires examination of sectioned or cleared parasite and host tissues (Figure 17.6b). As the hyphal cells of the haustoria enter the host, they will grow through the cortex until they reach vascular tissue. If haustoria are on the leaf, they will be attached to the veins, whereas if haustoria are on the stem, they will force their way through the cortical tissue and be connected to the host vascular cylinder. Although more details can be seen from slides prepared by sectioning the tissue with a microtome and then staining, free-hand sections are quick and easy and can show the overall structure of the haustorium. Procedure 17.1 gives the protocol for examining the *Cuscuta* haustorium. You should be able to see the haustorium in the host cortex. This portion of the haustorium that is inside the host is called the **endophyte**. If the haustorium is sufficiently mature, you can also see the tip of the endophyte branching. These cells are called **searching hyphae**. When they reach host vascular tissue, they connect with xylem and phloem cells. The *Cuscuta* vascular system originates at the parasite stele, traverses the haustorium and endophyte, and connects to the host vascular system. This appears as a thin strand composed of **tracheids** in the haustorium, but in the vicinity of the host stele, it appears as several branched strands. The host–parasite vascular system can be seen in great detail using free-hand sections as well as clearings (see Procedure 17.1).

Materials

Each student or team of students will need the following items:

- *Cuscuta* plants attached to the host
- Stereo and compound microscopes
- Glass slides and coverslips
- Double-edged razor blades snapped in half (these are sharper than single-edged razor blades)
- Forceps (fine tipped ones are best)
- Toluidine blue stain (1% in 50% ethanol/water, in dropper bottle)
- Wash bottle with distilled water
- 5% sodium hydroxide solution
- Safranin G stain (1% in 50% ethanol/water)
- Paper towels

Follow the protocol in Procedure 17.1 to complete this exercise.

Anticipated Results

Observations made in this experiment document the detailed structure and development of the haustorium of a parasitic flowering plant. The processes of haustorial initiation, host attachment, host penetration, and the establishment of xylem to xylem link-up between parasite and host can all be seen. Attachment of the searching hyphae to host phloem cannot be seen with free-hand sections; this requires light or electron microscopic methods. The genus *Cuscuta* obtains the contents of host xylem (water and minerals) and phloem (sugars, amino acids, etc.), thus it is a holoparasite. The morphological and physiological link between the two organisms is extremely intimate, and in effect, the two function as one organism.

Procedure 17.1
Morphology of the *Cuscuta* Haustorium

Step	Instructions and Comments
1	Locate a portion of the host plant that has coiled *Cuscuta* stems and attached prehaustoria. It is useful to find early (young) attachments as well as mature (older) attachments for comparative purposes. The next steps are easier if you use smaller host stems and leaf petioles (less than 4 mm in diameter).
2	With the razor blade, remove a 5-mm long section of host stem infected with dodder and place on a microscope slide. Add a drop of water.
3	Working under the stereo microscope, hold the host stem with haustoria with the forceps and carefully make a transverse section (also called a cross section). From the cut surface, move inward along the host stem and make another cut producing as thin a section as possible. Try to make a slice that tapers in thickness. Let this section drop into the water on the microscope slide. Take care that sections do not dry. If your first attempts are not very good, keep going to get a variety of sections to examine.
4	Remove all plant materials that are not a section. Add a drop of toluidine blue stain. Place a coverslip over the sections. Place the slide on a paper towel and, using the wash bottle, gently add water to one side of the coverslip and absorb it out from the other. This will wash away excess stain.
5	Place the slide on the stage of the compound microscope and view your sections at various magnifications.
6	To visualize just the vascular tissue of the host and parasite, a technique called clearing is used. The tissue is placed in a capped vial and covered with 5% NaOH in a 50°C–60°C oven overnight (or at room temperature for 2–3 days). The length of time in NaOH varies with different tissues; do not let the process proceed too far or else the plant material will dissolve! As the NaOH solution becomes discolored, change it with fresh NaOH. Rinse the tissue with water several times. Replace water with 50% ethanol. Stain with Safranin G (1% in 50% ethanol) for 2–24 h. Destain with 70% ethanol. The vascular tissue and other lignified cells should be red and other tissues light pink. The material can now be viewed using the compound microscope.

Questions

- Do prehaustoria form only on living hosts? You can test this by placing various objects such as glass rods, wooden dowels, dead tree branches, and so on in the path of the tendril. You may notice that the dodder will coil around vertical objects, but not horizontal ones? Why do you think this might be a useful adaptation?
- You probably noticed that *Cuscuta* forms attachments to its own stems. Do you think these haustoria are structurally and functionally the same as those attached to a host plant? Section some of these and compare to help develop your answer.
- The term used to describe the *Cuscuta* endophyte cells is "hyphae." This is the same term used to refer to the growing strands of fungi. Are these homologous structures? Why?
- The manner in which the haustorium grows through the prehaustorial tissue from the parasite stele is reminiscent of the development of branch roots in nonparasitic plants. What features are the same and what features are different between the two? Do you think *Cuscuta* haustoria are modified roots?
- The formation of a haustorial endophyte (from inside the prehaustorium) is apparently controlled by the presence of the plant growth regulator cytokinin (Tsivion, 1978). How would you design an experiment to test this?

Experiment 2. Seed Germination and Seedling Behavior

Dodder produces numerous flowers in dense cymose clusters (Figure 17.7a and b). These flowers have four to five sepals, petals, and stamens. The gynoecium is composed of two fused carpels. One of the most distinguishing features of the *Cuscuta* flower is the presence of fringed or cleft scales alternating with the stamens. The fruit is a capsule (Figure 17.7c), but it may be either dehiscent or indehiscent, depending upon the species. Each capsule contains up to four seeds that have very hard seed coats (Figure 17.7d). Because of this, dodder seeds need scarification to germinate (Hutchison and Ashton, 1979). This can be accomplished using chemical or mechanical means.

When the seed of *C. gronovii* germinates, a primary root emerges first from the seed followed by the stem (Truscott, 1966). This vestigial root does not have a root cap nor does it elongate or participate in providing water

FIGURE 17.7 (a) *Cuscuta campestris* (scaldweed) flowering stem twining around its host stem. (b) Inflorescence (cyme) removed to show flower details. (c) Fruits (capsules). (d) Seeds. (Courtesy of D. Nickrent.)

and nutrients as one sees in nonparasitic plants. In fact, this root and the base of the seedling will wither and will die as the shoot system grows and eventually attaches to a host. At this stage the dodder is not rooted in the soil and is an aerial parasite.

In *C. campestris*, maximum seed germination occurs between 30°C and 33°C, whereas no germination occurs above 39°C or below 15°C (Hutchison and Ashton, 1979). Germination does not appear to be dependent upon light; however, further development of the seedling is strongly influenced by light. *Cuscuta* seedlings respond to red light by the straightening of the hypocotyl hook. Once this hook straightens, the seedling stem grows actively and circumnutates. Seedlings that encounter no host will die, whereas those that make a haustorial connection will continue growing and developing. The growth rate (length increase per unit of time) and circumnutation rate (number of revolutions or arcs per unit of time) can be calculated from a time-lapse video.

Materials

Each student or team of students will need the following items:

Germination

- *Cuscuta* seeds
- Gloves
- Fume hood
- 50-mL beaker
- Concentrated sulfuric acid
- Petri dishes
- Filter paper circles
- Wood blocks (2) with attached fine-grade sandpaper

Cultivation of dodder seeds on greenhouse plants

- Germinated dodder seedlings
- 1.7-mL microfuge tubes
- Microfuge tube racks
- Pipettes and tips
- Potted coleus (*Solenostemon scutellarioides*) host plants

Time-lapse photography of circumnutation

- Germinated dodder seedlings in tubes in racks.
- Macintosh iPhone or iPad.
- Support stand for iPhone or iPad (many versions sold on the Internet).
- OSnap, free application. Excellent tutorial on how to use it here: http://www.osnapphotoapp.com/tutorials.php. This substitutes for a more expensive intervalometer, plus camera setup to obtain time-lapse video.

- Fluorescent lamps to provide supplemental lighting.
- Centimeter ruler.

Follow the instructions in Procedure 17.2 to complete this experiment.

Anticipated Results

It is expected that some, but not all, of the dodder seeds will germinate. With their hard seed coats, dodder can remain viable, but dormant in the soil for many years (Dawson et al., 1994). It is this viability and longevity that makes some species of *Cuscuta* formidable pests in agricultural situations. The time-lapse video should document that dodder seedlings (and actually all plants) are not stationary, but in fact are in constant motion, albeit very slow. If you conducted your time-lapse experiment in the laboratory without any host plants nearby, the circumnutation of the seedlings will show no directionality. Another study you can do is a time lapse, but this time put a potential host plant (e.g., coleus) near the seedling. Recent work has shown that *C. campestris* senses volatile compounds given-off by host plants and that its growth is then oriented in that direction (Runyon et al., 2006).

Questions

- The seeds of most *Cuscuta* are dormant and require scarification for germination to occur. An exception seems to be seen in *C. pedicellata* where no dormancy is documented (Lyshede,

Procedure 17.2

Seed Germination and Seedling Behavior in *Cuscuta*

Step	Instructions and Comments
1	Obtain about 40 dodder seeds. Two methods of scarification of the seed coat can be tried: chemical and mechanical.
2	While working in the fume hood, wear gloves, place 20 seeds in a beaker, and cover with concentrated sulfuric acid which will quickly turn dark. Allow to scarify for 15 min.
3	Decant the acid and thoroughly wash the seeds with distilled water. Do not let the seeds dry. Place 10 scarified seeds in each of two Petri dishes with moistened filter paper. Keep both the dishes at room temperature (about 25°C), but put one in the dark (a drawer) and the other in light on the benchtop. Within 5 days, the seeds will germinate producing the characteristic hook-shaped seedling.
4	Place the other 20 seeds between the two wood blocks with sandpaper. Rub the blocks against each other to scarify the seeds. Place the seeds in a beaker, rinse well with water, and place in Petri dishes as described in Step 3.
5	After 5 days, record the percent germination and appearance of the light and dark grown seedlings.
6	All germinated seeds (now seedlings) can be used in the next part of the experiment. Pipette 100 μL of water into as many microfuge tubes as there are germinated seeds. Place the seedlings, root end down, into the tubes. Note that the roots will remain undeveloped; the stem of the seedling will grow and circumnutate, "searching" for a host. Only by forming a haustorial connection to the host can this seedling survive.
7	Place half the seedlings (tubes in racks) in the greenhouse adjacent to a potted coleus host plant. Seedlings in nature may initially attach to small plants near the soil surface, and only later they are sufficiently robust to parasitize larger hosts. For this reason, it might be best to start the seedlings on small coleus plants (rooted from cuttings). At these early stages, it is also important to not allow the seedlings to dry out. Once the parasite is attached, it will become more vigorous and the tendrils can attach to larger plants.
8	The other half of the seedlings can be kept at ambient conditions (typical temperature, moisture, and light). For these, the circumnutation rate of the seedling stem can be documented using time-lapse photography. The instructions for using the OSnap application with your iPhone™ or iPad™ are clear and straightforward. Adjusting the timer interval, that is, the time between each photo that is taken, is the critical factor for recording movement in the seedlings. Start with one photo per minute and record the seedling's movement over at least 2 h. Be sure to plug in the power cord to your iPhone™ or iPad™ so that it does not run out of power during the experiment. A solid, neutral color background with a centimeter ruler should be placed behind the seedling. Illuminate the seedling with a cool light source (not incandescent bulbs that will cause heating).

1984). What are some possible reasons for this difference? Could there be an adaptive advantage to dormancy versus nondormancy in seeds from different species?

- It appears that in dodder, seed germination is entirely independent of any influence from the host plant. Do you know of any parasitic plants where the opposite is the case? A bit of Internet inquiry should help you answer this question.
- In *C. pedicellata*, seedlings undergo circumnutation for a few days, but after this they become prostrate and creep along the soil (Lyshede, 1985). Why do you think this happens? Could such behavior be an adaptation?
- Many dodder species have honeycombed seed coats when they are dry. When water is absorbed, the seed coat cells bulge out in fingerlike extensions. What possible role can you propose for the function of such structures?

Experiment 3. Growth and Development of Dodder on Different Hosts

One of the most common misconceptions about parasitic flowering plants is that most of them only grow on particular hosts, that is, they are host specific. Indeed, specialist parasites have evolved in all the major haustorial parasite clades (Table 17.1), but the majority of species are generalists. The two species of *Cuscuta* that are used in these exercises, *C. campestris* and *C. gronovii*, are generalists. That said, it is important to point out that the range of hosts parasitized in nature may differ from what is seen when the parasite is presented with hosts under controlled experimental conditions. Frequently, plants in which the parasite would never encounter in nature make perfectly suitable hosts. An example of this is the coleus plant (*S. scutellarioides*), which is native to Southeast Asia and Malaysia, outside the distributional range of *C. gronovii*.

Dodder is an excellent experimental organism because it is relatively easy to grow, plus it can be cloned. Having genetically identical individuals allows control in experiments where otherwise differences between individuals might interfere with the effects being measured. Moreover, hosts such as coleus are easy to propagate by cuttings, thus host clones are genetically uniform as well. A single dodder plant attached to one plant will spread to any available neighbors. Once attachment to the next plant is established, the dodder stem can be severed and the attached parasite will continue growing even though separated from the original host plant.

Although the dodder species can grow on many different host species, given a "choice," they will prefer some hosts over others. The research to date has not fully determined the cause for such host choice, but it has been suggested that dodder will preferentially parasitize hosts with higher nutritional quality. As mentioned in Experiment 2, *Cuscuta* seedlings can detect and discriminate among volatile compounds emanating from various host plants (Runyon et al., 2006).

The purpose of this exercise is to demonstrate how dodder can differentiate among different potential hosts and preferentially parasitize certain species. This experiment could be carried out with dodder seedlings such as the ones produced in Experiment 2, but as mentioned, the initial haustorial connection of the seedling is critical and dependent upon a number of environmental and host factors. For this reason, host preference studies will be conducted using dodder shoots already established on a host plant.

Materials

Each student or team of students will need the following items:

- A coleus plant that has previously been infected with dodder
- At least 10 species of plants representing a range of taxonomic groups and/or life forms
- Paper tags with string, permanent markers
- Single-edged razor blades

Follow the protocols in Procedure 17.3 to complete this experiment.

Anticipated Results

It is expected that the host range of *C. campestris* and *C. gronovii* will be quite large, parasitizing a wide taxonomic diversity of plants. In contrast, it might be expected that plants with dense pubescence or a very thick **cuticle** will present greater obstacles to parasitism. Another observation that is quite amazing is that a relatively small segment of dodder, excised from the mother plant, can survive for days under fairly severe (desiccating) conditions.

Questions

- In nature, dodder is often simultaneously attached to many different host plant species at one time. Each of these species has different cell surface features, different vascular anatomy, different defense chemistry, and so on, yet the single dodder individual can successfully form haustorial connections to all of them. What sorts of adaptations do you think dodder has evolved to allow it to be such a generalist parasite?
- Dodder stems are frequently seen attached to themselves; that is, they are self-parasitic. However, studies using one *Cuscuta reflexa* plant

Procedure 17.3

Growth and Development of *Cuscuta* on Different Host Plants

Step	Instructions and Comments
1	This exercise requires a coleus plant, grown to a fairly large size that has also been previously infected with dodder. Under most greenhouse conditions, a coleus cutting will be 0.5 m tall in 5 months. There should be sufficient numbers of dodder stems to allow segments to be removed and placed on other host plants. It should be pointed out that in *C. campestris* seedling attachment and normal twining will not occur if fluorescent light under greenhouse conditions in winter is given (Dawson et al., 1994); hence, natural sunlight supplemented with incandescent light should be used.
2	The number of potential hosts used here depends upon the diversity of plants available in your greenhouse. We suggest you present the dodder with both taxonomically diverse plants and plants with various growth forms. For the former, try ferns, gymnosperms, monocots, and dicots. For the latter, try glabrous versus pubescent plants, succulents, plants with milky latex, and so on.
3	The methodology for transferring dodder from one plant to another is straightforward. Using a razor blade, cut a segment of dodder stem approximately 5 cm long and secure it to another plant. Tags with strings are useful in that the plant number can be recorded and the strings are used to gently tie the dodder to the host plant. Be sure to make your greenhouse personnel aware of this experiment so that the dodder is not purposely or inadvertently removed (e.g., during watering).
4	Monitor the dodder cuttings at least once a day. Record any dodder plants that have died, been lost, are present, but not attached, and attached by haustoria. After about 2 weeks (or more), make your final observations, including information on which dodder cuttings have done well and which have not. Measurements of the size (length) of the dodder stems on these different hosts could be made.

attached to another injected with radioactively labeled sucrose showed that no labeled sugar moved to the parasitizing plant. Do you think there is any adaptive function to such haustoria?

- Dodder can detect volatile compounds coming from different host species and thus discriminate among hosts prior to forming haustorial connections. How do you think such a complex detection system has evolved?

Experiment 4. Dodder Is a Physiological Sink

In green plants, the photosynthates (sugars) are produced in leaf mesophyll cells, loaded into the phloem, translocated to other parts of the plant where the sugar is unloaded and used for metabolism (e.g., actively growing shoots) or stored (e.g., roots and fruits). This is accomplished by a process called **mass flow**, which explains the movement from **source cells** to **sink cells**. This experiment will show that *Cuscuta* attached to a host plant represents a very powerful physiological sink.

In some ways, the haustorium of *Cuscuta* functions like a combination of a mature leaf and a root. Like a root, it is involved in taking up water and minerals from the host xylem, moving these materials through the haustorium into the parasite vascular system. This water movement takes place within the tracheids, part of the **apoplast**. Inside the dodder haustorium, when a searching hypha contacts a host **sieve tube element**, it grasps, but does not penetrate, it like a hand with fingers. The sieve tube element is part of the **symplast**. Later, the contact hypha differentiates into an absorbing hypha which is functionally a specialized sieve tube element (with no companion cell). The ultrastructure of the parasite sieve element shows many features typical of transfer cells (Pate and Gunning, 1972), such as highly convoluted portions of the cell walls that are in contact with the host cell. Early radiolabeling and ultrastructural studies (Israel et al., 1980) showed the continuity between host and parasite phloem, and more recently interspecific plasmodesmata have been documented (Birschwilks et al., 2007). Thus, in terms of movement of sugars and amino acids in the phloem, the dodder haustorium functions like a leaf, moving these materials to parasite sinks (meristems, fruits, etc.).

Materials

Each student or team of students will need the following supplies:

- *Arabidopsis thaliana* (e.g., cv. Columbia)
- Coleus plants grown as described in Procedure 17.3 with a well-developed growth of dodder
- Paper envelopes

Procedure 17.4
Dodder as a Physiological Sink

Step	Instructions and Comments
1	The model organism *Arabidopsis thaliana* is a good host for *Cuscuta* (Birschwilks et al., 2007). Obtain seeds of *A. thaliana* (e.g., cv Columbia) and grow them in a growth chamber under short-day conditions to rosette stage. Next, transfer the potted plants to the greenhouse under long day (16 h light) to initiate flowering. Half of the *Arabidopsis* plants will be used for treatments and the others as controls.
2	Inflorescences in *Arabidopsis* typically initiate at about 28 days. When the inflorescence axis is a few centimeters above the rosette of leaves, inoculate half of the *Arabidopsis* host plants with cuttings of dodder. Be sure that each of the cuttings is equivalent in size (about 2–5 cm long). The dodder stems are obtained from an infected coleus plant as described in Procedure 17.3. Leave the other half of the *Arabidopsis* plants without dodder.
3	Let the experiment run for at least 3 weeks. From seed sowing, flowering in *Arabidopsis* typically lasts from day 32 to 49 and fruiting from day 48 to 61 (Boyes et al., 2001).
4	At the end of the experiment, record in a qualitative (descriptive) way how much dodder growth exists on your host plants. Also, count the number of developed fruits on the treatment and control plants. Finally, harvest the fruits (keeping those from each plant separate), place them in paper envelopes, dry them in an oven, and record the dry weight per plant of all fruits from the treatment and control plants.

- Drying oven
- Analytical balance

Follow the instructions provided in Procedure 17.4 to complete this exercise.

Anticipated Results

In studies with faba bean (*Vicia faba*), *C. campestris* was such an efficient parasite that it could divert the movement of nearly all phloem contents away from fruits (the typical sinks) and into its own tissues (Wolswinkel, 1974). This experiment should document the same phenomenon in *Arabidopsis*. It is expected that fruit number and fruit weight in the plants inoculated with dodder will be lower than those in the controls with no dodder.

Questions

- In green plants, the mass flow hypothesis requires the active transport of sucrose into the phloem at the source and active transport of sugars out of the phloem at the sink. The difference in sucrose concentration at the source and sink locations sets up an osmotic gradient and hydrostatic pressure, thus water (and sucrose) movement from source to sink. In Cuscuta, excising the parasite leaving only its coiled stem with haustoria attached to the host does not stop sucrose movement into the haustorium. How is this system different than what is seen in a green plant?
- Lateral buds that have actively growing meristems are considered sinks in typical green plants. As you may know, removing the apical meristem in a plant (decapitation) will change the concentration of auxin arriving at lateral meristems below and "release them" to elongate. Design an experiment using coleus plants to test which sink is stronger, lateral meristems or dodder.
- Until recently, it was not clear whether the symplast of dodder is connected to the symplast of its host. Past work has shown that host DNA and even viruses can be translocated from host to the dodder. More recent work has shown that dyes can move from host to parasite, but not a green fluorescent protein (GFP)—ubiquitin fusion (molecular size of 36 kDa). How are the two symplasts connected? What features of this connective structure might be involved in limiting the types of molecules moving between the two organisms?

REFERENCES

Baldev, B. 1962. In vitro studies of floral induction on stem apices of *Cuscuta reflexa* Roxb, a short day plant. *Ann. Bot.* 26:173–180.

Birschwilks, M., N. Sauer, D. Scheel, and S. Neumann. 2007. *Arabidopsis thaliana* is a susceptible host plant for the holoparasite *Cuscuta* spec. *Planta* 226:1231–1241.

Boyes, D., A. Zayed, R. Ascenzi, A. McCaskill, N. Hoffman, K. Davis, and J. Görach. 2001. Growth stage-based phenotypic analysis of *Arabidopsis*: A model for high throughput functional genomics in plants. *The Plant Cell* 13:1499–1510.

Costea, M. and F. J. Tardif. 2006. The biology of Canadian weeds. 133. *Cuscuta campestris* Yuncker, *C. gronovii* Willd. ex Schult., *C. umbrosa* Beyr. ex Hook., *C. epithymum* (L.) L. and *C. epilinum* Weihe. *Can. J. Plant Sci.* 86:293–316.

Dawson, J. H., L. J. Musselman, P. Wolswinkel, and I. Dörr. 1994. Biology and control of *Cuscuta*. *Rev. Weed Sci.* 6:265–317.

Fratianne, D. G. 1965. The interrelationship between the flowering of dodder and the flowering of some long and short day plants. *Amer. J. Bot.* 52:556–562.

Haidar, M. A., G. L. Orr, and P. Westra. 1997. Effects of light and mechanical stimulation on coiling and prehaustoria formation in *Cuscuta* spp. *Weed Res.* (Oxford) 37:219–228.

Hutchison, J. M., and F. M. Ashton. 1979. Effect of desiccation and scarification on the permeability and structure of the seed coat of *Cuscuta campestris*. *Amer. J. Bot.* 66:40–46.

Israel, S., I. Dörr, and R. Kollmann. 1980. Das Phloem der Haustorien von *Cuscuta*. *Protoplasma* 103:309–321.

Lane, H. C. and M. J. Kasperbauer. 1965. Photomorphogenic responses of dodder seedlings. *Plant Physiol.* 40:109–116.

Li, D. X., L. J. Wang, X. Yang, G. Zhang, and L. Chen. 2010. Proteomic analysis of blue light induced twining response in *Cuscuta australis*. *Plant Molec. Biol.* 72: 205–213.

Lyshede, O. B. 1984. Seed structure and germination in *Cuscuta pedicellata* with some notes on *C. campestris*. *Nord. J. Bot.* 4:669–674.

Lyshede, O. B. 1985. Morphological and anatomical features of *Cuscuta pedicellata* and *C. campestris*. *Nord. J. Bot.* 5:65–77.

Mathiasen, R. M., D. L. Nickrent, D. C. Shaw, and D. M. Watson. 2008. Mistletoes: Pathology, systematics, ecology, and management. *Plant Dis.* 92:988–1006.

Orr, G. L., M. A. Haidar, and D. A. Orr. 1996. Smallseed dodder (*Cuscuta planiflora*) phototropism toward far-red when in white light. *Weed Sci.* 44:233–240.

Pate, J. S. and E. S. Gunning. 1972. Transfer cells. *Annu. Rev. Plant Physiol.* 23:173–196.

Runyon, J. B., M. C. Mescher, and C. M. De Moraes. 2006. Volatile chemical cues guide host location and host selection by parasitic plants. *Science* 313:1964–1967.

Truscott, F. H. 1966. Some aspects of morphogenesis in *Cuscuta gronovii*. *Amer. J. Bot.* 53:739–750.

Tsivion, Y. 1978. Possible role of cytokinins in nonspecific recognition of a host and in early growth of haustoria in the parasitic plant, *Cuscuta campestris*. *Bot. Gaz.* 139:27–31.

Vaughn, K. C. 2002. Attachment of the parasitic weed dodder to the host. *Protoplasma* 219:227–237.

Wolswinkel, P. 1974. Complete inhibition of setting and growth of fruits of *Vicia faba* L., resulting from the draining of the phloem system by *Cuscuta* species. *Acta Bot. Neerl.* 23:48–60.

ONLINE RESOURCES

The Parasitic Plant Connection: http://www.parasiticplants.siu.edu/ (accessed May 9, 2014).

The Digital Atlas of Cuscuta: http://www.wlu.ca/page.php?grp_id=2147&p=8968 (accessed May 9, 2014).

The Angiosperm Phylogeny Website, Version 13: http://www.mobot.org/MOBOT/research/APweb/ (accessed May 9, 2014).

18 Abiotic Plant Disorders

Robert E. Schutzki, Bert Cregg, Tom Creswell, and Gail Ruhl

Concept Box

- Abiotic disorders are not caused by pathogens, insects, mites, or other biological agents.
- Abiotic disorders are often grouped by type of causal factor.
- The genetics of plants may make them inherently susceptible to certain disorders and damage.
- Environmental extremes are the most common contributor to abiotic disorders.
- Abiotic disorders may predispose plants to infection by pathogens or insect attack.
- Symptoms of abiotic disorders may closely mimic those caused by biotic agents.
- Diagnosis of the cause of abiotic disorders may be complex since causal factors may no longer be present by the time symptoms appear.
- Diagnosis of the cause of abiotic disorders requires collecting several types of information including:
    - Normal appearance of the plant involved
    - Symptoms present and time of appearance
    - Pattern of occurrence and whether other plants are affected
    - History of the plant
    - Site characteristics and site history
    - Documentation of cultural practices

The term "**abiotic** disorders" refers to a wide array of plant problems. We use the word "abiotic" to indicate that the symptom is not caused by a biological agent such as an insect, a mite, or a pathogen. Although not technically "abiotic," damage by mammals and birds is included here as another example of essentially mechanical injury that could be confused with disease symptoms. Abiotic disorders are associated with nonliving causal factors like weather, soils, chemicals, mechanical injuries, cultural practices, and, in some cases, a genetic predisposition within the plant itself. Abiotic disorders may be caused by a single, extreme environmental event (e.g., a sudden springtime freeze), by a complex of interrelated factors, or by chronic conditions such as a prolonged drought or planting an acid-loving species in an alkaline soil.

Abiotic plant problems are sometimes termed "physiological disorders," which reflects the fact that the injury or symptom we see, such as reduced growth or crown dieback, is ultimately due to the cumulative effects of the causal factors on the physiological processes needed for plant growth and development. When a tree is affected by severe drought, for example, water stress will result in the closure of the pores, or stomata, on the leaf. This conserves water in the leaf but also reduces the rate of photosynthesis and the ability of the plant to produce sugars for growth. If the drought stress occurs during hot weather, stomatal closure also limits the cooling effect of transpiration, and leaf scorch may occur. Nutritional imbalances also limit growth by reducing photosynthetic rate and other physiological and metabolic processes. Some plants, such as pin oak, have a limited ability to take up iron under alkaline soil conditions. Iron is essential for the synthesis of chlorophyll, so pin oaks on alkaline soils frequently develop severe leaf yellowing or chlorosis and have reduced rates of photosynthesis.

The cumulative and subtle nature of many physiological disorders can often make them difficult to diagnose. An extreme event such as a severe late freeze or a misapplied herbicide is an obvious "smoking gun" to indicate the underlying cause of injury. More often, however, diagnosing abiotic problems requires careful consideration of plant and site factors through a process of elimination to determine the source and potential remedy for the problem.

Abiotic disorders are usually classified by causal factor or symptom. In this chapter, we present abiotic disorders whose origins are due to biological/botanical factors, environmental (climate and weather) conditions, soil conditions, chemical applications, mechanical damage, and cultural practices. We also provide a framework for diagnosing problems and suggest steps to mitigate abiotic injuries before or after they occur.

BIOLOGICAL/BOTANICAL ABNORMALITIES

The genetic makeup of a species is expressed through its physiological and morphological characteristics. Whether these characteristics are perceived as normal or abnormal depends on the individual circumstances and location where they are found. Understanding the basis for these characteristics can be useful in determining whether they need to be addressed and can aid in outlining a course of action. In many cases, these characteristics are a fact of nature and simply need to be appreciated for what they are; in other cases, they can be altered or corrected by simple measures. Here are some common abiotic abnormalities that are biological/botanical in origin.

Genetic Mutations and Reversions

Genetic anomalies are prevalent in the plant world and often garner much attention. For example, seedling growers often find mutants in their nursery beds. Albino seedlings are common seedling mutants. Because they lack chlorophyll, these seedlings quickly die once they exhaust energy reserves from the seed. However, many mutants are stable and sustain growth. These stable mutants can exhibit unique form, foliage, flower, stem, and other characteristics that warrant their development and introduction into the trade.

For unknown reasons, some cultivars that originate as mutations exhibit a tendency to produce shoots that revert to their species type. Genetic reversions are common on dwarf Alberta spruce and Harlequin Norway maple (Figure 18.1). To maintain the cultivar characteristics, genetic reversions need to be pruned out quickly and completely when they are observed.

Chimeras

Chimeras are botanical abnormalities that are often confused with nutritional or chemical disorders. "Chimera" is a term used to describe a single plant with two genetically different tissue types. Leaf variegation is the most common example of chimeras in plant species. The difference in foliage color and the distribution of those colors are due to cell mutations in the meristematic tissue layers. Chimeras can be stable, in that the genetic differences are consistent and reproducible. These chimeras have spawned numerous variegated plant cultivars. Chimeras can also be unstable and unpredictable, surfacing sporadically on either shoots or individual leaves. The Bumald spireas are known for the sporadic appearance of unstable leaf chimeras (Figure 18.2). The stability of the chimera depends on the tissue layer where the mutation occurs. If you discover variegation on an individual stem and not over the entire plant, it is possible that it is a chimera and not an abiotic disorder related to plant nutrition or chemical injury.

FIGURE 18.1 Genetic reversion on "Harlequin" maple.

FIGURE 18.2 Chimera on "Anthony Waterer" spirea.

Leaf Abscission and Retention

Deciduous leaves emerge in the spring, function during the growing season, turn colors, and drop in the fall. There are times when leaf drop or retention is normal and times when they may be related to a plant problem. Leaf **abscission** or drop is triggered in the fall by seasonal changes in day length. Chlorophyll degrades and the plant hormones abscisic acid and ethylene cause an abscission layer to be formed at the base of the petiole. If a stem or leaf is injured or dies before these normal seasonal changes occur, abscission layers will not form and leaves will remain on the stem. Leaves remaining on isolated stems in the fall or winter may point to problems such as stems girdled by an insect borer, mechanical injury, or an early killing frost. In some deciduous plant species, retention of dead leaves (**marcescence**) is a normal characteristic. Typical examples of genera exhibiting marcescent leaves are oaks (Figure 18.3) and beeches. Species within these genera do not complete the formation of the abscission layer until the following spring. Leaves will fall when bud break occurs and stems resume growth. If leaf retention is of concern, identify the plant species, check for evidence of stem injury, and review recent weather conditions.

FIGURE 18.3 Leaf retention on oak.

Leaf drop also occurs on evergreens; it just happens with different aged leaves or needles, which usually function for two or more years, depending on the species. Normal leaf/needle yellowing and abscission occur on the interior of the plant, but yellowing or browning of needles on the shoot tips and in needles less than 3 years old is a sign of problems. Prolonged heat or drought stress, nutritional imbalance, and poor soil conditions are among the abiotic factors that can lead to early loss of evergreen leaves. Because evergreen needles have a waxy coating and normally function for 3 or more years, symptom appearance may be delayed by months following an injury or stress.

Graft Incompatibility

Many cultivars of woody ornamentals are propagated by grafting. Graft incompatibility is due to the failure of the bud or graft union between a scion and an understock. This may be due to a variety of causes including poor grafting technique resulting in a misalignment of scion and understock, viral or fungal infections, or large differences in growth rate between scion and understock. The failure of the graft union disrupts translocation in both the xylem and phloem and influences plant performance with age. In production systems, signs of graft incompatibility may be visible as an uncharacteristic intensity of the fall color. Disruption of translocation leads to a build-up of sugars during leaf senescence that intensifies fall foliage color. This is especially the case on *Acer rubrum* cultivars with known graft incompatibility problems, which eventually led to production on their own roots. Graft incompatibility remains a concern in fruit tree production. Incompatibility is not easily detected and can surface at various times in a plant's life. Unfortunately, for many, it surfaces when the trees are mature and affected by wind or some other environmental condition. Graft incompatibility can result in semimature trees snapping off at the base in windstorms. Failure of the tissue union is visible at the trunk/root collar. On plant species with known problems, check on the budding or grafting practices during production. Another form of incompatibility occurs when there are differences in growth rates between the scion and the understock. These differences can lead to disproportional growth at the junction of the two tissues.

Galls and Burls

Galls and burls are abnormal growths on woody plants. They can occur on branches, trunks, and, in some cases, on roots of woody plants. Galls may be caused by an insect or disease; however, the origin or causal factors for burls and other callus outgrowths are unknown. Abnormal growths form through a proliferation of cells. These cells continue to develop and do not differentiate into normal tissue types. In some cases, the outgrowth results from a cluster of shoots (Figure 18.4). Each year, new shoots are initiated, and the layers of the mass increase.

Branch Architecture

Genetic make-up contributes to branch architecture and the angle of attachment in plant species. Branch angle is a predictor of wood strength and a plant's ability to withstand compromising environmental conditions. Even though the loss of branches may be due to wind or ice load, the genetic predisposition for weak branch or wood strength is the inherent basis for the problem. Common knowledge tells us that vigorous upright branches with narrow attachment angles are prone to storm damage. Horizontal branch angles offer potentially greater wood strength. To ensure the stability and longevity of species with narrow branch angles or multistemmed upright crowns, corrective pruning should be performed on a semiregular schedule. Consider cabling or bracing on plants too old for corrective pruning.

ENVIRONMENTAL (CLIMATE AND WEATHER)

Climate and weather are certainly unpredictable as are their impacts on plants. Climate and weather present a complex of environmental conditions and encompass a range of potential injuries. Fortunately, experience and our ability to learn from the past aid our preparation. Understanding the impacts of climate and weather on plants involves plant science, a little meteorology, a dash of physics, and maybe some chemistry. These disciplines set the foundation for dealing with the interrelated environmental factors of temperature, light, precipitation, and airflow and their potential direct and indirect effects.

FIGURE 18.4 Trunk spouts leading to callus knob formation.

Low-Temperature Injury

Low-temperature impacts on plants can be classified as chilling or freezing. **Chilling** injury or chilling stress, as it is often termed, is associated with temperatures above freezing, but low enough to cause injury. Chilling injury is due to a sudden drop in temperature during an active period of plant growth or development. The greater the drop in temperature from normal, the more likely injury will occur. The extent of the injury may cause a reduction in photosynthesis or other metabolic processes, an alteration in growth or death of the exposed plant tissue. For example, a substantial drop in temperature during new shoot growth and leaf expansion may result in damage to the terminal meristem and a loss in shoot growth. Other signs of chilling injury may include wilting, desiccation, and/or a physical distortion of plant parts. Chilling injury is less serious than freezing injury on most herbaceous and woody plants, and plants can usually grow out of the damage.

Freezing injury, which is caused by subfreezing temperatures, is easier to identify and much more damaging than chilling injury. The impact of subfreezing temperatures may be felt from as little as one overnight episode or from a more sustained exposure episode. Plants subjected to subfreezing temperatures may exhibit ice formation in the vascular vessels, in the spaces between the cells and/or within the cells. The extent and pattern of ice formation vary with the plant tissue and its condition. Ice formation in the spaces between cells is usually not fatal and causes water to leave the cells, leading to dehydration or desiccation. If the freezing episode is cold and long enough, it can result in freezing within the cells. Internal or intracellular freezing ruptures membranes and leads to cell death. Freezing damage begins with dehydration and ultimately ends in plant tissue death. Typical signs of freezing injury are a blackened/brownish discoloration or bleaching of plant tissue (Figure 18.5). The severity of the freezing damage depends on the type of frost injury and the plant parts affected. If the freezing injury kills a significant number of buds or cambial tissue, the plant may die outright or suffer so much crown dieback that it becomes unusable or unacceptable in the landscape. More commonly, however, freezing injury may be limited to flower buds and minor shoot dieback. This damage may require corrective pruning and time to allow the plant to grow out of the damage.

Understanding low-temperature injury begins with an examination of cold hardiness. Cold hardiness refers to a plant's ability to withstand cold temperatures without sustaining injury. The level of hardiness that a plant can attain is based on genetics, preseason conditioning, and its current condition. The genetic origin of a plant contributes to its ability to "harden." Plants of a given species originating from warmer climates may not be as cold hardy when moved into northern regions as the same

FIGURE 18.5 Late spring freeze injury on yew.

species originating from northern regions. The same is true for heat tolerance when northern plants are moved to the South. Origin, not production location, is the governing factor in determining hardiness capability.

Preseason conditioning refers to the physiologic and metabolic changes that occur within a plant as it begins hardening. The process leading to cold hardening is generally considered a three-stage process. The first stage occurs at the end of the growing season. Growth has ceased; plants have formed terminal buds; and carbohydrates are accumulating in stem and root tissues. Short days and cool night temperatures of early autumn begin the acclimation process. During acclimation, metabolic changes occur within the plant, which permit it to withstand lower temperatures. During the initial stages of acclimation, however, hardiness levels may be lost as easily as they were obtained if temperatures rise or other environmental factors promote the resumption of growth. Acclimation continues and hardiness tolerance increases with consistent exposure to temperatures at or slightly below freezing. The final stage, the deepest level of cold hardiness, is achieved after prolonged exposure to subfreezing temperatures. Plants are said to be at midwinter levels when they reach their maximum cold tolerance for a given season. Maximum levels of hardiness are usually achieved by early January in the Northern Hemisphere. As spring approaches, plants begin to de-harden or lose cold tolerance with increasing temperatures. Late frosts, especially those following unseasonably warm early spring weather, can cause injury during this de-hardening phase when plants begin to break bud and initiate shoot growth. Remember that plant parts differ in their cold tolerance. Flower buds usually do not achieve the hardiness levels attained by vegetative buds and stem tissue.

The last factor contributing to a plant's ability to achieve maximum cold hardiness is plant health. Healthy plants achieve their cold hardiness potential; stressed plants may not. It is important to note that plants stressed by drought, flood, nutrient deficiencies, transplant shock, or pest problems may not become fully hardy because they may not be able to produce or store enough carbohydrates to attain levels necessary to avoid low-temperature injury. The best insurance for hardiness is plant health. In summary, genetics set the foundation level for low-temperature tolerance of a plant. Preseason conditioning and plant health determine whether a plant can approach these levels.

Sunscald and Frost Cracking

Sunscald and frost cracking are the results of the interaction of light and temperature (Figure 18.6). They are caused by thawing and freezing due to a rapid fluctuation in stem temperatures. Exposure to afternoon sun causes an increase in stem temperature and the subsequent thawing of stem/trunk moisture. As the sun sets, temperatures drop rapidly. If the temperature drops below freezing, ice crystals rupture internal tissue. Sunscald and frost cracking occur from the same causal conditions. In sunscald, the cambium, phloem, and xylem are damaged, a sunken area appears on the trunk, but the bark is not split. Frost cracking exhibits the same internal damage with vertical splits in the bark. The splits can reopen and close with changes in air temperature. As temperatures increase in spring, the bark tissue dries and the cracks remain open. Thin-barked plants are prone to sunscald and frost cracking especially when subjected to southwest exposures. Tree wraps can minimize sunscald and frost cracking, but plants with thin bark may always be susceptible to this problem if environmental conditions favor temperature fluctuations.

Winter Desiccation

Winter desiccation is not considered a direct result of low temperature, but originates from the interaction of temperature and wind. As the name implies,

FIGURE 18.6 Frost cracking of tree bark.

winter desiccation is actually a form of drought stress. Desiccation occurs when water absorption by the roots cannot replace water loss through the foliage, buds, and stems. Transpiration and evaporation from plant parts increase when temperatures rise above freezing. This is especially true on bright, sunny, windy days. If soils are frozen, water absorption cannot replenish plant tissue moisture, and dehydration and eventual desiccation injury occur. Signs of desiccation begin with leaf curl or wilting and progress to a browning of leaf margins and/or bud scales. If conditions are severe or persistent, browning will consume the entire plant part. If snow cover is present, a telltale snowline above which damage occurred can be observed on the plants. Desiccation usually affects the foliage of broad- and narrow-leaved evergreens (Figure 18.7); buds and stems for the most part are not affected unless adverse conditions are prolonged or compounded by freezing injury.

Desiccation injury varies with plant species, plant part, soil moisture content, depth of frozen soil, snow cover, and wind velocity. Desiccation is common on narrow- and broad-leaved evergreens. In colder climates, it occurs for the most part when the depth of frozen soil increases in mid- to late-winter. Desiccation injury can be minimized or prevented by protecting plants with screening or applying an antitranspirant. Wrapping or screening not only protects from wind but also helps minimize the degree of temperature fluctuations on the plant parts. Wrapping protects the entire plant; screening is usually positioned to intercept the prevailing winds. Wrapping may be the smart horticultural alternative on prized ornamentals or in high-profile landscapes. Antitranspirants are another viable means of protection, but keep in mind that they must be applied and allowed to dry at temperatures above freezing. Application should be applied before anticipated injury, usually by early January in the Northern Hemisphere. Degradation of the antitranspirant due to weather may require reapplication to extend the protection through the critical period.

FIGURE 18.7 Winter desiccation of broad-leaved evergreens.

Frost Heaving

Alternate freezing and thawing of soils and the resulting expansion and contraction can result in soil heaving and exposing the roots to cold and desiccation. This damage is especially a problem on newly planted or shallow-rooted plants. Fall-planted ground covers, perennials, and small container shrubs are highly vulnerable. Ground cover plugs and small container plants should be fully planted in soil. This may sound obvious, but with a tendency to not plant deep enough, these plants are often found planted in the mulch rather than surrounded by soil. In addition, a uniform mulch layer aids in preventing rapid soil temperature fluctuations. In larger container plants, planting high in the mulch layer may expose the upper portion of the root system enough to influence plant quality the following spring. Proper planting and mulching techniques are the best defense against frost heaving.

Snow and Ice

Ice and snow accumulation on branches can cause internal splits or cracks, bark tearing, and/or breakage. Plant architecture, branching structure, and wood strength can be predictors of damage from excessive snow or ice load. Unfortunately, hidden cavities in the wood or flaws in branch attachment can increase the damage potential. Plants with multiple stems can be tied to provide support.

This may be especially helpful on upright narrow-leaved evergreens. Excessive amounts of snow can be carefully removed to alleviate the stress on lateral branches. Ice should be allowed to melt naturally. Branches become extremely brittle when laden with ice. Mechanically removing the ice may increase the amount of injury. Once the ice has melted, the branches will usually return to their normal positions (Figure 18.8).

Drought and Heat

Moisture stress and extreme heat can greatly reduce plant growth, especially caliper growth. Heat can directly injure plants in extreme conditions, but the principal effect of heat is increased water loss and plant moisture stress. Moisture stress can cause significant mortality, particularly soon after transplanting. Wilting is the most common initial symptom of drought stress. Hardwood leaves and conifer leaders will curl or droop. As drought progresses, hardwood trees may begin to shed leaves, and conifers may begin to drop needles. Drought stress, however, reduces growth before visible symptoms such as wilting or leaf shedding become apparent. Even moderate stress can cause the stomata on the leaves to close, reducing photosynthetic production. Caliper growth is highly sensitive to water stress because cell turgor is required for radial expansion of new wood cells formed by the cambium. In fact, the high sensitivity of radial growth to moisture forms the basis of dendrochronology, the science of using tree rings to reconstruct past climates. Trees and shrubs with existing infections of *Botryosphaeria* species and similar dieback pathogens may decline very rapidly during drought stress due to a combination of the disease-causing disruption to the vascular system and a more rapid spread of the pathogen in the stressed plant.

Drought and heat stress can also cause some plant species to enter into an imposed dormancy. This mild state of dormancy can satisfy the chilling requirements of some species, causing isolated out-of-season flowering when the stress is relieved. Sunburn may also result when leaves that developed and acclimated in shaded conditions are exposed to bright sunlight by pruning or removal of nearby trees or as the upper leaves of a plant wilt during hot, dry periods (Figure 18.9).

To understand plant–water relations, it is important to remember that water forms a continuous path from the soil through the plant and into the atmosphere. This is often referred to as the soil–plant–atmosphere continuum. Any factor that reduces water uptake from the soil such as low soil moisture, loss of roots during transplanting, or frozen soil, or that increases evaporative demand and transpiration to the atmosphere such as high temperature or low humidity, can increase plant moisture stress.

Flooding

Although drought is a more common stress factor for plants, excess water can also be a problem. Plants vary widely in their tolerance of flooding. Many bottomland species can survive for months with their roots underwater, whereas other species may be killed by only a few days of inundation. Damage from winter flooding when soils are frozen may be minimal compared with similar water levels during active growth. Flooding causes several problems for plants that are not flood-tolerant. Without oxygen, roots begin to undergo anaerobic respiration, which results in the production of toxic compounds in the plant. Plants undergoing water stress may exhibit symptoms such as

FIGURE 18.8 Birch is often temporarily bent by snow and ice loads on branches.

FIGURE 18.9 Sunburn on summer foliage, which occurred after the viburnum was pruned in mid-summer.

wilting that are similar to symptoms of drought stress. Premature fall color is typically a sign of moisture stress, either too much or not enough. Excessive rain, standing water, and poor drainage in late summer can initiate premature fall color on affected plants. The same symptom exists for plants in severe drought stress. One of the prime indicator species for these types of summer moisture stress is *Euonymus alatus* "Compactus," commonly known as burning bush.

Lightning and Hail

Summer storms often bring lightning and, in some cases, hail. Trees in particular can be killed outright or severely damaged by lightning strikes. When lightning strikes a tree, it can fracture and splinter the wood or travel down from the entry point to the ground, carving a groove in the conductive tissue and bark (Figure 18.10). The extent of the injury depends on the intensity of the lightning, where it enters, the moisture level in the tissue, and the wood characteristics. Recommendations for treating lightning strikes are relatively simple: clean the damaged areas, remove splintered wood, cut loose bark back to firmly seated bark, and provide cultural practices that invigorate and promote plant growth, such as irrigation and balanced fertilization. Hail shreds and tears susceptible plant parts. It affects leaves, young stems, and branches, and in severe storms with large hail, it can damage or bruise bark. Damage from hail is usually immediate and short term. No treatment is necessary unless light pruning is needed to remove broken stems and branches.

FIGURE 18.10 Lightning strikes can result in extensive damage to underlying cambium.

SOILS

An understanding of plant–soil relationships is key not only to solving problems, but avoiding them in the first place.

Soil Type

"Soil texture" refers to the mixture of sand, silt, and clay particles in soil, and influences aeration, water retention, drainage, and nutrient-holding capacity. Soil texture and its relationship to water movement through soils is one of the most important factors leading to abiotic plant problems. We often hear the phrase "moist, well-drained soils" when referring to plant preferences. Unfortunately, this condition is not the rule in most planting situations. Sandy soils need amendments to increase water and nutrient-holding capacity. Clay soils need alterations to facilitate drainage. A thorough examination of physical and chemical properties of soil can aid in avoiding common problems associated with water deficits, flooding, and nutrient disorders.

Nutrient Problems

Nutrient problems rarely kill plants outright, but proper nutrient management is essential to optimizing growth and maintaining high-quality plants. Most nutrient problems are related to nutrient deficiencies (Figure 18.11), although nutrient toxicities may occur, especially in acidic soils. The extent and nature of nutrient deficiencies can depend on several factors, including soil type or planting media in production systems, site conditions in the landscape, and the plant.

Seventeen elements have been identified as essential for plant growth. Plants obtain carbon, hydrogen, and oxygen from air and water. The remaining elements are derived from soil or from a fertilization program. All of these are involved in one way or another with key physiological functions in the plant. For example, nitrogen and sulfur are components of essential amino acids needed to build proteins. Magnesium is a key element in chlorophyll molecules, and iron is involved in chlorophyll synthesis. In theory, any essential element may be limiting for growth. As a practical matter, only a handful of elements are commonly limiting in field nursery production and landscape systems.

Among the most common elemental deficiencies observed are nitrogen, phosphorus, potassium, iron, magnesium, and manganese. Iron and manganese deficiencies are often linked to specific plants and to alkaline soil. Nutrient deficiency symptoms include overall loss of vigor, reduced shoot growth, general yellowing or chlorosis of the leaves, interveinal chlorosis, marginal necrosis, and in severe cases, total leaf necrosis (Table 18.1).

Nutrient toxicities are less common in plants than nutrient deficiencies, but they can occur, particularly with some micronutrients. With elements such as boron or copper, the difference between sufficiency and toxicity is relatively small. Caution should be taken when using micronutrients in fertilization programs. If a soil test or foliar analysis indicates these elements are needed, one strategy is to apply them as split applications to reduce the possibility of toxicity. Nutrient deficiencies and toxicities need to be confirmed by a soil and leaf tissue analysis.

(a)

(b)

FIGURE 18.11 (a) Interveinal chlorosis of pin oak due to iron deficiency. (b) Manganese deficiency of red maple.

Salinity is the concentration of water-soluble salts in the soil or production medium. These salts are necessary for plant growth and development, but soluble salt levels in high concentrations can cause toxicity problems. Typical salts causing problems on plants are chlorides, sulfates, and ammonium. Soil or medium salinity is determined through electrical conductivity measurements. The most effective way to alleviate the negative impact of high soluble salts is through leaching. Leaching is effective in container production systems, but on field soils and in the landscape, the effectiveness of leaching depends on soil drainage.

Soil pH refers to the acidity or alkalinity of the soil. Soil pH influences the availability of nutrients in soil, the solubility of some nutrient elements, and the activity of soil microorganisms. It is generally accepted that the optimum pH for plant growth and the availability of a broad spectrum of nutrients is between 5.5 and 6.5 (slightly acid). As pH drops below or rises above this range, the solubility/availability of some nutrient elements may become limiting or toxic. Iron and manganese are known to be limiting at soil pH levels above 7 (levels above 7 are alkaline). Aluminum, manganese, and copper can be toxic at soil pH levels below 5.5 (acidic). Knowing the pH of soils and container media is essential for maintaining adequate growth and development as well as for diagnosing abiotic disorders.

CHEMICALS

Exposure to various chemicals can cause either chronic or acute plant injury, depending on the type and duration of the exposure.

TABLE 18.1
Symptoms of Nutrient Deficiencies in Plants

Nutrient	Symptoms
Nitrogen (N)	Chlorosis in older leaves, leaves smaller-than-normal, stunted plants
Phosphorus (P)	Purple-to-red leaves, smaller-than-normal leaves, limited root growth
Potassium (K)	Chlorosis in leaves, marginal chlorosis in older leaves
Magnesium (Mg)	Chlorosis in older leaves first, chlorosis may be interveinal
Calcium (Ca)	Leaf distortion such as cupping of leaves; fruit of some plants may rot on blossom end
Sulfur (S)	Chlorosis in young leaves
Iron (Fe)	New growth chlorotic, chlorosis often interveinal, major veins may be intensely green
Zinc (Zn)	Alternating bands of chlorosis in corn leaves, rosette, or "little leaf"
Manganese (Mn)	Interveinal chlorosis may progress to necrosis
Boron (B)	Stunting, distorted growth, meristem necrosis
Copper (Cu)	Chlorosis of tips of turfgrass, rosette of woody plants such as azalea

Salt Injury

Winter brings snow and ice and subsequently the need for deicing materials. Salt used for deicing roads and sidewalks affects plants in the landscape when it accumulates on stem tissue by airborne sprays and/or in soils because of runoff from treated areas. Salt buildup in soils is especially a problem in landscape beds adjacent to sidewalks or curbs where snow accumulates through plowing and shoveling. Deicing salts damage plants through direct sodium chloride toxicity, dehydration due to osmotic stress, reduced cold hardiness due to salt buildup on buds and stems, and/or its influence on soil nutrients. Salt injury usually expresses itself as terminal shoot dieback, bronze foliage on evergreens (Figure 18.12), bleaching of ground covers and turf, or total plant death in the case of ground covers and perennials.

Successive years of dieback on woody trees and shrubs could result in a "witches, broom" appearance on the outer branches. In regions with harsh winter conditions, it is essential to deal with deicing in the landscape design phase. Deicing practices must be considered in plant selection and landscape bed design in close proximity to paved surfaces. For existing landscapes, salt damage can be reduced by screening against salt spray and by using more plant-friendly deicing materials such as calcium magnesium acetate.

Herbicides

Herbicide injury symptoms vary with the plant and the herbicide applied. Understanding the mode of action of herbicides and their effects on nontarget plants will aid in recognizing injury symptoms and distinguishing herbicide injury from other abiotic and biotic problems. Three widely used types of herbicides are growth regulators, photosynthetic inhibitors, and plant enzyme inhibitors. Growth regulator-type herbicides such as 2,4-D are synthetic plant hormones that act by causing irregular growth. Essentially, weeds grow themselves to death.

FIGURE 18.12 Salt spray injury on white pine.

Symptoms of plant growth regulator herbicides' action on weeds include leaf cupping and twisted, distorted growth. These same symptoms are found on nontarget broadleaf plants. Photosynthetic inhibitors such as atrazine and simazine act by interrupting normal photosynthesis. These compounds block the mechanism that transforms sunlight into energy. Photosynthesis inhibitors require sunlight to work. Injury symptoms that appear on older leaves first include leaf yellowing and interveinal chlorosis. Plant enzyme inhibitors such as glyphosate are translocated in the plant and act upon enzymes that synthesize amino acids, protein, and other plant compounds. Injury symptoms including a yellowing and browning of leaves do not appear immediately. Some herbicides will cause root swelling and stunted shoot growth.

If herbicide injury is suspected as a causal factor, identify the compound applied, review its mode of action, and determine its persistence in the soil. This information will be useful in identifying proper treatment. Herbicide injury produces many of the same symptoms as nutrient deficiencies or toxicities (Figure 18.13). Nutrient deficiencies and toxicities can be eliminated as causes by leaf tissue and soil analysis.

Pesticides

Insecticides, fungicides, and other chemicals used in the control of plant pests can cause problems on both targeted and nontargeted plants. Several factors can influence whether the pesticide application is the solution or becomes part of the problem.

FIGURE 18.13 Growth regulator herbicide injury on grape. (Courtesy of Bruce Bordelon, Purdue University.)

As with any chemical application, injury will depend on the type, concentration, and application rate of the compound; plant species, its stage of growth, and physiological condition; and atmospheric conditions including light, temperature, humidity, and wind. Finding the perfect time to apply pesticides is always a challenge. Proper handling of the pesticides and close monitoring of weather conditions can avoid application-related injuries. Application injuries exhibit the same symptoms as nutrient disorders and other chemical applications. In most instances, leaf chlorosis, marginal and/or spotted necrosis, and total leaf necrosis are the visible symptoms.

Air Pollution

In some parts of the country, air pollutants can cause acute or chronic injury to plants, with symptoms usually appearing on the foliage. Injury symptoms include interveinal necrosis, marginal or tip necrosis, white or brown flecking or stippling on the leaf surface, and various degrees of chlorosis. The extent of the injury is very much dependent on other environmental and atmospheric conditions. Other factors include the type, concentration, and the length of exposure to the pollutant; the plant species, its stage of growth, and physiological condition; and atmospheric conditions.

Ozone is one of the most common pollutants that can damage plants. Ozone is produced in the atmosphere through a photochemical reaction between volatile hydrocarbons and nitrogen oxides in the presence of light. Ozone injury, if it is an issue, will generally occur in the summer, when temperatures support reaction rates. Sulfur dioxide is another pollutant that can injure plants. Sulfur dioxide emissions result from the burning of coal and other petroleum products. While improved filtration technology and emission controls have substantially reduced sulfur dioxide injury, isolated cases of volatile chemical spills or emissions may subject plants to injury. Such cases are easily diagnosed because of the close proximity of the plants to the isolated event.

MECHANICAL DAMAGE

While direct mechanical injury caused by birds, mammals, and humans may be technically considered "biotic," it is helpful to group them with other mechanical damage of abiotic origin for diagnostic purposes. Damage to the base of trees and shrubs is a common problem where string trimmers are used incorrectly. These wounds may open the plant to decay organisms and further damage. Trees planted too near driveways often suffer damage to branches from cars as they spread and obstruct the drive.

Many of our ornamentals become prime targets for deer, rabbits, squirrels, mice, and other rodents. Some ornamental species such as tulips, hostas, and taxus are preferred over natural forage. Deer browse is obvious on taxus, arborvitae, and other evergreens. Squirrels may feed on the petioles of newly expanded leaves or gnaw on the upper sides of branches within tree canopies (Figure 18.14). Rabbits, mice, and other rodents focus on the ground-level portions of deciduous trees and shrubs. Habitat reduction is the first-line of defense against small rodents. Unfortunately, many of our prized plant compositions also make for ideal cover. Minimizing cover may deter feeding but does not eliminate it. Caging plant crowns, wrapping wire mesh around the base of the plants, applying feeding repellents, and using poison baits work reasonably well. Quite often, our best bet is to apply multiple methods in various combinations. When repellents are used, keep in mind that they degrade quickly and need to be reapplied throughout the season. Poison baits are often used for mice and should be positioned to avoid feeding by other animals.

CULTURAL PRACTICES

Cultural practices are designed to promote and maintain plant growth and development. The same practices designed to promote can at times be the cause of abiotic problems.

Planting Process

The planting process involves preplanting examination and evaluation of the plant stock type (bare-root, container, balled and burlapped [B&B], or mechanical tree spade), the actual physical process of planting and the follow-up maintenance. Problems related to the planting process can originate within each of these. Consider the idiosyncrasies with each of the previously mentioned stock types, for example, the depth of the trunk/root collar on balled and burlaped (B&B) trees, encircling roots in container grown plants, root desiccation on bare-root plants, and glazing in mechanical tree spade plantings. Each one of these factors can contribute to the success or failure of the plants. Improper handling and planting procedures contribute

FIGURE 18.14 Squirrel gnawing leads to extensive stem injury that mimics a biotic canker.

significantly to abiotic consequences, especially when they are combined with soil or environmental limitations at the planting site. Symptoms related to poor planting procedures are similar to those of drought and flooding: shoot dieback, reduced leaf size, minimal shoot growth, root injury, and poor root regeneration. If plant excavation is possible, examine the root system for white root tips and signs of root regeneration.

Under drought or water deficits, white root tips will be absent, and existing roots will be dried and shriveled. Under excessive moisture and poor drainage, the root system will also lack white root tips and exhibit signs of anaerobic conditions. The blackened outer surface of the roots will slough off, exposing inner gray and water-soaked stained tissue. In addition to the obvious plant symptoms, evidence of twine around the base of the trunk, scars from staking, or other signs of mechanical injury may lead to conclusions on the causal factors and required treatment. Problems caused by poor planting procedures can be due to marginal plant stock quality; poor soil ball or container media moisture prior to planting or during establishment; improper planting depth, either too deep or too shallow; compacted planting sites and poor drainage through the soil profile; improper irrigation scheduling following planting; and improper mulching practices.

Girdling Roots

Girdling roots have long been recognized as an abiotic problem in both production and landscape systems (Figure 18.15). Encircling roots due to production methods, poor soil conditions, excessive mulch, and narrow planting sites have contributed in one form or another to the problem. A distinction is sometimes made between two forms of girdling roots. "Girdling roots" refers to the condition that occurs when roots encircle upon themselves. Stem-girdling roots encircle the tree stem above the trunk/root collar. Both of these conditions affect the structural stability and anchoring of the plant and restrict the roots system's ability to mine the soils adequately. Stem-girdling roots compress the conductive tissue in the trunk, restricting translocation and eventually leading to trunk decay. Each situation causes a slow but progressive decline in plant performance.

FIGURE 18.15 Girdling roots on red oak.

Treatment for girdling and stem-girdling roots consists of the selective removal of root sections. The extent of the removal varies with the condition and the length of time that the plant has been in place. Root removal may span several seasons to minimize stress to the plant. To eliminate or reduce the incidence of this problem, use proper planting procedures and long-term mulching practices. While planting, encircling roots should be cut or removed to ensure proper growth of new and existing roots into the surrounding soils. Excessive mulch layers around the bases of plants cause new roots to work their way upward to capitalize on optimal aeration, moisture, and nutrient levels. Roots remain in the mulch layers and encircle as continued mulching maintains the preferred environment. Mulch layers should be removed periodically and problem roots cut and redirected.

Mulch Practice

The well-documented benefits of mulch support its widespread use. Mulch conserves soil moisture, reduces soil erosion, minimizes weed growth, moderates soil temperatures, and contributes to soil fertility following decomposition. However, abiotic disorders can result from its improper or overuse. Excessive mulch and improper application lead to crown decay and plant decline. Problems associated with improper mulching include excessive moisture buildup on trunk collars and trunk decay, negative impacts on rooting depth, promotion of girdling roots, and nitrogen deficiencies in ground covers and annuals plantings.

Irrigation Management

Proper installation and operation of irrigation systems is essential for effective and productive plant growth and development. Proper timing, frequency, and rate of application not only ensure plant health and vigor but also provide for the efficient use of water resources. Supplemental irrigation is necessary during plant establishment. Demands for water during leaf emergence and expansion, shoot expansion, and root initiation require water at regular intervals. Once plants are established, however, irrigation can be adjusted accordingly. Abiotic problems related to irrigation operation can be due to moisture deficits resulting from improper scheduling and delivery or to flooding due to the failure of valves to shut off appropriately. Problems can be avoided by periodic checks to ensure that the system is operating correctly, and delivery is appropriate for the plant, season, and soil type (Figure 18.16).

- Plant selection: Choose plants that are reliably hardy in your area. Match the species to site conditions.
- Know environmental conditions: Analyze site conditions for seasonal patterns of light, wind, and moisture, noting microclimate variations.
- Understand soil conditions: Keep good records of soil test results and update them routinely.
- Understand water quality: Have water sources tested for alkalinity, calcium/magnesium ratio, soluble salts, and sodium.
- Manage irrigation properly: Consult local experts to develop a water management plan to give adequate water without over irrigating.
- Manage nutrition properly: Establish a routine of fertilization appropriate to the plant, site, and soil test results.

FIGURE 18.16 Avoiding Abiotic Disorders.

DIAGNOSIS

The cumulative and subtle nature of many physiological disorders can often make them difficult to diagnose. An extreme event such as a severe late freeze or a misapplied herbicide is an obvious "smoking gun" to indicate the underlying cause of injury. More often, however, diagnosing abiotic problems requires careful consideration of plant and site factors through a process of elimination to determine the source and potential remedy for the problem. Here are questions to remind you of the types of information you will need to collect in order to diagnose an abiotic disorder.

What Is the Plant?

It is essential to know the affected plant species and/or cultivar to know what is normal for that plant and to understand what stress factors should be considered. Certain plants are known for their susceptibility to soil conditions, for example, iron chlorosis in pin oak and premature fall color in burning bush due to soil moisture stress. Cultivars, varieties, and hybrids may differ from the species or parent species in susceptibility to disorders.

What Are the Symptoms?

Deviations from the normal appearance will be the first indication of a problem. Symptoms are changes in plant appearance initiated by the causal agents or factors. Either biotic agents or abiotic factors or conditions can initiate symptoms. A systematic evaluation of the plant from top to bottom can reveal several clues to the cause.

Leaves are often the first general indicator of plant problems. Discoloration or chlorosis may be a symptom of nutrient deficiency, or it could be related to a pest problem. Premature fall color is usually a symptom of environmental or soil-related stress. Leaf disfiguration or leaves with holes or shredded margins may indicate an insect, disease, or severe weather event. If a biological pest is suspected, look for direct signs of a pest, such as cankers, fungal fruiting bodies, insect feeding damage, or insect excrement. Twisted or malformed leaves may be associated with chemical injury from either a herbicide or a pesticide. Sudden blackening or browning of leaves usually indicates a change in temperature. High- or low-temperature extremes can kill shoot tips and cause a rapid decline in foliage appearance.

Trunk and branches are next in the inspection. We are looking for obvious damage to the bark. Bark damage may be in the form of holes from insect or bird injury, cracks or fissures from environmental fluctuations, mechanical damage from lawn mowers, or weed whips, and girdling by rabbits or other rodents. Bark bleeding or slime flux is a symptom of internal wood injury. The slime is caused by a bacterium and is sticky and foul smelling, and it often stains the outer bark. Wounds exuding slime flux need to be drained of the fermented fluids.

Roots are the most difficult part of the plant to examine, so they are the most overlooked in diagnostics. Root injury interferes with water and mineral uptake and is expressed in the plant parts farthest from the roots. Leaf scorch or dieback is the typical symptom of root-related problems. Keep in mind that symptoms of root injury are the same as those of a wide variety of causal factors. Symptoms are usually a result of poor uptake. Flooding, drought, or mechanical root damage will reduce root uptake. If you suspect root injury, dig deeper in the investigation. Flower and subsequent fruit displays may be subject to biotic and abiotic stress, resulting in a less than desirable ornamental display.

What Has Happened in the Surrounding Landscape?

It is important to determine the extent of the problem. Is it isolated on a single plant, on a single species, or in a single area, or does it exist on several species throughout the surrounding area? Isolated problems are usually related to a specific event and could be mechanical or chemical causes. Insects or diseases will usually be on plants of the same genera in the surrounding area. Environmental problems related to temperature, water, or wind will usually influence broader areas. It is also important to note when the problem appeared and whether it intensified with time.

What Is the History of the Plant?

The first basic question is how long has the plant been in its present location? The objective is to consider or rule-out any problems related to newly established plant materials. Transplant shock or site limitations during establishment may influence plant performance for several years following installation. Plant history should also include information on recent or routine cultural practices. Some plant problems may develop over time, and the visual change in appearance may not occur until the following growing season. Asking detailed questions will aid in identifying a cause and prescribing corrective recommendations.

What Are the Site Characteristics and Site History?

Reviewing site characteristics involves a comprehensive inventory of the site. We begin with documenting the characteristics of the plant's immediate location. Problems may surface because of its relationship to the area's exposure to light, wind, and compass orientation. Both chemical and physical characteristics of soil are important considerations, especially when the soil is either extremely light or heavy. In addition, soil pH may influence plant performance. Site history examines the impact of construction or disturbance on the site's soils and their subsequent impact on a plant's overall appearance. Weather patterns and/or specific events can trigger plant problems. Long-term weather patterns that result in marginal conditions such as prolonged drought or unseasonably high or low temperatures can predispose plants to other biotic or abiotic stresses. Weather events such as hail, late frost, and ice storms can have a direct impact on plant health and appearance.

What Routine or Periodic Cultural Practices Have Been Used?

Improper landscape management technique can lead to plant health problems. Timing and rates of chemical inputs such as fertilization and pesticide or herbicide applications can influence plant performance.

Planting records can provide a wealth of information about a plant's history and potential predisposition to problems. Information including the original stock type (bare-root, B&B, container, or mechanically spaded), planting date, maintenance during the warranty period, and any subsequent care can be useful in determining the extent of transplant shock on the establishment of newly planted landscape plants. Each of these may contribute to abiotic problems or increase a plant's susceptibility to biotic agents. Irrigation timing, frequency, and rates can influence plant health also. Water is essential for plant growth and development; too much of which, however, can lead to plant decline.

As mentioned earlier, the key to solving many of your plant problems will be keen observation of the physical environment, background information on the plant, the site, and cultural practices: knowledge of plant growth and development; and experience.

SUGGESTED READINGS

Costello, L. R., E. J. Perry, N. P. Matheny, M. J. Henry, and P. M. Geisel. 2003. *Abiotic Disorders of Landscape Plants*. Oakland, CA: University of California.

Kennelly, M., J. O'Mara, C. Rivard, G. L. Miller, and D. Smith. 2012. Introduction to abiotic disorders in plants. *The Plant Health Instructor*. DOI: 10.1094/PHI-I-2012-10-29-01. http://www.apsnet.org/edcenter/intropp/PathogenGroups/Pages/Abiotic.aspx.

Plant Health Care for Woody Ornamentals. 1997. *International Society of Arboriculture, Savoy, Ill.: Cooperative Extension Service, College of Agricultural, Consumer, and Environmental Science*. Champaign, IL: University of Illinois, Urbana-Champaign.

Part III

Plant–Pathogen Interactions

19 Virulence Factors Produced by Plant Pathogenic Bacteria

Rebecca Ann Melanson and Jong Hyun Ham

Concept Box

- Plant pathogenic bacteria produce a variety of virulence factors that contribute to the successful infection and colonization of host plants.
- Different types of virulence factors act as primary pathogenic determinants based on the nature of disease.
- Extracellular enzymes secreted through T1SS and T2SS are the primary virulence factors used by soft-rot–causing bacteria.
- Type III effector proteins secreted through T3SSs play important roles in the pathogenesis by plant pathogenic bacteria that have a narrow host range.
- Toxins, hormones, and extracellular saccharides (EPS) are other virulence factors of some plant pathogenic bacteria.
- The production of virulence factors should be properly regulated by regulatory systems for successful pathogenesis.

Plants possess complex defense systems in order to defend themselves against a diverse array of plant pathogens. In order to overcome these barriers of defense, plant pathogenic bacteria must produce a diverse array of **virulence factors**, factors that determine the level of disease. Since the beginning of plant pathology, various types of virulence factors have been identified and studied, and a number of secretion mechanisms that are responsible for the secretion and delivery of virulence factors to the host site have been identified. As technology continues to improve, additional virulence factors will be identified, and the mechanism(s) by which these and known virulence factors work will be elucidated. This knowledge will be useful in developing methods to combat these pathogens and prevent disease. This chapter presents a review of different kinds of virulence factors that are produced and utilized by plant pathogenic bacteria and discusses the functional characteristics of each type as well as the major secretion systems by which these virulence factors are secreted.

PROTEINACEOUS VIRULENCE FACTORS

Extracellular enzymes and type III effectors are proteinaceous virulence factors produced by plant pathogenic bacteria. Most of these virulence factors, whether they are secreted from the bacterium into the extracellular matrix or directly into the host, have enzymatic functions.

Extracellular Enzymes

Extracellular enzymes (see also Chapter 32) are important virulence factors of bacteria, such as *Pectobacterium chrysanthemi* and *P. carotovorum* that cause soft rots in their hosts. The function of known extracellular enzymes—cellulases, xylanases, pectinases, proteases, and many others—is to damage plant tissues and kill host cells. These enzymes differ in both substrate specificity and their mechanism of action. With the exception of the proteases, most of these known extracellular enzymes are secreted via type II secretion systems.

Pectinases are the most common degradative enzymes, and those produced by the members of the genus *Pectobacterium* are the best studied in regard to substrate specificity and mode of action. They can be classified according to the type of substrate they affect, the mode of cleavage or the point of attack. Pectinases can degrade **pectic acid**, a polymer composed of α-1,4 galacturonate chain interspersed with α-1,2-linked rhamnosyl residues for branching, or **pectin**, an acetylated or methylated pectic acid. They can also cleave their substrates through either hydrolysis or transelimination. Pectinases that cleave their substrates and produce saturated products through **hydrolysis** are polygalacturonases. Those that cleave their substrates and produce unsaturated products through **transelimination** (also called β-elimination) are pectate lyases. Pectinases can also be endo- or exopectinases and can cleave polymers randomly within the substrate or from the end of the substrate, respectively. The enzymatic activities and substrates of various pectinases are shown in Figure 19.1.

A single bacterium can produce multiple pectinases. For example, *Pectobacterium chrysanthemi* 3937 produces two acetyltransferases, PaeY and PaeX; two methyltransfereases, PemA and PemB; eight endopectate lyases, PelA, PelB, PelC, PelD, PelE, PelI, PelL, and PelZ; two exopectate lyases, PelX and PelW; four galacturonases, PehX, PehV, PehW, and PehN; and one rhamnogalaturonate lyase, RhiE; among others (Kazemi-Pour et al., 2004).

There are two phases in the induction of pectolytic enzyme activity. The first phase is an extracellular digestion phase in which basal levels of pectinases degrade natural galacturonase into oligogalacturonates that are taken up by the bacterial cell. Within the cell, the unsaturated digalacturonate is catabolized by the ketodeoxyuronate pathway to pyruvate and glyceraldehyde-3-phosphate. The digalacturonate intermediates and products of this pathway all induce genes for pectinase production, which is regulated by multiple environmental factors in a very complicated genetic system of interactions between plant and pathogen. In addition to inducing genes for pectinase production, some of the products of the pectolytic enzymes, mostly short-chain oligogalacturonides, trigger responses on the part of the plant, such as phytoalexin synthesis and protease inhibitor production that may be involved in host defense.

Type III Effectors

Type III effectors are proteins that are secreted directly from the bacterium into the host via a type III secretion system (T3SS). In general, type III effectors play important roles in the pathogenesis by pathogens that have a narrow host range, including *Pseudomonas syringae* pathovars and *Xanthomonas* spp., and a single bacterium produces multiple type III effectors. For example, over 28 type III effectors have been identified in the whole genome of *P. syringae* pv. *tomato* DC3000 (Lindeberg et al., 2012). The identification of these effectors and the genes that encode for them is difficult with conventional genetic analyses since the virulence phenotypes of individual type III effectors are often negligible and do not produce any obvious signals or hallmarks. Avirulence

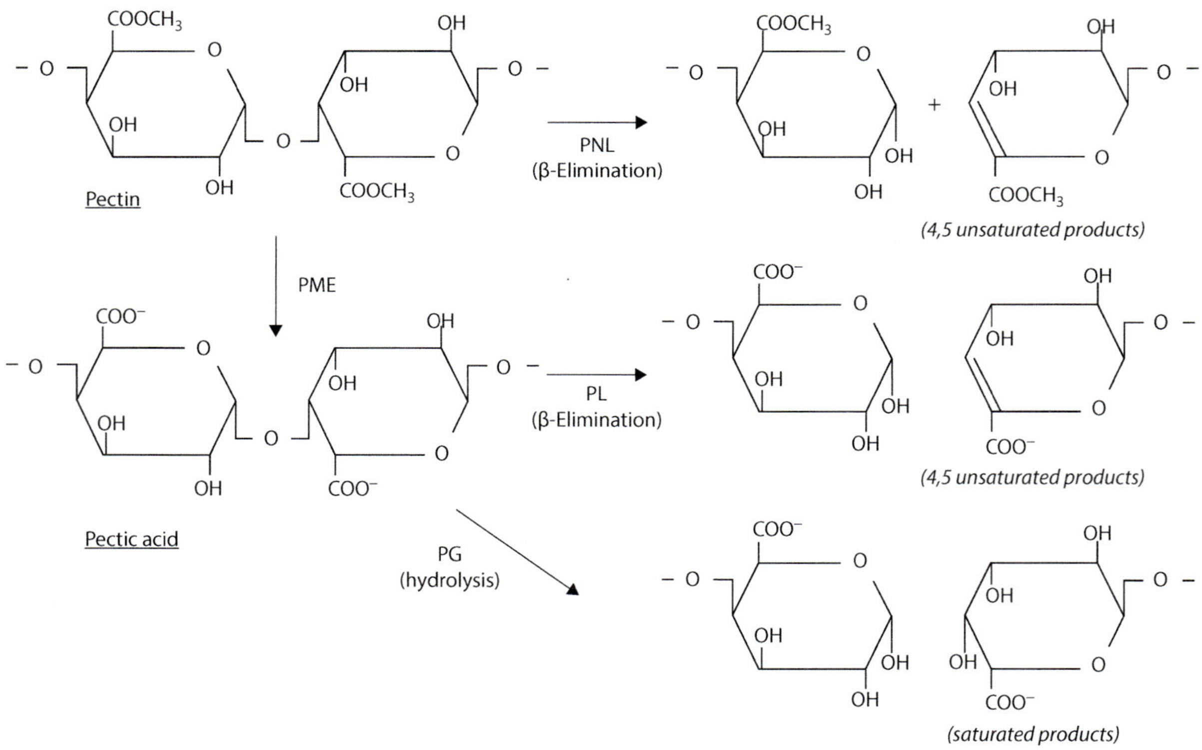

FIGURE 19.1 Cleavage of α-1,4 glycosidic bonds by various pectinases. PG, polygalacturonase; PME, pectin methylesterase; PL, pectic acid lyase; PNL, pectin lyase.

proteins, which are based on gene-for-gene interactions, were the first type III effectors identified because they produced a dominant genetic phenotype, specifically the hypersensitive response (HR), which is a hallmark of gene-for-gene interactions. Several experimental approaches are used to validate secretion by the T3SS. These approaches include Western blots of the supernatant fractions (Petnicki-Ocwieja et al., 2002), the Cya (*Bordetella pertussis* adenylate cyclase) reporter system (Schechter et al., 2006), and AvrRpt2 fusion experiments (Guttman et al., 2002).

The avirulence function of type III effectors is a form of self-betrayal to the bacterium because it allows for bacterial recognition by the host that initiates host defense systems. Individual type III effectors have biological and biochemical functions other than avirulence that contribute to the survival of the bacterium in the host as well as the onset of disease symptoms in the host. These functions include the suppression of plant defense systems, either the inhibition of the HR or the suppression of host basal resistance, host cell death, phosphorylation, and degradation of host cell components (Table 19.1).

Secretion Systems for Proteinaceous Virulence Factors

A number of secretion systems are utilized for the secretion of proteinaceous virulence factors in plant pathogenic bacteria. Just as a single bacterium can produce multiple virulence factors, it can also have multiple

TABLE 19.1

Selected Type III Effectors Identified in Plant Pathogenic Bacteria and Their Biological or Biochemical Functions within the Host

Type III Effector(s)	Pathogen(s)	Biological or Biochemical Activity (Known or Predicted Function)
VirPphA, AvrPphF, AvrPphC	*Pseudomonas syringae* pv. *phaseolicola*	Suppression of the hypersensitive response
AvrPtoB	*P. syringae* pv. *tomato*	Suppression of the hypersensitive response; E3 ubiquitin ligase activity
AvrPphE$_{Pto}$, AvrPpiB1$_{Pto}$, AvrRpt2, HopPtoD2, HopPtoE, HopPtoF, and HopPtoN	*P. syringae* pv. *tomato*	Suppression of the hypersensitive response
AvrPto, AvrRpt2	*P. syringae* pv. *tomato*	Suppression of host basal defenses
AvrRpm1	*P. syringae* pv. *maculicola*	Suppression of host basal defenses
DspA/E	*Erwinia amylovora*	Induction of host cell death; suppression of host defense
AvrE1, HopM1	*P. syringae* pv. *tomato*	Induction of host cell death; suppression of host defense
WtsE	*Pantoea stewartii* subsp. *stewartii*	Induction of host cell death; suppression of host defense
AvrPpiG1, HopPmaD, and HopPsyV	*P. syringae* pathovars	Cysteine proteases (degradation of host defense components)
AvrPphB, AvrPpiC2, HopPtoC, HopPtoN	*P. syringae* pathovars	Cysteine proteases (degradation of host defense components)
AvrRxv, AvrBsT, AvrXv4, XopD, and XopJ	*Xanthomonas campestris*	Cysteine proteases (degradation of host defense components)
Eop1	*E. amylovora*	Cysteine protease (degradation of host defense component)
Pop1, Pop2	*R. solanacearum*	Cysteine protease (degradation of host defense component)
HopPtoD2	*P. syringae* pv. *tomato*	Tyrosine phosphatase (suppression of the host defense pathway)
AvrB	*P. syringae* pv. *glycinea*	Phosphorylation (suppression of the host defense pathway)
AvrRpm1	*P. syringae* pv. *maculicola*	Phosphorylation (suppression of the host defense pathway)
HopPtoS1, HopPtoS2, HopPtoS3	*P. syringae* pv. *tomato*	ADP ribosyltransferase (suppression of the host defense pathway)

secretion systems. Each secretion system, named type I, type II, and so on through type VII, has its own machinery and mechanism for the secretion of particular virulence factors. In some cases, the secretion of proteins through one of these systems is a single-step process; in other cases, secretion through one of these systems is a two-step process and first requires proteins to be exported into the **periplasm** prior to secretion across the outer membrane. These bacterial secretion systems may also function dependently or independently of a general export or *sec*-pathway that is responsible for the secretion of proteins across the plasma membrane of Gram-positive bacteria and across the inner membrane of Gram-negative bacteria. Here, we discuss the most common secretion systems, T1SS, T2SS, and T3SS, that are used for the secretion of proteinaceous virulence factors in plant pathogenic bacteria. Figure 19.2 illustrates the general features of T1SS, T2SS, and T3SSs.

Type I Secretion Systems

The T1SS is a *sec*-independent secretion system that secretes exoenzymes across the bacterial plasma membrane(s) in a single step. The T1SS is composed of the following three major components: a pore-forming outer membrane protein (OMP), a membrane fusion protein (MFP), and an inner membrane adenosine triphosphate (ATP)-binding cassette (ABC) protein that hydrolyzes ATP to provide energy for translocation. Proteins secreted by the T1SS do not have the classical N-terminal signaling sequences. Rather, signaling occurs at the C-terminal end of the proteins. Cleavage of these signals does not occur during translocation, and specific sequences for signaling have not been identified.

The prototypical example of the T1SS is the *Escherichia coli* secretion system for the toxin, α-hemolysin (HlyA) in human pathogenic strains. In this pathogen, the T1SS is made up of the TolC, HlyD, and HlyB proteins, which are OMP, MFP, and ABC protein, respectively. An example of a T1SS in a plant pathogenic bacterium, *P. chrysanthemi*, which causes bacterial soft rots in vegetables, is the system composed of the proteins PrtF, PrtE, and PrtD. These proteins, OMP, MFP, and ABC protein, respectively, are responsible for the secretion of metalloproteases in this pathogen. Despite the presence of this T1SS in *P. chrysanthemi*, its importance in virulence however is very low as the proteases it secretes play only minor roles in virulence. T1SSs are also found in the plant pathogens *Agrobacterium tumefaciens*, *P. syringae* pv. *tomato*, *Ralstonia solanacearum*, *Xanthomonas axonopodis* pv. *citri*, and *Xylella fastidiosa* (Tseng et al., 2009).

Type II Secretion Systems

T2SS is a *sec*-dependent secretion system that secretes exoenzymes in a two-step process. It is a multi-protein complex that spans the inner and outer bacterial membranes. The major components of T2SS are proteins C, D, E, and O. Protein C is a "gatekeeper" involved in the recognition of proteins to be secreted. Protein D belongs to a family of proteins termed **secretins** that are integral OMPs. These proteins are predicted to consist largely of transmembrane β-strands that form a β-barrel structure with 12–14 subunits. Protein E is a cytoplasmic ABC protein, a traffic ATPase that provides energy for some aspect of the secretion process. Protein O is a peptidase for the N-terminal signals of proteins G, H, I, J, and K that likely form a pilus-like structure.

The first step in enzyme secretion through the T2SS is the export of the target protein through the *sec* pathway into the periplasm of the bacterial cell. When en route to the periplasm, the N-terminal signal sequence, the first 2 to less than 15 residues of a signal peptide, is

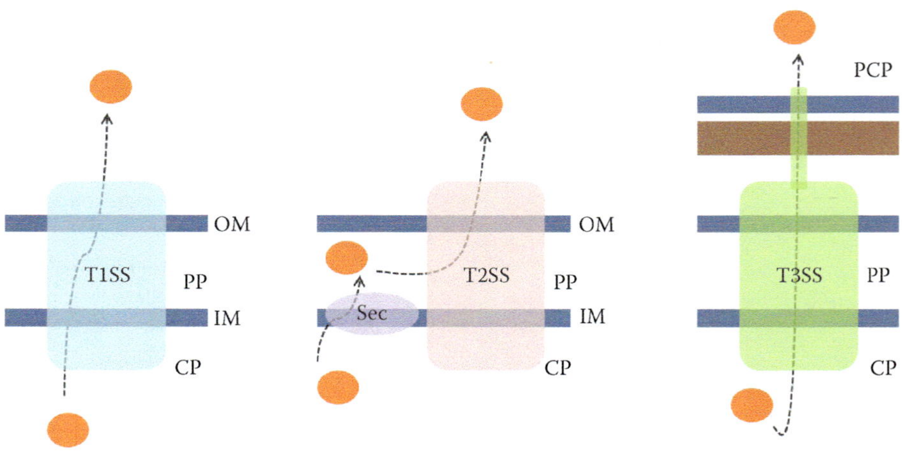

FIGURE 19.2 A schematic overview of protein secretion via the type I, type II, or type III secretion systems. CP, cytoplasm; IM, inner membrane; OM, outer membrane; PP, periplasm; PCM, plant cell membrane; PCW, plant cell wall; PCP, plant cytoplasm; Sec, *sec* pathway; T1SS, type I secretion system; T2SS, type II secretion system; T3SS, type III secretion system.

processed by signal peptidase. There are exceptions to this, however, as some proteins are processed and folded into mature proteins in the periplasm. From the periplasm, proteins are then secreted across the outer membrane. Secretion across the outer membrane requires the use of 12–16 accessory proteins, collectively referred to as the **secreton**. It has been suggested that this transport via the secreton be renamed the secreton-dependent pathway (Douzi et al., 2012).

Because a diverse array of proteins is secreted by the T2SS, this pathway is also called the **general secretory pathway**. Many proteins that secreted through the T2SS are putative virulence factors of both plant and animal pathogens. The main substrates of the T2SS are exoenzymes, such as **isozymes** of pectate lyases, polygalacturonases, and cellulases, which attack plant cell walls. Thus, the T2SS plays a major role in soft-rot causing bacteria that rely on exoenzymes for virulence. Plant pathogens with known T2SS include *P. chrysanthemi*, *P. carotovorum*, *R. solanacearum*, and *X. campestris*.

There is some level of specificity in the secretion of virulence factors via the T2SS. Even though the components of the T2SS and the enzymes secreted via this system are similar, enzymes from one bacterium are not always compatible with the secretion system of another bacterium. For example, *P. carotovorum* cannot secrete enzymes produced by *P. chrysanthemi* and vice versa; however, the enzymes and secretion systems from *P. carotovorum* and *P. atrosepticum* are compatible with each other. Studies with *Klebsiella* and *Pectobacterium* spp. suggest that the specificity of these interactions and the transport of specific substrate proteins may be determined by protein C in these pathogens and the OutD protein in *P. chrysanthemi* (Lindeberg et al., 1996).

The organization of the loci encoding the T2SS components is relatively conserved. Most of the genes are transcribed from a single operon that contains several overlapping genes. Although some gene clusters that encode the T2SS in some pathogens are constitutively expressed, others are regulated. The major operon of the *out* genes that encode the T2SS in *P. chrysanthemi*, for example, is negatively regulated by the protein KdgR, which is also a negative regulator of multiple pectinases.

Type III Secretion Systems

T3SSs are *sec*-independent secretion systems whose characteristic feature is the ability to secrete virulence factors directly from the bacterial cell into the host cell. This phenomenon is triggered, in some cases, upon contact with the host cell. Thus, secretion through the T3SS has been called "contact dependent" secretion. However, not all T3SSs are contact dependent, and some effector molecules exported via this system can be released into the external environment rather than into the host cell. T3SSs are made up of multiple components and are homologous to the Gram-negative flagellar export apparatus. The archetypal T3SS was first identified in pathogenic *Yersinia* spp. for the secretion of Yop proteins. In this system, the secretin-like OMP, YscC, was demonstrated to form ring-shaped structures with large central pores that facilitate the secretion of effector molecules and stabilize the needle-like complex that is formed by other components of the T3SS (Koster et al., 1997).

Signals of type III effectors recognized for secretion via a T3SS are not clearly defined like the N-terminal signal peptides for the *sec*-pathway. Yet, it is generally accepted that the first 50 amino acids of a type III effector are required for the secretion via a T3SS. Analysis of the type III effectors produced by *P. syringae* pv. *tomato* revealed the following six predictive rules for secretion-associated patterns: (1) isoleucine, leucine, alanine, or proline are present in position 3 or 4 after proline or a polar or basic amino acid, but not in both; (2) methionine, isoleucine, leucine, phenylalanine, tyrosine, or tryptophan are rarely present in position 5; (3) aspartic acid or glutamic acid is not present within the first 12 amino acids; (4) cysteine is rare after position 5; (5) the N-terminal 50 amino acids are rich in polar residues, especially serine and glutamine, comprising an amphipathic structure; and 6) more than three consecutive residues consisting of methionine, isoleucine, leucine, valine, phenylalanine, tyrosine, or tryptophan do not occur in the first 50 N-terminal amino acids (Petnicki-Ocwieja et al., 2002).

The genes that encode the T3SS are collectively called the *hypersensitive response and pathogenicity* genes, or ***hrp* genes**, because functional T3SSs are required in many plant pathogenic bacteria for eliciting an HR in nonhosts or resistant hosts and causing disease in susceptible hosts. Plant pathogenic bacteria can be divided into two separate groups based on the regulation of these *hrp* genes (Alfano and Collmer, 1997). In the first group, which includes *Erwinia amylovora*, *P. carotovorum*, *P. chrysanthemi* and *Pseudomonas* spp., an alternate sigma factor, HrpL, controls the expression of most of the *hrp* and *avr* (avirulence) genes. This factor binds to a Hrp-box, identified by the sequence GGAACCNA-N14-CCACNNA, which was originally identified in the promoter region of *avr* genes. Expression of *hrpL* requires a sigma-54 factor, encoded by *rpoN*, and another protein, HrpS, an NtrC-family transcriptional activator protein that functions upstream of HrpL in the regulatory pathway for the expression of *hrp* genes. In the second group, which includes *R. solanacearum* and *X. campestris*, *hrp* gene expression is regulated by HrpG, a protein similar to the OmpR subclass of the two-component regulatory system, and a second regulatory protein. In *X. campestris* pv. *vesicatoria*, this second regulatory protein is HrpX; in *R. solanacearum*, this protein, HrpB, is homologous to and shares 40% identity to HrpX. Unlike the first group, a sigma-54 factor is not required for the binding of HrpB or HrpX to the promoter of the *hrp* genes. HrpB or HrpX binding also occurs at a PIP-box, identified by the sequence, TTCGC-N15-TTCGC, in the promoter region of *hrp* genes rather than a Hrp-box. In *R. solanacearum*, an additional four regulatory components located upstream of HrpG in this regulatory pathway, PrhA, PrhI, PrhR, and PrhJ, have been identified for the regulation of the *hrp* genes for the T3SS (Brito et al., 1999).

NONPROTEINACEOUS VIRULENCE FACTORS

A number of nonproteinaceous virulence factors, including **extracellular polysaccharides (EPS),** toxins, and hormones, are also important virulence factors for many plant pathogenic bacteria. These virulence factors are responsible for many diseases-causing wilts, chlorosis, necrosis, and **hypertrophy** or **hyperplasia** in hosts. An overview of these virulence factors with examples of the pathogens that use them and how they induce disease within the host is presented as well as an overview of how bacterial pathogens protect themselves from their own toxins.

Extracellular Polysaccharides

Several bacteria utilize EPS as a virulence factor. (Table 19.2). They not only induce disease symptoms in the host but also serve to protect the bacterium within the host. EPS have the capacity to hold water and, thus, help to prevent bacterial desiccation. They also help to immobilize bacteria on host cell walls and, in doing so, may help avoid recognition by host cells. In addition, EPS can induce and maintain water soaking in the plant and can restrict water transport when they accumulate in xylem cells.

Many of the pathogens that use EPS as virulence factors produce wilting diseases in their hosts. In some bacteria, such as *Pantoea stewartii* subsp. *stewartii*, the cause of Stewart's wilt and leaf blight in corn, EPS is a virulence factor but is also essential for pathogenicity. In general, there tends to be a correlation between EPS formation and virulence with more EPS being produced by bacteria that are more virulent. This correlation is particularly strong in *R. solanacearum* and *E. amylovora*.

Although EPS can be detrimental to their plant hosts and can affect the production of crops, we have made use of the ability of *X. campestris* to produce the EPS xanthan gum. This EPS is now industrially produced and is used in many foods and household products, including ice cream, toothpaste, and cosmetics. Other EPS produced by plant pathogenic bacteria include levan and amylovoran from *E. amylovora*, stewartan from *P. stewartii*, and alginate from various **pathovars** of *P. syringae*.

Toxins

Toxins, microbial metabolites that are harmful to the host in small doses, also serve as virulence factors of many plant pathogenic bacteria and were the first bacterial virulence factors to be identified. Toxins can be harmful to host plants in a number of ways. They can interfere with the metabolism of the host causing increased or decreased levels of other metabolites, change the permeability of host cell membranes resulting in reduced host resistance, or interfere with active transport across host cell membranes creating environments in intracellular spaces that are conducive for bacterial reproduction. All known bacterial toxins are not host-specific, and most toxins are virulence determinants (not **pathogenicity factors**), so do not account for host specificity of the bacterial pathogen. Classical toxins include glycopeptides, peptides, and small metabolites. Some of the major toxins produced by plant pathogenic bacteria are included in Table 19.3.

Chlorosis-inducing toxins refer to those toxins that cause a yellowing of the plant tissue due to the decomposition of chlorophyll or the inhibition of its formation. A number of chlorosis-inducing toxins, produced by different pathovars of *Pseudomonas syringae*, have been identified, and the mechanisms by which they induce chlorosis are determined. Tabtoxin, produced by *P. syringae* pv. *tabaci*, is a dipeptide compound that is composed of threonine and tabtoxinine-β-lactam. The active form of this toxin, tabtoxinine-β-lactam, inactivates the enzyme glutamine synthetase, which leads to depleted glutamine levels, and consequently, the accumulation of ammonia. The buildup of toxic levels of ammonia uncouples **photosynthesis** and **photorespiration** and destroys the thylakoid membrane of the chloroplast causing chlorosis as well as necrosis. These effects suppress

TABLE 19.2
Extracellular Polysaccharides Produced by Bacterial Plant Pathogens

Extracellular Polysaccharide	Pathogen
Alginate	*Pseudomonas syringae* pathovars
Amylovoran, levan	*Erwinia amylovora*
Stewartan	*Pantoea stewartii*
Xanthan gum	*Xanthomonas campestris*

TABLE 19.3
Major Toxins Produced by Plant Pathogenic Bacteria

Toxin(s)	Pathogen(s)	Major Host(s)
Coronatine	*P. syringae* pvs. *coronafaciens*, *glycinea*, *tomato*, others	*Arabidopsis thaliana*, beans, tomato
Phaseolotoxin	*P. syringae* pv. *phaseolicola*	Beans
Tabtoxin	*P. syringae* pv. *tabaci*	Tobacco, soybean
Tagetitoxin	*P. syringae* pv. *tagetis*	Sunflower
Thaxtomin A, Thaxtomin B	*Streptomyces scabies*	Potato
Thaxtomin C	*Streptomyces ipomoeae*	Sweet potato
Toxoflavin, Fervenulin	*Burkholderia glumae*	Rice
Tolaasin	*Pseudomonas tolaasii*	Mushroom

the host's ability to actively respond to the pathogen. Coronatine, produced by a number of *P. syringae* pathovars, including pathovars *coronofaciens*, *glycinea*, and *tomato*, is a polyketide that causing chlorosis as well as stunting. Coronatine mimics jasmonic acid (JA) and leads to the activation of the JA-dependent defense pathway in the host, which is typically induced in response to wounds. The JA-dependent defense pathway and the salicylic acid (SA)-dependent defense pathway, which is typically induced by pathogen infection, antagonize each other. Thus, the activation of the JA-dependent defense pathway leads to the suppression of the SA-dependent defense pathway and the suppression of host defense mechanisms against plant pathogens. Phaseolotoxin, produced by *P. syringae* pv. *phaseolicola*, is a tripeptide compound that is composed of a modified ornithine-alanine-arginine that contains a phosphosulfinyl group. Phaseolotoxin itself is not toxic to the host, but cleavage of the toxin by host peptidases produces a toxic moiety, octicidin, which inhibits ornithine carbamoyl transferase (OCTase). Another toxin, tagetitoxin, produced by *P. syringae* pv. *tagetis*, contains a hemithioketal ring structure and may have a general role in the repression of chloroplast genes. Because of this role, only developing leaf tissues are affected and become chlorotic.

Necrosis-inducing toxins cause lysis of the plant cell membranes resulting in the leakage of cellular nutrients into the intracellular or extracellular spaces. Two groups of cyclic lipodepsipeptides necrosis-inducing toxins are produced by *P. syringae* pv. *syringae*. The first group consists of those toxins that are made up of nine peptides and includes syringomycins, syringotoxins, and syringostatins. The second group consists of toxins that are made up of 22–25 peptides and includes syringopeptins. Another necrosis-inducing toxin, tolaasin, is produced by the mushroom pathogen *P. tolaasii*.

Thaxtomin A, B, and C produced by *Streptomyces* spp. infecting potato and sweet potato are scab-inducing toxins that contribute to scab diseases of these hosts. Thaxtomin A and B are cyclic dipeptides, produced by the potato pathogen *S. scabiei*, which can induce necrosis, hypertrophy, and abnormalities in mitosis and cell division in root cells. Both the toxins are thought to be determinants of pathogenicity. Thaxtomin C, which has not been extensively studied, is produced by the sweet potato pathogen *Streptomyces ipomoeae*.

Other toxins produced by bacterial plant pathogens include glycoproteins and toxoflavin and its derivative fervenulin. *Clavibacter michiganensis* subsp. *michiganensis* and subsp. *sepedonicus* produce glycoproteins that are utilized as toxins. Glycoproteins produced by *C. michiganensis* subsp. *michiganensis*, the cause of bacterial wilt and canker in tomato, cause white necrotic spots on leaves. *Burkholderia* spp., including the rice pathogens *B. gladioli* and *Burkholderia glumae* (Case Study 19.1), produce the toxins toxoflavin and fervenulin. Toxoflavin, a major virulence factor of *B. glumae*, is the major causal agent of bacterial panicle blight of rice (Ham et al., 2011). It functions as an active electron carrier between nicotinamide adenine dinucleotide—reduced form (NADH) and oxygen. This toxin allows for electron transport in the plant to bypass the cytochrome system and leads to the buildup of hydrogen peroxide. Toxoflavin causes a discoloration of rice panicles, probably due to cell death caused by high levels of hydrogen peroxide, which are harmful to the host.

In most instances, toxins that are harmful to the host can also be harmful to the pathogen. Toxin-producing bacteria, therefore, must have some mechanism(s) in place to avoid poisoning themselves with their toxic metabolites. Several different mechanisms that prevent self-toxicity by bacterial pathogens have been identified. One mechanism involves selective transport and/or the inward inhibition of the toxin. *B. glumae*, for example, has a toxoflavin transport system, encoded by the *toxFGHI* operon, to actively transport toxoflavin outside the cell. Another mechanism involves the production of isozymes that are insensitive to toxin inhibition. *P. syringae* pv. *phaseolicola*, for example, produces two isoforms of OCTase. One isoform is insensitive to phaseolotoxin and is produced when the toxin is

CASE STUDY 19.1

BURKHOLDERIA GLUMAE—AN EMERGING THREAT TO GLOBAL RICE PRODUCTION

- *Burkholderia glumae* is the major causal agent of bacterial panicle blight (BPB) of rice (also referred to as bacterial grain rot in Asian countries).
- BPB affects rice production in the major rice-producing areas of the world and has been shown to reduce yield by as much as 75%.
- Major epidemics of BPB occurred in 1995, 1998, 2000, and 2010 in the rice-producing areas of the southeastern United States and were associated with high humidity and high temperatures.
- Increasing global temperatures increase the risk of rice yield loss by this disease.
- Rice varieties that are completely resistant to this pathogen have not yet been identified.
- Oxolinic acid applied as a seed treatment or foliar spray is effective at controlling BPB.
  - Oxolinic acid-resistant strains have been isolated from fields treated with this chemical.
  - Oxolinic acid is not commercially available in many countries.
- Biological control with avirulent *B. glumae* strains or with *Bacillus* spp., *Pseudomonas* spp. or *Saccharomyces* spp. is promising, but needs to be evaluated in the field.
- *Burkholderia glumae* produces a number of virulence factors, including toxoflavin, lipase, flagella, and type III effectors.
- A **quorum-sensing** system mediated by the *N*-acyl homoserine lactone (AHL) synthase, TofI, and the AHL receptor, TofR, is the central regulatory system that controls the production of the major virulence factors in *B. glumae*, including toxoflavin, lipase, and flagella.
- The only method of management for bacterial panicle blight for rice producers in areas where oxolinic acid cannot be used is the use of pathogen-free seed and rice varieties that show some level of resistance.

being produced. The second isoform is sensitive to phaseolotoxin and is produced when the toxin in not being produced. In some bacteria, the form of the toxin produced by the bacterium in the cytoplasm is not toxic or is in an inactive form. Tabtoxin, for example, is inactive in the bacterium and is altered to its active form, tabtoxinine-β-lactam, by plant hydrolases or by bacterial enzymes located in the periplasm. Phaseolotoxin is also inactive, but is converted to its active, toxic form, octicidin by the plant host. In some bacteria, enzymes in the cytoplasm detoxify the toxin by converting it from its active form to its inactive form. For example, β-lactamase in *P. syringae* pv. *tabaci* detoxifies the toxic tabtoxinine-β-lactam to the inactive form tabtoxinine.

As these examples indicate, a bacterium may have more than one mechanism in place to prevent the harmful effects of the toxin it produces. Additional examples of bacterial self-protection against toxins will undoubtedly be discovered through the continued study of bacterial pathogenesis and the genome sequencing of toxin-producing plant pathogenic bacteria.

HORMONES

Plant growth **hormones**, such as indole acetic acid (IAA), cytokinins, and auxin, which stimulate plant growth and cell elongation, and are responsible for the production of galls and excrescences in plant hosts may also be virulence factors of some plant pathogenic bacteria. In *Pseudomonas savastanoi* pv. *savastanoi*, the cause of olive knot on olive, for example, loss of the ability to produce auxin may be associated with the loss of virulence. The genes that encode for these plant growth hormones may be located on the bacterial chromosome or on bacterial **plasmids**.

One of the most well-known gall-forming plant pathogens, *Agrobacterium tumefaciens*, the cause of crown gall in a number of hosts, causes gall formation by transferring genes for hormone synthesis into host cells. This gene transfer from an *A. tumefaciens* plasmid has been thoroughly studied and is accomplished through a specialized secretion system, the T4SS. Other plant pathogenic bacteria that utilize hormones as virulence factors include several pathovars of *Pantoea agglomerans*, including pathovars *betae*, *gypsophilae*, and *milletiae*.

LABORATORY EXERCISES

Numerous assay systems have been developed to determine the production of virulence factors by plant pathogenic bacteria. These systems allow for the characterization of virulence factors that are produced by a given pathogen. Some systems for specific virulence factors are important tools for various areas of phytobacteriology, including etiology, taxonomy, and genetics. Most importantly, however, convenient and high-throughput systems that can clearly indicate the production of virulence factors are useful for molecular genetic studies of plant pathogenic bacteria, including the identification of genetic elements involved in virulence. For example, random mutants of a pathogen generated through transposon mutagenesis can be screened in these assay systems to identify mutants that exhibit altered phenotypes in the production of the particular virulence factor being studied. Mutants deficient in the production of virulence factors would have mutations in the genes for biosynthesis or secretion/export of those virulence factors or for the positive regulation of these virulence genes. In contrast, mutants showing increased production of virulence factors would have mutations in the genes for the negative regulation of the corresponding virulence genes. Mutated genes in these mutants can be easily identified through sequencing of the flanking regions of the transposon inserted in the genome, using the known sequences of the transposon.

In this chapter, two laboratory exercises for determining two distinctive virulence factors produced by various plant pathogenic bacteria are described. The first exercise uses an assay system to detect pectolytic activity of plant pathogenic bacteria. This assay also indicates the functionality of a T2SS, which secretes pectinases and other extracellular enzymes. The second exercise uses an assay system to detect the ability of bacterial pathogens to elicit an HR on tobacco leaves. Because HR elicitation by bacterial pathogens is dependent on type III effectors secreted via a T3SS, this assay indicates the functionality of a T3SS in the pathogen.

Experiment 1. Detection of Pectolytic Activity on Pectate Semi-Solid Agar

This experiment is on the basis of the protocol developed by Starr et al. (1977). Pectic polymers are primary components of plant cell walls. Many plant pathogenic bacteria, particularly soft rot-causing bacteria, produce a diverse array of pectinases. Pectolytic activities of plant pathogenic bacteria can be detected on pectate semisolid agar (PSSA), which contains polygalacturonic acid. Pitting zones form on PSSA around bacterial colonies that produce and secrete pectinases (Figure 19.3). The purposes of this experiment are to test various bacterial pathogens for their ability to produce pectinases and break down PSSA and to determine the effect that pH has on pectolytic activity.

Materials

Each student or team of students will require the following materials:

- Bacteria
  - Plant pathogenic bacteria that cause soft rots, i.e., *Pectobacterium carotovora* subsp. *carotovora*
  - Plant pathogenic bacteria that do not cause soft rots, that is, *Pseudomonas* spp. or *Xanthomonas* spp.
  - Nonpathogenic bacteria, i.e., *Pseudomonas fluorescens* or *E. coli*
- Bromothymol blue (BTB)
- $CaCl_2 \cdot 2H_2O$ (for 10% solution in water)
- NaOH (for 1 N solution and for 0.1 % BTB solution)
- HCl (for 1 N solution)
- Polygalacturonic acid sodium salt (Sigma-Aldrich, Saint Louis, MO; Product # P3850)
- Yeast extract
- Agar
- pH paper
- Micropipettes and tips
- Magnetic stirrer with heating function
- Magnetic stirring bar
- 5-mL and 10-mL pipettes
- Incubator (28°C–30°C)
- Petri plates

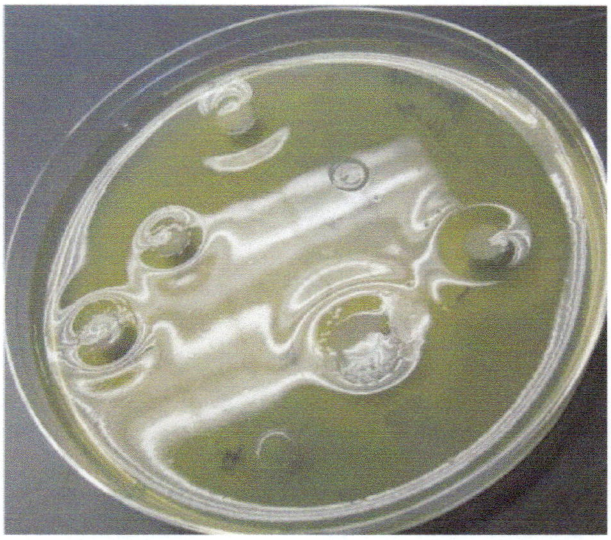

FIGURE 19.3 Pitting zones around bacterial spots on a pectate semisolid agar plate indicate bacterial pectinase activity.

Follow the protocol in Procedure 19.1 to complete this experiment.

Anticipated Results

Pitting zones will be seen on the surface of PSSA around the bacterial spots of bacterial species or strains that produce and secrete pectinases (Figure 19.3). This is due to the degradation of the polygalacturonic acid in the media by the polygalacturonases produced by these bacteria. Furthermore, the pectolytic activities of a bacterial species or strain will vary depending on the

Procedure 19.1

Detection of Pectolytic Activity on Pectate Semisolid Agar

Step	Instructions and Comments
1	Add 200 mL of distilled water in a 1-L Erlenmeyer flask and stir the water with a magnetic stirrer.
2	While stirring the water, add: a. 1.2 mL of 10% $CaCl_2{\bullet}2H_2O$ solution (in water), b. 2.0 mL of 0.1% BTB solution (in 6.4×10^{-4} N NaOH), and c. 1.0 g of yeast extract.
3	Begin to heat the solution and slowly add: a. 6.0 g of polygalacturonic acid and b. 0.6 g of agar.
4	Continue heating and stirring the solution until the polygalacturonic acid and agar are completely melted. **Caution:** Never let the solution boil.
5	Adjust the pH of the solution to the desired level while the solution is still hot and being stirred. a. For acidic PSSA (pH 4.5–4.7), add 1 N HCl prior to autoclaving at 121°C for 15 min. b. For neutral PSSA (pH 6.9–7.1), add 1 N HCl or 1 N NaOH prior to autoclaving at 121°C for 15 min. c. For basic PSSA (pH 8.1–8.3), add 1 N NaOH *after* autoclaving at 121°C for 15 min. **Caution:** Polygalacturonic acid is labile at a high pH, so the pH of basic PSSA should be adjusted *after* it is autoclaved.
Note:	The pH of PSSA typically decreases slightly (approximately 0.3–0.5 units) after it is autoclaved; therefore, the pH of the acidic and neutral PSSA should be adjusted again after autoclaving.
6	Pour the autoclaved PSSA, adjusted to the proper pH, into Petri dishes when it is still hot (approximately 65°C). **Caution:** Polygalacturonic acid solidifies at higher temperatures than agar and cannot be melted again. Use safety gloves when pouring PSSA.
7	Replace the lids of the Petri dishes with new ones when the PSSA is cooled (approximately 40°C) to prevent the formation of moisture under the lids.
8	Spot 10 µL of an overnight culture of each bacterial species on the acidic, neutral, and basic PSSA plates and incubate plates in the dark at 28°C–30°C.
9	Observe PSSA plates after 2–3 days of incubation or pitting zones (sunken areas) produced around the bacterial spots on each plate.

pH (acidic, neutral, or basic pH) of the PSSA. See also Chapter 32 for other experiments concerning enzymatic activities.

Questions

- Which bacterial strain showed the strongest or weakest pectolytic activity?
- At which pH (acidic, neutral, or basic) did each bacterial species or strain show the strongest or weakest pectolytic activity? Why does pH affect pectolytic activity?
- If a given plant pathogen does not show pectolytic activity on PSSA, does it mean that this pathogen does not have genes for pectinases production and secretion? Why?
- How can we use PSSA to identify new bacterial genes involved in pectolytic activity?

Experiment 2. Elicitation of the Hypersensitive Response on Tobacco Leaves

The HR is a form of rapid, programmed cell death, which is often accompanied by gene-for-gene resistance derived from incompatible plant–pathogen interactions (Figure 19.4). Gene-for-gene resistance, which is usually mediated by the interaction between a plant's resistance gene and a pathogen's avirulence gene, is generally robust and effectively defends a host plant from pathogen invasion. This resistance, however, can easily be abolished by the occurrence of pathogens that lack avirulence genes.

In many Gram-negative plant pathogenic bacteria, the collective actions of multiple type III effectors are responsible for pathogenicity in susceptible hosts, while the action of a single type III effector acting as an avirulence protein in resistant hosts causes gene-for-gene resistance. This means that T3SSs, which are responsible for the secretion of type III effectors, are required for both gene-for-gene resistance and pathogenesis by bacterial pathogens. T3SSs also secrete **harpin** proteins, which cause HR elicitation in many plants, including tobacco, by physically binding to the plant cell wall rather than by acting through the gene-for-gene resistance pathway. This requirement of a T3SS for pathogenesis and the HR is the reason that genes encoding the components of a T3SS were originally named *hrp* genes for HR and pathogenicity genes, and the reason that the HR phenotype of Gram-negative plant pathogenic bacteria in nonhost plants or in resistant-host plants is considered to be an indicator of the presence of a functional T3SS in a pathogen.

The HR is normally confined to plant cells that come in direct contact with avirulent bacteria (Figure 19.4). One bacterial cell carrying an avirulence gene can elicit an HR in one plant cell. Thus, the HR cannot be observed macroscopically if low concentrations of a bacterial suspension ($<10^5$ CFU/mL) are infiltrated into a nonhost plant. In this case, the death of plant cells due to HR can be observed with a microscope after cell staining with trypan blue. An HR is visible as a confluent necrotic lesion to the unaided eye, however, when high concentrations of bacterial cells (generally $> 5 \times 10^6$ CFU/mL) are infiltrated into a nonhost plant. In addition, HR elicitation can be suppressed by various chemicals that inhibit plant metabolism, revealing that HR elicitation is a consequence of the active cellular processes of a plant cell (programmed cell death).

The purposes of this experiment are to observe the development and nature of the HR caused by the tomato pathogen *P. syringae* pv. *tomato* in tobacco leaves and to test the effects of cycloheximide and sodium vanadate on HR elicitation.

FIGURE 19.4 A typical hypersensitive response elicited by a bacterial plant pathogen on a tobacco leaf.

Materials

Each student or team of students will require the following supplies:

- Bacteria:
  - *P. syringae* pv. *tomato* (not pathogenic to tobacco). *Note*: If this pathogen is not available, other pathovars of *P. syringae*, *Xanthomonas* spp. or *E. amylovora* can be used.
  - *P. syringae* pv. *tabaci* (tobacco pathogen).
  - An *hrp*⁻ derivative of *P. syringae* pv. *tomato* or of the pathogen chosen in #1 that is defective in the T3SS (optional).
  - Nonpathogenic bacteria that do not have a T3SS, i.e., *E. coli* or *P. fluorescens*.
- Luria Bertani (LB) or King's B (KB) agar media
- Disposable 1-mL needleless syringes
- Razorblades
- $MgCl_2$
- Cycloheximide
- Sodium orthovanadate (Na_3VO_4)
- Six-well cell culture plates
- Sterile cotton swabs (or a metal inoculation loop that can be sterilized by flaming)
- Spectrophotometer
- Drying oven (for incubation at 95°C)
- Incubator (28°C–30°C)
- Glass slides and coverslips
- Compound microscope
- Permanent markers

Follow the instructions in Procedure 19.2 to complete this experiment.

Anticipated Results

Tobacco leaves infiltrated with *P. syringae* pv. *tomato* (or any other plant pathogen suggested as alternatives) at a high bacterial concentration (approximately 10^8 CFU/mL) will show tissue collapse and will form a necrotic lesion, limited to the infiltrated areas, within 8–18 h. The HR elicited by these pathogens will be suppressed by the addition of the cycloheximide or sodium orthovanadate. Infiltration of the tobacco pathogen, *P. syringae* pv. *tabaci*, at a high cell concentration will also result in tissue collapse along with necrosis and chlorosis, but this symptom will neither be confined to the infiltrated areas nor suppressed by cycloheximide or sodium orthovanadate. Infiltration of type III secretion mutants or saprophytes at a high bacterial concentration

Procedure 19.2

Elicitation of the hypersensitive response on tobacco leaves

Step	Instructions and Comments
	Part I Elicitation of the HR on Tobacco Leaves
1	Use sterile cotton swabs (or an inoculation loop) to suspend bacterial cells from 24- to 48-h-old bacterial cultures grown on LB or KB agar plates in approximately 10–20 mL of sterile 10 mM $MgCl_2$.
	a. For a high concentration bacterial suspension (approximately 10^8 CFU/mL), adjust the concentration to $OD_{600} = 0.2$ by adding more bacteria to the suspension.
	For a low concentration bacterial suspension (approximately 10^5 CFU/mL), dilute the high concentration suspension 1000 times using 10 mM $MgCl_2$ solution.
2	Use a needleless syringe to infiltrate each bacterial suspension and 10 mM $MgCl_2$ control into a tobacco leaf.
	a. Make a pinhole in the center of the leaf panel to be infiltrated with a syringe needle.
	b. Fill the needleless syringe with one of the bacterial suspensions to be tested or with buffer.
	c. Place the tip of the syringe over the pinhole on the abaxial side of the leaf and support the point of infiltration with your finger on the adaxial side of the leaf.
	d. Slowly infiltrate the leaf with the bacterial suspension or buffer until approximately 20 cm^2 of leaf area has been infiltrated.
3	Mark the infiltrated (water-soaked) area with a permanent marker to determine if the HR is confined to the infiltrated area later.
	Part II Cycloheximide and Sodium Orthovanadate Effects on HR Elicitation
4	Repeat Steps 2 and 3 on a new tobacco leaf.
5	After 10 min, infiltrate 50 μg/mL of cycloheximide or 100 μM of sodium orthovanadate into the areas previously infiltrated with bacterial suspensions.
	Caution: Cycloheximide and sodium orthovanadate are also toxic to humans! Wear chemically resistant nitrile gloves!
6	Infiltrate cycloheximide and sodium orthovanadate alone as controls.
7	Observe symptoms on tobacco leaves infiltrated in Part I and Part II at 8–18 h after infiltration.
	Part III Trypan Blue Staining of Tobacco Leaves (Koch and Slusarenko, 1990)
1	Prepare the trypan blue staining solution by dissolving 10 mg of trypan blue in 40 mL of lactophenol solution (10 mL of lactic acid, 10 mL of saturated phenol, 10 mL of glycerol, and 10 mL of distilled water).
2	Use a razorblade to cut the leaf areas previously infiltrated with a low concentration of bacterial cells (approximately 10^5 CFU/mL) into 2 × 2 cm pieces.
3	Place three to four leaf samples in each well of a six-well culture plate.
4	Add 6 mL of trypan blue staining solution to each well containing leaf samples.
5	Incubate the six-well culture plate at 95°C for 5 min.
6	When the plate is cooled, use a pipette to remove the trypan blue staining solution and destain the leaf samples in chloral hydrate solution (250 g of chloral hydrate dissolved in 100 mL of distilled water) for at least 30 min at room temperature. In the United States, research use of chloral hydrate requires a drug license because it is a schedule IV controlled substance. Consult your university compliance officers.
7	Mount each leaf sample in the chloral hydrate solution on a glass slide, cover the leaf sample with a coverslip, and observe the stained plant cells using a compound microscope.

will not cause visible tissue collapse or the formation of a necrotic lesion. In leaf tissues infiltrated with suspensions of low bacterial concentration, similar patterns of HR elicitation will also be observed if dead cells in leaf tissues are stained with trypan blue and observed under a microscope.

Questions

- What is the difference between the cell death caused by *P. syringae* pv. *tomato* and that caused by *P. syringae* pv. *tabaci* in tobacco leaves?
- How does cycloheximide and sodium orthovanadate inhibit plant metabolism? What is their mode of action?
- What experiment could be performed to prove that the suppression of HR elicitation by cycloheximide or sodium orthovanadate is due to the inhibition of the metabolism of the plant and not to a toxic effect on the infiltrated bacteria?
- Why do some plant pathogenic bacteria not elicit a HR in nonhost plants, including tobacco? Does this mean that these pathogens do not have a functional T3SS? Why?
- What is the role of the HR in the gene-for-gene resistance of plants to bacteria?

REFERENCES

Alfano, J. R., and A. Collmer. 1997. The type III (Hrp) secretion pathway of plant pathogenic bacteria: Trafficking harpins, Avr proteins and death. *J. Bacteriol.* 179:5655–5662.

Brito, B., M. Marenda, P. Barberis, C. Boucher, and S. Genin. 1999. *prhJ* and *hrpG*, two new components of the plant signal-dependent regulatory cascade controlled by PrhA in *Ralstonia solanacearum*. *Mol. Microbiol.* 31:237–251.

Douzi, B., A. Filloux, and R. Voulhoux. 2012. On the path to uncover the bacterial type II secretion system. *Philos. Trans. R. Soc. Lond. B. Biol. Sci.* 367:1059–1072.

Guttman, D. S., B. A. Vinatzer, S. F. Sarkar, M. V. Ranall, G. Kettler, and J. T. Greenberg. 2002. A functional screen for the type III (Hrp) secretome of the plant pathogen *Pseudomonas syringae*. *Science* 295:1722–1726.

Ham, J. H., R. A. Melanson, and M. C. Rush. 2011. *Burkholderia glumae*: Next major pathogen of rice? *Mol. Plant Pathol.* 12:329–339.

Kazemi-Pour, N., G. Condemine, and N. Hugouvieux-Cotte-Pattat. 2004. The secretome of the plant pathogenic bacterium *Erwinia chrysanthemi*. *Proteomics* 4:3177–3186.

Koch, E., and A. Slusarenko. 1990. Arabidopsis is susceptible to infection by a downy mildew fungus. *Plant Cell* 2:437–445.

Koster, M., W. Bitter, H. de Cock, A. Allaoui, G. R. Cornelis, and J. Tommassen. 1997. The outer membrane component, YscC, of the Yop secretion machinery of *Yersinia enterocolitica* forms a ring-shaped multimeric complex. *Mol. Microbiol.* 26:789–797.

Lindeberg, M., S. Cunnac, and A. Collmer. 2012. *Pseudomonas syringae* type III effector repertoires: Last words in endless arguments. *Trends Microbiol.* 20:199–208.

Lindeberg, M., G. P. Salmond, and A. Collmer. 1996. Complementation of deletion mutaitons in a cloned functional cluster of *Erwinia chrysanthemi out* genes with *Erwinia carotovora out* homologues reveals OutC and OutD as candidate gatekeepers of species-specific secretion of proteins via the type II pathway. *Mol. Microbiol.* 20:175–190.

Petnicki-Ocwieja, T., D. J. Schneider, V. C. Tam, S. T. Chancey, L. Shan, Y. Jamir, Y., L. M. Schechter, M. D. Janes, C. R. Buell, X. Tang, A. Collmer, and J. R. Alfano. 2002. Genomewide identification of proteins secreted by the Hrp type III protein secretion system of Pseudomonas syringae pv. tomato DC3000. *Proc. Natl. Acad. Sci. U. S. A.* 99:7652–7657.

Schechter, L. M., M. Vencato, K. L. Jordan, S. E. Schneider, D. J. Schneider, and A. Collmer. 2006. Multiple approaches to a complete inventory of *Pseudomonas syringae* pv. *tomato* DC3000 type III secretion system effector proteins. *Mol. Plant-Microbe Interact.* 19:1180–1192.

Starr, M. P., A. K. Chatterjee, P. B. Starr, and G. E. Buchanan. 1977. Enzymatic degradation of polygalacturonic acid by *Yersinia* and *Klebsiella* species in relation to clinical laboratory procedures. *J. Clin. Microbiol.* 6:379–386.

Tseng, T. T., B. M. Tyler, and J. C. Setubal. 2009. Protein secretion systems in bacterial-host associations, and their description in the gene ontology. *BMC Microbiol.* 9(Suppl 1):S2.

SUGGESTED READINGS

Bent, A. F., and D. Mackey. 2007. Elicitors, effectors, and *R* genes: The new paradigm and a lifetime supply of questions. *Annu. Rev. Phytopathol.* 45:399–436.

Charkowski, A., C. Blanco, G. Condemine, D. Expert, T. Franza, C. Hayes, N. Hugouvieux-Cotte-Pattat, E. Lopez Solanilla, D. Low, L. Moleleki, M. Pirhonen, A. Pitman, N. Perna, S. Reverchon, P. Rodriguez Palenzuela, M. San Francisco, I. Toth, S. Tsuyumu, J. van der Walls, J. van der Wolf, F. Van Gijsegem, C. H. Yang, and I. Yedidia. 2012. The role of secretion systems and small molecules in soft-rot Enterobacteriaceae pathogenicity. *Anuu. Rev. Phytopathol.* 50:425–449.

Denny, T. P. 1995. Involvement of bacterial exopolysaccharides in plant pathogenesis. *Annu. Rev. Phytopathol.* 33:173–197.

Durbin, R. D., and P. J. Langston-Unkefer. 1988. The mechanisms for self-protection against bacterial phytotoxins. *Annu. Rev. Phytopathol.* 26:313–329.

Gross, D. C. 1991. Molecular and genetic analysis of toxin production by pathovars of *Pseudomonas syringae*. *Annu. Rev. Phytopathol.* 29:247–278.

Hann, D. R., and J. P. Rathjen. 2010. The long and winding road: Virulence effector proteins of plant pathogenic bacteria. *Cell Mol. Life Sci.* 67:3425–3434.

Salmond, G. P. C. 1994. Secretion of extracellular virulence factors by plant pathogenic bacteria. *Annu. Rev. Phytopathol.* 32:181–200.

20 Physical and Physiological Host Defenses

Kimberly D. Gwinn and David I. Yates

Concept Box

- Most plants are resistant to most pathogens.
- Resistant plants produce passive barriers that the pathogen cannot overcome or they must be able to activate successful defense(s) that arrest pathogen development.
- Although both active and passive defenses can be important at any stage of the disease cycle, passive defense is usually more important during prepenetration and penetration, and active defense is usually most important in the infection stage.
- Active disease responses require signal recognition/transduction followed by gene activation.
- Examples of passive defenses are inhibitory plant surface chemicals, thick cuticles, lignified tissues, and phytoanticipins. Examples of active defenses are antimicrobial proteins, phytoalexins, the hypersensitive response (HR), and systemic resistance.

Healthy plants grow in an atmosphere crowded with fungal spores, bacterial cells, and viruses. Despite the high numbers of fungal spores, bacterial cells, and nematodes that thrive in the rhizosphere (soil immediately around the roots of the plant), healthy roots predominate. In the face of this onslaught of potential pathogens, plants defend themselves with an arsenal of weapons, and as a result, most plants are resistant to most pathogens. Plants have developed defensive strategies that successful pathogens must overcome. Although plants defend themselves against potential pathogens in different ways, strategies are classified into the following two basic categories: **passive defense** (physical barriers or chemical reservoirs present before pathogen recognition) and **active defense** (activated after recognition of the pathogen as nonself). Knowledge and exploitation of host defenses can lead to new pathogen control strategies such as "fungicides" that turn on resistance; transgenic plants that can silence viruses; plants that overproduce bioactive natural products; and plants that produce antimicrobial proteins. Successful host defenses disrupt the disease cycle (Chapter 1), primarily in the prepenetration, penetration, or infection phases. In general, passive defenses against pathogen attack are more prevalent in the prepenetration and penetration phases, and active defenses are more important in the infection phase.

PREPENETRATION

Passive barriers are most important for preventing disease establishment during prepenetration. Pathogens must attach to the surface and out-compete existing nonpathogenic organisms that have colonized the plant surface. Specific chemistry and spatial architecture of the plant surface control the suitability of the microenvironment and, thus, determine if the pathogen can survive, recognize the host, and colonize the plant surface.

Surfaces of the plant contain both nutrients and compounds that inhibit pathogen establishment. Pathogens must compete with other organisms on the surface for nutrients including sugars, nitrogen, and essential inorganic molecules. The nature and quantity of available nutrients may favor the growth of the pathogen or the other organisms on the surface. Populations of microorganisms are also controlled by the presence of antimicrobial compounds on the plant surface. Successful pathogens have evolved methods to detoxify or escape these compounds. For example, lipopolysaccharides (LPS), which are cell surface components of Gram-negative bacteria, act as barriers that exclude antimicrobial compounds and allow the bacterium to grow under unfavorable environmental conditions (see Case

CASE STUDY 20.1

Bacterial Wilt

- Bacterial pathogen—*Ralstonia solanacearum*. This bacterium causes bacterial wilt of over 200 plant species (Figure 20.1) in more than 50 families, and it has been identified as one of the top 10 plant pathogenic bacteria in scientific and economic importance (Mansfield et al., 2012). The vascular tissues of stems, roots, and tubers turn brown and, when cut, ooze a stream of bacteria. *Ralstonia solanacearum* colonizes both the soil, which is nutrient-poor, and the inside of a plant, which is nutrient-rich but well-defended.
- Bacterial cells attach to the plant roots and form micro-colonies at sites on the root which are rich in nutrients and are vulnerable (e.g., lateral root emergence and the root elongation zone), and bacteria rapidly develop within the intercellular spaces of the inner cortex.
- While entering the plant cortex and the vascular system, the pathogen causes minimal tissue/cell damage and either avoids or suppresses host recognition. **Exopolysaccharides** may be responsible for masking bacterial structures that are targets of host recognition because strains of the bacterium that do not produce the polysaccharide simply agglutinate and degenerate in the cortex, perhaps due to defense responses. Exopolysaccharides trigger enhanced expression of the ethylene and salicylic acid defense response pathways in wilt-resistant cultivars but not in susceptible ones.
- Bacteria sense that they have arrived in the plant cell and induce genes for host infection, the so-called *hrp* genes. Inactivation of the *hrp* genes causes nearly complete loss of virulence, and bacteria that lack these induce a hypersensitive response (HR) on resistant plants. The *hrp* genes encode proteins (harpins) that allow translocation of avirulence factors across bacterial membrane for delivery to the host cell. PopA and PopW, two *hrp* genes, induce HR-like responses in plant tissues. When harpin genes are constitutively expressed in plant cells or harpins are sprayed on plants, they do not cause HR cell death, but they do enhance defense responses to diverse plant pathogens, including fungi, oomycetes, bacteria, and viruses (Choi et al., 2013). PopW can be used to induce systemic acquired resistance (SAR) for the control of *Tobacco mosaic virus* (TMV) (Li et al., 2011) (Case Study 20.2).
- Plant resistance based on effector-triggered immunity (ETI), and polygenic resistance to bacterial wilt have been identified (Peeters et al., 2013).

FIGURE 20.1 Bacterial wilt of tomato. (Courtesy of the American Phytopathological Society.)

Study 20.1). Bacteria that lack **lipopolysaccharides (LPS)** are generally less-effective pathogens. It is important to note, however, that in plants resistant to these pathogens, LPS can function as signals to turn on active resistance mechanisms (Silipo et al., 2010).

Plant surface architecture and microenvironment can determine if the plant will be a suitable host for a pathogen. Pathogens must compete with other plant surface-colonizing organisms for sites with high available nutrients and water such as depressions on the leaf surfaces where abundant nutrients are available due to localized leakage or injury. Availability of free water is essential for bacterial pathogens that must multiply, for fungal spores that must germinate, and for nematodes that must hatch from eggs on the plant surface. The physical attributes of the plant surface can affect these processes and limit pathogens. Plant hairs or trichomes protrude from leaves and stems, disrupt continuity of the surface, and reduce the availability of free water. For some fungi, formation of penetration structures is induced by physical contact with the ridges and valleys of the host surface; in others, recognition of host chemicals can induce the formation of structures for attachment and penetration. In prepenetration, waxes on the surface of many plant parts limit the availability of the free water. The thickness of physical barriers also plays an important role in host defense during prepenetration, penetration, and infection. The role of cuticle and cell wall thickness are discussed in the section on penetration. Even small changes in the structural architecture on the plant surface may alter the ability of a pathogen to infect a plant (Łaźniewska et al., 2012).

PENETRATION—PASSIVE

Physical barriers can limit pathogen penetration. The cuticle, a waxy layer made primarily of cutin, determines the hydrophobicity of plant leaves and stems and is generally the first point of contact between pathogen and plant (see Case Study 20.2). The thickness of the cuticle plays an essential role in host defense against some pathogens—fungi penetrate a thin cuticle layer more easily than a thicker cuticle. The thickness of the cuticle may also control the release of bioactive compounds that inhibit the growth pathogens. Suberin, a compound similar to cutin that is associated with cork cells, can be a passive defense or formed in an active response. Cell wall thickness plays a role both in the epidermal cells and in the cells being invaded by a pathogen. Host plant cell walls can be breached by pathogen-produced enzymes that break down cuticles or components of the host cell walls (Chapter 32). Enzymes produced by pathogens that degrade plant cell walls provide insight into the molecular arms races between pathogens and hosts. Some plants foil this strategy by producing compounds that inhibit the cuticle and cell wall-degrading enzymes. Successful pathogens that can breach the host defense often have enzymes that are not sensitive to inhibitors produced by the plant (Łaźniewska et al., 2012).

CASE STUDY 20.2

Powdery Mildew of Barley

- Fungal pathogen—*Blumeria graminis* f. sp. *hordei (Bgh)*. This obligate fungal pathogen is the causal agent of powdery mildew of barley and ranks as one of the top 10 most important plant pathogenic fungi (Dean et al., 2012).
- Although host cuticle plays an important role in resistance to many pathogens, *Bgh* effectively degrades the barley cuticle using esterase enzymes that break the ester bonds that hold the cutin molecules together. Esterases are produced within 2 h of the conidia landing on the host.
- In susceptible cultivars, some cells resist penetration and produce papillae. Gene expression differed in infected and resistant cells. In general, resistant cells had greater expression of genes associated with resistance including signaling, reactive oxygen production, secondary metabolism, and cell wall degrading enzymes than cells that were infected (Gjetting et al., 2007).
- Host genes govern resistance at different stages of the interaction (Figure 20.2). In plants containing the *Mla* gene, papillae formed by the attacked cell arrest growth of *Bgh*, and an hypersensitive response (HR) is formed in the attacked cell (Figure 20.2a). If the fungus breaches the papilla defense, cells immediately subjacent undergo an HR (Figure 20.2b). *Arabidopsis* plants that lack genes controlling nonhost immunity can serve as compatible hosts for some isolates of *Bgh*; effector triggered-immunity (ETI)-based resistance can be added by insertion of the barley *Mla* gene into this host background. Plants with the resistant response differ in transcriptional activities from plants with a compatible reaction. Transcriptional changes resulting in resistance or compatibility were similar in both barley and *Arabidopsis* (Hacquard et al., 2013).
- In plants containing the *mlo* gene, penetration of the fungus is stopped by formation of effective papillae (Figure 20.2c); this type of resistance is effective against all races of the pathogen. Occasionally, a few mildew colonies appear in *mlo*-resistant barley. Silicon-rich halos are observed in wheat and barley infected with *Bgh*. Silicon levels are higher in regions of failed penetration attempts than in sites of successful penetration. Peroxidases and hydrolytic enzymes are also found in the halos.
- In susceptible plants, cell wall penetration is followed by the formation of a haustorium and elongated secondary hyphae (Figure 20.2d).
- Barley infected with a mycorrhizal fungus, *Piroformospora indica*, has enhanced active defense to *Bgh* consistent with induced systemic resistance (ISR). Transcription of defense-related responses (e.g., apoptosis and pathogenesis-related protein) produced in response to *Bgh* was greater and more rapid in mycorrhizal than nonmycorrhizal plants (Molitor et al., 2011).

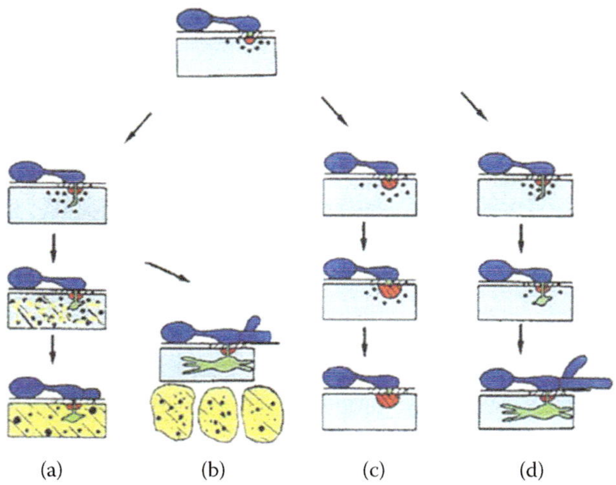

FIGURE 20.2 Reactions of barley to *Blumeria graminis* f. sp. *hordei* (*Bgh*) mediated by the resistance genes, *Mla* (a and b) and *mlo* (c), and in susceptible plants (d). (a) In plants with *Mla*, papillae (red) impede penetration of the fungus, *Bgh* (dark blue), into the plant cell (light blue,) and a hypersensitive reaction (HR) (yellow) occurs in the attacked cell. (b) In a few cells of plants with *Mla* genes, *Bgh* is able to penetrate the papilla and form a haustorium (green); further disease development is limited by an hypersensitive response (HR) in cells beneath the attacked epidermal cell. (c) In plants with the recessive *mlo* gene, papilla formation is sufficient for the protection of the plant cell against *Bgh*, and no HR occurs. (d) In plants without resistance genes, papilla formation is insufficient to inhibit cell wall penetration, and *Bgh* forms a haustorium, does not stimulate an HR, and disease progression is not impeded. (Courtesy of the American Society of Plant Biologists.)

PENETRATION—ACTIVE

Plants have developed a number of active mechanisms to limit the entry of pathogens that overcome preestablished barriers. The key to active defense is host recognition of the pathogen. Plants react to the presence of a pathogen by signal recognition, followed by signal transduction, and finally gene activation. Gene activation results in end products that contain, inhibit, or kill the invader. Plants react to invasion by pathogens with quick and efficacious defense responses to penetration. The following three categories of active defense have been described: primary, secondary, and systemic. Primary responses are those limited to the cell in contact with the pathogen. Secondary responses are induced in cells adjacent to the cell in contact with the pathogen, and systemic responses are induced throughout the entire plant.

An understanding of **signal recognition/transduction** is essential to understand active defense mechanisms. A signal recognition/transduction pathway is activated in response to either **effectors**, compounds that serve to facilitate infection in virulent pathogens but trigger immunity in plants with corresponding **resistance genes (R-genes)**, or molecules recognized by the innate (nonhost) immune system, the so-called pathogen-associated molecular patterns (**PAMPs**). The pathway involves ion fluxes, oxidative bursts, signaling molecules, and activation of mitogen-activated protein kinases (MAPKs). These early events signal defense responses, including the activation of diverse defense genes, cell wall strengthening, phytoalexin biosynthesis, hypersensitive response (HR), and induced resistance (Meng and Zhang, 2013). An ion flux is an efflux of Cl^- and K^+ ions coupled with an influx of H^+ and Ca^{2+}. Currently, Ca^{2+} is believed to be a central mediating component of early plant defense responses involving elaborate signal decoding systems (Kudla et al., 2010). An oxidative burst, the rapid production of hydrogen peroxide (H_2O_2), is believed to signal cell death, overwhelm antioxidant cell protection, lead to rapid cell wall reinforcement, and induce gene expression. Signal molecules are the active defense systems of all plants studied thus far, and signals are induced by small-molecular-weight compounds, such as salicylic acid, jasmonic acid, ethylene (Figure 20.3), glycerol-3-phosphate, and abietane diterpenoid hydroabietinal. Jasmonic acid and its methyl ester, methyl jasmonate, increase in response to pathogen attack, both locally and systemically, and are both preexisting and induced compounds. These molecules are produced at penetration for some fungal and nematode diseases and in early infection for bacterial, viral, and some fungal and nematode diseases. *Arabidopsis* mutants impaired in the jasmonic acid response have increased susceptibility to necrotrophic fungi, but not to biotrophic fungi. Ethylene, a gaseous plant growth regulator, is synthesized in both compatible and incompatible reactions. Ethylene appears to mediate resistance against necrotrophic fungal pathogens and nonhost resistance.

Treatment of plants with salicylic acid, the active component in aspirin, induces systemic resistance in many plants and has been implicated as an important signal for the activation of both **PAMP-triggered immunity (PTI)** and **effector-triggered immunity (ETI)** (Vlot et al., 2009). Some chemical signals move through the plant and initiate protection of distant plant tissues (e.g., **systemic acquired resistance, SAR**). The nature of the mobile signal is not known, but several possible mobile signals (methyl salicylic acid, jasmonic acid, azelaic acid, glycerol-3-phosphate, and abietane diterpenoid hydroabietinal) have been identified (Fu and Dong, 2013). In response to

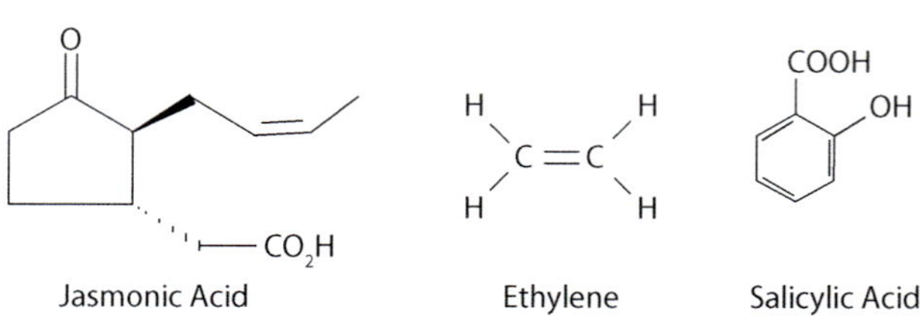

FIGURE 20.3 Examples of signal compounds; jasmonic acid, ethylene, and salicylic acid. The active defense systems of all plants studied thus far are induced by these molecules.

these signals, genes are activated, resulting in increased translation and transcription of proteins that are directly or indirectly involved in host defense.

Plants can limit penetration of many pathogens to the first cell that is attacked or to cells in the immediate vicinity via primary and secondary defensive responses. Early responses to pathogens that directly penetrate into the cells include aggregation of the cytoplasm and rapid formation of cell wall thickenings or papillae in the cells under attack. Papillae are cell wall appositions deposited between the plasma membrane and the cell wall (Figure 20.2). They consist primarily of callose, a polymer of β-1,3 D-glucose residues, but also may contain cellulose, phenolic compounds, and lignin. The growth and intrusion of the fungal penetration peg across the plant cell wall may be stopped by papilla or the fungus may successfully transverse it. Halos, locally modified regions of the host cell wall around a penetration sites, are rich in phenolics, silicon, lipids, proteins (peroxidases), and lignin. Halos are often larger at unsuccessful penetration sites than at successful penetration sites. Epidermal cells adjacent to papillae or appressoria are often lignified. In addition to papilla, there are many other structural barriers. Suberization, deposition of insoluble polymers at the cell wall, also increases resistance to some fungal pathogens. Growth of pathogens may be stopped temporarily in unripe fruit, but may resume as the fruit tissue ripens, causing postharvest diseases. For example, in some fruit diseases caused by *Colletotrichum gloeosporioides*, conidia germinate within hours of landing on the surface and penetrate the cuticle to form swollen hyphal structures that remain quiescent. Mechanisms that activate the hyphae to grow as the fruit matures are mostly unknown, but may be related to changes in the plant cell wall that renders it ineffective as a barrier (Prusky et al., 2013). For pathogens that enter the plant through stomatal openings rather than directly penetrate into the cells, plants have evolved mechanisms to regulate the stomatal aperture as defense strategy (Zeng et al., 2010).

INFECTION—PASSIVE

Plants produce a vast array of low-molecular-weight secondary metabolic compounds that are present at concentrations that affect pathogens. General classes of compounds in the defensive arsenal of plants include the following: flavonoids, phenolics, glucosinolates, terpenoids, and alkaloids. Antimicrobial plant defense compounds are grouped into the following two classes: the **phytoanticipins** and **phytoalexins**. Phytoanticipins are preestablished, small-molecular-weight compounds stored in the plant cell or released from a glucoside (see Case Study 20.3).

CASE STUDY 20.3

Antifungal Phytoanticipins in Mango Fruit

- Fungal pathogen—*Colletotrichum gloeosporioides*. This pathogen causes mango anthracnose. Losses due to mango anthracnose can approach 100% in fruit produced under wet or humid conditions.
- Unripe fruit actively developing mango fruit is more resistant to mango anthracnose than ripe fruit. Unripe fruit typically contains more of the plant defense chemicals that inhibit fungal penetration and infection. Mechanisms that activate the change to an active infection are mostly unknown, but may be related to changes in the cell wall that render it ineffective as a barrier (Prusky et al., 2013) or to timely production and concentrations of secondary metabolites (e.g., gallotannins and resorcinols) and antimicrobial proteins (e.g., chitinase).
- On unripe fruit, conidia germinate and form infection hyphae that remain quiescent (dormant). Unripe mango has high concentration gallotannins and resorcinols in the outer peel (exocarp). Gallotannin and resorcinol concentrations are higher in resistant cultivars of mango.
- Gallotannin concentrations decline as fruit ripens and hyphae are no longer inhibited. In resistant cultivars, gallaotannin concentration does not decline as rapidly as that in susceptible cultivars. As fruit over-ripens, anthracnose lesions begin to expand, and hyphae penetrates the peel (Figure 20.4).
- Mango fruits also contain substantial latex early in development that decreases in volume and in concentration as fruits ripen. The latex is contained in the exocarp and in the outer regions of the mesocarp. Resorcinol concentration is several times greater in latex than the fruit peel. Latex also contains chitinase, an enzyme that has the ability to rapidly digest the wall of conidia of *C. gloeosporioides*. Fruits drained of latex develop anthracnose to a greater extent than fruits with latex (summarized from Karunanayake et al., 2011).
- Resistance of unripe fruits to and the quiescence of *C. gloeosporioides* appears to be due to a combined effect of gallotannins and resorcinols in the outer exocarp and the resorcinols and chitinase in the latex.

FIGURE 20.4 Symptoms of mango anthracnose caused by *Colletotrichum gloeosporiodes* on ripe fruit. (Courtesy of Scott Nelson, University of Hawai'i at Mānoa.)

Phytoalexins are formed in response to the pathogen and so are discussed in the section on active defense responses. Some compounds are phytoalexins in one species and phytoanticipins in another. Most antimicrobial natural products are broad spectrum, and specificity, if it exists, is determined by the pathogen's ability to break down the compounds. In a few studies, plants with lower amounts of antimicrobial compounds are more susceptible to disease. Crops may be susceptible to many pathogens because selective breeding for other characteristics has decreased fitness by reducing numbers and concentrations of natural products (Großkinsky et al., 2012). Many phytoanticipins are stored within vesicles, whereas others are stored in the glycosylated form until cells are damaged. Glycosylation or chemically bonding to a glucose molecule converts a reactive and toxic phytoanticipin to a stable, nonreactive storage form that is more likely to be water soluble. Many transgenic plants that have been engineered to make glycosidic phytoanticipins have increased resistance to disease-causing organisms.

INFECTION—ACTIVE

Phytoalexins

Phytoalexins (Figure 20.5) are antimicrobial natural products that are produced after infection or elicitation by abiotic agents. These antimicrobial molecules are chemically diverse because they are manufactured by a number or combination of biosynthetic pathways. However, plant species in the same family tend to make phytoalexins that are derived from the same pathway and, therefore, are very similar in structure. The bioactivity of phytoalexins is not limited to their role in disease resistance. They can be important medicinal and nutraceutical compounds. For example, resveratrol is important in the resistance to *Botrytis cinerea*,

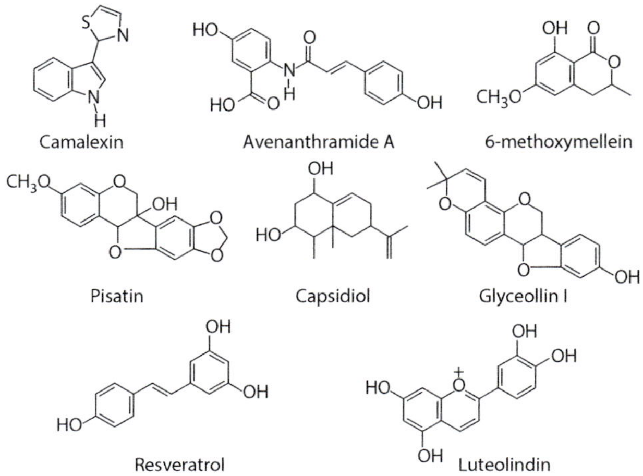

FIGURE 20.5 Structures of selected phytoalexins.

the causal agent of gray mold of grapes. Plants genetically engineered to overproduce resveratrol are more resistant to pathogens that cannot detoxify the compound (Großkinsky et al., 2012). Because it has cardioprotective, antitumor, neuroprotective, and antioxidant properties, plants that have high concentrations of resveratrol or its phytoanticipin glycoside (piceid) are highly valued.

Although it is generally accepted that phytoalexins play a role in host defense, they are likely only one part of an overall defense strategy. Several lines of evidence support a role for phytoalexins in host defense and include the following: (1) in ETI, resistance is often associated with phytoalexin production; (2) phytoalexins accumulate rapidly to inhibitory concentrations at the site of pathogen development; (3) pathogens can overcome host resistance by phytoalexin detoxification; and (4) plants genetically transformed to overproduce phytoalexins are more resistant to disease than nontransformed plants (Hammerschmidt, 1999; Großkinsky et al., 2012). Phytoalexins are produced by the cell under attack as a primary response and in adjacent cells as a secondary response.

Antimicrobial Proteins

Antimicrobial proteins have been detected in many plant species and tissues. The widespread localization of antimicrobial proteins, such as chitinases and β-glucanases in plants, coupled with their activity against pathogens *in vitro* suggests that these enzymes may serve a protective role. Plants that are transformed with genes that code for these antimicrobial proteins can be more resistant to pathogens.

Hypersensitive Response

The active responses described above are nonspecific; they occur in response to pathogens and other organisms. During initial contact of plant and pathogen, the plant cells recognize the PAMPs and turn on responses that result in resistance (PTI). ETI, however, is highly specific and occurs only when the product of a pathogen avirulence gene (effector) interacts with the product of a plant resistance gene (R-gene). Pathogen effectors are injected into the host cell to suppress PTI host defense and cause disease. Plants that have R-genes recognize effectors and turn on specific resistance (ETI). Activation of this gene-for-gene resistance results in a cascade of reactions within the cell. The HR, rapid death of a few host cells that limits the progression of the infection, is a manifestation of recognition of the effector (Figure 20.6). Typically, a HR includes signal transduction, programmed cell death, increased activation of defense-related genes (e.g., synthesis of phytoalexins, salicylic acid, and antimicrobial proteins), and a distant induction of general defense mechanisms that serve to protect the plant (i.e., SAR).

Elicitation of the primary responses results from the recognition of a pathogen effector protein by a host receptor protein. Most effector proteins have no apparent enzymatic activity, but are capable of binding to the receptor protein, which is a product of the R-gene. When the two proteins interact, a signal transduction pathway is activated, and the primary response is initiated. A rapid burst of oxidative metabolism leads to the production of superoxide and subsequent production of hydrogen peroxide that precedes the development of visual symptoms of a HR.

Plants appear to have adapted **programmed cell death** or **apoptosis** a general process commonly associated with reproductive and xylem tissue development, as a host defense response. The attacked cell and several cells around it die in response to chemical signals. The sacrifice of these cells isolates the pathogen and is a successful resistance mechanism against biotrophic pathogens. The HR is not limited to biotrophic pathogens. Other classes of pathogens may also be killed by the reactive oxygen species generated by the oxidative burst or by the antimicrobial compounds and enzymes formed in response to gene activation.

FIGURE 20.6 Hypersensitive response in tobacco. (Courtesy of the American Phytopathological Society.)

Systemic Resistance

Although plant systems do not truly mimic the immune system of mammals, they do have the ability to better resist pathogens after exposure to other organisms. Infection or colonization of the plant by one organism can induce host resistance to other pathogens. The two major types of systemic resistance are SAR and **induced systemic resistance (ISR)**. In SAR, the attacking organism causes plant cell damage (usually HR-like), which in turn elicits the production of signaling compounds by adjacent cells. Signals include salicylic acid, methyl salicylic acid, azelaic acid, glycerol-3-phosphate, and abietane diterpenoid dehydroabietinal. These signals then lead to systemic expression of the antimicrobial enzymes such as chitinases and glucanases in the unchallenged cells/tissues that protect the rest of the plant from infection. Methyl salicylate, which is volatile, may serve as a signal to neighboring plants as well. Application of SA or its analogs (2,6-dichloroisonicotinic acid [INA] and benzothiadiazole S-methyl ester [BTH]) can induce SAR in plants that have not been attacked by pathogens (see Case Study 20.4). Bacteria and fungi that colonize the roots, but do not cause disease, may induce ISR. In ISR, both jasmonic acid and ethylene signal other portions of the plant, but antimicrobial enzymes are not produced.

LABORATORY EXERCISE

The following laboratory exercise is designed to demonstrate the effects of essential oils (commonly used as candy flavorings) on the growth of plant pathogenic fungi and to demonstrate the impact of wound healing on disease.

Experiment 1. Effect of Volatile Compounds from Candy Flavorings on the Growth of Plant Pathogenic Fungi

Essential oils are highly volatile substances isolated from an odiferous plant; the term essential was used because these oils were thought to contain the essence of odor and flavor. The oil bears the name of the genus or common name of the plant from which it is derived and can be somewhat misleading because chemistry can be highly variable

CASE STUDY 20.4

Tobacco Mosaic Virus

- Viral pathogen—*Tobacco mosaic virus* (TMV). Plant virologists rank TMV as the most important plant virus in terms of scientific and economic importance (Scholthof, 2011; Scholthof et al., 2011). All viruses are obligate biotrophic organisms (Figure 20.7).
- In some cultivars of solanaceous crops, symptoms of TMV range from mild to severe mottling, chlorosis, and dwarfing of leaves. Tobacco cultivars with the N-gene produce a hypersensitive response (HR) in response to TMV.
- Methyl salicylate, released from necrotic lesions on tobacco caused by TMV, increases the resistance of uninfected neighboring plants. The Pathogenesis related 1 (*PR1*) gene is also induced in neighboring plants.
- Concentration of a protein located in the plastid can control host reaction to TMV. Overexpression of this protein increases the rapidity of the HR; low concentrations lead to a suppression of the HR.
- Application of PopW (a harpin protein produced by *Ralstonia solanacearum*—see Case Study 20.1) stimulated an Systemic acquired resistance (SAR) response in tobacco plants and effectively suppressed TMV infection in the greenhouse and in the field up to 100 days after treatment (Li et al., 2011).
- Rapid cell death occurs in plants with an N-gene except when plants are transformed with viral genes that slow down the HR and result in susceptible interaction.
- At least three protein products have been demonstrated to function as avirulence determinants: the replicase protein (in lines carrying the N-gene), the coat protein (in tobacco lines carrying the N'-gene), and the movement protein (in tomato lines carrying Tm-2 and Tm-22).

(a) (b)

FIGURE 20.7 *Tobacco mosaic virus* of tobacco in plant without (a) and with (b) N-gene resistance. (Courtesy of Karen-Beth Scholthof, Texas A&M University.)

within a genus or species (Gwinn et al., 2010). Since antiquity, essential oils have been used as perfumes, medicines, and flavorings. Essential oils are well known for their antibacterial, antifungal, antiherbivore, and antioxidant activities and have been proposed as natural and safe pesticides. Essential oils of many plants, such as sage, oregano, citrus, and various mints, contain antifungal phytochemicals, which are compounds produced by plants. Effects of essential oils on postharvest and grain spoilage pathogens (or similar species) have been studied most often.

Candy flavorings or aromatherapy oils are easily manipulated, nontoxic sources of phytochemicals. Many (clove, lemon, orange, peppermint, and spearmint) are extracted without solvents as pure oils. These contain several antifungal phytochemicals, most notably various monoterpenes.

General Considerations

Candy flavorings or aromatherapy oils that are pure essential oils can be purchased at local grocery or specialty shops, but care should be taken to avoid flavorings that contain alcohols or other solvents because these will confound the results. Clove, lemon, orange, peppermint, and spearmint oils produced by Lorann Gourmet (Lorann Oils, Inc., 4518 Aurelius, Lansing, MI 48910) can be used with good results.

Any pathogen that is easily cultured on solid media can be used in these experiments. For example, *Alternaria*, *Fusarium*, and *Sclerotinia* species are inhibited by compounds found in the candy flavorings. These experiments are designed for teams of students. Each student should have an opportunity to transfer the mycelial plug to the culture medium, pipette oils onto the filter paper, and measure fungal colony diameter.

Materials

The following items are needed for each team of students:

- For each pathogen to be tested, 14 Petri dishes (10-cm-diameter) containing a fungal growth medium are needed. This is based on three oils at two concentrations run in duplicate, plus dishes for nonamended controls. The choice of medium should be dictated by the choice of pathogen. Potato dextrose agar works well for most fungal pathogens.
- Two actively growing cultures of each pathogen.
- Cork borers or sterile plastic straws (5–7 mm diameter) for cutting similar size mycelial plugs.
- Bunsen burner and 70% ethanol.
- Dissecting probe.
- 10–100 μL pipette and tips.
- Parafilm.
- Filter paper.
- Plastic ruler.

Follow the protocol listed in Procedure 20.1 to complete this experiment.

Anticipated Results

Treatment with essential oils should reduce the growth of all fungi, that is, comparative growth index (CGI) should be less than zero for all compounds. Growth should be inhibited more by the higher concentration (50 μL) of the oil than the lower concentration (5 μL). Some oils will reduce growth more than others. For example, fungal growth in an atmosphere of lemon oil should be less than growth in an atmosphere of orange oil; both should be less than controls.

Questions

- What role might antimicrobial phytochemicals play in plant defense?
- Would a pathogen of orange rind, such as *Penicillium* species, be more or less sensitive or have the same sensitivity to volatile compounds produced by orange than *Fusarium*, which is usually considered a nonpathogen of orange? Why or why not?

Procedure 20.1

Effect of Essential Oils on Fungal Growth

Step	Instructions and Comments
1	Label dishes with your name, genus of pathogen, amount and type of flavoring, and date.
2	Cut several disks (~5 mm) from the edge of an actively growing culture using a cork borer or sterile plastic straw (Figure 20.8a).
3	Place one agar disk, mycelium side down, in the center of the culture medium. If available, place a single sclerotium at the center of the dish (Figure 20.8b). Inoculate all dishes in the same manner.
4	Layer two pieces of filter paper inside the lid of each dish. They should fit snugly and the bottom layer should be in contact with the inside of the lid.
5	Pipette a drop (0, 5, or 50 μL) of candy flavoring at the center of the filter paper. Invert the lid and fit the bottom of the dish into the lid. The Petri dish should be upside-down with the mycelium plug above the filter paper (Figure 20.8c and d). Each treatment should be repeated in a separate dish.
6	Incubate the cultures at room temperature for 1 week. Dishes containing different flavorings should be stored in separate areas. Observe growth of the fungi daily. Control treatments (0 μL) should not be allowed to grow to the edge of the Petri dish. Data should be recorded when the mycelium of the control nears the edge of the dishes or after 1 week.
7	Measure and record the diameter of each fungal colony. Determine the mean diameter for each treatment. Calculate CGI using the following equation: $$CGI = \frac{Dt}{Dc} - 1$$ where Dc is the mean colony diameter for controls and Dt is the mean colony diameter for each treatment. Values will be greater than 0 if mean growth in the treatment exceeds mean growth of control. Values will be less than 0 if mean growth in the treatment is less than the control.

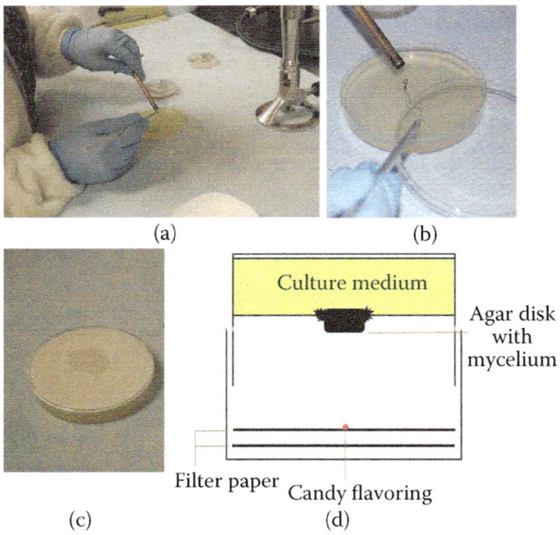

FIGURE 20.8 Procedures for laboratory exercise on the effect of essential oils on fungal growth. (a) Disks are cut from the edge of an actively growing fungal culture using a cork borer. (b) Agar disk is placed with mycelium side down at the center of the culture dish filled with appropriate medium. (c) Two pieces of filter paper are fitted in the lid of the culture dish and a drop of essential oil pipetted on the filter paper. (d) Schematic for the upside-down culture dish showing the mycelial plug above the filter paper. (Courtesy of MiKayla Goodman and Sharena Domingo, University of Tennessee.)

REFERENCES

Choi, M.-S., W. Kim, C. Lee, and C.-S. Oh. 2013. Harpins, multifunctional proteins secreted by gram-negative plant-pathogenic bacteria. *Molec. Plant Microbe Interact.* 26:1115–1122.

Dean, R., J. A. L. van Kan, Z. A. Pretorius, K. E. Hammond-Kosack, A. Di Pietro, P. D. Spanu, J. J. Rudd, M. Dickman, R. Kahmann, J. Ellis, G. D. and Foster. 2012. The top 10 fungal pathogens in molecular plant pathology. *Mol. Plant Pathol.* 13:414–430.

Fu, Z.Q., and X. Dong. 2013. Systemic acquired resistance: Turning local infection into global defense. *Annu. Rev. Plant Biol.* 64:839–863.

Gjetting, T., P. H. Hagedorn, P. Schweizer, H. Thordal-Christensen, T. L. W. Carver, M. F. Lyngkjaer. 2007. Single-cell transcript profiling of barley attacked by the powdery mildew fungus. *Molec. Plant Microbe Interact.* 20:235–246.

Großkinsky, D. K., E. van der Graaff, and T. Roitsch. 2012. Phytoalexin transgenics in crop protection- fairy tale with a happy end? *Plant Sci.* 195:54–70.

Gwinn, K. D., B. H. Ownley, S. E. Greene, M. M. Clark, C. L. Taylor, T. N. Springfield, D. J. Trently, J. F. Green, A. Reed, and S. L. Hamilton. 2010. Role of essential oils in control of Rhizoctonia damping-off in tomato with bioactive monarda herbage. *Phytopathology* 100:493–501.

Hacquard, S., B. Kracher, T. Maekawa, S. Vernaldi, P. Schulze-Lefert, and E. Ver Loren van Themaat. 2013. Mosaic genome structure of the barley powdery mildew pathogen and conservation of transcriptional programs in divergent hosts. *Proc. Natl. Acad. Sci.* 110:E2219–E2228 DOI: 10.1073/pnas.1306807110.

Hammerschmidt, R. 1999. Phytoalexins: what have we learned after 60 years? *Annu. Rev. Phytopathol.* 37:285–306.

Karunanayake, L. C., N. Adikaram, B. M. M. Kumarihamy, B. M. R. Bandara, and C. Abayasekara. 2011. Role of antifungal gallotannins, resorcinols and chitinases in the constitutive defence of immature mango (*Mangifera indica* L.) against *Colletotrichum gloeosporioides*. *J. Phytopathol.* 159:657–656

Kudla, J., O. Batistič, and K. Hashimoto. 2010. Calcium signals: The lead currency of plant information processing. *Plant Cell* 22:541–563.

Łaźniewska, J., V. K. Macioszek, and A. K. Kononowicz. 2012. Plant-fungus interface: The role of surface structure in plant resistance and susceptibility to plant pathogenic fungi. Physiol. Molec. *Plant Pathol.* 78:24–30.

Li, J.-G., J. Cao, F.-F Sun, D.-D. Niu, F. Yan, H.-X. Liu, J.-H. Guo. 2011. Control of *Tobacco mosaic virus* by PopW as a result of induced resistance in tobacco under greenhouse and field conditions. *Phytopathology* 101:1202–1208.

Mansfield, J., S. Genin, S. Magori, V. Citovsky, M. Sriariyanum, P. Ronald, M. Dow, V. Verdier, S. V. Beer, M. A. Machado, I. Toth, G. Salmond, and G. D. Foster. 2012. Top 10 plant pathogenic bacteria in molecular plant pathology. *Molec. Plant Pathol.* 13:614–629.

Meng, X., and S. Zhang. 2013. MAPK cascades in plant disease resistance signaling. *Annu. Rev. Phytopathol.* 51:245–266.

Molitor, A., D. Zajic, L. M. Voll, J. Pons-Kühnemann, B. Samans, K.-H. Kogel, and F. Waller. 2011. Barley leaf transcriptome and metabolite analysis reveals new aspects of compatibility and *Piriformospora indica*-mediated systemic induced resistance to powdery mildew. *Molec. Plant Microbe Interact.* 24:1427–1439.

Peeters, N., A. Guidot, F. Vailleau, and M. Valls. 2013. *Ralstonia solanacearum*, a widespread bacterial plant pathogen in the post-genomic era. *Molec. Plant Pathol.* 14:651–662.

Prusky, D., N. Alkan, T. Mengiste, and R. Fluhr. 2013. Quiescent and necrotrophic lifestyle choice during postharvest disease development. *Annu. Rev. Phytopathol.* 51:155–176.

Scholthof, K.-B.G. 2011. TMV in 1930: Francis O. Holmes and the local lesion assay. *Microbe* 6:221–225.

Scholthof, K.-B. G., S. Adkins, H. Czosnek, P. Palukaitis, E. Jacquot, T. Hohn, B. Hohn, K. Saunders, T. Candresse, P. Ahlquist, C. Hemenway, and G. D. Foster. 2011. Top 10 plant viruses in molecular plant pathology. *Mol. Plant Pathol.* 12:938–954.

Silipo, A., G. Erbs, T. Shinya, J. M. Dow, M. Parrilli, R. Lanzetta, N. Shibuya, M.-A. Newman, and A. Molinaro. 2010. Glycoconjugates as elicitors or suppressors of plant innate immunity. *Glycobiology* 20:406–419.

Vlot, A. C., D. A. Dempsey, and D. F. Klessig. 2009. Salicylic acid, a multifaceted hormone to combat disease. *Annu. Rev. Phytopathol.* 47:177–206.

Zeng, W., M. Maeli, and S. Y. He. 2010. Plant stomata: A checkpoint of host immunity and pathogen virulence. *Curr. Opin. Biotechnol.* 21:599–603.

21 Disruption of Plant Function

Melissa B. Riley

Concept Box

- Plant disease symptoms often result from plant pathogens affecting the normal physiological activities of the plant.
- Effects of a pathogen on the normal physiology of the plant can be at the macroscopic level (i.e., production of galls) or at the microscopic level (i.e., alternation of cell membrane permeability).
- Major plant physiological activities affected by plant pathogens include photosynthesis, respiration, production of plant growth hormones, absorption/translocation of water and nutrients, and transcription and translation.
- Study of plant pathogens led to the discovery of important compounds associated with normal plant activity—Gibberellins, which are plant growth hormones, are produced by some fungal plant pathogens.
- Reduced photosynthesis and increased respiration induced by plant pathogens ultimately result in reduced plant growth and yield.
- Disruptions of absorption/translocation within a plant can result in mineral deficiency symptoms.
- One of the first responses of a plant-to-plant pathogen attack is alteration in cell membrane permeability.

Since the beginning of time, humans have observed the effects of plant pathogens on plants, although they did not know about plant pathogens or how they caused the observed effects. In many cases, the processes that result in symptom development are not completely understood. Plant disease has been defined as alteration in the normal **physiology** of a plant, but this definition obviously requires extensive knowledge of normal plant physiology. Normal plant physiology includes many processes such as **photosynthesis**, **respiration**, **absorption**, and **translocation** of water, transport of photosynthetic products, production of compounds, such as plant growth **hormones**, **enzymes**, proteins, carbohydrates, lipids, and nucleic acids by the plant; and movement of materials between individual cells. Changes in any of these processes may have an effect on the overall appearance of the plant, which is observed as symptoms of disease. How do plant pathogens interrupt these processes? Responses may be something as small as the alteration of the metabolism within a cell or the movement of materials across a cell membrane to an overall plant response such as reduction in crop yield. We have learned much about the effects of plant pathogens on the normal physiology of plants, but still many questions remain, especially related to the exact sequence of events that result in the expression of symptoms.

Some common symptoms observed in response to plant pathogens and their possible relationship to normal plant physiology and examples of plant pathogens are provided in Table 21.1. This chapter examines how plant pathogens can disrupt the normal functions of a plant. Specifically, the effects of these pathogens on the major activities of the plant, including photosynthesis; respiration; production of plant growth hormones; absorption and translocation of water and nutrients; protein, carbohydrate, and lipid production; and cell **permeability** will be considered.

PHOTOSYNTHESIS

Essentially all organisms on earth depend on the process of photosynthesis in which plants absorb solar energy and convert it into carbohydrates that can be further utilized as energy sources. Photosynthesis occurs in the **chloroplasts** (Figure 21.1) of plant cells where carbon dioxide and water in the presence of solar energy and chlorophyll

TABLE 21.1
Common Plant Disease Symptoms, the Physiological Processes Affected, and Examples of a Disease and the Associated Plant Pathogen Affecting the Process

Symptom	Physiological Function	Example Disease/Pathogen
Chlorosis	Photosynthesis	*Tobacco mosaic virus*
Wilting	Xylem transport	Bacterial wilt/tomato and tobacco—*Ralstonia solanacearum*
Hyperplasia-cell division	Growth hormone regulation	Crown gall—*Agrobacterium tumefaciens* Black knot/plum—*Dibotryon morbosum*
Necrosis	Many different functions	Fire blight/apple—*Erwinia amylovora*
Hypertrophy-cell enlargement	Growth hormone regulation	Root knots—*Meloidogyne incognita* (root-knot nematode)
Leaf abscission	Growth hormone regulation	Coffee rust—*Hemileia vastatrix*
Etiolation	Growth hormone regulation	Bakanae "foolish seedling" disease of rice—*Gibberella fujikuroi*
Stunting	Many different functions	Many different viral diseases
Abnormal leaf formation	Growth hormone regulation, respiration	*Cucumber mosaic virus*—ornamentals

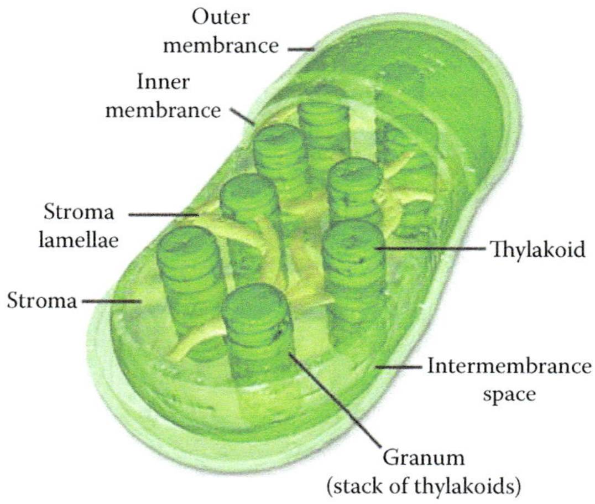

FIGURE 21.1 Plant chloroplast structure. (Courtesy of Michael W. Davidson and Florida State University.)

are ultimately converted to carbohydrates. This process is generally expressed by the following formula:

$$6CO_2 + 6H_2O + \text{Light} + \text{Chlorophyll} \rightarrow C_6H_{12}O_6 + 6O_2$$

The process can be divided into a light and a dark portion. During the light phase, solar energy produces reduced chemical compounds in the form of reduced nicotinamide adenine dinucleotide phosphate (NADPH) and adenosine 5′ triphosphate (ATP). During the dark phase, energy captured in the NADPH and ATP is utilized to convert carbon dioxide into carbohydrates. These reactions can occur through the following two major pathways: (1) the Calvin cycle in C_3 plants and (2) the C_4 cycle in C_4 plants. Alteration or inhibition in the overall photosynthetic activity of the plant can result in obvious gross symptoms in the plant, such as chlorosis (Figure 21.2), or reduced growth and yield. Reduction in the photosynthetic activity of the plant can be accomplished in many different ways. One way is from a reduction in total leaf area due to leaf necrosis and thereby destruction of photosynthetic tissue. A study of the effects of rust and anthracnose diseases on the photosynthetic competence of diseased bean leaves revealed that rust infections reduced photosynthetic rates in direct proportion to visible lesion area, whereas the reduction in photosynthetic activity in anthracnose-infected leaves was greater than could be attributed to the visibly infected area (Lopes and Berger, 2001). Photosynthesis in the green areas beyond the necrotic symptoms was severely impaired in anthracnose-infected leaves, indicating that the pathogen was affecting photosynthesis beyond the development of necrotic areas. Chlorosis is a common symptom observed in response to plant pathogens and can result from the inhibition of chlorophyll synthesis (Almási et al., 2000), increases in the rate of chlorophyll degradation (Strelkov et al., 1998), and reduction in the chloroplast size and number (Kema et al., 1996).

Blumeriella jaapii, the causal agent of cherry leaf spot, seems to mainly interfere with the enzymatic process of the Calvin cycle, and these effects were observed prior to the occurrence of visible symptoms (Niederleitner and Knoppik, 1997). The activity of three specific enzymes of the Calvin cycle, including ribulose-1,5-bisphosphate carboxylase/oxygenase (RUBISCO), decreased in powdery mildew-infected wheat leaves (Wright et al., 1995). RUBISCO activity was also decreased in *Arabidopsis thaliana* in response to *Albugo candida* (white blister rust; Chapter 8; Tang et al., 1996), in sugar beet infected with *Beet curly top virus* (Swiech et al., 2001), and *Parthenocissus quinquefolia* infected with *Xylella*

FIGURE 21.2 Foliar chlorosis and wilt of tobacco seedlings caused by *Pythium* sp. (From Mila, M. and Gutierrez, W.A., Pythium root rot in Tobacco Greenhouses, Plant Disease Fact Sheets, NC State University, https://www.cals.ncsu.edu/plantpath/extension/clinic/fact_sheets/index.php?do=disease&id=15. Courtesy of Walter A. Gutierrez, USDA APHIS.)

fastidiosa (McElrone and Forseth, 2004) indicating that different types of pathogens can have similar effects on their hosts. An interesting report to note is that carbon fixation associated with the dark phase of photosynthesis in uninfected bean leaves on plants infected with *Uromyces phaseoli* (bean rust; Chapter 14) is increased compared with that in bean leaves of uninfected controls (Murray and Walters, 1992). These results indicated that the pathogen can affect the photosynthetic activity in an area of the plant that is not infected. This can be a result of the plant trying to compensate for the loss of activity in infected leaves. In studies of the wheat cultivar Miriam, which is susceptible, but tolerant to *Septoria tritici* blotch (STB), the rate of carbon fixation per unit chlorophyll and per green leaf area was higher than that observed in healthy plants. Some plants exhibit enhanced photosynthesis in the remaining green tissue to compensate for loss of photosynthetic tissue due to the pathogen (Zuckerman et al., 1997).

The photosynthetic system of plants can be affected in many ways by pathogens. Cypress infected with *Seiridium cardinale* contained less total chlorophyll and carotenoids, reduced RUBSCO and nitrate activity, but more importantly, the infection resulted in an inactivation of the donor side of photosystem II (Muthuchelian et al., 2005). Examples such as this illustrate how the application of functional genomics techniques can permit more complete descriptions of disrupted photosynthetic systems. Microarrays and advanced protein analysis (Chapter 33) can be utilized to assess the effects of a pathogen on each of the enzymes in the photosynthetic process. Regardless of the details of disease process, the overall effect of plant pathogens that interfere with the photosynthetic potential is less carbohydrate production. Ultimately disease is manifested in reduced growth and yield. In a study on the effects of widespread viral infections of orchids, specifically *Oncidium*, healthy plants following virus elimination had a 17% increase in plant height, 65% increase in the inflorescence size, and 21% increase in photosynthetic capacity (Chia and He, 1999).

RESPIRATION

Respiration in plants involves oxidative processes in which complex molecules, such as carbohydrates produced during photosynthesis, are broken down into carbon dioxide, water, and energy. Energy is transferred and available to life processes when ADP is converted into ATP. The conversion reaction of ADP to ATP is referred to as **oxidative phosphorylation** and ATP serves as the energy component needed for almost all the operations in the plant cell. There are numerous metabolic pathways that serve to generate the substrates needed for oxidative phosphorylation and include the following: **glycolysis** (often referred to as Embden–Meyerhof pathway), tricarboxylic acid cycle (TCA), and oxidation of lipids. Another pathway that may be important, but that is not directly connected to oxidative phosphorylation, is the oxidative pentose phosphate pathway in which NADPH is generated along with pentose phosphates that can enter the glycolysis pathway. Respiration in plant pathogens (except viruses) has metabolic pathways similar to those of the plant host. Because of this it is difficult to differentiate between increases in respiration in plants in response to a plant pathogen or to determine if increases in respiration are due to the activity of the pathogen. Increases in respiration have generally been measured by determining increases in oxygen consumption. Because viral pathogens do not have respiratory pathways, they have been used to determine the effect of plant pathogens on respiration. All oxygen consumption in a viral infection can be directly correlated with plant host respiration.

Respiration usually increases in response to an invasion of a plant by a pathogen. Respiration increases more rapidly in plants exhibiting a resistance reaction due to the requirement for energy to rapidly produce and mobilize defense mechanisms. The rapid rise in respiration is followed by a decrease back to normal rates. In plants exhibiting a susceptible response to the pathogen, respiration levels increase more slowly, but continue to rise. Elevated respiration rates can be the result of increased activity of many different enzymes, faster breakdown of carbohydrates, including starch, and an uncoupling of oxidative phosphorylation. Uncoupling (ADP is not converted to ATP) results in increased concentration of ADP, which stimulates respiration. The activity of six enzymes associated with glycolysis and mitochondrial respiration in *Cucurbita pepo* infected with *Cucumber mosaic virus* (CMV) increased within lesions, whereas levels of RUBISCO involved in photosynthesis decreased only slightly (Técsi et al., 1996). Instead of using viruses as a nonrespiring pathogen, heat-killed bacteria have been employed as the elicitor of host responses. Oxygen uptake in tobacco cells treated with dead bacteria increased within 4 min and lasted for 10 min; thereafter, respiratory rates returned to a steady state that was approximately twice the initial rate (Baker et al., 2000).

ABSORPTION AND TRANSLOCATION OF WATER AND NUTRIENTS

Many pathogens, including viruses, fungi, bacteria, and nematodes, disrupt the absorption and translocation of nutrients in plants. In some cases, this may be on a small scale, such as within a single leaf, but in other cases the pathogen can affect the absorption and translocation of water and nutrients throughout the entire plant. This disruption can result from destruction of the root system, blockage of movement of water and nutrients in the xylem or phloem, or the redirection of host nutrients.

Some of the most obvious symptoms noted in response to plant pathogens that attack the plant roots are wilting, chlorosis, and general decline of the host plant (Figure 21.2). Root systems of plants infected with *Phytophthora*, *Pythium*, or *Thielaviopsis* are often nonfunctional. Rotting and destruction of roots make it difficult, if not impossible, for the plant to absorb and transport water and nutrients normally obtained by the uninfected root system. Many nematodes (Chapter 6) cause disruption of absorption and translocation of water and nutrients in the host plant through the destruction of the small feeder roots. Ectoparasitic nematodes, such as ring nematodes (*Criconemella* sp.; Chapter 6) in turf grasses and peach trees, can lead to a significant reduction of feeder roots that can lead to chlorosis as well as plant mineral deficiency symptoms.

Other plant pathogens are able to affect the absorption and translocation of water and nutrients within the plant because of their presence in the xylem and phloem tissues. *Ralstonia solanacearum*, causal agent of bacterial wilt in tobacco, tomato, eggplant, brown rot of potato, and Moko disease in banana, invades the xylem through wounds. The bacteria spread through the plant, multiply rapidly within the xylem, and produce copious amounts of exopolysaccharides (EPSs) that essentially dam or block water flow (Figure 21.3). Loss of water translocation results in rapid wilting of young plants, wilting of a portion of more mature plants, chlorosis, and general decline. Bacterial streaming can be observed when the stem of a young infected tobacco or tomato plant exhibiting wilt symptoms due to *R. solanacearum* is cut and immediately placed in water. Fungal agents such as *Fusarium oxysporum* f. sp. *lycopersici* (wilt in tomato) and *Ophiostoma novo-ulmi* (Dutch elm disease) decrease water flow through the xylem. This decrease in water flow can be attributed to physical obstruction of vessels with hyphae, secretion of polysaccharides and pectolytic enzymes, production of gums/mucilages, and the formation of **tyloses** in the xylem by the host plant. The most obvious result in blockage of the vascular elements is the reduction in water movement, but it also diminishes the transport of essential minerals throughout the plant.

The presence of various plant pathogens can also result in the redirection of nutrients and resources within the plant. Pathogens divert resources such as carbohydrates and amino acids of the host for their own for growth and reproduction. Leaves of *Arabidopsis thaliana* infected with *Albugo candida* showed increased levels of both soluble carbohydrates and starch when compared with uninfected leaves. Activities of both wall-bound and soluble invertases, enzymes involved in sucrose hydrolysis, were higher in infected leaves compared with those in control leaves. Additionally, an invertase isozyme was present in infected leaves and not found in healthy tissues (Tang et al., 1996). In studies of CMV infection in cotyledons of marrow plants, virus replication and synthesis of viral protein created a strong sink within lesions, resulting in increased photosynthesis and starch accumulation. There was more than twice as much starch hydrolase activity within lesions when compared

FIGURE 21.3 Gray brown discoloration of vascular tissues and bacterial ooze in potato tuber infected by *Ralstonia solanacearum* race 3 biovar 2 (http://www.apsnet.org/publications/imageresources/Pages/IW000109.aspx). Potato tuber infected by *R. solanacearum* race 3 biovar 2. (Courtesy of Patrice Champoiseau, University of Florida—Institute of Food and Agricultural Services. APS publication number: IW000109.)

with that in areas outside the lesions and to healthy leaves (Técsi et al., 1996).

Healthy melon plants have stachyose as a major component of sugars in phloem. Absolute levels of stachyose in phloem sap were similar in CMV-infected leaves, but sucrose levels dramatically increased and the sucrose to stachyose ratio was 15- to 40-fold higher compared with that of healthy plants. Elevated sucrose concentrations did not appear to result from stachyose hydrolysis because there was no corresponding increase in galactose. Therefore, it is possible that CMV infection affected the movement of sucrose and was not due to metabolism (Shalitin and Wolf, 2000).

PLANT GROWTH HORMONES

Normal plant growth is controlled by various plant growth hormones and regulators, which have numerous effects and interactions within the plant especially related to plant: plant pathogen interactions (Table 21.2). Alterations in the levels of these compounds can have significant effects on the plant's growth. Many different plant pathogens cause the host to change the hormone concentration in the plant, or in some cases, they actually produce these growth-controlling substances themselves. Symptoms observed in response to plant pathogens such as galls or tumor growth, excessive branching, leaf **abscission** or **epinasty**, abnormal leaf shape, and abnormal fruit ripening can be attributed to alterations in the levels of various hormones.

The pathogen *Gibberella fujikuroi*, causal agent of Bakanae or "foolish seedling" disease in rice, produces gibberellins in the host plant and causes elongation of the internodes. The disease results in spindly plants that often **lodge**. Prior to the identification of gibberellins as an actual plant growth hormone, a Japanese scientist identified gibberellins from the culture filtrates of *G. fujikuroi*. This is an example of how the study of plant pathogens and their activities has been important to other areas of plant research. Conversely, *Fusarium* wilt of oil palm, which results in stunting, may be caused by an inhibition of gibberellin synthesis, because similar symptoms were observed following the application of an inhibitor of gibberellin synthesis. Application of gibberellin to infected palms resulted in a partial, but not complete, elimination of symptoms (Mepsted et al., 1995).

Some of the best and most complete evidence to demonstrate the production of plant growth hormones by a plant pathogen is with *Agrobacterium tumefaciens*, the causal agent of crown gall. This bacterium has a self-replicating circular piece of DNA referred to as the Ti (tumor inducing) **plasmid**. The plasmid has the ability to transfer a section of its DNA, referred to as the T-DNA, into the host cell where it becomes incorporated into a host cell chromosome within the nucleus. T-DNA contains genes for the production of opines, cytokinin, and auxin. Opines (amino acids) are an unusual nutrient source that cannot be utilized by the plant cell or by many other organisms except *Agrobacterium*, which has genes for opine metabolism. The production of cytokinin and auxin resulting from the T-DNA incorporation into the plant cell is not under the normal control mechanisms of the plant. The overproduction of these hormones results in gall development due to uncontrolled cell division (**hyperplasia**) and enlargement (**hypertrophy**). The

TABLE 21.2
Major Groups of Plant Growth Hormones and Their Associated Physiological Activities within the Plant

Hormone Class	Physiological Activity
Auxin (indole-3-acetic acid or IAA)	Promote cell elongation Induce high levels of ethylene formation resulting in growth inhibition Induce cambial cell division Initiate root formation
Gibberellins (GA: many forms)	Promote stem elongation by cell elongation Stimulate α-amylase production in seeds Stimulate flower production Retard leaf and fruit senescence
Cytokinins (Zeatin (Z), Z riboside (ZR), and ZR phosphate)	Promote cell division and differentiation Inhibit senescence of plant organs Induce stomatal opening Suppress auxin-induced apical dominance
Ethylene	Stimulate fruit ripening Stimulate leaf and flower abscission
Abscisic acid (ABA)	Induce maturation of seeds including development of desiccation tolerance Induces water stress-associated stomatal closure

ultimate result is the observed symptom, which gives the disease its common name, crown gall (Figure 21.4).

Pseudomonas savastanoi, the causal agent of olive knot disease, contains genes for auxin production on a plasmid as well as on its bacterial chromosome. In this case, the bacteria rather than the plant host produce the auxin that induces gall formation. In studies of *Erwinia herbicola* pv. *gypsophilae*, cytokinin and auxin genes are present in a pathogenicity-associated plasmid, which induces the production of a gall. The size of root galls observed in response to *Plasmodiophora brassicae*, the causal agent of clubroot of cabbage (Chapter 9), was correlated with the free indole acetic acid content in the clubs, indicating the role of auxins in symptom development (Grsic-Rausch et al., 2000). Increases in cytokinins and the expression of four or five putative cytokinin synthase genes from *Brassica rapa* were also correlated with the development of clubroot (Ando et al., 2005).

Ethylene is a gaseous plant growth hormone. It has many different effects in the plant as shown in Table 21.2. Early in the 1900s it was noted that oranges shipped with bananas promoted banana ripening. Ethylene, which caused fruit ripening, was not produced by the oranges, but by the fungus *Penicillium digitatum*, the cause of green mold on oranges. A non-ethylene-producing mutant showed that ethylene production has no significant role in the pathogenicity of *P. digitatum* even though low levels of ethylene predispose fruit to postharvest diseases. Many plant pathogens are known to induce ethylene production in the host as a response to infection. Ethylene then induces many other responses within the plant such as the production of various pathogenesis-related (PR) proteins (Chapters 20 and 24) and defense-related compounds.

FIGURE 21.4 Crown gall on a Euonymus stem (http://www.ipm.iastate.edu/ipm/hortnews/files/images/crown-gall-Bob-Dodds.jpg). (Courtesy of Bob Dodds. From Christine, E., *Horticulture & Home Pest News*, IC-497 (2), February 7, 2007, http://www.ipm.iastate.edu/ipm/hortnews/2007/2-7/crowngall.html.)

Ethylene production by the host or pathogen can result in premature ripening of fruit, leaf epinasty (downward bending), leaf abscission, and chlorosis.

Abscisic acid (ABA) in general is viewed as a growth inhibitor, and increases in its production have been noted in some plant diseases. ABA production associated with these plant–pathogen interactions resulted from a plant host response to the plant pathogen rather than being produced by the plant pathogen. ABA-producing strains of plant pathogenic fungi have been identified, but the involvement of ABA in pathogenesis has not been thoroughly investigated.

Many times plant pathogens can cause multiple changes in plant growth regulators. In a study of premature fruit drop in citrus caused by *Colletotrichum acutatum*, ethylene evolution increased threefold, indole-3-acetic acid (IAA) increased 140-fold, ABA showed no change, both *trans*- and *cis*-12-oxo-phytodienoic acid (precursors of jasmonic acid [JA]) increased 8- to 10-fold, *trans*-JA was unchanged, *cis*-JA increased fivefold, and salicylic acid (a signaling compound; Chapter 20) increased twofold. Many of the enzymes associated with the production of ethylene, IAA, and JA were also shown to be upregulated in infected plants (Lahey et al., 2004). Viroids, the smallest known infectious agents specifically, *potato spindle tuber viroid* (PSTVd) infection, have been shown to have numerous effects on hormone biosynthesis in tomatoes both positively and negatively (Owens et al., 2012) with resulting effects on the physiology of the tomato plant.

ALTERATION OF CELL PERMEABILITY

The plant cell membrane consists of a bilayer of phospholipids with embedded proteins. Membranes exist within the cell wall and surrounding all plant organelles such as chloroplasts and mitochondria. Proteins within the cell membrane are important in the regulation and movement of materials across the membrane and are extremely important to the integrity of the cell. When this integrity is altered, the membrane is unable to control the movement of materials across the membrane and substances may be lost from the cell. When pathogens attack plants, one of the first plant responses that can generally be detected is the alteration of membrane **permeability**. Membrane integrity is measured by determining electrolyte (charged ions, such as K^+ or Cl^-) leakage from cells. Several toxins associated with plant pathogens have been shown to affect cell membrane permeability or integrity (Figure 21.5). Examples of these toxins include T-toxin (*Cochliobolus heterostrophus*: southern corn leaf blight), HC-toxin (*Cochliobolus carbonum*: leaf spot disease in maize), AM-toxin (*Alternaria alternata*: alternaria leaf blotch of apple), and syringomycin E, syringotoxin, and syringopeptin

FIGURE 21.5 The pathogenic potential of *Alternaria* species can be related to the production of toxins (https://cropphotoupdate.files.wordpress.com/2014/02/screen-shot-26022014-1440-2.jpeg, 2014_02_26 Devon: Oilseed-Rape disease Alternaria ©2014 CropShots). (Courtesy of Crop Photo Update UK. http://cropphotoupdate.com/category/oilseed-rape/page/2/. February 26, 2014.)

(*Pseudomonas syringae* pv. *syringae*: necrotic lesions in a broad range of monocot and dicot species, resulting in tip dieback, bud and flower blast, spots and blisters on fruit, stem canker, and leaf blight). Many of these toxins are cytotoxic in plant cells at nanomole (10^{-9} M) concentrations and cause necrosis by forming ion channels, which freely allow the passage of divalent cations (e.g., calcium and magnesium). In a study of the cultivar-specific necrosis toxin produced by *Pyrenophora tritici-repentis* (tan spot, a widespread foliar disease of wheat), electrolyte leakage was observed in a toxin-sensitive cultivar after 4 h, but was not observed in a toxin-insensitive cultivar (Kwon et al., 1996).

TRANSCRIPTION AND TRANSLATION

The processes of **transcription** and **translation** are often affected by plant pathogens, but are most clearly observed with plant viruses. Transcription is the process whereby information associated with a DNA sequence or gene is copied into a complementary piece of RNA referred to as messenger RNA (mRNA) (Figure 21.6). RNA polymerase recognizes the start of a gene on the DNA, combines with the DNA separating the two nucleotide strands and then progresses down the DNA section making the complementary strand of RNA until a stop code is encountered. Translation is the process by which this piece of mRNA is copied into a protein whose amino acid sequence is determined by the sequence of nucleotides in the mRNA. This process involves a ribosome binding to the mRNA and essentially reading three nucleotides at a time that correspond to specific amino acids, which are brought to the ribosome active site by a transfer RNA (tRNA). The amino acids are

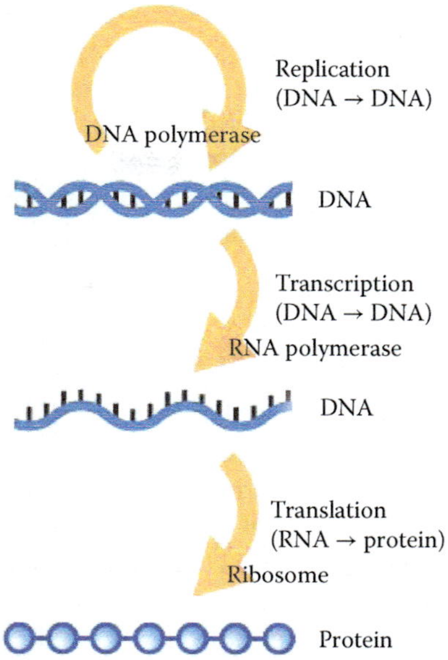

FIGURE 21.6 Central dogma of molecular biology (https://upload.wikimedia.org/wikipedia/commons/6/68/Central_Dogma_of_Molecular_Biochemistry_with_Enzymes.jpg; Author Daniel Horspool).

connected by peptide bonds forming a protein. Some processing of the protein may be required after the amino acid sequence is complete. The enzymes that are responsible for the production of other materials within the plant cell such as the polysaccharides and lipids are included in the proteins produced by this process.

Virus particles that infect plants consist of genetic information in the form of ssDNA, dsRNA, or ssRNA surrounded by a protein coat (Chapter 4). Viruses do not have the ability to self-replicate, but instead utilize the cellular machinery of the host. The DNA or RNA from the virus moves into the host cell, where it undergoes transcription and translation, allowing the genetic material to be produced along with the protein coat. The viral genome serves as the template for transcription for the production of the nucleic acid component of the virus, and translation serves for the production of the components of the viral protein coat. The total protein content of 20 cultivars of alfalfa infected with *Phoma medicaginis* was determined following infection. Infected leaves had lower protein content in 19 of 20 cultivars with an average reduction of 22%. Five cultivars had protein levels reduced from 35% to 65% indicating a significant variability associated with different cultivars (Hwang et al., 2006).

Much of the previous work concerning the effects of plant pathogens on the physiology of its host has involved the detection of changes in the production of specific enzymes, such as the enzymes involved in glycolysis or chlorophyll biosynthesis and degradation, the measurement of chlorophyll levels, hormone concentrations, and membrane permeability. With the advancement of techniques to measure changes in gene expression (Golem and Culver, 2003; Li et al., 2006; Whitman et al., 2006) and the resulting protein production as well as protein identification (Devos et al., 2006), a better and more complete understanding of the effects of a pathogen on the host physiology is becoming possible and is advancing rapidly.

LABORATORY EXERCISES

"Seeing is believing" and "Prove it to me" are common phrases that are used in many situations. These ideas are particularly important in research—we always want to know what the effect of the plant pathogen is on the plant. The overall objective of the following exercises is to illustrate some of the effects that plant pathogens have on the physiological processes of the plant. We see the results when we observe symptoms, but in these exercises we will be looking at the effects of plant pathogens on specific physiological activities of the plant. The first exercise can be conducted with many different plants. The only requirement is that the plant pathogen causes symptoms resulting in a chlorotic or mosaic appearance on leaves. Healthy plants with uninfected leaves are used in comparison with the concentration of chlorophyll present in the infected leaf. The second exercise was developed to be used with the bacterial pathogen *Ralstonia solanacearum* and tomato plants. This exercise illustrates the blockage of the xylem system by bacteria and how these bacteria can also be isolated from the xylem stream on nutrient agar plates. The final exercise was developed to illustrate how a plant pathogen, *Agrobacterium tumefaciens*, can alter the normal growth patterns of plant tissue.

Experiment 1. Effects of a Plant Pathogen on Chlorophyll

Chlorophyll is probably one of the most important molecules on earth. In combination with other compounds and enzymes in the plant, chlorophyll is able to harvest light energy and transfer it into a usable form of chemical energy that the plant utilizes for growth and reproduction. This stored chemical energy also supplies energy for many other forms of life on earth, including humans and plant pathogens. Any time when we eat food, we are utilizing this energy either directly or indirectly. The amount of chlorophyll associated with cells has a direct effect on the amount of the stored energy that a plant ultimately produces. A reduction in the chlorophyll can lead to effects such as stunting due to the loss of the energy conversion potential. Chlorophyll extractions such as the one outlined here (Moran, 1982) can be easily done and provide a quick measurement of the effects of plant pathogens on chlorophyll content in plants.

Materials

Each student or team of students will require the following items:

- Leaves from a plant exhibiting chlorotic symptoms or mosaic symptoms. The specific plant is not that important, but it is vital to have a healthy plant that can be used for comparison. Examples of plants that can be used include tobacco plants infected with *Tobacco mosaic virus* or cucumber plants infected with *Cucumber mosaic virus*.
- Balance to determine the weight of individual leaves.
- #5 cork borer.
- Ruler.
- Test tubes.
- Dimethylformamide (DMF). *Caution*: Use in a chemical hood only and wear rubber gloves to prevent absorption through skin. Obtain information on material safety data sheet prior to use.
- Disposable rubber gloves.
- Spectrophotometer to measure absorbance at 664 and 647 nm.
- Disposable or quartz cuvettes for spectrophotometer.
- Pipettes for transferring DMF.
- Incubator set at 10°C without lighting.
- Calculator.

Follow the instructions in Procedure 21.1 to complete the experiment.

Anticipated Results

The results of this experiment provide a basis for suggesting that one pathogen or strain of the pathogen causes more severe reactions in the plant than another. Reductions in chlorophyll have a direct effect on plant growth. If the chlorophyll is reduced by 10% by one pathogen and 40% by a second pathogen, it is clear that this second pathogen is more severely affecting the plant based on the effect on chlorophyll content. In many cases, however, reductions in chlorophyll are not the only effect a plant pathogen has, so other tests may be needed to quantify the total effect that the plant pathogen has on the plant. This experiment should have definite results. An additional component to the experiment may be to determine the variability of results for a specific group illustrating the purpose of having replicates in experiments.

Procedure 21.1

Determination of Total Chlorophyll Present in Leaf Area in Comparison with Healthy and Infected Leaf Areas and to Compare the Effects of Different Pathogens

Step	Instructions and Comments
1	Choose leaves from a plant exhibiting symptoms such as chlorosis or mosaic patterns. Nonsymptomatic leaves must also be obtained from the same plant or from the same general area of a healthy plant. Leaves associated with different diseased plants can be compared in order to compare the effects of different plant pathogens.
2	Cut 10 disks from each type of leaf area (normal and symptomatic) with a #5 cork borer (1-cm-diameter). Calculate the area of the leaf disk with the following formula: $$\text{Area} = \pi r^2 \times 10$$ where $\pi = 3.14$ and $r = 0.5 \times$ diameter, or in this case 0.5 cm. Obtain weight of leaf disks prior to proceeding to the next step.
3	Weigh the leaf disks and place them in 5 mL cold DMF. Incubate in the dark at 10°C for a minimum of 48 h.
4	Determine absorbance of the DMF at 664 nm and 647 nm and calculate total chlorophyll with the following formula: $$\text{Total chlorophyll} = 7.04\ \text{Abs}_{664} + 20.27\ \text{Abs}_{647} = \mu\text{g chlorophyll / mL}$$
5	Convert μg/mL to μg/cm^2 by multiplying the total chlorophyll by 5 mL (DMF solution volume) and dividing by the total leaf area associated with the disks as calculated in Step 2. Convert μg/mL to μg/gm fresh weight by multiplying the chlorophyll total by 5 mL and dividing by the weight in grams obtained in Step 3.

Questions

- What is the percent reduction in the levels of chlorophyll associated with the infected leaves compared with the healthy leaves?
- What is the variation in chlorophyll values within the laboratory for the infected leaves and healthy leaves? Minimum value? Maximum value? Are there similar levels of variation for the healthy and infected groups or is there more variation associated with the infected leaves? How can you explain the variation within these groups?
- Why would it be important to know the variation in chlorophyll levels of plants prior to setting up a research project investigating the effects of a plant pathogen on chlorophyll levels of a plant?
- How can plant pathogens affect the total level of chlorophyll in a plant?

Experiment 2. Disruption of Water Translocation

Disruption of the absorption and translocation of water within the plant can occur when plant pathogens cause a destruction of the root system, block the movement of water and nutrients in the xylem and phloem, or redirect the movement of host nutrients. The blockage of the water flow within plants is a common mechanism for plant pathogens that results in wilting of the plant and ultimately may kill the plant. Many of the pathogens that cause these symptoms are very good saprophytes in the soil and survive for many years even in the absence of a host. *Ralstonia solanacearum* utilized in this exercise is common in tropical and subtropical areas and causes severe diseases in tobacco, tomato, potato, and eggplant in warm areas outside the tropics. Several races of this organism have various host ranges, and a race that attacks tomato plants is utilized in this exercise.

Follow the instructions in Procedure 21.2 to complete the experiment.

Materials

Each student or team of students will require the following items:

- 2- to 4-week-old tomato plants ("Marion" or "Rutgers" cultivars that are susceptible and various other cultivars may be included to compare susceptibility of different cultivars)

Procedure 21.2

Disruption of Vascular Flow in Tomato Infected with ***Ralstonia solanacearum***

Step	Instructions and Comments
1	Obtain tomato plants approximately 2–4 weeks old. "Marion" and "Rutgers" cultivars are extremely susceptible, but other cultivars can be included in comparison with resistance to *R. solanacearum*.
2	Grow a virulent strain of *R. solanacearum* in nutrient broth or tryptic soy broth on the shaker for 24 h or until medium is turbid.
3	Using a 1-mL sterile syringe with 1/2-inch needle (smallest diameter possible), draw up 0.2 mL of bacterial suspension.
4	Inject the bacterial suspension just below the surface and into the lower stem of the tomato. Incubate the inoculated plants at approximately 30°C until wilt symptoms develop. This usually occurs within 2–7 days. Generally, the older the tomato plant, the longer it will take for wilt to occur.
5	After the plants wilt, place a couple of drops of sterile distilled water on a glass slide. Cut the tomato stem across the area above the site of inoculation and immediately place the cut stem in the water on the slide to observe bacteria streaming from the cut surface. Water on slide should become turbid due to the presence of bacteria.
6	Take a sterile loop and streak the water–bacteria suspension from the slide onto either nutrient agar or tryptic soy agar in Petri dishes and seal with Parafilm®. Incubate the sealed dishes at 28°C–30°C for 24 h and observe bacterial growth.
7	Touch loop to the bacteria–water mix on the glass slide prepared in Step 5 and transfer a drop to a clean slide. Spread the suspension and air-dry.
8	Heat-fix the bacteria to the slide by passing it quickly through the flame of an alcohol burner. Be careful not to get the slide too hot.
9	Add a couple of drops of safranin stain to the slide for 30–60 s; then rinse with water. Wear gloves for this step.
10	Blot, do not rub, the slide dry with a paper towel. Rubbing will remove the bacteria.
11	Observe the stained bacteria by using an oil immersion lens on a compound microscope.

- 24- to 48-h-old culture of *R. solanacearum* (race pathogenic to tomato plants) grown in nutrient broth or tryptic soy broth (Difco, Lansing, MI)
- 1-mL sterile syringes with 1/2-inch needles
- Tryptic soy broth or nutrient broth
- 30°C growth chamber for tomato plants
- Glass microscope slides and coverslips
- Sterile distilled water in dropper bottle
- Nutrient agar or tryptic soy agar in Petri dishes (10-cm-diameter)
- Shaker
- 28°C–30°C bacterial culture incubator
- Bacterial transfer loop that can be sterilized by flaming
- Alcohol burner
- Safranin O (0.25% solution in 10% ethanol, 90% water—should be made up initially in ethanol)
- Disposable nitrile gloves
- Paper towels
- Sink with water
- Compound microscope with oil immersion lens

Anticipated Results

Tomato plants infected with *R. solanacearum* will generally wilt within 2–4 days, depending on the strain and the age of the tomato plant. Older plants will take longer to wilt. The temperature of the growth chamber is vital. If the temperature is less than 30°C the wilting of the plant may be delayed or may not occur. Bacterial colonies should appear within 48 h on the nutrient agar plates on which the bacteria were transferred from the slide. Bacteria should be obvious on the stained slide made from the xylem exudates—water mix.

Questions

- Is the wilting associated with *R. solanacearum* infection of tomato strictly due to the presence of bacteria in the xylem stream blocking the passage of water?
- What compounds do some bacteria produce that may also be important in the blockage of water flow in a plant?

- How are bacteria such as *R. solanacearum* disseminated in nature? What can be done to manage diseases such as *R. solanacearum*? What are the limitations?
- Were there any observable differences between cultivars of tomato used in the experiment?
- How many different types of bacterial colonies were isolated from the xylem exudate? Are these other organisms involved in the wilting observed? What might you need to do to determine their involvement?

Experiment 3. Effect of *Agrobacterium* Species on Normal Plant Growth

Plant pathogens are known to either produce normal plant growth hormones or have the ability to increase or alter the production of plant growth hormones by the plant. Virulent stains of *Agrobacterium* contain plasmids that contain genes for the production of various plant growth hormones. The genes associated with the plant growth hormones are transferred to the plant during the infection process, and as a result, more plant hormones are produced that interferes with normal plant function and development. The presence of these extra levels of hormones causes various abnormalities in the normal growth of the plant. Tomato plants serve as experimental subjects and can rapidly produce callus or root growth following inoculation with *Agrobacterium*.

Materials

Each student or team of students requires the following items:

- Tomato plants 4–6 weeks old (various cultivars can be used in comparison with "Marion" and "Rutgers" cultivars, which are susceptible)
- 24- to 48-h-old culture of virulent strain of *A. tumefaciens* (induces gall formation) and *A. rhizogenes* (induces root formation) growing on nutrient agar or tryptic soy agar
- Sterile scalpel with #10 blade
- Shaker
- Parafilm®
- Balance
- Dissecting microscope
- Growth chamber, greenhouse, or window where tomato plants can be maintained for several weeks

Follow the instructions in Procedure 21.3 to complete the experiment.

Anticipated Results

The tomato plants should be observed on a weekly basis. The plants should begin to show symptoms of gall or root formation within approximately 2–3 weeks. The first indication of abnormal growth will be a roughness on the surface of the stem where the plant was inoculated. Plant height should be measured for all plants prior to inoculation and at the end of the experiment. Control plants

Procedure 21.3
Induction of Gall or Root Formation by Inoculation of Tomato Plants with ***Agrobacterium tumefaciens*** or ***A. rhizogenes***

Step	Instructions and Comments
1	Obtain tomato plants approximately 4–6 weeks old. ("Marion" and "Rutgers" cultivars are susceptible, but other cultivars can be included in comparison with gall or root formation following inoculation.)
2	Grow a virulent strain of *Agrobacterium tumefaciens* (for the induction of gall formation) and *A. rhizogenes* (for the induction of root formation) on nutrient agar or tryptic soy agar for 24 h or until growth is observable on the dish.
3	Take a sterile scalpel and touch it to the bacterial growth in the Petri dish.
4	Cut the surface of the tomato stem with the scalpel coated with the bacteria and then wrap a piece of Parafilm® around the cut. Be careful not to cut completely through the tomato stem.
5	For the control treatment take a sterile scalpel and touch it to the surface of sterile agar dish and then cut the surface of a tomato stem with the scalpel followed by wrapping the Parafilm®.
6	Remove the Parafilm® after 1 day. Observe the tomato plants weekly for the production of callus tissue (galls) or roots. Abnormal growth should be observed after approximately 2–3 weeks. Plants can be incubated in a window, greenhouse, or growth chamber.
7	Compare callus/root growth in different tomato cultivars by measuring the size/number of abnormal growths. Cut the abnormal growth associated with the plant and weigh the tumor/root growth.

that are cut with the scalpel but are not inoculated with *Agrobacterium* should be included. The total weight of abnormal callus or root growth at the end of the experiment (after approximately 6 weeks) should be determined and should be related to the resistance of various tomato cultivars to *Agrobacterium*.

Questions

- What is the average callus or abnormal root growth for each of the cultivars of tomatoes used in the experiment? Does the average weight of callus or root growth correspond to the resistance to *Agrobacterium*? Why or why not?
- Is there any response on the control plants? If so, what is that response? Why does it occur?
- Does the infection with *Agrobacterium* have any effect on the overall height increase of the infected plant compared to that of the control? Will you expect it to have an effect? Why or why not?
- Why is Parafilm® placed around the plant after it is inoculated?
- What are the methods used for the management of *Agrobacterium* in nature?

REFERENCES

Almási, A., D. Apatini, K. Bóka, B. Böddi, and R. Gáborjányi. 2000. BSMV infection inhibits chlorophyll biosynthesis in barley plants. *Physiol. Mol. Plant Pathol.* 56:227–233.

Ando, S., T. Asano, S. Tsushima, S. Kamachi, T. Hagio, and Y. Tabei. 2005. Changes in gene expression of putative isopentenyltransferase during clubroot development in Chinese cabbage (*Brassica rapa* L.). *Physiol. Mol. Plant Pathol.* 67:59–67.

Baker, C. J., E. W. Orlandi, and K. L. Deahl. 2000. Oxygen metabolism in plant/bacteria interactions: Characterization of the oxygen uptake response of plant suspension cells. *Physiol. Mol. Plant Pathol.* 57:159–167.

Chia, T. -F., and J. He. 1999. Photosynthetic capacity in *Oncidium* (Orchidaceae) plants after virus eradication. *Environ. Exp. Bot.* 42:11–16.

Devos, S., K. Laukens, P. Deckers, D. Van Der Straeten, T. Beeckman, D. Inzé, H. Van Onckelen, E. Witters, and E. Prinsen. 2006. A hormone and proteome approach to picturing the initial metabolic events during *Plasmodiophora brassicae* infection on *Arabidopsis*. *Mol. Plant Microbe Interact.* 19:1431–1443.

Golem, S., and J. N. Culver. 2003. *Tobacco mosaic virus* induced alteration in the gene expression profile of *Arabidopsis thaliana*. *Mol. Plant Microbe Interact.* 16:681–688.

Grsic-Rausch, S., P. Kobelt, J. M. Siemens, M. Bischoff, and J. Ludwig-Müller. 2000. Expression and localization of nitrilase during symptom development of the clubroot disease in *Arabidopsis*. *Plant Physiol.* 122:369–378.

Hwang, S. -F., H. Wang, B. D. Gossen, K. -F. Chang, G. D. Thurbill, and R. J. Howard. 2006. Impact of foliar diseases on photosynthesis, protein content and seed yield of alfalfa and efficacy of fungicide application. *Eur. J. Plant Path.* 115:389–399.

Kema, G. H. J., D. Yu, F. H. J. Rijkenberg, M. W. Shaw, and R. P. Baayen. 1996. Histology of the pathogenesis of *Mycosphaerella graminicola* in wheat. *Phytopathology* 86:777–786.

Kwon, C. Y., J. B. Rasmussen, L. J. Fancl, and S. W. Meinhardt. 1996. A quantitative bioassay for necrosis toxin from *Pyrenophora tritici-repentis* based on electrolyte leakage. *Phytopathology* 86:1360–1363.

Lahey, K. A., R. Yuan, J. K. Burns, P. P. Ueng, L. W. Timmer, and K.-R. Chung. 2004. Induction of phytohormones and differential gene expression in citrus flowers infected by the fungus *Colletotrichum acutatum*. *Mol. Plant Microbe Interact.* 17:1394–1401.

Li, C. Y. Bai, E. Jacobsen, R. Visser, P. Lindhout, and G. Bonnema. 2006. Tomato defense to the powdery mildew fungus: differences in expression of genes in susceptible, monogenic-and polygenic resistance responses are mainly in timing. *Plant Mol. Biol.* 62:127–140.

Lopes, D. B., and R. D. Berger. 2001. The effects of rust and anthracnose on the photosynthetic competence of diseased bean leaves. *Phytopathology* 91:212–220.

McElrone, A. J., and I. N. Forseth. 2004. Photosynthetic responses of a temperate liana to *Xylella fastidiosa* infection and water stress. *J. Phytopathol.* 152:9–20.

Mepsted, R., J. Flood, and R. M. Cooper. 1995. Fusarium wilt of oil palm I. Possible causes of stunting. *Physiol. Mol. Plant Pathol.* 46:361–372.

Moran, R. 1982. Formulae for determination of chlorophyllous pigments extracted with N,N-dimethylformamide. *Plant Physiol.* 69:1376–1381.

Murray, D. C., and D. R. Walters. 1992. Increased photosynthesis and resistance to rust infection in upper, uninfected leaves of rusted broad bean (*Vicia faba* L.). *New Phytol.* 120:235–242.

Muthuchelian, K., N. La Porta, M. Bertamini, and N. Nedunchezhian. 2005. Cypress canker induced inhibition of photosynthesis in field grown cypress (*Cupressus sempervirens* L.) needles. *Physiol. Mol. Plant Pathol.* 67:33–39.

Niederleitner, S. and D. Knoppik. 1997. Effects of the cherry leaf spot pathogen *Blumeriella jaapii* on gas exchange before and after expression of symptoms on cherry leaves. *Physiol. Mol. Plant Pathol.* 51:145–153.

Owens, R. A., K. B. Tech, J. Y Shao, T. Sano., and C. J. Baker. 2012. Global analysis of tomato gene expression during *Potato spindle tuber viroid* infection reveals a complex array of changes affecting hormone signaling. *Mol. Plant Microbe Interact.* 25:582–598.

Shalitin, D., and S. Wolf. 2000. Cucumber mosaic virus infection affects sugar transport in melon plants. *Plant Physiol.* 123:597–604.

Strelkov, S. E., L. Lamari, and G. M. Ballance. 1998. Induced chlorophyll degradation by a chlorosis toxin from *Pyrenophora tritici-repentis*. *Can. J. Plant Pathol.* 20:428–435.

Swiech, R., S. Browning, D. Molsen, D. C. Stenger, and G. P. Holbrook. 2001. Photosynthetic responses of sugar beet and *Nicotiana benthamiana* Domin, infected with beet curly top virus. *Physiol. Mol. Plant Pathol.* 58:43–52.

Tang, X., S. A. Rolfe, and J. D. Scholes. 1996. The effect of *Albugo candida* (white blister rust) on the photosynthetic and carbohydrate metabolism of leaves of *Arabidopsis thaliana*. *Plant Cell Environ.* 19:967–975.

Técsi, L. I., A. M. Smith, A. J. Maule, and R. C. Leegood. 1996. A spacial analysis of physiological changes associated with infection of cotyledons of marrow plants with cucumber mosaic virus. *Plant Physiol.* 111:975–985.

Whitman, S. A., C. Yang, and M. M. Goodin. 2006. Global impact: Elucidating plant responses to viral infection. *Mol. Plant Microbe Interact.* 19:1207–1215.

Wright, D. P., B. C. Baldwin, M. C. Shephard, and J. D. Scholes. 1995. Source-sink relationships in wheat leaves infected with powdery mildew. II. Changes in the regulation of the Calvin cycle. *Physiol. Mol. Plant Pathol.* 47:255–267.

Zuckerman, E., A. E. Eshel, and Z. Eyal. 1997. Physiological aspects related to tolerance of spring wheat cultivars to *Septoria tritici* blotch. *Phytopathology* 87:60–65.

Part IV

Epidemiology and Disease Control

22 Plant Disease Epidemiology

Kira L. Bowen

Concept Box

- Epidemiology is the study of properties of pathogens, hosts, and the environment that lead to an increase in disease in a population.
- Polycyclic pathogens that can readily reproduce and disperse usually cause the most damaging epidemics.
- Temperature, moisture, wind, soil properties, radiation, and other components can comprise a conducive environment.
- Host developmental stage and population uniformity can affect epidemic development.
- Disease spreads in space as it increases in incidence (numbers of infected plants).
- Mathematical models that allow disease predictions also contribute to disease management.

Epidemiology is the study of factors that lead to an increase in disease in a population. Understanding the reasons why diseases increase in populations of plants, and what can affect those increases, can contribute to decisions concerning plant disease management. Man's ability to produce food and fiber is, in part, limited by his ability to manage plant diseases. Thus, plant disease epidemiology has contributed to the highly technological culture in which we currently live. Concepts related to plant disease epidemiology are discussed in this chapter.

COMPONENTS OF AN EPIDEMIC

Disease occurs on a plant when a virulent pathogen and a susceptible host interact in a conducive environment. In order for disease to increase, three components—pathogen, host, and environment—must continue to interact over time. The interactions of these components might be thought of as a three-sided pyramid, with each side representing each component of the epidemic and the height of the pyramid representing the disease level. Thus, as the susceptibility of the host population increases or as the environment becomes more conducive, the resultant disease level is also greater or higher on the pyramid. In a population of plants, disease becomes important when the damage caused by that disease increases to the extent that there are social and economic impacts. There are numerous examples of plant disease epidemics that have had profound impacts on human history, and these illustrate the interactions needed over time between the pathogen, host, and environment. Well-known examples include the potato late blight epidemic of 1845, chestnut blight of the early twentieth century, and the southern corn leaf blight epidemic of 1970. Some more recent examples of epidemics affecting U.S. citizens include sudden oak death, citrus canker, and head blight of wheat and barley.

Potato Late Blight

The potato, *Solanum tuberosum* L., native to the American continent did well as a crop in the cool damp climate of Ireland. Prior to the 1840s, farmers in Ireland had become dependent upon potato for food, and the crop was widely grown across this island country. Therefore, two of the three necessary components for a disease epidemic were present—the susceptible crop and an environment that was conducive to disease. *Phytophthora infestans* (Mont.) De Bary, the fungus-like oomycete that causes potato late blight, was introduced into Ireland by early 1844, providing the third element for an epidemic. It is probable that this pathogen had become widely established by the end of the 1844 growing season (Andrivon, 1996), but had caused only low levels of disease. In 1845, excessively cool and wet weather prevailed throughout Europe. This weather, which was highly conducive to the development of *P. infestans*, allowed rapid increase in the severity of late blight disease and resulted in the loss of the entire potato crop by the end of the 1845 growing season. Because farmers were almost entirely reliant on potato for food, this loss led to the Irish potato famine, with many of the Irish people dying or emigrating to other countries.

Chestnut Blight

The American chestnut, *Castanea dentata* (Marsh.) Borkh., was an important and predominant tree species in native forests of the eastern United States through the eighteenth century. The rot-resistant wood was important for the timber industry, and the bark was a source of tannins for the leather industry. American chestnut also provided food in the form of nuts for wildlife as well as humans. In 1904, American chestnut trees in the New York Zoological Park began wilting and suddenly dying apparently from a canker disease. During the next year, the disease, called chestnut blight, was noticed on American chestnuts in other areas of New York. By 1911, chestnut blight had spread through New Jersey, south to Virginia, and north into Massachusetts. Communities in the Appalachians depended on the American chestnut tree for their livelihood, so the spread of this disease caused great concern. Attempts to stop the spread of chestnut blight were unsuccessful. Mature chestnuts trees continued to die throughout the eastern United States with devastating effects on Appalachian communities.

Chestnut blight is caused by the pathogen, *Cryphonectria parasitica* (Murrill) Barr (previously *Endothia parasitica*), and is spread when sticky conidia exuding from cankers adhere to birds and insects. This pathogen was brought to the United States with botanical specimens from China. In China, *C. parasitica* is **endemic** where it affects native Chinese chestnuts causing low levels of the disease. American chestnuts had no resistance to *C. parasitica*, and there were no elements of the environment in the United States that effectively limited its spread. The chestnut blight epidemic changed the makeup of American forests in less than 50 years, but only after the introduction of the virulent pathogen into the United States.

Southern Corn Leaf Blight

Southern corn leaf blight, caused by *Bipolaris maydis* (Y. Nisik. and C. Miyake) Shoemaker (anamorph of *Cochliobolus heterostrophus*), is endemic to the United States. Through the 1960s, hybrid field corn cultivars (*Zea mays* L.) were generally resistant to this disease. However, this reaction changed due to a change in the way hybrid seed was produced. Hybrid seed can be produced by planting several rows of a desired "female parent" corn inbred (A) between one or two rows of the "male parent" corn inbred (B). Pollen-producing tassels are removed from the female parents in these hybrid seed production fields. Because all pollens in the field are from the male parent, and all seeds produced on the female parent are a result of the cross (A × B) and are hybrid seeds. Removing tassels from female corn parents was extremely labor intensive, so when cytoplasmic male sterility (i.e., corn with sterile pollen) was discovered, it became widely used in inbred lines serving as female parents. One type of cytoplasmic male sterility, known as *Tcms*, was incorporated into numerous corn-breeding lines used as female parents, and the hybrid plants produced retained traces of *Tcms* cytoplasm.

By 1970, 80% of hybrid corn in the United States was produced using *Tcms*, creating a genetic uniformity in corn throughout corn-producing regions. About that time, corn cultivars that had previously been resistant to southern corn leaf blight were becoming diseased. As *Tcms* had been increasingly incorporated into corn germplasm, a new race of *B. maydis* had developed. This race, dubbed Race T, was more aggressive than the former dominant Race O of the pathogen and was highly virulent to *Tcms* corn. *B. maydis* Race T developed quickly on the susceptible host populations of *Tcms* plants in the hot moist conditions prevailing in the southern corn-producing regions of the United States (Figure 22.1). Losses due to this disease were 100% in some fields in the South where the epidemic apparently initiated. Although the moist weather of the south contributed to disease development, the southern corn leaf blight epidemic of 1970 was a result of man's creation of the susceptible host population.

These examples of historically important plant disease epidemics illustrate the interaction of each of the components of the disease pyramid over time. American chestnut trees were an important part of eastern U.S. forests until the introduction of the virulent pathogen *C. parasitica* to the United States where the environment was suitable for disease development. If the weather had not been highly conducive to the rapid development of *P. infestans* in Ireland in 1845, the Irish potato famine may never have happened. The environment, host, and pathogen interaction allowing southern corn leaf blight did exist before the 1970s, but had been adequately managed with plant resistance. This disease only became a problem when the host population was changed to a susceptible population. These plant disease epidemics also illustrate additional concepts relating to plant disease development and epidemiology.

PATHOGEN

Chestnut blight arose in the United States after the importation of infected botanical specimens for a collection prior to 1907. Similarly, potato late blight in Ireland

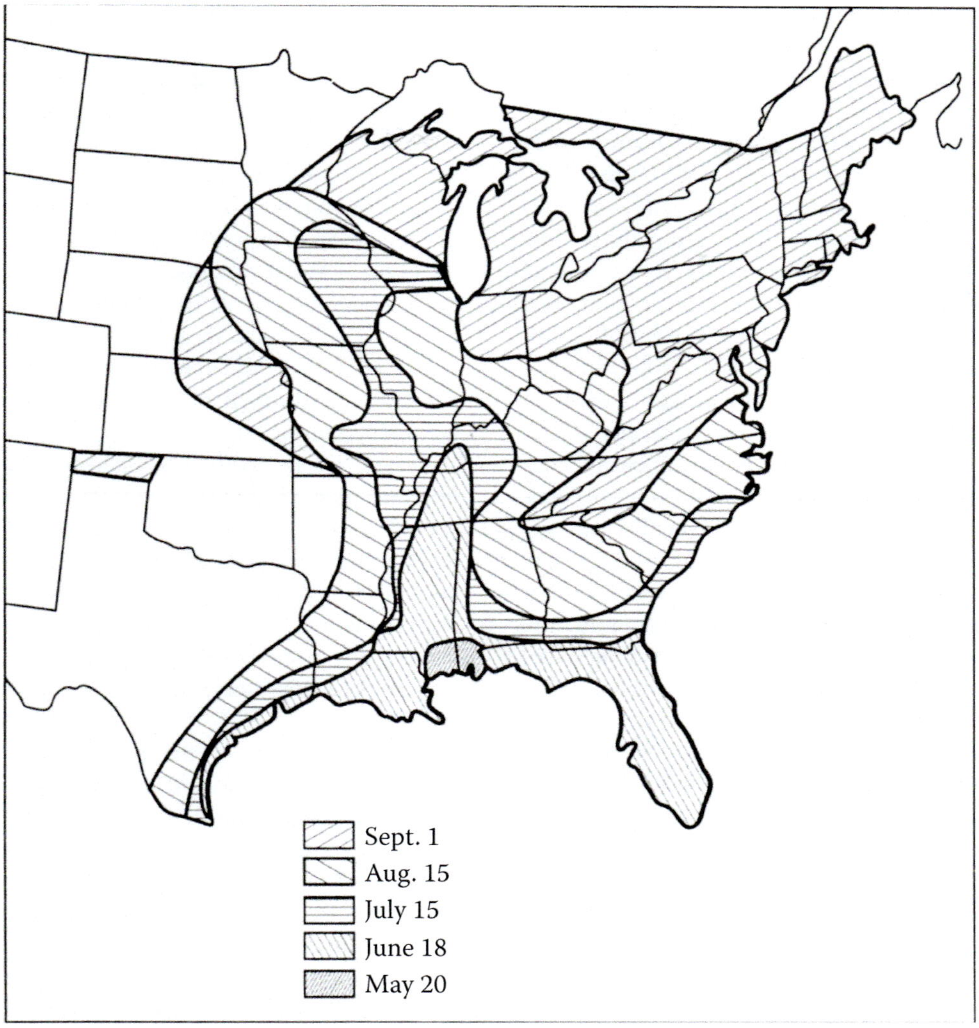

FIGURE 22.1 Disease progress of southern corn leaf blight during the 1970 growing season. (Adapted from Moore, M. F., *Plant Dis. Rept.*, 54, 1970.)

probably traces back to a shipment of tubers, with at least one infected tuber, into Belgium from the Americas. These infected specimens and tubers provided the **initial inoculum** of virulent pathogens for these plant disease epidemics. Given that the favorable environment and susceptible hosts were present upon introduction of the initial inoculum, the pathogen increased, causing more disease.

In the recent years, several pathogens have been newly found in the United States, including *Phytophthora ramorum*, the cause of sudden oak death (Davidson et al., 2003), *Phakopsora pachyrhizi* Syd., causing Asian soybean rust (Schneider et al., 2005), and *Candidatus* Liberibacter asiaticus bacteria, the cause of Huanglongbing disease, otherwise known as citrus greening (USDA, 2005). Initial inoculum of these newly introduced pathogens came to the United States in varying ways. For example, it is possible that *Phytophthora ramorum* arose spontaneously from other *Phytophthora* spp., which are genetically flexible organisms. Hurricane winds may have carried spores of *P. pachyrhizi* to U.S. Gulf Coast states (Pan et al., 2006) from South America. The pathogen responsible for Huanglongbing disease is vectored by specific species of psyllid insects that have been found in Florida since 1998.

While plant disease epidemics can be initiated by the introduction of an exotic pathogen to a new location, more commonly, initial inoculum for an epidemic is the quantity of the pathogen that survives a period without a host, as through a winter (also called primary inoculum) or the quantity that arrives at a location after a dispersal event. The initial inoculum for a plant disease epidemic may come from infected seed or may be the viable propagules of a soilborne pathogen at planting time. Initial inoculum can be from a single infected plant brought into a greenhouse or even from a vector for many viral and bacterial diseases of plants. Of course, the greater the quantity of initial inoculum, the greater the disease level at the start of an epidemic.

Potato late blight and southern corn leaf blight disease levels increased in a matter of days or weeks, whereas the increase in chestnut blight was observed over years. Chestnut blight increased over years, not because the tree has a long life span, but because of inherent characteristics of the pathogen *C. parasitica*. *C. parasitica* is considered **monocyclic**; that is, this pathogen reproduces only once in a growing season. There are pathogens of annual plants that are considered monocyclic, such as species of *Sclerotium* that cause diseases on numerous plants including stem rot of peanut, white rot of onion, and southern stem blight of tomato. *Sclerotium rolfsii* survives as sclerotia, and these germinate and cause plant infections during a growing season. At the end of the growing season, new sclerotia are produced on infected plants to serve as inoculum for the following growing season. Like chestnut blight, many canker diseases are monocyclic. Azalea gall, caused by *Exobasidium vaccinii* (Fuckel) Woronin, common in the spring in southern landscapes, is also a monocyclic disease. Certain microcyclic rusts that have no repeating urediniospore stage, such as cedar-apple rust caused by *Gymnosporagnium juniperi-virginianae* Schwein, are also monocyclic diseases. Plant disease epidemics that are due to monocyclic diseases will only increase over years, not over a few months.

The pathogens that cause southern corn leaf blight and potato late blight are both **polycyclic** because they multiply or reproduce several times in a growing season. It is likely that only one or a few potato tubers with a viable infection of *P. infestans* survived the trip from the Americas, to Europe, and then to Ireland by early 1844. That single infected tuber produced a plant with foliage infections from which sporangia and/or zoospores were produced. Sporangia are easily spread on wind currents, infecting more plants in moist conditions and producing new sporangia in as little as 5 days when weather conditions are optimum. Thus, in a 90-day growing season, this pathogen could have reproduced 17 times, resulting in ever greater disease intensity with each cycle. Most plant disease epidemics that occur unexpectedly are polycyclic because of the rapidity with which they can develop.

Polycyclic diseases occur on perennial plants, just as they do on annual crops such as potatoes or corn. Apple scab, caused by *Venturia inaequalis* (Cooke) G. Wint., for example, is a polycyclic disease of apple trees, as is fire blight. *Erwinia amylovora* (Burrill, 1882) Winslow et al., 1920, the bacterial cause of fire blight, will reproduce and cause increasing numbers of infections as long as moist conditions and appropriate temperatures prevail. Many rust diseases that have the repeating urediniospore stage, including zoysia rust and leaf and stem rusts of cereals, are polycyclic, as are powdery mildew diseases. Conidia of *Microsphaera pulchra* and *Uncinula necator* (Schwein.) Burrill, causing powdery mildew of dogwood and grape, respectively, can infect their hosts and produce more conidia every 5–8 days under favorable conditions. Diseases of plants caused by viruses are also often polycyclic, especially when caused by viruses that are vectored by insects and as long as the insects are moving among plants. Cucumber mosaic and tomato spotted wilt are examples of polycyclic diseases caused by viruses. Diseases caused by nematodes can also be polycyclic. Root-knot nematodes (*Meloidogyne* spp.) complete a life cycle in 3–4 weeks when soil temperatures are 25°C–30°C. In the southern United States, these nematodes go through three to four reproductive cycles in a growing season.

There are also pathogens that cause diseases that are **polyetic**. Polyetic diseases are those that take several years from the time of infection until symptoms develop and the pathogen reproduces. Because *Cronartium ribicola* Fisch. can take several years to grow into the main stem of the tree before sporulating, white pine blister rust is a polyetic disease. Mistletoe parasites are also polyetic because they reproduce only after a few years of plant growth.

Another aspect of the pathogen that affects epidemic development is the diversity within its population. Different individuals or isolates of the same pathogen may vary in the host cultivar that they can infect, and these are called **races** of a pathogen. Races have been identified in a number of plant pathogens, including *Phytophthora sojae* M. J. Kaufman & J.W. Gerdemann, cause of soybean root rot, and *Xanthomonas campestris* pv. *vesicatoria*, cause of bacterial leaf spot of pepper. Differences in a pathogen population may also affect the aggressiveness of the pathogen, as seen with Race T of *B. maydis*. When a new race of a pathogen develops, plant epidemics can become severe in crops previously thought to be resistant to that pathogen. This is what happened in wheat with *Puccinia graminis* Pers.:Pers. f. sp. *tritici* Eriks. and E. Henn. through the early part of the twentieth century. Wheat cultivars were developed and released to growers with improved resistance to stem rust, but then, within a few years, the disease once again became severe on the crop.

ENVIRONMENT

Many of the most destructive diseases of plants are polycyclic diseases, and each of them differs in some way. While the disease cycles for southern corn leaf blight and potato leaf blight can reoccur very rapidly when conditions are optimum, many disease cycles are somewhat longer. Disease cycles, from infection by inoculum through production of more inoculum, lengthen in time with less than optimum conditions. Temperature and moisture, and the interaction of these two components of weather, are often the most important factors affecting the length of a disease cycle. For example, the southern corn leaf blight disease cycle can be completed in 3 days when moist conditions prevail and temperatures are between 20°C and 30°C. If drying occurs, especially if relative humidity (RH) remains below 90% for 18 or more hours in a 24-h period, the disease cycle lengthens. Urediniospores of *P. graminis* f. sp. *tritici*, causing stem rust of wheat, cause infection and produce more urediniospores in as little as 7 days in ideal conditions at 30°C. However, under normal field conditions, this cycle repeats every 14–21 days (Roelfs, 1985).

In the field, temperature and moisture fluctuations detract from ideal conditions for pathogen development. Such fluctuations may be second-by-second as wind currents vary, but we are usually most aware of those that occur diurnally, in a 24-h cycle or due to weather systems moving through a region. Unfavorable weather conditions often limit disease development substantially, as can be seen with decreasing levels of powdery mildew in winter wheat as daytime temperatures warm or with minimal black spot on roses during dry seasons. Diseases that seem to disappear during the summer, such as brown patch of warm-season turfgrasses caused by *Rhizoctonia* spp., are actually limited by high temperatures and reinitiate in late summer as temperatures decrease.

Temperature and moisture are readily understood as influences on pathogen development, but there are other aspects of the environment that can affect pathogen development. Wind, for example, plays several important roles. Wind currents are one of the most important means by which plant pathogen inoculum is **dispersed** or becomes spread over a geographical area. It is easy to imagine how a fungal spore can be carried on the wind from one plant to another in a field, but it is possible for inoculum to move even greater distances as might have been the case with hurricanes carrying *Phakopsora pachyrhizi* (causing Asian soybean rust) to the United States. Generally, faster wind currents carry inoculum propagules further, and smaller propagules are carried further than larger propagules. Urediniospores of *P. graminis* f. sp. *tritici*, which are relatively small spores, have been carried by wind currents up to 680 km (>420 miles) from their source. Wind currents also have direct effects on temperature and moisture because they can cool and dry surfaces, which could affect the length of a disease cycle.

Solar radiation, the energy from the sun, is another aspect of the environment that influences disease development. Radiation can inhibit or stimulate germination of a number of fungal spores. Most bacterial cells quickly lose viability with exposure to radiation. Dogwood anthracnose lesions develop more slowly and have reduced sporulation on foliage in full sun than in full shade. Conversely, spores of *Botrytis* sp., causing gray mold of flowering plants, have been found to germinate only with exposure to particular UV wavelengths. Thus, the prevalence of overcast versus clear skies can contribute to the development of an epidemic.

Other aspects of the environment can affect disease development. Soil, for example, can have physical and chemical characteristics that profoundly affect disease development, particularly diseases by soilborne pathogens. Nematodes develop and move most readily in moist, but not in waterlogged soil and near the soil surface where oxygen availability is not limiting. Most fungi need oxygen for growth and development; hence, root rots by *Sclerotium* spp. tend to begin near the soil surface. Soil acidity is one of the chemical properties that can affect disease development, as seen with take-all of wheat and turfgrasses by *Gaeumannomyces graminis* (Sacc.) Arx & D. Olivier that is favored in more neutral (higher pH) soils. Soil nutrients can also affect the development of both soilborne and airborne pathogens in causing foliar diseases. For example, nitrogen fertilization has been associated with increased potato late blight severity. High N is also conducive to brown patch development in turf caused by *Rhizoctonia* spp., and excessive fertilization can initiate a severe outbreak of this disease. In a very broad sense, the economic and political environment that led to reduced tillage practices over substantial acreage in the North Central States may be responsible for recent severe epidemics of Fusarium scab of wheat (McMullen et al., 1997).

HOST

In addition to influences of the environment, the genetic makeup of a host population can affect an epidemic. A higher degree of genetic uniformity among plants is likely to allow more rapid and more severe disease development than would greater genetic diversity. In any population of plants, there will be some differences between individual plants that can affect the amount of disease resulting from an infection. However, plants that are self-pollinated, such as wheat, will have less genetic variability between individuals in a population than those that are cross-pollinated, such as corn. Plants that are vegetatively propagated, such as potatoes and many ornamentals, are more genetically uniform than even self-pollinated plants. Thus, epidemics would be expected to develop most rapidly across a population of clonally or vegetatively propagated plants, less rapidly in self-pollinating plants, and more slowly in cross-pollinated populations. In wild or natural plant populations, where plant species are intermingled, and different ages and origins of plants exist together, disease epidemics are considered rare. This rarity of plant disease epidemics in mixed species populations is primarily due to the probability of pathogen propagules landing on an appropriate host.

Humans, of course, are responsible for creating massive populations of identical plants that can readily succumb to disease problems. We have also discovered that certain plants have resistance or the inherent ability to withstand pathogen infection or disease development. Two types of resistance are recognized, one is easy to select for or breed into plants because it is usually encoded for by a single gene. In addition, this **single gene resistance** frequently prevents the development of any symptoms of disease. The disadvantage of this type of resistance is that a change in the pathogen can overcome that single gene and allow as much disease as if the plant were susceptible. Thus, adaptation of the pathogen to this type of resistance can result in the development of a severe plant disease epidemic. In wheat, for example, numerous genes have been identified that encode for resistance to stem rust, and these genes can be found in different cultivars of wheat. Because races of *P. graminis* f. sp. *tritici* shift from year to year, different cultivars need to be planted from one year to the next to avoid a devastating disease epidemic. Another type of resistance is **multigenic resistance**, and this is encoded by several to many genes. Multigenic resistance is difficult to incorporate into a single cultivar, but is not as readily overcome as resistance due to a single gene. However, multigene resistance usually does allow some disease development, which is why this is also called partial resistance. An advantage to this type of resistance is that it is effective against all races of a pathogen. Epidemics do develop in crops with multigenic resistance, but disease severity generally stays low (Vanderplank, 1963).

Tissue age or age of host plants can also influence the rate of disease development. Many pathogens, such as *V. inaequalis*, the cause of apple scab, will infect young tissues more readily than older tissues, so disease develops more quickly early in the growing season. There are also plant diseases, such as Sclerotinia blight of peanut caused by *Sclerotinia minor* Jagger, charcoal rot of soybean (caused by *Macrophomina phaseolina* [Tassi] Goidanich), and chestnut blight, that do not develop until plants have entered their reproductive stages. Most often, plants are more susceptible to diseases when they are young or when the plant or particular plant parts are actively growing. Young tissue can be more susceptible to pathogen infection because natural barriers (e.g., cuticular coats) have yet to develop.

As plants grow, they produce more leaf and stem tissues, and all these tissues have the potential to become diseased. If disease is not increasing as rapidly as the plant is growing, the proportion of diseased tissue on that plant will appear to decrease. Similarly, if defoliation occurs due to disease, as happens with peanut leaf spot diseases and black spot of rose, the proportion of the plant that is diseased may appear to decrease. Many plants have limited or finite growth during a season, so that disease can eventually affect the entire plant if conditions remain appropriate. Thus, cultivars of plants that have longer growing seasons may suffer heavier damage due to a particular disease than other, shorter-season cultivars because an epidemic developed for a longer period of time.

As plants grow, they also go through various developmental stages that involve physiological changes, and these changes can influence disease development. Although it might be a tedious work, it is easy to understand that various aspects of plant size can be measured. Plant height, numbers of leaves, total area of foliage, root depth, root volume, and root mass, if monitored over time, would reflect plant growth. Host development, too, can be monitored, and for many plants distinct **host growth stage** keys or diagrams are available. In corn, for example, distinct stages were delimited by Hanway (1963) and include emergence, tassel formation, silk development, and several stages in kernel maturation. Each of these developmental stages is separated in time, but not necessarily by equal time increments. Growth stage keys have been developed for most agronomic crops, including cereals, soybeans, and peanuts.

DISEASE

Prior to the complete devastation of their crops in 1845, Irish farmers did not notice late blight on their potatoes even though it had likely occurred in the preceding year. Late blight that may have affected potato plants in 1844 would have been easy to overlook, in part because the concept of plant disease was not known. However, even today, it is difficult to find the disease in any plant population when damage levels are low. Most people who work regularly with plants will not notice symptoms of a disease until it has increased to about 1% intensity. This initial observation of disease is called **onset**, and onset is often preceded by several to many reproductive cycles by the pathogen following the introduction of the initial inoculum. Yet the difference between onset at ~1% disease level and severely diseased plants at 80% damage might only be another one or few reproductive cycles. Thus, it appears that disease development is slow at the beginning of an epidemic. This can be illustrated with a disease that may occur, for example, in an acre of 20,000 plants. A single lesion of this disease might occupy 0.1% of the foliage of a single plant and account for about 0.000005% of all foliage in the acre of plants. If that single lesion produces inoculum that results in a 20-fold increase in numbers of lesions, then the second "generation" of disease lesions would occupy 0.0001% of foliage in the acre, and so on. It would take more than four disease cycles before disease at ~1% is readily noticed, and only 1.6 more cycles before 100% of foliage is diseased if that disease increase continued at this rate (Figure 22.2).

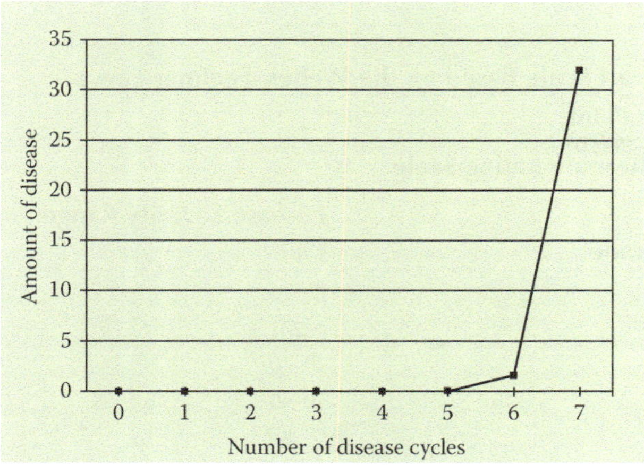

FIGURE 22.2 Increases in disease that starts from a single lesion affecting 0.000005% of foliage on one plant among 20,000 plants, when each disease cycle allows the development of 20 new lesions from each older lesion.

Epidemics are not always rapid developments of severe disease; epidemics occur when there is *any* increase in disease over time. Thus, even the increase of one diseased plant to two disease plants reflects an epidemic. An increase in disease **incidence**, the number or proportion of plants in a population that are affected by disease is due to spread or dispersal of disease-causing inoculum. When formerly healthy plants become diseased due to inoculum spread from another plant, this is termed **alloinfection**. Some inoculum spreads no further than to leaves of the originally diseased plant. However, even this **autoinfection** of the same plant leads to a greater disease level through an increase in severity. Disease **severity** represents the proportion of tissue of a single plant that is affected by disease. It is probable that potato late blight was widely established through Ireland at the end of the 1844 growing season and that disease incidence was high, with more than 50% of individual plants affected. However, the level of disease on individual plants was probably fairly low, perhaps less than 5% disease severity.

Disease severity and disease incidence are usually positively related to one another. That is, as disease severity increases, so does disease incidence. However, there are a number of diseases that increase only in incidence because a single infection systemically affects the entire plant, frequently causing plant death. Bacterial wilt on tomato by *Ralstonia solanacearum* (Smith) Yabuuchi et al. and Verticillium wilts of many plants including redbud, snapdragon, and most vegetables are just some of the diseases that only increase in incidence over time in a plant population. While there may be differences in the stage of disease progress at some point in time among systemically infected plants in a population, these differences are more likely due to when infection occurred or genetic differences between the infected plants.

Disease Rating Scales

Disease severity, or the proportion of the plant that is affected by disease, can range from 0% to 100%. However, the entire range of disease severity is difficult to distinguish with precision, especially when disease levels are moderate or between 20% and 80%. The Weber–Fechner law of visual discrimination relates to this ability to distinguish moderate levels of plant disease severity. This law states that an individual's ability to see difference decreases by the logarithm of the intensity of the stimulus. In other words, when the stimulus is a disease lesion on a leaf amidst healthy tissue, it is easier to distinguish 1% from 3% or 5% disease severity than to tell 20% from 40% severity. Similarly, when a leaf is nearly completely diseased, as happens with southern corn leaf blight, it is relatively easy to distinguish 95% from 99% disease severity, but more difficult to determine 65% from 85% disease. It is easier to distinguish between high levels of disease as severity approaches 100% because the visual stimulus becomes the healthy tissue amidst diseased tissue.

Horsfall and Barratt developed a disease rating scale based on the Weber–Fechner law and the difficulty of distinguishing among moderate levels of disease severity. This rating scale consists of 12 levels or grades, each of which includes disease severity ranges that decrease as disease approaches 0% or 100% (Table 22.1). In addition, disease severity ranges are symmetrical around 50%. The Horsfall–Barratt disease rating scale is only one example of many aids that can be used in rating disease severity in order to monitor increase in disease over time. For example, in evaluating disease on a plant, it is relatively easy to think of whether the severity encompasses 1/5, 2/5, 3/5, etc., etc., of the plant, and this consideration can be implemented as a rating scale with 0 for no disease and 5 for additional levels (Table 22.1). One advantage of the Horsfall–Barratt scale over one based on fifths of a plant is that there are more grades or levels in the scale. The more restricted ranges of disease at low and high levels allow greater accuracy in keeping track of an increase in disease over time. Rating systems with fewer grades may be more useful for comparing the effectiveness of fungicidal products or the susceptibility of a number of cultivars.

Disease Diagrams

As an aid in assessing disease severity on plants, disease diagrams have been developed for a number of diseases on several important crops (e.g., James, 1971). Such diagrams provide pictorial aids for determining particular disease levels (Figure 22.3). While frequently used in training, such pictorial aids improve

TABLE 22.1
Two Rating Scales for Assessing Disease: The Horsfall–Barratt Scale Based on the Weber–Fechner Law of Visual Acuity and an Alternate Scale Based on 1/5th of the Plant

Horsfall–Barratt		Alternate Rating Scale	
Grade	Disease Ranges (% Severity or Incidence)	Grade	Disease Severity Range (%)
0	0	0	0
1	0–3	1	0–20
2	3–6	2	20–40
3	6–12	3	40–60
4	12–25	4	60–80
5	25–50	5	80–100
6	50–75		
7	75–88		
8	88–94		
9	94–97		
10	97–100		
11	100		

FIGURE 22.3 Disease severity diagram of rating scale for hypothetical foliar disease of a small grain.

the precision of disease ratings, even by experienced personnel. In addition, these aids provide consistency in disease evaluations made at different times and by different evaluators. Current technology allows us to easily develop customized diagrams for specific diseases on any host.

Complications

Disease assessment, especially with the help of rating scales and disease diagrams, is not always straightforward. Often, disease symptoms vary in color, and these colors may not be distinct from one another. Similarly, more than one disease may be affecting a single leaf or plant at any point in time.

Disease in Space

As illustrated by the southern corn leaf blight and chestnut blight epidemics, increases in disease incidence over time can also lead to an increase in the size of the area in which a disease can be found (Figure 22.1). At a single point in time, severity on individual plants is expected to decrease as distance increases from the original infection. This is called a **disease gradient**. Gradients can be due to gradual changes in the microclimate or soil type, but are most often seen with increasing distances from the first infection and are due to dispersal. Those pathogens that are most easily dispersed, like powdery mildew fungi and *Phytophthora infestans*, will infect plants further and more quickly from an original infection site than pathogens with larger propagules that are dispersed with more difficulty. *P. infestans* is said to have a "flat" dispersal gradient because of its ease of spread (Figure 22.4). Conversely, nematodes and *Gaeumannomyces graminis*, causal organism of take-all of grasses, are considered to have "steep" dispersal gradients because their spread is due to their own movement and growth through soil. Actually, dispersal gradients are not usually applicable to diseases caused by nematodes because groups of plants affected by nematodes tend to be clustered in a field. Diseases that occur in patches, such as those caused by nematodes and other soilborne pathogens are considered "focal diseases."

A number of diseases not only cause tissue damage but also cause defoliation or premature loss of foliage. In addition to defoliation, disease development can cause lodging

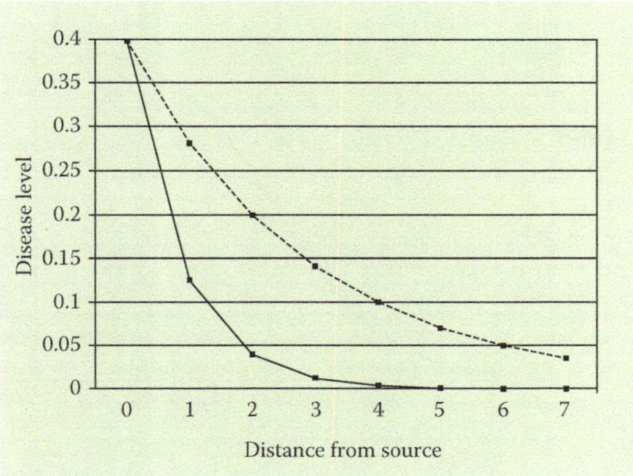

FIGURE 22.4 Graphical representation of disease gradients or the decrease of disease with increasing distance from source. Solid line represents "steeper" disease gradient.

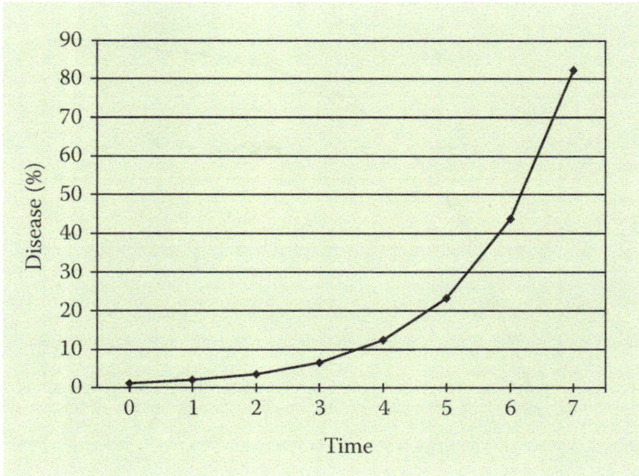

FIGURE 22.5 Exponential model of disease development over time.

or the falling over of plants, stunting or a decrease in size, wilting, and diminished quality (sugar, protein content, or toxin contamination). Any of these disease characteristics can increase in intensity over time.

DISEASE ANALYSIS OR MODELS FOR EPIDEMICS

In order to understand plant disease epidemics, mathematical models have been used as a means of simplification. These mathematical models provide a quantitative way to represent the sometimes complicated processes of an epidemic. Mathematical models have also allowed comparisons of epidemics from different times or places and provide a framework for forecasting future disease levels.

EXPONENTIAL MODEL

One model that has been applied to plant disease development over time is the exponential model (sometimes called the logarithmic model) (Figure 22.5):

$$y_t = y_0 e^{rt} \quad (22.1)$$

where y_t is the amount of affected tissue after time intervals t, given a starting disease level (onset) of y_0 and a rate parameter (r), where e is the mathematical constant 2.718. Note that the graph of this model looks similar to one presented previously (Figure 22.2) in that disease stays low during the first few cycles, then increases rapidly. The linear form of this model is often easier to understand:

$$\ln y_t = \ln y_0 + rt \quad (22.2)$$

TERMINOLOGY

While applying this and similar models to plant diseases, several things should be noted. For example, the variable t can represent different time intervals, depending on the pathosystem. As noted earlier in this chapter, the time interval applicable to late blight of potato would be a day, whereas for chestnut blight months or years might be more applicable. Similarly, the infection rate r differs among various pathosystems. Calculations done on a number of plant disease epidemics indicate that $0.10 < r < 0.56$ for diseases of annual plants. It should also be noted that r is not translatable to units such as conidia/day; the value of r reflects many processes that affect disease development over time. The starting disease level is y_0 and represents the initial inoculum or initial disease level.

MONOMOLECULAR MODEL

While the exponential model allows for infinite increase in the value of y, usually there is some finite limit to the amount of plant tissue that can become diseased. This is especially true when dealing with disease severity, when the proportion of disease cannot become greater than 100% or when considering the proportion of plants in a finite population. Therefore, the exponential model is not often applicable to disease increase in plant populations. The monomolecular model is another model for simplifying disease increase over time and with this mode (Figure 22.6), disease levels are limited to 100%:

$$y_t = 1 - (1 - y_0)e^{-rt} \quad (22.3)$$

The linearized form of the monomolecular model is

$$\ln y_t = \ln(1/[1 \quad y_0]) \quad (22.4)$$

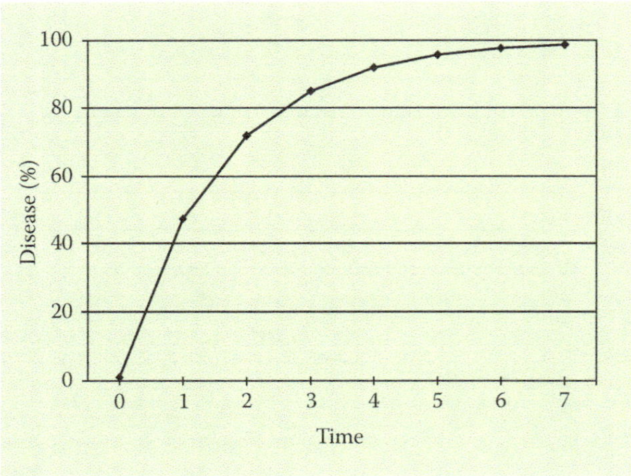

FIGURE 22.6 Increase of disease over time according to the monomolecular model.

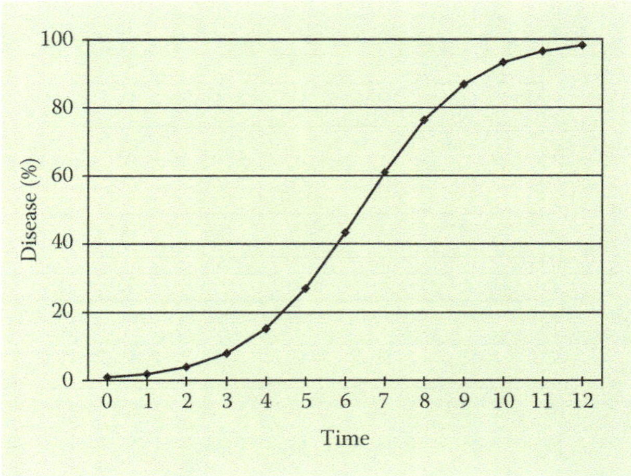

FIGURE 22.7 Increase of disease in time as described by the logistic model.

Logistic Model

A third model for simplifying disease increase in time is the logistic model (Figure 22.7). This model also limits disease to a maximum of 100%:

$$y_t = 1/\left(1+e^{-\left[\left(\ln y_0/(1-y_0)\right)+rt\right]}\right) \quad (22.5)$$

The linearized form of the logistic model is

$$\ln\left[y_t/(1-y_t)\right] = \ln\left[y_0/(1-y_0)\right] + rt \quad (22.6)$$

It should be noted that when disease levels are low, that is, less than 10%, the exponential model is very similar to the logistic model. Thus, either the exponential model or the logistic model will apply during early stages of an epidemic. When disease levels are high, approaching 100%, the monomolecular model is very similar to the logistic model. Thus, either the monomolecular or the logistic model will apply during later stages of many epidemics.

Simple Assumptions

These models idealize disease progress since they assume constant and uniform values for each of the variables. They also simplify the processes of disease development and have been fit to numerous plant disease epidemics, from potato late blight development to the development of dollar spot by *Sclerotinia homeocarpa* F.T. Bennett on creeping bentgrass. Vanderplank, in his 1963 book *Plant Diseases: Epidemics and Control*, was the first to quantitatively analyze plant disease epidemics. These initial analyses were criticized and built upon and ultimately led to the plant disease epidemiology as an autonomous area of study.

The infection rate of these models is where most of the simplification comes in relative to understanding epidemics. As described earlier in this chapter, there are numerous factors that influence the rate of epidemic development, and these are incorporated into the value of r. In real life, an increase in plant disease over time appears to be somewhat erratic—sometimes increasing rapidly, sometimes seeming to decrease. Quantitative models smooth out these apparent stops and starts of disease increase, as seen in the difference between Figures 22.2 and 22.5. In addition, these models for plant disease increase over time and do not account for any latent period, but assume that new infections are immediately visible and instantaneously infectious. All of these fluctuations and stops and starts in disease development are incorporated into the infection rate r.

Control Applications

Using the linearized forms (Equations 22.2, 22.4, and 22.6) of the models for disease increase in time, it is easy to see how changes on the right side of each equation affect the left side. For example, a decrease in the value of y_0 will decrease y_t if the values of other parameters stay the same. Thus, these equations help in understanding the effects of many control measures.

Effects on y_0

The amount of viable inoculum of a soilborne pathogen often decreases with rotation to a nonhost crop and, for use in the models presented, can be represented by a lower value for y_0. Initial inoculum is also decreased through sanitation, which is the removal of inoculum, and through the use of protectant fungicides, which are fatal to inoculum. Single gene resistance is also considered to have a detrimental effect on y_0. A decrease in initial inoculum, if substantial enough, can result in lower disease at the end of a finite time period, such as a growing season, compared with no decrease in y_0 (Figure

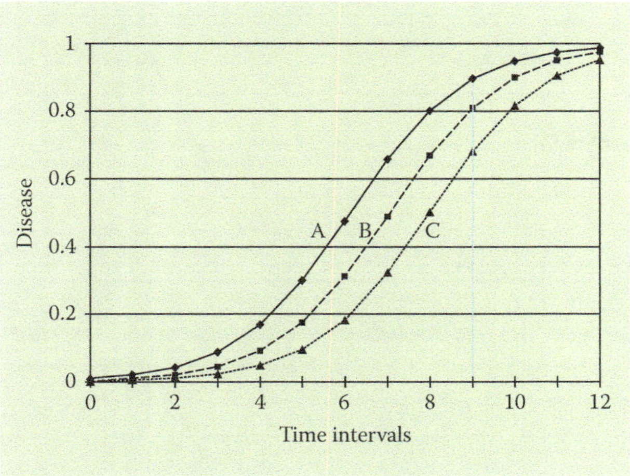

FIGURE 22.8 Disease development in time according to the logistic model with three levels of initial inoculum. (A) $y_0 = 0.01$; (B) $y_0 = 0.005$ (50% of A); and (C) $y_0 = 0.0025$ (50% of B). The vertical line is at $t = 9$.

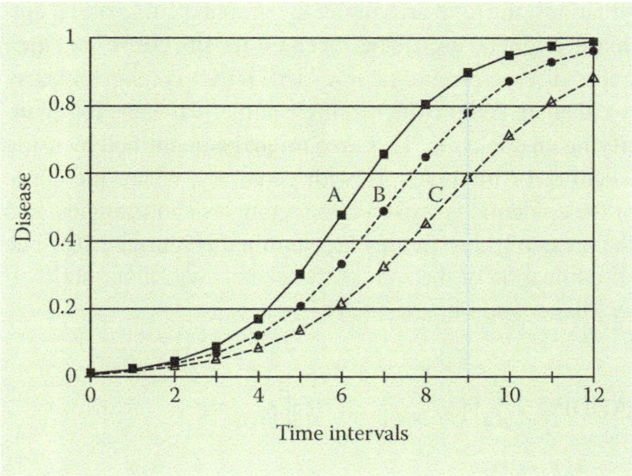

FIGURE 22.9 Graphs of disease in time according to the logistic model with three levels for the rate parameter. (A) $r = 0.75$; (B) $r = 0.65$; and (C) $r = 0.55$. The vertical line is at $t = 9$.

22.8). The figure Logistic disease increase beginning at $y_0 = 0.01$ will reach 90% at $t = 9$ (Line A). However, when initial inoculum is decreased such that $y_0 = 0.005$, disease at $t = 9$ is 81% (Line B). A further decrease to $y_0 = 0.0025$ results in disease at 68% at $t = 9$ (Line C). Lower disease levels are desirable because less disease is generally associated with higher yield.

Effects on *t*

Control practices can also affect the *t* variable of models for disease increase. For example, early harvest or a short-season cultivar would end an epidemic earlier, perhaps at $t = 8$ rather than $t = 9$. For example, with Line C in Figure 22.8, an earlier harvest at $t = 8$ rather than at $t = 9$ could mean disease severity of 50% instead of 68%.

Effects on *r*

Control measures can also affect the infection rate, *r*, of these models. For example, multigene resistance in a plant is considered to reduce *r* compared with plant susceptibility. The use of sterol biosynthesis-inhibiting fungicides, which delay fungal development and reduce inoculum production, reduces the rate of disease increase. Cultural methods such as pruning a tree or shrub to increase air flow, thereby decreasing the duration of moist periods, and use of drip irrigation instead of sprinkler irrigation, also reduce the value of *r* in these models (Figure 22.9). The figure shows that disease increase, according to the logistic model with $r = 0.75$, will reach 90% at $t = 9$ (Line A). When *r* is decreased to 0.65 then 0.55, disease at $t = 9$ will be 78% (Line B) then 59% (Line C).

Each of the components of these models act together and distinctively for each pathosystem. In general, polycyclic diseases have low y_0 but high *r* values. Because polycyclic diseases can develop very quickly, the level of y_0 has less influence on the epidemic than does *r*. Thus, control strategies for diseases with high *r* values should be aimed at decreasing *r*. Black spot disease of rose is a polycyclic disease with a high *r* value in the southern United States. Reduction of y_0 through sanitation by the removal of fallen infected foliage is recommended for minimizing black spot. However, on many rose plants, fungicide applications are still needed in order to manage this disease. Similarly, peanut leaf spot diseases, caused by *Cercospora arachidicola* Hori and *Mycosphaerella berkeleyi* Jenkins (formerly, *Cercosporidium personatum*), are polycyclic diseases with high values for *r*. Initial inoculum for these leaf spot diseases is believed to originate from plant debris remaining in the field. Because these fungal pathogens only infect peanut, the only source of inoculum is from previously infected tissue. However, if peanuts are grown in a field not previously planted to this crop, leaf spot diseases still develop rapidly from the small quantities of inoculum that are carried to these new fields by wind. Powdery mildew diseases of many plants are additional examples of polycyclic diseases with high *r* values.

Conversely, most monocyclic diseases have low *r* values, but relatively high y_0 values. Most effective control strategies for monocyclic diseases would be to reduce y_0. As presented earlier, southern blight of tomato is a monocyclic disease. If planting is done in a field that had not been previously cropped to tomato, the incidence of southern blight is likely to remain low through the growing season. Thus, crop rotation is a recommended and effective strategy for minimizing southern blight of tomato, a monocyclic disease.

While increase of disease severity over time may be fit to the logistic model, a simpler means of comparing different epidemics is frequently desired. In these situations, the total area under the graphed curve of the epidemic might be used. The "**area under the curve**" or "area under disease progress curve" (**AUDPC**) is relatively easy to calculate and provides a single numerical value for quantifying an epidemic. This area might be quantified by using calculus, or more simply with geometry, where the curve of the epidemic is broken into rectangles and triangles, and the areas of these figures are summed (Figure 22.10). The simplification of the sum of these areas was first published by Shaner and Finney (1977):

$$\text{AUDPC} = \sum_{(i=1)}^{n} \left[\left(y_{(i+n1)} + y_i\right)/2\right]\left[t_{(i+1)} - t_i\right] \quad (22.7)$$

Several epidemics can be compared using AUDPC values, most easily when the monitored duration of those epidemics is the same.

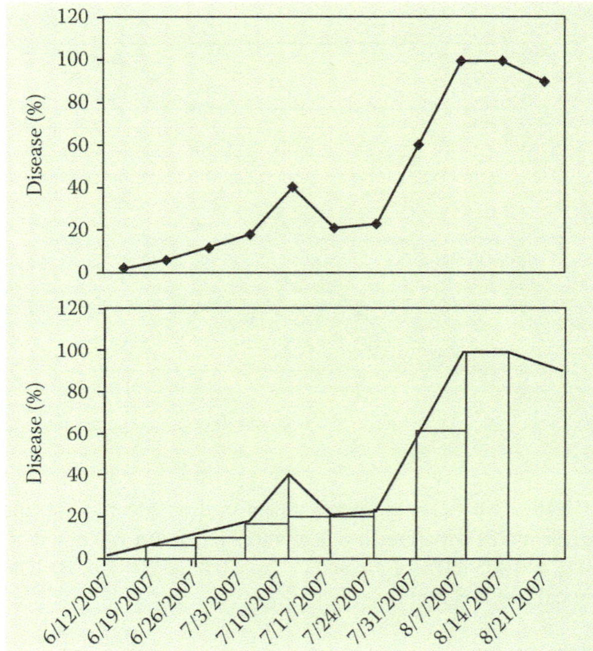

FIGURE 22.10 Geometric approach to calculating area under the disease progress curve (AUDPC) by summing all areas of rectangles and triangles that fit under the disease curve. Area of rectangle = (width × height) or (difference between adjacent time intervals × difference between adjacent disease levels). Area of triangle = ½ (width × height) or ½ (difference between adjacent time intervals × difference between adjacent disease levels).

Spatial Modeling

As has been mentioned, when there is disease increase in a population, the increase occurs over time as well as over geographical area. Spatial aspects of epidemics can also be described using mathematical models. Such spatial models can be used to determine how far a disease can spread. One such model is an exponential model with different parameters than the preceding model:

$$y = a\mathrm{e}^{-bx} \quad (22.8)$$

The linear form being:

$$\log y = \log a - b\left(\log x\right) \quad (22.9)$$

where y is disease at some distance x from an infection source, a is the amount of viable inoculum, and b is the ease (or difficulty) by which the pathogen is dispersed. As the value of b approaches 0, the pathogen spreads easily. Thus, diseases caused by pathogens that disperse readily, such as *Botrytis* spp. with its small conidia, have small values for b. Values for b for soilborne diseases are high, approaching 1, since the pathogens causing many of these diseases spread primarily by means of their own growth or movement.

Spatial analyses of plant diseases can contribute to improved management of disease problems. For example, spatial analyses have been used with geostatistics for the documentation of the role of weeds as alternate hosts in tomato virus management in the Del Fuerte Valley of Mexico (Nelson et al., 1999). These analyses have also reinforced the need for management strategies that reduce insect vectors and pathogen reservoirs when dealing with Pierce's disease of grape (Tubajika et al., 2004).

CROP LOSSES

In each of the three examples at the beginning of this chapter, plants and crops were killed or yields were substantially reduced due to plant disease epidemics. Potato late blight killed potato plants, and any tubers that did develop and were eventually harvested were likely to rot in storage due to infection by *Phytophthora infestans*. Similarly, mature American chestnut trees died from infection with *Cryphonectria parasitica* throughout eastern U.S. forests. While not all plant pathogens kill their hosts, any disease will detrimentally affect plant growth and development. In the vast majority of diseases, leaf spots or even a few rotten roots will not negatively impact human lives. However, when disease incidence or severity increases to the point that most of a plant's foliage becomes necrotic or limbs of trees do not produce fruit, this is a cause for concern.

One of the primary goals in gaining knowledge about plant diseases is to reduce crop losses to an acceptable level. An acceptable level of loss may be that which cannot be controlled with affordable management strategies. As the value of the yield of a crop or an ornamental plant increases, more can be done to manage any disease that is detrimental to the yield or the aesthetics of plants such as ornamentals. However, not all levels of disease have noticeably detrimental effects on yield or plant development. So, first the relationship between disease and yield must be determined and understood.

The relationship between disease and yield (or losses in yield) has been determined for numerous pathosystems. With some plants, such as wheat, specific growth stages affect different components of the productivity of that plant, and any disease occurring during that growth stage can subsequently affect yield. For example, the flag leaf, or the leaf just below the grain head of wheat, has been documented as contributing more to grain fill than other leaves. For this reason, yield loss models for foliar diseases of cereals often use the percent disease severity on the flag leaf (e.g., Tillman et al., 1999) for estimating losses due to disease. Of course, in order to use such estimates of loss, disease severity must be determined with some accuracy that justifies the use of standardized rating scales as previously presented in this chapter. Another type of loss relationship can be illustrated with southern stem rot in peanuts where each locus of a *Sclerotium rolfsii* infection at plant maturity, up to 30 cm in length of 30.5-m row, causes 0.9%–2.9% loss in yield quantity (Bowen et al., 1992). Knowledge of this relationship can aid a grower in determining how yields could change if disease incidence increased or decreased and may provide justification for the implementation of control measures. Perhaps a means of estimating crop loss due to disease before the occurrence of disease would be more useful. Such knowledge would allow the use of control measures before damage and losses have developed.

DISEASE FORECASTING AND MANAGEMENT

One of the uses of the mathematical models that describe plant disease development over time is for predicting or forecasting future disease levels. Knowledge about a future disease level can assist in making disease management decisions.

With monocyclic diseases, disease levels might be forecast by a direct assessment of the amount of initial inoculum at the beginning of the season. Indeed, this has been done with white rot of onion, caused by *Sclerotium cepivorum* Berk. in New Jersey (Adams, 1981) and for sclerotium rot of sugar beets, caused by *S. rolfsii* in sugar beets in Uruguay (Backman et al., 1981). With both of these diseases, sclerotia in the soil provide the initial inoculum. The number of sclerotia found in soil can be used to predict the amount of disease at harvest or yield loss. Similarly, nematode populations are determined prior to planting to estimate the potential damage in a crop. When initial inoculum levels are high, substantial disease is predicted, and the recommendation is made to plant the crop elsewhere or to take protective measures. Another disease for which disease is predicted from initial inoculum is Stewart's wilt of corn. The assessment of initial inoculum, however, is indirect because it is the corn flea beetle vector of the pathogen, *Erwinia stewartii* (Smith) Dye that is considered rather than the pathogen itself. The corn flea beetle population is adversely affected by cold temperatures during winter months. If the average winter temperature is at or below freezing, few flea beetles survive, and the risk of high disease is substantially reduced. Knowledge about the corn flea beetle survival helps growers determine which cultivar they should plant, especially when cultivars differ in their resistance to Stewart's wilt (Pataky et al., 1995).

Initial inoculum also plays a role in the forecast for apple scab. However, the actual amount of initial inoculum is not quantified, rather the weather that is conducive for release of spores and infection by initial inoculum is monitored and analyzed. When infection is favored, the forecast recommends initiation of a fungicidal spray program. Thus, regular fungicidal sprays are used to manage this polycyclic disease, but the forecast is largely based on weather that is favorable for initial inoculum.

Most polycyclic diseases are best controlled by reducing the infection rate. As presented earlier, weather is one of the primary influences on the rate of development of an epidemic. Thus, some of the best forecasts of polycyclic diseases are based on weather conditions that are favorable for pathogen development. In Alabama, for example, the AU-Pnut leaf spot advisory provides a recommendation for fungicide applications on peanuts when weather is favorable for infection by the pathogens causing peanut leaf spots. This advisory is based on the chance for rain and the accumulation of rain events beginning at planting or following a period of protection after a fungicide spray. Further north, in Virginia and North Carolina, temperature can be a limiting factor for the development of peanut leaf spot diseases, so the Virginia peanut leaf spot advisory system includes temperature as well as moisture in predicting favorable conditions for disease increase. These two forecast systems, based on reducing the infection rate of an epidemic, have been shown to save growers one to two fungicide applications per year compared with applications made every 14 days on a calendar schedule through a growing season.

Moisture and temperature are also used in predicting potato late blight using BLITECAST. This forecasting system computes daily disease severity values that are based on numbers of hours during which RH > 90% and the average temperature during moist periods. Greater disease severity values are predicted with longer periods of high RH, and higher temperatures allow some shortening of high RH periods. Disease severity values for BLITECAST are accumulated over a 7-day interval and, along with numbers of rain days, are used in recommending whether to apply a fungicide to the crop. In addition to the recommendation to spray or not spray, BLITECAST suggests spray intervals depending on the accumulated disease severity values.

A similar forecast system that uses moisture and temperature for determining disease severity values is TOM-CAST, a management aid for control of early blight (caused by *Alternaria solani* Sorauer), Septoria leaf spot, and fruit anthracnose (caused by *Colletotrichum coccodes* [Wallr.] S. Hughes) in tomatoes. Disease severity values in TOM-CAST are greatest with wet weather and temperatures between 21°C and 27°C (70°F and 80°F). Fungicide applications are recommended by TOM-CAST when disease severity values have accumulated to specific levels. Each of these forecasting systems provide recommendations for applying fungicides and are useful for decreasing the number of fungicide applications when the environment is not conducive for disease development.

In addition to the plant disease forecasting systems presented here, numerous additional systems have been developed. Many of these systems have been implemented with some success. The reasons that forecasting systems for plant diseases have not been more widely used are varied. One reason that growers do not adapt to using a plant disease forecast or advisory system is that they consider the possible reduction in pesticide applications as too risky. Adding to the perception of risk is that pesticide labels may not address the possible use of forecasting systems. Another reason for lack of implementation of disease forecasts has been the limited availability of appropriate weather data, particularly for information such as hours of leaf wetness. However, this is changing with increased access to the Internet and ease of obtaining customized local and accurate weather information.

CONCLUSION

Plant disease forecasts and advisory systems are tools that have been developed through epidemiological studies. Plant disease epidemiology, or the study of the factors that allow disease increase, provides information that can improve the efficiency of plant and crop production. Numerous opportunities exist for continued application of knowledge gained through epidemiology for improving our ability to manage plant diseases.

LABORATORY EXERCISES

Experiment 1. Disease Assessment

Materials

The following materials will be needed for each student or team of students:

- Diseased leaves or fruit or other items (e.g., potato tubers, peanut pods) that show a range of disease damage. Such a range of disease can develop over time, so these materials can be collected at different points in time from the same plant population. It is recommended that at least five specimens with varying disease/damage be available for each student and that each student have similar disease levels in their collection. Alternatively, copies of disease diagrams can be used (e.g., James, 1971).

Follow the instructions in Procedure 22.1 to complete this exercise.

Anticipated Results

Students will gain an appreciation for differential disease levels and for the "art" of assessing disease severity. Ideally, using *in vivo* materials, students can see how disease severity might develop over time if materials were collected as such. If used, copies of disease diagrams can be organized by the instructor to simulate disease development over time or to compare student improvement with practice.

Procedure 22.1

Disease Assessment

Step	Instructions or Comments
1	Using Table 22.1, assign a rating to each sample based on the Horsfall–Barratt scale; then rate each sample using the alternative scale. Do this for each specimen.
2	Calculate a mean disease rating for each student using both the rating scales.
3	Compare disease ratings among students.

Questions

- What levels of disease were the most difficult to rate?
- Were there complicating factors affecting the symptoms that were assessed?
- What was taken into consideration when deciding a value to assign to each sample?

Experiment 2. Disease Progress

In order to understand plant disease epidemics, mathematical models have been used as a means of simplification. These mathematical models provide a quantitative way to represent the sometimes complicated processes that comprise an epidemic. The logistic model is often used to depict plant disease development:

$$y_t = 1/\left(1 + e^{-[(\ln y_0/(1-y_0)) + rt]}\right)$$

In this model, y_t is the amount of affected tissue after t time intervals, given a starting disease level y_0 and a rate parameter (r), where e is the mathematical constant = 2.718.

Materials

Each student or team of students will require the following items:

- Calculator
- Graph paper

Follow the instructions provided in Procedure 22.2 to complete this part of the exercise.

Questions

- How did the plotted lines differ?
- What plant disease management method might have been taken to reduce y_0, as shown in differences between Lines 1 and 3?
- What approach might be taken to reduce r, as demonstrated by differences between Lines 1 and 2?
- What management strategies might be more effective to reduce the theoretical disease that is modeled?

Experiment 3. Disease Advisory System

Disease forecasting or advisory systems can help growers reduce fungicide use while maintaining an adequate level of disease control. Pathogen infection and subsequent disease development requires specific environmental conditions that can be monitored and used for these systems. For example, AU-Pnuts advisory for leaf spots on peanuts works in the most southeastern United States states where temperatures are always favorable for leaf spot diseases. This advisory uses rain forecasts and rainfall events. Five-day rain forecasts are used, and the chance of rain is averaged over 5 days. There are two sets of rules for use in this advisory; one set applies only to the first fungicide application of the season while somewhat different rules apply to the second and subsequent fungicide applications (Table 22.2). When the 5-day average for a chance of rain and the number of rain events that have occurred align as shown in the table, a fungicide application is recommended. Daily counts are made of rain events, and the 5-day rain forecast start when seedlings emerge or 10 days following any fungicide application. This 10-day period is set because a fungicide application provides 10 days of fungal pathogen control.

Follow the instructions in Procedure 22.3 to complete this exercise.

Questions

- Identify days on which to apply fungicides according to AU-Pnuts for
  - An initial spray and subsequent sprays, if peanut plants emerged on Day 0, and
  - Subsequent sprays assuming that Day 0 is the end of a 10-day fungicidal efficacy period.

Procedure 22.2
Disease Progress

Step	Instructions and Comments
1	By using Equation 22.5 or 22.6, and $y_0 = 0.01$, $r = 0.48$, calculate y_t values for $t = 0, 1, 2, \ldots, 20$.
2	Graph the data, calculated in Step 1, on graph paper. Connect data points to create Line 1.
3	Repeat Steps 1 and 2, using $y_0 = 0.01$ and $r = 0.24$ for Line 2.
4	Repeat Steps 1 and 2, using $y_0 = 0.005$ and $r = 0.48$ for Line 3.
5	Compare plotted lines.

TABLE 22.2
According to the AU-Pnuts Advisory System for Managing Leaf Spots of Peanut, Fungicide Applications Are Recommended Based on Rain Events

Five-Day Average Chance for Rain	Number of Rain Events (≥0.10 inch in 24 h)	
	First Spray of Season	Subsequent Sprays
≥50%	4	0
≥40%	5	1
≥20%	–	2
Any or none	6	3

Procedure 22.3
Fungicide Application Timing Based on Weather Conditions

Step	Instructions and Comments
1	Given the following weather information (Table 22.3), calculate the average probability for rain over all possible 5-day intervals starting with Day 1.
2	Identify "rain events" of >0.1 in.

TABLE 22.3
Weather Information for Forecasting Fungicide Applications

Day	Average Temperature	Rainfall (in.)	Forecast (Rain Probability)	Day	Average Temperature	Rainfall (in.)	Forecast
1	62	0.01	0	16	81	0	0
2	67	0	0	17	79	0.09	30
3	73	0	0	18	76	1.33	100
4	74	0.06	30	19	82	0.26	90
5	73	0.64	50	20	78	0	50
6	75	0.29	50	21	81	0	0
7	74	0.61	100	22	82	0	0
8	75	0.54	100	23	83	0	0
9	74	0.55	60	24	85	0	0
10	79	0.01	10	25	84	0	0
11	80	0.1	10	26	83	0.05	20
12	76	0	0	27	82	0	60
13	79	0	0	28	83	0.02	30
14	81	0	0	29	86	0	0
15	83	0	0	30	87	0	0

- Which variable, average forecast or rain events, seems to be more important for the first spray of the season? Which of these two variables is more important for subsequent sprays? If the important variable seems to differ based on first or subsequent application, how can this be explained?
- If the forecast for next 3 days (days 31, 32, and 33) is 100%, are any additional fungicide applications recommended during this 30-day period? If so, what day?
- If average rain chances were not part of this advisory, on what days would fungicide applications be made assuming that Day 0 is the end of the 10-day fungicidal effectiveness period?

REFERENCES

Adams, P. B. 1981. Forecasting onion white rot disease. *Phytopathology* 71:1178–1181.

Andrivon, D. 1996. The origin of *Phytophthora infestans* populations present in Europe in the 1840s: A critical review of historical and scientific evidence. *Plant Pathol.* 45:1027–1035.

Backman, P. A., R. Rodriguez-Kabana, M. C. Caulin, E. Beltramini, and N. Ziliani. 1981.Using the soil-tray technique to predict the incidence of *Sclerotium* rot in sugarbeets. *Plant Dis.* 65:419–421.

Bowen, K. L., A. K. Hagan, and R. Weeks. 1992. Seven years of *Sclerotium rolfsii* in peanut fields: Yield losses and means of minimization. *Plant Dis.* 76:982–985.

Davidson, J. M., S. Werres, M. Garbelotto, E. M. Hansen, and D. M. Rizzo. 2003. Sudden oak death and associated diseases caused by *Phytophthora ramorum. Plant Health Progress*. doi:10.1094/PHP-2003-0707-01-DG.

Hanway, J. J. 1963. Growth stages of corn (*Zea mays* L.). *Agron. J.* 55:487–492.

James, C. 1971. *A Manual of Assessment for Plant Diseases*. St. Paul, MN: American Phytopathological Society.

McMullen, M., R. Jones, and D. Gallenberg. 1997. Scab of wheat and barley: A re-emerging disease of devastating impact. *Plant Dis.* 81:1340–1348.

Moore, W. F. 1970. Origin and spread of southern corn leaf blight in 1970. *Plant Dis. Rept.* 54:1104–1108.

Nelson, M. R., T. V. Orum, and R. Jaime-Garcia. 1999. Applications of geographic information systems and geostatistics in plant disease epidemiology and management. *Plant Dis.* 83:308–319.

Pan, Z, X. B. Yang, S. Pivonia, L. Xue, R. Pasken, and J. Roads. 2006. Long-term prediction of soybean rust entry into the continental United States. *Plant Dis.* 90:840–846.

Pataky, J. K., J. A. Hawk, T. Weldekidan, and P. Fallah Moghaddam. 1995. Incidence and severity of Stewart's bacterial wilt on sequential plants of resistant and susceptible sweet corn hybrids. *Plant Dis.* 79:1202–1207.

Roelfs, A. P. 1985. Wheat and Rye Stem Rust. Pages 3–37 in: *The Cereal Rusts*, Vol. II. A. P. Roelfs, and W. R. Bushnell (eds). Orlando, FL: Academic Press.

Schneider, R. W., C. A. Hollier, H. K. Whitam, J. M. McKemy, L. Levy, and R. DeVries-Paterson. 2005. First report of soybean rust caused by *Phakopsora pachyrhizi* in the continental United States. *Plant Dis.* 89:774.

Shaner, G., and R. E. Finney. 1977. The effect of nitrogen fertilization on the expression of slow-mildewing resistance in Knox wheat. *Phytopathology* 67:1051–1056.

Tillman, B. L., W. S. Kursell, S. A. Harrison, and J. S. Russin. 1999. Yield loss caused by bacterial streak in winter wheat. *Plant Dis.* 83:609–614.

Tubajika, K. M., E. L. Civerolo, M. A. Ciomperlik, D. A. Luvisi, and J. M. Hashim. 2004. Analysis of the spatial patterns of Pierce's disease incidence in the lower San Joaquin Valley in California. *Phytopathology* 94:1136–1144.

USDA. 2005. U.S. Department of Agriculture and Florida Department of Agriculture confirm detection of citrus greening. *Dept. Press Release*. Online publication.

Vanderplank, J. E. 1963. *Plant Diseases: Epidemics and Control*. New York, NY: Academic Press.

23 Host Resistance

Peter Balint-Kurti, H. David Shew, and Christina Cowger

Concept Box

- Disease resistance is a crucial trait for any crop plant.
- The degree of disease resistance varies within plant populations. Much of this variation has a genetic basis.
- Plant disease resistance can be broadly categorized into several classes including qualitative and quantitative resistance, nonhost resistance, and tolerance.
- Some resistance is dependent on environment or growth stage.
- In practice, all these different forms of resistance interact with each other in complex ways.
- Disease resistance can be assessed in different ways. Accurate assessment is vital for correct characterization of resistance and for making selections during breeding.
- Breeding for resistance utilizes traditional as well as modern methods such as genomic selection and makes use of primary, secondary, and tertiary gene pools.
- Management of genetic resistance in the field is important for improving its effectiveness and durability. Reducing or offsetting genetic monoculture can help prolong the durability of resistance.

Just as humans vary in their susceptibility to any disease, so we can observe similar variation in susceptibility/resistance within a natural population of any plant species. A portion of this variation is caused by the environment, that is to say, by variation in things like the amount of **inoculum**, the particular microclimates, and the growth conditions encountered by each plant. However, a significant—in most cases the most significant—cause of this variation is the genetic variation among individuals.

Plants possess a diverse set of disease resistance **genes** that confer resistance to different diseases in a variety of ways. Within any natural plant population, the complement of genes and **alleles** carried by each plant varies. This genetic variation underlies the variation in disease resistance that is the main subject matter of this chapter.

It is helpful to bear in mind when thinking about genetic resistance that resistance and susceptibility are two sides of the same coin and that one is always defined in relation to the other. If a gene conferring resistance to a disease exists in the same form in every single member of a plant species, then we generally cannot detect its presence, since the effect of any gene can only be measured by comparing plants differing for its presence/absence. Thus, there are likely to be many genes that are very important for disease resistance which remain undetected and uncharacterized because there is no **functional variation** (i.e., variation that causes a change in phenotype) in them within the populations that we have observed. When we talk about resistance genes, we are really limiting ourselves to resistance genes for which we can observe functional variation. For these genes, there are at least two alleles within the observed population: a resistance allele and a susceptibility allele. When we talk about genetic resistance, we could equally well talk about genetic susceptibility. The presence of a **dominant** resistance allele implies the presence of a **recessive** susceptibility allele, and vice-versa.

TYPES OF DISEASE RESISTANCE

Each plant–pathogen interaction and each case of resistance are unique. However, there are many features and mechanisms that are shared across host species and diseases. Different terms are used to characterize and classify the types of resistance we observe. It is becoming increasingly clear that these classifications are somewhat artificial and tend to represent extremes of what is actually a continuum, but nevertheless, they are useful to organize our thinking, understanding the mechanisms involved in resistance, and determining how they might best be exploited and deployed in agriculture. Various texts have classified plant disease resistance in different ways and have called certain phenomena by different names. Here we attempt to employ the classifications that are in most general use, while making reference to alternative names and classifications as appropriate.

Non-Host Resistance

Most plants are resistant to most diseases. To put it another way, most plant pathogens have severely restricted **host ranges**. For example, the maize rust pathogen *Puccinia sorghi* infects maize and its close relative teosinte, but is unable to infect other monocotyledonous hosts such as wheat or barley. Similarly the potato late blight pathogen, *Phytophthora infestans*, can infect potato and its close relatives, tomato and nightshade, but is generally not found on other hosts. In general, most plant-pathogenic microorganisms have host ranges restricted to one or a few closely related hosts, though there are exceptions; for example, *Botrytis cinerea* infects over 200 hosts (Jarvis, 1977). The general resistance of entire plant species to specific pathogens of other species is known as **nonhost resistance**.

The genetic basis of nonhost resistance is not entirely understood. It is clear that there are many different genetic and physiological mechanisms at work, depending on the specific plant–pathogen interaction being considered. In recent years, a mechanism known as microbe-associated molecular pattern (**MAMP**)-triggered immunity (**MTI**, also known as pathogen-associated molecular pattern [**PAMP**]-triggered immunity, **PTI**) has been elucidated, which is believed to be the basis of much of the nonhost resistance we observe. Briefly, plants have a set of genes that encode proteins capable of detecting molecular features and patterns generally associated with microbes. These features and patterns are known as **MAMPs** or **PAMPS** (Uma et al., 2011). Typical MAMPs include flagellin (the main protein making up the flagellum in most bacteria) and chitin (the main constituent of fungal cell walls). When these MAMPs are detected by the plant, a low-level, barely-observable defense response is triggered, sufficient to protect the plant against most microorganisms to which it is a nonhost. In order to infect the host plant, pathogen species have evolved ways of suppressing or evading MTI. Because MTI mechanisms vary between host species, pathogen adaptations to evade or suppress MTI on one host rarely work on hosts of a different species. It is this host-specific suppression/evasion of MTI that is believed to underlie restricted host ranges of many if not most plant pathogens.

Generally, nonhost resistance has a **multigenic** or **oligogenic** basis; that is, it is controlled by two or more genes. Other forms of resistance, e.g., qualitative, have a **monogenic** basis, which means that resistance is conditioned by a single gene. For further discussion of MAMPs and nonhost resistance, the reader is directed to several excellent recent reviews (Bent and Mackey, 2007; Uma et al., 2011) and Chapters 20 and 21 in this book.

Qualitative Resistance

This type of resistance usually confers a very high level of resistance. Figure 23.1 illustrates a hypothetical case

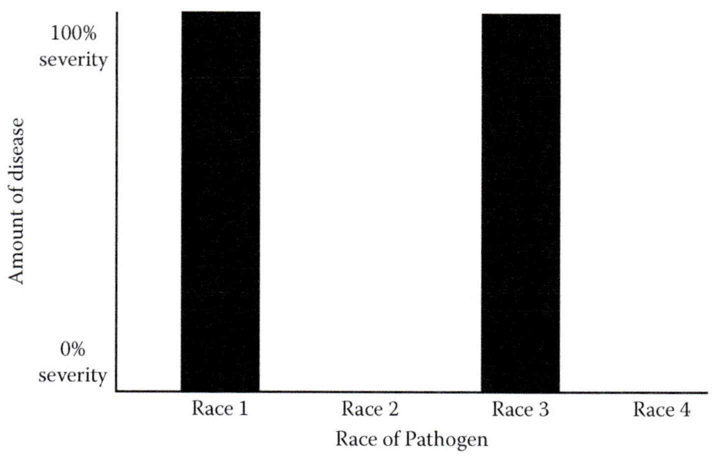

FIGURE 23.1 A hypothetical case of qualitative resistance for a host plant species with respect to four races of a pathogen species. The plant is susceptible to races 1 and 3 and resistant to races 2 and 4.

of qualitative resistance with respect to four races of a pathogen species. Two typical features of qualitative resistance are apparent: first, the high amount of resistance conferred (see Race 2, Race 4); and second, the fact that resistance is not effective against all races of the pathogen. In cases like this where the host plant is susceptible to some races of a pathogen and resistant to others, the resistance is said to be **race-specific**.

The genetic basis of qualitative resistance is generally **monogenic**; that is, there are single genes or alleles in the host whose presence or absence determines whether susceptibility or resistance is observed. From the 1930s to the 1950s, H. H. Flor conducted a long series of pioneering experiments on the interaction between flax and the pathogen that causes flax rust, looking at the genetics of both resistance in the host and virulence in the pathogen. He determined that "the pathogenic range of each physiological race of the pathogen is conditioned by pairs of factors that are specific for each different resistant or immune factor possessed by the host variety" (for a summary of Flor's work see Loegering and Ellingboe, 1987). To put it another way, for each dominant resistance gene (**R-gene**) in the host, there is a corresponding dominant **avirulence** gene (**Avr gene**) in the pathogen. Resistance is only conferred if both the resistance gene and the corresponding avirulence gene are present in the same interaction. This is known as the gene-for-gene interaction and is illustrated in Figure 23.2. Gene-for-gene interactions have subsequently been shown to be very widespread, being found in practically every plant species that has been studied. Gene-for-gene interactions are involved in resistance to bacterial, fungal, viral, and nematode diseases and are even important for resistance to aphids and other plant-parasitic insects.

Since, in most interactions, multiple R-genes are **segregating** in the host population, things can get complicated. The potential number of **pathogen races** = 2^N, where N is the number of host resistance genes. So with three R-genes, there may be up to eight different pathogen races, and so on. Figure 23.3 illustrates a case in which two different R-genes are segregating in the host population and two corresponding Avr genes are segregating in the pathogen population. As can be seen, all that is required for resistance is that a single corresponding R/Avr-gene pair is present in an interaction.

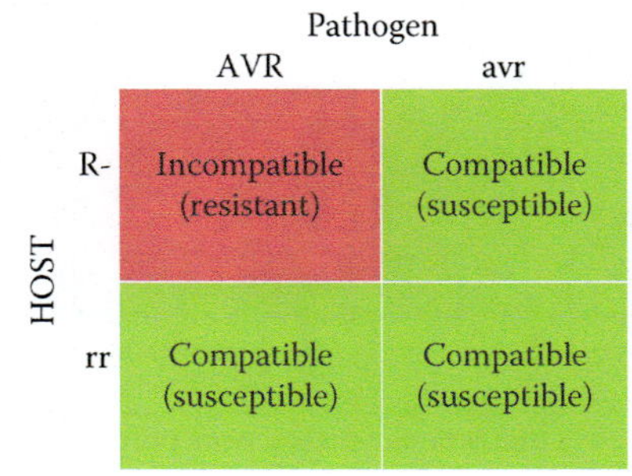

FIGURE 23.2 The gene-for-gene interaction first identified by H.H. Flor. If the pathogen possesses the dominant avirulence gene Avr, and the host possesses a dominant resistance gene R, the interaction is "incompatible"—a resistant interaction occurs. If either the host or the pathogen lacks that dominant gene, however, the interaction is termed "compatible" and disease occurs.

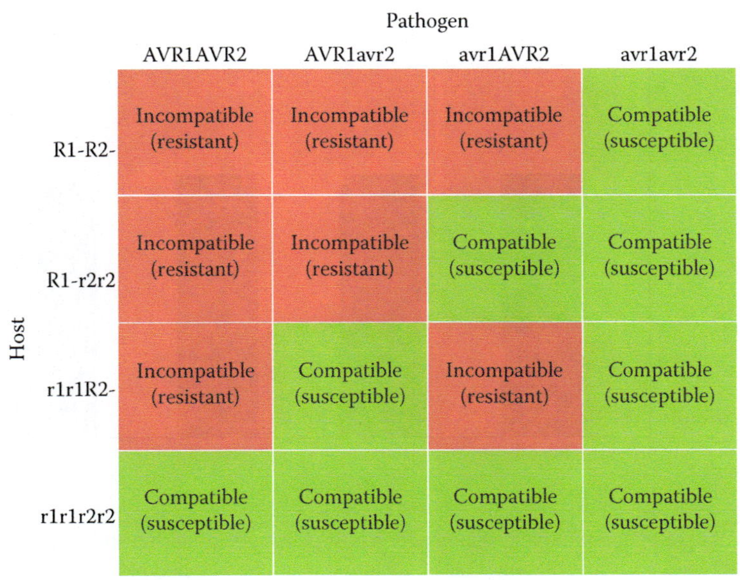

FIGURE 23.3 The case in which a host plant–pathogen interaction is governed by two different loci for resistance and virulence. The dominant genes are R1 and R2 in the host, matched respectively by AVR1 and AVR2 in the pathogen.

Qualitative resistance is often (though not always) associated with an observable **hypersensitive response (HR)**, a rapid and localized death of host cells around the point of pathogen penetration and associated biochemical and histological changes (Figure 23.4). The association of qualitative resistance with HR is likely the main reason that qualitative, gene-for-gene type resistance is largely (though not always) associated with resistance to **biotrophic** and **hemibiotrophic** rather than **necrotrophic** pathogens, since encircling dead tissue provides an effective barrier to the former, but not to the latter.

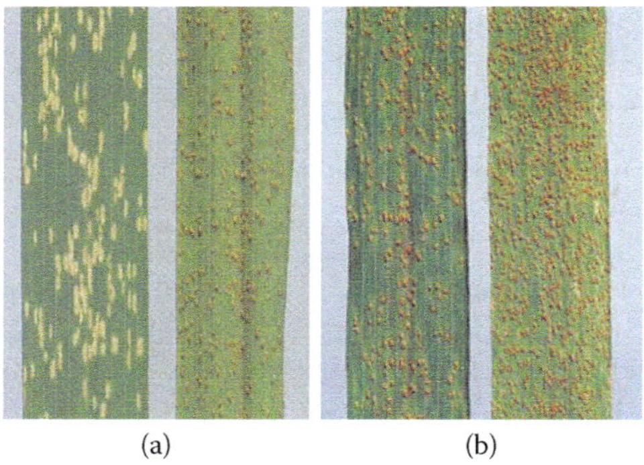

FIGURE 23.4 Phenotypic distinction between quantitative and qualitative resistance in the wheat leaf rust (*Puccinia recondita*, formerly named *Puccinia triticina*) pathosystem. (a) Contrast between incompatible (resistant) reaction on the left, due to presence of the *Lr37* resistance gene, and compatible (susceptible) disease interaction on the right. (b) Quantitative differences in the severity of leaf rust in two cultivars differing in levels of partial resistance. Note the common reaction type in (b) (pustule size, absence of chlorosis are the same on both) while amount of disease is different. (From http://www.globalrust.org/traction/permalink/multimedia325.)

Quantitative Resistance

Quantitative resistance in many ways is the opposite of qualitative resistance. It is partial as opposed to complete; it confers resistance to biotrophic and necrotrophic diseases, and it usually has a multigenic rather than a monogenic basis. Figure 23.5 illustrates another difference: quantitative resistance is generally assumed to be race nonspecific, that is, to be equally effective against all races or strains of the pathogen. However, in some cases, this assumption has been tested and found to be untrue. In these cases, partial resistance is more effective against some pathogen strains than others (Poland et al., 2009).

Quantitative resistance tends to be based on many genes, each with a minor or moderate effect and each acting **additively**. In some cases, a large number of genes are at work. For example, a study of quantitative resistance to two fungal diseases of maize, southern leaf blight and northern leaf blight, identified about 30 **genetic loci** in each case, with each locus responsible for less than 5% of the total variation (Kump et al., 2011; Poland et al., 2011). In many cases, the presence or absence of quantitative resistance genes in an individual plant may have a barely perceptible effect. However, 5 or 10 genes in combination can confer substantial levels of resistance, as illustrated in Figure 23.6. This figure also demonstrates how quantitative resistance varies in a continuous way compared with the discrete resistant and susceptible classes typical of qualitative resistance. Unlike qualitative resistance, quantitative resistance tends to be somewhat environmentally dependent, but it is generally more **durable** than qualitative resistance.

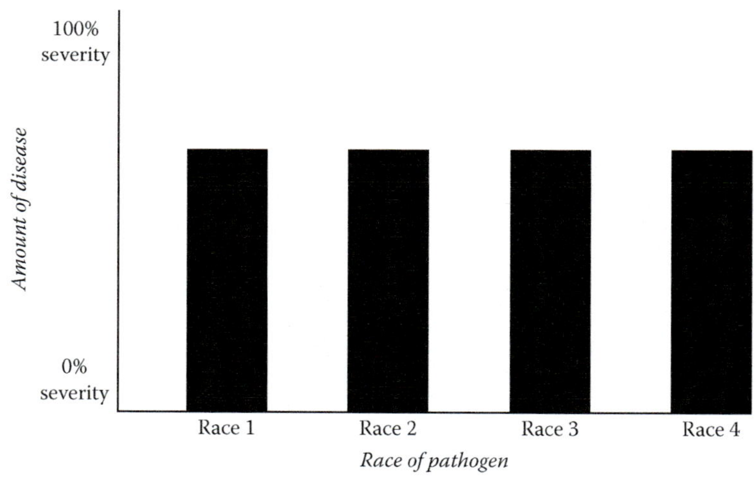

FIGURE 23.5 A hypothetical case of quantitative resistance for a host plant species with respect to four races of a pathogen species; contrast with Figure 23.1. The plant is susceptible to all races, meaning it can be infected by all of them, yet partially resistant. Note that in this case the levels of partial resistance to each race are the same; that is, resistance is not race-specific.

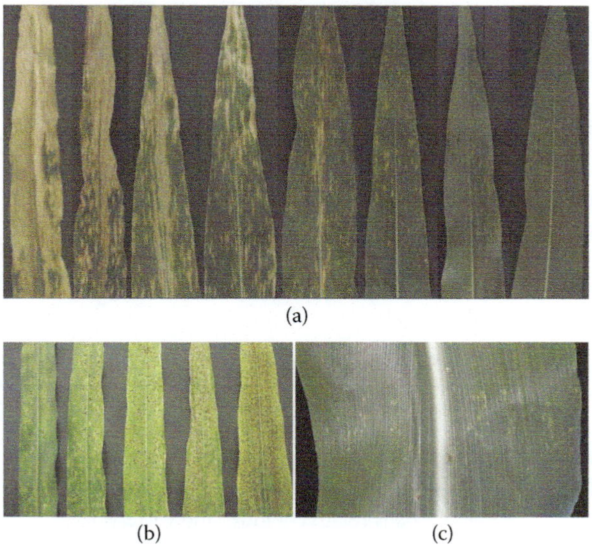

FIGURE 23.6 Examples of quantitative disease in maize. Quantitative variation for resistance to (a) *Bipolaris maydis*, the causal agent of southern corn leaf blight, and (b) *Puccinia sorghi*, the causal agent of maize common rust in a segregating maize population. Contrast to (c), which shows a hypersensitive resistance reaction to *P. sorghi* due to the R-gene *Rp1*. Although the resistance provided by R-genes is highly effective, it is often subject to breakdown resulting from pathogen evolution. (Courtesy of Chi-Ren Shyu and Jason Green (a); Jesse Poland (b); and Jerald Pataky (c). Originally appeared in Poland et al., *Trends Plant Sci.*, 14, 2009.)

Alternative Terminology

Qualitative and quantitative resistances are known by a number of essentially synonymous names. In particular, Vanderplank (1984) coined the terms "**vertical resistance**" and "**horizontal resistance**" for qualitative and quantitative resistance, respectively, and these terms are still in common use. A comparison of Figures 23.1 and 23.5 indicates the reasons for this nomenclature. Qualitative resistance is alternatively known as gene-for-gene resistance, R-gene resistance, major-gene resistance, hypersensitive resistance, and effector-triggered immunity (ETI), a recently coined term that refers to the mechanism of pathogen perception, and is discussed in Chapter 20 and in a recent review (Bent and Mackey, 2007). As well as the term horizontal resistance, alternative terms for quantitative resistance include minor-gene resistance, additive resistance, incomplete resistance, partial resistance, race-nonspecific resistance and multigenic resistance.

Resistance That Depends on Growth Stage

An interesting feature of plant resistance to disease is that its expression may vary greatly, depending on the plant's growth stage. The most common kind of growth-stage-specific resistance is called "**adult-plant" resistance (APR)**. The gene or genes that condition APR are expressed primarily or more fully when the plant has reached a certain stage of maturity. A good example is APR in wheat to powdery mildew, caused by the fungus *Blumeria graminis* f. sp. *tritici*. This trait is also sometimes called "slow-mildewing." APR can be recognized when, for example, two wheat genotypes are scored as equally susceptible early in a powdery mildew epidemic, but later in the epidemic one genotype is significantly more diseased than the other (Figure 23.7). As can be seen from this example, recognizing APR generally involves multiple observations of multiple genotypes experiencing an epidemic over the entire life cycle of the plant.

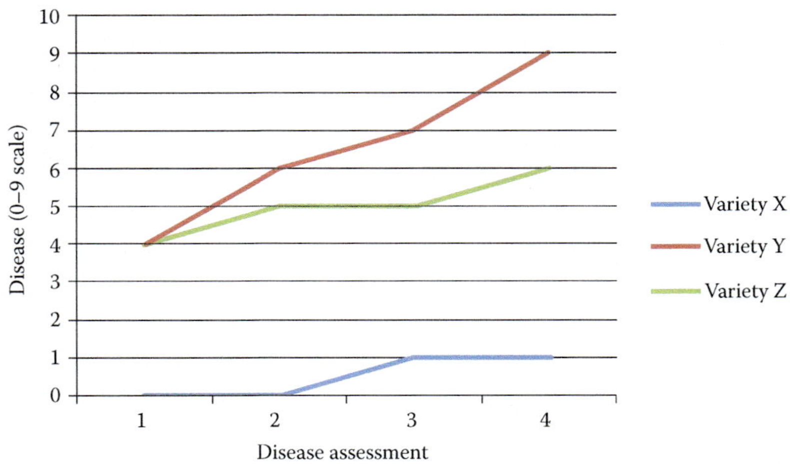

FIGURE 23.7 Adult-plant resistance: Variety X is classified as resistant at all growth stages; Variety Y is susceptible; and Variety Z has adult-plant resistance. The resistance of Variety Z is partial as compared with that of Variety X, and manifests itself as the crop matures and in comparison with Variety Y.

Resistance That Depends on the Environment

The degree of resistance expressed by a plant can also vary in response to environmental conditions. One of the best known examples is high-temperature adult plant (HTAP) resistance in wheat and barley to the stripe rust fungal pathogen, *Puccinia striiformis*. HTAP is particularly effective after cereal stems elongate—in other words, when the hitherto grasslike plants send up seed stalks. This HTAP resistance increases as plants get older and temperatures remain above 10°C at night and between 25°C and 30°C during the day. This type of resistance is usually not specific to certain pathogen races and is thought to be durable.

TOLERANCE

For the most part, disease only matters in agriculture in as much as it affects yield (exceptions include cosmetic damage to vegetables, fruits and ornamental crops, and mycotoxins produced by fungi in field crops). **Tolerance** is the ability of a variety to yield more than a less-tolerant variety, given equally severe disease symptoms in both cases (Figure 23.8). Tolerance should not be confused with quantitative resistance (though it often is!). Ironically, although tolerance is arguably the most important disease-related trait a variety should possess, almost nothing is known of its genetic basis or mechanism beyond the fact that it exists and that it has a genetic basis (Schafer, 1971). This is largely because it is a very hard trait to measure and calculate compared with simple measures of disease symptoms.

THE REAL WORLD

As discussed at the beginning of this section, each plant–pathogen interaction is unique, and in many cases, the resistance we observe does not conform exactly to the dogma of qualitative or quantitative resistance, but may lie somewhere in between. For instance, qualitative resistance is often not complete—some pathogen growth is allowed. Moreover, partial resistance may sometimes be conferred by a single gene instead of many genes acting in an additive fashion. Similarly, quantitative resistance is sometimes race-specific, whereas in some cases qualitative resistance is not. This is discussed in more detail in a recent review (Poland et al., 2009). Furthermore, quantitative and qualitative resistance mechanisms often function simultaneously in the same interaction. An example of this is given in Figure 23.9 (adapted from Figure 7.1 in Vanderplank 1984). It details the interaction of two potato varieties, Kennebec and Maritta, with a set of races of *Phytophthora infestans*. Both the varieties have the same major resistance gene, R1, and so have complete (i.e., qualitative) resistance to races avirulent to R1, but

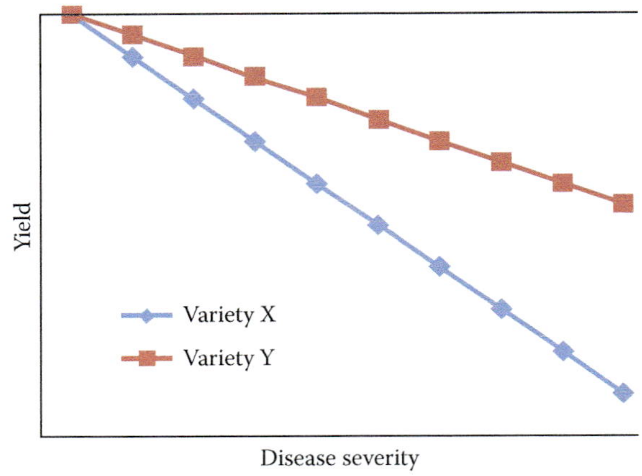

FIGURE 23.8 Disease tolerance: Variety Y shows greater tolerance than Variety X because under a given level of disease pressure, the yield loss is relatively lower for Variety Y.

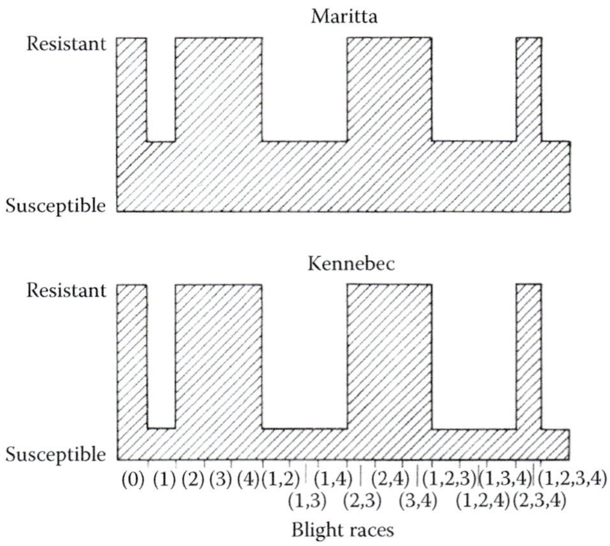

FIGURE 23.9 Quantitative and qualitative resistance at work simultaneously. Two potato varieties have the same major gene conferring resistance to several races, but have different degrees of partial resistance to other races. (From Vanderplank, J. E., *Disease Resistance in Plants,* Academic Press, New York, 1984. With permission.)

they differ in their levels of quantitative resistance to the other races.

MEASUREMENT AND DESCRIPTION OF DISEASE RESISTANCE

Inoculation

Disease resistance is classically measured by growing plants in the field or in controlled environments and assessing their performance under disease pressure. Controlled environments are usually greenhouses or growth chambers. Plants can become infected either by naturally occurring epidemics in the field or by artificial inoculation. Inoculation with fungal pathogens is often done by directly applying the pathogen in a water suspension, sometimes with a surfactant and/or a "sticker" (something to make the spores adhere to the leaves) such as agar. Alternatively, sometimes the inoculum is provided by allowing the pathogen to grow on another substrate, such as the host species itself or a medium such as sterilized grain, and then spreading the colonized plant material in the field around the crop to be infected. Susceptible plants, either previously inoculated or not, may be planted into and around field nurseries as "spreaders" to encourage disease development.

Rating

Many methods have been developed over the years to describe the levels of resistance. Often, plant pathologists will record two separate measures of disease: **incidence**, or percentage of plants that exhibit symptoms, and **severity**, or percentage of host tissue that is symptomatic. The simplest method is to make a direct visual estimate of the percentages. Diagrams can be used as an aid to estimation (example in Figure 23.10). It is straightforward to analyze data collected as percentages; for example, a genotype with 20% disease severity is twice as diseased as a plant with 10% disease severity.

Rating scales in which estimates are made within certain ranges of disease percentages are also utilized. Such scales can be preferable because it is unnecessarily time-consuming to assign precise percentage estimates to large numbers of lines, and because human estimators are better at detecting small differences at the extremes of a scale (where there is either very little disease or a large amount of disease). An example of a disease scale with unequal class widths is the Horsfall–Barratt scale, which consists of 12 classes or grades of disease (see Table 22.1 in the previous chapter). These types of scales are popular with plant breeders who usually have to speedily evaluate large numbers of lines, and so often look for ways to simplify the rating process.

Another rating method is to collect **ordinal** data. In an **ordinal scale**, a relevant set of numbered categories is devised based on a ranking of symptom levels, and the disease symptoms are assigned to a particular category. Importantly, in an ordinal scale, although the rankings are ordered from least disease to most disease, relative levels of difference between consecutive categories are not necessarily constant; instead, qualitative criteria such as leaf wilting, color, and so on are often used to distinguish levels (Figure 23.11).

Statistical analysis of disease data recorded on scales with unequal class widths or on ordinal scales can be a little more complicated than analysis of straight percentage data. In general, these types of data are best analyzed using either **nonparametric** or **parametric** generalized linear models for multiple-category data, rather than conventional analysis of variance (Madden et al., 2007).

Measurements at separate time points during an epidemic provide a picture of how disease progresses on different genotypes, which is another way of assessing the degree of host plant resistance. This allows the calculation of **area under the disease progress curve (AUDPC)**, which is a good way to compare levels of quantitative resistance (Figure 23.12). (Note that Figure 23.12 also illustrates the adult-plant resistance described earlier; the APR of Variety 1 can be detected in comparison with the full susceptibility of Variety 2.)

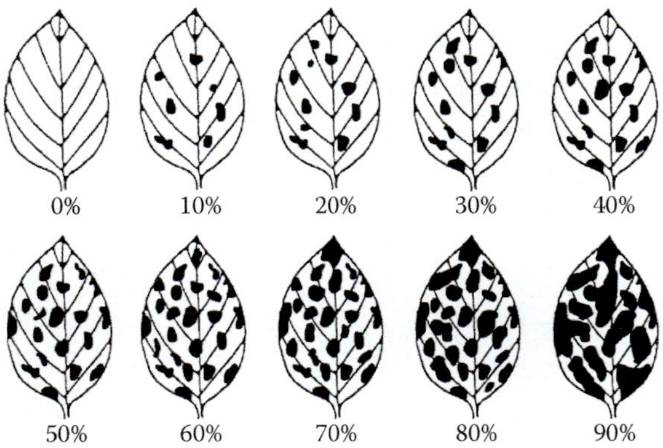

FIGURE 23.10 A diagram to aid in rating the severity of soybean rust, caused by *Phakopsora pachyrhizi*. (Based on a figure in *Kentucky Integrated Crop Manual for Soybeans*, IPM-3. 2009. pg. 3. Found in the University of Kentucky Cooperative Extension Service, Plant Pathology Fact Sheet PPFS-MISC-06, http://www.ca.uky.edu/agcollege/plantpathology/ext_files/PPFShtml/ PPFS-MISC-6.pdf.)

For some diseases, specialized scales have been devised to take into account both the severity of disease symptoms and also the phenotypes of partial resistance. One example is the cereal rusts, caused by different species of the fungal genus *Puccinia* that infect wheat, barley, and other small grains. The varying host responses are called "infection types," or ITs. Different ITs are described by the presence or absence of necrosis or chlorosis and differences in the size of the uredia, which are the infectious pustules (Table 23.1).

FIGURE 23.11 Ordinal 0–6 scale used to rate shoot symptoms of sugar beet caused by Rhizoctonia crown and root rot on sugar beet. (From Hillnhütter et al., *Nematology*, 13, 2011.)

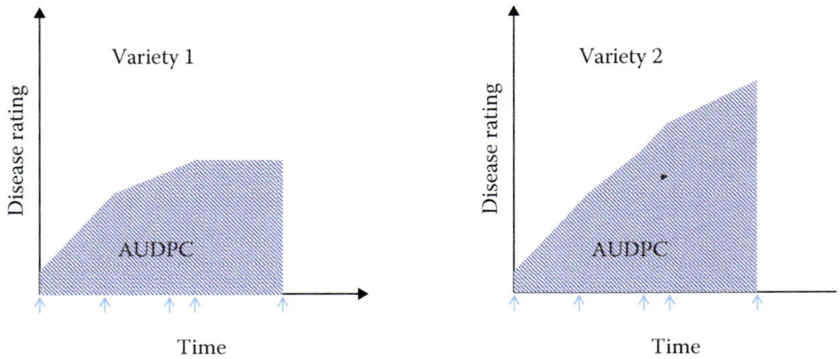

FIGURE 23.12 An illustration of "area under the disease progress curve" (AUDPC) represented by the shaded area of each graph calculated for two varieties. Each variety was scored at the same five time points (indicated by arrows on the x-axis). Although the initial two ratings were the same, the disease subsequently progressed much more quickly in the variety on the right. AUDPC measurements act as summary scores that take into account the results of multiple ratings taken over the growing season at irregular intervals.

TABLE 23.1
Infection Types Used in Scoring Reactions of Small-Grain Seedlings to Rusts Caused by *Puccinia* Species

Host response (class)	Infection type	Symptoms
Immune (resistant)	0—low	No uredia or macroscopic sign of infection
Nearly immune (resistant)	0—low	No uredia but necrotic or chlorotic flecks
Very resistant	1—low	Small uredia with necrotic border
Moderately resistant	2—low	Small to medium uredia with chlorosis or necrosis
Heterogeneous	X—low	Random distribution of variable-sized uredia
Heterogeneous	Y—low	Variable-sized uredia, decreasing in size with distance from the leaf tip
Heterogeneous	Z—low	Variable-sized uredia, decreasing in size with distance from the leaf base
Moderately susceptible	3—low	Medium-sized uredia
Susceptible	4—high	Large uredia without chlorosis or necrosis

Source: Adapted from Table 1 in Roelfs (1988).

Another type of rating scale takes into account how high disease has progressed on the plant, as well as how much disease the plant has. As an example, the Saari–Prescott scale is especially designed to draw out the differences in cultivar response to splash-dispersed wheat diseases (Saari and Prescott, 1975). It is a double-digit scale (00–99) in which the first digit indicates vertical disease progress on the plant, and the second digit refers to severity measured as diseased leaf area on a 0-to-9 scale.

Sometimes, relative levels of host plant resistance are inferred by estimating the quantity of the pathogen in the respective host genotypes. A traditional method of doing this was the enzyme-linked immunosorbent assay (ELISA), which uses a specific antibody to react with the antigen (in this case proteins produced by the pathogen) in a plant sample. The intensity of the color change produced in the reaction can be taken to reflect the amount of pathogen in the sample. More recently, the real-time polymerase chain reaction (PCR, also called quantitative real-time PCR or **qPCR**) has been used for this purpose. The PCR quantifies the pathogen DNA present in a sample of the infected host, which is assumed to correspond to overall pathogen biomass and, therefore, to the level of resistance in the host.

Ultimately, most methods of rating disease involve some compromise between speed and accuracy. For instance, diseased area percentages might be estimated by eye rather than by using more accurate (but much more time-consuming) image analysis methods. Easily observed symptoms are rated rather than measuring the amount of pathogen present in the plant or (best of all but hardest to measure) measuring the ultimate yield loss that can be attributed to the disease. For breeding purposes, a constant relationship between symptom severity and ultimate yield loss is assumed (although tolerance could in fact explain part of a cultivar's yield performance). To confirm the effects of a certain disease-resistance genotype on yield, though, it is necessary to compare yields from diseased and nondiseased plots of the cultivar with the absence of disease often assured through fungicide applications.

BREEDING FOR DISEASE RESISTANCE

General Strategies

Practical plant breeding programs seek to incorporate disease resistance genes into plant genotypes that are otherwise appealing to those who will grow them. That is, the lines should have high yield, desirable end-use qualities, and/or attractive ornamental features. Disease resistance is of little use if no one wants to grow the resistant line. Thus, the general approach is to **introgress** genes for resistance to the major diseases of interest into disease-susceptible, but otherwise agronomically or horticulturally superior cultivars.

A single gene for resistance to a given disease is likely to be readily overcome by the evolution of that pathogen population, as discussed later in text. So breeders often seek to "pyramid" (stack together) multiple genes for resistance to a particular disease in a single line. **Gene pyramids** are thought to be most effective when each of the genes is still effective at the time of pyramiding. Another strategy is putting major-effect genes into a background of quantitative resistance. This is useful because if the pathogen population evolves to overcome the single gene, there will still be quantitative resistance to "fall back on."

Gene Pools

The most straightforward source of new disease resistance genes is the **primary gene pool** of a given plant species, which consists of individuals in the same or a very closely related species with which crossing is easiest. For example, wheat (*Triticum aestivum*) is a hexaploid species (six sets of each chromosome) with three genomes (A, B, and D), so the primary gene pool includes **landraces** that are also hexaploid. It also includes species that are closely related to cultivated wheat, such as the diploids *T. urartu* (donor of the A genome) and *Aegilops tauschii* (donor of the D genome), and the tetraploid *T. turgidum* (durum wheat). Crossing wheat with these species is relatively simple, because chromosomes are **homologous**, meaning that they pair properly during meiosis, and hybrids can be recovered easily.

Introgressing genes is somewhat trickier when using the **secondary gene pool**, which primarily consists of species closely related to the crop of interest (See Case Study 23.1). In the case of wheat, that includes tetraploid *Triticum* and *Aegilops* species, such as *Triticum timopheevii* and *Triticum araraticum*, which share one homologous and one **homoeologous** genome with *T. aestivum*. More distantly related species make up the **tertiary gene pool**. For wheat, this includes goat grasses (*Aegilops* spp.) and even more remote relatives such as rye (*Secale cereale*) and wheatgrass (*Thinopyrum intermedium*). What makes crossing with species from the secondary and tertiary gene pools tricky is that homoeologous chromosomes may pair poorly or not at all during meiosis. Hybrids may be weak or sterile when crossing a crop species with its more distant wild relatives. To transfer new resistance genes from secondary and tertiary gene pool members, special procedures are often needed to create viable recombinant progeny.

Molecular Approaches in Resistance Breeding

In breeding populations, DNA-based markers known as molecular markers can be used to characterize the genetic constitution of individual plants; that is, they can be used to determine which regions of the genome have been inherited from which parent. New methodologies are making the generation of extremely large amounts of molecular-marker (or genotypic) data very rapid and very cheap. The major types of molecular markers (single nucleotide polymorphisms [SNPs], simple sequence repeats [SSRs], and amplified fragment length polymorphism [AFLPs], etc.) are discussed in Chapter 33. For our purposes, it is enough to understand that in modern agricultural research, the generation of genotypic information is not, or soon will not be, a limiting factor.

Molecular markers have added a new dimension to breeding for disease resistance. It can often be difficult to tell if a plant genotype has resistance to a certain disease. For example, a soilborne pathogen may not be spread evenly throughout the soil in a nursery, but rather scattered here and there in clumps. So all genotypes planted in that nursery will not be evenly exposed to the pathogen. "Escapes," or lines in plots without soilborne inoculum, could be mistakenly rated as resistant. The same applies even with some airborne or rain-splashed diseases if disease pressure occurs inconsistently over a field. In cases like this, molecular markers that have been previously determined to be closely linked to a known disease-resistance gene can be used to indicate the likely presence or absence of the gene. This is especially useful where accurate assessments of the resistance phenotype are difficult to obtain.

Marker-assisted selection (MAS) describes the use of molecular markers associated with specific genes, rather than or in addition to the phenotypes those genes confer, to make selections in a breeding program. MAS can be very useful for the introgression of qualitative resistance genes. When the goal is to introgress from the donor parent just the resistance gene and nothing else, MAS, using markers located throughout the genome, can be used to select both *for* the presence of the resistance gene and *against* the presence of other parts of the donor genome. MAS is also very useful for gene pyramiding because if one resistance gene is present, it becomes harder to observe the phenotypic effect of adding a second resistance gene. However, MAS is less effective for complex disease resistance traits that are controlled by many genes of small effect. In such a case, it is hard to find and select for markers closely linked to all the loci that contribute to the phenotype.

In response to this limitation, and taking advantage of the fact that so much genotypic data can now be generated very cheaply, new techniques such as **genomic selection (GS)** are rapidly gaining ground. In GS, a "training population," usually consisting of a large diverse set of germplasm drawn from available breeding lines, must first be characterized. This training population is assessed phenotypically over several environments and is genotyped with a (large) genome-wide set of markers. **Marker effects** are then calculated for each marker for each trait. In essence, a marker effect is the degree of association a marker has with the trait of interest. For instance, if a particular marker allele is disproportionately found in lines with above-average disease resistance, then a significant effect on disease resistance is attributed to that marker allele. The key with GS is that specific loci associated with a trait are not identified and selected for. Instead, the effect of every marker over the whole genome is fitted into a complicated statistical model, and then analysis based on this model is used to make selections. The process of using statistical associations across a genome to predict disease resistance levels is especially powerful where the resistant phenotype is the result of the small, additive effects of genes at many loci. However, the effectiveness of GS ultimately depends upon the accuracy and relevance of the data collected from the training population.

Transgenic Approaches

Transgenesis, also known as "genetic engineering," refers to the insertion of foreign DNA into the genome of an organism of interest using modern techniques for **genetic transformation**. Because genetic transformation uses DNA sequences independent from their species of origin, and since the genetic code is largely consistent throughout the entire living world, genes from any living creature (the so-called quaternary gene pool) can, in theory, be utilized for the enhancement of plant disease resistance by this approach.

A thorough analysis of the theory and practice of plant genetic engineering and its use to create plants with enhanced disease resistance requires much more detail than can be given here. For a comprehensive treatment, see some recent texts and reviews such as Hammond-Kosack and Parker (2003), Tourte (2003), Gurr and Rushto (2005), and Wulff et al. (2011).

Since plants were first transformed with foreign DNA in the early 1980s (Chilton 2001), the potential for manipulating and improving agronomic traits has been clear, and much effort has been devoted in this direction. Some significant (and rather lucrative) successes have been achieved, specifically in the areas of herbicide tolerance (e.g., "Roundup Ready" lines) and insect resistance using Bt genes and their derivatives (Sanchis and Bourguet, 2008). On the other hand, there have been significant disappointments. Perhaps the largest of these is in the area of transgenically conferred disease resistance. At the time of writing, 30 years after the first successful plant transformation and despite much research effort, and with the exception of a few virus-resistant lines, no transgenically conferred plant disease resistance traits are commercially available.

Virus resistance is often cited as a genetic engineering success story, specifically its use to confer papaya ringspot resistance, which has been credited with saving the Hawaiian papaya industry (Gonsalves, 1998). The intimate host–virus association, involving the exchange of DNA and the expression of pathogen genes by the host, enabled the use of methodologies based on **RNA-interference (RNAi)** to achieve effective and durable resistance (http://www.apsnet.org/publications/apsnetfeatures/Pages/papayaringspot.aspx). Several virus-resistant cucurbit varieties have also been made commercially available, but these remain as relatively low-volume niche products.

Commercially viable transgenic resistance to bacteria and fungi has so far proved elusive. This is due to a number of factors including the following:

- The trait must save the farmer money or increase yield sufficiently to justify the increased cost of the seed. Financial savings usually derive from decreased application of pesticides which control multiple diseases. Often, therefore, transgenic traits must confer durable resistance to multiple diseases in order to be commercially viable. This has been difficult to achieve.
- Numerous intellectual property and technical issues often limit the scope of what can be achieved.
- Negative public perception of genetic engineering and the associated high regulatory hurdles have made it financially prohibitive to release transgenic traits commercially, with the exception of "blockbuster" traits such as "Roundup-Ready."
- Perhaps the biggest hurdle is the continued success of traditional breeding techniques, augmented by the use of molecular markers (see earlier), in producing disease-resistant cultivars much more cheaply and effectively than has so far been achieved using transgenic approaches.

Continued rapid progress in our understanding of the molecular mechanisms associated with plant disease and plant gene expression offers hope that transgenic techniques will eventually be useful for conferring disease resistance, especially in important crops such as banana, which are very difficult to breed in a traditional way. In particular, breakthroughs in the understanding of mechanisms associated with MTI, RNAi, and **TAL effectors** and their targets provide promising future avenues (discussed in Wulff et al., 2011).

HOW DO PLANT PATHOGEN POPULATIONS RESPOND TO HOST RESISTANCE?

Adaptation to Resistance

In modern, industrialized agriculture, deployment of plant disease resistance calls forth powerful "countermeasures" from the organisms it is aimed at defeating. As soon as a resistance gene is utilized in a crop species, it exerts **selection pressure** on the pathogens of that crop. Although the resistance gene may initially be completely effective, the more it is used, the more it favors the multiplication of pathogen strains with mutations conferring virulence to it. These mutant individuals may be initially rare, but they have such an advantage over avirulent strains that they attain high frequencies in the pathogen population in just a few years.

Genetic uniformity, having large areas planted with varieties with the same resistance gene or genes, poses great risks by maximizing the likelihood of appearance and buildup of virulent pathogen populations (see Case Studies 23.1 and 23.2). This leads to the "boom-and-bust" dynamic (Figure 23.13): the increasing frequency of a host resistance allele is matched, with a time lag, by the increasing frequency of a pathogen virulence allele that can overcome it. The defeated resistance allele is replaced by a new, effective allele, which in turn soon selects for the matching virulence in the pathogen population, and so on.

Pathologists have devoted considerable effort to understand the forces that increase or decrease the ability of a pathogen population to evolve to overcome resistance. Taken together, such factors determine the durability of a particular source of resistance and can shorten or prolong the life of commercially important cultivars.

One factor is the nature of the resistance, whether it is conferred by a single gene or by multiple smaller-effect genetic loci. Qualitative resistance tends to more rapidly select for virulent mutant strains. For example, a major gene for resistance to wheat powdery mildew, *Pm17*, was highly effective when it was initially deployed in the eastern United States. Commercial wheat varieties bearing *Pm17* were first planted in the mildew-prone mid-Atlantic states of Virginia and North Carolina in 2002. Those varieties were immune to powdery mildew, and by 2004, they were grown in large areas. In 2009, mildew infections were found throughout plots of lines with *Pm17*, and by 2011, the gene was completely ineffective in the mid-Atlantic region. It is typical for popular eastern U.S. wheat cultivars to lose market share after a few years; in this case, an important effective R-gene was lost to producers as well.

Quantitative resistance can also select for more aggressive pathogen strains, although we have less empirical evidence about this relationship than about

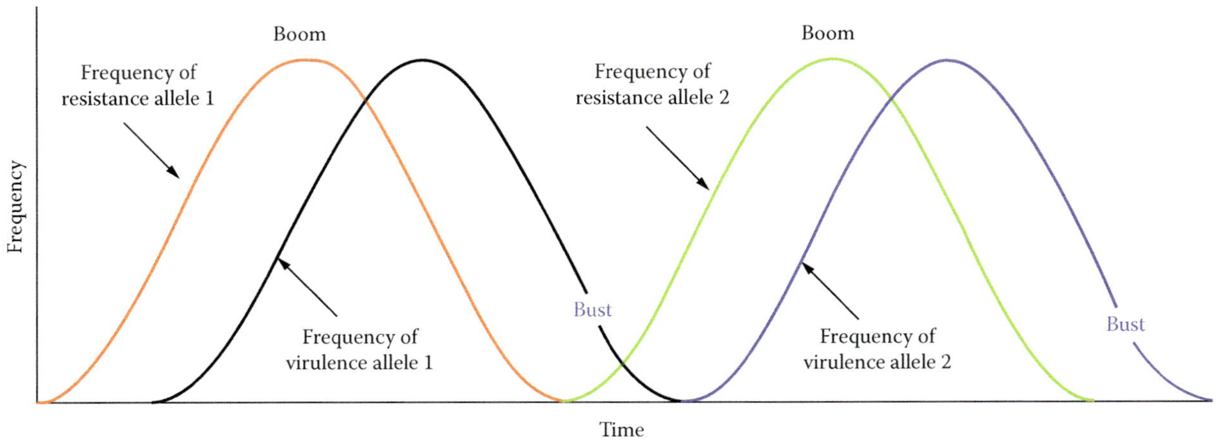

FIGURE 23.13 Boom and bust cycle undergone by host resistance genes and pathogen virulence in modern industrial agriculture. When they are first commercially deployed, resistance alleles are highly effective, but as they occupy more planted area, the frequency of the matching virulence increases accordingly in the pathogen population, leading to sequential defeat of the resistance alleles. Note that although the frequency of the defeated resistance alleles may return to zero due to producers' choices, it is unlikely that the thus-unnecessary pathogen virulence alleles also return to a frequency of zero.

the effect of qualitative resistance. In one of a small number of field experiments addressing this question, wheat varieties with different levels of partial resistance to *Mycosphaerella graminicola*, a leaf blotch fungus, were planted in field plots (Cowger and Mundt, 2002). Epidemics of *M. graminicola* are initiated by a shower of windborne ascospores on small wheat plants and proceed via multiple cycles of splash-dispersed conidia during the growing season. Researchers collected samples of the pathogen at the beginning of the season and then again at the end, after there had been time for host resistance level to exercise a selective effect on the pathogen populations in the plots. Aggressiveness of the early and late pathogen samples was compared by inoculating them on wheat seedlings in the greenhouse. On average, the pathogen had shifted toward greater aggressiveness in the field plots of partially resistant cultivars than in plots of susceptible cultivars. A similar Dutch experiment with potato cyst nematodes had the same result. Thus, higher levels of quantitative or partial resistance appear to select for greater aggressiveness in the pathogen. Nevertheless, the breakdown of quantitative resistance appears to be slower than the breakdown of major genes, allowing breeders more time to come up with new resistant lines.

Mutations to virulence against qualitative resistance are changes in the **effectors** used by the pathogen to attack the plant and cause disease. Obviously, such a mutation can lead to a reduction in disease-causing ability that may be severe. However, it appears that such fitness costs can be offset via compensatory mutations in the effector itself, or in other genes. Fitness can also be restored through a shift to reliance on alternative biochemical pathways. Pathologists have devoted a great deal of effort to investigating whether mutation to virulence decreases pathogen fitness, and whether pathogen strains that carry multiple virulence mutations are on average less fit than those that carry fewer virulence mutations. It is clear that an individual strain or race of a pathogen can carry many virulence alleles simultaneously and still be at a high frequency in the population. This can be true even when the matching resistance genes are not common in the host plant population.

So is the presumed fitness cost of mutation to virulence illusory or insignificant? Once a defeated resistance gene is withdrawn from commercial production, will the frequency of virulence to it in the pathogen population decrease? This "**stabilizing selection**" would be operative if virulence carried a fitness penalty; once there was no advantage to having an unnecessary virulence mutation, a strain carrying it would be outcompeted. Whether this occurs turns out to be difficult to determine, because strains of a given pathogen differ in many ways, which add up to their background fitness, and it is hard to hold everything else constant and just compare the effects of differences at one virulence locus. There is some evidence of such stabilizing selection among plant viruses. Because of their small genomes, these viruses may pay higher fitness penalties for evolving to pathogenicity (Fraile et al., 2011).

The Bottom Line?

When qualitative resistance is withdrawn from widespread commercial production, the frequency of virulence to it may decrease, but likely not to zero, so the virulence can "flare up" again if the same gene is redeployed. However, the magnitude of the fitness cost of a particular mutation to virulence can vary, depending on the function of the gene in question. Mutations to virulence against some R-genes appear to be particularly costly. This probably helps explain why, in practice, the durability of R-genes seems to vary. In a few cases, the effectiveness of a resistance gene has remained constant for decades, although sadly such cases are notable for being rare.

On the pathogen side, various features of an organism's biology help predict whether it will overcome host resistance rapidly, slowly, or not at all. One important factor is the amount of genetic diversity in the pathogen population. A population with low levels of diversity will be slower to evolve to virulence than a population with high levels of diversity, where chances are better of a favorable allele winding up in a good genetic background. This relates to "effective population size": how many different genotypes contribute to reproduction of the population, providing their respective genotypes to be selected by host plant diversity (see Case Study 23.3)?

Mode of reproduction is a related and important factor. Pathogens that undergo regular recombination pose higher risks than pathogens that undergo little or no recombination. Among microorganisms, recombination occurs through sexual reproduction but can result from other processes as well, such as bacterial conjugation, recombination between viral genomes in plants with mixed infections, and hyphal **anastomosis** and/or **parasexual recombination** in fungi.

Pathogens with mixed reproductive systems (both sexual and asexual reproduction) pose the highest risk of evolution. Sexual recombination allows many new combinations of alleles to come together and be tested against host resistance. The fittest genotypes can then undergo asexual reproduction, often in multiple cycles per growing season, which allows advantageous allele combinations to remain together and attain high frequencies in the population. These asexually reproduced individuals are also often dispersed over long distances as windblown conidia, spreading the fittest genotypes from one growing area to many others.

Gene flow and genotype flow also greatly affect pathogen evolution to virulence. If there is migration and mixing among dispersed subgroups of a pathogen population, it is more likely that the specific recombination events needed to place useful virulence mutations in highly fit backgrounds will occur. Conversely, a primarily soilborne pathogen that does not travel long distances has less opportunity for such recombination.

MAXIMIZING DURABILITY OF RESISTANCE

Overall, the **durability** of a resistance gene is likely to be optimized by avoiding the widespread and uniform cultivation of a cultivar solely or primarily protected by that gene. Breeders generally seek to introgress R-genes into backgrounds containing quantitative resistance, both because the cultivar will have "something to fall back" on if the R-gene is defeated, and because quantitative resistance prolongs the effective life of qualitative resistance.

Several measures that depart from resistance **monoculture** have proved beneficial both for reducing disease epidemics and for increasing the durability of individual qualitative resistance genes. These include rotating cultivars with different resistant genes; growing a higher number of different crops per unit geographic area; and intercropping, or interspersing rows or plots of different, complementary crop species within a single field. All these measures reduce the continuous spatial area of genetically uniform host tissue.

However, even within the basic industrial agriculture model of heavy reliance on a few crops and one crop species per field, there are strategies to increase host heterogeneity. One such approach, most commonly employed with small-grain cereals, is called **multilines**. A multiline is composed of several **isolines**, or lines that are genetically identical except for one or more resistance genes. For example, a barley multiline could consist of four or five isolines of a cultivar, into each of which a different R-gene or pair of R-genes has been backcrossed. Multilines that result in some (rather than no) disease, the so-called dirty crop approach, are believed to increase the durability of specific resistance genes by lessening the selective advantage of the corresponding virulence mutations.

A related strategy, **variety mixtures**, gets around the main disadvantage of multilines, which is the resource-intensiveness of generating the backcrossed isolines. These mixtures, sometimes also called "blends," usually consist of two to five released commercial cultivars of the same crop species (Figure 23.14). Commonly, the cultivars complement each other for various important traits, including disease resistance and yield. For example, a high-yielding, but powdery mildew-susceptible, barley variety is blended with a lower-yielding, but mildew-resistant, variety. Disease in such mixtures is reduced by three basic mechanisms. First, the inoculum for each cycle of pathogen growth is diluted because a portion of it falls upon resistant tissue, thus reducing the overall rate of epidemic development. Second, resistant plants simply operate as a barrier between one susceptible plant and another, preventing a portion of the spores that are released from reaching a susceptible target. Third is the benefit from induced resistance: lines have their resistance machinery activated by pathogen strains that may be virulent to other components of the mixture, but not to them, and thus subsequent pathogen attacks on them are less effective (Calonnec et al., 1996). Due to the ease of mixing seeds prior to planting, variety of mixtures have been most widely used in small grains, but the strategy has also been explored with larger plants such as potato,

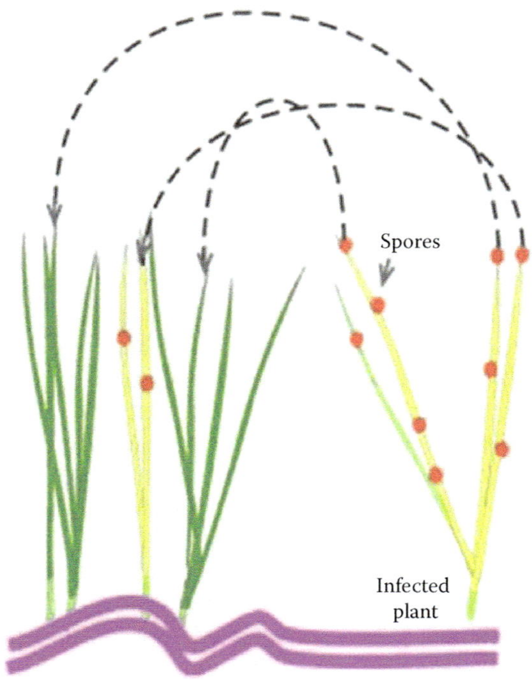

FIGURE 23.14 A 50:50 genotype mixture of yellow (susceptible) and green (resistant) plants. Half of the inoculum from the susceptible plants is ineffective because it lands on resistant tissue. (From Rosalie Bliss, Blended wheat varieties show their strength, May/June 2007 issue of Agricultural Research magazine, http://www.ars.usda.gov/is/ar/archive/may07/wheat0507.htm)

coffee, willows for fiber, and apple. For more on multilines and mixtures, the reader is directed to a recent review (Mundt, 2002).

In the case of qualitative resistance to plant-parasitic insects, such as Hessian fly of wheat, gene pyramids have been widely used. This is also true where cotton or corn has been engineered to carry individual *Bacillus thuringiensis* (*Bt*) bacterial genes that kill insect pests. This strategy is based on the assumption of a low probability that an insect can assemble the mutations needed to render it insensitive to multiple *Bt* toxins with different modes of action. In the case of *Bt*-resistant corn and cotton, farmers are required to plant **refuges** consisting of a small area of susceptible plants. The refuge works because it favors the majority insect population that is vulnerable to the *Bt* toxins and increases the likelihood that rare insensitive mutant individuals will mate with sensitive ones, slowing the evolution of homozygous insensitive bugs.

It is interesting to compare durability of disease resistance in agricultural systems with durability in natural communities of mixed plant populations and their pathogens. In contrast to human-managed systems, wild plant-pathogen systems are said to be coevolving: each party exerts selection pressure on the other. Disease is a feature of wild plant populations as well, but as a general rule epidemics are less uniform, widespread, and intense than in agricultural systems. The relative rarity of damaging and large-scale epidemics has led observers to characterize wild plants' relationship with their parasites as "trench warfare." This is because pathogen incidence and frequency ebbs and flows but host populations are usually not devastated. The diversity of both virulence and resistance alleles is high, with no one allele sweeping a population (attaining a 100% frequency). Such natural systems contain features that are of interest to humans trying to replace the "boom and bust" dynamic with more sustainable agricultural approaches.

LABORATORY EXERCISES

Experiment 1. Use of the Hypersensitive Reaction (HR) to Determine the Presence of Complete Resistance Genes by Inoculation of Tobacco Cultivars with *Tobacco Mosaic Virus*

Tobacco mosaic virus (TMV) is an important pathogen of tobacco in many areas of production. The virus is very easily transmitted mechanically from plant to plant, and no chemical control measures are effective once a plant is infected. Therefore, management is focused on preventing the virus from coming in contact with plants and developing resistant varieties that prevent the virus replication if introduced into the plant.

The most effective genetic resistance is based on the presence of a single resistance gene that stops the virus from replicating and moving out of the initially infected cells by inducing HR at the point of infection. Because all viruses require living cells to replicate, and because without replication the virus cannot move to other cells, the HR effectively stops disease development. Some cultivars possess resistance genes. In this experiment, you will inoculate susceptible and resistant cultivars of tobacco and observe symptom development over a 2-week period. This experiment has the following objectives:

- Learn methods used to transmit viruses.
- Observe susceptible and hypersensitive resistance responses to a virus disease.

Materials

Students should work in teams of 2–4. Each team will require the following materials:

- Phosphate buffer
- SEB1 sample extraction buffer

- TMV Immunostrips (Agdia Inc.)
- Tobacco leaves infected with TMV
- 2 tobacco plants (6–8 weeks old) of each variety—Hicks and Ky 14 x L8
- Mortar and pestle
- Carborundum—a very fine and hard industrial abrasive (silicon carbide)
- Cotton swaps, latex gloves, paper towels, pot labels, sharpie, scissors

Follow the protocol found in Procedure 23.1 to complete this experiment.

Anticipated Results

Refer to Table 23.2 for typical symptoms on leaves of resistant and susceptible tobacco plants.

Questions

- What is the importance of phosphate buffer in the inoculation procedure?
- What is the importance of the carborundum in the inoculation procedure?
- What type of symptom developed on the Ky 14 x L8 plants? How long did it take for the symptoms to develop?
- Do you think that the type of resistance observed in this experiment would be highly effective? Why?
- Did you detect the virus in the inoculated leaf of each host? Briefly discuss why or why not.
- Did you detect systemic movement of the virus? In which variety? Briefly discuss why or why not.

Procedure 23.1

Resistant and Susceptible Cultivars of Tobacco to *Tobacco Mosaic Virus* (TMV)

Step	Instructions and Comments
1	Work in teams of 2 to 4. Place 8–10 mL of buffer into the mortar and add 1–2 small tobacco leaves with obvious symptoms of TMV. Grind the leaves in the buffer until the tissue is macerated.
2	Sprinkle a small amount of carborundum onto the surface of—two to three leaves of each of the two tobacco plants. Choose leaves that are young to moderately young. Do not use the bud leaves or the oldest lower leaves.
3	Dip the tip of the cotton swap into the TMV buffer solution and then gently rub the areas of the leaves that have the carborundum, frequently adding fresh inoculum to the swab. Do not rub too hard as this will injure or kill cells!
4	Label the pots with your section and table number and take to the greenhouse. Observe the symptoms over a 2-week period after 2–3 days, 7 days, and 14 days and compare with the symptoms listed in Table 23.2.
5	Use TMV Immunostrips (Agdia Inc.) for detecting the presence of TMV in tissue (follow manufacturer's instructions). Assay both the inoculated leaves and leaves above and below them.

TABLE 23.2

Symptoms Observed on Seedlings Inoculated with *Tobacco mosaic virus*

	Days after Inoculation					
	2–3 Days		7 Days		14 Days	
Variety	**Control**	**Inoculated**	**Control**	**Inoculated**	**Control**	**Inoculated**
Hicks	No symptoms	No symptoms	No symptoms	No symptoms or slight mosaic in newest leaves	No symptoms	Mosaic in new leaves; no necrosis
KY14 x L8	No symptoms	Necrotic spots, 4–5 mm in diameter	No symptoms	Browning of original spots, no new spots, and no mosaic	No symptoms	No further lesion development; no mosaic

Experiment 2. Inoculation of Susceptible and Resistant Corn Inbreds with the Southern Corn Leaf Blight Pathogen, *Bipolaris maydis*

Southern corn leaf blight (SCLB) is caused by the fungus *Bipolaris maydis*. This was a minor disease of corn until 1970. Control of SCLB is primarily based on the use of resistant hybrids that have high levels of quantitative resistance. In this experiment, you will inoculate two corn inbreds (you can also use two varieties of sweet corn if inbreds are not available) that differ in level of partial resistance to SCLB. You will also use two inoculation methods to investigate the level of resistance present. One method is based on inoculation of detached leaf parts and the other uses of growing seedlings. This experiment has the following two objectives:

- Learn methods used to inoculate a foliar fungal pathogen.
- Observe the effectiveness of partial resistance by observing the response of corn inbreds to a common leaf spot pathogen.

Materials

Each student or team of students will need the following items:

- Spore suspension of *B. maydis*; 5000 conidia/mL
- Seedlings of two corn inbreds, B-73 and Mo17
- Preval sprayer with reservoir (pressurized paint applicator available from any home supply store)
- 1-mL pipettes (sterile)
- Incubation box
- Paper towels, labels, sharpie

Follow the instructions in Procedure 23.2 to complete this experiment.

Anticipated Results

Lesions should appear on the leaves of susceptible plants after 2 days and after 7 days on plants with partial resistance.

Procedure 23.2

Partial Resistance of Corn Inbreds to a Common Leaf Spot Pathogen

Step	Instructions and Comments
	Whole Plant Inoculation
1	Swirl the spore suspension and pour it into the reservoir at the bottom of the sprayer.
2	Aiming away from the corn plant, spray several seconds until a fine spray is coming from the sprayer. Next, carefully spray the entire plant for about 3–4 s or until you see numerous small drops on all of the leaves. Do not over wet the leaves. Carefully cover the plant with a large plastic bag for 24 h.
	Whorl Inoculation
1	Drop 1 or 2 mL of spore suspension into the whorl of a plant. Cover plant with a large plastic bag for 24 h.
2	Describe the symptoms observed on whole plants and on whorls inoculated with *Bipolaris maydis* 2 and 7 days after inoculation. Rate for number, size, and position of lesions.

CASE STUDY 23.1

Chestnut Blight Disease in North America—*Exotic Disease Devastates Continental Ecosystem*

- At the turn of the twentieth century, American chestnut trees made up around a quarter of the hardwood trees in the forests of the eastern U.S. seaboard and the Appalachian Mountains.
- In the early twentieth century, chestnut blight, caused by the accidental introduction of the fungus *Cryphonectria parasitica* from Asia, almost completely wiped out the chestnut.
- Chinese chestnut varieties have good resistance to chestnut blight. Efforts are underway to breed blight-resistant American chestnut varieties, using crosses with the Chinese chestnut. This is a long-term undertaking, as each generation of testing takes 5–8 years.
- This is an example of the use of exotic germplasm, from the point of origin of the pathogen, to confer resistance to introduced diseases.

CASE STUDY 23.2

Disease in Central American Banana Plantations—*The Perils of Monoculture*

- The Asian banana cultivar Gros Michel was grown over most of the plantation acreage in Central America and the Caribbean from the late nineteenth century until the 1950s.
- Panama disease, a root disease caused by *Fusarium oxysporum*, swept through the plantations in the 1950s and devastated production.
- Production was switched to the Cavendish variety, which was resistant to Panama disease.
- "Cavendish" is still by far the predominant variety grown in plantations today.
- The following major diseases now threaten and limit the production of the "Cavendish":
  - Black Sigatoka disease, a foliar fungal disease for which plantations are sprayed with a fungicide up to once a week.
  - A new form of Panama disease—subtropical race 4—to which "Cavendish" is susceptible. This new race has appeared in Asia, and it is feared that it will make its way to Central America and the Caribbean.
- The extreme genetic uniformity of banana plantations has predisposed them to the threat of devastating epidemics.
- The fact that it is extremely hard to breed bananas makes germplasm improvement through traditional methods difficult.
  - Genetic engineering may be a reasonable alternative approach to breeding disease resistant varieties.

CASE STUDY 23.3

Ug99 Race of Wheat Stem Rust—Rapid Mutation and Migration Threaten a Crop Globally

- Wheat stem rust was a devastating problem in North American wheat production in the first half of the twentieth century. In some years up through the 1950s, stem rust caused near-complete crop losses in the United States and Canada.
- By the 1970s, the wheat stem rust problem was almost completely resolved in North America. This was for two main reasons. First, a government-coordinated campaign to eradicate the barberry, the alternate host of stem rust, prevented the fungus from reproducing sexually, greatly reducing its ability to evolve to virulence. Second, the time for stem rust reproduction on its primary host, wheat, was shortened due to breeding of earlier-maturing wheat varieties.
- Thanks to those measures, a small set of stem rust resistance genes remained effective for decades. This was essentially true worldwide throughout the latter part of the twentieth century, with very few exceptions.
- However, in 1999, the effective genes were suddenly overcome in Uganda, where the new virulent stem rust race was christened Ug99.
- Ug99 and its descendants continued to mutate, leading to the emergence of a family of Ug99-derived races. This virulent family spread rapidly to neighboring countries in eastern Africa, and also to Zimbabwe, South Africa, Sudan, Yemen, and Iran.
- In 2011, it was estimated that only 5%–10% of the area of wheat planted in Africa and Asia had adequate resistance to Ug99 and its derivatives. Fears were raised that the Ug99 family could devastate wheat production throughout Africa, the Middle East, and South Asia.
- This is an example of how a plant pathogen can both rapidly evolve and migrate, so that in short order it stops being a historical footnote and becomes a threat to global food security.

Questions

- Were there any differences observed for the two cultivars in the whole plant inoculation? Describe the differences.
- Describe the distribution of symptoms on the whole plant and leaf whorl inoculations? Why did the lesions develop in these patterns?
- Based on your results, does the resistance affect lesion size? Is this an important component of partial resistance? Why?
- Which inoculation method would you use to quantify the level of partial resistance in the varieties?

REFERENCES AND SUGGESTED READINGS

Bent, A. F., and D. Mackey. 2007. Elicitors, effectors, and R Genes: The new paradigm and a lifetime supply of questions. *Annu. Rev. Phytopath.* 45:399–436.

Calonnec, A., H. Goyeau, C. de Vallavieille-Pope. 1996. Effects of induced resistance on infection efficiency and sporulation of *Puccinia striiformis* on seedlings in varietal mixtures and on field epidemics in pure stands. *Eur. J. Plant Pathol.* 102:733–741.

Chilton, M.-D. 2001. Agrobacterium. A Memoir. *Plant Physiol.* 125:9–14.

Cowger, C., and C. C. Mundt. 2002. Aggressiveness of *Mycosphaerella graminicola* isolates from susceptible and partially resistant wheat cultivars. *Phytopathology* 92:624–630.

Fraile, A., I. Pagán, G. Anastasio, E. Sáez, and F. García-Arenal. 2011. Rapid genetic diversification and high fitness penalties associated with pathogenicity evolution in a plant virus. *Molec. Biol. Evol.* 28:1425–1437.

Gonsalves, D. 1998. Control of papaya ringspot virus in papaya: A case study. *Annu. Rev. Phytopath.* 36:415–437.

Gurr, S. J., and P. J. Rushton. 2005. Engineering plants with increased disease resistance: What are we going to express? *Trends Biotech.* 23:275–282.

Hammond-Kosack, K. E., and J. E. Parker. 2003. Deciphering plant-pathogen communication: Fresh perspectives for molecular resistance breeding. *Curr. Opin. Biotechnol.* 14:177–193.

Hillnhütter, C., R. A. Sikora, and E.-C. Oerke. 2011. Influence of different levels of resistance or tolerance in sugar beet cultivars on complex interactions between *Heterodera schachtii* and *Rhizoctonia solani*. *Nematology* 13:319–332.

Jarvis, W. 1977. *Botryotinia* and *Botrytis* species. Taxonomy and pathogenicity. *Can. Dep. Agric. Monogr.* 15:195.

Kump, K. L., P. J. Bradbury, R. J. Wisser, E. S. Buckler, A. R. Belcher, M. A. Oropeza-Rosas, J. C. Zwonitzer, S. Kresovich, M. D. McMullen, D. Ware, P. J. Balint-Kurti, and J. B. Holland. 2011. Genome-wide association study of quantitative resistance to southern leaf blight in the maize nested association mapping population. *Nat. Genet.* 43:163–168.

Loegering, W. Q., and A. H. Ellingboe. 1987. Flor, H.H. - pioneer in phytopathology. *Annu. Rev. Phytopath.* 25:59–66.

Madden, L.V., G. Hughes, and F. Van den Bosch. 2007. *The Study of Plant Disease Epidemics*. St. Paul, MN: American Phytopathological Society.

Mundt, C. C. 2002. Use of multiline cultivars and cultivar mixtures for disease management. *Annu. Rev. Phytopath.* 40:381–410.

Poland, J. A., P. J. Balint-Kurti, R. J. Wisser, R. C. Pratt, and R. J. Nelson. 2009. Shades of gray: The world of quantitative disease resistance. *Trends Plant Sci.* 14:21–29.

Poland, J. A., P. J. Bradbury, E. S. Buckler, and R. J. Nelson. 2011. Genome-wide nested association mapping of quantitative resistance to northern leaf blight in maize. *Proc. Nat. Acad. Sci.* (USA) 108:6893–6898.

Roelfs, A. P. 1988. Genetic control of phenotypes in wheat stem rust. *Annu. Rev. Phytopathol.* 26:351–367.

Saari, E. E., and J. M. Prescott. 1975. A scale for appraising the foliar intensity of wheat diseases. *Plant Dis. Rep.* 59:377–380.

Sanchis, V., and D. Bourguet. 2008. *Bacillus thuringiensis*: Applications in agriculture and insect resistance management. A review. *Agron. Sustain. Dev.* 28:11–20.

Schafer, J. F. 1971. Tolerance to Plant Disease. Annu. Rev. Phytopath. 9:235–252.

Tourte, Y. 2003. *Genetically Modified Organisms: Transgenesis in Plants*. Enfield, NH: Science Publishers.

Uma, B., T. Swaroopa Rani, and A. R. Podile. 2011. Warriors at the gate that never sleep: Non-host resistance in plants. *J. Plant Physiol.* 168:2141–2152.

Vanderplank, J. E. 1984. *Disease Resistance in Plants.* 2nd Edition. New York, NY: Academic Press.

Wulff, B. B. H., D. M. Horvath, and E. R. Ward. 2011. Improving immunity in crops: New tactics in an old game. *Curr. Opin. Plant Biol.* 14:468–476.

24 Plant–Fungal Interactions at the Molecular Level
The Biological Approach to Fungal Pathogen Control

Ricardo Manuel de Seixas Boavida Ferreira and Sara Alexandra Valadas da Silva Monteiro

Concept Box

- Fungal pathogens cause great losses to global agriculture every year.
- Plant–pathogen interactions may be considered as biochemical warfare, the outcome of which determines the establishment of resistance or disease.
- Increasing public concern over the use of toxic fungicides, together with more restrictive legislation associated with huge costs of discovering novel fungicides has led to a move from chemical to biological input strategies to control plant pathogens.
- Control of pathogens based on the minute amounts of plant volatile organic compounds (PVOCs) emitted and liberated into the atmosphere when a plant is challenged by a pathogen involves detection, identification, and quantification of the precise blend of chemicals. The technology required to accomplish this task is not yet available to use at an agricultural scale.
- Unveiling the complex molecular interactions that take place between host and pathogens will open the way to a wide range of possibilities of controlling pathogens. Research in plant pathogen control may be regarded as follows: the past was chemical, the present the "-omics," and the future whole organisms (e.g., microbiomes and neurologically controlled insects), plus the emitted blend of the volatiles they induce.

FUNGAL PATHOGENS

Fungi are an extremely diverse group of organisms with about 250,000 species widely distributed in essentially every ecosystem. The total weight of fungi on earth has been estimated as 4×10^{14} g, thus exceeding that of humans. In contrast to the situation in animals, in which fungal diseases are less frequent, many plant diseases are caused by fungi. Approximately 120 genera of fungi, 30 types of viruses, and 8 genera of bacteria are responsible for more than 11,000 diseases that have been described so far in plants.

Plants constitute an excellent ecosystem for microorganisms because they offer a wide diversity of habitats, including the **phyllosphere** (aerial plant part), the **rhizosphere** (zone of influence of the root system), and the **endosphere** (internal transport system). Therefore, plant organs, both above and below ground, are continuously and permanently exposed to a vast range of potential pathogens. However, because the normal state of a plant is healthy, development of disease requires the coincidence of a susceptible host, a virulent pathogen, and a favorable environment (Chapter 1). In other words, not all pathogens can attack all plants, and a single plant is not susceptible to the plethora of **phytopathogens**. In fact, only a very small proportion of these pathogens are capable of invading any plant host successfully and causing disease. In addition, pathogens and host plants change during their life cycles according to their stage of development, which affects pathogen virulence and host susceptibility. Disease is also strongly dependent on environmental conditions, such as water availability, temperature, and plant surface wetness. For these reasons, disease development is usually less frequent than one would expect (see disease triangle Figure 1.9).

The huge economic importance derived from pests and diseases in modern agriculture hardly requires any mention. Estimates from 2005 indicate that the use of chemicals amounts to 3×10^{12} g per year, at a value of $40,000,000,000 per year, with pests and diseases still causing crop losses on the order of 30%–40%. It is now clear that pathogen control in crops will assume increasing importance in human and animal nutrition.

The increasing world population in the foreseeable future and the permanent decrease in global arable land (at least, when taken as a proportion of the total planet land area) indicate that we are heading for an inevitable critical point, caused by a worldwide shortage in food supply, whose consequences remain to be properly evaluated. Our apparent inability to change the course of these two phenomena will limit the way we manage our natural resources in the near future, which will determine whether the critical point will occur in the near future, or be delayed. One of the most sensible ways to delay the critical point is by improving crop yield and disease control without collateral damage to the agricultural environment and to organisms other than pathogens.

The overall approach/strategy generally adapted to tackle fungal infections has been hampered by considering phytopathogens and human pathogens as belonging to separate scientific discourses. From the pathogen point of view, there is no reason whatsoever for making such a distinction. In fact, there are several fungal species (e.g., *Alternaria alternata*) that infect both humans and plants. The major differences are the hosts, especially in what concerns their responses, defenses, and sensitivities. Surprisingly, an increasing number of features are shared by both human and plant organisms in their everyday inevitable encounter with fungal pathogens.

THE BATTLE FOR FUNGAL PATHOGENESIS

The attempted infection of a plant by a pathogen, such as a fungus or an oomycete, may be regarded as a fierce war, composed of many battles. The major weapons in the battles are proteins and smaller chemical compounds produced by both organisms. The outcome of this complex confrontation determines the failure or success of the attempted pathogenesis, that is, establishment of resistance or disease, or the evolution to mutualism (Figure 24.1).

In their long coevolution with pathogens, plants evolved an intricate and elaborate array of defensive compounds that are being discovered at a rapid pace. At the same time, those very same pathogens developed means to overcome plant resistance mechanisms in what must have been a multimillion-year, ping-pong like coevolution, in which plant and pathogen successively added new chemical weapons in this perpetual war. As each defensive innovation was established in the host, new ways to circumvent it evolved in the pathogen. For example, the tobacco pathogen, *A. alternata*, synthesizes and secretes the reactive oxygen quencher mannitol as a means of suppressing reactive oxygen-mediated plant defenses. Interestingly, the non-mannitol-containing host tobacco plants respond by expressing a pathogen-induced mannitol dehydrogenase that catabolizes mannitol of fungal origin. The constitutive expression of a celery (*Apium graveolens*) mannitol dehydrogenase cDNA in tobacco plants conferred enhanced resistance to *A. alternata*, but not to the non-mannitol-producing fungal pathogen *Cercospora nicotianae*.

More recently, phytopathogens overcame plant resistance through acquisition of virulence factors. These newly evolved pathogen race-specific factors (Chapter 23) drove the coevolution of plant resistance genes and the development of phylogenetically related, pathogen race/plant cultivar–specific disease resistance. Over time, the coevolutionary struggles between would-be pathogens and their erstwhile hosts have generated some of the most complex and interesting interactions known to biology. This complex coevolution process has probably led to redundancy or ineffectiveness of some defensive genes, which often encode numerous enzymes with overlapping activities, and to the exquisite specificity observed between many pathogens and their hosts.

The pathogen must therefore successfully overcome host obstacles before it succeeds in causing disease. To this end, pathogens must evade, suppress, or manipulate host defenses by microbial effectors that act inside the plant cell, which often requires extensive modification of plant gene expression and metabolism to its own benefit and/or to abolish defense reactions.

Plant cell walls, which consist mainly of polysaccharides (i.e., cellulose, hemicelluloses, and pectins) and proteins, play an important role in defending plants against pathogens (Figure 24.1). Many pathogenic fungi release an array of cell wall degrading, hydrolytic enzymes that fragment plant cell wall polymers, including proteases and **glycanases** (e.g., **galacturonases**, **xylanases**, and **glucanases**), thus facilitating colonization of the host cells. The importance of the secreted proteases in pathogenesis is highlighted by the observation that protease-deficient fungal mutants lose the ability to induce lesions in plants. Also, during appressorium formation, several fungal proteases exhibit specificity toward fibrous **hydroxyproline-rich proteins** found in plant cell walls.

Pectin, a major component of plant cell walls, is cleaved by fungal **endopolygalacturonases (EPGs)** with the transient formation of elicitor-active **oligogalacturonides (OGAs)** with degrees of polymerization between 9 and 15. Thus, the oligosaccharides generated by glycanases

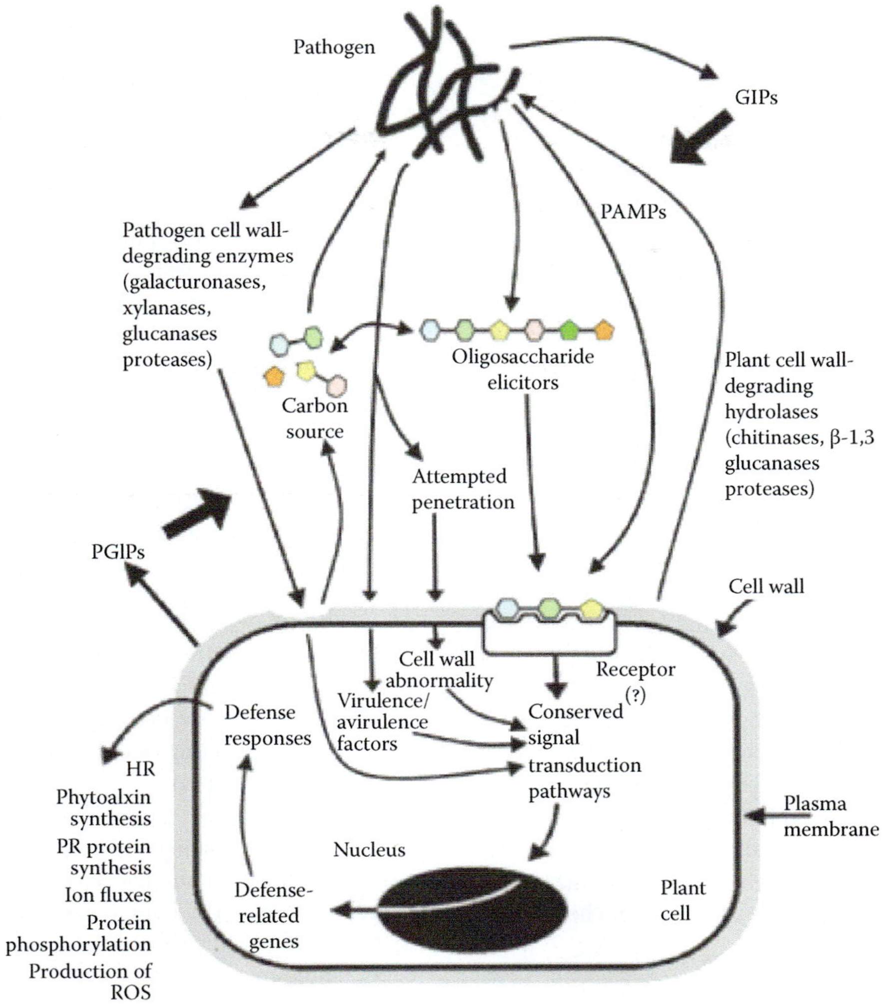

FIGURE 24.1 Molecular chain of events that typically underlies a host–pathogen interaction: the protein warfare going on between plant host and attacking pathogen during attempted infection and the path leading to plant defense response. Pathogen refers to fungal or oomycete pathogens. Abbreviations: PAMP, pathogen-associated molecular pattern; GIP, glucanase inhibitor protein; PGIP, polygalacturonase-inhibiting protein; HR, hypersensitive response; PR, pathogenesis-related; ROS, reactive oxygen species. (From Ferreira, R. B. et al., *Plant Pathology Concepts and Laboratory Exercises*, Boca Raton, FL: CRC Press, 2008.)

not only provide the fungus with a carbon source, but are perceived by and elicit plant defense mechanisms. For this reason, OGAs are rapidly converted to smaller, biologically inactive fragments by the EPGs. For laboratory exercises involving measuring and detecting extracellular enzymes produced by pathogens, see Chapter 31.

To increase the lifetime of the biologically active oligosaccharides, plants release inhibitors of the fungal glycanases. These include inhibitors of pectin-degrading enzymes such as **polygalacturonases**, **pectin methyl esterases**, and **pectin lyases** and cross-linking glycan (known earlier as hemicelluloses)-degrading enzymes such as **endoxylanases** and xyloglucan endoglucanases. For example, plant polygalacturonase-inhibiting proteins (PGIPs) are glycoproteins present in the **apoplast** of many plants that form reversible high-affinity complexes with fungal EPGs, reducing their catalytic activity by one to two orders of magnitude. By limiting EPG activity, the lifetime and concentration of OGAs are increased, prolonging and/or enhancing plant defense responses. PGIPs limit the growth of plant pathogens and also elicit defense responses in plants. They belong to the superfamily of leucine-rich repeat (LRR) proteins that also include the products of several plant resistance genes. Fragmentation of the fungal cell wall by plant derived **chitinases**, and β-1,3-glucanases also generates oligosaccharides that induce plant defense responses. In response, fungi produce **glucanase inhibitor proteins (GIPs)** that prevent degradation of their own cell wall, thus limiting their perception by plants.

Plants evolved complex, multilevel series of structural and chemical barriers that are both constitutive or preformed and inducible. These defenses may involve strengthening of the cell wall, hypersensitive response (HR), oxidative burst, **phytoalexins**, and pathogenesis-related

(PR) proteins. The pathogen must successfully overcome each of these obstacles before succeeding in causing disease. In some cases, the pathogen needs to modulate or modify plant cell metabolism to its own benefit and/or to abolish plant defense reactions. Central to the activation of plant responses is the timely perception of the pathogen by the plant. A crucial role is played by elicitors, which, depending on their mode of action, are broadly classified into nonspecific elicitors and highly specific elicitors or virulence effector/**avirulence** factors. A protein battle for penetration is then initiated, marking the attempted transition of the pathogen from extracellular to invasive growth, before disease can be established.

THE GREAT COMPLEXITY OF HOST–PARASITE INTERACTIONS

The early view of a host–pathogen interaction involving the sequence "evolution of a new defensive innovation established in the host," followed by "a new way to circumvent it in the pathogen," is simplistic in view of the extreme complexity of some interactions whose biochemical details have come to light at the molecular level. Rather than the new weapon/new defense sequence, it now seems well established that the pathogen requires access to modulate or control host gene expression to its own benefit and/or to abolish defense reactions in order to establish pathogenesis. This has been demonstrated by the green island effect (Figure 24.2). A detached barley leaf was inoculated with *Blumeria graminis* f. sp. *hordei* (responsible for powdery mildew in barley) on one side and then placed in darkness to accelerate senescence of the leaf. Only those cells underneath the fungal colony on the uninoculated side remained green, which indicated the fungus was capable of actively suppressing senescence of host tissue.

(a) (b)

FIGURE 24.2 Modulation of host gene expression by fungal pathogens as demonstrated by the green island effect. A detached barley leaf was inoculated with *Blumeria graminis* f. sp. *hordei* on one side and then placed in darkness to accelerate senescence of the leaf. The uninoculated (a) and inoculated (b) sides are shown. The cells underneath the colony remained green, indicating that the fungus actively suppressed senescence of host tissue. (From Schulze-Lefert, P, and Vogel, J., *Trends Plant Sci*., 5, 343–348, 2000. With permission.)

In summary, host–parasite interactions are essentially biochemical warfare. Sometimes, the intricate complexity underlying the attempted parasitism evolved to almost unimaginable levels, so did the host defense mechanisms and the parasite ways to circumvent them. In some cases, it is as if the fungal pathogen could think for itself!

CONTROL OF FUNGAL PATHOGENS: THE SHIFT FROM CHEMICAL TO BIOLOGICAL

A comprehensive definition of plant pathology encompasses the following three concepts: (i) the study of microorganisms and environmental factors that cause diseases in plants; (ii) the study of the mechanisms by which these factors induce disease in plants; and (iii) development of methods for preventing or controlling disease and reducing the damage caused.

The largest manufacturers of **phytopharmaceuticals** are chemical companies. Increasing public concern over the use of chemical fungicides and residues on foods and decreasing efficacy of compounds are severely impacting manufacturers. Current estimates are that 100,000 new chemical compounds are tested for each new fungicide that enters the market, a figure that significantly raises development costs of new compounds. Unfortunately, the low-toxicity products currently on the market often exhibit efficacy rates well below those of traditional, more toxic products. The additional capacity of fungal pathogens, bacteria, and other infectious organisms to develop resistance mechanisms against some top pesticides has led some prominent companies in the field to abandon the extremely high cost of development of new chemical fungicides.

The era of chemical fungicides seems to be heading to an end, with the large fungicide manufacturers slowing down on the release of novel active ingredients. This section could have been most appropriately titled "The worldwide drama of fungal pathogens." A door is definitely opening for biological control, which will certainly assume a leading role in the future of phytopathology. The search is based on the intricate molecular interactions that take place between host and pathogen. The use of **transcriptomic**, **proteomic**, **glycomic**, and **metabolomic** techniques will facilitate the identification of biological molecules that are essential for the establishment of pathogenesis. Each one of these molecules will constitute a potential target for pathogen control.

BIOLOGICAL, WHOLE ORGANISM APPROACH TO CONTROL PLANT–PATHOGEN INTERACTIONS

The use of living organisms in successful control of crop pests and diseases has long been known. Most people have heard of releasing ladybirds (bugs) for aphid control as an attempt to reestablish/reintroduce the natural homeostatic and self-regulated equilibrium between predators and prey, typical of many natural environments. Such balance has been disrupted (i.e., biologically deregulated) in many instances by successive years of application of toxic chemical pesticides.

The recent trend of introducing **mycopathogenic** bacteria or fungi (e.g., specific strains of *Bacillus subtilis*, *B. pumilus*, *Trichoderma gamsii*, and *T. asperellum*) to control phytopathogenic fungi results in the release of zillions of living microorganisms into the environment. At least at first thought, this procedure seems potentially far more dangerous than using transgenic crops. The medium-term overall effect on global environmental biology remains to be elucidated, but may induce dramatic imbalances in the microbiota of many ecosystems, whose importance is still far from understood. In addition, some of the microbial genera in use contain species that are pathogenic to man, for example, *Trichoderma longibrachiatum*, *T. pseudokoningii*, *T. citrinoviride*, *T. koningii*, *T. harzianum*, *T. viride*, and *T. inhamatum*. The bacterial species in use include *Bacillus anthracis*, *B. cereus* (a common cause of food poisoning), *B. thuringiensis*, *B. larvae*, *B. lentimorbus*, *B. papillae*, and strains of *B. sphaericus* (pathogens of insects). It should be noted that infectious agents have crossed species barriers and infected the human population, occasionally with devastating effects. A highly promising alternative for research in disease control is to exploit the biochemical warfare that occurs between parasite and host for the establishment of pathogenesis.

THE MICROBIOME

The term "**microbiome**" was coined by Joshua Lederberg to describe microorganisms inhabiting the human body that should arguably be included as part of the human genome because of their influence on human physiology. The term is now defined as the complete set of microorganisms, including their genomes and environmental interactions, which inhabit a particular location, be it a human, plant, pond, or rock.

In addition to their known beneficial effects, residential microbiomes are receiving increasing interest in different areas of the biology of higher organisms, the real importance of which is only starting to be appreciated. The large gaps in our knowledge result from the great complexity of the studies and the difficulty in identifying the organisms comprising a microbiome, which may vary from host to host within a particular species and/or change during the life cycle of a single individual. In this sense, a given disease may be regarded as a specific imbalance in the resident microbiome, with one or more pathogenic microbes assuming a predominant role in the microbiome. The pathogen may or may not be a common member of the normal microbiome.

To adapt to a wide range of diverse habitats across environmental gradients, all plants exhibit **phenotypic plasticity**, that is, the capacity, via controlled changes in the transcriptome, proteome, and metabolome, to originate multiple phenotypes from a single genotype, depending on the environmental conditions. The adaptation of plants to stress conditions involves a combination of phenotypic plasticity and **genetic adaptation**, both of which involve processes exclusively derived from the plant genome. However, there is compelling evidence for a role of the microbiome on plant stress tolerance and fitness. There are at least three classes of fungal symbionts involved in these interactions, including the following:

1. Mycorrhizal fungi, which are associated with plant roots and share nutrients with their plant hosts
2. Class 1 **endophytes**, for example, **clavicipitaceous fastidious fungi** that infect cool season grasses
3. Class 2 endophytes, whose ecological roles have only recently begun to be elucidated

Plants use the root microbiome, that is, the consortium of fungi, bacteria, and viruses that live in their root systems, to survive stressful conditions. For example, mycorrhizal fungi increase root and shoot biomass, improve plant resistance to biotic stresses such as pathogens and herbivores, improve tolerance to abiotic stresses such as heat, salt, and drought, and enhance nutrient absorption (e.g., phosphorous and nitrogen), especially under the conditions of nutrient deficiency. Class 1 endophytes produce chemicals that deter herbivory. Class 2 endophytes are the largest group of fungal symbionts. They are readily culturable on artificial media, are thought to colonize all plants in natural ecosystems, and confer stress tolerance (e.g., heat, drought, and pathogens) to host species. For example, the grass, *Dichanthelium lanuginosum* grows at 70°C in the geothermal hot springs of Yellowstone National Park in the United States. However, the grass fails to grow at high temperatures if its seeds are treated to kill their fungal endophytes.

This observation prompted the following hypothesis: is it possible to improve growth of a normal plant under conditions of water stress by transferring the microbiome from a drought-tolerant plant? The answer to this question came when fungal spores from *D. lanuginosum* previously sprayed on wheat seeds allowed wheat to grow to temperatures up to 70°C, instead of the normal 38°C and required 50% less water.

Different microbiomes can extend the adaptability of plants in a number of unsuitable environments. Spraying rice seeds with endophytes from a salt-adapted dune grass (*Leymus mollis*) and from a low temperature-adapted strawberry plant (*Fragaria vesca*) allowed rice plants to tolerate salt and cold, respectively, to grow five times larger and require half the water under those conditions. These rice plants produced a greater number of larger roots within 24 h of being sprayed with the endophytes and expressed genes involved in stress resistance and drought tolerance.

Practical application of endophytes under real agricultural conditions still causes an imbalance in the natural ecosystem microbiome, but shows a number of advantages over genetically modified (GM) plants. For example, unlike endophyte-treated plants, drought-tolerant GM plants grow poorly when water is abundant. In addition, farmers can decide whether or not to spray their seeds with endophytes at the beginning of the growing season. These observations suggest the possibility of using microbiomes to control fungal pathogens, especially when soilborne diseases are concerned.

BIOLOGICAL, MOLECULAR APPROACH IN PLANT–PATHOGEN INTERACTIONS CONTROL

For any given compatible host–pathogen interaction, each biomolecule (RNA, protein, carbohydrate, lipid, or metabolite) that is essential for pathogen virulence or host defense may be a potential target for disease control. In addition to the whole organism approach, two additional areas of biological control may be considered, which are properly addressed by any combination of transcriptomics, proteomics, **glycomics**, **lipidomics**, and metabolomics, that is, the "-omics" era of pathogen control. The first approach involves airborne chemical signaling, in which volatile compounds most often released by the host are perceived by the target organism, typically located elsewhere; and second, the other operating through bioactive nonvolatile biomolecules or related biomolecules that operate near or at the site of physical contact between pathogen/herbivore and host, including the interior of the cell.

AIRBORNE CHEMICAL SIGNALING IN PLANT–PATHOGEN INTERACTIONS

One of the areas with a huge potential in the future to control plant pathogens involves the complex and intricate airborne chemical communications that take place among organisms. The volatile compounds involved function as chemical signals in a language that evolved and became established over many millions of years, among members of the same species or among organisms of different species within or from distinct kingdoms. These airborne volatile chemical signals are released by one organism, are carried by the wind, and perceived by the target organism that is typically in a different location.

All plants produce and emit substantial amounts of volatile metabolites, known as **phytogenic volatile organic compounds (PVOCs)**, into the atmosphere, which may account for up to 20% of their photosynthetically assimilated CO_2 (Figure 24.3). On a global scale, the carbon emitted in PVOCs exceeds by several fold that emitted from anthropogenic sources. PVOCs are produced by a range of physiological processes in many different plant tissues and are themselves extremely diverse (~30,000 compounds), including alkanes, alkenes, alcohols, aldehydes, ethers, esters, and carboxylic acids. The most abundant are volatile isoprenoids, which play important roles in atmospheric, ecological, and physiological sciences. Chemical communication with other organisms is one of the main roles played by PVOCs. Plants alter the pattern of emitted PVOCs when subjected to biotic or abiotic stresses, the former including predation by herbivores and infection by parasites, most of which are fungi, prokaryotes (bacteria and **mollicutes**), parasitic plants, viruses and viroids, nematodes, and protozoa.

Figure 24.3 highlights the delicate equilibrium established among the plant metabolic processes that consume photosynthetically assimilated carbon, the plant metabolic routes that release carbon dioxide, and the biomass of the corresponding plant, ecosystem, and so on. The net outcome results in gain or loss of biomass by the plant ecosystem.

Knowledge of pathogen-induced PVOC emission by hosts is still scarce. Fungi- and oomycete-induced PVOCs encompassing aerial and root infections have been found in many plant host: pathogen interactions, including the following: peanut: *Sclerotium rolfsii*, silver birch: *Marssonina betulae*, oil palm: *Ganoderma boninense*, willow: *Melampsora epitea*, tomato: *Erysiphe orontii* and *Botrytis cinerea*, potato: *Phytophthora infestans*, mustard: *Alternaria brassicae*, and cucumber: *Pythium aphanidermatum*. In some cases, host-derived volatiles are recognized as signals by plant pathogenic fungi to regulate certain steps of their development. The rust fungus, *Uromyces fable*, an obligate

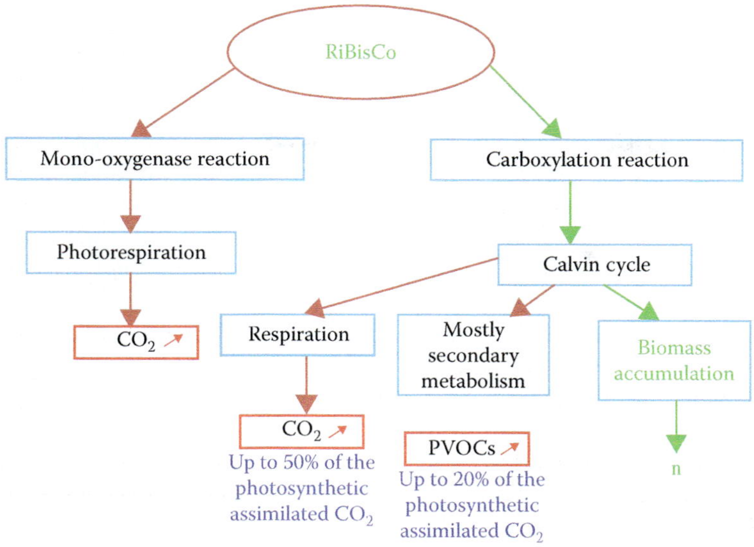

FIGURE 24.3 Diagram highlighting nature's delicate balance of biomass accumulation: catalytic activities of ribulose bisphosphate carboxylase/oxygenase (the enzyme responsible for the photosynthetic carbon assimilation; RuBisCo in short) versus biomass accumulation.

biotrophic pathogen, stimulates volatile emission by its host, broad bean. Three of the PVOCs (nonanal, decanal, and hexenyl acetate) promote development of haustoria.

Herbivorous insects use olfactory information about volatile substances to assess the health/infection status of their potential hosts. The attraction and oviposition response of the grapevine moth, *Lobesia botrana*, to healthy grapes were compared with the response to grapes infected with a phytopathogenic fungus, *B. cinerea*. Fungal infection elicited substantial reductions in both attraction from a distance and oviposition on the host. The active volatiles emanating from the infected grapes, including 3-methyl-1-butanol, were the signal that elicited the observed behavior.

Plants can vary considerably in the volatile blends they emit, in terms of both chemical composition (quality) and quantity of volatile compounds. A key issue involves determining whether the emission of a particular blend of PVOCs is specific for a given pathogen, host, or **pathosystem**. There is evidence for the existence and absence of specificity, depending on the pathosystem under analysis. The absence of specificity in the emission of particular PVOCs is well established. Indeed, emission of the same PVOCs has been induced with different infectious and noninfectious diseases or when different plant species were challenged with a similar infection. For example, most of the PVOCs detected with fungal infection are also detected with ozone treatment of tomato plants, whereas *Tobacco mosaic virus* (TMV) infection in both tobacco and tomato induce an increase in the emission of methyl salicylate. PVOCs that are often detected following the onset of disease, regardless of the disease and independent of the plant species, include (*Z*)-3-hexenol, methyl salicylate, (*E*)-*β*-ocimene, linalool, (*E*)-*β*-farnesene, (*E*)-4,8-dimethylnona-1,3,7-triene (DMNT), and (*E*,*E*)-4,8,12-trimethyl-1,3,7,11-tridecatetraene (TMTT). Apparently, no chemical substance has ever been exclusively associated with one specific disease.

Different diseases attack different plant parts in various ways. As a result, depending on the disease, specific plant structures or tissues will be involved. As a consequence, the chemical substances associated with the particular type of infected plant part might be used to characterize the disease that harms the plant, but not to differentiate between different diseases that attack the same tissue of the plant. For example, upon infection, there is a correlation between certain PVOCs emitted and the plant tissue that emits them. Thus, pathogen-induced damage of host cell membranes results in the local emission of several **lipoxygenase (LOX) products** at the site of damage, whereas damaged glandular trichomes will result in the local emission of stored **terpenes** or other secondary metabolites. These emissions will therefore characterize diseases in which damage of cell membranes or of **glandular trichomes**, respectively, play an important role. In addition, the precise blend of PVOCs per trichome depends on its position on the plant. For instance, in the case of tomato, the amount of β-caryophyllene in stem trichomes is much larger than in leaf trichomes. Although a certain plant species may emit similar PVOCs upon induction by different diseases, and different plant species may emit the same PVOCs after being challenged with a similar disease, it remains to be seen whether the precise composition of a PVOC blend may be specific for a certain plant–pathogen interaction.

Detailed analyses allowed for the detection of qualitative differences among similar PVOC blends, suggesting that it may be possible to discriminate between different plant diseases based on PVOC emission. Although containing the same major compounds (e.g., methyl salicylate, numerous LOX products, DMNT, TMTT, (*E*)-β-ocimene, and α-farnesene), potato plants exposed to different stressors, namely a pathogen (*Phytophthora infestans*) and four types of herbivores (mites [*Tetranychus urticae*], thrips [*Frankliniella occidentalis*], aphids [*Myzus persicae*], and caterpillars [*Spodoptera exigua*]), released PVOC blends whose precise compositions were quite characteristic for the applied organism.

Most research on plant–pathogen interactions has focused on pathogen-induced plant defenses or on the response of insect vectors to infected versus noninfected plants. However, studies that highlight complex multiple interactions between the pathogen and herbivore, the insect (vector, predator) and the host plant are emerging. Transmission of plant pathogens by insect vectors is a complex biological process involving interactions between the plant, insect, and pathogen. Pathogens can induce changes in the traits of their primary hosts (e.g., morphology, primary and secondary metabolites, emitted volatiles, and nutrients) as well as their vectors (e.g., fecundity, survival, and behavior) to affect the frequency and nature of interactions between hosts and vector. Pathogens can also induce the release of chemicals from plants that benefit their survival and spread. For example, infection of apple with "*Candidatus* Phytoplasma mali" (causal agent of apple proliferation) induces release of β-caryophyllene, which attracts apple psyllids (*Cacopsylla picta*) to infected trees and facilitates pathogen spread. Similarly, plants infected with *Cucumber mosaic virus* (CMV), *Potato leaf roll virus* (PLRV), or *Barley yellow dwarf virus* (BYDV) are more attractive to their aphid vectors (*Aphis gossypii*, *Myzus persicae*, *Rhopalosiphum padi*, or *Schizaphis graminum*), than noninfected plants due to induced production of volatiles as a result of infection. Another example of how a pathogen induces host responses that modify the behavior of its insect vector is provided by "*Candidatus* Liberibacter asiaticus," a fastidious, phloem-limited bacterium responsible for causing huanglongbing disease of citrus. The bacterium induces the host to release methyl salicylate, which increases attractiveness of infected plants to its insect vector, *Diaphorina citri*. The duration of initial feeding on infected plants is sufficiently long for the vectors to acquire the pathogen. Subsequently the vector, which initially preferred infected plants, is dispersed to noninfected plants as their preferred location of prolonged settling likely because of suboptimum nutritional content of infected plants. Apparently, the bacterial pathogen manipulates the behavior of its insect vector to promote its own proliferation.

A surprising example illustrates both our lack of knowledge and the importance of airborne communication between organisms. A simple experiment was recently performed, which may settle the long-held dispute about whether baker's yeast (*Saccharomyces cerevisiae*) is naturally present on grapes or not, and about widespread claims associating specific *S. cerevisiae* strains to specific **terroirs**. Strange as it may seem, the life cycle of baker's yeast under natural conditions remains to be elucidated, despite a number of recent reports that tentatively identified habitats in which the yeast apparently overwinters. The yeast has been used by humans over millennia and has economical, technological, ecological, and biological (as a model organism in biology) significance in modern society. Mechanical injuries were induced in mature grapes before harvest. One batch of damaged grapes was wrapped in mesh with a pore size sufficiently small to prevent access of small insects (ranging from fruit flies to bees and wasps), whereas the other batch was left untouched. Interestingly, the wound healed in the covered group, whereas the uncovered group became infected by a wide array of microorganisms. Possibly, the wounded grapes released volatiles (e.g., phenylacetic acid) that were recognized by insects as a signal for food. As they feed, the insects not only maintain the "open" wound but also act as vectors for microbes, including *S. cerevisiae* and saprophytic fungi. This study supports the nonexistence of terroir-specific strains of *S. cerevisiae*, and the view that *S. cerevisiae* is not normally encountered on healthy grapes. Accordingly, collecting one bunch of grapes, or a group of bunches carrying a single damaged grape provides evidence for the presence of *S. cerevisiae*. The claim that *S. cerevisiae* is detected in healthy grapes may therefore derive from improper sampling methods.

Sensing the presence of diseases in plants is a challenging task, mostly derived from their inability to communicate with us. Much remains to be uncovered for a number of reasons including the following:

1. Techniques available to detect volatiles are inherently difficult due to the gaseous nature of the volatiles.
2. Most volatiles act at extremely low concentrations, with some known to function at the level of single molecules.
3. Many signaling compounds are released by hosts and/or pathogens in a situation that results from an incompatible or unsuccessful compatible interaction, that is, disease not established.

4. Extreme sensitivity and specificity of the corresponding receptors make their detection difficult, especially under natural environments where myriad volatiles from a large number of organisms, undergoing all sorts of chemical interactions, occur simultaneously.

Such experiments are most often conducted under confined environments and highly controlled conditions. In addition to the "-omics," this interdisciplinary approach requires a deep understanding of plant pathology, plant physiology, and sensor technology. Nevertheless, a large number of experiments have been reported on airborne biological control/chemical signaling.

The possibility of recognizing the initial phases of a compatible interaction between host and herbivore/pathogen by the specific mixture of volatiles released by the host remains a major challenge. Dynamic sampling coupled with gas chromatography, followed by an appropriate detector, is considered the most appropriate method for application under conditions of open-air agriculture. However, practical application of such instrumentation in agriculture remains a challenge, mainly due to high costs.

As referred to previously, when plants are subjected to herbivory or attack by parasites, they typically release a blend of volatile chemicals, most often terpenes. The capacity for early detection and identification of those compounds may provide a way to control the agent at the initial stages of attack/infection, possibly avoiding at the same time the preventive widespread application of pesticides currently associated with intensive farming. This approach is certainly more accurate, covering a wider scope of applications than old procedures that are still in use. For example, rose plants are planted at selected edges in vineyards as an indicator of powdery mildew infection, based on the higher susceptibility of the former to *Sphaerotheca pannosa* than of the latter to *Erysiphe necator*. However, this approach suffers from a series of drawbacks among which is that the released volatiles are present in minute concentrations, making their detection, identification, and origin difficult to determine.

A major goal of disease control measures is recognition of the pathogen/herbivore through the detection of specific chemical substances, the characterization (identification and quantification) of the specific blend of emitted PVOCs upon the onset of disease, or the specific time course of the disease-induced PVOC emission. The situation is complicated under the real conditions that prevail in the wild or under open-air agriculture, which are far more complex than those reported in studies involving relatively simple interactions. Simultaneously, plants face attempts of pathogen infections and herbivore predation, perceive chemical signals emanated from surrounding plants or pathogen/insect/host systems, and/or are exposed to abiotic stresses, which are also known to alter PVOC emissions.

NONVOLATILE MOLECULAR INTERACTIONS IN PLANT–PATHOGEN RELATIONS

The door is definitely open for biological control, based on the complex molecular interactions that take place between host and pathogen during attempted pathogenesis and disease. Two most critical episodes occur at the molecular level and precede the establishment of pathogenesis: perception of the pathogen by the host and the molecular war that ensues (Figure 24.1). Each episode comprises an intricate molecular chain of events, where each step may be regarded as a potential source of targets, which may subsequently be developed to control the disease. Screening for potential targets in either host or pathogen is most appropriately addressed by any

TABLE 24.1
Major Levels and Selected Sublevels of "-Omics" Relevant to Plant–Pathogen Interactions

Basic Unit	Organizational Unit	Omics
Gene	Genome	Genomics
Transcript	Transcriptome	Transcriptomics
Protein	Proteome	Proteomics
Protein–protein interaction	Interactome	Interactomics
Lectin	Lectinome	Lectinomics
Xylem	Xylome	Xylomics
Phloem	Phloome	Phloomics
Carbohydrates	Glycome	Glycomics
Oligosaccharide side chains projected outwards from a cell membrane	Exoglycome	Exoglycomics
Metabolite	Metabolome	Metabolomics

Note: Any active interaction among cells or between a molecule and a cell will result in an "-omics" alteration. Major levels are indicated in dark blue. Selected sublevels are in light blue.

combination of transcriptomics, proteomics, glycomics, and metabolomics (Table 24.1), taking into consideration that any biological response is based on altered gene expression and carried out as a result of protein synthesis. A schematic diagram of the major factors controlling protein synthesis is presented in Figure 24.4.

The Age of "-Omes"

The **genome** is the complete DNA sequence of an organism, that is, the full complement of genetic information in the organism, including both coding and noncoding regions, with genomics encompassing the study of the genes and their functions, as well as their interactions within a genome. First cited in PubMed in 1932, the word was coined in analogy with chromosome (from the Greek *chroma* = color and *soma* = body, due to its property of being very strongly stained by certain dyes). It is a static concept because at least in what concerns the nuclear genome, it is common to all eukaryotic cells in a multicellular organism and is not altered by environmental or physiological conditions. The huge success of large-scale quantitative biological processes, such as those relating to genome sequencing, explains the subsequent widespread use of the suffix "-ome" in many areas of the biological sciences. All "-omes" included in Table 24.1, other than the genome, correspond to dynamic concepts since they inherently vary from tissue to tissue and with the environmental, physiological, and pathological conditions.

Altered gene expression in both plant and parasite is at the very heart of every host–pathogen interaction. Efficiency has been improved one step further in some prokaryotes, where transcription and translation occur simultaneously in the same macromolecular complex in the cytoplasm.

Evidence points to a potentially important role of alternative splicing in plant–pathogen interactions. *Pseudoperonospora cubensis*, an obligate pathogen, is the causal agent of cucurbit downy mildew, a foliar disease of global economic importance. This oomycete has a series of RXLR (Arginine-X-Leucine-Arginine, where X is any amino acid) and RXLR-like effector proteins, which likely function as virulence or avirulence determinants during the course of host infection. One such effector protein, RXLR protein 1 (PscRXLR1), arises as a product of alternative splicing, making this the first example of an alternative splicing event in plant pathogenic oomycetes transforming a noneffector gene to a functional effector protein. Another example is that the stress response genes are overrepresented among transcripts that show intron retention. Thus, alternative processing in *Phytophthora sojae* family 5 endoglucanases revealed the generation of both coding and noncoding RNA isoforms. On the other hand, several mechanisms of alternative splicing have been reported for plant resistance (*R*) genes belonging to the toll/interleukin-1 receptor (TIR)-nucleotide-binding site (NBS)-LRR class.

Several other "-omes" deserve a dedicated and special mention when host–pathogen interactions are considered (Table 24.1). **Interactome** is defined as the whole set of molecular interactions in cells. However, in a restricted sense, the term is frequently used to refer to protein–protein interactions (PPIs), although

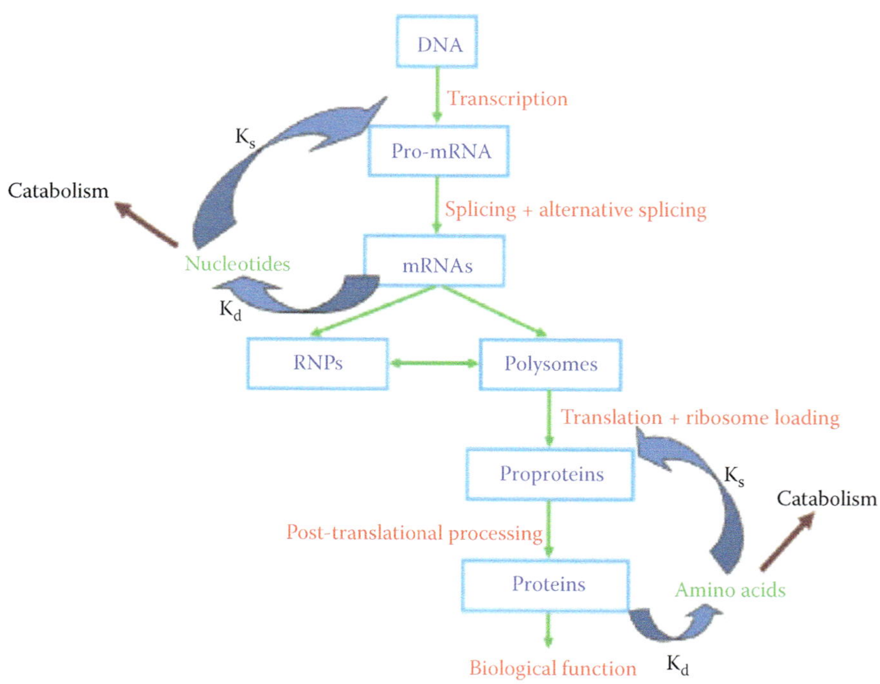

FIGURE 24.4 Schematic diagram of the major factors controlling protein synthesis.

proteins often establish important interactions with other molecules (e.g., protein–DNA, protein–RNA, and protein–ligand interactions). The first plant interactome (*Arabidopsis thaliana*) was published in 2011, and 6200 PPIs were identified among 2700 proteins, 40% of which involved plant-specific proteins (Centre for Cancer Systems Biology; http://ccsb.dfci.harvard.edu/web/www/ccsb/). Surprisingly, many of these interactions involve a small set of highly connected proteins referred to as hubs, which are an appealing target for plant pathogens. Subsequently, the interactions between *A. thaliana* proteins and virulence proteins produced by two very different pathogens were mapped. Pathogen effector proteins that manipulate the plant's immune system tend to interact with highly connected plant proteins, and effectors from different pathogens target the same hub proteins. Confirmation that these highly connected proteins are involved in immune defense was achieved by exposing mutant plants with impaired protein-coding genes to pathogens. It seems that the two pathogens under analysis independently evolved different modes of attack that converge on the highly connected proteins, an observation that suggests that the pathogens may have evolved to target proteins that participate in more than one immune response or pathogen recognition pathway.

The basic notions that xylem and phloem are the circulatory systems within plants, with the xylem sap containing water and mineral salts, and the phloem sap translocating specific **photoassimilates**, such as sucrose and asparagine, remained unchallenged for many years. It is now known that xylem contains RNAs and proteins in addition to transporting water, nutrients, and metabolites and is also involved in long-distance signaling in response to pathogens, symbionts, and environmental stresses. Phloem serves as a major trafficking pathway for assimilates (including a wide range of compounds like water, minerals, amino acids, organic acids, sugars, and sugar alcohols), viruses, RNA, plant hormones, metabolites, and proteins with functions ranging from synthesis and metabolism to signaling. The occurrence of proteins, mRNAs, and other macromolecules within the plant vasculature are surprising since these elements lack the capability for mRNA and protein synthesis. However, signal sequences were subsequently identified that direct and deliver the "**xylome**" and "**phloome**" proteins to their respective sites. These macromolecules not only occur sporadically, but a large number of RNAs and soluble proteins are present on a permanent basis. Proteins can even be regarded as a major component of phloem sap, given that, for example, cucurbit exudates contain high concentrations of up to 100 mg protein/mL.

The concepts of xylome and phloome may therefore be used to specify those subsets of the plant proteome that are found within the two long-distance transport systems of plants. Both xylem and phloem play a very important role in phytopathology, especially in signaling (e.g., salicylic acid, **jasmonic acid**, and ethylene) and translocation of defensive biomolecules (e.g., lectins, chitinases, and proteases). PR proteins, peroxidases, chitinases, and proteases are abundantly present in both the xylem and phloem saps of several plant species. The composition of the xylome, which is relatively constant from plant to plant, can be altered considerably by pathogen infection. In addition, these proteins can be detected in xylem sap at some distance from the infection site. For example, a stress-induced protein (FISP 17) showing homology to PR proteins was identified in the stem sap of soybean seedlings infected with *Fusarium virguliforme* (formerly known as *F. solani* f. sp. *glycines*), the causal agent of sudden death syndrome of soybean.

When a plant–pathogen interaction is addressed at the molecular level, proteins are usually first considered, since they are responsible for the vast majority of tasks executed by cells in living organisms. Thus, during a plant–pathogen interaction, perception of the parasite and the war for pathogenesis that subsequently develops are carried by the proteins themselves and/or the products that result from their action. The most common molecular approach to first address a plant–pathogen interaction is to analyze (with two-dimensional [2D] electrophoresis [isoelectric focusing followed by denaturing, reducing polyacrylamide gel electrophoresis, IEF/SDS-PAGE]) the evolution of the plant and/or pathogen proteomes during attempted pathogenesis, followed by polypeptide spot identification, typically performed by mass spectrometry. However, there is a big constraint in this approach. Figure 24.5 illustrates the mathematical relationship between **translatome** and proteome in a typical mammalian cell—we may reasonably assume a similar relationship for a plant or fungal cell. Thus, a typical eukaryotic cell may contain approximately 8000 different proteins, of which ~10% accounts for 90% of the cell total protein and ~90% of which correspond to approximately 3% of the cell protein (Figure 24.5), meaning that most cellular proteins (and certainly those involved in metabolic regulation) occur in minute amounts. Assuming that a typical 2D gel allows identification of 800 spots, 90% of the proteins present in the pathosystem will pass unnoticed. Of course, there are ways to minimize this effect, but the vast majority of the cell proteins will remain undetectable. One of the approaches utilized is labeling (radioactive isotopes, fluorescent probes, etc., which are far more sensitive to detect in electrophoresed polypeptides than conventional protein-staining methods), the proteins that are being synthesized during the plant–pathogen interaction. Alternatively, it is possible to employ two or more different fluorescent probes (that do not affect in a significant way of protein mobility during electrophoresis) and use each one to label the proteins present in any of the protein extracts under analysis. Protein extract from

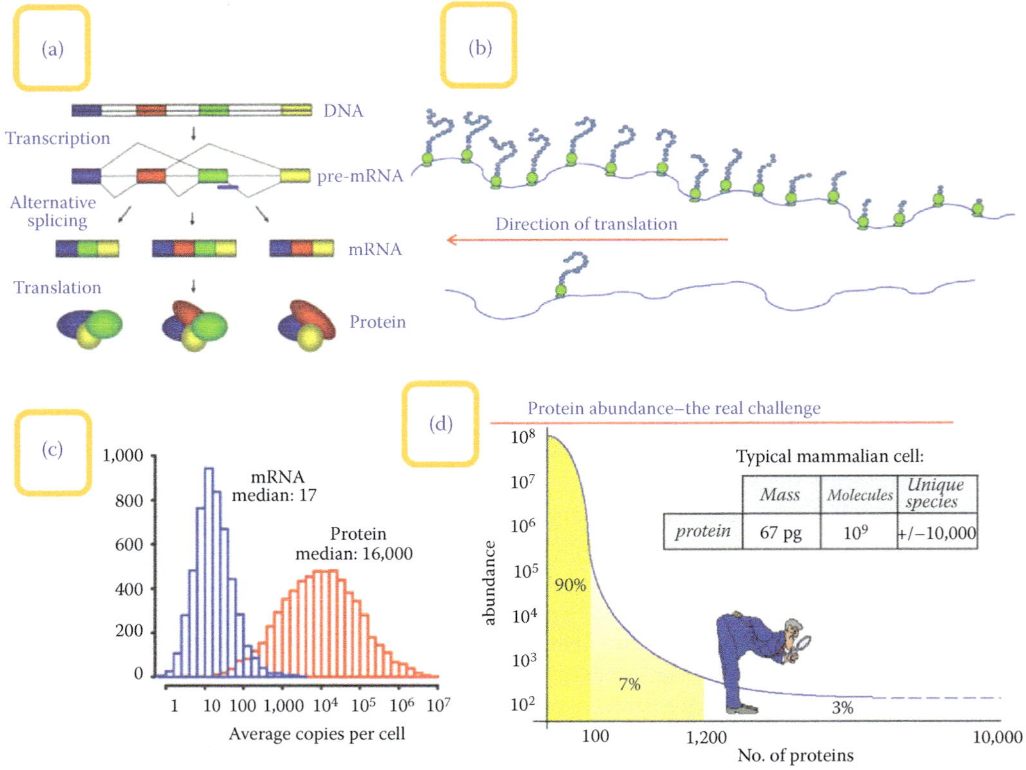

FIGURE 24.5 Protein synthesis and the proteome. (a) Schematic illustration of alternative splicing. Alternative splicing of pre-mRNA drives the generation of multiple protein isoforms through the assembly of different combinations of splice sites within a single gene. (b) Ribosome loading, illustrating a heavy loaded polysome (top) and a monosome (bottom). (c) Distribution of distinct RNA (blue) and protein (red) molecules plotted as a function of the number of identical copies. (d) The mathematical relationship between translatome and proteome in a typical mammalian cell.

a healthy plant leaf and protein extract prepared from a plant leaf challenged with a pathogenic fungus is an example. The two samples are mixed and fractionated by 2D electrophoresis. Analysis of the proportion of two fluorescent probes present in each polypeptide spot will reveal whether that polypeptide has been up-regulated, down-regulated, or not affected by the pathogen challenge. Such results require confirmation by RT^2-PCR (real-time, reverse transcriptase-polymerase chain reaction). Yet other approaches may involve organelle isolation (e.g., mitochondria or chloroplasts) prior to 2D electrophoresis when our interests are centered on an organelle protein.

Nucleic acids are a useful alternative to the aforementioned constraints for the following reasons: unlike proteins, they are not easily denatured in an irreversible manner; and using an appropriate sets of primers, they may be amplified exponentially by RT-PCR, meaning that in theory it is possible to start with a single DNA fragment or mRNA molecule and end up with millions of identical copies. The only way to modestly increase the number of molecules of a given protein is via the mechanism of protein synthesis, either using *in vitro* translation systems or *in vivo* recombinant expression, both of which require the mRNA encoding the required protein. Therefore, transcriptomics is the method of choice to study any process leading to metabolic changes because this method can pinpoint those RNAs that are present in minute amounts and may undergo substantial up- or down-regulation during the course of a plant–pathogen interaction. Nevertheless, transcriptomic methodologies such as microarray technologies, subtractive suppressive hybridization, and so on are not free from pitfalls. Thus, some mRNAs are not translated into protein, and the level of a particular mRNA does not always match with the concentration of the corresponding protein (Figure 24.5). In addition, transcriptomic approaches require confirmation by RT^2-PC and do not avoid the need to study the encoded protein, which becomes absolutely essential at subsequent stages. However, these methods provide a way to obtain the protein by recombinant technology, eventually followed by the production of either polyclonal- or monoclonal-specific antibodies.

Identification of a potential target opens the way to any of a number of alternatives including recombinant approaches. However, due to their myriad implications, GM plants are outside the scope of this chapter. Nevertheless, three of its applications deserve a brief mention: (i) immunomodulation (see later); (ii) transient expression using viral vectors; and (iii) RNA interference (see later).

IMMUNOMODULATION

Immunomodulation can be defined as a molecular technique that allows interference with cellular metabolism, signal transduction, or pathogen infectivity by the ectopic expression of genes encoding recombinant antibodies or antibody fragments, thus modulating the function of a corresponding antigen. Antibodies or antibody fragments produced in plants by the expression of corresponding encoding genes or gene fragments are often referred to as "**plantibodies**." Therefore, immunomodulation can be used to "immunize" the plant against pathogen infection if the plantibody recognizes, binds, and inactivates a pathogen virulence product, such as an enzyme, toxin, or other pathogen factor involved in disease development (Figure 24.6). Immunomodulation in subcellular environments, such as the cytosol and nucleus, should turn immunomodulation into a powerful and attractive tool for gene inactivation, complementary to the classical antisense and cosuppression approaches. Using appropriate signal sequences, plantibodies can be directed to specific sites including the xylem.

Countless studies have been reported on the transcriptomics, proteomics, and metabolomics of host–pathogen interactions. The vast majority of proteomic studies, including those performed on PR proteins and antifungal proteins, deal with the 10% or so "visible" proteins (Figure 24.5) of either host or pathogen, whereas detailed transcriptomic studies identify large numbers of up- and down-regulated proteins making it difficult, at the present state of knowledge, to draw overall conclusions and to select one or a group of proteins as targets for disease control. As a result, the focus is changing to emerging areas such as **RNA interference** and the **exoglycome**. Many metabolomic studies have also been reported in phytopathology with those addressing phytoalexins and elicitors assuming a central position.

RNA INTERFERENCE (RNAI)

Transcript technology involving RNA transcripts to control pathogenesis is rapidly assuming an increasing importance as a tool in phytopathology. RNA interference is a highly evolutionarily conserved, endogenous pathway for negative post-transcriptional regulation that controls

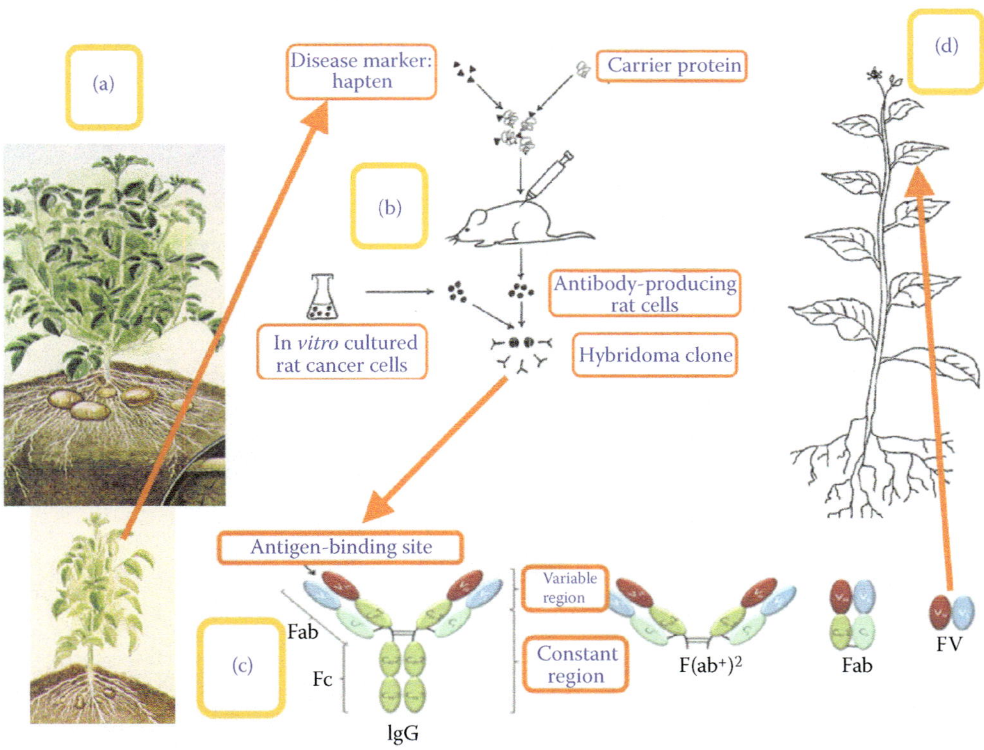

FIGURE 24.6 Immunomodulation. (a) Healthy and pathogen-infected plant. (b) Production of monoclonal antibodies specific for a pathogen protein or hapten. (c) Identification of the nucleotide sequences corresponding to the hypervariable region of the monoclonal antibody. (d) Recombinant production of the selected monoclonal antibody fragment within the plant tissues (plantibody), capable of recognizing and binding to the pathogen protein or hapten. *Note*: Hybridoma is a hybrid cell resulting from the fusion of a lymphocyte and a tumor cell and used to culture specific monoclonal antibodies.

the expression of protein-coding genes. Discovered in the nematode, *Caenorhabditis elegans*, RNAi was later found in a wide range of other eukaryotic organisms, including plants, animals (from nematodes to mammals), and fungi (e.g., *Neurospora crassa*, *Cladosporium fulvum*, *Cryptococcus neoformans*, *Magnaporthe oryzae*, and *Mucor circinelloides*). It is a natural mechanism for sequence-specific gene silencing, based on the observation that small noncoding RNAs play a central role in RNA silencing by a mechanism of cosuppression. RNAi refers collectively to diverse RNA-based processes that result in sequence-specific inhibition of gene expression at the transcription, and mRNA stability or translational levels (Figure 24.7). It has most likely evolved as a mechanism for cells to eliminate foreign genes since it is triggered by double-stranded RNA (dsRNA) derived from endogenous parasitic or exogenous pathogenic sources of nucleic acids. The unifying features of RNAi are the production of small RNAs (21–24 nucleotides long) that act as specific determinants for down-regulating gene expression and the requirement for one or more members of the **Argonaute protein family**.

The possibility of synthesizing an siRNA outside cells and then introducing it into them has emerged as a method of choice for gene targeting in fungi, viruses, bacteria, and plants. RNAi has been used to manipulate gene expression in plants, animals, and fungi. For example, transformation constructs that produce RNAs with the ability to form duplexes, such as hairpins, are highly effective in inducing gene silencing. One of the biggest

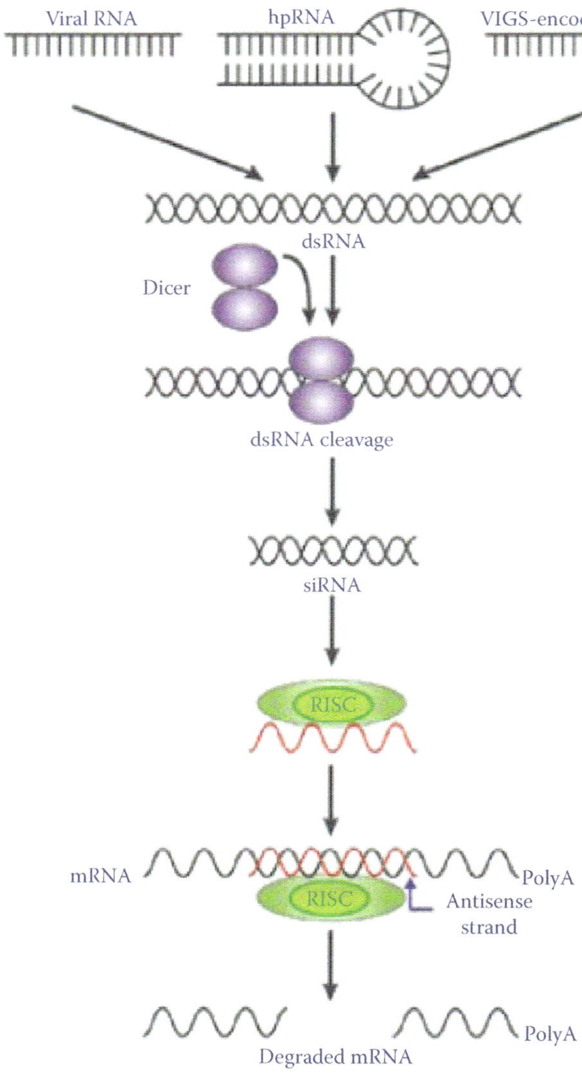

FIGURE 24.7 Double-stranded RNA (dsRNA) from replicating viral RNA, viral-vector-derived (VIGS or virus-induced gene silencing) RNA or hairpin RNA (hpRNA) transcribed from a transgene, is processed by a Dicer-containing complex to generate siRNAs. An endonuclease-containing complex (called the RNAi silencing complex, RISC), is guided by the antisense strand of the siRNA to cleave specific mRNAs, thus promoting their degradation. (Adapted from Waterhouse, P. M, and Helliwell C. A., *Nat. Rev. Genet.*, 4, 29–38, 2003.)

challenges in RNAi research is the delivery of the active molecules that will trigger the RNAi pathway in plants. Several methods exist for delivery of dsRNA or siRNA into different cells and tissues.

Simultaneous silencing of several unrelated genes by introducing a single chimeric construct has been demonstrated for several fungi, for example, *Venturia inaequalis*, the causal agent of apple scab, the most economically important disease of apples worldwide. The first effort toward the systematic silencing of *Magnaporthe grisea*, a causal organism of rice blast, used the enhanced green florescent protein gene as a model. A protocol was developed later for silencing *mpg*1 and polyketide synthase-like genes. *Mpg*1 gene is a **hydrophobin gene** that is essential for pathogenicity as it acts as a cellular relay for adhesion and triggers development of the **appressorium**. The work on this host–pathogen system revealed silencing of the above genes at varying degrees in 70%–90% of the resulting transformants. Ten to fifteen percent of the silenced transformants exhibited almost a "null phenotype." Multiple gene silencing has been achieved in *Cryptococcus neoformans* using chimeric hairpin constructs and in plants using partial sense constructs.

RNAi has great potential as a gene-silencing technology, promising to revolutionize experimental biology. It may have important practical applications in functional genomics, therapeutic intervention, agriculture, and other areas. It has possibilities for creating custom "knockdowns" of gene activity, allowing silencing of genes of interest. This is especially useful for genes for which the success of silencing is difficult to establish *in vitro*, for example, genes with a low expression level, or an intricate developmental regulation, or in the case of a pathogen, those that are exclusively expressed during infection. Gene silencing may occur throughout an organism or in specific tissues, works in both cultured cells and whole organisms, and can selectively silence genes at particular stages of the organism's life cycle. Therefore, this technique has great potential in agriculture specifically for nutritional improvement of plants and the management of plant diseases.

THE FUNGAL EXOGLYCOME

Glycomics—The Sugar Code

There are three major classes of biological macromolecules, all of which encode information essential for the life of the organism: nucleic acids, proteins, and carbohydrates. The original concept of the Central Dogma of Molecular Biology, first articulated by Francis Crick in 1958 and restated in a *Nature* paper published in 1970, ignored posttranslational modifications (such as protein glycosylation) that greatly magnify the functions of a single protein encoded by a particular gene. The apparatus required for glycan assembly is unique. The biosynthesis of nucleic acids and proteins is carried out by relatively simple template mechanisms, whereas glycans require a complicated nontemplate assembly line, where many different workers and machines (membranous organelles and vesicles, enzymes, transporters, and structural proteins) function in harmony to manufacture the final product from a complex collection of primary building materials.

Nucleic acids and proteins are linear molecules in which the individual subunits (nucleotides and amino acids, respectively) are linked by identical phosphate and peptide bonds, respectively (Figure 24.8). Compared with nucleotides and amino acids, monosaccharide versatility for isomer formation (code words) is unsurpassed (Figure 24.9) for a number of reasons. A considerable number of candidate monosaccharides and monosaccharide derivatives exist in both the furanose or pyranose forms. In addition, the linkages between individual subunits are not identical; the anomeric carbon atom of one monosaccharide may be linked via an α or β bond to any other carbon atom containing an hydroxyl group on the adjoining monosaccharide. The resulting

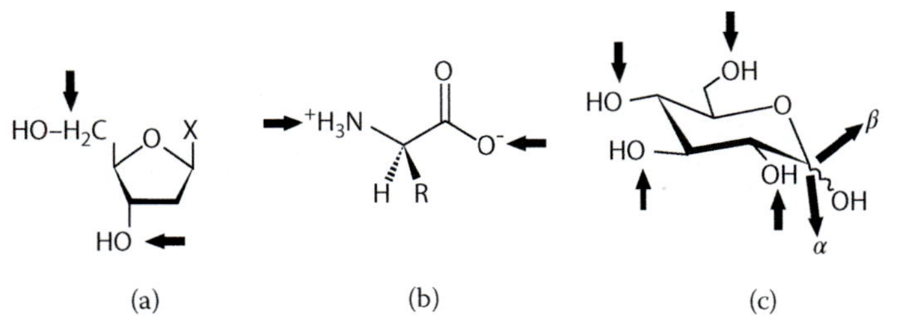

FIGURE 24.8 Illustration of the linkage points (arrows) for oligomer formation in biomolecules. In comparison with the phosphodiester and peptide bonds, strictly constrained in their capacity for oligomer formation, a large panel of glycosidic linkages can be formed. The phosphodiester bond in nucleic acid biosynthesis (a) and the peptide bond in protein biosynthesis (b) yield linear oligomers. In contrast, the glycosidic linkage in oligosaccharides (c) can involve any hydroxyl group on the second monomer, opening the way to both linear and branched structures. This unique property is symbolized by arrows directed toward the hydroxyl groups. The anomeric configuration (α or β) in chain elongation can vary, as symbolized by two bold arrows pointing away from the molecule. (Extracted from Ferreira, R. et al., *Eur. J. Plant Pathol.*, 133, 117–140, 2012; Originally from Gabius H.-J., *The Sugar Code: Fundamentals of Glycosciences*, Wiley-VCH Verlag GmbH & Co. KGaA, Weinheim, 2009.)

- The sequence of monosaccharides, originated by the variety of different monomer building blocks
- The linear or branched oligo- or polymer structures and, in the latter case, the branching patterns
- The different molecular forms that monosaccharides can adopt, such as ring opened or closed, different ring sizes (furanose or pyranose), and conformations
- The isomeric diversity, including the two possible stereochemical linkages between units, i.e., the α or β configuration assumed by the anomeric carbon
- The linkage points: the anomeric carbon atom of one monosaccharide may be linked to any carbon atom containing a hydroxyl group on the adjoining monosaccharide
- The diversity of secondary modifications of monomers (e.g., methylation, sulfation, acetylation, and phosphorylation)
- The different modes of attachment for cell-surface oligosaccharides (including glycolipids and *N*- or *O*-linked glycoproteins)
- Their indirect relationship to the genome
- The range of molecular contexts in which the modifications are found
- The fact that most carbohydrates lack chromophores or fluorophores, a property that makes their detection difficult

FIGURE 24.9 Characteristics of oligosaccharides that contribute to structural complexity. *Note*: The structural complexity and diversity of carbohydrates underlies the large coding capacity of the sugar code, but is also responsible for difficulties of analysis and/or synthesis. (Adapted from Ferreira, R. et al., *Eur. J. Plant Pathol*., 133, 117–140, 2012.)

molecule may be linear or branched, may be covalently attached to a protein or lipid, or may undergo a wide variety of other chemical modifications. Carbohydrates and oligosaccharides in particular are therefore highly efficient vehicles for information storage. This remarkable coding capacity of carbohydrates has been referred to as the sugar code.

Following nucleic acids and proteins, carbohydrates, or more specifically oligosaccharides, have recently been recognized as the third code/alphabet of life, with a coding capacity, which far exceeds those of the other two polymers (Figure 24.10); this is clearly illustrated by the following example in actual numbers. Considering a degree of polymerization of DP = 6, only 4096 (4^6) hexanucleotides are possible with the four letters in the DNA language, 6.4×10^7 (20^6) hexapeptides from the 20 classical protein amino acids, but a staggering number of 1.44×10^{15} hexasaccharides from 20 monosaccharides. As with nucleic acids and proteins, only a tiny fraction of all possible carbohydrate structures occur naturally.

With such a wide diversity of structures and the associated coding capacity, it comes as no surprise the diversity of functions that have been attributed to carbohydrates. Carbohydrates are much more than simple biochemical fuels or (as polymers) the molecular concrete to convey stability to plants or insects, with the potential of sugars stretching far beyond energy metabolism and cell wall stability. An obvious sign for a wide range of physiological roles is the observation that oligosaccharide chains are frequently presented by proteins and lipids. Their significance is to impart a discrete recognition role on the carrier, the essence of the sugar code concept. Protein–carbohydrate and carbohydrate–carbohydrate interactions control salient aspects of intra- and intercellular communication and trafficking and are at the basis of a variety of essential biological phenomena. They play an essential and direct role in cell–cell recognition, in the distinction between self from nonself, and in the warfare that takes place between host and pathogen before infection is established. They are also involved in adhesion of infectious agents to host cells, and cell adhesion in the immune system, malignancy, and metastasis.

Glycosylation represents one of the most frequently occurring post-translational modifications of proteins. About 0.5%–1% of human genes encode proteins related to biosynthesis, function, and degradation of sugar chains. When compared with genomics and proteomics, glycosciences have received far less attention than they deserve. An obvious explanation why research in glycosciences (structural and functional glycomics and **lectinomics**) has lagged behind the fields of genomics and proteomics is that oligo- and polysaccharides are much more complex and difficult to study than nucleic acids and proteins, which results primarily from the complexity of their structures (Figure 24.9).

The Fungal Cell Membrane and the Fungal Exoglycome

All cells, from bacteria to mammals, as well as viruses, possess numerous oligosaccharide side chains on the outer surface of their cell membrane that are projected into the extracellular milieu (Figure 24.11). This dense layer of oligosaccharides that coats cells explains the importance of glycobiology in multicellular life and serves as identification molecules to the surrounding world. They mark cells and tissues as "self," allowing discrimination from "non-self," and send signals to the immune system when tissues are injured. Analogous to the term glycome, which

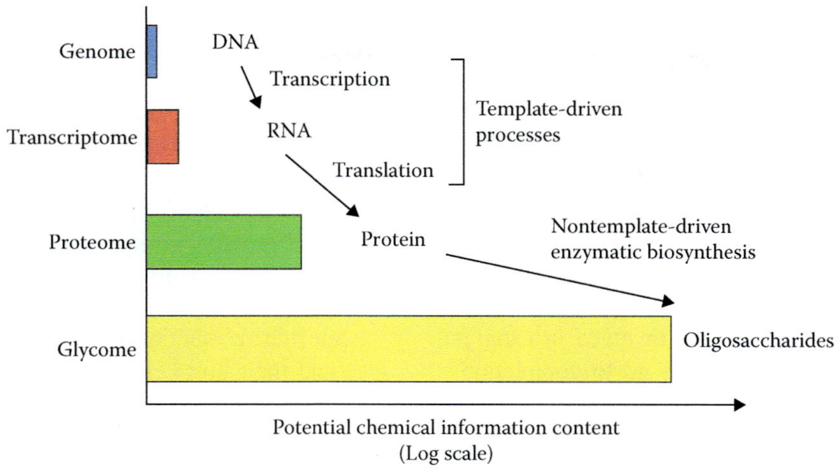

FIGURE 24.10 Glycome enhancement of the molecular and functional diversity of the proteome. Protein expression is based on a genetically encoded template, but post-translational modifications of proteins dramatically enhance their functional diversity. The glycome represents the main class of post-translational modifications, providing biological access to a vast information space at minimum genetic cost. Note that in the information flow from the genome to the glycome, the biosynthesis of oligosaccharides is not encoded via a template-driven system. Note also the log scale of potential chemical information. (Adapted from Turnbull, J. E., and Field R. A., *Nat. Chem. Biol.*, 3, 74–77, 2007.)

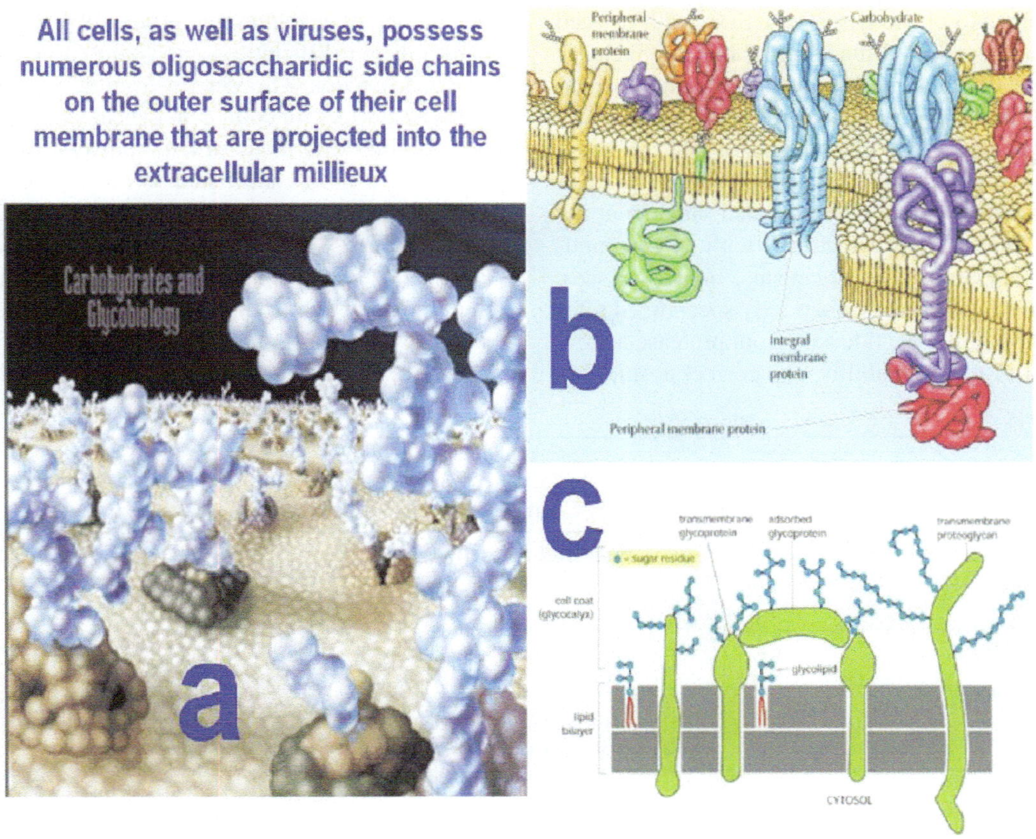

FIGURE 24.11 Three illustrations showing the typical oligosaccharide side chains, which compose the exoglycome on the external surface of cell membranes. (Extracted from Ferreira, R. et al., *Eur. J. Plant Pathol.* 133, 117–140, 2012; Originals from (a) front cover of *Science*, 291, 2001; (b) Cooper, G. M., *The Cell: A Molecular Approach*, Sinauer Associates, Sunderland, 2000; and (c) Alberts, B. et al., *Molecular Biology of the Cell*, Garland Science, New York, 1994.)

is the carbohydrate analog to proteome and genome, the word exoglycome was proposed to collectively represent the oligosaccharide side chains presented at the outer surface of the cell membrane, thus excluding the cell wall carbohydrates, which tend to have a passive role in mediating cell communication with the outside world. The exoglycome may therefore be defined as the collection of carbohydrate moieties present in *N*-linked glycoproteins, *O*-linked glycoproteins, and glycolipids that protrude outwards from the cell membrane.

Because many of the reactions involved in exoglycome carbohydrate synthesis do not go to completion, oligosaccharides modifying an individual site are typically expressed as a mixture of **glycoforms** elaborating a core structure. This diversity multiplies the structural and functional complexity of proteins and lipids by orders of magnitude (Figure 24.10). The mature structures are cell-type specific and depend on multiple factors, including enzyme and nucleotide sugar availability and kinetics of **glycoconjugate** transport. As a result, glycoconjugate oligosaccharides are mixtures of variants (glycoforms) on a core structure. This serves to diversify the biological functions of the limited number of genes in higher animal genomes to create tissue structures with defined boundaries and controlled functions. Table 24.2 includes a proposed list of functions displayed by the oligosaccharides present in the outer surface of cells in the form of *N*- or *O*-glycoproteins and glycolipids.

It is now well established that the importance of the exoglycome cannot be overestimated, namely at the level of the interactions of each cell with its surrounding milieu, including other organisms, organs, tissues, cells, and molecules. Apart from poorly specialized roles (e.g., protection of the peptide axis from protease action, improvement of water solubility, and correct positioning of glycoprotein), highly specialized roles are attributed to the exoglycome of host and pathogen cells due to the extremely high coding capacity of the sugar code. These specialized roles include regulation of cell–cell interactions (discrimination between self and nonself), regulation of cell–matrix interactions, regulation of cell–molecule interactions (oligosaccharide–lectin interactions), quality control of proteins critical to the development and function of complex multicellular organisms, and receptors for viruses, bacteria, fungi, and other organisms.

Specific recognition processes between fungal parasites and their host cell targets may be mediated by the interaction of carbohydrate-binding proteins (see later) on the surface of one type of cell that combine with complementary sugars on the surface of another. The selectivity of such proteins for microbial surface sugars has been demonstrated for the genus *Candida*, where the adherence of *C. albicans* and *C. parapsilosis* to epithelial cells correlates with fungal cell surface carbohydrates.

Difficulties in Studying the Exoglycome

The emerging -omics domain of glycomics has lagged behind that of genomics and proteomics, mainly because of the inherent difficulties in analyzing carbohydrate structure and function. This derives from the huge capacity of oligosaccharides for coding biological information, which is due to their enormous structural diversity. Several lines of evidence have contributed to the lack of knowledge currently existing in what the exoglycomes of most cells are concerned—the absence of appropriate techniques to analyze them. It was not until recently that the first tools required for such complex analyses became available and are still being developed. Their importance is highlighted by the large number of very recent review articles

TABLE 24.2
Proposed Functions for the Oligosaccharides that Compose the Exoglycome

Nonspecific roles
Protection of the peptide axis from protease action
Improvement of water solubility
Correct positioning of the glycoprotein
Highly specialized roles
Regulation of cell–cell interactions (discrimination between self and nonself)
Regulation of cell–matrix interactions
Regulation of cell–molecule interactions (oligosaccharide–lectin interactions)
Quality control of proteins critical to the development and function of complex multicellular organisms
Receptors for viruses, bacteria, and possibly fungi

Source: Ferreira, R. et al., *Eur. J. Plant Pathol.*, 133, 117–140, 2012.

published on this subject, with new automated methods promising a dramatic change and many compounds based on complex carbohydrates now in clinical trials for many diseases. In fact, recombinant glycoprotein drug products represent a multibillion dollar market. Effective analytical methods are needed to speed up the identification of new targets and the development of industrial glycoprotein products, both new and biosimilar.

The Changing Exoglycome

The exoglycome oligosaccharides are expressed in a cell-type-specific and temporally regulated manner to allow cell phenotypes to change dynamically in response to environmental stimuli. For example, oligosaccharide structures on cell surfaces vary according to spatial location in tissue and temporal factors. Thus, cells respond to environmental stimuli by altering the carbohydrate structures on their surfaces. They are synthesized in a nontemplate controlled manner, and mature structures arise by the coordinated expression of numerous genes that code for glycosyltransferases, glycosidases, and other enzymes that synthesize and remodel oligosaccharide chains, as well as accessory enzymes involved in the synthesis and transport of nucleotide sugars. Their structures are regulated dynamically. Indeed, the basic principle of glycosylation, that is, the lack of template, makes the exoglycome inherently sensitive to all changes within the cell (and possibly outside as well), and the oligosaccharide structures that are being produced at any moment actually mirror all relevant past events in the cell.

A changing exoglycome is also possibly used by pathogenic fungi, which may alter their surface molecular architecture to elude, circumvent, or avoid host defense mechanisms. Studies at the level of the cell membrane are an area of research that has been neglected in plant fungal pathogens. It is therefore surprising that cell membrane proteomes such as the corresponding exoglycomes of the human and plant pathogens *Alternaria alternata* and *Aspergillus fumigatus* have not been studied and characterized. The putative alteration of such "pathogen -omes" in response to the presence of the host may well contribute to identify polypeptide or oligosaccharide targets as disease markers that may eventually lead to ways to control the disease, for example, to save human lives and/or improve plant productivity.

LECTINOMICS AND THE OLIGOSACCHARIDE RECEPTORS

To encode the complex, biomolecular recognition ability established between carbohydrates and proteins, nature has at its disposal a powerful information tool, the sugar code. The role of reading the sugar-encoded messages is mainly played by a specific class of carbohydrate-binding proteins or receptors. Indeed, the role attributed to oligosaccharides as the third code of life implies the existence of the corresponding receptors. Such receptors, termed **lectins**, may be defined as proteins of nonimmune origin (thus excluding immunoglobulins) that bind in a stable manner (thus excluding enzymes and carbohydrate sensor/transport proteins) to carbohydrates.

Lectins are proteins with a ubiquitous distribution in nature, and plants are the richest source of lectins, namely legume seeds and the storage organs of plants in general. Many lectins are believed to have evolved to recognize and bind specific oligosaccharide structures. Lectins exhibit specificity for sugars, but they typically display a much higher binding affinity for oligosaccharides and glycoproteins, having the capacity to precipitate glycoproteins and multivalent branched oligosaccharides.

The biological functions of lectins are multiple and have been reviewed recently. Some authors consider seed lectins as reserve proteins with a passive function as they undergo patterns of synthesis, accumulation, and degradation that parallel those of typical storage proteins. They may be present in high levels and exhibit no other known functions, especially in what concerns their activity. The presence of large quantities of lectins in legume seeds that seem to fulfill a storage role remains a mystery.

Some authors consider lectins as biologically active proteins because most plant lectins may not only play a role in the plant itself, for instance, as a reserve of organic carbon, nitrogen, and sulfur, or as a specific recognition factor, but they also interact with glycoconjugates of other organisms, thus playing a role in plant defense. There is a compelling evidence that endogenous lectins and glycoconjugates may contribute to metastasis via attachment of malignant cells to vascular endothelial cells or subendothelial matrix components and subsequent adherence to them. These reports prompted great interest in the search for naturally occurring lectins from plants and other sources that would selectively bind tumor-specific glycoconjugates.

The interaction between the fungal pathogen exoglycome and its host is an area of research that has been neglected. The significance of some fungal pathogens (e.g., *Candida albicans*, *Aspergillus fumigatus*, and Alternaria *alternata*) in human lives and of a very large number of fungal phytopathogens suggests that the fungal exoglycome is an especially important research topic, particularly in what concerns the potential host-induced changes in the fungal exoglycome and its analysis to identify potential targets for disease control.

To avoid recognition by host defense mechanisms, it is conceivable for some pathogens to selectively alter their exoglycome to establish pathogenesis. Should they occur, such precise alterations may prove extremely useful, in both diagnosis and treatment, allowing development of a number of tools for their specific recognition. A number of additional potential applications for exoglycome-based approaches for fungal control are the expression of xylem-directed lectins or the expression of plantibodies in plant tissues, specific for any PAMP or other specific features exhibited by one or a group of pathogens.

FURTHER READING

Agrios, G. N. 2005. *Plant Pathology*, 5th edition. Amsterdam, the Netherlands: Academic Press.

De Jaeger, G., C. De Wilde, D. Eeckhout, E. Fiers, and A. Depicker. 2000. The plantibody approach: Expression of antibody genes in plants to modulate plant metabolism or to obtain pathogen resistance. *Plant Molecular Biology* 43:419–428.

Ferreira, R., R. Freitas, and S. Monteiro. 2012. Targeting carbohydrates: A novel paradigm for fungal control. *European Journal of Plant Pathology* 133:117–140.

Ferreira R. B., S. Monteiro, R. Freitas, C. N. Santos, Z. Chen, L. M. Batista, J. Duarte, A. Borges, and A. R. Teixeira. 2006. Fungal pathogens: The battle for plant infection. *Critical Reviews in Plant Sciences* 25:505–524.

Ferreira, R. B., S. Monteiro, R. Freitas, C. N. Santos, Z. Chen, L. M. Batista, J. Duarte, A. Borges, and A. R. Teixeira. 2007. The role of plant defence proteins in fungal pathogenesis. *Molecular Plant Pathology* 8:677–700.

Gabius, H.-J. (ed). 2009. *The Sugar Code: Fundamentals of Glycosciences*. Weinheim, Germany: Wiley-VCH Verlag GmbH & Co. KGaA.

Jansen, R. M. C., J. Wildt, I. F. Kappers, H. J. Bouwmeester, J. W. Hofstee, and E. J. van Henten. 2011. Detection of diseased plants by analysis of volatile organic compound emission. *Annual Reviews of Phytopathology* 49:157–174.

Lis, H., and N. Sharon. 1998. Lectins: Carbohydrate-specific proteins that mediate cellular recognition. *Chemical Reviews* 98:637–674.

Loureiro V., M. M. Ferreira, S. Monteiro, and R. B. Ferreira. 2012. The microbial community of grape berry. Pages 241–268 in: H. Gerós, M. M. Chaves, and S. Delrot (eds). *The Biochemistry of the Grape Berry.* United Arab Emirates: Bentham Science Pub.

Ribeiro A., S. Catarino, and R. B. Ferreira. 2012. Multiple lectin detection by cell membrane affinity binding. *Carbohydrate Research* 352:206–210.

25 Cultural Management of Plant Diseases

Craig S. Rothrock and Terry N. Spurlock

Concept Box

- Cultural control of plant diseases is one of the most important management tools that growers, homeowners, and landscapers have to limit diseases and disease losses.
- Cultural practices apply several disease control strategies including eradication, reducing the amount of inoculum of a pathogen, avoidance, and providing an environment for the production that limits the likelihood of disease.
- Disease control by the use of cultural practices requires a thorough understanding of the plant's development, the biology of the pathogen, and an understanding of how location and management of the crop alter environmental conditions.

Cultural control of plant diseases is one of the most important management tools for growers, homeowners, and landscapers to manage diseases. Anyone who grows plants knows that there are numerous steps to be successful in growing a crop for appearance or to harvest for food or fiber. These steps may include site selection, land or site preparation, selecting a time to establish the crop, deciding on how many seeds or plants to place in an area, managing plant growth through fertility, irrigation, training, or pruning, and timing harvest. Growers and homeowners may not appreciate how these production practices and recommendations for a crop or plant were selected for their area. Many of these are selected to maximize plant growth and yield. However, within many of these production recommendations are strategies to minimize losses due to plant diseases or other pests in your location. In addition, if diseases are not taken into consideration when recommendations are developed or new production practices are implemented, the incidence and severity of disease may change dramatically.

Disease control principles utilized in cultural practices to limit disease development include **avoidance** and **eradication**. Avoidance involves creating a situation or environment to reduce the likelihood of disease development by implementing a production system that minimizes the potential for disease losses, maximizes yields, and limits the need for applications of pesticides later in the season. Eradication is the principle of eliminating or reducing inoculum of a pathogen available for the initiation of disease. A number of cultural practices that affect disease development are discussed along with some examples demonstrating their impact on plant diseases. Many of the cultural practices discussed require advanced planning prior to establishing the crop.

SITE SELECTION

Often the first step in producing a crop or planning a landscape is **site selection**. Considerations may include selecting crops adapted to your region. If you are at the limit of latitude or environment for a crop or landscape plant, these plants will grow poorly. A plant under stress is often more susceptible to disease. Within an area adapted for a crop, considerations for a site include soil texture. For example, soils with high clay content do not allow water to percolate easily into the soil and thus may result in standing water or saturated conditions. This creates an anaerobic environment that limits oxygen for root growth. In addition, these sites are often more prone to problems from diseases caused by the oomycetes *Phytophthora* and *Pythium* (Chapter 8 and Figure 25.1). Sites that have very sandy soils may drain rapidly and increase the risk of limited water for growth. A number of diseases are associated with drought stress including diseases caused by many *Fusarium* spp. and *Macrophomina phaseolina*, the causal agent of charcoal rot. In addition, nematode damage (Chapter 6) may increase with soils having greater sand content. Site preparation may be done to alter site characteristics or soil properties. Organic amendments may be added to improve water retention and aeration. Sand may be added to clay soils to improve drainage and aeration. Other considerations for site selection may be shading or hours of direct sunlight or topography. Sites may also be selected to avoid inoculum of specific pathogens or the presence of vectors, which may introduce a pathogen into a crop.

An extreme example of site selection is growing plants inside greenhouses. In the greenhouse, the environment can be controlled for conditions favorable to the plant. Soil has been replaced with soilless potting media to improve the properties for container-grown plants, such as water-holding capacity, decreased bulk density, and increased pore space. In addition, the absence of soil minimizes the likelihood of introducing soilborne pathogens (Chapter 16) with the growing medium.

TILLAGE

Tillage is the one of the primary methods of site preparation for many crops. Historically, energy-intensive tillage practices like **moldboard plowing** were the primary tillage practice that was used. This tillage reduced soil compaction and improved aeration in the layer being disturbed, 6–12 inches deep, and inverted the soil to bury **crop residue** and incorporate fertilizer. This practice had a number of other benefits including tillage-reduced weed problems and increased the decomposition of crop residue, reducing the survival of inoculum of many pathogens, especially foliar pathogens. The dramatic increase in the incidence and severity of gray leaf spot in corn has been directly associated with the adoption of **conservation-tillage** practices (Figure 25.2). Burying crop residue also reduces the spread of spores of the pathogen by wind or splashing water because residue is not left on the soil surface. However, with many soilborne pathogens, movement of soil and crop residue in soil dramatically increases the movement of inoculum of a pathogen in soil (Figure 25.3). Soil carried on equipment may also spread the pathogen from field to field or on a smaller scale from one area of a garden or landscape to a new area

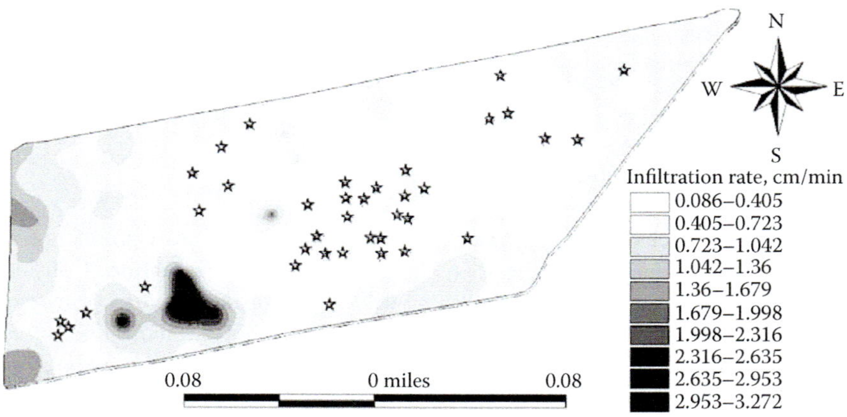

FIGURE 25.1 Map of infiltration rates with stars showing areas of Phytophthora root rot symptoms on cranberry. (From Pozdnyakova et al., *Comp. Elec. Agric.*, 37, 2002).

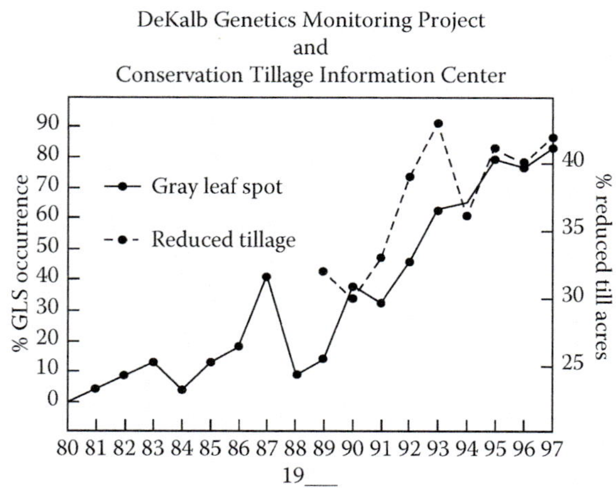

FIGURE 25.2 Association of the increase in gray leaf spot in corn in relation to the adoption of conservation tillage practices. (From Lipps, P.E., *APSnet Features,* 1998).

FIGURE 25.3 Take-all disease symptoms (light-colored patches in the field due to premature heads termed whiteheads), showing the spread of inoculum of the take-all pathogen, *Gaeumannomyces graminis* var. *tritici*, in wheat with the direction of tillage operations.

where the pathogen was absent. In recent years, tillage practices are less frequent and shallower, which reduces the depth and degree of residue burial and decomposition. The benefits of residues left on the soil surface for reducing soil erosion have been stressed with the adoption of conservation tillage and no tillage. In addition, time, money, and energy are saved from the reduced use of heavy equipment and the number of trips across a field. Weed suppression now often relies on the use of pre- and post-emergent herbicides.

Tillage operations may also involve raising the row or area where the crop is planted, called bedding. Bedding may be important to allow drainage, avoid saturated soils near the seed or base of plants, and assist in irrigation and warming of soils. Planting on beds is often used for crops including cotton, soybean, fruit crops like blueberry, many vegetable crops, and landscape plants. Tillage and bedding operations, in addition to changing the location of the pathogen in the soil and limiting survival of some pathogens, provide an environment that improves emergence and growth of crops. Disease losses are minimized through avoiding conditions that favor the pathogen or placing the plant under stress.

CROP ROTATION

Plant pathogens often have a limited number of hosts that the pathogen can parasitize, allowing survival and reproduction of the organism. The host range of a pathogen may be very specific, having only one or a few plant species affected, as with some obligate parasites, or a very broad host range that include a range of plant species and families. For many obligate pathogens, the lack of a susceptible crop or living host tissue may mean that inoculum of the pathogen will die. For other pathogens, if they have no long-term means of survival (e.g., **chlamydospores, oospores,** or **sclerotia**) or the ability to grow **saprophytically**, once host residue is decomposed, the population of the pathogen is eliminated. Crop rotation or crop sequences or leaving an area **fallow** takes advantage of this knowledge of pathogen biology to grow crops in a sequence to allow inoculum of the pathogen to be reduced or eradicated. Rotation may be effective for some pathogens if used by home gardeners to minimize disease problems. Many recommendations for row crops and vegetable production highlight the benefit of rotation for fertility management, but also for disease control. Fusarium head blight can be a serious disease of wheat and barley, causing yield losses and also affecting grain quality by decreasing grain test weight and increasing the level of mycotoxins found in the grain. The disease is difficult to control, but a rotation of wheat or barley with a leguminous crop, such as soybean, instead of corn, a host, will decrease the inoculum for the next season because the legume is a nonhost for the fungus (Table 25.1). Plant parasitic nematodes are obligate parasites, and thus rotation with a nonhost crop will have a dramatic impact on the population of the nematode (Table 25.2). However, in some cases, for example, Fusarium wilt of cotton (caused by *F. oxysporum* f. sp. *vasinfectum*), although crop rotation may prevent an increase in the pathogen population, it may not significantly reduce the number of *Fusarium* spores in soil so that disease cannot be eliminated by crop rotation alone (see Case Study 25.1). Consideration of the host range and survival strategies of a pathogen will help ensure that crop rotation is a successful and economical disease control practice.

TABLE 25.1
Residue Cover and Fusarium Head Blight Severity on Wheat for Previous Crop and Tillage Combinations

Treatment	Corn	Wheat	Soybean	Average
			Residue Cover (%)	
Moldboard plow	12 bc[z]	9 ab	5 a	9 r
Chisel plow	42 e	34 d	17 c	31 s
No tillage	67 f	83 g	46 e	65 t
Average	41 y	42 y	22 z	
			Disease Severity (%)	
Moldboard plow	17 b–e	16 a–d	14 a–c	16 r
Chisel plow	26 f	19 c–e	16 a–e	20 s
No tillage	26 f	20 de	17 b–e	21 s
Average	23 z	18 y	16 x	

Source: Adapted from Dill-Macky and Jones, R. K., *Plant Dis.*, 84, 2000.

[z] Means followed by the same letter are not significantly different at $P \leq 0.05$, and the main effects were tested separately (previous crop treatment = r, s, t; tillage practices = x, y, z) from means of previous crop–tillage combinations (a through g). Interaction for previous crop × tillage was significant.

TABLE 25.2

Effect of Crop Rotation with Peanut (P) and Cotton (C) on End-of-Season Juvenile Populations of *Meloidogyne arenaria* in a Field Experiment at the Wiregrass Substation, Near Headland, Alabama

Cropping Sequence and Year						Juveniles per 100 cm^3 Soil					
1985	1986	1987	1988	1989	1990	1985	1986	1987	1988	1989	1990
P	P	P	P	P	P	579	72	144	97	155	225
C	P	C	P	C	P	5	41	1	59	2	128
C	C	P	C	C	P	15	16	24	8	6	88

Source: Adapted from Rodriguez-Kabana et al., *Suppl. J. Nematol.* 23, 4S, 1991.

CASE STUDY 25.1

Fusarium Wilt of Cotton—Managing a Disease Interaction

- Fusarium wilt of cotton is caused by *F. oxysporum* f. sp. *vasinfectum*.
- The disease is most prevalent when temperatures are between 30°C and 32°C (86°F and 90°F).
- Fusarium wilt of cotton is a disease complex in most areas with symptoms and yield loss being substantially greater when cotton is also infected by the root-knot nematode *(Meloidogyne incognita)*.
- Root-knot nematode activity is affected by soil texture. Damage is typically more severe in soils with higher sand content.
- Symptoms of Fusarium wilt include vascular discoloration, yellow chlorotic and necrotic foliage, wilting, and plant death. Symptoms of the root-knot nematode include galling of infected roots and subsequent root dysfunction resulting in nutrient deficiencies and water stress.
- **Resistance** is being emphasized in areas where an interaction with the nematode is not important. These cultivars typically show wilt symptoms in the presence of the nematode.
- **Crop rotation** with nonhost crops is a good way to manage root-knot nematode populations. Nonhost crops include resistant soybean cultivars, rice, grain sorghum, and peanut. Crop rotation is not effective for *F. oxysporum* f. sp. *vasinfectum*.
- Chemical **nematicides** or **soil fumigants** to control root-knot populations can reduce Fusarium wilt.

PLANTING DATE

Planting date is one of the most important decisions made by a farmer, nursery producer, or gardener. What determines when to plant or transplant a crop? Among the considerations of when to plant are length of growing season and weather conditions, including rainfall, soil, and air temperatures. Calendar date is less important than these environmental conditions at or shortly after planting. Soils warm slowly in the spring, and warming is determined by a number of factors including slope, soil texture, and soil moisture. With a southern slope in the northern hemisphere, soils warm earlier in the spring. Dark-colored soils absorb more sunlight and are warmer than light-colored soils. Soils with more water warm slower, but transfer the heat deeper. Soils that are covered with residue or mulches or plant cover warm more slowly than bare soil. There may be minimal or critical temperatures for germination or avoiding cold temperature injury for each plant species. However, when planting in soils above these limiting temperatures, growers often encounter problems with emergence resulting in poor stand or seedlings that are low in vigor and are delayed in development. In many situations, this is due to seedling diseases. For a long time, it has been appreciated that when a plant is under stress, it is more susceptible to these pathogens. This is associated with slower growth and development and increased loss of root **exudates**. These exudates may be cues to the dormant pathogen that a host is present and allows germination of the pathogen and growth toward and infection of the plant. For cotton, a warm season crop, planting in cool soil conditions with rainfall soon after planting greatly reduced plant stand and increased seedling diseases caused by the seedling disease pathogens *Rhizoctonia solani* and *Pythium* spp. (Figure 25.4). Similar responses can be seen with numerous crops depending on their

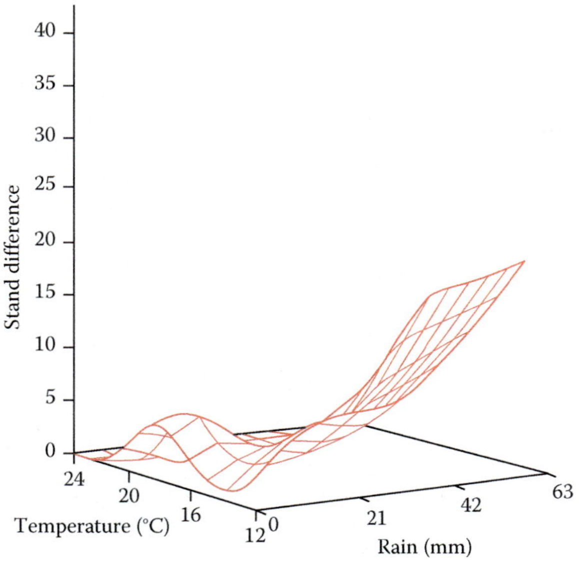

FIGURE 25.4 Stand improvement from the control of *Pythium* spp. causing seedling disease on cotton, with use of metalaxyl seed treatment, in relation to planting environment the first three days after planting. The vertical axis is the difference in percent stand of metalaxyl-treated seeds compared with no seed treatment. (From Rothrock et al., *Plant Dis.*, 96, 2012.)

susceptibility to these seedling disease pathogens and the stress that an unfavorable planting environment places on that crop. Along with planting date, considerations such as preparation of the seedbed and planting depth may increase or decrease disease severity.

IRRIGATION

Water is a limiting factor for plant growth. Frequency, amount, and duration of irrigation have a large impact on plant growth and yield and may also impact pathogens and the severity of disease. There are many irrigation practices used for growing plants such as flooding, furrow irrigation, allowing the water to run between the rows, and overhead irrigation (sprinklers or pivot systems). As technology changes and with concerns over use of water for agricultural production, new methods to use water more efficiently have been developed. For example, drip or trickle irrigation systems deliver small amounts of water to the base of the plant, thus enabling more efficient water use. In greenhouse production, advanced systems may include **ebb and flow** or **hydroponic systems**. Pathogens often require free water to germinate or break dormancy, grow to or on the host, and infect. *Botrytis cinerea*, which causes rots of many types of fruit, requires as little as 4 h of free water for infection in grapes with the proportion of grapes infected increasing as duration of free water increases (Figure 25.5). Pivot irrigation or sprinklers increase the likelihood of keeping the foliage wet. Time of watering may also be important.

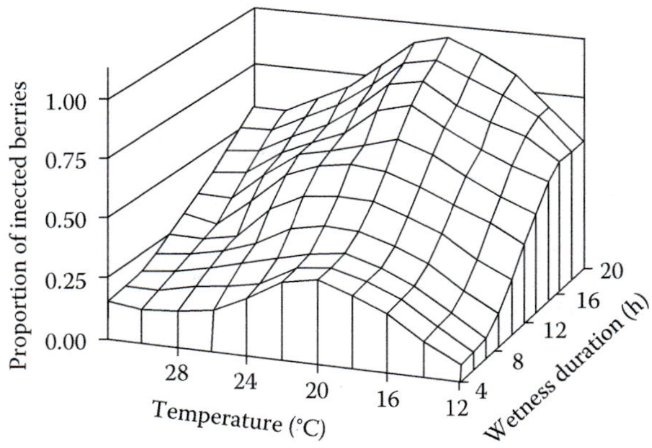

FIGURE 25.5 The proportion of grape berries infected by *Botrytis cinerea* is affected by wetness duration and temperature. (From Broome et al., *Phytopathology*, 85, 1995.)

Watering late in the day when the foliage may remain wet through the night may increase disease progression dramatically. Overhead irrigation may also increase the spread of foliar pathogens by splashing spores from crop residue or neighboring diseased leaves to healthy leaves. For soilborne pathogens, water may be most important to allow conditions favorable for the movement of the pathogen and infection. Flooded or saturated soils increase infection by *Pythium* and *Phytophthora* species, which produce motile zoospores. *Phytophthora capsici* caused much greater damage on pepper as irrigation frequency increased from every 21 days to every 7 days (Figure 25.6). Water movement in the field may be important for the dissemination of pathogens in the field with the movement of soil or infested residue (Figure 25.7). At the other extreme, too little water may cause stress on the plant allowing for disease to progress and disease symptoms to be more severe.

PLANT POPULATION

The number of seeds planted in a row or how far plants are spaced apart is another important consideration when planting. A grower tries to maximize yields or in a landscape, make the most attractive landscape. However, plant disease is also a consideration. Obviously, a **monoculture** often used for crop production, or to give a uniform appearance in the landscape, increases the likelihood of disease, thus increasing disease incidence and severity. A healthy plant is closer to a diseased plant at high plant populations. For a soilborne pathogen, spread may be limited by plants separated by a few centimeters, whereas with airborne pathogens, plant distance may not limit dispersal. In a landscape or garden, an increase in the number and density of immune plants or plants that are not hosts may limit

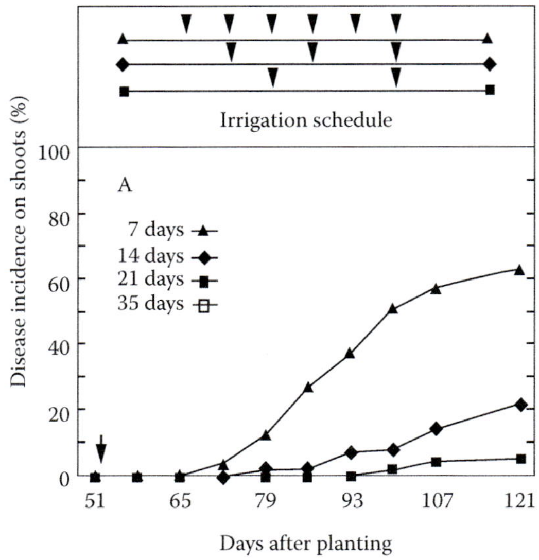

FIGURE 25.6 Disease progression of Phytophthora root rot of pepper in soil infested with *Phytophthora capsici*. Arrows indicate when irrigation was applied. (From Café-Filho, A.C, and Duniway, J.M., *Plant Dis.*, 79, 1995.)

FIGURE 25.7 Spread of inoculum of *Fusarium oxysporum* f. sp. *vasinfectum* as a result of furrow irrigation of cotton in Australia.

the disease by decreasing the production of inoculum and the likelihood of encountering a susceptible plant. When intercropped with corn or cowpea, cassava was less affected by cassava mosaic disease caused by whitefly transmitted viruses in the genus *Begomovirus* than when grown in a monoculture (Figure 25.8). However, for other diseases, including some virus diseases, due to the changing behavior of the vector and verticillium wilt, increasing plant populations may actually decrease the percentage of plants infected.

Plant populations may also affect the microenvironment. Decreasing plant spacing may result in a denser

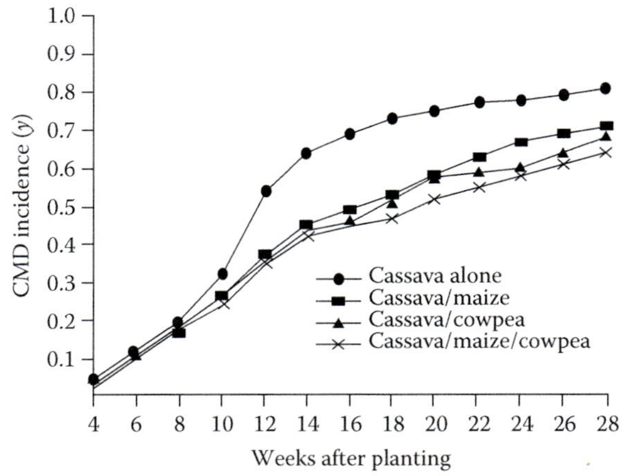

FIGURE 25.8 Cassava mosaic disease (CMD) progress in cassava alone and cassava intercropped with maize and/or cowpea. (From Fondong et al., *J. Phytopathology,* 150, 2002.)

canopy, increasing the likelihood of sufficient humidity or duration of leaf wetness for infection due to lower light intensity and reduced airflow, as discussed earlier, whether plants are hosts or not. Selecting plants with more open plant architecture may improve light infiltration and air movement, reducing the duration of leaf wetness.

PLANT FERTILITY

Soil fertility is a critical part of any recommendation for the production of plants and is the reason that soil tests prior to establishing a crop or landscape are encouraged. Soil pH is important for plant production, but also for managing diseases. Club root of cabbage (Chapter 9) (caused by *Plasmodiophora brassicae*) is effectively controlled by maintaining an alkaline soil pH (Table 25.3). Diseases favored by neutral to alkaline soil pH include potato scab (caused by *Streptomyces scabiei*), take-all of small grains (caused by *Gaeumannomyces graminis* var. tritici; see Case Study 25.2), and black root rot (caused by *Thielaviopsis basicola*). Diseases favored by acid soil pH include Fusarium wilt. Soil pH also affects the availability of many nutrients in the soil. As pH increases, some mineral elements, including Fe, Mn, Zn, Cu, and Co, are less available. Thus, iron deficiency symptoms are often seen on crops or landscape plants grown in neutral to alkaline soils because iron is less available to the plant. As soil pH decreases below 5.5, N, Ca, Mg, K, P, S, and Mo are less available. Soil pH also influences the composition of soil microflora.

A balanced fertility program is critical to good crop appearance and yield. Excessive nitrogen fertility will often lead to more succulent and excessive growth and greater disease susceptibility. Excessive growth may limit air movement and increase relative humidity in the canopy and hours of leaf wetness, as discussed earlier,

TABLE 25.3
Clubroot of Chinese Cabbage in Field Miniplots Following the Incorporation of Lime or Uniform Mixing of Lime[y]

Soil treatment	Soil pH	Diseased plants (%)
No lime	6.0	91 a[z]
No lime, mixed	6.2	96 a
Lime	7.0	34 b
Lime, mixed	7.0	6 c

Source: Adapted from Dobson et al., *Plant Dis.*, 67, 1983.

[y] *Hydrated lime applied at 4483 kg/ha in a silt loam soil and rotovated twice for field application or mixing soil after passing through a 2-mm sieve. $CaCl_2$—pH determinations.*

[z] *Means of 2 experiments, 4 replications per experiment, and 15 plants per experimental unit. Means followed by the same letter are not significantly different based on Duncan's multiple range test ($P = 0.05$).*

favoring disease. Excessive nitrogen increases rice blast caused by *Magnaporthe oryzae*. Deficiency of nutrients may also increase disease. For example, K deficiency increases foliar diseases on cotton. There are many examples of how excesses or deficiencies of macro- or micro-elements affect diseases on numerous crops. In addition, nutrient deficiencies are considered abiotic diseases, and pictures of disease symptoms can be found in most books on diseases of crops.

SANITATION

Sanitation is an important practice in plant production in greenhouses, landscapes, gardens, and production fields. These practices are primarily directed at eradicating or reducing inoculum of the pathogen. Removing diseased plant material, leaves, or more often plants showing disease symptoms is important to control diseases caused by viruses, bacteria, fungi, oomycetes, and nematodes. For example, in greenhouse production, this may require removing pots containing dead or diseased plants or cleaning-up leaves and other plant material from the production area during the production of a crop or prior to establishing a new planting (see Case Study 25.3). Chemicals to disinfest the area may also be used. For an orchard or landscape, this may involve removing entire trees that have died and removing some of the soil surrounding the plant. Raking-up and disposing diseased leaves, flowers, fruit, or branches may also reduce the inoculum of the pathogen. In other situations, diseased branches should be pruned out of an affected plant. This is a common recommendation for fire blight control on pear and apple caused by *Erwinia amylovora* to reduce the inoculum in the following year and eliminate the disease from the affected tree. Sanitation may also involve eliminating weeds that may serve as a reservoir of inoculum for a pathogen or vector.

Sanitation is also important when handling plants. When pruning-out the diseased material from a tree or removing plants, tools may introduce the pathogen to a healthy plant. Disinfesting tools is an important practice to limit the spread of disease. Many pathogens may be present, but do not cause disease symptoms at the time of pruning. This is frequently the case with diseases caused by viruses. On a larger scale, producers may choose to disinfest the equipment to limit the movement of soil to another field or prior to the equipment coming onto a farm.

CASE STUDY 25.2

Take-All of Wheat

- Take-all of wheat is caused by the fungus *Gaeumannomyces graminis* var. *tritici*.
- In the field, diseased plants die prematurely causing the "white head" symptom, characteristic of the disease.
- The fungus colonizes the roots and crown of the wheat plant as dark ectotrophic hyphae on the surface, often called runner hyphae, and the vascular tissue of roots producing black lesions.
- The pathogen survives saprophytically on dead host tissue or parasitically on other hosts, such as weeds, in the field.
- **Crop rotation** to a nonhost crop will allow time for decomposition of the host residue, and pathogen death before the next susceptible crop is planted.
- **Soil pH** above 6.5 favors disease, and application of acidifying fertilizers is a management option.
- **Fertility** should be maintained in fields with a history of take-all to minimize disease losses.
- **Fungicides** are not effective for the management of the disease.
- **Resistance** is not found.

CASE STUDY 25.3

Downy Mildew on Impatiens—A Limiting Factor for Impatiens in the Landscape

- Downy mildew, caused by the oomycete *Plasmopara obducens*, has become an increasingly important disease on Impatiens (*Impatiens walleriana*) in the United States and around the world.
- Symptoms begin as yellow to pale-green foliage and can advance to stunting of plants, malformation of foliage and flowers, and severe defoliation prior to plant death.
- A diagnostic sign of the pathogen includes blue, gray, or silver growth, and sporangia on the underside of leaves.
- The disease is favored by cool wet conditions.
- This polycyclic disease is likely introduced into ornamental planting by purchasing infected plant material from nurseries.
- No resistant cultivars are available although the New Guinea Impatiens (*Impatiens hawker*) has good resistance.
- **Crop rotation** of a nonsusceptible plant within the year and the following year is recommended.
- **Chemical control** for homeowners is available, but the products typically purchased are less effective than those used by commercial growers.
- **Irrigation** such as overhead irrigation that wets the leaves increases the disease progress. A more efficient drip system may limit over watering and avoid the foliage remaining wet for long periods of time.
- **Sanitation** could be effective if the planting beds are scouted regularly for disease. If one or two plants appear to be diseased, removing the plants and surrounding soil, placing the removed material in a plastic bag, and disposing of it away from any beds utilized for Impatiens planting could limit disease progress.

Sanitation may also involve trying to eliminate the pathogen on a larger scale. For example, crop residues should be composted if the material is to be used later as a soil amendment or mulch. It is important that the composting process heats the residue to 150°F (55°C) to eliminate the pathogens by **pasteurization**. This heating process is similar to pasteurization of milk and juices. Soils may also be pasteurized by solarization. Solarization takes place when moist soil is covered with clear plastic. Sunlight heats the soil by radiant heat similar to a greenhouse. With sufficient days and weeks of sunlight, soil temperatures in the top few inches of the soil are pasteurized. Temperatures may not reach those of composting or steam pasteurization of potting media, but due to the increase in the duration of these elevated temperatures, often 40°C–45°C, pathogens are eliminated. Another soil sanitation technique is summer flooding. A water level is established above the soil for a long enough duration during the summer for the soil to become anaerobic, and after sufficient time many soilborne pathogens will be eliminated due to lack of oxygen in soil. This process takes place during rice production where fields are flooded soon after rice is established and continues until harvest.

OTHER CULTURAL PRACTICES

Adequate air flow in a plant canopy is important to maintain leaf temperature near air temperature and reduce humidity and the formation of dew. As discussed earlier in the chapter, many of these pathogens require free water to germinate, grow, and infect the plant. Air movement may be improved in a landscape or orchard by pruning or thinning the vegetation. This increases air movement and may also reduce shading. Pruning leaves in the fruit zone for grapes is done in vineyards to reduce botrytis bunch rot and to decrease the likelihood of leaf wetness periods favorable for infection, as discussed earlier. Fans are frequently used in greenhouses to manage temperature, but are also used to help control disease by reducing differences between leaf and air temperature and reducing relative humidity and free water in the greenhouse. On golf courses, fans may be used to minimize the periods of leaf wetness and reduce relative humidity especially where cool-season turfgrasses are being grown as putting greens in more temperate climates (Figure 25.9).

Mulches may be used to maintain soil moisture. These may include straw, grass clippings, composted bark, newspaper, many inorganic materials, and landscape fabric. Mulches prevent evaporation of water, as well as suppress weeds, and may limit splashing of inoculum from soil or host residue onto the plant. Clear or black plastic is often used in vegetable production as part of the production system, to suppress weeds, reduce evaporation and water fluctuations, and influence soil temperature.

FIGURE 25.9 Oscillating fans used to cool the canopy of a cool-season turfgrass putting green during summer.

Soil amendments have historically been added to improve soil properties or fertility for plant growth. Amendments may include common materials like grass clippings, leaves, crop residues or waste, animal waste or green manures, and plant material grown for the purpose of amending soils. Incorporation of these materials into the soil may have beneficial and detrimental effects depending on the amount and timing of application. Increasingly, soil organic amendments are being emphasized for improving soil health, and increasing the numbers and diversity of soil microflora and fauna. In addition, amendments may impact soilborne plant pathogens and influence disease incidence and severity. The nature of the organic amendment and the amount needed to maximize plant growth and suppress diseases may not be obvious, and may require considerable research. One example of successful use of organic amendments is the use of amendments from brassica crops like canola, broccoli, and mustard, as seed meal or green plant material. There are a number of pathogens that have been suppressed by these amendments, in part through a process known as **biofumigation**. Plants in the Brassicaceae produce compounds known as **glucosinolates**, which are sulfur-containing secondary metabolites that are hydrolyzed into **isothiocyanates** and other biologically active compounds when plant tissue is destroyed. Brassica cover crops have been shown to significantly reduce disease caused by *Phytophthora capsici* in squash (Figure 25.10).

In summary, from the examples used in this chapter, it should be clear that many cultural practices chosen when producing a crop have an impact on plant diseases. The effectiveness of any of these practices requires an understanding of the biology of the important pathogens on a crop and the interaction between the host and the pathogen. By selecting cultural practices that minimize disease pressure, the need for more drastic and expensive inputs may often be avoided.

LABORATORY EXCERCISES

The experiments are designed to demonstrate the importance of cultural practices in managing plant diseases. The experiments can be completed in 2–3 weeks and focus on seedling diseases because of the ease of assessment. In addition, the experiments can easily be adapted to many crops, including field and vegetable crops, and seedling pathogens are present in most soils. Experiments are designed to demonstrate the importance of planting date (soil temperature), inoculum placement (tillage), and eradicating inoculum by heating soils (pasteurization, by solarization or composting) on disease.

FIGURE 25.10 Reduction of disease caused by *Phytophthora capsici* on squash in field trials in 2010 using various cover crops as soil amendments. Rye, mefenoxam (Ridomil Gold)-treated soil, or nontreated soil served as controls. Bars with the same letters are not significantly different at $P = 0.05$. (From Ji et al., *Pest Manag. Sci.*, 68, 2012.)

Experiment 1. Effects of Planting Date and Soil Temperature on Seedling Disease

Materials (Experiment 1 and 2 should be done simultaneously and the materials list can be used for both experiments)

Each lab group will require the following materials:

- Seed not treated with fungicides.
  - Field crops: corn, cotton, soybean, wheat, or rice
  - Vegetable crops: corn, pea, or bean
- A culture of *Pythium* spp., such as *Pythium aphanidermatum*, *P. ultimum*, or *P. irregulare.* As an alternative, the seedling pathogen *Thielaviopsis basicola* works well for planting date studies. Cultures can be obtained from the American Type Culture Collection (ATCC). However, an Animal and Plant Health Inspection Service (APHIS)—Plant Protection and Quarantine (PPQ) permit must be obtained in order for the ATCC to ship the culture. A plant pathology department at a land-grant university in your state may have cultures, which are not under permit because they were collected locally, and they would be able to share.
  - Alternatively, *Pythium* species may be baited-out from almost any soil used for field crop or vegetable production. To recover the pathogen, take fresh soil and place seed in the soil and water to allow imbibition and germination of the seed. After 24–72 h wash seed thoroughly and place on water agar. Most *Pythium* species will be the fastest growing colonies coming from the seed. Preliminary tests of pathogenicity can be done by growing a culture of the isolate on water agar and placing surface-sterilized seed on the culture and examining for seed or seedling rot as described by Broders et al. (2007).
  - Alternatively, *T. basicola* may be isolated from soil used for cotton production in some states or potting medium growing pansies at many nurseries or retailers. To obtain a culture of *T. basicola*, place carrot slices in a Petri dish containing wet filter paper and sprinkle soil or potting medium over the carrots. The fungus should colonize the carrots, if present, in 7–10 days.
- Plastic pots (10-cm-diameter), styrofoam cups, or soup bowls.
- Commercial vermiculite, sand, or potting media work for these short-term experiments. Vermiculite holds moisture well and allows roots to be recovered easily. Be aware that commercially purchased potting media may not be free of pathogens.
- Growth chamber facilities or other facilities to maintain constant growing temperatures.

Follow the protocol in Procedure 25.1 to complete these experiments.

Procedure 25.1
Effect of Planting Temperature on Seedling Disease Severity

Step	Instructions and Comments
1A	Infest potting medium with pathogen or pathogens. Infestation can be done by cutting up a culture of *Pythium* species growing on a medium such as cornmeal agar and mixing approximately one culture per pot. A more precise technique is to produce your inoculum in sand-cornmeal medium. This can be prepared by adding 100 mL of sand, 5.6 mL of cornmeal, and 40 mL of deionized water in a 250-mL flask. Cap the flask with a cotton or foam plug and aluminum foil and autoclave for 40 min on two consecutive days. After cooling, add several pieces of an actively growing culture to the medium and incubate for 9 days; shake periodically to assure uniform colonization. About 8 g of inoculum per pot should provide adequate seedling disease pressure. Prepare pots of potting medium that are not infested as controls.
1B	**Alternative pathogen for procedure**: *Thielaviopsis basicola* causes black cortical root rot on seedlings, rather than acute symptoms of seed rot or seedling death and generally does not reduce stand, but does produce distinctive symptoms. For the production of *Thielaviopsis* inoculum, see Procedures 16.8 and 16.9. Infest potting medium with the pathogen by collecting endoconidia and chlamydospores of *T. basicola*. Add spores to the

(Continued)

potting medium, from 100 to 1000 spores/g should give good disease levels. Prepare pots of potting medium that are not infested as controls.

2 Plant 10 seeds at an appropriate depth for the crop in a pot of 10-cm-diameter and water.

3 Place one pot for each student group in each of the temperatures selected. Three temperatures are adequate for the procedure. A range of temperatures at 4-degree increments should be selected. For example, for most of the crops 20°C, 24°C, and 28°C will give a range of temperatures favorable and unfavorable for germination and emergence. Pea, soybean, and cotton are effective for this assay. Wheat or other cool season crops may need to be grown at lower temperatures.

4 Water plants periodically to maintain favorable conditions for seedling disease.

5 Record emergence (number of plants that are above the soil surface after 7 and 14 days). After 14 or 21 days, record final stand depending on the crop.

6 Seedlings can be removed from soil to examine for root or hypocotyl (coleoptile) symptoms at the time of the final stand count. Rate the roots for percent discoloration. Assess plant growth, growth stage, or weight. However, often stand data are sufficient.

7 Isolations may be done from plants onto one of several culture media including water agar, cornmeal agar, or potato dextrose agar for *Pythium*, or carrot juice agar or potato dextrose agar for *Thielaviopsis*. See Table 16.2 for agar medium recipes.

8 Record data for the emergence and final stand of viable plants, root discoloration, and plant growth for infested and noninfested pots for each temperature. The data for each group can be used as replicates for statistical analysis.

Anticipated Results

In the noninfested pots, students should see good stands of seedlings, assuming that the temperatures selected for growth were not limiting for the crop. Growth differences of seedlings among the temperatures may also be evident. For the infested pots, the importance of planting date (soil temperature) on seedling disease severity and establishing a crop should be evident over the range of temperatures chosen. Seedling disease will be evident in a reduced number of seedlings that emerged from the 10 seeds planted. In addition, difference between emergence and final stand should indicate seedlings that died after emergence from postemergence damping-off.

Questions

- What is the importance of planting date on establishing a crop?
- Does the temperature response observed for a crop reflect temperatures unfavorable to the crop or temperatures that increase virulence of the pathogen?
- What can be done to minimize damage from seedling diseases caused by *Pythium* prior to planting?
- What options does a grower have for the crop if seedling disease occurs?

Experiment 2. Effects of Tillage or Inoculum Location on Disease

Materials—see list for Experiment 1

Follow the instructions in Procedure 25.2 to complete this experiment.

Anticipated Results

Good stands of seedlings should be evident in the noninfested pots. For the infested pots in Experiment 2, the importance of tillage (inoculum location) on seedling disease severity and crop establishment should be evident by comparing the emergence from the infested pots in Experiment 1. Higher rates of emergence should be evident in pots where the inoculum was covered with soil (placed further away from the seed) before planting. Placing inoculum away from contact with the seed should limit the effects of the pathogen on stand losses from seed rot, just as tillage displaces inoculum and creates environmental conditions that enhance decomposition of infested plant material created by diseased plants from the previous crop.

Question

- Is the location of pathogen inoculum in soil important for the development of seedling diseases?

Experiment 3. Effect of Soil Pasteurization, Solarization, or Composting on Inoculum Survival

Materials

Each lab group will require the following items:

- Six capped disposable centrifuge tubes, screw-cap vials, or other closed containers
- Hot water baths of temperature of 45°C, 55°C, and 70°C
- Infested soil or potting medium or cultures to infest soil or potting medium
- Appropriate selective media

Follow the instructions in Procedure 25.3 to complete this experiment. Refer to instructions related to the production of the pathogen (*Pythium* and *Thielaviopsis*) inoculum in Procedure 25.1.

Anticipated Results

Counts of pathogen colonies should be lower with increased soil pasteurization temperature due to the reduction in viability of pathogen inoculum. The importance of temperature in composting of residue or pasteurization of soil should be evident.

Questions

- Is heat treatment an effective method of treating soil, potting media, or crop residue to eradicate pathogens?
- Do you think duration of heat treatment would affect the temperature at which the treatment is effective?

Procedure 25.2

Effect of Tillage or Inoculum Location on Seedling Disease

Step	Instructions and Comments
1	Infest potting medium with pathogen or pathogens. This can be accomplished by using a culture on agar or sand-cornmeal inoculum as outlined in Procedure 25.1. Place 0.5–1 cm of noninfested potting medium above the infested medium to separate the seed from the pathogen. Place seed on the surface of this noninfested layer and cover the seed with noninfested potting medium to a similar depth as in Procedure 25.1. This procedure should be done in combination with Procedure 25.1 to compare stand for noninfested pots and uniformly infested pots. Steps 2 through 8 are completed as outlined in Procedure 25.1.

Procedure 25.3

Effect of Soil Pasteurization, Solarization, or Composting on Inoculum Survival

Step	Instructions and Comments
1	Infest soil or potting medium with spores, sclerotia, or hyphae of selected pathogen. Make sure that the medium is moist. If you have been working with *Pythium* or *Thielaviopsis*, these pathogens will work for this assay. This may also be done with naturally infested soil or crop residue mixed with soil or potting medium.
2	Place 10 g of soil in each of six capped disposable centrifuge tubes, screw-cap vials, or other closed container.
3	Place two tubes in each of three hot water baths at a temperature of 45°C, 55°C, or 70°C for 30 or 60 min. Include time for the soil to come up to temperature.
4	Place infested medium or soil into 100 mL of molten (40°C–45°C) agar of an appropriate selective medium to detect the pathogen. Swirl and dispense into several Petri dishes.
5	Count colonies after 2–10 days as appropriate for the pathogen.
6	Graph colony-forming units on the vertical axis and time on the horizontal axis for each duration and temperature treatment.

REFERENCES

Broders, K. D., P. E. Lipps, P. A. Paul, and A. E. Dorrance. 2007. Characterization of *Pythium* spp. associated with corn and soybean seed and seedling disease in Ohio. *Plant Dis*. 91:727–735.

Broome, J. C., J. T. English, J. J. Marois, B. A. Latorre, and J. C. Aviles. 1995. Development of an infection model for Botrytis bunch rot of grapes based on wetness duration and temperature. *Phytopathology* 85:97–102.

Café-Filho, A. C., and J. M. Duniway. 1995. Effects of furrow irrigation schedules and host genotype on Phytophthora root rot of pepper. *Plant Dis*. 79:39–43.

Dill-Macky, R., and R. K. Jones. 2000. The effect of previous crop residues and tillage on Fusarium head blight of wheat. *Plant Dis*. 84:71–76.

Dobson, R. L., R. L. Gabrielson, S. A. Baker, and L. Bennett. 1983. Effects of lime particle size and distribution and fertilizer formulation on clubroot disease caused by *Plasmodiophora brassicae*. *Plant Dis*. 67:50–52.

Fondong, V. N., J. M. Thresh, and S. Zok. 2002. Spatial and temporal spread of cassava mosaic virus disease in cassava grown alone and when intercropped with maize and/or cowpea. *J. Phytopathology* 150:365–374.

Ji, P., D. Kone, Y. Jingfang, K. L. Jackson, and A. S. Csinos. 2012. Soil amendments with *Brasssica* cover crops for management of Phytophthora blight on squash. *Pest Manag. Sci*. 68:639–644.

Lipps, P. E. 1998. Gray leaf spot: A global threat to corn production. *APSnet Features*. Online DOI:10:1094/APSnetFeature-1998-0598.

Pozdnyakova, L., P. V. Oudemans, M. G. Hughes, and D. Gimenez. 2002. Estimation of spatial and spectral properties of phytophthora root rot and its effects on cranberry yield. *Comp. Elec. Agric*. 37:57–70.

Rodriguez-Kabana, R., D. G. Robertson, L. Wells, C. F. Weaver, and P. S. King. 1991. Cotton as a rotation crop for the management of *Meloidogyne arenaria* and *Sclerotium rolfsii* in peanut. *Suppl. J. Nematol*. 23(4S):652–657.

Rothrock, C. S., S. A. Winters, P. K. Miller, E. Gbur, L. M. Verhalen, B. E. Greenhagen, T. S. Isakeit, W. E. Batson, Jr., F. M. Bourland, P. D. Colyer, T. A. Wheeler, H. W. Kaufman, G. L. Sciumbato, P. M. Thaxton, K. S. Lawrence, W. S. Gazaway, A. Y. Chambers, M. A. Newman, T. L. Kirkpatrick, J. D. Barham, P. M. Phipps, F. M. Shokes, L. J. Littlefield, G. B. Padgett, R. B. Hutmacher, R. M. Davis, R. C. Kemerait, D. R. Sumner, K. W. Seebold, Jr., J. D. Mueller, and R. H. Garber. 2012. Importance of fungicide seed treatment and environment on seedling diseases of cotton. *Plant Dis*. 96:1805–1817.

SUGGESTED READINGS

Allen, T. W., A. Martinez, and L. L. Burpee. 2004. Pythium blight of turfgrass. *The Plant Health Instructor*. DOI:10.1094/PHI-I-2004-0929-01

Bockus, W. W., and N. A. Tisserat. 2000. Take-all root rot. *The Plant Health Instructor.* DOI:10.1094/PHI-I-2000-1020-01. Updated 2005.

Caitlin, N., and M. Daughtrey. 2013. Impatiens downy mildew in the landscape. Cornell Univ. Fact Sheet. https://s3.amazonaws.com/assets.cce.cornell.edu/attachments/2559/DM-landscape.pdf?1409319690.

Davis, R. M., P. D. Colyer, C. S. Rothrock, and J. K. Kochman. 2006. Fusarium wilt of cotton: Populations diversity and implication for management. *Plant Dis*. 90:692–703.

Greer, A., G. Wilson, and T. Kirkpatrick. 2006. *Management of Economically Important Nematodes of Arkansas Cotton*. University of Arkansas Division of Agriculture. FSA7567.

Johnson, K. B. 2000. Fire blight of apple and pear. *The Plant Health Instructor*. DOI:10.1094/PHI-I-2000-0726-01 Updated 2005.

Long, D. H., F. N. Lee, and D. O. TeBeest. 2000. Effect of nitrogen fertilization on disease progress of rice blast on susceptible and resistant cultivars. *Plant Dis*. 84:403–409.

Milus, G., R. Cartwright, C. Rothrock, and C. Parsons. 2009. *Management of Take-All Disease of Wheat in Arkansas*. University of Arkansas Division of Agriculture. FSA7526.

Schmale III, D. G., and G. C. Bergstrom. 2003. Fusarium head blight in wheat. *The Plant Health Instructor*. DOI:10.1094/PHI-I-2003-0612-01. Updated 2010.

Schubert, T. 2012. Downy mildew of impatiens caused by *Plasmopara obducens*. Thurston, H. D. 1990. Plant disease management practices of traditional farmers. *Plant Dis*. 74:96–101.

26 Chemical Control of Plant Diseases Caused by Fungi

Jason E. Woodward and Alan S. Windham

Concept Box

- Fungicides are often the primary tool used in the management of fungal diseases of plants.
- In general, there are two types of fungicides: protectants and systemics.
- Fungicides can be identified based on their product or trade name, as well as an active ingredient(s), which often doubles as the common name of a product. Furthermore, fungicides can be categorized by chemical group, mode of action, or breadth of activity.
- Major commercial fungicide formulations include dusts, granules, wettable powders, emulsifiable concentrates, and flowable formulations.
- The site of activity is used to differentiate fungicides that are active against a single point in one metabolic pathway (single-site) versus multi-site fungicides, which affect a number of different metabolic sites within a fungus.

Plants are susceptible to infection caused by a wide range of organisms including bacteria, fungi, parasitic plants, nematodes, and viruses. Fungal diseases represent approximately 70% of plant diseases worldwide. Fungicides account for the greatest chemical expenditure to manage diseases of plants. Disease management is best achieved through the integration of different control tactics including, but not limited to, the selection of resistant or tolerant cultivars (Chapter 23), crop rotation, planting date, sanitation (Chapter 25), and the application of fungicides. Selection of the proper fungicide is critical in the development of effective disease management strategies. Efficacy of most fungicides used in the management of plant diseases is enhanced when preventive applications are made.

The first record of chemical control of diseases occurred in the middle of the seventeenth century with the brining of grain with salt water followed by liming. Such practices were conducted to control bunt diseases (caused by *Tilletia* spp.; Chapter 14) and were implemented following observations that seed reclaimed from shipwrecked vessels did not develop disease symptoms. Inorganic compounds such as arsenic, cadmium, copper, mercury, and sulfur comprised some of the earliest fungicides. For example, the **Bordeaux mixture** (named for the grape-producing region of France where it was used in the management of downy mildew; Chapter 8) is a product of copper sulfate and hydrated lime and is still used extensively today for the management of various foliar pathogens. Fungicides containing arsenic, cadmium, or mercury are no longer commercially available due to issues with mammalian toxicity, as well as environmental hazards; however, both copper- and sulfur-based fungicides are used in the management of a wide range of fungal diseases. The nature of fungicides requires that multiple applications be made so that emerging tissues are protected.

FUNGICIDE MOVEMENT AND SYSTEMICITY

Fungicides are chemical or physical agents that kill or inhibit the growth and/or development of fungi. In general, there are two types of fungicides: protectants and systemics. **Protectant fungicides** are chemicals that are applied to plant surfaces prior to inoculation that interact with a fungus to prevent infections from occurring, often through the inhibition of spore germination. Protectant fungicides are also referred to as contact fungicides due to the lack of postinfection activity. In contrast, a **systemic fungicide** is a chemical that is absorbed into and transported throughout a plant and may provide some curative or postinfection activity.

Movement of fungicides within a plant can vary greatly among different classes. Some fungicides are locally systemic and move only a short distance within the plant, whereas others are moderately systemic. Others are highly mobile and readily move through the plant vascular system. Most systemic fungicides move upward (acropetally) in the xylem; however, the fungicide, fosetyl-Al, is capable of moving both upward and downward (bidirectionally). Many of the fungicides used for disease control today are synthetic compounds that vary either in how they function or in their mode of action.

CATEGORIZING FUNGICIDES

Fungicides can be identified based on their product or trade name, as well as an active ingredient(s), which often doubles as the common name of a product. Furthermore, fungicides can be categorized by chemical group, mode of action, or breadth of activity (Table 26.1). Chemical grouping is based on the considerations of the molecule(s) in accordance with chemical nomenclature and may or may not have a similar chemical structure. Classification via biochemical mode of action distinguishes fungicide groups according to

TABLE 26.1

Classification of Fungicide Active Ingredients Based on Fungicide Resistance Action Committee (FRAC) Group, Mode of Action and Chemical Family Group

FRAC Group	Mode of Action	Chemical Family (Group)	Active Ingredients
1	Mitosis and cell division	Benzimidazoles	Thiabendazole
1		Thiophanates	Thiophanate-methyl
2	Respiration		iprodione Vinclozolin
3	Sterol synthesis	Imidazoles	Imazilil
3		Piperazines	Triforine
3		Pyrimidines	Fenarimol
3		Triazoles	Bitertanol Cyproconazole Difenoconazole Fenbuconazole Flusilazole Ipconazole Metconazole Myclobutanil Propiconazole Prothioconazole Tebuconazole Tetraconazole Triadimefon Triadimenol Triticonazole
4	Nucleic acid synthesis		Metalaxyl Mefenoxam
7	Respiration		Boscalid Carboxin Flutolanil
9	Protein synthesis		Cyprodinil
11	Respiration	Methoxyacrylates	Azoxystrobin Picoxystrobin
11		Methoxycarbamates	Pyraclostrobin
11		Oximino acetates	Kresoxim methyl Trifloxystrobin
11		Oxazolidine-diones	Famoxadone
11		Dihydro-dioxazines	Fluoxastrobin
11		Imidazolinones	Fenamidone

(Continued)

TABLE 26.1 *(CONTINUED)*
Classification of Fungicide Active Ingredients Based on Fungicide Resistance Action Committee (FRAC) Group, Mode of Action and Chemical Family Group

FRAC Group	Mode of Action	Chemical Family (Group)	Active Ingredients
12	Signaling		Fludioxonil
13	Signaling		Quinoxyfen
14	Lipids and membranes		Chloroneb Dicloran PCNB
14		1,2,4-Thiadiazoles	Etridiazole
17	Sterol synthesis		Fenhexamid
19	Cell wall synthesis	Peptidyl pyrimidine nucleoside	Polyoxin
21	Respiration	Cyanoimidazole	Cyazofamid
22	Cell division		Zoxamide
24	Protein synthesis		Kasugamycin
25	Protein synthesis		Streptomycin
27	Unknown		Cymoxanil
28	Cell membrane permeability		Propamocarb
29	Respiration	2,6-Dinitro-anilines	Fluazinam
30	Respiration	Tri phenyl tin compounds	Fentin hydroxide
33	Unknown	Ethyl phosphonates	Fosetyl-Al Phosphorous acid and salts
40	Cell wall synthesis	Cinnamic acid amides	Dimethomorph
40		Mandelic acid amides	Mandipropamid
41	Protein synthesis		Oxytetracycline
P	Host plant defense induction	Benzo-thiadiazole BTH	Scibenzolar-S-methyl
M	Multisite contact activity	Inorganic	Copper
		Inorganic	Sulfur
		Dithiocarbamates and relatives	Ferbam Mancozeb Maeb Metiram Thiram Ziram
		Phthalimides	Captan
		Chloronitriles Phthalonitriles	Chlorothalonil
		Guanidines	Dodine
NC	Not classified	Diverse	Mineral oils Organic oils Potassium bicarbonate

which biosynthetic pathways are affected. Examples of specific modes of action include damaging of cell membranes, inactivating critical enzymes or proteins, interfering with energy production or respiration, and inhibiting sterols or chitin. Breadth of activity is used to differentiate fungicides that are active against a single point in one metabolic pathway (single-site) versus multi-site fungicides, which affect a number of different metabolic sites within a fungus. Many of the fungicides that affect a single site have an increased risk of resistance developing within the pathogen population. In response to this, an international group of scientists, comprising the Fungicide Resistance Action Committee (FRAC), work together to recommend use patterns to try and avoid the development of resistance in high-risk fungicide groups.

FUNGICIDE FORMULATIONS

Other attributes of fungicides include how the marketed products are formulated. The major commercial fungicide formulations include dusts, granules, wettable powders, emulsifiable concentrates, and flowable formulations. Dusts represent some of the oldest agricultural pest control products that are used and are commonly used in various homeowner scenarios. Granule formulations are variations of dusts with considerably lower potential for drift. One difference between the two is that granules consist of considerably larger particles. This attribute has been exploited in using granular fungicides for the management of diseases caused by soilborne pathogens as they have the ability to better penetrate the plant canopy and be delivered to the soil surface. Wettable powders (also referred to as dispersible powders) were developed as an alternative to dusts, to circumvent issues associated with drift and adhesion problems. An emulsifiable concentrate consists of an active ingredient dissolved in a solvent that when mixed and agitated with water forms an emulsion. In general, emulsifiable concentrates have less of a problem with the clogging of nozzles in application equipment. Unlike other formulations, where active ingredients are specified as a percentage of the total volume, the amount of active ingredient in an emulsifiable concentrate is expressed as pounds per gallon of the formulation. Liquid flowable formulations combine the characteristics of emulsifiable concentrates and wettable powders, especially when the solid state of an active ingredient does not dissolve in water or oil.

LABORATORY EXERCISES

Due to the significant economic impact of fungal diseases on yield and quality of plants, chemical control measures are often warranted. Plants are susceptible to infection by fungal pathogens throughout much of their lifecycle; however, chemical control measures are primarily used most readily during the establishment and development of a crop, to maintain productivity and quality, as well as improving storage quality and longevity. The overall objective of the following exercises is to introduce students to various aspects of disease management using fungicides. The first exercise will familiarize students with pesticide labels and material safety data sheets (MSDS), which provides essential information for proper usage of pesticides and safety information. The second experiment demonstrates the effect of increasing concentrations of the fungicide pentachloronitrobenzene (PCNB) on hyphal growth of *Sclerotium rolfsii*. This information makes it possible to determine fungicide efficacy, as well as monitor the effect of the fungicide on the fungus with regard to the development of fungicide resistance. The third experiment was developed to evaluate the constituents of seed treatment fungicides on different causal agents associated with the seedling disease complex. Most seed treatment products contain combinations of protectant and systemic compounds to broaden the spectrum of activity. Adjustments to use rates can be made to increase activity; however, precautions must be made to ensure that the viability of the seed is not adversely affected. The fourth exercise will be conducted to illustrate how fungicide application timing can affect the development of postharvest diseases. The fifth and last exercise is designed to give students experience in calibrating sprayers and field application of materials.

Experiment 1. Reading and Comprehending a Pesticide Label and Material Safety Data Sheet

Pesticide labels contain essential information about the safe use of pesticides. Additional facts about mixing, application, and storage may be found on the label. Information on the label should be reviewed before purchasing, mixing, applying, storing, or disposing of the pesticide. Pesticide applicators or farm workers who are exposed to pesticides should be familiar with information on how to handle an accidental poisoning or spill prior to the incident. The goal of this exercise is to familiarize students with the safety and use information found on a pesticide label and MSDS.

Materials

- Provide each student with a sample pesticide label from a product guide or agrichemical company website.
- MSDS may be found at the back of product guides or agrichemical company website.

Follow the instructions in Procedure 26.1 to complete the exercise.

Questions

- Whom would you call for questions about pesticide use or safety?
- What would you do if a pesticide spill occurred at your business?
- How should pesticides be stored?
- What does it mean if a pesticide is a carcinogen?
- What does it mean if a pesticide is mutagenic?

Experiment 2. Biological Activity of Fungicides

Several techniques have been developed to screen fungicides for activity toward fungal pathogens. Studies have shown that testing of fungicides under artificial

Procedure 26.1
Reading Pesticide Labels and Material Data Safety Sheets

Step	Instructions and Comments
1	Find the following information on a sample pesticide label: • Trade name • Manufacturer • Common name • Formulation • Percentage active ingredient • EPA registration number • Signal word • Precautionary statements (Are there any hazards to humans or wildlife?) • Does the pesticide list Worker Protection Standards (WPS)? • What is the reentry interval (REI)? • What personal protective equipment (PPE) (Figure 26.1) is required? • List any environmental hazards • List any restrictions • How is the pesticide to be applied? • List emergency telephone numbers
2	Find the following information on the MSDS: • Pesticide trade name • Common name • Toxicological information • Acute effects of exposure • Chronic effects of exposure • Carcinogen (yes or no) • Teratogenicity (birth defects) • Reproductive effects • Neurotoxicity • Mutagenicity (genetic effects) • Accidental spill or leak information • Ecological information • Handling and storage requirements

FIGURE 26.1 Worker using a backpack sprayer with personal protection equipment (spray suit and gloves).

conditions is comparable to natural situations. *In vitro* assays are routinely used to evaluate the biological activity of fungicides on certain fungal species. Results from such testing provide information on the toxicity of new molecules as well as the chemical mode of action. The procedures used to screen fungicides should be quick and simple, and the results reliable and reproducible. The measurement of mycelial growth in an agar medium is especially useful when determining fungicide toxicity toward fungi that fail to produce asexual spores. Results from *in vitro* sensitivity assays are used to predict field performance of fungicides.

Materials

Each group of students will require the following items:

- Two- to three-day-old cultures of *Sclerotium rolfsii* growing on potato dextrose agar
- Petri dishes (n = 3) containing approximately 20 mL of potato dextrose agar amended with PCNB at concentrations of 0, 50, 20, 10, 2, 1, and 0.5 µg/mL
- Disposable nitrile gloves
- A #4 cork borer
- Alcohol lamp with matches
- Parafilm®
- Ruler or caliper
- Data sheet and graphing paper
- Calculator

Follow the protocol in Procedure 26.2 to complete the experiment.

Anticipated Results

Radial growth can be observed by 12 h after inoculation. Growth will be suppressed at lower temperatures (26°C–28°C is optimum) or if temperatures fluctuate dramatically. Ensure that Petri dishes remain in the dark to minimize the impact that light may have on the fungicide (photolysis). On the nonamended medium, the fungus will cover the surface of the medium within 4–5 days; therefore, measurements should be taken prior to the mycelium reaching the edge. Increased concentrations of the fungicide should adversely affect fungal growth compared with the nonamended controls. Furthermore, the number of sclerotia (survival structures of the fungus) may be reduced following 10–14 days of incubation.

Questions

- How did increasing concentrations of PCNB affect radial growth of *S. rolfsii*? What was your calculated EC_{50} value?
- Why were measurements made in two directions on multiple replicates?
- How was the production of sclerotia affected at higher concentrations of PCNB?
- What mode of action does PCNB exhibit when interfering with growth and development of *S. rolfsii*? What metabolic pathway is interrupted by PCNB?

Experiment 3. Effect of Seed Treatment on Disease of Cotton

Seedling diseases are a major factor affecting cotton (*Gossypium hirsutum* L.) production worldwide. Several microorganisms, including *Pythium* spp., *Rhizoctonia solani*, and *Thielaviopsis basicola*, are involved in the disease complex. Collectively, such pathogens negatively affect the germination of seed, as well as emergence, growth, and survival of seedlings. The appearance of

Procedure 26.2

In vitro Sensitivity of *Sclerotium rolfsii* to the Fungicide PCNB

Step	Instructions and Comments
1	Amend quarter-strength potato dextrose agar with technical grade PCNB (CAS: 82-68-8) to achieve final concentrations of 0.5 1, 2, 10, 20, and 50 µg/mL of medium. The fungicide should be dissolved in acetone (0.05% vol/vol) before mixing with agar that has been cooled to 60°C. Nonamended media will serve as the control. Label each Petri dish with the appropriate concentration. (Petri dishes containing amended concentrations can be prepared 1–2 days prior to use.)
2	Surface sterilize a #4 cork borer and obtain hyphal plugs of 1-cm-diameter from the edge of actively growing *S. rolfsii* colonies. Inoculate Petri dishes by placing the plug hyphal side down on the center of each of the six different concentrations and nonamended control media. Three replicates should be used for each concentration.
3	Wrap Petri dishes with Parafilm®, place them on the desktop, and cover with a box. Incubate at room temperature in the dark for 3 days.
4	Measure mycelial growth (in mm) from the edge of the inoculum plug at two locations perpendicular to one another, record your data, and calculate the mean mycelial growth for each concentration.
5	Using logarithm (base 10) graphing paper, plot the means of growth on each of the concentrations. Determine the effective concentration to inhibit growth by 50% (EC_{50} value). This can be done by regressing the percent inhibition (100 – [colony diameter on amended medium / colony diameter on the control × 100]) against the log (base 10) of the fungicide concentration.
6	Count the number of sclerotia produced by the fungus on each of the fungicide-amended media after 10–14 days.

seedling diseases in the field occur as uniform stands resulting from rotted seed or dead seedlings. Under extreme circumstances, significant yield losses may be experienced and replanting of infected areas may be required. More often, stand losses associated with seedling diseases negatively affect crop maturity, which may ultimately delay harvest. Cool temperatures and moist soil favor the development of seedling disease. Such conditions are often experienced, as producers typically plant early to maximize the growing season. Furthermore, supplemental irrigation is often applied to ensure moisture is available for germination. Currently, all cotton seed sold commercially contain fungicide constituents with activity toward *Pythium* spp., *R. solani*, and *T. basicola*.

Materials

Each group of students will require the following items:

- Premixed soils containing fungal inocula
- Twenty Styrofoam cups
- Dissecting needle
- Alcohol lamp
- Nitrile gloves
- Cotton seed treated with the appropriate fungicides and nontreated control
- Data sheets
- Dissecting kits including microscope slides and coverslips

Follow the instructions in Procedure 26.3 to complete this experiment.

Anticipated Results

Results from this exercise will document the importance of fungicide seed treatments on the emergence and survival of cotton seedlings. Fungicide seed treatments should provide increased stand establishment over seed not treated with fungicides. Stands for Apron XL-treated seed will be improved over other fungicide treatments when planted into soil containing *Pythium ultimum*. Likewise, Vitavax-PCNB will provide better stands than other treatments when grown in soil infested

Procedure 26.3

Effect of Fungicide Seed Treatments on Cotton Seedling Disease Complex

Step	Instructions and Comments
1	Inocula of *Pythium ultimum* and *Thielaviopsis basicola* can be prepared on potato dextrose agar, whereas inoculum of *Rhizoctonia solani* AG-4 should be prepared on potato dextrose agar amended with 3.5 g of wheat bran per liter. Incubate inoculated Petri dishes for 2 weeks. Blend media containing *P. ultimum* and *R. solani* in water for 3 min in a commercial blender at low speed. Chlamydospores of *T. basicola* can be harvested from flooded Petri dishes and counted with a hemacytometer (inoculum should be produced prior to meeting and made available to students beforehand).
2	Fungal inocula should be thoroughly mixed with potting mix (1:1 peat : sand). Adjust inoculum densities of *P. ultimum* to 150 CFU per gram of potting mix, 5 CFU of *R. solani* per 100 g of potting mix and 150 conidia/chlamydospores of *T. basicola* per gram of potting mix. Noninfested potting mix will serve as a control.
3	Each group will receive twelve 20-oz Styrofoam cups. Make perforations at the bottom of each cup using a sterile dissecting needle to allow for drainage. Fill cups approximately three-fourths with each of the aforementioned infested potting mixes and label with a marker.
4	Obtain cotton seeds treated with 0.32 oz/cwt metalaxyl (Apron XL), 6.0 oz/cwt carboxin PCNB (Vitavax-PCNB), and 0.84 oz/cwt myclobutanil (Systhane 40WP). Nontreated seed (black seed) should be included as a control.
5	Plant seeds ($n = 3$) of each treatment (three fungicides plus one nontreated) with three replications in 20-oz cups filled with *P. ultimum*, *R. solani*, and *T. basicola* infested and noninfested potting mixes. Seeds should be planted to a depth of 0.5 to 1.0 in.
6	Cups should be arranged in trays, well-watered, placed in a growth chamber set at 18°C–20°C and incubated for 1–3 weeks.
7	Record pre- and post-emergence damping-off and final stands of seedlings for all treatments. Plant heights should be measured and root discoloration (percent root necrosis) scored on plants recovered from potting mix infested with *T. basicola*.
8	Microscopic examinations and/or isolations of infected seedlings should be performed to identify the causal agent(s).

with *R. solani*. Finally, root necrosis will be decreased and plant height increased for seedlings emerging from seed treated with Systhane 40WP when challenged with *T. basicola*. Microscopic examinations or isolations from infected tissues will reveal the causal agent associated with seedling disease symptoms.

Questions

- What types of symptoms are associated with seedling disease? Which of these symptoms were observed in this experiment?
- What are the different morphological characteristics associated with the pathogens involved in the seedling disease complex? Why is it important that the causal agent be properly diagnosed when making management decisions?
- What effect does maintaining the experiment at cooler temperatures have on disease development? How would conducting the experiment at higher temperatures (25°C–30°C) impact your results?
- Were there observable differences in plant establishment among the different pathogen/fungicide combinations?
- Why is it important to have noninoculated and nontreated controls?
- What other fungal genera are capable of causing seedling disease?

Experiment 4. Effect of Fungicide Application Timing on Disease Development

Many fruit crops receive applications of fungicides during the course of the growing season to minimize the damaging effects of fungal diseases. Historically, there have been fewer products available for the management of postharvest diseases, such as gray mold or Botrytis rot, caused by *Botrytis cinerea*. More recently, fungicides that are considered to have reduced risk have been commercialized. Such products are considered to have reduced risk based on low mammalian toxicity, which would pose fewer risks to human health; low toxicity toward nontarget organisms, such as bees, birds, and fish; low potential for groundwater contamination; lower use rates; and compatibility with integrated pest management systems (Adaskaveg et al., 2002). Applications of postharvest fungicides are typically made on the same day fruits are harvested. These applications are made in order to protect the fruit prior to colonization of wounds that may occur during harvest procedures. Fungicide applications are made during some stage of packaging with low-volume or ultra-low-volume applicators that limit fungicide runoff. This step is often preceded by a surface washing station, which is used to remove pathogen inoculum from the surface. Without the application of a postharvest fungicide, usually fruits cannot be shipped to distant markets or displayed on the shelf as ambient temperatures increase the risk of decay.

Materials

Each group of students will require the following items:

- Actively sporulating cultures of *B. cinerea*
- Sterile water (100 mL)
- Sterile 10-mL pipette with bulb
- Rubber spatula (to harvest spores)
- Sterile, small glass funnel
- Cheesecloth
- 10-mL sterile conical tubes
- Hemacytometer
- Tween 20
- Disposable nitrile gloves
- Aerosol spray bottle
- Containers of detached strawberries (for appropriate application timing)
- Lab tape and marker
- Access to growth chamber or covered enclosure with humidifier
- Data sheets

Follow the protocol in Procedure 26.4 to complete this experiment.

Anticipated Results

Wound inoculations will facilitate the development of gray mold on nontreated fruit. Applications of Scala SC made 24 and 48 h prior to inoculation with *B. cinerea* will provide similar levels of disease suppression. Greater levels of disease should be observed on fruits that were inoculated 24–48 h prior to the fungicide being applied. Considerable disease will develop on nontreated fruit (positive control), whereas relatively low amounts of gray mold should be observed on noninoculated fruit (negative control).

Questions

- How did the application timing of the fungicide affect disease development?
- How would substituting a contact fungicide affect your results and why?

Experiment 5. Effect of Nozzle Size on Water Output of a Small, Hand Sprayer

Sprayer calibration is essential for the safe, economical, and effective application of pesticides. Hand sprayers are used to spray small areas of lawns, shrubs, or small

Procedure 26.4

Inhibition of a Postharvest Fruit Rot in Strawberry with a Systemic Fungicide

Step	Instructions and Comments
1	Grow a virulent strain of *Botrytis cinerea* on clarified V-8 agar or potato dextrose agar for 2 weeks or until sporulation occurs.
2	Flood harvest *B. cinerea* colonies ($n = 3$ per group) by pipetting 15 mL of sterile water on top of the medium. Gently rub a rubber spatula over the surface of the culture to dislodge the conidia. Combine the contents in a 50-mL conical tube by passing the water through a sterile glass funnel lined with five layers of sterile cheesecloth.
3	Count the number of conidia using a hemacytometer and dilute to $1–3 \times 10^5$ /mL in 0.05% Tween 20.
4	Wash detached strawberry fruit with tap water and allow to air dry. Separate into four groups to receive the following treatments: (a) noninoculated control, (b) curative applications (made 24 and 48 h after fruits are inoculated), and (c) preventive application (made prior to fruit being inoculated).
5	The fungicide pyrimethanil (Scala SC, Bayer CropScience, Research Triangle Park, NC) should be applied to detached fruit until run-off using a garden spray bottle at a rate of 290 µg/mL. Applications should be made 24 and 48 h prior to and after inoculation with *B. cinerea*. Nontreated and noninoculated fruits will serve as positive and negative controls, respectively. Uniformly spray fruits with conidial suspensions using an aerosol spray bottle (applications should be made by a licensed pesticide applicator and made available to students during the lab period).
6	Labeled containers of strawberries (corresponding to when fruits were treated) should be obtained from the instructor.
7	Position containers on trays and place in a dew chamber set at 15°C, >90% relative humidity (RH), and a 12-h photoperiod. If a dew chamber is not available, cover a $1.3 \times 0.7 \times 0.6$ enclosure constructed from 2.54-cm PVC pipe with clear plastic. Maintain RH by placing a humidifier inside the enclosure.
8	Incubate the containers for 24–48 h and move to a lab bench under ambient environmental conditions.
9	Continue to incubate at room temperature for 7–14 days monitoring for disease development. Record disease incidence (number of infected fruits) and severity (percentage of fruit surface exhibiting gray mold symptoms) for each of the different treatments.

trees and may be used to apply fungicides, insecticides, and herbicides. The volume of water applied to a given area is often determined by the pesticide and target pest. For example, many turf herbicides are applied in 1 gal water/1000 ft^2, whereas most turf fungicides are applied in 3 gal water/1000 ft^2 since good coverage is necessary to protect the foliage.

The amount of water output of most agricultural sprayers over a given area is determined by travel time (in the case of hand sprayers, walking speed), pressure, and nozzle type and size. With most hand sprayers, output is controlled by nozzle size, as pressure is constant if the sprayer is continually pumped. Constant pressure is easier to maintain with backpack sprayers that can be pumped during application. Calibrate the sprayer in a parking lot or other area where the spray pattern of the nozzle on the ground is easily observed.

Materials

The following items are required for each student or team of students:

- Carboy of water or easily accessible source of water
- 1- to 3-gal compressed air hand sprayer or a solo backpack sprayer
- Flat fan nozzles (Delevan D1, D3, D5 or Spraying Systems TK1, TK3, TK5)
- 1-L graduated cylinder

Follow the protocols outlined in Procedure 26.5 to complete this exercise.

Anticipated Results

Flat fan nozzles should give uniform coverage. Water output should have been increased as nozzles with larger

Procedure 26.5

Effect of Nozzle Size on Output of a Small, Hand Sprayer

Step	Instructions and Comments
1	Measure a test area of 500 ft^2 (e.g., 20 × 25 ft) on a parking lot (where it is easy to see the coverage and spray pattern) or a grassy area.
2	Fill sprayer with a measured volume of water or to a marked level on the sprayer.
3	Uniformly spray the test area. Make sure that the surface is evenly wet.
4	Release compressed air and determine the amount of water used by measuring the water remaining in the sprayer or the amount of water needed to raise the water level to the initial level.
5	Calculate the application rate of the sprayer with Method 2:128th of an acre calibration method.
6	Measure a test area of 342 ft^2 (approximately 18.5 × 18.5 ft).
7	Repeat Steps 2–4.
8	Another method for determining water output is to record the amount of time that it takes to spray the test plot. Spray the test plot three times and average the spray times.
9	Fill sprayer with water, pressurize and catch the output in the 1-lL graduated cylinder for the average time required to spray the test plot. The output in fluid ounces is equal to the number of gallons the sprayer would apply to an acre.

orifice sizes were tested. If adequate coverage is not provided at a normal walking speed, then nozzle size should be increased or walking speed decreased.

Questions

- How would you adjust water output of a compressed air hand sprayer? A tractor-mounted hydraulic sprayer?
- What types of nozzles are used to apply pesticides?

REFERENCES

Adaskaveg, J. E., H. Förster, N. F. Sommer. 2002. Principles of postharvest pathology and management of decays of edible horticultural crops. In: A. Kader (ed). *Postharvest Technology of Horticultural Crops*, 4th edition. Oakland, CA: UC DANR Pub 3311, pp. 163–195.

27 Management of Plant Pathogens and Pests by Microbial Biological Control Agents

Dmitri V. Mavrodi, Olga V. Mavrodi, Leonardo De La Fuente, Blanca B. Landa, Linda S. Thomashow, and David M. Weller

CONCEPT BOX

- Plant diseases, pests, and weeds can be controlled by antagonistic organisms and/or their metabolites.
- Mechanisms of biological control of plant pathogens and pests include antibiosis/toxin production, parasitism/predation, competition/niche exclusion, and induced systemic resistance (ISR).
- Biological control agents often control pathogens via multiple biocontrol mechanisms.
- Biological control agents may also promote plant growth directly by producing growth-promoting compounds that mobilize nutrients or reduce the effects of ethylene (ET) produced by the plant during stress.
- Commercial production of biopesticides is a rapidly growing industry driven by growing environmental concerns and restrictions on the use of traditional chemical pesticides.

Biological control is the use of beneficial living organisms called **biocontrol agents (BCAs)** to reduce the numbers or effects of undesirable pests. BCAs include natural or genetically modified pathogens, parasites, or predators that are deliberately introduced in the environment or, if already present, are encouraged to multiply with the purpose of reducing the survival and/or activity of pest organisms. This chapter focuses on microorganisms that are defined in legal and commercial terms as "**microbial biopesticides**" and are commercially produced for the purpose of controlling plant diseases, pests, and weeds or for the manipulation of plant physiology and productivity (Glare et al., 2012; http://www.epa.gov). Microbial biopesticides may consist of live or killed bacteria, fungi, yeasts, protozoa, viruses, and/or the metabolites they produce. In addition to these, the U.S. Environmental Protection Agency (EPA) also defines biopesticides as "naturally occurring substances that control pests (biochemical pesticides)" and "pesticidal substances produced by plants containing added genetic material (plant-incorporated protectants)," which are beyond the scope of this chapter and are not discussed here.

ANTAGONISM AS A MAJOR BIOCONTROL MECHANISM

Biological control is a complex phenomenon, and pathogen suppression often results from the synergistic interaction of different types of biocontrol mechanisms. Most often, the biocontrol results from antagonism, in which a BCA adversely affects the population dynamics and/or activity of a pathogen. This antagonism may result from a direct interaction, which involves physical contact and either (i) destruction of the pathogen by antibiotics, **toxins**, and lytic enzymes produced by the BCA or (ii) parasitism/predation of the pathogen by the BCA. Alternatively, antagonism may result from indirect interactions where the BCA suppresses the pathogen by competing for essential nutrients or space. The underlying basis for each of these types of interactions can be complex and multifactorial.

ANTIBIOSIS AND PRODUCTION OF TOXINS

Antibiosis is a major mechanism underlying biological control and involves the inhibition or destruction of a pathogen by a metabolite produced by a BCA (Baker and Cook, 1974). These inhibitory metabolites, or antibiotics, are low-molecular-weight organic compounds that, even at low concentrations, suppress the growth or metabolic activities of pathogens (Thomashow and Weller, 1995). The key role of antibiotics in the suppression of plant pathogens has been documented in numerous studies published over the past four decades. Mining of recently sequenced genomes of BCAs has revealed that most BCAs have the capacity to produce multiple antibiotics and antagonistic metabolites, often with broad and overlapping spectra of activity. To date, however, only a small proportion of antibiotics produced by BCAs have been studied as is evidenced from the recent flood of genomic-based discoveries of new antimicrobial and insecticidal compounds (Gross and Loper, 2009).

The biochemistry and genetics of antibiotic production is best studied in biocontrol strains that belong to *Bacillus* (see Case Study 27.1) and *Pseudomonas* (see Case Study 27.2) genera. The types of antibiotics synthesized by these BCAs are very diverse and include peptides and polyketides and hybrids thereof, cyclic lipopeptides, derivatives of chorismate and amino acids, macrolides and aminoglycosides. Another well-studied

CASE STUDY 27.1

Bacillus subtilis var. *Amyloliquefaciens* Strain FZB24

Biocontrol agent—*Bacillus subtilis* var. *amyloliquefaciens*, a Gram-positive bacterium

- Target pathogens—*Rhizoctonia* spp. and *Fusarium* spp.
- Disease—Soilborne diseases caused by *Rhizoctonia* and *Fusarium*, on fruit and leafy vegetables, cucurbits, vines, herbs and spices, ornamentals, shrubs, fruit, shade and forest trees, and turf. Also can be used on plants grown outdoors, indoors, and in greenhouses.
- Mechanisms of biocontrol—*Bacillus subtilis* bacteria produce a class of lipopeptide antibiotics including iturins. Iturins help *B. subtilis* bacteria outcompete other microorganisms by either killing them (inhibits plant pathogen spore germination, disrupts germ tube growth) or reducing their growth rate (interferes with the attachment of the pathogen to the plant) and have direct fungicidal activity on pathogens. *B. subtilis* induces SAR against bacterial pathogens.
- Application and formulation—Taegro® (Novozymes A/S). Contains 13% active ingredient with a minimum of 1.0×10^{10} colony forming units (CFU)/g and 87% other ingredients. Taegro® must be premixed thoroughly with water and applied within a few hours of mixing. Taegro® can be applied to transplants by dipping or drenching and to seedlings or to newly rooted cuttings by drenching (http://www.syngenta-us.com/thrive/product/taegro-eco-fights-diseases.html [accessed 07/12/16]; Fravel, 2005).

CASE STUDY 27.2

Pseudomonas aureofaciens Strain BS1393

Biocontrol agent—*Pseudomonas aureofaciens*, a Gram-negative bacterium

- Target pathogens—*Botryotinia, Fusarium, Helminthosporium, Pseudocercosporella, Pythium, Erysiphe, Septoria, Pyrenophora, Puccinia, Pseudomonas, Xanthomonas, Rhizoctonia, Cladosporium,* and *Cercospora.*
- Disease—Gray mold, Cercospora leaf spot, black rot, Fusarium head blight, Pythium damping-off, Septoria leaf spot, rust diseases in cereal, vegetable, and fruit crops.
- Mechanisms of biocontrol—*P. aureofaciens* BS1393 produces phenazine compounds and stimulates root development and growth of plants, improves phosphorus nutrition, strengthens crop immunity, activates microbiological activity in soil microflora, and increases crop yields by 10%–15% and their quality.
- Application and formulation—Pseudobacterin®2 (Biona Group; Sibbiopharm) is registered and used in Russia and Ukraine as a foliar spray. It contains live cells of *P. aureofaciens* BS1393 (5.0×10^9 CFU/mL) and a set of biologically active substances. The application rate of Pseudobacterin®2 is 2–3 L/ha mixed with 200–300 L of water for cereal crops and 800–1200 L of water for vegetable and fruit crops (http://biona.ua/en/catalog/rastenievodstvo/ [accessed Jul 14, 2016]).

example of the involvement of antibiotics in biological control is provided by the commercially produced biocontrol bacterium *Agrobacterium radiobacter* K84 (see Case Study 27.3). This BCA produces the antibiotic agrocin 84, which kills the closely related plant pathogenic bacterium *Agrobacterium tumefaciens*, a causative agent of gall disease in a wide range of dicotyledonous plants (Kim et al., 2006). Agrocin 84 is very selective and kills only certain strains of *A. tumefaciens* that carry nopaline/agrocinopine gall-inducing plasmids. Although agrocin 84 is a low-molecular-weight antibiotic, its high selectivity closely resembles the action of bacteriocins, a class of bacterial proteinaceous toxins that specifically inhibit the growth of sensitive strains of the same or closely related species.

In order to suppress plant pathogens, antibiotic compounds must be produced by BCAs in sufficient amounts at or near the site of infection. The quantification of antibiotics *in situ* is technically very challenging, but nevertheless possible with the use of modern molecular and analytical techniques. The application of such tools to BCAs has revealed that antibiotic biosynthesis genes are indeed expressed *in situ* and that antibiotics may accumulate, at least transiently, in amounts sufficient for the inhibition of plant pathogens. Finally, recent studies suggest that antibiotics play several ecological roles that are of crucial importance for the successful establishment and performance of BCAs. In addition to pathogen suppression, antibiotics may protect BCAs from predation by protozoa and nematodes. At subinhibitory concentrations, antibiotics act as molecular signals that regulate colony morphology and biofilm formation. Certain classes of antibiotics play important roles in surface motility and regulate redox homeostasis and the acquisition of micronutrients by BCAs (Raaijmakers and Mazzola, 2012).

In addition to antibiotics, diverse toxins, volatile metabolites, and bacteriocins also have been implicated as important antagonistic factors in BCAs. The best known of these are Cry (crystal) toxins of the entomopathogenic bacterium *Bacillus thuringiensis* (Bt) (Bravoa et al., 2011). Bt represents the most widely manufactured biopesticide and is used worldwide to control larvae of pest butterflies, moths, beetles, mosquitoes, and blackflies. The commercial biopesticide preparations contain a mixture of Bt spores and insecticidal toxins in the form of crystal-like parasporal inclusion bodies. The parasporal bodies are made of protoxin (an inactive form of toxin) that requires ingestion and exposure to the high pH of the larval midgut for activation. The active toxin possesses membrane-disrupting activity and forms pores in the midgut epithelium, which allows Bt to invade the larval hemocoel (body cavity) and ultimately kill the infected host. Bt strains and their toxins exhibit high selectivity toward particular groups of insects and are completely safe for nonhost organisms. Genes encoding Cry toxins have been stably introduced into the genomes of different crops. Such genetically engineered "Bt crops" express toxins and kill insects that feed on their tissues.

Parasitism

Some BCAs directly attack and kill phytopathogens and pests and use them as a source of food. Such **hyperparasites** encompass many highly effective commercialized BCAs including the mycoparasitic fungi *Trichoderma* and *Paraphaeosphaeria* (formerly *Coniothyrium*). *Trichoderma* is a **rhizocompetent** filamentous fungus that attacks fungal pathogens by sensing their hyphae, coiling along them, and ultimately penetrating them with the help of powerful lytic enzymes

CASE STUDY 27.3

Agrobacterium radiobacter Strain K84

Biocontrol agent—*Agrobacterium radiobacter*, a Gram-negative bacterium

- Target pathogens—*A. tumefaciens* and *A. rhizogenes*, Gram-negative soilborne bacteria.
- Disease—Crown gall on certain fruit, nut, and ornamental nursery stock (including bare-root seedlings, trays of very young plants, and planting stock).
- Mechanisms of biocontrol—A nonpathogenic *A. radiobacter* strain K84 produces an antiagrobacterial antibiotic compound, agrocin 84, and competes for niche and nutrients with pathogenic *A. tumefaciens* and *A. rhizogenes*.
- Application and formulation—Galltrol-A® (AgBioChem, Inc.). Sold as a bacterial culture in Petri dishes containing 50 mL of agar culture medium with 1.2×10^{11} colony-forming units (CFUs) of bacteria (equivalent to 1 L of 1.2×10^{8} cells/mL). Bacterial cells are washed from one Petri plate into 1 gallon of water and agitated well before adding to a sprayer. Plants are treated (maximum 12 h) after wounding (http://agbiochem.com/products/ [Accessed Jul 14, 2016]; Fravel 2005).

(Druzhinina et al., 2011). In addition to being a mycoparasite, *Trichoderma* competes with fungal pathogens for nutrients and space and produces volatile and nonvolatile antimicrobials. Finally, *Trichoderma* can also act as an opportunistic plant symbiont by colonizing root surfaces and synthesizing phytohormones and defense response elicitors that enhance plant growth and disease resistance (see Case Studies 27.4 and 27.5). In addition to attacking living hyphae, some hyperparasitic BCAs can also attack resting structures, which are produced by some fungal pathogens, and serve as an important source of inoculum in the field. For example, the obligate mycoparasite, *Paraphaeosphaeria minitans* (formerly *Coniothyrium minitans*), is used to manage *Sclerotinia sclerotiorum*, a destructive pathogen of many economically important crops (Paulitz and Belanger, 2001). *Paraphaeosphaeria* controls *Sclerotinia* by destroying its melanized sclerotia that otherwise are highly resistant to chemical and biological degradation. Although details of this interaction are poorly understood, it is known that *Paraphaeosphaeria* destroys sclerotia of *Sclerotinia* by penetrating them via the use of cell-wall-degrading enzymes and physical pressure.

Mycoparasitic fungi are not the only kind of hyperparasites that are commercially exploited for the control of plant pathogens and pests. *Purpureocillium lilacinum* (formerly *Paecilomyces lilacinus*) is a naturally occurring soil fungus that has been commercially formulated as a bionematicide for the control of sedentary plant parasitic nematodes. *P. lilacinum* attacks eggs, juvenile, and female nematodes by secreting hydrolyzing enzymes and has been used successfully to control a range of nematode

CASE STUDY 27.4

TRICHODERMA ASPERELLUM T11 AND *TRICHODERMA ATROVIRIDE* T25

Biocontrol agent—*Trichoderma*, a mycoparasitic ascomycete fungus

- Target pathogens—*Fusarium, Phoma, Phytophthora, Pythium, Rhizoctonia* and *Sclerotinia species,* and *Polymyxa betae.*
- Disease—Damping-off diseases, Rhizomania disease of sugar beet, and drop of lettuce.
- Mechanisms of biocontrol—Broad spectrum of mechanisms of action including strengthening of plant defense mechanisms and nutrient solubilization, general reduction of plant pathogen populations in soil mainly through direct and indirect mycoparasitim, and competition for nutrients and soil substrates.
- Application and formulation—Tusal WG® (Certis Europe) is sold as water dispersible granules containing 1×10^8 CFU/g of each *T. harzianum* and *T. atroviride*. It can be applied by seed pelleting or through drip irrigation systems. It is certified for use in greenhouses and open fields and in organic farming (for more information visit http://www.certiseurope.es/ and http://www.nbt.es/NBTBiocontrol).

CASE STUDY 27.5

TRICHODERMA ASPERELLUM STRAIN ICC012 AND *TRICHODERMA GAMSII* STRAIN ICC080

Biocontrol agent—*Trichoderma*, a mycoparasitic ascomycete fungus

- Target pathogens—*Armillaria, Phytophthora, Pythium, Rhizoctonia, Rosellinia, Sclerotinia, Sclerotium rolfsii, T. basicola*, and *Verticillium dahliae*.
- Disease—Root and stem rots, Sclerotium root rot, Verticillium wilt, and white mold.
- Mechanisms of biocontrol—Direct and indirect mycoparasitism of soilborne plant pathogens and competition for nutrients.
- Application and formulation—Registered and used in the United States and Europe with different names BioTam® (AgraQuest), Radix® (Certis Europe), Tenet WT® (Agrian, Inc.) sold as wettable powder containing 5×10^7 CFU/g of *T. asperellum* and of *T. gamsii*. Products can be applied by seed pelleting or through drip irrigation systems and are certified for use in greenhouses and open fields and in organic farming (for more information see the biopesticides registration action document "decision_PC-119208_4-Mar-10.pdf" at http://www.epa.gov).

species on different crops. Several species of entomopathogenic fungi, and in particular those belonging to the genera *Metarhizium, Beauveria, Lecanicillium,* and *Isaria*, are produced as biopesticides for the management of pest insects (Vega et al., 2009). These fungi have evolved specialized mechanisms for the enzymatic degradation of insect cuticle and for overcoming insect immune defenses and are capable of rapidly killing insect pests. Interestingly, recent studies suggest that many fungal entomopathogens also are capable of colonizing plant tissues and providing phytostimulatory effects, and even protecting their hosts against phytopathogenic fungi. Finally, several types of entomopathogenic viruses, particularly those belonging to the baculovirus group, are produced as commercial bioinsecticides for the control of pest butterfly and moth caterpillars. Like other types of viruses, baculoviruses are obligate parasites. They infect larvae upon ingestion and then use the host cellular machinery to multiply within the infected insect, ultimately killing it.

Competition

Plant-associated microbial communities are tremendously diverse. For example, recent studies based on culture-independent techniques such as deep metagenomic and microbial tag sequencing estimated that the **rhizosphere** of plants grown in agricultural soils may harbor tens of thousands of different bacterial species (Mendes et al., 2011). However, although plant surfaces and rhizosphere soil have more available nutrients than bulk soil, plant-associated microbes still have to constantly compete for the same pool of resources. Such **competition** among species sharing an ecological niche and physiological requirements often results in the dominance of the competitor that is more efficient in obtaining nutrients and/or has higher reproduction or growth rates.

Competition is an important mechanism underlying the capacity of some BCAs to control pathogens and reduce disease incidence and severity. In some cases, the BCA can actively restrict or remove a certain nutrient from a pathogen, as happens with beneficial rhizosphere-dwelling fluorescent *Pseudomonas* spp. that control the soilborne pathogen *Fusarium oxysporum.* Chlamydospores of *F. oxysporum* require an exogenous supply of iron for efficient germination. In calcareous soils, which have particularly low amounts of soluble iron, the sequestration of iron by **siderophores** secreted by fluorescent *Pseudomonas* spp. leads to the inhibition of germination of *Fusarium* chlamydospores (Lemanceau and Alabouvette, 1993). In other cases, a BCA may rapidly colonize plant surfaces and exhaust the limited available space and resources essential for growth and activity of the pathogen. Such site exclusion is attributed to the ability of the nontoxigenic *Aspergillus flavus* (see Case Study 27.6) to protect corn, peanuts, and cottonseed from closely related strains that produce hepatotoxic and carcinogenic aflatoxins. Similarly, *Pseudomonas fluorescens* A506 protects apples and pears from fire blight by colonizing flower tissues and outcompeting the pathogen *Erwinia amylovora* (Stockwell et al., 2010). Yet another example of pathogen exclusion via competition is provided by the BCAs *Yarrowia lipolytica* (formerly *Candida oleophila*) strain O and *Pseudomonas syringae* strains ESC-10 and ESC-11, which are used for the prevention of postharvest decay caused by the gray mold *Botrytis cinerea* and the blue mold *Penicillium expansum* on apples, pears, and other fruits (Mercier and Marrone, 2006).

CASE STUDY 27.6

Aspergillus flavus Strain NRRL 21882

Biocontrol agent—*Aspergillus flavus*, an aflatoxin nonproducing ascomycete fungus

- Target pathogens—*A. flavus* producing toxin and a carcinogenic compound, aflatoxin, in peanuts and corn (field, sweet, and popcorn).
- Disease—*A. flavus* is responsible for infection of corn and peanuts, and aflatoxin contamination prior to harvest or during storage. *A. flavus* also causes diseases of animals. When processed, aflatoxins get into the general food supply of pet and human foods as well as feedstocks for agricultural animals.
- Mechanisms of biocontrol—Niche competition is the primary mechanism. Afla-Guard®GR (*A. flavus* NRRL 21882) displaces aflatoxin-producing *A. flavus.*
- Application and formulation—Afla-Guard®GR (Syngenta Crop Protection, Inc.) contains 0.0094% active ingredient with a minimum of 1.2×10^8 CFU/lb of *A. flavus* NRRL 21882 and inert ingredients. Afla-Guard®GR can be delivered as soil or aerial applications (http://www.syngentacropprotection.com/afla-guard-gr-biocontrol-agent [accessed Jul 14, 2016]; Fravel, 2005).

INDUCED SYSTEMIC RESISTANCE

Plants possess multiple constitutive and inducible defense mechanisms involved in counteracting pathogen attacks. Constitutive defense is provided by the cell wall, which acts as a physical and chemical barrier against invading pathogens. Inducible defense mechanisms are triggered by contact with the pathogen and include two types of responses (Van Wees et al., 2008). The first type of inducible defense response is called systemic acquired resistance (SAR). It is mediated by the plant hormone salicylic acid (SA) and pathogenesis-related (PR) proteins and results in deposition of callose, production of antimicrobials and reactive oxygen species, and programmed cell death in tissues surrounding the site of infection. The second type of inducible defense response is known as **induced systemic resistance** (ISR), which involves exposure of the plant to a BCA prior to infection. This perception of a resistance-inducing agent does not result in the immediate activation of PR proteins and other defense mechanisms. Rather, it primes the plant's innate immunity for a faster and/or stronger response to subsequent pathogen attack. Early experiments with specific *Arabidopsis* mutants revealed that canonical ISR is mediated by signaling via the plant hormones jasmonic acid (JA) and ethylene (ET). However, studies conducted during the past decade have revealed that signaling pathways involved in ISR induction are very diverse and depend on the nature of the BCA and the plant–pathogen system (Chapter 20).

Many BCAs can suppress phytopathogens via induction of ISR, and this defense mechanism has been proven effective in the control of diverse phytopathogens and pests including viruses, bacteria, fungi, oomycetes, nematodes, and herbivorous insects (De Vleesschauwer and Höfte, 2009). The essential feature of ISR is spatial separation of the BCA and the pathogen, which therefore excludes the possibility of direct antagonism. A classic example is the induction of resistance to a foliar pathogen in aboveground plant tissues upon treatment of plant roots by an ISR-inducing BCA. Molecular determinants that trigger ISR are best studied in biocontrol bacteria and include flagella, lipopolysaccharide, exopolysaccharides, biosurfactants, N-acyl-L-homoserine lactones, siderophores, antibiotics, and volatiles.

PLANT GROWTH PROMOTION

In addition to suppression of plant pathogens and induction of ISR, many BCAs also improve plant health and vigor. The application of such beneficial microorganisms does not inhibit pathogens directly, but rather increases plant tolerance to biotic and abiotic stresses by stimulating nutrient uptake and organ development. The promotion of plant growth by microorganisms is a multifactorial phenotype that includes solubilization of inorganic phosphate, nitrogen fixation, production of iron-chelating siderophores, synthesis of phytohormones, and modulation of plant ET levels or any combination thereof (Kim et al., 2011).

Phosphorus (P) is one of the essential macronutrients for plant growth. Despite the overall abundance of P in soil, the majority of it is present in the form of highly insoluble mineral and organic phosphates that are not suitable for assimilation by plants. Many types of BCAs are known to render P available to plants by solubilizing the soil phosphates via production of organic acids (gluconic and citric), acid phosphatases, and the release of protons.

Certain groups of plant growth-promoting and biocontrol bacteria are capable of supplying plants with another essential macronutrient, nitrogen (N). Such beneficial bacteria, also called **diazotrophs**, can fix atmospheric N via the nitrogenase enzyme complex. The largest amount of N is provided to plants by diazotrophs called rhizobia, which form symbiotic associations with leguminous plants and are not generally considered BCAs. BCAs with confirmed diazotrophic activity include some species of *Burkholderia* and *Azospirillum*, which live in close proximity to plant roots and form associative relationships with their plant hosts. Such bacteria provide only a modest amount of fixed N to their host plants. However, the process of nitrogen fixation is important for the establishment, survival, and performance of these BCAs in the environment.

In addition to macronutrients, plants require several microelements for proper growth and development. One of these, iron, is abundant in the environment, but mostly in the insoluble oxidized form, which is biologically inaccessible. To cope with the limited supply of iron, microorganisms scavenge it by producing low-molecular weight metabolites with high affinity for Fe^{3+} called siderophores. Siderophores are secreted in the environment, where they bind iron, and the resultant Fe-siderophore complexes are taken up by dedicated membrane receptors. Siderophores produced by BCAs can efficiently supply iron to plants and improve their growth under conditions of iron limitation or heavy metal pollution.

Many BCAs are known to stimulate germination of seed and tubers, promote stem and root growth, and alleviate the negative effects of various biotic and abiotic stresses. This beneficial activity is attributable to the capacity of plant-associated microorganisms to modulate host phytohormone levels. Treatment of plants with strains producing auxins, cytokinins, and/or gibberellins commonly results in better formation of root hairs, increased root growth and branching, and, as a result, improved mineral and nutrient uptake. Certain groups of BCAs also produce an enzyme called 1-aminocyclopropane-1-carboxylate (ACC) deaminase, which cleaves the immediate precursor of the plant hormone ET. Lower ET levels can result in longer roots and less inhibition of ET-sensitive plant growth following environmental or pathogen-induced stress.

DISEASE SUPPRESSIVE SOILS

Soilborne plant pathogens are major yield constraints in crop production and are difficult to control (Chapter 16). Plants lack genetic resistance to most soilborne pathogens and instead rely on the stimulation and support of populations of antagonistic rhizosphere microorganisms. The special type of soils that are called "disease suppressive" provides the best example of this symbiosis. In **disease suppressive soils** "the pathogen does not establish or persist, establishes but causes little or no damage, or establishes and causes disease for a while, but thereafter the disease is less important, although the pathogen may persist in the soil" (Baker and Cook, 1974). In contrast, conducive (nonsuppressive) soils are soils in which disease readily occurs. The capacity of a suppressive soil to restrict the growth and/or activity of a pathogen is either associated with the natural soil properties or is initiated and sustained by crop monoculture.

Suppressive soils owe their activity to both general and specific types of suppression (Weller et al., 2002). General suppression results from the ability of the total microbial biomass in any soil to compete with the soilborne pathogen and suppress its growth and/or activity to a limited extent. In contrast, specific suppression results from the activity of a select group(s) of microorganisms. It is highly effective and is transferable between soils. Suppressive soils are known for many pathogens, but the basis of the suppressiveness in most cases is unknown. Soils in which the mechanism of specific suppressiveness is at least partially understood include those suppressive to take-all disease of wheat caused by *Gaeumannomyces graminis* var. *tritici*; to Fusarium wilt caused by *F. oxysporum*; to Rhizoctonia root rot caused by *Rhizoctonia solani*; to black root rot of tobacco caused by *Thielaviopsis basicola*; to potato scab caused by *Streptomyces scabiei*; and to the plant parasitic nematode *Heterodera schachtii*.

For example, the suppressiveness to take-all, also known as take-all decline (TAD), develops in fields with monoculture of wheat after one or more severe outbreaks of the disease (Weller et al., 2002). TAD manifests itself as a spontaneous decrease in take-all incidence and severity. The primary mechanism responsible for this phenomenon is microbiological changes in the soil, resulting in antagonism of the pathogen, *G. graminis* var. *tritici*. In particular, TAD is associated with the buildup of large populations of a distinct type of beneficial *Pseudomonas* bacteria. These bacteria actively colonize roots of wheat and suppress *G. graminis* var. *tritici* via production of the polyketide antibiotic 2,4-diacetylphloroglucinol (2,4-DAPG). The antibiotic produced by bacteria in the plant rhizosphere plays a key role in the suppression of disease, as the take-all pathogen is highly sensitive to 2,4-DAPG. TAD is a field phenomenon that occurs globally, and 2,4-DAPG-producing pseudomonads have been recovered from TAD soils studied in different parts of the world. Interestingly, rhizosphere-dwelling *Pseudomonas* that produce other types of antibiotics have recently been implicated as key antagonistic components of microbial communities from a French soil suppressive to Fusarium wilt of melon (Mazurier et al. 2009) and a Dutch soil suppressive to Rhizoctonia root rot of sugar beet (Mendes et al. 2011).

COMMERCIAL IMPLEMENTATION OF BIOLOGICAL CONTROL

Microbial BCAs have been around for a long time. As early as in 1835, the Italian entomologist Agostino Bassi demonstrated that white muscadine disease in silkworms is caused by the fungus *Beauveria bassiana*. Another early discovered biocontrol microorganism is the entomopathogenic bacterium *Bacillus thuringiensis*, which was first isolated from infected silkworms by the Japanese biologist Shigetane Ishiwatari and rediscovered a decade later in Germany by Ernst Berliner. *B. thuringiensis* was also the earliest commercially manufactured microbial BCA. The first commercial Bt products appeared in France in the late 1930s, and the widespread production and use of *B. thuringiensis* in the United States started in the 1950s. Many more microorganisms—antagonists or hyperparasites of plant pathogens and pest insects—were identified during the first half of the twentieth century, and a number of them were successfully tested in field-scale trials. However, the rates of adoption of microbial biopesticides remained low because of the rapid rise of cheaper and more efficient, but at the same time more toxic, synthetic chemical pesticides.

In the 1980s and 1990s, the situation started to change. The drawbacks of the overuse of synthetic pesticides led to the resurgence in academic and industrial research related to new biopesticide development. A number of new microbial BCAs were successfully commercialized including *Agrobacterium radiobacter* for the control of crown gall disease of woody crops and *Pseudomonas fluorescens* for the prevention of fire blight in orchards.

In the past decade, growing public awareness of the long-term impact of synthetic pesticides on human health, growing environmental concerns, and restrictions on the use of traditional chemical pesticides have led to consumer demands for organically grown food (Glare et al., 2012). Farmers have become more receptive to the incorporation of biopesticides in both organic and conventional systems. Finally, biopesticide technology has matured, which has resulted in improved formulation techniques, large-scale manufacturing of biopesticides, improved application methods, and better storage and shelf life. All these factors have led to resurgence in the development and registration of new microbial biopesticides. In fact, since 1995, the U.S. EPA has registered more than 100 new biopesticide-active ingredients (http://www.epa.gov).

Currently, microbial biopesticides are used on specialty and orchard crops, field crops such as corn and soybeans, and certain forage crops, as well as in the areas of public health and forestry. Recent estimates offered for the size of the biopesticide market vary dramatically due to the incredible diversity in biopesticide products and areas of their application. Despite these differences, most recent studies agree that the biopesticide market is growing (Glare et al., 2012). For example, in 2014, the global biopesticide market was valued at $2.09 billion, and is expected to reach $5.11 billion by 2020 (http://www.researchandmarkets.com/research/qdjgpv/global_bio).

COMMERCIALLY AVAILABLE MICROBIAL BIOPESTICIDES

Microbial biopesticides are commercially produced in the form of dusts, dry and wettable powders, granules, or liquids and are applied via incorporation into soil or as foliar sprays, root dips, or seed or fruit coatings or drenches. Live BCAs are often formulated in a dormant state with a supply of nutrients, which prolongs shelf life of the product and aids quick establishment in the environment upon application. Products based on bacteria belonging to Firmicutes (*Bacillus*) are the most well-known and widely used of all biopesticides. Insecticides based on *B. thuringiensis* and some closely related bacterial species are used worldwide to control caterpillar pests, or beetle, fly, and mosquito larvae. Other species of *Bacillus* (i.e., *B. subtilis, B. pumilus*, and *B. amyloliquefaciens*) and some Proteobacteria (i.e., *Agrobacterium, Pseudomonas*, and *Azospirillum*) are used to control various plant diseases, insects, nematodes, and weeds and to promote plant and root growth.

Fungal biopesticides are based on diverse members of the Ascomycetes (*Trichoderma, Beauveria, Metarhizium, Paraphaeosphaeria, Gliocladium, Purpureocillium*, nonpathogenic *Fusarium*) and include species that are parasitic to various eukaryotes. The most popular commercial fungal biopesticides are those based on *Trichoderma* spp. and *B. bassiana*. Beneficial *Trichoderma* (see Case Studies 27.4, 25.5, and 27.7) has the ability to antagonize phytopathogenic fungi and to stimulate plant growth and defense responses. It is used in the field, on vegetable and ornamental crops, and in forestry industries to control a variety of diseases and pests. *B. bassiana* (see Case Study 27.7) is a broad-host-range parasite of many insect species. It has been proven to be effective in controlling economically important crop pests such as thrips, aphids, and whitefly. Other fungal biopesticides are used to manage various plant pathogenic fungi or nematodes and to control different weeds or postharvest damage on fruits and vegetables.

Commercially produced viral biopesticides are mostly based on baculoviruses that infect and kill several species of pest *Lepidoptera* larvae (caterpillars). The viral insecticides help to minimize chemical residues and reduce the risk of resistance to chemical insecticides. The viruses are also very specific in their action, which allows managing crop pests with minimal off-target

CASE STUDY 27.7

Beauveria bassiana Strain Atcc 74040

Biocontrol agent—*Beauveria bassiana*, an entomopathogenic ascomycete fungus

- Target insects—Adult and larval stages of whitefly species, which belong to the family Aleyrodidae, *Bactrocera oleae, Ceratitis capitata, Frankliniella occidentalis, Rhagoletis cerasi*, and *Trialeurodes vaporariorum*.
- Pests—White flies, tetranychid mites, thrips, wireworms, and fruit flies.
- Mechanisms of biocontrol—The entomopathogenic fungus acts primarily by contact: once attached to the insect's cuticle, the conidia germinate producing penetration hyphae, which enter and proliferate inside the insect's body. The fungus invades and feeds on its host, causing its death due to dehydration and/or depletion of nutrients.
- Application and formulation—Naturalis-L® (Belchim Crop Protection Ltd.; Troy Biosciences, Inc.) is a bioinsecticide based on an oil dispersion formulation containing 7.16% w/w of living conidia of the naturally occurring *B. bassiana* strain ATCC 74040. The formulated product contains at least 2.3×10^7 viable spores/mL. The product can be used in food crops and ornamental plant production. Since the bioinsecticide is a dispersion of living conidia in vegetable oil, it can also be used in organic farming (for more information visit http://orgprints.org/13651).

effects. The granulovirus of the codling moth *Cydia pomonella* (CpGV) represents an example of a commercially successful viral insecticide. It is accepted for use in both conventional and organic farming and is employed worldwide to control populations of codling moth and limit damage in pome fruits. Other commercial viral insecticides contain the nuclear polyhedrosis virus and are used for the control of larvae of the beet armyworm (*Spodoptera exigua*), corn earworm (*Helicoverpa zea*), tobacco budworm (*Heliothis virescens*), and other important pest insects. A nonexhaustive list of biocontrol species and brands of biopesticides that are commercially produced in the United States and some other countries is given in Table 27.1.

TABLE 27.1
Biocontrol Species and Brands of Biopesticides That Are Commercially Produced in the United States and Other Countries

Pathogen Group	
	Bactericides
Bacteria:	*Agrobacterium radiobacter* (Galltrol, Nogall, Diegall, Norbac 84C)
	Pseudomonas fluorescens A506 (BlightBan)
	Fungicides
Bacteria:	*Bacillus amyloliquefaciens* (Quantum 400, Taegro, DoubleNickel 55)
	Bacillus subtilis (BioYield, Kodiak, Serenade, Companion, Cease, etc.)
	Mixes of *Bacillus* strains (Voodoo Juice, GH Subculture M)
	Pseudomonas chlororaphis (Pseudobacterin)
	Pseudomonas syringae (Bio-Save)
	Streptomyces griseoviridis (Mycostop)
Fungi:	*Aspergillus flavus* (Afla-Guard)
	Aureobasidium pullulans (Boni-Protect)
	Candida oleophila O
	Coniothyrium minitans (ContansWG)
	Trichoderma spp. (SoilGuard, T22, RootShield, PlantShield, TrichoderMax, Fungistop)
	Herbicides
Bacteria:	*Streptomyces acidiscabies* (MBI-005)
Fungi:	*Phoma macrostoma* (94–44B)
	Sclerotinia minor (Sarritor)
	Insecticides
Bacteria:	*Bacillus thuringiensis* and other *Bacillus* species (many commercial formulations)
Fungi:	*Beauveria bassiana* (BoveMax)
	Metarhizium anisopliae (MethaMax, Met52)
	Paecilomyces fumosoroseus (PRF-97)
Protozoa:	*Nosema locustae* (NoLo Bait)
Viruses:	*Cydia pomonella granulosus virus* (CYD-X)
	Nuclear polyhedrosis virus (GemStar; Spod-X)
	Nematicides
Bacteria:	*Pasteuria usage* (Econem)
Fungi:	*Myrothecium verrucaria* (DiTera)
	Paecilomyces lilacinus (MeloConWG)
	Plant Growth Stimulants and Biofertilizers
Bacteria:	*Azospirillum* spp. (Azo-Green, Azo-Nit, AzoMax)
	Pseudomonas fluorescens (Planrhyz)
	Pseudomonas chlororaphis (NIVA 2B, BioZlac, Respecta)
	Rhizobium spp. and closely related species (i.e., *Mesorhizobium, Bradyrhizobium*, and *Sinorhizobium* spp.) (numerous manufacturers and commercial formulations)
Fungi:	*Clonostachys rosea* (Endofine)
	Fusarium moniliforme (Gybbersib); *Penicillium biaii* (JumpStart)
	Mixtures of mycorrhizal fungi (RhizoMyco, RhizoMyx, RhizoPlex)

LABORATORY EXERCISES

General Considerations

These experiments are designed to demonstrate the ubiquity in nature of fluorescent *Pseudomonas* with the potential as BCAs. Students will isolate bacterial strains from plant rhizospheres and compare them with a known biocontrol strain. The exercises are designed for teams of two to three students each. Remember to maintain aseptic conditions during the course of these exercises. Aseptic conditions are especially important for Experiments 2 and 3. Use latex or acetonitrile gloves and laboratory coats during these experiments. The microbes used in these experiments can be purchased at the American Type Culture Collection (ATCC). An approved permit from the Plant Protection Quarantine section of the U.S. Department of Agriculture is required to purchase or transport plant pathogens such as *Rhizoctonia solani* and *Pythium ultimum*. To apply for this permit, go to http://www.aphis.usda.gov/permits/ppq_epermits.shtml and http://www.aphis.usda.gov/forms/. Permits are not required for nonpathogenic organisms such as *Pseudomonas fluorescens* Pf-5 and *P. fluorescens* 2-79.

Experiment 1. Isolation of Fluorescent *Pseudomonas* from the Rhizospheres of Different Plants

The goal of this experiment is to demonstrate that fluorescent *Pseudomonas* spp. are widespread in nature and can be found in the rhizospheres of most plants. During this exercise, students will understand the difference between bulk soil (loose) versus rhizosphere soil (closely attached to the root system). Isolates obtained by each group (Experiment 1) should be stored and used for testing the antagonistic activity against plant pathogenic fungi (Experiment 2), and screening for the presence of genes coding for antibiotics (Experiment 3).

Materials

Each group of students will require the following items:

- Vortex mixer.
- Incubator at 28°C; if incubator is not available, keep plates at room temperature and extend incubation times.
- Ultraviolet transilluminator; be careful to wear protective gear (lab coat, gloves, and face shield) to avoid irradiation.
- Bunsen burner.
- Glass "hockey stick."
- 70% ethanol solution.
- Scissors.
- Paper towels.
- Sharpie markers.
- Pen and paper
- Calculator.
- Several strips of Parafilm® (5 × 2 cm).
- Pipette and corresponding tips to measure 100 μL (P100 or P200).
- 1-mL and 10-mL serological pipettes.
- Sterile solution of NaCl (9 g/L) in distilled water (total of ~100 mL/group) and six test tubes for aliquoting 9 mL of sterile NaCl.
- Pseudomonas agar F (PAF; Difco) Petri plates. Prepare medium following manufacturer's instructions. Each team will need at least eight plates of PAF and at least six plates of PAF with antibiotics added. For the latter, add the following antibiotics at the corresponding final concentrations: cycloheximide (Chex) 100 μg/mL, ampicillin (Amp) 50 μg/mL, and chloramphenicol (Cm) 12.5 μg/mL. Chex is a eukaryotic antibiotic and will be used to prevent the growth of fungi and oomycetes common in the soil. Since fluorescent *Pseudomonas* spp. are naturally resistant to Amp and Cm, these antibiotics will make the medium semiselective for this group of bacteria. Antibiotics should be added after the medium is autoclaved and cooled to ~45°C in a water bath. Antibiotic stock solutions should be prepared and kept in a freezer at –20°C (Sambrook and Russell, 2001). Prepare plates at least 1 day prior to this experiment to let them solidify and dry. Plates can be stored in sealed container or plastic bag at 4°C to keep them from drying out for later use.

Follow the instructions outlined in Procedure 27.1 to complete the experiment.

Anticipated Results

Each team will identify and isolate fluorescent *Pseudomonas* from the rhizosphere of their chosen plant.

Questions

- What is rhizosphere? Why is rhizosphere important for plant growth?
- Why are fluorescent *Pseudomonas* spp. easy to find everywhere? What do you think is their function in nature?
- What is the effect of the antibiotics that you added to the media? Were you able to kill the growth of a few or many of the microbes in the soil by adding antibiotics to the medium?

Procedure 27.1
Isolation of Fluorescent *Pseudomonas* from Rhizospheres of Different Plants

Step	Instructions and Comments
1	Carefully dig one plant (grass, weed, and ornamental) and keep as much as possible of the root system. Different groups should pick different types of plants.
2	Remove bulk soil adhered to the roots by gently shaking and separating big pieces of soil with your hands. If necessary, submerge the roots in water momentarily to detach bulk soil.
3	Using scissors cut three to four pieces of the root (each 1–2 in. long) and place them in a test tube containing 10 mL of a sterile solution of NaCl (9 g/L). Secure tube cap with Parafilm®.
4	Vortex for 10 min to detach rhizosphere soil.
5	Prepare six 1:10 serial dilutions, transferring 1 mL of rhizosphere soil suspension into tubes containing 9 mL of NaCl, mixing very well after each transfer.
6	From each dilution, collect 100 µL and spread with a sterile hockey stick onto plates of PAF medium containing the following antibiotics: Chex 100 µg/mL, Amp 50 µg/mL, and Cm 12.5 µg/mL. Label Petri plates with group number and dilution factor. Plate—one to two dilutions in media without antibiotics to observe the effect of antibiotics on suppression of unwanted microflora.
7	Incubate plates at 28°C for 24–48 h.
8	With forceps, take the cut pieces of the root system from the tube, rinse them in tap water, and place roots between paper towels. Weigh roots within 3 min and record numbers.
9	When bacterial colonies are detected on the agar plates (24–48 h), observe them under ultraviolet illumination. Circle fluorescent colonies (fluorescent *Pseudomonas* candidates) with a Sharpie, and count how many fluorescent bacteria are found on a dilution plate with 3 to 300 colonies.
10	Transfer five to six fluorescent colonies from your plates onto new PAF plates without antibiotics. Incubate plates for 24–48 h and seal each plate with Parafilm to prevent moisture loss and maintain sterility. Keep plates at +4°C for Experiments 2 and 3.
11	Calculate how many fluorescent *Pseudomonas* are obtained per gram of root of your chosen plant (take into consideration the dilution factor).

Experiment 2. Suppression of Plant Pathogens by Metabolites Produced by Fluorescent *Pseudomonas*

This experiment is designed to test the antagonistic ability of the isolates obtained in Experiment 1 *in vitro* against common soilborne plant pathogens, such as the fungus *Rhizoctonia solani* and the oomycete *Pythium ultimum*. Strain *P. fluorescens* Pf-5 will be used as a control. Pf-5 produces several antibiotics including 2,4-DAPG, pyrrolnitrin, and pyoluteorin. Pf-5 antagonizes *R. solani* and *P. ultimum in vitro* and protects cotton plants against infection by these pathogens. Besides detecting antagonism, some simple experiments will be conducted to explore the nature of the antagonistic factors produced by fluorescent *Pseudomonas*. Procedure 27.2 describes approaches to determine whether or not the active factors are secreted by the bacteria into the medium, and if these factors are regulated by iron.

Materials

Students will need the following items:

- Incubator at 28°C; if incubator is not available, keep plates at room temperature and extend incubation times.
- UV light (wavelength, 360 nm).
- Cultures of isolates obtained from Experiment 1, *P. fluorescens* Pf-5 (ATCC BAA-477), *R. solani* (ATCC number 14011 or other), and *P. ultimum* (ATCC number 32939 or other).
- Bunsen burner.
- Inoculation loop.
- Forceps.
- Ethanol solution.
- Sharpie marker.
- Pen and paper.
- Metric ruler.
- Parafilm® (8–10 pieces, 10 × 2 cm).

- Sterile blue pipette tips (for P1000 or similar).
- PAF (Difco) and diluted potato dextrose agar (PDA) Petri plates. Prepare media following manufacturer's instructions. If potato dextrose broth (PDB, Difco) is used for the preparation of PDA plates, make 1:10 dilution of PDB diluting 1:10 media component, and add 20 g/L of agar. Each team will need at least six plates of PAF and two plates of PDA. Plates can be stored at 4°C for later use.
- Sterile cellophane discs (9-cm-diameter); cut cellophane disks, wrap in aluminum foil and sterilize by autoclaving.
- Filter-sterilized stock solution of 37 mM $FeCl_3 \cdot 6H_2O$.

Follow the instructions outlined in Procedure 27.2 to complete Experiment 2.

Anticipated Results

At least one of the isolates obtained in Experiment 1 will show an inhibition halo against one or both of the plant pathogens tested. *P. fluorescens* Pf-5, used as control, should show an inhibition halo against *R. solani* and *P. ultimum*. Since the most common mechanism of antagonism involves production of antibiotics secreted into the medium, it is anticipated that the cellophane experiments (Step 9 in Procedure 27.2) will reveal antagonism even in the absence of bacterial cells.

Questions

- How many bacterial isolates that are tested show some kind of antagonism against plant pathogens? Is this a common or a rare trait?
- What do you need to do to prove that selected isolates are active *in planta* and could be used as commercial BCAs? Describe each step in a protocol.

Experiment 3. Detection of Genes Involved in Biosynthesis of Antibiotics by PCR

This experiment requires trained personnel and equipment that is more sophisticated than the previous ones. If resources are available, divide the class into teams of three members; otherwise turn this experiment into a demonstration conducted by the professor with the help of a few students. The goal of this experiment is to demonstrate the utility of the polymerase chain reaction (PCR) to identify genes involved in the biosynthesis of antibiotics (phenazines, 2,4-DAPG) by fluorescent *Pseudomonas*. These antibiotics contribute to the biocontrol activity of many fluorescent *Pseudomonas* isolates worldwide. It is important for the students to follow the process of isolation of fluorescent *Pseudomonas* from the rhizosphere (Experiment 1); to assess their biocontrol activity against plant pathogens *in vitro* (Experiment 2); and to detect genes involved in the biosynthesis of antibiotics by PCR (Experiment 3). For this experiment, the control strains *P. fluorescens* Pf-5 will be used to amplify the *phlD* gene of the 2,4-DAPG biosynthesis locus and *P. fluorescens* 2-79 will be used to amplify the *phzF* gene of the phenazine biosynthesis locus.

Materials

Each team of students will require the following items:

- Thermal cycler (DNA Engine PTC-200 from Bio-Rad or similar)
- Horizontal gel electrophoresis unit with power supply
- Gel documentation system with transilluminator (with light appropriate for DNA stain selected) and camera
- Centrifuge for Eppendorf tubes
- Freezers (–20°C and –80°C)
- Set of pipettes, clean and dedicated to molecular biology work (P2, P20, P200, and P1000 from Gilson, or similar)
- PAF (Difco)
- Cultures of isolates from Experiment 1 and *P. fluorescens* Pf-5 (ATCC BAA-477) and *P. fluorescens* 2-79 (NRRL B-15132)
- Agarose
- 250-mL Erlenmeyer flask
- Fast Blast DNA stain (Bio-Rad) or other DNA stain
- Taq DNA polymerase with 5× enzyme buffer and 25 mM stock of $MgCl_2$ (Taq Flexi DNA polymerase from Promega or similar)
- 2 mM stock of deoxynucleotide triphosphates (dATP, dCTP, dGTP, and dTTP; Promega, or similar)
- 1-kb DNA ladder molecular weight marker (New England Biolabs or similar)
- Loading dye
- Primers specific for each antibiotic biosynthesis gene (Table 27.2)
- Sterile deionized H_2O for PCR
- Sterile pipette tips (to fit all pipettes used)
- Petri dishes
- Microcentrifuge (Eppendorf) tubes
- PCR tubes
- Ice
- Cold box for handling enzymes chilled to –20°C

Procedure 27.2
Suppression of Plant Pathogens by Metabolites Produced by Fluorescent *Pseudomonas*

Step	Instructions and Comments
1	Five to seven days prior to the experiment, start growing cultures of *R. solani* (Rs) and *P. ultimum* (Pu) in separate PDA plates and incubate them at 28°C.
2	Three to five days prior to the experiment, start growing fluorescent *Pseudomonas* isolates (one or two per team) obtained in Experiment 27.1 and control bacteria *P. fluorescens* Pf-5 on PAF plates and incubate them at 28°C.
3	One day before the experiment, start the "antagonism" plates. With a sterile inoculation loop, transfer bacteria from Step 2 to new PDA plates as a ~1 cm diameter circle close to the edge of the plates (see Figure 27.1). Incubate bacterial plates for 24 h at 28°C.
4	On the day of the experiment, using a sterile blue pipette tip (hold from the tip end upside down), and cut an agar plug from the actively growing mycelium of Rs and Pu. With the help of a sterile loop, transfer the agar plug to the center of PAF plates (the number of plates will depend on the number of bacterial isolates to be tested by each team). One of the plates will be used as a "control" for Rs and Pu growth, and another one will be used to test "antagonistic" activity of fluorescent *Pseudomonas* isolates (up to four isolates per plate, see Figure 27.1).
5	*Optional step*: Duplicate "antagonism" plates using PAF + Fe (addition of filter-sterilized $FeCl_3 \cdot 6H_2O$ after autoclaving the medium; final concentration 37 μM). When analyzing antagonism results (Step 7), compare PAF and PAF + Fe to determine whether or not the antagonism of the bacteria against the plant pathogen is influenced by iron concentration (possibly due to siderophore production).
6	Seal each Petri plate with Parafilm® to prevent moisture loss and maintain sterility. Incubate "control" and "antagonism" plates at 28°C for 4–5 days.
7	Observe plates daily, and once the mycelia of Rs and Pu fill the whole surface of the control plates, measure inhibition haloes for each isolate and Pf-5 (Figure 27.1). Use a ruler to measure and record the distance between the edge of the bacterial colony facing the fungus (Rs) or oomycete (Pu), and the edge of the mycelial growth of Rs or Pu. If both edges touch each other, record as "0," and consider these isolates to be nonantagonistic to the plant pathogen.
8	Observe if, in the inhibition halo, you can detect any color diffusing in the agar. Orange may be indicative of antibiotics such as phenazine, whereas fluorescent yellow may be indicative of siderophore production (visible under UV light at 360 nm wavelength).
9	*Optional*: To follow-up on your above observations, select isolates that show antagonism against Rs or Pu, and some that did not. Follow Steps 9–11 to elucidate whether secreted metabolites are responsible for the antagonistic activity.
10	Repeat Steps 1–7, but this time the "antagonism" plates will be prepared differently. For this, a circle of sterile cellophane will be placed on top of the PAF plates covering the whole surface. Streak the fluorescent pseudomonads as a ~1 cm diameter circle on top of the cellophane, and incubate for 48 h.
11	At the bottom of the Petri plate, mark the spots where bacteria were growing with a Sharpie. Aseptically and with the help of sterile forceps, remove cellophane with bacterial growth. You may be able to observe a colored pigment diffused into the agar below spots where bacteria were growing, indicative of secreted metabolites. Inoculate with Rs or Pu as described in Step 4 and incubate as in Step 6.
12	Repeat observations of antagonism against plant pathogens (Step 7), but this time you will be able to describe whether only the material diffusing into the agar (metabolites produced and secreted by the bacterium) is sufficient to stop the growth of the plant pathogen, or if the whole bacterial cell is needed (compare with the results in Step 7).

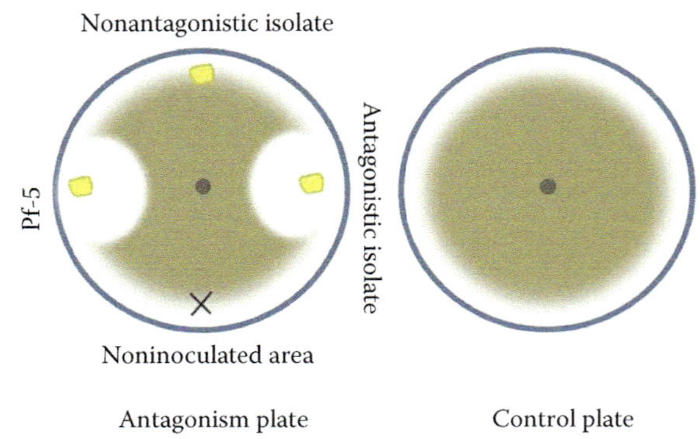

FIGURE 27.1 Diagram of experimental setting for antagonistic activity. Small black circles in the center of each Petri dish represent the initial inoculum of *Rhizoctonia solani*. Brown area represents mycelial growth by this fungus (control plate is completely covered with mycelia). In the example, *Pseudomonas fluorescens* Pf-5 and the "antagonistic isolate" create a zone of inhibition between the bacterial colony and fungal mycelia, while the "nonantagonistic" isolate did not.

Follow the instructions outlined in Procedures 27.3 and 27.4 to complete Experiment 3.

Anticipated Results

After performing PCRs and running the agarose gel, bands of the specific sizes corresponding to biosynthesis genes of the two different antibiotics will be visualized. PCR with *P. fluorescens* Pf-5 should amplify a band of 649 base pairs (bp), indicating the presence of the 2,4-DAPG biosynthesis locus, while PCR with phenazine primers should be negative. PCR with *P. fluorescens* 2-79 should amplify a band of 427 bp, indicating the presence of the phenazine biosynthesis locus, whereas PCR with phloroglucinol primers should be negative. Negative control containing master mix and no added DNA should not have any amplified product in PCRs with both sets of primers. If negative control has some amplified product, use fresh aliquots of all PCR reagents and repeat PCR.

TABLE 27.2

PCR Primers and Conditions for the Detection of Biosynthesis Genes for Common Antibiotics Produced by Fluorescent *Pseudomonas*

Target Pathway	Primer Name	Sequence(5′ to 3′)	Annealing Temperature (°C)	Amplicon Size (bp)
Phenazine	Ps_low1	CCR TAG GCC GGT GAG AAC	57	427
	Ps_up1	ATC TTC ACC CCG GTC AAC G		
2,4-DAPG	B2BF	ACC CAC CGC AGC ATC GTT TAT GAG C	60	629
	BPR4	CCG GTA TGG AAG ATG AAA AAG TC		

Sources: McSpadden Gardener et al., *Phytopathology* 91:44–54, 2001; and Mavrodi et al. *Appl. Environ. Microbiol.* 76:866–879, 2010.

Procedure 27.3

Detection by PCR of Genes Involved in Biosynthesis of Antibiotics

Step	Instructions and Comments
1	Prepare solutions as described in Procedure 27.4
2	One day before experiment, streak cultures of *Pseudomonas fluorescens* Pf-5, *P. fluorescens* 2-79, and isolates from Experiment 1 (with recorded biocontrol activity) onto fresh PAF plates and incubate at 28°C for 24 h.
3	General consideration: keep all PCR reagents stored at –20°C; thaw them before preparing the PCR reactions; vortex and centrifuge briefly; and keep all reagents on ice while setting up the PCR mixtures. Be extremely careful while handling the *Taq* DNA polymerase, which should be the last reagent to be added, and should be kept in a cold container (kept at –20°C) at all times.
4	Mix PCRs containing 1 × *Taq* reaction buffer, 1.5 mM $MgCl_2$, 200 µM of each dNTPs, 20 pmol of each primer, 1.2 units of *Taq* DNA polymerase, and the DNA template (see Procedure 27.4).

5 PCR amplification should be performed with a thermal cycler. The cycling program for *phlD* consists of an initial denaturing step at 95°C for 3 min, followed by 35 cycles of melting at 94°C for 60 s, annealing at 60°C for 60 s, extension at 72°C for 60 s, and a final extension at 72°C for 5 min. The cycling program for *phzF* consists of an initial denaturing step at 94°C for 3 min followed by 30 cycles of melting at 94°C for 20 s, annealing at 57°C for 15 s, extension at 72°C for 40 s, and a final extension at 72°C for 5 min.

6 Use an aliquot of the PCR product (typically 5–10 µL, mixed with a loading dye) for electrophoresis in 0.8% agarose gels (containing a DNA stain) and visualized under UV light following standard procedures (Sambrook and Russell, 2001).

7 Determine the amplicon size in comparison with a 1-kb DNA ladder. The control strain Pf-5 should be positive for *phlD*, but not for *phzF*. PCR with *P. fluorescens* 2-79 should be positive for *phzF*, but negative for *phlD*. Negative control containing master mix and no added DNA should not have any amplified product(s) in PCRs with either set of primers.

Procedure 27.4

Solutions That Must Be Prepared in Advance

Step	Instructions and Comments
1	*0.8% Agarose gels* containing Fast Blast DNA stain (Bio-Rad) or a fluorescent DNA stain, such as SYBR Safe (Bio-Rad) for greater sensitivity—follow manufacturer's instructions for the concentration of stain to add to gels. Fluorescent stains require UV light for visualization. Place the necessary agarose and 0.5× Tris-borate-ethylenediaminetetraacetic acid (EDTA) buffer in a 250-mL Erlenmeyer flask covered with plastic wrap. Punch holes in the plastic wrap to let air flow, and place in the microwave for consecutive intervals of 30 s, carefully swirl thoroughly between cycles until agarose is completely dissolved in the buffer. Cool down the agarose to 45°C–50°C; add required DNA stain; swirl thoroughly, avoid generating a lot of bubbles; and pour onto electrophoresis tray. For agarose gel staining, wear nitrile gloves and lab coat. Use a face shield and protective clothing if you used a fluorescent stain and will be taking pictures of the agarose gel with a UV light source.
2	*0.5× Tris-borate-EDTA buffer*: Prepare a 5× concentrated stock solution using 54 g Tris base, 27.5 g boric acid, and 20 mL 0.5 M EDTA (pH 8.0) in deionized water. Store the stock solution in a glass bottle at room temperature. The working solution is made by 1:10 dilution of the stock solution in deionized water.
3	*2 mM Stock of deoxynucleotide triphosphates (dNTPs)*. This means that the final concentration of each dNTP (i.e., dATP, dCTP, dGTP, and dTTP) is 2 mM. Note that all dNTPs together make 2 mM. Store at –20°C.
4	*DNA template*: DNA for PCR amplification will be obtained from single colonies of bacterial isolates grown on PAF for 1 day. Pick one colony with a pipette tip by lightly touching the colony, dip into the PCR mixture, and mix thoroughly. It is important to keep a minimal amount of cells in the PCR; a larger amount of cells can inhibit PCR.

Questions

- How does PCR work? What are the three different steps in PCR? What is the master mix?
- How will you prove that the antibiotic is being produced by the bacterium if you know that PCR for genes involved in antibiotic production was positive?
- What are the advantages and risks of the use of antibiotics in humans, animals, and plants?

REFERENCES

Baker, K. F., and R. J. Cook. 1974. *Biological Control of Plant Pathogens*. San Francisco, CA: WH Freeman.

Bravoa, A., S. Likitvivatanavong, S. S. Gill, and M. Soberon. 2011. *Bacillus thuringiensis*: A story of a successful bioinsecticide. *Insect Biochem. Mol. Biol.* 41:423–431.

De Vleesschauwer, D., and M. Höfte. 2009. Rhizobacteria-induced systemic resistance. *Adv. Botanical Res.* 51:223–281.

Druzhinina, I. S., V. Seidl-Seiboth, A. Herrera-Estrella, B. A. Horwitz, C. M. Kenerley, E. Monte, P. K. Mukherjee, S. Zeilinger, I. V. Grigoriev, and C. P. Kubicek. 2011. *Trichoderma*: The genomics of opportunistic success. *Nat. Rev. Microbiol.* 16:749–759.

Fravel, D. R. 2005. Commercialization and implementation of biocontrol. *Annu. Rev. Phytopathol.* 43:337–359.

Glare, T., J. Caradus, W. Gelernter, T. Jackson, N. Keyhani, J. Köhl, P. Marrone, L. Morin, and A. Stewart. 2012. Have biopesticides come of age? *Trends Biotechnol.* 30:250–258.

Gross, H., and J. E. Loper. 2009. Genomics of secondary metabolite production by *Pseudomonas* spp. *Nat. Prod. Chem. Biol.* 14:53–63.

Kim, J.-G., B. K. Park, S.-U. Kim, D. Choi, B. H. Nahm, J. S. Moon, J. S. Reader, S. K. Farrand, and I. Hwang. 2006. Bases of biocontrol: Sequence predicts synthesis and mode of action of agrocin 84, the Trojan Horse antibiotic that controls crown gall. *Proc. Natl. Acad. Sci.* (USA) 103:8846–8851.

Kim, Y. C., J. Leveau, B. B. McSpadden Gardener, E. A. Pierson, L. S. Pierson, and C. M. Ryu. 2011. The multifactorial basis for plant health promotion by plant-associated bacteria. *Appl. Environ. Microbiol.* 77:1548–1555.

Lemanceau, P., and C. Alabouvette. 1993. Suppression of Fusarium wilts by fluorescent pseudomonads: Mechanisms and applications. *Biocontrol Sci. Technol.* 3:219–234.

Mavrodi, D. V., T. L. Peever, O. V. Mavrodi, J. A. Parejko, J. Raaijmakers, P. Lemanceau, S. Mazurier, L. Heide, W. Blankenfeldt, D. M. Weller, and L. S. Thomashow. 2010. Diversity and evolution of the phenazine biosynthesis pathway. *Appl. Environ. Microbiol.* 76:866–879.

Mazurier, S., T. Corberand, P. Lemanceau, and J. M. Raaijmakers. 2009. Phenazine antibiotics produced by fluorescent pseudomonads contribute to natural soil suppressiveness to Fusarium wilt. *ISME J.* 3:977-991.

McSpadden Gardener, B. B., D. V. Mavrodi, L. S. Thomashow, and D. M. Weller. 2001. A rapid polymerase chain reaction-based assay characterizing rhizosphere populations of 2,4-diacetylphloroglucinol-producing bacteria. *Phytopathology* 91:44–54.

Mendes, R., M. Kruijt, I. de Bruijn, E. Dekkers, M. van der Voort, J. H. Schneider, Y. M. Piceno, T. Z. DeSantis, G. L. Andersen, P. A. Bakker and J. M. Raaijmakers. 2011. Deciphering the rhizosphere microbiome for disease-suppressive bacteria. *Science* 332(6033):1097-1100.

Mercier, J., and P. G. Marrone. 2006. Biological control of microbial spoilage of fresh produce. Pages 523–539 in: *Microbiology of Fruits and Vegetables*. J. R. Gorny, A. E. Yousef, and G. M. Sapers (eds). Boca Raton, FL: CRC Press.

Paulitz, T. C., and R. R. Belanger. 2001. Biological control in greenhouse systems. *Annu. Rev. Phytopathol.* 39:103–133.

Raaijmakers, J. M., and M. Mazzola. 2012. Diversity and natural functions of antibiotics produced by beneficial and plant pathogenic bacteria. *Annu. Rev. Phytopathol.* 50:20.1–20.22.

Sambrook, J., and D. W. Russell. 2001. *Molecular Cloning: A Laboratory Manual.* 3rd edition. Cold Spring Harbor, NY: Cold Spring Harbor Laboratory Press.

Stockwell, V. O., K. B. Johnson, D. Sugar, and J. E. Loper. 2010. Control of fire blight by *Pseudomonas fluorescens* A506 and *Pantoea vagans* C9-1 applied as single strains and mixed inocula. Phytopathology 100:1330–1339.

Thomashow, L. S., R. F. Bonsall, and D. M. Weller. 1997. Antibiotic production by soil and rhizosphere microbes in situ. Pages 493–499 in: *Manual of Environmental Microbiology*, C. J. Hurst, G. R. Knudsen, M. J. McInerney, L. D. Stetzenbach, and M. V. Walter (eds). ASM Press, Washington, D.C.

Thomashow, L. S., and D. M. Weller. 1995. Current concepts in the use of introduced bacteria for biological disease control: Mechanisms and antifungal metabolites. Pages 187–235 in: Plant-Microbe *Interactions*, Vol. 1. G. Stacey, and N. Keen (eds.), New York, NY: Chapman & Hall.

Van Wees, S. C., S. Van der Ent, and C. M. Pieterse. 2008. Plant immune responses triggered by beneficial microbes. *Curr. Opin. Plant Biol.* 11:443–448.

Vega, F. E., M. S. Goettel, M. Blackwell, D. Chandler, M. A. Jackson, S. Keller, M. Koike, N. K. Maniania, A. Monzon, B. H. Ownley, J. K. Pell, D. E. N. Rangel, and H. E. Roy. 2009. Fungal entomopathogens: new insights on their ecology. *Fungal Ecol.* 2:149–159.

Weller, D. M., J. M. Raaijmakers, B. B. McSpadden Gardener, and L. S. Thomashow. 2002. Microbial populations responsible for specific soil suppressiveness to plant pathogens. *Annu. Rev. Microbiol.* 40:309–348.

28 Integrated Pest Management

Anton Baudoin

Concept Box

- Integrated pest management (IPM) originated largely as a reaction to the drawbacks and limitations of insecticide misuse and overuse, and to a lesser extent that of other pesticides.
- IPM is defined as a sustainable approach to managing pests by combining biological, cultural, physical, and chemical tools in a way that minimizes economic, health, and environmental risks (Jacobsen, 1997).
- Integrated disease management relies on exclusion, plant disease resistance, cultural practices, and the use of chemicals to the extent that they are available and economical.
- The economic damage threshold concept is at the core of IPM philosophy and refers to the pest level at which control should be applied to avoid reaching the economic injury level (EIL).
- The EIL is pathogen population or disease level where the cost of an additional increment of control is balanced by the resulting increment in damage reduction.

HISTORY

As a concept, **integrated pest management** (IPM) was named and defined starting in the late 1950s and the 1960s. As a practice, it existed before the development of modern pesticides, but its emergence as a "new" concept resulted from a period of overuse of pesticides, leading to growing recognition of their drawbacks and limitations. The problems were most acute with insecticides, and entomologists led the way in the development of "IPM."

The years that followed the Second World War have been called "The Golden Age of Pesticides." Sulfur, copper, arsenic, and other inorganic pesticides had been used for decades, but the development of organic chemicals as pesticides, such as DDT and its cousins among insecticides, and 2,4-D and relatives among the herbicides, revolutionized the attitudes of both scientists and farmers about pest management. What had formerly been backbreaking work or a difficult undertaking meeting with limited success, keeping weeds, insect pests, and diseases under control, now became easy and as simple as popping a pill: apply the appropriate pesticide and the problem was solved! The development of these new tools coincided with the increased use of chemical fertilizers, rapid mechanization, and migration of labor away from farms. Because of the availability of the new pesticides, public and grower expectations for blemish-free produce increased. All of these developments increased the dependence on and use of pesticides.

However, limitations and drawbacks emerged as the use of pesticides expanded over the years. These include the following:

- Some pests became **resistant** to pesticides. A prominent example was mosquitoes in areas where malarial eradication campaigns relied heavily on DDT in the 1950s. Many other important pests, including the Colorado potato beetle, houseflies, human body lice, apple maggot, and coddling moth also developed DDT resistance.
- **Secondary pests** developed. A pesticide application may lead to the development of a different pest that would not have been a problem otherwise. Spider mite populations commonly flare up when broad-spectrum insecticides are used because elimination of natural enemies permits unchecked development of the new pest.
- The destruction of natural enemies can also lead to **target pest resurgence**. A pesticide application may reduce the population of the target pest as well as the natural enemies, but the pest may bounce back more quickly if it can reproduce or migrate away from the site.

- Pesticides can damage other nontarget organisms, such as birds, fish, amphibians, and the food organisms on which they depend, in the environment.
- Some pesticides may cause cancer and other diseases in humans.

Initially, these problems were most acutely experienced with insecticides. Pest resistance to organic insecticides became an important issue in the 1950s and 1960s, but resistance to herbicides and fungicides were uncommon until the 1970s and 1980s. Secondary pest development and target pest resurgence can occur in response to fungicides, but documented examples are rare. Insecticides tend to be more toxic to humans and to wildlife than most herbicides and fungicides. Health and environmental risk from the latter two groups certainly exist. For examples, mercury-containing fungicides that were once commonly used as seed treatments are now banned; copper fungicides damaged aquatic organisms; and the herbicide paraquat is highly toxic to humans. However, such cases are less common compared with the deleterious effects of some insecticides, nematicides, and rodenticides.

As scientists became increasingly aware of the drawbacks mentioned, they started devising improvements, for which the umbrella term "integrated pest management" became the most commonly used term. The publication of Rachel Carson's book *Silent Spring* in 1962 is often credited with bringing problems due to use, misuse, and overuse of pesticides to the attention of the public and governments, and providing a major push for IPM's implementation.

DEFINITION AND CHARACTERISTICS

Many and sometimes slightly different definitions of IPM exist. In this chapter, we will use the following definition by the U.S. National Coalition for IPM: "A sustainable approach to managing pests by combining biological, cultural, physical and chemical tools in a way that minimizes economic, health and environmental risks" (Jacobsen, 1997).

The term "**pest**" has both a narrow (insects and mites) and a broad meaning (insects and mites plus pathogens, nematodes, weeds, rodents, birds, etc.). The broader meaning is widely cited and recognized in scientific and policy-making circles, but entomologists still commonly use "IPM" to refer to insects and mite pests only, scarcely mentioning pathogens and weeds. Because of this historical association with insects, many plant pathologists have been somewhat reluctant to adopt the term IPM, and some have preferred alternatives such as plant health management, which covers the broad range of pests, but also abiotic crop production limitations.

IPM, as originally developed with a focus on insects and mites, was understood to include the following tenants:

1. Reliance on biological control by natural enemies.
2. Monitoring of pest populations, and making spray decisions based on comparing population size with **economic thresholds (ETs)**, because complete prevention or eradication is usually unrealistic and expensive.
3. Using pesticides only when needed, and adjusting pesticide types or spray timing to minimize harm to natural enemies.
4. Minimizing the need for pesticide use by implementing other tactics, such as resistant cultivars and cultural practices including crop rotation and alternative methods such as disruption of insect mating, and so on.

Plant disease management developed along somewhat different lines than insect pest management. Disease-resistant varieties of many crops had already become commonly used as a mainstay of plant disease control several decades before the Golden Age of Pesticides, whereas with some exceptions, they are a more limited and more recent addition for insect pest management. In addition, the fungicides available until the mid-1960s were protectants (Chapter 26) and although very useful, they were not the "silver bullets" that could completely eliminate disease problems. When highly effective systemic fungicides became available in the late 1960s, the optimism that the new tools would revolutionize fungal disease control did not last long. Within a few years, it became apparent that pathogen resistance would be a major impediment to heavy and sole reliance on systemic fungicides. Thus, integrated plant disease management has tools and emphases that differ somewhat from insect pest management. The IPM concept has evolved to accommodate such variations.

We use the term "pest" in the broad sense, but focus the discussion on plant disease management. Let us take a more detailed look at the characteristics of the IPM approach as it applies to plant diseases.

The base layer (Figure 28.1) or first line of defense, in a plant disease management program (Case Study 28.1) consists of efforts to keep pathogens out of the crop, the region, or the country, through quarantine and other exclusionary practices. The next layer, and the major foundation of many programs, is largely composed of disease-resistant varieties (Chapter 23) and cultural practices (Chapter 25). Biological and physical control practices are important, but usually play a much smaller role than other management strategies. Only those pathogens that still make it through these barriers need to be treated with chemical applications to the extent they are available

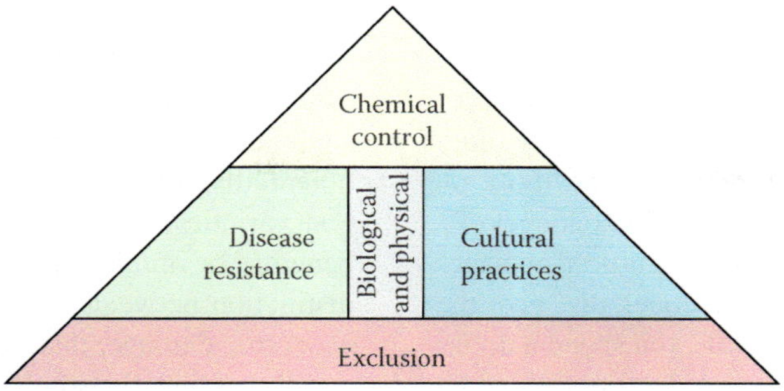

FIGURE 28.1 The disease management "pyramid."

CASE STUDY 28.1

Home Garden IPM practices

Before or at planting:

- Rotate annual plants. Do not rotate tomatoes with the peppers (they have too many diseases in common), but rotate the tomatoes with, for example, sweet corn.
- Sanitation (removal of inoculum sources), including debris from last year and weeds.
- Choose disease-resistant cultivars or disease-resistant crops.
- Purchase high-quality pathogen-free seeds or transplants.
- Create a good planting bed, with well-drained soil, add compost to improve air spaces, organic matter, drainage, and microbial activity can enhance biological control.
- Plant each vegetable at its optimum time. Planting a warm-season plant in cold soil will greatly increase the likelihood of damping-off diseases.

During growing season:

- Check regularly for any disease or physiological problems.
- Mulching: retains, reduces erosion, and minimizes splashing of soil/inoculum.
- Sanitation (removal of diseased material):—difficult to accomplish.
- Timing of watering to minimize leaf wetness.
- Chemical use as needed: "softer" chemistry (oil, soap, sulfur, and bicarbonate), timing optimized.

Preparation for next year:

- Cover crop
- Sanitation
- Tillage

and economical. The relative importance of these management tool categories varies with the crop.

IPM practices aim to minimize harmful effects on environment and human health when chemicals are used by selecting the least toxic pesticides that will be effective, and to adjust application timing to minimize hazards. Reading the information on the pesticide label is the first step in determining potential hazards (Chapter 26). "Signal words" indicate degree of risk to the user: danger (highly toxic), warning (moderately toxic), or caution (less toxic). Statements and specific directions about risk to aquatic life, birds, and so on are often included as well. "Soft" chemicals, such as paraffinic or vegetable oils, bicarbonates, even milk, all of these have some efficacy against, for example, powdery mildews; they are not always as effective as the most effective commercial products, but may be sufficient in some circumstances. Efforts have been made to provide a more comprehensive accounting of risks, such as the environmental impact quotient (EIQ; Kovach et al., 1992, http://www.nysipm.cornell.edu/publications/eiq/; try the EIQ calculator), which calculates a single number based on estimated impacts on farm workers, consumers (including pesticide residues in food commodities as well as drinking water), and ecology (aquatic life, wildlife, and beneficial organisms such as bees, and predators and parasites of pests).

The **economic threshold** concept is at the core of IPM philosophy, although ETs for plant diseases are often difficult to quantify. The idea is do not spend more on additional control practices when the cost is no longer offset by additional revenue. When putting a pest management plan together, include those practices that give the best results for minimum costs. Would you spend a dollar if the resulting disease prevention or reduction yielded $8 worth of additional revenue due to protection of yield and quality? Probably yes. If the next dollar spent on a second control method or a second application provided an extra $4 in revenue, that expenditure is easily justified as well. If we continue adding additional effort and expense, then there comes a time where an additional dollar spent (marginal cost of pest control, in economic terms) will only yield an additional dollar in revenue. Any additional disease control expenditures beyond this balance point are expected to be economically unsound. At this balance point, some population of the pathogen or level of disease is probably still present in the field. This pathogen population or disease level is known as the **economic injury level** (**EIL**) or the economic damage threshold (Figure 28.2). Any damage less than the economic damage threshold does not justify additional management expenses.

Is it a best management practice to leave a little pathogen or a little disease in the field, which may rebound later to cause economic loss? One must indeed take into account all costs and losses, including those incurred in the long term, such as the survival or buildup of inoculum for future seasons. This makes it more difficult to determine valid EILs, but the reasoning is still valid. A disease that is well established in a region will tend to come back sooner or later, despite control measures.

Generally, we should not wait for disease to build up to the EIL to take action. Many management tools are preventive but require lead time (Table 28.2). Some, such as planting resistant varieties or implementing crop rotation, need to be implemented before the growing season starts. There may be little cost associated with some of those practices (e.g., planting a resistant rather than a susceptible variety), but if there is, the costs (including, potentially, lower yield or quality of some disease-resistant varieties) have to be weighed against the expected benefits. In order to take the lead time into account, a distinction between the EIL (pest level at which we get economic damage, that is, exceeding the cost of control) and the **action threshold** or ET (the pest level at which control should be applied to avoid reaching the EIL) must be made.

An ET is not a static number. If crop prices decline, protecting from a certain percentage loss (e.g., 5%) is worth less than if crop value is high, and so the ET will increase (more disease should be allowed before action is taken). If the cost or efficacy of management practices changes, the ET will change as well. For example, if a disease is currently controlled by several chemical applications, but a new cultivar becomes available with good resistance against the disease, the ET may decline dramatically. ETs have to be verified and adjusted based on local production and management practices as well as economic conditions.

Basing disease management action on scouting of disease levels can be effective for diseases where detection of the disease or pathogen leaves enough time to take action. A good example is the monitoring of soilborne nematode populations by taking soil samples commonly at the end of the growing season of an annual crop. The resulting action may be the planting of a resistant variety or applying a nematicide at planting time of the next season's crop. Application of a fungicide for control of rust and powdery mildew of small grains can be based on severity of symptoms as detected through regular scouting. For many other diseases, basing action on disease development may be less suitable if the time of detection is likely too late to take the requisite action (Chapter 22). Fungicide applications, especially applications of protectant fungicides, are often not effective unless applied before infection takes

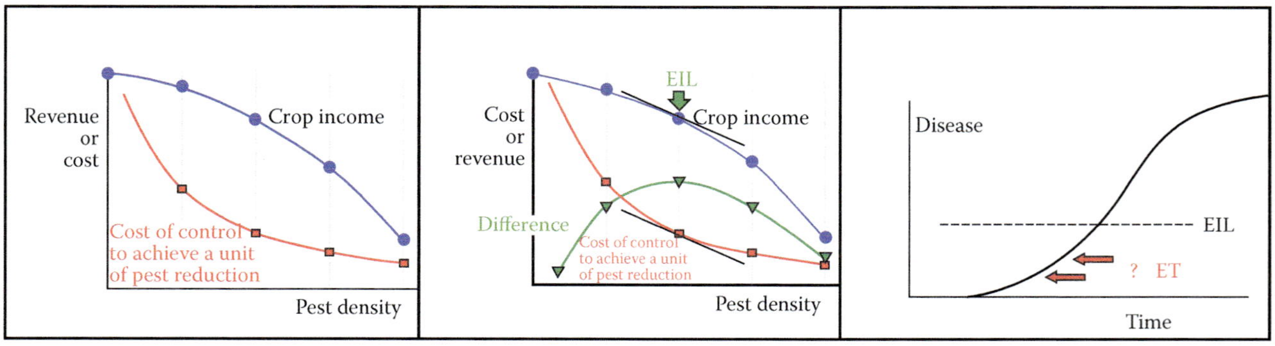

FIGURE 28.2 The economic injury level (EIL) is the pest density where the cost of an additional increment of control is balanced by the resulting increment in damage reduction. The economic threshold or action threshold is the pest level at which control should be applied to avoid reaching the EIL. ET can be variable, as indicated by '?'.

place or in the early rather than the mid-stages of an epidemic. Diseases such as rusts, powdery, and downy mildews of many crops can develop explosively, and delaying fungicide applications for a certain amount of symptom development can lead to unmanageable disease outbreaks. Therefore, for many diseases, scouting for disease severity is replaced by monitoring for conditions that are conducive for disease development. These may be expressed as either infection periods or risk indices. A few examples are listed in Table 28.1.

Conservation of natural enemies, another major component of insect IPM programs, plays a much smaller role in disease management programs. The introduction or application of natural enemies for biological control of pathogens has had some successes (Chapter 27), but overall has not reached nearly the pervasiveness that it has in insect pest management. Interference by fungicide/pesticide application with natural biological control of plant diseases can certainly occur, but is not a common phenomenon.

INTEGRATION

Just as "Pest" can have a narrow or a broad meaning, something similar is true for "Integrated." How integrated should a pest management program be, before it can be called "IPM?" This can be a matter of importance when crop insurance or government programs mandate use of IPM. There is no single correct answer. Combining several control methods (partially resistant varieties with reduced chemical use) for one pest or pest group is integration at one level. Programs that coordinate and combine methods for several pest groups, for example, weeds and diseases or diseases and insects, constitute the next

TABLE 28.1
Forecasting Programs and Disease-Risk Indices

Pathosystem (Crop/Disease)	Basis of Forecasting (Factors Affecting Disease Development)	Reference(s)
Peanut/tomato spotted wilt	Disease history Insecticide application Peanut variety Plant population Planting date Row pattern Tillage	http://www.apsnet.org/edcenter/intropp/lessons/viruses/Pages/TomatoSpottedWilt.aspx
Peanut/sclerotinia blight and other fungal diseases	Air and soil temperature Crop growth Duration of high humidity Rainfall	Phipps et al. (1997)
Grape/powdery mildew	Air temperature	http://www.apsnet.org/publications/apsnetfeatures/Pages/UCDavisRisk.aspx
Corn/Stewart's wilt	Winter temperature	http://www.extension.iastate.edu/CropNews/2011/0407nutterhodgsonrobertson.htm http://www.apsnet.org/edcenter/intropp/lessons/prokaryotes/Pages/StewartWilt.aspx Esker et al. (2006)
Fusarium head scab/small grains	Weather conditions prior to flowering	http://www.wheatscab.psu.edu/De Wolf et al. (2003)
Soybean/soybean rust	Development of disease in source areas Regional "sentinel plots" Scouting Weather conditions Wind patterns that affect distance and direction of dispersal	http://sbr.ipmpipe.org/ (U.S. nationwide network)
Apple and pear/fire blight	Temperature Rainfall Wetness due to heavy dew or fog Tree bud development (hail, high winds)	http://www.caf.wvu.edu/kearneysville/maryblyt/

TABLE 28.2
IPM Preventive Practices

No IPM	Low	Medium	High or Biointensive IPM
• No preventive practices, monitoring or alternative, nonchemical control methods • Reliance on chemical controls	• Attractant baits/crops • Crop rotation • Cultivation • Disease-free seed/planting material • Edge treatment • Pest-resistant varieties • Scouting • Selective pesticides • Sprayer calibration	• Green manures/compost • Maintenance of refugia • Nutrient and water monitoring • Pest biotype monitoring • Precision agriculture-based treatments • Use of induced resistance activators • Weather-based forecasting models	• Biocontrols • Interactive pest/weather/crop models • Pheromones • Primary reliance on preventive nonchemical practices • Release of beneficials • Soil solarization • Trap crops

Source: Adapted from Jacobsen, B.J., *Annu. Rev. Phytopathol.* 35, 373–91, 1997.

level. Interactions are obvious where insects either serve as vectors of pathogens or create infection courts. For example, the citrus leaf miner creates leaf injuries that promote development of citrus canker. Fungicides, such as sulfur, captan, and mancozeb, can be harmful to the natural enemies of insect and mite pests, including fungi that parasitize pests, predators, and parasites, and overuse of these fungicides can lead to secondary outbreaks of pests.

Interactions of plant disease and weeds are also common. Weeds may serve as alternative, reservoir, or overwintering hosts, and dense weed stands may increase canopy humidity making it more favorable to disease development. In contrast, weeds may provide some benefits by harboring natural enemies of insects or mites. The use of herbicides for weed control makes no-till agriculture possible. Omitting tillage has many advantages, including reductions in fuel use and soil erosion, but affects pathogen survival on crop residues, insect survival, and soil warm-up in spring and cool-down in the fall. One example of a three-way interaction is when herbicides became available that could be used in growing corn to selectively control Johnson grass. Controlling Johnson grass after the corn seedlings had emerged sometimes led to an increase in virus diseases in the corn. Apparently, the insect vectors, which served as overwintering hosts for the corn viruses, moved from the dying Johnson grass to corn (King and Hagood, 2003).

Beyond integrating pest management practices with each other, a grower has to integrate pest management with agronomic or horticultural practices, labor availability, and marketing opportunities at the farm or production unit level. Some considerations transcend even this level and should be considered in the context of the entire **agro-ecosystem** (Lewis et al., 1997). Agriculture is more than crop and livestock production. Other aspects are watershed management, pollution control, native species preservation, and recreational values. These may not always be valued or priced properly in our current economic systems (Robertson and Swinton, 2005). Additionally, increasing biodiversity and crop diversification may increase long-term resilience of agricultural systems (Altierri, 1999; Lin, 2011).

IMPLEMENTATION AND ECONOMIC PERSPECTIVE

One trend in the implementation of IPM programs is that "simple" solutions (such as pesticide applications) are replaced by more knowledge-based solutions. Not only does the pest or disease have to be properly identified in order to choose the correct pesticide or chemical control, but also behavior and ecology of the control agents need to be understood in relation to the environment and to a combination of practices, which each might contribute to pest suppression and prevention (CAST, 2003). Collecting the information needed for ecology-based decisions, such as monitoring development of diseases on a regular basis (scouting) can be time-consuming and require training and experience, thus entailing considerable cost.

A grower who sees a risk of losing the crop and livelihood will naturally have a different perspective on pesticide use than a consumer who only considers the risks of pesticides in the food they buy. Although many agricultural producers and managers are strongly committed to safe and sustainable practices, they need to make a living in a challenging economic environment where weather often presents big risks and crop prices may be low. Ideally, minimizing unnecessary chemical applications and replacing chemicals with alternative practices will be economically beneficial. Where this is not always the case, governments and other agencies can stimulate IPM by requiring IPM practices in return for farm support payments. Some experimentation has been done with **IPM labeling** of fruit or vegetable commodities, where sale opportunities and the prices growers receive may increase, along similar lines as for organically produced produce. Some examples of such labeling programs may be found at http://www.ipminstitute.org/. Practices that reward the bottom line will naturally be more likely to be implemented than those that only cost money.

Criticisms

Environmentalists and a health- and safety-conscious public may feel that IPM still allows too much pesticide use, whereas farm chemical suppliers may feel that attempts to reduce pesticide use raises unnecessary barriers to cost-effective farming while not necessarily reducing harm or increasing safety. IPM may have inadvertently increased some specific risks. The search for more selective, narrow-spectrum pesticides, and the practice of delaying pesticide application until thresholds have been reached, both may have enhanced the risk of pesticide resistance development. These problems and possible misjudgments/questions are not due to the IPM philosophy, but to the difficulty of estimating long-term risks of different approaches that can be quantified only in hindsight.

IPM EXAMPLE—WHEAT DISEASE MANAGEMENT IN THE SOUTHEASTERN UNITED STATES

Crop management decisions can be divided into choices set in place at the beginning of a season (crop and cultivar selection, crop rotation, method of tillage/seedbed preparation, seed quality, seed treatment, planting date, seeding method and rate, fall fertility, and planting time) and those that can still be applied or modified in the course of a season (added fertility, insecticides, herbicides, and foliar fungicides) (Cook and Veseth, 1991). Disease management is a key component of high-yielding wheat, and most diseases are best managed through the use of multiple tactics. Wheat producers have a significant portion of their total disease management program in place once the seeds are sown.

Variety Selection

Decisions relating to variety selection are, perhaps, the most important decisions that can be made in managing diseases. Every commercially available wheat variety has a unique range of reactions to diseases common in the region. Selecting two or three varieties with the greatest amount of available resistance to the diseases is important. To do this, some idea about the disease history on the farm or in the region is necessary. There are two good reasons to plant more than one variety. First, it spreads risk because a single variety may be damaged by an unexpected outbreak of a disease (Chapter 22). Second, planting several varieties with different maturities can help with the logistics of harvesting and, if this crop is used in rotation, soybean planting (double cropping).

Crop Rotation

Crop rotation helps to manage wheat pathogens that survive between crops in residue. When a crop other than wheat is grown in a field, levels of pathogens specific to wheat decline as the residue of the wheat crop deteriorates. Crop rotation is helpful in the management of hidden diseases, such as pythium root rot and take-all. Rotating fields out of wheat is the only practical means of controlling take-all disease. Rotation can also reduce infections by foliar fungi such as *Stagonospora*, *Septoria*, and *Pyrenophora*, which survive on wheat stubble. However, spores of pathogens from neighboring fields can negate the beneficial effects of rotation on these diseases.

As an example, in some areas of the soft red winter wheat-growing region in the eastern United States, wheat is planted in the fall following corn harvest. For many wheat diseases, corn is a good nonhost rotation crop. However, the fungi that cause Fusarium head blight (FHB) of wheat, also attack corn, causing stalk and ear rots, and corn residue provide a substrate for the FHB fungi to survive between seasons. On an individual field basis, planting wheat behind a noncorn crop, even in a conventionally tilled environment, does not greatly reduce the potential for FHB. This is because FHB inoculum is released and blown around readily by the wind, so the disease will be a regional problem regardless of an individual farm's rotation scheme and the tillage method used. In-field inoculum and residue management can be helpful in areas where regional cropping and tillage practices do not favor FHB inoculum production and dispersion.

Tillage

Tillage hastens the breakdown of wheat stubble that harbors certain disease organisms. This can help reduce levels of take-all and foliar diseases, such as Septoria leaf blotch and tan spot. "Help" is the operative word here because it is unlikely that tillage will suffice in the absence of other management methods. For fields in a wheat/double-crop soybean/corn rotation, tillage prior to planting corn results in a significant decline in surviving wheat stubble. The year between wheat crops in this rotation also helps, except where high levels of the take-all fungus exist. In those cases, two or more years between wheat crops may be required to reduce inoculum levels.

Seed Quality, Seed Fungicides, Seeding Rate, and Planting Method

Seed quality, seed fungicides, seeding rate, and planting method influence stand establishment and seedling development. Sufficient stands are necessary to achieve the highest possible yields. There must be an excellent seed germination rate followed by the emergence and development of seedlings to ensure the desired stands. Using high-quality (i.e., certified) seed treated with a broad-spectrum fungicide and good planting techniques fosters good stand establishment. Excessively dense stands are undesirable because they encourage foliar and head diseases by reducing air circulation and light penetration into the canopy later in the season and may promote lodging of plants because of weakened stems. Certified seed is also important because it minimizes the risk of bringing in seedborne diseases such as loose smut.

Planting Date

The trend in recent years has been to plant wheat earlier. An important wheat pest is the Hessian fly, an insect that can destroy wheat plants, but this insect ceases activity in the fall, which is known as the "fly-free date." Early-planted wheat, defined as wheat planted before the Hessian fly-free planting date, is at greater risk of damage caused by *Barley yellow dwarf virus* (BYDV), take-all disease, and Hessian fly than a crop planted later. If logistic considerations cause a grower to plant some wheat acres before the fly-free date for that area, those acres should have a crop rotation plan and be planted with a variety that can tolerate some BYDV. The use of seed treatment or fall-applied foliar insecticides may also be appropriate in these situations. These pesticides can help to reduce populations of the aphids that transmit BYDV to wheat, barley, and oats. However, cost of treatment is a consideration, and, thus, treatment is usually targeted at high-risk, early-planted acres. Planting all wheat acreage before the fly-free date is extremely risky.

Nitrogen Fertility

Too much nitrogen fertilizer applied in the fall can encourage excessive growth that could increase problems with BYDV, and most foliar diseases caused by fungi. Increased problems with BYDV are associated with an extended period of activity by aphids that transmit the virus when stands are lush in the fall. The same situation encourages infection and overwintering of pathogens causing foliar diseases, such as leaf rust, powdery mildew, and leaf blotch complex. Excessive spring nitrogen results in lush stands that promote disease in a manner similar to that associated with excessive seeding rates. Furthermore, lodging may increase, making harvest difficult and resulting in higher moisture within the canopy of fallen plants, which creates favorable conditions for fungal diseases. Excessive nitrogen also leads to runoff, leaching, and water pollution.

Fungicide Seed Treatments

Obtaining and keeping a good stand of wheat is a key component of high yields. Wheat producers use the management strategy of treating seed with a fungicide to obtain excellent stands. Wheat seed treatment fungicides affect the following in stand establishment and disease control.

Encourage Good Stand Establishment

At the time of wheat planting, the soil may be excessively cool, wet, or dry, the seed may be planted too deep or shallow, or seed-to-soil contact may be poor in no-till plantings; all slow the germination process and predispose developing seedlings to infection by seedborne and soilborne fungi. Species of *Pythium* are probably the primary pathogens in excessively wet and warm soils. The other conditions mentioned favor infection of seedlings by species of *Fusarium*, *Rhizoctonia*, *Septoria*, and *Stagonospora*. Infection can result in fewer emerged seedlings, and reduced vigor of the seedlings that do emerge.

If high-quality, high-germination rate seed is used and the seed is treated, most producers will have little difficulty achieving dense, vigorous stands of wheat seedlings. However, treating high-quality seed with a fungicide only rarely results in stand or yield increases. Nonetheless, seed treatment with a good general-use fungicide is still advisable as a form of low-cost insurance to protect seedlings from adverse soil conditions that may develop after planting. Seed treatment fungicides often have minimal environmental impact because of their low toxicity, low per-acre use rates, rapid breakdown, and targeted application method.

If seed quality is marginal because of fungi such as *Fusarium*, seed treatment can be used to bring seed with less than optimum germination to acceptable levels. Seed testing laboratories can provide tests that indicate whether or not fungicides would be likely to enhance germination. Poor response of low-to-moderate germination seed lots to fungicides is indicative of a high percentage of dead seed, mechanical damage, or some factor apart from disease. Seed lots of low germination are not likely to be helped by any seed treatment fungicide and should not be used where high yield is a goal.

Control of Loose Smut

Starting in the 1960s, the use of seed treatment fungicides, such as carboxin, allowed seed lots to be "cleaned up" by eliminating loose smut from infected seed. Carboxin is still considered by many to be the standard for loose smut control. It is inexpensive and usually highly effective. Newer seed treatment materials are also highly effective against loose smut and are extremely active at low rates. Because of the very low use rates of these products, application can often be done only by seed conditioners with proper experience and equipment. This eliminates the option of on-farm seed treatment and may increase the cost. Some on-farm hopper-box treatments are still available and may be inexpensive. However, complete coverage of all seeds is essential and can be accomplished on farm only with considerable effort and planning. Poor coverage equals poor results and a waste of money. Having seed treated by a professional eliminates potential problems of poor fungicide distribution on the seeds.

Control Foliar Diseases

The new generation of systemic, sterol-inhibiting seed treatment fungicides (Chapter 26) are absorbed by the roots, migrated to the leaves, and can provide fall protection from several fungal diseases including Septoria and Stagonospora leaf blotches (leaf blotch complex), leaf rust, and powdery mildew. Occasionally, reduction of these diseases extends into early spring as a result of reduced inoculum levels of the causal fungi in fields planted with treated seed. This activity can be quite substantial, as is the case with triadimenol and powdery mildew, where control of powdery mildew until head emergence in the following spring is not uncommon. Nonetheless, seed treatment fungicides should not be considered as a total replacement for spring-applied foliar fungicides because no seed treatment provides season-long control of foliar diseases.

In general, sterol-inhibiting fungicide seed treatments, because of their specific mode of action, are not highly effective against many common soilborne pathogens. For this reason, most are marketed as a mixture with either thiram or captan, both of which provide at least moderate activity against a wide range of soilborne fungi. Formulations may also include metalaxyl to control soilborne *Pythium* species.

Costs of fungicides for wheat seed treatment and the degree of disease control vary widely, so it is critical to assess cost–benefit ratios of the various fungicides. The specific purpose of a seed treatment fungicide is an important consideration. Fungicide labels (Chapter 26) list the diseases controlled, which should be read and followed completely. As one example, triadimenol seed treatment is relatively expensive. If powdery mildew is a widespread problem on a farm, triadimenol may negate the need for an early spring foliar fungicide application on a mildew-susceptible variety, making the economics of triadimenol seed treatment more favorable. If loose smut or general soilborne pathogens are the main concern, and the risk of powdery mildew is minimal, triadimenol is not the most economical choice because less expensive materials are as good as triadimenol at managing these diseases.

Considerations in the Application of Foliar Fungicides

A grower will need to decide how important wheat is to the total farming operation. If wheat is important to the profitability of the farm, it is advisable to make both time and monetary commitments to produce the best crop possible. If wheat is of only secondary importance, scouting and fungicide applications may not be worthwhile and management of diseases using only resistance and cultural practices should be considered.

Scouting

Scouting for diseases and other problems is time-consuming, but important for two reasons. First, it helps to determine if and when to apply fungicides (as well as insecticides). Second, and more generally, regular scouting ensures early problem detection and helps to build an on-farm information database that can be used to select appropriate disease management tactics for future crops. However, scouting takes time and is expensive if we have to hire personnel or a consultant. Fungicide application decisions should not be based on what is found along the edges of a field or what is seen from the seat of a vehicle. A good decision is based on the average disease situation in a field, which requires assessing disease levels at 8–10 sites spread over the field, and this takes time.

Crop Yield Potential

Does the field have sufficient yield potential to justify a foliar fungicide application? Spraying with fungicides protects only yield already "built" into a crop; fungicides do not increase yield. Although various techniques can be used to estimate yield potential, experienced producers can look at a crop after green-up and know intuitively if the crop is worth protecting. Depending on grain prices and chemical and application costs, an additional three to eight bushels per acre will need to be harvested to offset the cost of a fungicide application. The higher the yield potential of a crop, the more likely an economic benefit is to be realized from applying a foliar fungicide if disease becomes a problem. The yield potential of a crop can be reduced by inclement weather conditions, insect pest outbreaks, and poor weed control or by diseases that do not respond to foliar fungicides.

What Diseases Are Expected?

Fungicides manage a relatively small number of fungal diseases. Fortunately, the ones controlled, such as leaf rust, powdery mildew, leaf blotch complex, glume blotch, and tan spot, are those that commonly reduce yields of soft red winter wheat in the southern and eastern United States. Diseases that are not affected by foliar fungicides include all virus and bacterial diseases and other fungal diseases including take-all and loose smut. Fusarium head blight is reduced, but not controlled, by the application of a foliar fungicide. Thus, proper identification of the disease is critical in developing a control strategy.

Disease Reactions of the Wheat Varieties Planted

Typically, foliar fungicides are not necessary against a particular fungal disease on a wheat variety rated as resistant or moderately resistant to that disease. Many varieties have resistance to some, but not all relevant diseases. Additionally, leaf rust and powdery mildew can adapt to and attack a formerly resistant variety. This can happen in a single season, so growers and consultants need to be vigilant and still scout those crops.

Choice of Fungicide

Extension or manufacturer literature may be used to compare the options with respect to price, activity spectrum, and other pertinent details. Product labels provide detailed use instructions and product limitations. Apply all pesticides according to label specifications.

How Many Fungicide Applications Are There?

Nearly every producer says they will only make one, and few indicate two applications. The answer determines the approach to fungicide use. If a grower is going to make only one fungicide application, timing of the application is critical. Two fungicide applications reduce the risk of miss-timing, and crops receiving two fungicide applications often yield significantly more than crops getting even the best-timed single application. However, the question is whether or not the economic benefit from the additional treatment is greater than its cost. The extra treatment is most likely to pay for itself if crop prices are high and early disease pressure is moderate to heavy.

Timing—Growth Stage

The growth stage of the crop is important for the following two reasons: (1) all fungicides must be applied within specific growth-stage restrictions. Propiconazole products (e.g., Tilt), for example, cannot be legally applied once the crop is flowering. All fungicides have well-defined days-to-harvest restrictions to ensure that residues at harvest will be minimized; (2) Fungicides provide the greatest benefit when plants are protected from disease between flag leaf emergence and soft dough stage. In much of the southeastern United States, the stage in which fungicide applications are most beneficial is typically from mid-head emergence through flowering. The top leaves, that is, the flag leaf (F) and the second leaf down (F-1), as well as the head are the most important to protect because they have the greatest effect on yield and grain quality. However, a fungicide application against FHB should be applied at flowering, a timing that differs from that against foliar diseases.

Timing—Disease Development

To be effective, fungicides must be applied early in an epidemic. Applying fungicide earlier and disease development later can still cause severe damage. In addition, applying fungicides too early, before any disease is visible, may result in no economic benefit if disease pressure remains low. If applied too late, significant levels of disease may already be present on the flag leaf. For leaf blotch complex, if F-2 and lower leaves have symptoms and if it has rained recently, there is a good chance that the flag leaf and head are already infected, with first symptoms appearing 7–12 days later. The best way to limit this risk is to start scouting operations in the stages just before flag leaf extension and examine symptoms on lower leaves.

A single fungicide application made during heading generally performs at least well, and usually better, than a single application made at flag leaf emergence. A single application at flag leaf emergence (regardless of the fungicide used) frequently allows late-season disease pressure to build to excessive levels. As a result, the crop is damaged although early diseases may have been kept in check. Heading applications usually limit disease buildup on the upper leaves and the head, although disease is allowed to develop unchecked early in the season.

How much disease is enough to justify a foliar application of a fungicide? Guidelines have been determined as to what levels of disease at specific stages will translate into what degree of yield loss (see earlier discussion of ET). Many states have developed threshold guidelines to help producers make informed fungicide use decisions.

Timing—Other

Thresholds must be used with some common sense. For example, if a specific threshold is reached for powdery mildew, application of a fungicide would not be recommended if an extended period of hot, dry weather is predicted. The threshold indicates that yield loss caused by one or more of the aforementioned diseases is likely; however, they do not mean losses will definitely occur. Weather can always intervene and impact the development and progression of a disease epidemic. Weather may also severely reduce the yield potential of the crop due to non-disease factors such as a spring freeze, lodging, delayed harvest, and poor grain fill period. Unfortunately, the above situations are always a risk to the grower. There is no way to develop disease thresholds that are accurate in all situations.

Application

A period of rainy weather can interfere with timely application. If fields become too muddy to allow tractor-based applications, use of an airplane may still be possible, but application in the rain will not be effective in either case. You can make all the right fungicide-use decisions, but if inferior application technique is used, all can be lost!

Understand the Risks

A variety of factors determine whether fungicide applications will be worthwhile. In all instances where fungicides are used, check the response of the crop to the treatment by leaving a non-treated strip in the field for comparison (Hollier et al., 2001).

Questions for Thought and Discussion

- List the major components of an IPM program for a crop of your choice.
- Discuss IPM strategies and tactics based on the perspective of the following: consumer, producer (including turf or landscape manager), agricultural supplier, and agricultural consultant.
- Determine the EIQ of a fungicide.
- Explore IPM elements and guidelines by state or commodity.

REFERENCES

Altieri, M. A. 1999. The ecological role of biodiversity in agroecosystems. Agriculture, Ecosystems, and Environment 74:19-31.

Carson, R. L. 1962. *Silent Spring*. Boston, MA: Houghton Mifflin.

CAST. 2003. Integrated pest management: Current and future strategies. Task Force Report 140. Ames, IA: Council for Agricultural Sciences and Technology.

Cook, R. J., and R. J. Veseth. 1991. *Wheat Health Management*. St. Paul, MN: APS Press.

De Wolf, E. D., L. V. Madden, and P. E. Lipps. 2003. Risk assessment models for wheat Fusarium head blight epidemics based on within-season weather data. *Phytopathology* 93:428–435.

Esker, P. D., J. Harri, P. M. Dixon, and F. W. Nutter, Jr. 2006. Comparison of models for forecasting of Stewart's disease of corn in Iowa. *Plant Disease* 90:1353–1357.

Hollier, C. A., D. E. Hershman, C. Overstreet, and B. M. Cunfer. 2001. Management of Wheat Diseases in the Southeastern United States: An Integrated Pest Management Approach. Baton Rouge, LA: Louisiana State University.

Jacobsen, B. J. 1997. Role of plant pathology in integrated pest management. *Annual Review of Phytopathology* 35:373–391.

King, S. R., and E. D Hagood, Jr. 2003. Effect of application rate and timing of AE F130030 03 plus AE F107892 for control of italian ryegrass (Lolium multiflorum) in barley (Hordeum vulgare). Weed Technology 17:866-870.

Kovach, J., C. Petzoldt, J. Degnil, and J. Tette. 1992. *A Method to Measure the Environmental Impact of Pesticides*. New York's Food and Life Sciences Bulletin, No. 139. http://www.nysipm.cornell.edu/publications/eiq/.

Lewis, W. J., J. C. Lenteren, S. S. Phatak, and J. H. Tumlinson, III. 1997. A total system approach to sustainable pest management. Proceedings of the National Academy of Science (USA) 94:12243-12248.

Lin, B.B. 2011. Resilience in agriculture through crop diversification: Adaptive management for environmental change. BioScience 61:183-193.

Phipps, P. M., S. H. Deck, and D. R. Walker. 1997. Weather-based crop and disease advisories for peanut in Virginia. *Plant Disease* 81:236–244.

Robertson, G. P., and S. M. Swinton. 2005. Reconciling agricultural productivity and environmental integrity: A grand challenge for agriculture. Frontiers in Ecology and the Environment 3:38-46.

29 Organic Agriculture and Plant Disease

David M. Butler and Erin N. Rosskopf

Concept Box

- Organic crop production is a legally defined production system, which is managed through integration of cultural, biological, and mechanical practices.
- Organic standards and regulations differ slightly according to political governing body and certification agencies, but share a common philosophy and a number of broad, multifunctional goals.
- Management of plant disease in organic production systems relies primarily on the design of cropping systems that are less susceptible to the development of disease and secondarily on the use of biological, botanical, mineral, and approved synthetic substances to manage disease occurrences.
- Advancements in scientific knowledge in the fields of pathogen biology and agroecology will allow for the development of more effective management of plant diseases in organic cropping systems.

Organic agriculture refers to agricultural production systems that are managed according to a number of standards, which vary by governing body or political entity, but which share a common philosophy and a set of general management practices. In popular culture, organic crop production is generally understood to refer to crop production without the use of genetically modified organisms or synthetically derived pesticides, fertilizers, or growth regulators. This definition often leads conventional agricultural producers and the general public to the conclusion that modern organic farming is no different from farming systems that existed historically and prior to the wide availability of synthetically derived agricultural inputs. While this is largely true in a narrow sense, it fails to acknowledge advancements in scientific understanding as well as the underlying philosophy and ecological principles of modern organic farming systems; principles that are vital to understanding management of plant disease in organic crop production.

ORGANIC BACKGROUND

In the United States, the **Organic Foods Production Act** of 1990 (Title 21 of the Food, Agriculture, Conservation, and Trade Act or "Farm Bill") established the National Organic Program (NOP) within the U.S. Department of Agriculture (USDA) to "develop, implement, and administer" standards for the production of foods labeled as organic. The act was largely a response to organic producers, processors, and consumers who were concerned with inconsistent standards for the production and labeling of organic foods. When the final regulations were published in 2001, **organic production** was defined as "a system that is managed … to respond to site-specific conditions by integrating cultural, biological, and mechanical practices that foster cycling of resources, promote ecological balance, and conserve biodiversity." As we can see, the spirit of organic farming goes far beyond a prohibition on synthetic inputs to encompass design and management of crop production systems with a number of broad, multifunctional goals. This definition established by the NOP is very much in keeping with the mid-twentieth-century writings of the pioneers of modern organic farming systems, such as Albert Howard, Eve Balfour, Walter Northbourne, and Jerome Rodale. In fact, Northbourne is widely regarded to have been the first to use the term "organic" to refer to these farming systems. Northbourne, writing in his 1940 book, *Look to the Land*, expounded on farm management as an ecologically balanced and dynamic "organic whole," a concept that is echoed today in the growing and related field of **agroecology**. Agroecology seeks to apply ecological principles, or the relationships among and between living organisms and their environment, to the management of agricultural systems or agroecosystems. The principles of modern agroecology are relevant to all cropping systems, but even more so to management of organic cropping systems where producers must first rely on design and management of the agroecosystem or cropping system to limit disease pressure and then resort to approved inputs only when these practices are not sufficient.

This strategy of utilizing systems-based approaches is first outlined by the NOP in the brief standard published in the U.S. code of federal regulations for the management of plant diseases (as well as weeds and other pests). The regulations state that producers must, "use management practices to prevent crop pests, weeds, and diseases including but not limited to:"

1. "Crop rotation and soil and crop nutrient management practices..."
2. "Sanitation measures to remove disease vectors..."
3. "Cultural practices that enhance crop health, including selection of plant species and varieties with regard to suitability to site-specific conditions and resistance to prevalent pests, weeds, and diseases."

Title 7, Part 205.206 Code of U.S. Federal Regulations

More specific to plant diseases, organic producers are required to utilize "management practices that suppress the spread of disease organisms; or application of nonsynthetic biological, botanical, or mineral inputs." When the aforementioned practices fail, producers are permitted to resort to a limited number of synthetic substances approved for use in organic crop production that are discussed later in this chapter. Conditions for use of such substances must be clearly documented in a producer's **organic system plan** (Case Study 29.1), a document required for organic certification and annual organic inspections that detail farm management practices.

Many of the control methods and management practices described in this chapter are not unique to organic cropping systems. In fact, many of these practices are used by growers across the spectrum of production systems with management tactics differing in levels of synthetic inputs of pesticides and fertilizers, inputs of organic matter, cropping system diversity, and other management practices. In this chapter, we examine how these practices (e.g., biological control, cultural control, host resistance, and integrated pest management) are utilized in organic cropping systems; refer to the relevant chapters in this text for more detailed descriptions of these practices and general information on their use in other cropping systems. In organic systems, these tactics are often employed in concert to achieve acceptable control of plant diseases. This strategy was initially described by Liebman and Gallandt (1997) as a "many little hammers" approach for ecological management of weeds in cropping systems, but also has use for describing the management of plant pathogens. The use of many "little hammers" or tactics can be as effective as "big hammers" (i.e., synthetic pesticides) due to "additive, synergistic, and cumulative" effects of each tactic as well as the reduced risk of pathogen adaptation to a single tactic or stressor.

SYSTEMS-BASED APPROACHES

Crop Rotation

The prominent position of **crop rotation** within the NOP standards is apt considering its importance to the management of plant disease and to the management of organic farming systems in general. Crop rotation is a

CASE STUDY 29.1

Do not Forget the Organic System Plan

- Organic management of downy mildew (*Pseudoperonospora cubensis*) on cucurbits is difficult.
- Cultural practices can contribute to the management of this disease. A farmer can plant resistant varieties, space plantings to increase air circulation and reduce canopy humidity, and use drip irrigation to minimize sustained leaf wetness.
- Cultural practices should be listed in the farm plan, but downy mildew can reach epidemic proportions quickly. If a cucurbit grower observes small yellow leaf spots symptomatic of downy mildew, additional control measures should be employed immediately before sporangia are evident on the underside leaf surface.
- Some organically approved treatments for downy mildew include limited amounts of certain copper-based products described in this chapter and some biologically based fungicides like the commercial product Serenade® (*Bacillus subtilis*). These materials should be used as soon as symptoms are observed and applied in the morning to prevent phytotoxicity.
- If this disease was not expected and these materials were not listed in the organic system plan, then the farm plan must be amended and approved by the farm's organic certification agency prior to the emergency treatment.

planned system of growing a number of different crops on the same land over a period of time. In practice, crop rotation generally has both spatial and temporal dimensions on working organic farms; each crop is present in some area of the farm each year, but is only present in a given field in one segment of a several year crop rotation. Two aspects of crop rotation that are important for the management of plant disease include **rotation length**, the number of years within a rotation sequence before it is repeated, and **rotation diversity**, the number of different crop species and/or botanical families within a crop rotation sequence. Generally, crop rotations that are longer and more diverse will result in fewer incidences of many crop diseases. However, the biology and life cycle of the causal pathogens of problematic diseases for a given crop or region will be key factors in determining the impact that crop rotation may have on disease control. Pathogens that cannot survive in soil or on crop residues for long periods of time, which lack a large number of crop or weed species hosts, and that have a limited dispersal range will be more easily controlled by crop rotation. Conversely, pathogens with persistent survival structures (e.g., *Sclerotium rolfsii*), numerous crop and weed species hosts (e.g., *Pythium* spp.), a wide dispersal range (e.g., *Phytophthora infestans*) or that are often introduced on seeds or transplants (e.g., *Xanthomonas fragariae* on strawberry) are unlikely to be controlled by crop rotation alone. However, beneficial impacts of crop rotation on the cropping system can have a positive effect on the management of diseases caused by many of these pathogens as well. The spatial component of crop rotation is also an important consideration for pathogens with intermediate dispersal ranges (e.g., *Phytophthora capsici* and *Alternaria solani*), making sanitation an important consideration to prevent inoculum movement between fields at different points in the crop rotation.

In organic production systems, most organic certifiers (third-party organizations that verify producers are following USDA organic standards) require that a crop rotation contain at least three crops, although this is not explicitly stated in NOP regulations. Most organic producers, especially those growing horticultural crops for local markets or those who integrate livestock with crop production, will generally have a more diverse rotation than three crops that will include a number of cash crops as well as **cover crops** or perennial sod (hay and pasture) crops. Cover crops are plants that are grown primarily for the improvement of the agroecosystem (e.g., biological nitrogen fixation, increased soil organic matter, erosion prevention, and pest suppression) rather than for harvest and market. In addition to general functions of added biological diversity and increasing crop rotation diversity, a number of cover crops suppress certain soilborne plant pathogens. Examples of these pathosystems include the following: Sudan grass (*Sorghum bicolor* var. *sudanense*) and the wilt pathogen *Verticillium dahliae*; nonhosts or poor hosts (e.g., sunn hemp, *Crotalaria juncea*) to a number of plant parasitic nematodes (*Meloidogyne* species); and even trap crops, for example, arugula (*Eruca sativa*) for some plant-parasitic nematodes (*Meloidogyne hapla*) (Figure 29.1).

The rotation length that organic farmers must utilize is also not specified by the NOP, as rotations should ideally be developed to address cropping system management concerns at the farm- or even field-scale. In terms of management of plant diseases, it is imperative that rotation length be adequate to reduce inoculum levels of plant pathogens prior to replanting of a susceptible crop within the same field. For organic horticultural crops, a general recommendation of a 5- to 7-year rotation is often given. In practice, this may be difficult for many producers to implement due to land limitation, comparative economic value of the rotation crops, or lack of specialized equipment for producing a variety of crops. Since rotations for organic field crops are typically shorter than 5–7 years, these producers have the ability to implement **sod-based rotations** (perennial forage grasses and legumes). Sod-based rotations reduce the incidence of some diseases and function by either introducing a perennial crop that does not serve as a host for a given pathogen (e.g., bahiagrass, *Paspalam notatum*, and the peanut root-knot nematode [RKN], *Meloidogyne arenaria*) or by creating a cropping system environment that is unfavorable to disease organisms, which favor more frequent disturbances. Likewise, producers in more arid regions will often utilize a fallow period within their crop rotation. While the primary consideration for use of a fallow period in arid regions is often storage of rain water within the soil profile, fallow

FIGURE 29.1 Arugula as a cover crop can serve as a trap crop for some plant-parasitic nematodes. Here, arugula is grown prior to the production of tomato and bell pepper crops. (Courtesy of D. Butler.)

periods can also serve to manage plant disease due to the lack of hosts for a given time period.

In organic cropping system management, crop rotation is often thought of as a way to manage plant diseases (as well as soil quality, soil fertility, weeds, etc.); however, we can also think of rotation as a way to combat **yield decline**, which is a common problem in monoculture crops grown without rotation or in very simple rotations. Yield declines in the general range of 10%–50% have been widely reported for crops grown without rotation. Yield decline is typically thought to be caused by a number of factors, including increases in populations of plant pathogens or **deleterious rhizosphere microorganisms** (organisms that do not penetrate plant vascular tissue, but have a negative impact on crop growth), which readily thrive in and adapt to the consistent environmental conditions and resource availability of monoculture. Declines in soil quality due to reduced soil biodiversity or extraction of soil nutrients in similar patterns are also potential mechanisms of yield decline in monoculture that can be easily addressed by crop rotation.

SOIL MANAGEMENT PRACTICES

Soil management is often understood by organic practitioners to form the basis of organic crop production systems. Considering the importance that early organic pioneers placed on soil fertility and the "Law of Return," a concept that advocated returning all organic residues and wastes back to agricultural land following a managed composting process, is not surprising. While a major driver behind the application of composts and other organic residues to organic land is the maintenance of soil fertility, a number of other processes occur simultaneously that have both direct and indirect effects on the plants and plant pathogens within a cropping system. The organic residues help to increase soil organic matter, which has a number of beneficial effects on the physical, chemical, and biological properties of soil. These include increased soil water holding capacity, reduced bulk density, increased cation exchange capacity, and increased soil biological activity. Soil biological activity increases due to the added organic materials that act as substrates for soil fauna. The type of residues that are added can also be important, as a greater diversity of residues (e.g., plant species, stage of maturity, and stage of decomposition) tends to increase biodiversity of soil microorganisms. This increase in biodiversity impacts the levels of disease observed in organic cropping systems, likely due to increased competition to and predation of pathogenic organisms in the soil environment by nonpathogenic organisms.

The maintenance and accumulation of soil organic matter can also have positive effects on crop health in organic farming systems, largely due to indirect effects of improved soil properties. Improved soil chemical properties allow for a more consistent and adequate supply of plant nutrients mineralized from soil organic matter, creating crops that are less stressed and less vulnerable to infection by pathogens. Maintaining this consistent supply of nutrients in organic farming systems helps to reduce the incidence of many fungal, bacterial, and insect-vectored viral diseases, which can be exacerbated by lush growth created by high soil nitrogen fertility.

SANITATION

"**Sanitation** measures to reduce disease vectors" are required by NOP regulations as a disease management tactic for organic farmers. Sanitation measures can include such practices as removing diseased plants or infected parts of plants so as to reduce the spread of inoculum and potential for the development of disease within a field. In organic strawberry, early spring removal of leaves killed by winter weather can help to reduce the incidence of gray mold caused by *Botrytis cinerea*. Other examples include the timely removal of fruiting canes of blackberry after harvest so as to prevent overwintering of the cane blight pathogen (*Leptosphaeria coniothyrium*) and the removal of fallen fruit in orchard systems to manage various fruit rot diseases. Such practices are often too labor intensive for use on many conventional farms when diseases can be managed through use of pesticides, but are a key strategy for managing disease in certain organic crops. In the case of gray mold of strawberry, many nonorganic growers may also come to rely on sanitation practices as part of their management because of the recent development of resistance by the causal pathogen to a number of commonly used fungicides.

It is important to recognize that within the organic philosophy, there is a slight tension between many sanitation practices and "the Law of Return," especially when large amounts of crop residues are involved. Whereas the safest method of disposal of diseased plants would be through burning or disposal in a landfill or other deep burial, this is not in keeping with the practice of composting and utilizing all crop residues. There does not need to be a conflict between these practices if composting is followed according to NOP standards, which specify that composts must maintain temperatures of 55°C–77°C (131°F–170°F) for 3–15 days (depending on composting method) in order to destroy the inoculum of plant pathogens. Issues often arise with both weed seed and plant pathogen inoculum when appropriate composting procedures are not followed. Similarly, it is sometimes recommended that producers use deep tillage to remove residues and pathogen inoculum from the soil surface, thus reducing the incidence of disease in subsequent crops. This would generally be discouraged by most organic producers and certifiers because of potential negative effects on soil quality.

Additional sanitation measures include practices such as disinfection of field equipment, hand tools and pruning shears, transplant trays and pots or irrigation systems to prevent the spread of pathogens from plant to plant or from field to field. In organic production systems, a limited number of synthetic products can be used for this purpose in addition to natural products such as vinegar (acetic acid) and hot water. Synthetically derived products that are currently allowed for this purpose in organic production systems include alcohols, chlorine products (e.g., bleach), and hydrogen peroxide.

The procurement and use of disease-free planting stock is another sanitation measure that is imperative in organic cropping systems, which in general lack curative controls for introduced pathogens. For seedborne pathogens, organic growers can utilize hot water treatments or weak chlorine solutions to remove pathogens (although chlorine solutions may not be acceptable to all producers and certifiers). A number of protocols are available from state agricultural extension agencies for the treatment of seeds with hot water. Organic producers largely grow their own transplants, partially due to the lack of an organic transplant industry in much of the country, economic issues, as well as the ability to ensure that pathogens are not introduced to the farm.

The removal of alternate hosts of pathogens near organic crop production sites can be a key strategy for managing certain pathogens. For example, removal of eastern red cedar (*Juniperus virginiana*) near apple orchards can be a feasible strategy to manage cedar-apple rust (*Gymnosporangium juniperi-virginianae*) in the absence of host resistance. Likewise, removal of wild blackberry and raspberry (*Rubus* spp.) near domesticated production sites can help to control a number of fungal and viral diseases for which there are few economically feasible organic alternatives. There are a number of weeds that can serve as alternate hosts for pathogens of crops, which in turn could spread to growing crops and limit the benefits of a well-planned crop rotation. For example, many weeds in the Solanaceae (e.g., nightshades) can host a number of pathogens of Solanaceae crops (e.g., tomato, potato, pepper, eggplant, and tomatillo). Tomato spotted wilt virus infects a large number of weed species, which allows for transmission of the virus to susceptible crops by insects (i.e., thrips). The control of volunteer crops (crops that have reseeded from a previous production cycle) is also important as the presence of these plants can serve as hosts to pathogens and limit the benefits of crop rotation. Weeds and volunteer crops can also potentially impact the microclimate of crops, such as increased humidity in the crop canopy leading to more favorable conditions for the development of some foliar diseases.

FIGURE 29.2 Mulching is an example of a sanitation practice to reduce the spread of pathogen inoculum from soil. Here, both synthetic polyethylene mulch and living cover crop mulch are used in the production of strawberry. (Courtesy of D. Butler.)

Mulching is another practice widely used by organic growers to improve crop production, from in-row plastic mulches in vegetable production, organic mulches such as straw or mechanically killed cover crop residues, or living mulches such as between row cover crops (Figure 29.2). These mulches serve to maintain soil moisture, modify soil temperatures, maintain cleaner produce, and suppress weeds. Mulches also play a role in sanitation, by preventing the spread of pathogen inoculum by limiting contact of pathogen inoculum in the soil with crop leaves or fruit tissue.

CULTURAL PRACTICES TO ENHANCE CROP HEALTH

A number of management practices that address systems-level issues in cropping systems could be classified under what the NOP considers "cultural practices to enhance crop health." The only explicitly stated strategy is the selection of crop varieties for resistance to endemic diseases on a given farm, but there are several management practices used on organic farms to manage plant diseases that can be considered cultural practices, which enhance crop health.

HOST RESISTANCE

The selection of crop species resistant to diseases known to be problematic in a given location, termed host resistance, is a common strategy used by organic growers to manage disease. With a concerted effort in building disease resistance in modern cultivars over the past several decades, organic growers often have several disease-resistant varieties from which to choose that can be utilized in their given systems. However, these varieties may lack other characteristics that are appealing to organic growers as many modern varieties are developed and selected under conditions of high nitrogen fertility, lack of weed competition, and high levels of insect control; conditions that are not typically representative of those on organic farms. For this reason, some organic growers favor older varieties that were developed in agricultural systems more similar to modern organic farming systems and that may be perceived to have better flavor or better nutritional characteristics. However, the level of disease resistance in these varieties varies widely. Recent efforts are underway to develop modern varieties specifically bred for organic production with a recognition that the stresses on plants in organic systems often differ greatly from those in conventional systems.

Nevertheless, when acceptable resistant varieties are available to growers, they can greatly improve yields in organic systems. Advances in plant breeding for disease resistance are having a large impact in organic systems. The commercial availability of several quality apple cultivars resistant or tolerant to troublesome diseases, such as cedar-apple rust, fire blight (*Erwinia amylovora*), and apple scab (*Venturia inaequalis*), are making organic apple production in the humid eastern United States more feasible. The 2011 release of a late blight (*P. infestans*)-resistant tomato cultivar promised to reduce the risk of organic tomato production without the use of copper sprays. Because there are few cucumber varieties tolerant to bacterial wilt (*Erwinia tracheiphila*), the use of these varieties greatly simplifies organic management of the insect vectors (cucumber beetles) that transmit bacterial wilt.

Grafting is another technique that incorporates host resistance of a cultivar of a crop plant or closely related species with undesirable horticultural characteristics (such as yield, fruit size, or flavor) and combines it as a rootstock with a more desirable but less disease-resistant cultivar or scion. While most commonly used in the production of tree fruits and nuts, grafting of certain fruiting vegetables (e.g., tomato, pepper, eggplant, melon, and cucumber) for the management of soilborne disease is also gaining popularity in the United States. This approach has been widely used in Japan and Korea for decades and is now being used extensively in areas of Europe, the Middle East, and North Africa. Typically, the rootstocks are of a "wild type" with high resistance or tolerance of certain soilborne pathogens to which the scion (shoot) is not resistant.

CROP SEEDING METHODS

The prohibition of synthetic pesticides in organic production also limits the use of seed treatments such as thiram (a dimethyl dithiocarbamate fungicide), which is used in conventional systems to limit seed decay and seedling mortality caused by damping-off and other diseases caused by pathogens, such as *Pythium* and *Rhizoctonia* species. Organic producers currently lack any comparably effective organic seed treatments and thus must largely rely on cultural methods to achieve adequate plant densities. One method is to increase the seeding rate in organic systems as compared with conventional systems so that potentially increased seed decay and seedling mortality will not result in inadequate crop stand densities (Figure 29.3). Altering planting date can also be used in organic cropping systems to influence the environment that germinating seeds and seedlings are exposed to in the field. By seeding slightly later in the spring or earlier in the fall, producers can increase the chances that seeds will germinate and grow rapidly due to warmer soils, helping to prevent seedling diseases favored by cool, moist soils.

Similarly, transplanting of certain vegetable crops allows producers to place a plant in the field at a growth stage that is no longer susceptible to many seedling diseases. The use of pathogen-free transplant media and a controlled environment for transplant production also

FIGURE 29.3 Increased seeding rates for organic field crops can help compensate for potentially higher losses to seed and seedling diseases. Here, seeding rates for organic wheat production are evaluated. (Courtesy of D. Butler.)

limits seed decay during germination. On the negative side, transplanting does increase labor and supply costs, but is well worth it for many organic vegetables (e.g., tomato, pepper, eggplant, cucumber, melon, pumpkin, squash, and onion) as compared with direct seeding. Some organic producers are even beginning to transplant crops not traditionally transplanted, such as sweet corn, due to the advantages of earliness and control of seedling diseases facilitated by transplanting. In addition, transplanting also provides an opportunity to add inputs specifically to the rhizosphere of each plant (e.g., biological control organisms).

For many foliar diseases, increasing plant spacing can help to increase movement of air between plants, thus decreasing humidity within the crop canopy, which favors the development and spread of foliar diseases, such as early blight of tomato caused by *Alternaria solani*. Although this is a good strategy for growers of transplanted horticultural crops such as tomato where weeds can be easily controlled through the use of plastic or organic mulches, this strategy is rarely used for direct-seeded organic crops, such as field crops due to the greater concern of weed control. Growers of these crops often use close plant spacing and high seeding rates for these crops in order to permit the crop canopy to close more quickly to shade competing weeds. With weed control being considered the greatest threat to yields or organic field crops and many direct seeded horticultural crops, the management of weeds must be taken into consideration when changing cultural practices to manage plant diseases.

Another strategy promoted in the field of agroecology is the use of mixtures of crop species, termed a **polyculture**, due to the benefits of complimentary resource use that are possible with a mixture of species. This strategy is also useful in the management of plant disease in organic systems for several reasons. First, a well-designed mixture of crop species will typically not be susceptible to similar plant diseases. If an outbreak of disease occurs in one area of the field, it will not spread as easily to other areas of the field as a proportion of inoculum will be intercepted by nonsusceptible crop foliage (termed the "flypaper effect"). Second, if a devastating plant disease were to occur in one crop in the mixture, the other species would still be able to produce a crop and utilize some of the resources (e.g., nutrients, light, and water) left unused by the diseased crop (this phenomenon is termed "yield compensation"). Third, insects vectoring plant diseases are also less able to find target crop resources due to their less concentrated nature or due to greater abundance of natural enemies. Fourth and last, the polyculture can also impact the soil biodiversity through similar mechanisms as those that allow crop rotations and cover crops to increase soil biodiversity: greater diversity of residues and root exudates, leading to potential decreases in plant disease as discussed earlier.

In planting of woody perennial crops, such as orchards or vineyards, special care is taken in the layout of orchard rows in order to improve the environmental conditions (e.g. sunlight, air movement, and temperature inversion) that impact production. In organic systems, this becomes a key factor in managing plant diseases. Typically, rows in these systems will be oriented in a north to south direction if possible. This allows for maximum capture of sunlight and thus greater photosynthesis and also for more complete drying of the crop canopy during the day, which helps to control many foliar and fruit rot pathogens (e.g., black rot of grape, *Phyllosticta ampelicida* (formerly named *Guignardia bidwellii*]). In rows oriented east to west, the north side of the crop canopy can remain moist for extended periods of time due to shading, which favors development of disease. Similarly, these perennial crops are best planted on a gentle slope due to the natural flow of air on slopes caused by the downward movement of denser air from higher elevations. This air movement further dries the crop canopy and reduces humidity in the canopy.

IRRIGATION METHODS

The impact of irrigation methods on control of plant pathogens is described in the chapter on cultural control of plant diseases (Chapter 25). These concepts do not differ in organic production, but do become even more important as growers lack the chemical controls to control disease exacerbated by poor cultural practices.

PRUNING/TRAINING SYSTEMS

The use of appropriate training and pruning systems is another cultural measure that can be used in some organic production systems to manage plant disease. In organic fruit production systems, such as apple, proper pruning helps to open the crop canopy to allow for more sunlight and air penetration. This increase in light, ultraviolet radiation, and air movement helps to suppress the development of disease and spread of pathogen inoculum. Training systems, such as appropriate trellising in grapes, serve a similar function by creating a crop canopy that intercepts a maximum amount of sunlight but remain open enough to allow for air movement and light penetration.

PROTECTED CULTURE SYSTEMS

Protected culture systems, such as high tunnels where crops are grown in the soil, but protected by passively heated and ventilated greenhouse-like structures, are gaining popularity across the globe for the benefits they bring to crop producers such as lengthened growing seasons (Figure 29.4). Organic growers of horticultural

FIGURE 29.4 High tunnels and other protected culture systems influence plant disease development through altered environmental conditions. Here, romaine lettuce management is evaluated for organic high tunnel production systems. (Courtesy of D. Butler.)

crops have been especially keen to adopt the technology due to the benefits of disease control offered for many crops because of the modified environment inside a high tunnel. Because rainfall is excluded, crop foliage remains dry, which helps control many foliar fungal and bacterial diseases. The lack of rainfall also prevents soil particle (and associated pathogen inoculum) splash onto fruit and foliage. Conversely, diseases and insect vectors favored by dry conditions (e.g., powdery mildew and aphids) may be more severe. The high tunnel also acts to physically exclude many sources of pathogen inoculum and insects that may vector plant diseases (e.g., cucumber beetles and bacterial wilt). On the negative side, high tunnels may limit producers' crop rotations, as there are typically only a few high-value crops that are planted in these structures in which growers have made a substantial investment. One solution to this issue is the more recent development of simple, moveable high tunnels, so that producers can more easily rotate crops and prevent the buildup of pathogen inoculum within heavily used areas.

SOIL DISINFESTATION

To control soilborne pathogens and plant-parasitic nematodes, conventional growers have long resorted to broad-spectrum soil fumigants such as methyl bromide. Many soil fumigants allow growers to produce crops without crop rotation due to near-total elimination of soilborne pathogen inoculum, weed propagules, and nematodes immediately following fumigation. In organic systems, growers rely more on crop rotation, but for many growers with limited land or limited diversity of high-value crops, crop rotation is often limited. In these cases, there are several options that organic growers can use to "disinfest" soil. The term disinfestation is more appropriate than sterilization as these methods typically reduce the inoculum of plant pathogens and/or the incidence of disease through a number of mechanisms, but generally do not eliminate pathogenic and nonpathogenic soil microorganisms entirely. Mechanisms are often complex, and include a number of physical (e.g., thermal inactivation), chemical (e.g., increased nutrient availability, creation or release of pesticidal compounds), and biological mechanisms (e.g., creation of an environment favorable to beneficial soil fauna).

Two disinfestation methods include **solarization** and **steam disinfestation**, both of which use heat to reduce numbers of soilborne plant pathogens and plant-parasitic nematodes prior to planting a susceptible horticultural crop. Solarization has been practiced since the mid-1970s. The process involves the use of clear plastic mulch covering moist soils during several weeks of warm sunny weather to disinfest soils. Temperatures have been known to reach as high as 60°C, at shallow depths (5 cm), and 45°C, at deeper depths (20 cm); temperatures with sufficient time are lethal to many soilborne pathogens. Limitations of solarization include the length of time needed to be out of production, the limited seasonality of effectiveness, the lack of control deeper in the soil, and that usefulness of this disinfestation method is limited to warmer, sunnier regions. Solarization can be paired with several alternative soil treatment practices to increase effectiveness, as we discuss in the following paragraphs. Steam disinfestation is a method that has been practiced in the nursery industry for over a century. This practice typically involves heating of the soil to temperatures of 70°C for 20 min. This method has demonstrated acceptable control of many pathogens and nematodes, but degradation of soil quality is also possible with repeated use and could be considered a "big hammer" approach to pest management as discussed previously in this chapter.

Biofumigation is the practice of incorporating cover crops, crop residues, or seed meals of a number of glucosinolate-containing species in the Brassicaceae into the soil prior to crop planting. When these tissues are disrupted and incorporated into moist soil, the release of pesticidal isothiocyanates is triggered by the co-occurring enzyme myrosinase. Other release products are also thought to play a role (e.g., nitriles) as well as general effects of incorporated organic matter. Synthetic isothiocyanate-generating compounds are actually utilized as commercial soil fumigants. This method is effective in controlling a number of pathogens (e.g., *Rhizoctonia* and *Pythium* spp.), although there is some question to the relative effect of isothiocyanates versus the impact of the organic amendment on soil biological activity.

Limitations to biofumigation include the substantial expense and limited availability of the seed meal products and the fact that a large amount of nutrients is incorporated when the seed meals are used at effective rates.

Cover crops of mustards that work well in biofumigation are a more economical and culturally acceptable tactic for many organic growers. Similarly, there has also been a recent focus on the incorporation of a number of bioactive crop residues such as cover crops and many herb species for the control of soilborne plant pathogens. The effects of these treatments are improved at higher temperatures, which allows for solarization to improve treatment effects in addition to gaining solarization benefits, such as improved weed control.

A method gaining a great deal of recent research is referred to as biological or **anaerobic soil disinfestation**. This method relies on the soil incorporation of a labile carbon source to stimulate soil microbial growth and respiration, mulching with plastic to limit gas exchange, and irrigation of the topsoil to flush soil pore space, limit resupply of oxygen, and allow for transport of toxic (i.e., pesticidal) decomposition byproducts through the soil solution. As anaerobic conditions form in this soil environment, the anaerobic decomposition of the added carbon source during a few weeks of treatment creates toxic byproducts and volatile compounds that are believed to play a role in the control of many soilborne plant pathogens and plant-parasitic nematodes. Solarization has synergistic effects with this treatment.

ORGANIC AMENDMENTS

The heavy reliance in organic agriculture on building soil quality through increases in soil organic matter by conservation of crop and cover crop residues as well as the application of organic residues, such as animal manures and composts, also has an effect on the management of plant disease. The application of organic residues and other increases in soil organic matter improve control of plant disease. The mechanisms are complex, but likely relate to the release of a variety of bioactive compounds, as well as the improvement of soil quality and soil structure for improved plant growth and increased activity of nonpathogenic soil microorganisms. This is most notable in the ample data accumulated on the control of plant parasitic nematodes with organic amendments.

REACTIVE MANAGEMENT PRACTICES AND INPUT-BASED MANAGEMENT

Biological Controls

For more information on biological control, see Chapter 28. In organic production, biological controls are permitted as long as they adhere to all other NOP regulations. For example, they cannot contain unapproved inert ingredients nor can they be a product of genetic modification. Although there are several excellent biologically based insect control products for use in both organic and conventional production, there are not a large number of biological controls commonly used in organic production for the control of plant pathogens. Some of the more commonly used products include the following: *Bacillus subtilis*, a commercial product that is labeled for the control of fire blight, and several other bacterial and fungal pathogens; *Trichoderma harzianum*, which is labeled for the control of several root diseases caused by *Pythium*, *Rhizoctonia*, and *Fusarium* species; and *Bacillus pumilus*, labeled for the control of powdery mildew and other foliar diseases. *Trichoderma*-based products are examples of biological control agents that could be incorporated during transplant production.

BOTANICAL

While marketed and utilized primarily for insect control in organic production systems, oil extracted from the seeds of the neem tree (*Azadaracta indica*) has fungicidal activity, and several products are labeled for the control of various leaf spot, rust, mildew, and scab diseases. Garlic (*Allium sativum*) extract is labeled for the control of powdery and downy mildew as well as rusts. A number of plant extracts (e.g., rosemary, *Rosmarinus officinalis*; lavender, *Lavandula* species; and other common herbs and spices) have activity against certain pathogens, but at this point, commercialization and use by organic farmers have been limited. Research in this area continues to evolve and will likely impact how plant pathogens are managed organically in the future.

MINERAL INPUTS

The NOP guidelines specify that "mineral" inputs can be used as disease control measures. In this case, "mineral" refers to naturally occurring (often mined) materials used in the formulations for disease control. An example would be kaolin clay, which when sprayed on the canopy of apple trees helps to prevent powdery mildew (in addition to the control of some economically important arthropod pests of apple). While some may consider various formulations of copper, sulfur, and lime products to be "minerals," these are classified as synthetics by the NOP, and therefore, all restrictions imposed on synthetics must be followed by producers and certifiers. Although many of these products can be obtained from mined sources, most commercial products are byproducts of other industrial processes or contain inert ingredients that require restrictions on use according to the NOP.

APPLICATION OF APPROVED SYNTHETICS

Bordeaux mixture, aptly named for the wine-producing region of France where it was developed for the control of downy mildew in vineyards, is a mixture of copper sulfate and calcium hydroxide (slaked lime). It has activity against a number of bacterial and fungal pathogens common in temperate fruit production. A number of other copper products (copper hydroxide, copper oxide, and copper oxychloride) can also be used in organic production to manage bacterial and fungal diseases. Many organic tomato growers regularly use products such as copper hydroxide to manage fungal pathogens causing diseases such as early blight and late blight in the absence of resistant varieties. The NOP regulations require that these materials be applied in a way that minimizes the accumulation of copper in the soil, which could be problematic with repeated, long-term application. In general, these materials should only be used when other management practices described in this chapter fail to prevent disease and when environmental conditions favor development of disease. Other products used to manage diseases include preparations of sulfur, lime sulfur (elemental sulfur added to calcium hydroxide slurry), potassium, or sodium bicarbonate, paracetic acid, and horticultural oils. The antibiotic streptomycin has historically been used to manage fire blight in apple and pear, but its use is likely to be phased out by the NOP in the near future in order to maintain its use as an antibiotic for human and domesticated animal health. It is strongly recommended that organic growers always inform their organic certification agency before changing management practices to ensure that they are staying within NOP guidelines.

LABORATORY EXERCISES

Controlling soilborne plant pathogens is often a very difficult task even in conventional agricultural systems. For organic growers, there are fewer options available for the control of specific diseases that may develop even though overall management practices are designed to reduce the inoculum and threat of plant disease. Since organic growers of horticultural crops are unable to use tools like broad-spectrum biocides (soil fumigants) that are very often used to eliminate pathogens in chemically based systems, crop rotations, organic amendments, and attempts to create disease-suppressive soils are commonly employed, but may fail to control all potential pathogens. The first laboratory exercise can be employed using soils collected from farms with different management strategies. The goal is to determine if incorporating a soil-derived biological control agent can reduce plant disease and if the effect of the biological control agent differs depending on soil properties and past management. Pathogen inoculum can be added to the collected soils or the effects of adding the organically acceptable biological control product without the presence of a specific pathogen could be evaluated. The second exercise was developed to introduce the concept of grafting scions with desirable horticultural characteristics onto rootstocks with disease resistance characteristics. Students will perform the grafting procedure and evaluate the impacts of grafting in pathogen-infested and pathogen-non-infested soil.

Experiment 1. Effect of Biological Control Agents in Soils Under Contrasting Management Systems

The concept of biological control of plant pathogens is not a new one. As early as 1926, it was suggested that disease could be reduced by the addition of organic matter to soil, which would support antagonists of the potato scab agent, *Streptomyces scabiei* (Sanford, 1926). Since that time, major advances in the study of biological control led to the characterization of plant disease-suppressive soils, defined competition between pathogenic and non-pathogenic strains of organisms, and established the concepts of antibiosis and hyperparasitism as well as characterizing systems in which many different mechanisms are at play. While these mechanisms are attributed to the direct interaction between pathogen and biological control agent, this interaction is undoubtedly influenced by the physical, chemical, and biological properties of the soil to which the biological control agent is added. The purpose of this laboratory is to observe differences in the level of plant disease occurring when a biological control agent is added to varying soil types containing one of two different pathogens.

Materials

Each team of students will need the following items:

- Field soil from multiple locations, preferably differing in management (e.g., organic vs. conventional farm growing similar crops, compost-amended vs. unamended, or fumigated vs. nonfumigated soils)
- *Sclerotium rolfsii* cultured on wheat seeds
- *Pythium myriotylum* cultured on wheat seeds
- Planting containers (pots)
- Tomato or pepper transplants
- SoilGard™ (*Trichoderma virens*) or other OMRI-approved biological control material
- Labels and permanent marking pens
- Laboratory gloves
- Large plastic bags for mixing soil
- Scale

Follow the instructions outlined in Procedure 29.1 to complete the experiment.

Anticipated Results

The results of this experiment provide a basis for suggesting that an introduced biological control agent may

Procedure 29.1
Effect of Biological Control Agents in Soils under Contrasting Management Systems

Step	Instructions and Comments
1	Divide soil source into half and mix SoilGard™ into one-half of each soil at a rate of 0.9 kg per m^3 of soil. If using a 15-cm (6") pot, this is approximately 1.5 g of SoilGard™ per pot of soil. If using a different product, read and follow the directions on the product label. Remember to keep soil from different sources separately during the entire experiment.
2	Fill nine pots of each soil source with soil containing SoilGard™ and nine pots of each soil source with no SoilGard™ added. Allow pots to incubate for 24–48 h.
3	After 24–48 h, mix 2 g of *Pythium*-infested wheat into three pots of each treatment and 2 g of *Sclerotium*-infested wheat into three other pots of each treatment. Maintain three SoilGard™ amended pots with no pathogens added and three pots with no pathogens and no SoilGard™.
4	Label all pots.
5	Plant one plant into the center of each pot.
6	Observe the plants for growth and disease development for 2–4 weeks. Record results.

be more effective against one type of pathogen than another (a fungal plant pathogen vs. an oomycete). It is possible that pathogens that rapidly produce highly resistant survival structures, like sclerotia, could outgrow or survive in soil longer than the biological control agent. Previous soil management tactics can influence the effectiveness of a biological control agent because of inhibition or augmentation of the mechanisms through which the biological control agent might work. If the level of microbial diversity in an organic soil is already very high, there may be no additional benefit derived from adding a biological control agent. During the course of the experiment, it may be possible to observe the interaction of the pathogen and the biological control agent, either microscopically or with the naked eye. The use of inoculated and noninoculated control plants will illustrate the importance of using controls and the variability in results will demonstrate the need for replication in experiments.

Questions

- What did you observe in the treatments containing no biological control agent?
- What would you hypothesize to be the effect of the addition of the biological control agent?
- Did the biological control agent perform as the label indicated?
- How do you think using the field soil(s) affected the experiment? What was the impact of past soil management on the effectiveness of the added biological control agent? What are the implications for organic cropping system management?
- How might these results differ if you started with sterile potting medium?

Experiment 2. Vegetable Grafting for the Control of Root-Knot Nematode and Bacterial Wilt

Many organic growers are interested in targeting specialty markets with products that are not typically found in conventional production systems. Heirloom tomatoes are an example of this type of specialty crop. These varieties are typically open-pollinated tomatoes that have limited resistance to plant diseases, but have desirable fruit qualities. One way to address this issue is to produce grafted tomato plants for transplanting to the field. Organically produced seeds of many heirloom varieties are now commercially available, and disease-resistant rootstocks are also becoming more widely available. Rootstocks can be chosen for resistance to pests commonly found in organic production systems (Rivard and Louws, 2008), as well as for pathogens that are difficult to control in all production systems (McAvoy et al., 2012), particularly those that survive for long periods in soil without a host.

Materials

Each group of students will require the following items:

- Two-week-old tomato plants (two to four true leaves) of a bacterial wilt-resistant rootstock (e.g., “Cheong Gang”); an RKN (*Meloidogyne* species)-resistant rootstock (e.g., “Multifort”); and a susceptible heirloom tomato variety (e.g., “German Johnson” or “Yellow Brandywine”)
- 2-mm silicon tube-shaped grafting clips
- New sterile razor blades (warn students to take care in using blades)
- Laboratory gloves
- Organic potting medium

- Planting containers (pots)
- Labels and permanent marking pens
- 24- to 48 h-old culture of *Ralstonia solanacearum* (tomato race) grown on nutrient agar
- Sterile deionized water
- Sterile rods
- RKN-infested tomato roots or field soil naturally infested with RKN
- Spectrophotometer
- Test tubes
- Wooden handle probes

Follow the instructions outlined in Procedure 29.2 to complete the experiment.

Anticipated Results

If the tomato plants chosen for each rootstock–scion combination have stems that are similar in size and are cut at similar angles, successful grafts should occur 85%–90% of the time. Grafted tomato plants will often grow more vigorously than nongrafted plants of the same variety. Tomato plants affected by bacterial wilt will generally die rapidly when growing in greenhouses at 30°C or higher. Rootstocks resistant to the pathogens included in the study should provide protection for the scion of grafted plants.

Questions

- What types of symptoms were observed on each type of plant in each treatment?
- What other (if any) benefits of grafting were observed?
- If sterile conditions were not used during grafting, what types of problems could arise?
- If the rootstock were cut above the cotyledons, what might occur?

Procedure 29.2

Vegetable Grafting for the Control of Root-Knot Nematode and Bacterial Wilt

Step	Instructions and Comments
1	Select rootstock and scion plants that have stem diameters as closely matched as possible.
2	Wearing gloves and using a completely clean surface area, *carefully* cut the stem of the rootstock at a 45° or greater angle with a sterile razor blade. The cut should be made below the cotyledon of the rootstock (Figure 29.5).
3	Using sterile technique, cut the scion using the same angle and a new sterile blade.
4	Trim off the largest of the leaves from the scion.
5	Carefully join the scion with the rootstock using the silicon tube clip. Repeat for the desired number of replicates for each rootstock/scion combination needed.
6	Place plants in a "healing chamber" or growth chamber in the dark at 21°C–27°C and 95% relative humidity. Allow the plants to remain under these conditions for 3 days and begin introducing light for another 3 days. After 7 days in the chamber, move plants to full sunlight for another week before planting.
7	Prepare *Ralstonia solanacearum* inoculum by removing bacteria from plates using a sterile loop to transfer to sterile deionized water. Adjust the concentration spectrophotometrically ($OD_{600\ nm} = 10^8$ CFU/mL) and dilute to 10^7 CFU/mL.
8	Fill pots with organic potting mix and inoculate each of nine pots with *R. solanacearum* using 75 mL of bacterial suspension (10^7 CFU/mL). Mix soil with sterile rod.
9	Fill nine pots with organic potting medium and 20 g of chopped RKN-infested tomato roots or fill nine pots with naturally RKN-infested field soil.
10	Fill nine pots with organic potting medium alone. These pots will serve as non-inoculated controls.
11	Plant three non-grafted heirloom plants into three pots of each treatment (three in bacteria-inoculated pots, three RKN-inoculated pots, and three untreated pots). Repeat for each of the grafted plant types.
12	Observe plants in the greenhouse for symptoms of disease. Record the incidence of wilting.
13	Remove wilted plants, cut the stems, and immerse in a test tube of water to observe bacterial streaming.
14	After 4 weeks, carefully remove plants from the nematode-treated and -untreated pots and compare the root systems for presence of galling. Record results.

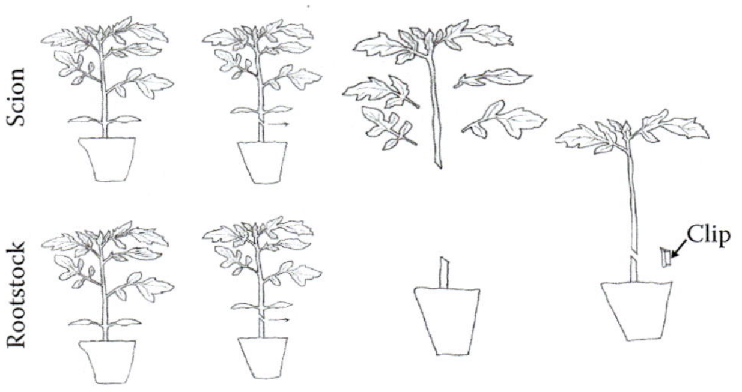

FIGURE 29.5 Tomato grafting procedure. (Illustration courtesy of Kate Rotindo, USDA-ARS.)

DISCLAIMER

The use of trade, firm, or corporation names in these methods is for the information and convenience of the reader. Such use does not constitute an official endorsement or approval by the USDA Agricultural Research Service (ARS), of any product or service to the exclusion of others that may be suitable. In addition, The USDA-ARS makes no warranties as to the merchantability or fitness of the methodologies described on these pages for any particular purpose, or any other warranties expressed or implied. These methodologies provide a guide and do not replace published work. The USDA-ARS is not liable for any damages resulting from the use or misuse of these methodologies.

REFERENCES

Liebman, M., and E. R. Gallandt. 1997. Many little hammers: Ecological management of crop-weed interactions. Pages 291–343 in: *Ecology in Agriculture,* L. E. Jackson (ed). San Diego, CA: Acad. Press.

McAvoy, T., J. H. Freeman, S. L. Rideout, S. M. Olson, and M. Paret. 2012. Evaluation of grafting using hybrid rootstocks for management of bacterial wilt in field tomato production. *HortSci.* 47:621–625.

Rivard, C. L., and F. J. Louws. 2008. Grafting to manage soilborne disease in heirloom tomato production. *HortScience*. 43:2104–2111.

Sanford, G. B. 1926. Some factors affecting the pathogenicity of *Actinomyces scabies*. *Phytopathology* 16:525–547.

SUGGESTED READINGS

Altieri, M. A. 1999. The ecological role of biodiversity in agroecosystems. *Agric. Ecosyst. Environ.* 74:19–31.

Bailey, K. L., and G. Lazarovits. 2003. Suppressing soil-borne diseases with residue management and organic amendments. *Soil Tillage Res.* 72:169–180.

Bennett, A. J., G. D. Bending, D. Chandler, S. Hilton, and P. Mills. 2012. Meeting the demand for crop production: The challenge of yield decline in crops grown in short rotations. *Biol. Rev.* 87:52–71.

Brussaard, L., P. C. de Ruiter, and G. G. Brown. 2007. Soil biodiversity for agricultural sustainability. *Agric. Ecosyst. Environ.* 121:233–244.

Butler, D. M., E. N. Rosskopf, N. Kokalis-Burelle, J. Albano, J. Muramoto, and C. Shennan. 2012. Exploring warm-season cover crops as carbon sources for anaerobic soil disinfestation (ASD). *Plant Soil* 355:149–165.

Davis, J. R., O. C. Huisman, D. T. Westermann, S. L. Hafez, D. O. Everson, L. H. Sorensen, and A. T. Schneider. 1996. Effects of green manures on Verticillium wilt of potato. *Phytopathol.* 86:444–453.

Fravel, D. R. 1988. Role of antibiosis in the biocontrol of plant diseases. *Annu. Rev. Phytopathol.* 26:75–91.

Heckman, J. 2006. A history of organic farming: Transitions from Sir Albert Howard's *War in the Soil* to USDA National Organic Program. *Renew. Agric. Food Syst.* 21:143–150.

Klein, E., J. Katan, and A. Gamliel. 2011. Combining residues of herb crops with soil heating for control of soilborne pathogens in a controlled laboratory system. *Crop Prot.* 30:368–374.

Lazarovits, G. 2001. Management of soil-borne plant pathogens with organic soil amendments: A disease control strategy salvaged from the past. *Can. J. Plant Pathol.* 23:1–7.

Matthiessen, J. N., and J. A. Kirkegaard. 2006. Biofumigation and enhanced biodegradation: Opportunity and challenge in soilborne pest and disease management. *Crit. Rev. Plant Sci.* 25:235–265.

McGrath, M. T. 2009. Managing plant diseases with crop rotation. Pages 32–40 in: *Crop Rotation on Organic Farms: A Planning Manual,* C. L. Mohler and S. E. Johnson (eds). Ithaca, NY: Natural Resource, Agriculture, and Engineering Service.

Melakeberhan, H., A. Xu, A. Kravchenko, S. Mennan, and E. Riga. 2006. Potential use of arugula (*Eruca sativa* L.) as a trap crop for *Meloidogyne hapla. Nematol.* 8:793–799.

Papavizas, G. C. 1985. Trichoderma and Gliocladium: Biology, ecology, and potential for biocontrol. *Annu Rev. Phytopathol.* 23:23–54.

Peck, G. M., and I. A. Merwin (eds.). 2009. *A Grower's Guide to Organic Apples*. Ithaca, NY: Cornell University Cooperative Extension, New York State Department of Agriculture and Markets, New York State Integrated Pest Management.

Scofield, A. M. 1986. Organic farming: The origin of the name. *Biol. Agric. Hort.* 4:1–5.

Shennan, C. 2008. Biotic interactions, ecological knowledge and agriculture. *Phil. Trans. R. Soc. B.* 363:717–739.

Stapleton, J. J., C. Summers, J. Mitchell, and T. Prather. 2010. Deleterious activity of cultivated grasses (Poaceae) and residues on soilborne fungal, nematode and weed pests. *Phytoparasitica* 38:61–69.

Tjamos, E. C., G. C. Papavizas, and R. J. Cook (eds). 1992. *Biological Control of Plant Diseases: Progress and Challenges for the Future*. New York, NY: Plenum Press.

Van Brugen, A. H. C. 1995. Plant disease severity in high-input compared to reduced-input and organic farming systems. *Plant Dis*. 79:976–984.

Wang, K.-H., B. H. Sipes, and D. P. Schmitt. 2002. *Crotalaria* as a cover crop for nematode management: A review. *Nematropica* 32:35–57.

Widmer, T., and N. Laurent. 2006. Plant extracts containing caffeic acid and rosmarinic acid inhibit zoospore germination of *Phytophthora* spp. pathogenic to *Theobroma cacao*. *Eur. J. Plant Pathol.* 115:377–388.

Part V

Special Topics

30 Plant Disease Diagnostics

Kevin L. Ong

Concept Box

- Plant disease diagnosis is the process of figuring out the problem uing a logical and calculative method.
- Skills needed for plant disease diagnosis include personal skills (observations, knowledge), and technical skills (ability to use analytical procedures and detection tools).
- Preliminary or initial diagnosis, based on observation and knowledge, will determine which confirmatory or specialized tests to use.
- Confirmatory or detection tests are developed from previous research with known pathogens and are not available for all pathogens or causal agents.
- The ultimate test to diagnose a plant disease problem is to show proof of pathogenicity, satisfying Koch's postulate.
- An accurate diagnosis is usually based on the result of more than one approach or procedure to derive an answer.

Plant disease diagnosis is the act or process of determining and identifying a disease problem. This act can range from a simple guess based on the observed symptoms to confirmatory testing where a healthy plant is challenged with the recovered suspect pathogen or agent. Within this continuum, there are a large number of methods and techniques that are available to the person who is diagnosing. Most people have performed some level of plant disease diagnosis. For example, a homeowner seeing a dead patch of grass in his/her lawn concludes that it appears to be similar to take-all patch that they read about in their local newspaper garden section.

Plant disease diagnosis presents many challenges. First, not all plant diseases exhibit symptoms that are easily visible or recognizable. In these situations, techniques other than just observation are needed to ascertain the causal agent of the problem. Second, plants may vary in their response to the same pathogen or causal agent; they may not all exhibit the same symptoms. In this situation, knowledge about the plant in question is extremely helpful to determine the reaction of the particular plant to the causal agent. Third, occasionally, symptoms of a disease problem may not be necessarily present in the same location as the pathogen. For example, symptoms of twig dieback may be a result of diminished root function caused by root rot. In this case, the location of symptoms (twig) is far removed from the cause (root rot). Fourth, there are times when the causal agent of a plant problem is not present or may be difficult to recover where symptoms are noticed. For example, the bacterium *Erwinia amylovora* causes fire blight on ornamental pear, but cannot be recovered from symptomatic dead branches. In this situation, one must recognize that the pathogen, most likely, will be found causing damage on tissue that is not yet dead. Fifth, there are also situations where disease arises from more than one causal agent, requiring the ability to discriminate among the various clues pointing to different causal agents. Finally and sixth, there are some disease problems where a

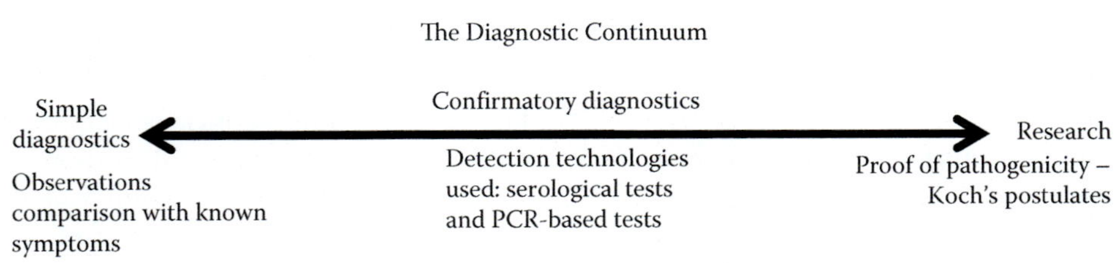

consistently recovered pathogen or identified causal agent has not been previously associated with the damage on the plant. In this situation, research (Koch's postulates—Chapters 1 and 31) must be conducted to demonstrate the pathogenicity of the causal agent.

A common issue in disease identification is the need for accuracy and speed of diagnosis. Accuracy in diagnosis is important for success in disease management. However, accuracy or confidence in the diagnosis is inversely related to the speed at which diagnosis can be made. The simple approach of matching the symptoms to a disease relies on the presence of distinct recognizable symptoms and the ability of the practitioner to recognize those symptoms. This method is very quick, but the possibility remains that similar symptoms can be caused by many different pathogens or causal agents and so erodes the certainty of the diagnosis. In these situations, we rely on the knowledge and experience of the person who is diagnosing to recognize the symptoms. The stakes are much higher for growers or farmers who rely on their crops for their livelihood. Both speed and accuracy are critical. Therefore, a balanced approach to produce a quality diagnosis quickly is necessary. To meet this need, there are practitioners who specialize in the problems of specific crops. Their training, knowledge, and experience in a focused fashion on a limited number of crops enable them to recognize the specific and various nuances of the problem to provide a reliable preliminary diagnosis. Many of these practitioners also have training in various tools and methods used in the process of diagnosis. Some work as specialists or consultants and perform much "field" or on-site diagnostics. Some work as diagnosticians in plant disease diagnostic clinics or centers. In general, successful practitioners have a wide range of knowledge in areas such as soil science, horticulture, agronomy, biology, crop production, entomology, and, most importantly, plant pathology.

A quality diagnosis relies on the following abilities of the practitioner to (1) distill information and/or observations of the plants, growing area, and environment; (2) apply knowledge in understanding potential disease-causing agents; and (3) utilize scientific tools and resources to identify the causal agent. In this chapter, we will explore the basic skills needed to perform plant disease diagnosis. We will also explore various tools that are typically used to help with the diagnosis of plant diseases.

BASIC SKILL NEEDS FOR DIAGNOSIS

Diagnosis begins with an innate curiosity of wanting to know what happened to the plant, typically precipitated by an observation that the plant is not normal. Knowledge of the plant and its growth habits is critical to identify whether the plant is normal or diseased.

KNOWLEDGE OF PLANTS

How do you know if the plant is diseased? People assume that if a plant is not green, then there might be a problem. However, there are ornamental plants with variegated leaves, such as pothos (*Epipremnum aureum*) "Marble Queen." This variegation is not the result of a disease or physiological condition, but is instead a quality that was bred for this particular plant.

There are characteristics of plants that may be novel, but to an unfamiliar observer, and these may be a source of fear and panic. For example, Texas live oak (*Quercus fusiformis*) is an evergreen through the winter months in Texas. However, they shed their leaves in the spring in advance of new foliage that will appear soon after shedding. People who are unaware of this characteristic of Texas live oak become very concerned about their trees when they notice the leaves dropping in the spring. Their reasoning is that the tree survived well in the winter and they wonder if the cold might have injured the tree to cause the defoliation. This is a logical thought process, and one ought to be concern if the plant is a holly (*Ilex* sp.) or *Photinia* sp., but not for Texas live oak where this is normal. Knowledge about plant characteristics are important in determining whether or not there is a disease problem.

There are many horticultural books on various plants and characteristics of different varieties. For example, there are about 150 species of roses and over 6500 different varieties, although only a few are available commercially. There are numerous books, bulletins, factsheets, and Web pages about the characteristics of different roses. Some are resistant to a fungal disease known as black spot, whereas others are highly susceptible. Similarly, corn growers have to decide on the variety of corn that they can obtain and will be planting for any given growing season. Each of these varieties has specific characteristics. For example, growers in areas where corn stalk rot (a fungal disease caused by *Gibberella*, *Fusarium*, or *Diplodia*) is a problem would benefit from planting a variety that has resistance to this disease.

OBSERVATION

To a certain degree, all of us observe our surroundings. To be proficient in diagnostics, one must be a keen observer. What should be observed? There are things that are obvious and things that are not so obvious. For example, a brittle branch is an obvious sign that the branch is dead. Fungal fruiting bodies in cracks or cankers on the dead branch may not be so obvious; yet, reveal themselves as a contributing factor to the demise of that branch. Take a logical approach to observation of the plant and its surrounding. The following order of observation is one technique that involves maintaining a chronological and systematic approach to collecting information.

1. Observe the diseased/problem plant. Note what is abnormal. What do you see that makes you think this plant is diseased? Identify the symptoms that are observed. Take care to observe if any patterns are associated with the symptoms, such as wilting leaves occurring on only one side of the plant, or tip **dieback** occurring on scattered branches throughout the plant. Look for evidence of the causal agent. Most plant diseases are caused by fungal pathogens. Are there **signs** of fungal growth (fuzzy mycelial growth) on **symptomatic** areas of the plant?
2. Observe the area where the plant is growing. Does the location of the plant contribute factors that might stress the plant? For example, does the affected plant appears to be located in a low-lying area where water might pool? Does the location of the plant contribute to symptom development by a causal agent? For example, the affected plant was planted too close to a south-facing brick or metal wall and exhibits scorching symptoms. Does the area surrounding the plant show possible disturbance? For example, evidence of trenching for an irrigation system suggests that roots of the affected plant may have been disturbed.
3. Observe other plants in the immediate area. Are there other plants of the same kind showing similar or progressive damage? Are there other types of plants exhibiting symptoms that are abnormal? Are these symptoms similar to the affected plant? This gathered information can provide clues about whether the plant problem is caused by a biological causal agent or one that is abiotic. For example, if multiple species of plants are showing somewhat similar symptoms of necrotic leaf spot, it is possible that the symptoms are the result of a nondiscriminatory causal agent such as chemicals (herbicide).
4. Observation of the plant over a period of time. Many times information on plant history may not be possible if you were consulted for the problem after it was observed. In order to gain such information, it is necessary to interview the owner of the plant or people who may have knowledge of the plant (witnesses). Examples of probing questions to gain historical information include the following: How old is the plant? How was it planted? Was it a containerized plant? When did you first notice the problem? What did you feed or treat the plant within the last 4–6 months? Did you take any action after first noticing the problem? The information collected may provide a progressive view of the onset of the problem. On most occasions, the problem actually began prior to symptom observation.

In recent years, tools to take images (photos) have become readily available. Digital cameras are relatively inexpensive and images can be transmitted almost instantly. Even "point and shoot" cameras can take good photographs of a subject. Furthermore, many cellular phones have an onboard camera. Some of these cameras can capture good-quality images. Photography is a great way to document what was observed (Case Study 30.1; Figure 30.1). The image also serves as a visual record.

CASE STUDY 30.1

A Picture Is Worth a Thousand Words

- Photography is a good method to capture observations and freeze the image at a particular time.
- The goal of photographing the plant subject is to convey the problem that you are observing:
  - The big picture: image of the affected plant in the garden shows any interaction between the affected plant and the rest of the garden.
  - The focus: image of the affected plant should show how the problem is manifested and what parts of the plant are damaged.
  - The details: close-up of the affected plant may reveal some characteristics that would help identify the pathogen.
- The limitation of photography for diagnosis is that some information may be missing or you would be unable to be describe key information with only a photograph:
  - A photograph is two-dimensional; therefore, there is no definitive way to show the texture of the damaged area on the plant or the topography of the land in relation to the diseased plant.
  - You must be deliberate to add or include things in the photograph that would infer a size description.
- Photography is useful in diagnostics to share with other experts to help determine the next course of action, such as what type of sample to collect.

FIGURE 30.1 Tools for sampling plant tissue and soil. Note magnifying glass and camera for documentation of symptoms and signs in the field.

All the information collected from the approaches noted above or the visual record can provide many clues and result in the development of a list of potential suspects.

ACCESS TO RESOURCE MATERIAL

Observations made about the affected plant and its surroundings would yield information that can be used to identify potential causal agents. Reference materials are important to complete this step in the process of diagnosis. In this age, extensive information and resources are accessible via the Internet. A search engine and several key words describing the symptoms and host plant may yield numerous references. A major challenge in applying information from the Internet is the reliability of that information. Look for information that is from reputable sources such as Land-Grant universities, Cooperative Extension Service (eXtension.org), professional organization (American Phytopathological Society or Entomological Society of America), or the U.S. Department of Agriculture (USDA). There are also many good printed materials that are useful as references in diagnosis. Disease compendia or disease host indices have images and information that are helpful to identify potential pathogens or causal agents. Examples of these books include the disease compendia series from APS Press and the Ortho Problem Solver book.

With these resources, a simple diagnosis can be made on the basis of the convergence of observed symptoms and known or reported symptoms. As previously mentioned, this may always be possible. In such cases, further investigation is needed to better ascertain the nature of the problem.

SAMPLING FOR FURTHER INVESTIGATION

In the event that a simple diagnosis cannot be made or confirmation of a particular pathogen or causal agent is needed, samples should be taken of the plant for further study/analysis in the laboratory, or to be submitted to a plant disease clinic/center. Help can be obtained from local county extension offices for the general public (farmers, growers, and homeowner) who need assistance with sampling for diagnosis. The Cooperative Extension Service is a nationwide education network in every state in the United States and is based at the state's Land-Grant university. The Extension Service of each state may have regional- or county-level offices throughout the state where service is provided locally. Extension agents at these local offices can provide guidance to collect the proper samples and assist in submitting them to a plant disease clinic. In any case, there are some considerations when collecting samples for diagnosis submission.

First, the sample collected should be representative of the observed problem. For example, if leaf spot and fruit rot are the only abnormal symptoms observed on an apple tree, then leaves and fruits showing early, middle, and late stages of the damage should be collected, if available. An accurate diagnosis depends on a sample that is representative of the situation. Second, the sample should be fresh. Ideally, specimens collected should be recently infected or in early stages of disease spread. This type of plant tissue provides the best opportunity to identify, isolate, and/or recover a pathogen because the pathogen is actively developing (growing and causing disease). For example, the disease that develops in rainy weather may be dependent on those wet conditions for disease development. Collecting the sample at a later time when conditions are much drier may result in the inability to identify the pathogen because the pathogen may no longer be active. Additionally, symptoms may have been altered, and contamination by secondary microorganisms may have developed to prevent an accurate diagnosis. Third, an adequate sample should be collected for a proper diagnosis. How large should an adequate sample be? The answer to this question varies depending on the type of problem and plant. For example, a branch with 10–20 symptomatic leaves may be sufficient when working with a leaf spot problem. For turf grass samples, divots measuring approximately 7" × 7" in area with sufficient depth to include the roots from the edge of damaged areas would be adequate for diagnosis. On the other hand, whole plants may be needed when there is suspicion that root problems may be contributing to foliar symptoms. In some cases, collecting the whole plant is not possible, such as the case with trees. In these situations, representative samples from symptomatic or suspected parts would be the appropriate approach for a sample submission. Avoid dead plants as samples. When the plant tissue is dead, (1) the plant pathogen may not have survived on dead tissue and thus cannot be detected, or (2) the plant part is overrun with saprophytic microorganisms that make diagnosis difficult or impossible. Sometimes soil samples around damaged and healthy plants should be collected when soil pH or nutrients is suspected to be the contributors of the plant problem. Some laboratories have the capability to test soils for fertility or nutrient content.

Once the sample is collected, it must be packaged to preserve the freshness and integrity of the sample (Figure 30.2). For example, samples can be placed in a plastic bag to prevent desiccation; care should be taken not to introduce excessive moisture. Sample parts containing soil, such as the roots, should be packaged to prevent contaminating other plant parts. The sample should be placed in a crush-proof box or container to preserve the integrity of the sample. Crushed plant parts could result in loss of visible symptoms. In addition, plant tissue that is crushed may release moisture and nutrients that allow secondary decay microorganisms to grow and interfere with the diagnosis.

If submitting a specimen to a plant disease clinic or diagnostic laboratory, the submitter is usually required to complete a form that accompanies the sample. These forms are used to collect the submitter's observations and are indispensable to the diagnostician. For example, a question asking a submitter to note their irrigation practices may reveal if overwatering was a potential contributor to the root rot occurring on the submitted plant sample.

If the sample is to be mailed to the diagnostic laboratory, overnight mail or courier service is suggested. The shorter transit allows fresh samples to be received by the diagnostic service and maximizes the likelihood of an accurate diagnosis. In addition to short transport times, keeping the sample cool and ensuring that sample integrity is sound are good practices to preserve the freshness of the sample when transporting samples to the laboratory for further investigation.

FIGURE 30.2 Packaging for sample transport. Sample and forms are packaged separately in a crush-proof box.

SOIL PROPERTIES

Occasionally, soil properties can contribute to poor growth, predisposition to diseases, or increase the likelihood of nutrient toxicity or deficiency response of the plant. If soil or growing medium is collected for analysis and there is reason to believe that the soil may contribute to the problem, the soil should be analyzed. The two simplest test typically done in a plant disease clinic are soil **pH** and **total soluble salts** determination.

Soil pH can be determined using a pH meter or litmus paper. Litmus paper was a method that was previously used to rapidly evaluate the pH of a solution. This method is no longer used extensively. Rather, a pH meter is favored to rapidly determine the pH of the soil or growing medium. In the laboratory, pH meters are normally available and relatively inexpensive. Benchtop units are used and may have more functions than a handheld version. Handheld or portable pH meters are small and easy to carry, making them the choice for work in the field. To determine the pH, a small amount of soil or growing medium is mixed with an equal amount of distilled or deionized water. This mixture is allowed to stand for 30 min to 1 h, which permits most of the soil or growing medium to settle. A pH reading of the liquid solution would reveal the pH of the soil. Alternatively, one can pour distilled or deionized water into a pot of soil or growing medium and allow the liquid to percolate through it. The pass-through liquid (leachate) is captured and a pH reading is made. The pH reading will provide information on whether the pH of the growing medium or soil was appropriate for the plant. For example, azalea thrives in acidic soil, but performs poorly in highly alkaline soils. In addition, extreme ranges of pH, too acidic or too alkaline, can be an indication that the symptom on the affected plant may be due to nutritional imbalances. For example, extremely acidic soils may cause some minor element/nutrients toxicities in some plants.

Total soluble salts can be measured using an electrical conductivity (EC) meter. This instrument is used much like the pH meter. The conductivity reading is typically given in milliSiemens per centimeter (mS/cm). Interpretation charts accompany specific procedures for various methods of taking an EC reading from soil or growing medium. These procedures refer to the different ways to obtain a reading, such as the **"pour-thru" method** where distilled deionized water is poured into the pot, the leachate is collected, and a reading is made from the leachate. A low reading such as 0–0.9 mS/cm may indicate that the soil or growing medium lacks mineral nutrients and could cause nutrient deficiency symptoms on the plants (Anon, 2015). On the other hand, high

EC values are an indicator of high salts and can contribute to symptoms such as leaf scorching.

Typically, these tests are conducted by a soil testing laboratory or facility. In addition to these two readings, a soil testing laboratory can determine various mineral nutrient levels in soil. This information can provide evidence to support or deny problems caused by soil nutrient imbalance (lack of or excess).

MICROSCOPY

In the field, a simple magnifier (5× to 20×) is useful to look at a lesion or damage on plant parts. Magnification of the plant subject or the causal agent of disease is helpful to determine the details of disease symptoms and pathogen structures. Microscopes, dissecting and compound, are magnification tools that are useful in examining fungi and bacteria in a laboratory setting (see Chapter 3).

A dissecting microscope typically allows for magnification up to 60×. This low magnification range is useful for the examination of specimens, especially to look at larger fungal structures, such as fruiting bodies and spore masses, in greater detail. This also allows for finer touches when manipulating those fungal structures. For example, with a dissecting scope the diagnostician can see and identify fungal structures and transfer them by forceps to a glass slide for further examination at much higher magnifications on the compound microscope.

A compound microscope has magnification ranging from 40× to 1000×. This range of magnification is sufficient to examine fungal fruiting bodies and spores in greater detail. Many fungal pathogens can be identified by the spore morphology and/or key hyphal structures. The compound microscope can also be used to visualize masses of bacteria as they ooze out of diseased plant tissue. When coupled with specific dyes, such as those used for the Gram stain, bacteria can be characterized.

Occasionally, viral diseases may be identified via a compound microscope. Infected plant tissue can be stained with a protein or nucleic acid stain to highlight the protein or nucleic component of the virus. Aggregations of virus particles in the plant cell are known as **viral inclusion bodies** (Chapter 4); these can be viewed at 450× and are specific to individual virus groups (Christie and Edwardson, 1986).

Electron microscopes utilize an electron beam to resolve and magnify an object better than the compound microscope. Modern electron microscopes can visually enlarge an object upwards of 10 million times. There are several types of electron microscopes that all use the electron beam to create the enlarged image. This technology has been used to study various plant pathogens, but is best known for use in the study of viruses (Chapter 4). It is a tool that has been used to detect viruses in diseased plant samples. Materials that are examined using electron microscopy have to be specially treated or processed. The samples may have to be fixed, chemically or with liquid nitrogen (cryofixation), which is a way to preserve the sample at a moment in time so as to prevent deterioration of tissue. The sample may need to be embedded, typically in plastic, so that thin layers of tissue can be sliced thinly, and examined using an electron microscope. The samples can also be stained, but staining for electron microscopy is usually done with heavy metals such as lead, uranium, and/or tungsten. The staining helps to produce better contrast among structures in the electron microscopy examination. The electron microscope is not used often for diagnostic or detection purposes because they are expensive to purchase and maintain. Furthermore, sample preparation and processing prior to electron microscopy examination are tedious. However, the electron microscope is instrumental in the identification and characterization of previously unknown viruses.

INCUBATION

Many plant diseases are caused by fungi. Incubation of a suspect sample in a moist chamber simulates high-humidity conditions and is conducive to fungal growth and/or continued fungal development (sporulation, development of resting or fruiting structures, e.g., **sclerotia** or **pycnidia**) or characteristic hyphal or mycelial structures. A moist chamber is usually set up after initial triage is done on the specimen and a fungal pathogen is suspected. A moist chamber is basically a closed container or plastic bag (not necessarily sealed from air exchange) with a wet sponge or paper towel as the source of moisture to maintain high relative humidity. The plant specimen is placed in this chamber and exposed to high relative humidity, not free water. Humid conditions are favored by many microbes for growth and sporulation. As such, limitations of the moist chamber are the nondiscriminatory growth of microorganisms that are **saprophytes** that may outcompete and overgrow the plant specimen, making it impossible to detect the pathogen spores and/or structures (Shurtleff and Averre, 1997). In addition, growth of many microorganisms leads to quicker breakdown or decay of the plant specimen. Samples in a moist chamber should be monitored daily to avoid some of these limitations. Because the moist chamber does not discriminate for pathogen growth, familiarity and knowledge of both plant pathogenic and saprophytic fungi are very helpful.

ISOLATION AND RECOVERY

Culturing to isolate and recover the suspect pathogen is a technique used on both fungal and bacterial pathogens. The specimen parts used for culturing typically consist of infected tissue bordering on healthy tissue (interphase/transitional area between healthy and bad tissues).

This piece of plant tissue is usually surface-sterilized to reduce surface contaminant or other saprophytes that are lurking on the plant tissue surface. The rationale is that the pathogen causing the damage would be present in the damaged internal tissue. **Surface sterilization** can be achieved in many different ways. Most common is soaking of the tissue in a diluted bleach solution, for example, 10% (v/v) bleach solution. Ethanol, peroxide, or products containing *n*-alkyl can be used as the sterilizing agent. Usual exposure time of the plant material to the sterilizing agent can range from a few seconds to several minutes, depending on the structure if the plant tissue (permeability of the plant tissue). Sterilized plant tissue should be handled **aseptically** throughout this process. After a period of exposure to these chemicals, the plant tissue is rinsed in sterile water and excess water blotted off with blotting paper or clean, sterile paper towels. Plant tissue is sectioned into small pieces and placed on sterile culture medium in Petri dishes. These dishes are maintained at room temperature or in an incubator at a specific temperature (such as 28°C–30°C for bacterial growth or ~25°C for fungal growth). For both fungal and bacterial agents, the culture dishes are typically maintained for 3–7 days and examined daily for microbial growth.

The type of culture medium is important to the type of microorganism one expects to isolate. For example, potato dextrose agar (PDA) is generally a nutritious medium that many fungal and bacterial agents can utilize. PDA is used for the growth of most fungal pathogens. Addition of several drops of concentrated lactic acid to PDA reduces the pH of the medium and suppresses the growth of bacteria. Acidified PDA is a medium that is typically used to suppress the growth of bacterial contaminants while allowing the fungi to grow (Chapter 2). For bacteria, a highly nutritious medium, such as nutrient agar, is a good general recovery culture medium. There are many other types of agar that have been described to recover and grow microorganisms. Selecting the correct medium for the correct application is an art. For example, water agar is a good choice for isolating *Pythium* species because this water mold can outgrow other fungi or bacteria on this medium—thus allowing for a possible "clean" recovery.

If surface sterilization was done correctly and if there is a pathogen in the plant material, we would expect to see fungal growth emanating from internal parts of the plated tissue. Another clue is recovered cultures that are consistent morphologically. These pure cultures would allow for the plant pathologist to identify the pathogen. Some fungi have distinctive growth in culture, whereas other fungi produce spores or fruiting bodies that are unique. Distinctive growth habit and characteristics of spores and fruiting bodies are useful in the identification of fungal agents.

If a bacterial agent is suspected, surface-sterilized plant material exhibiting damage is cut up or crushed in a drop of sterile water. This action helps to release the bacteria from the confines of the plant tissues. If an infection is severe, it very likely that bacterial oozing is visible. The macerated tissue is left in the sterile water for several (3–5) minutes; then a small amount of the liquid is plated on a suitable culture medium. If the surface sterilization procedure was done correctly, one would expect to see a consistent colony type on the culture medium. Bacteria, unlike fungi, do not grow unique structures that could be used for identification. Bacterial identification relies on specific physiological test/reaction characteristics or molecular characteristics.

Baiting

Baiting is a technique that is used to lure the pathogen to you. The bait can be an attractant or a food source for the pathogen. The bait, usually a desirable plant host, can stimulate pathogen growth. For example, the oomycete *Phytophthora* sporulates and produce zoospores that are motile. This characteristic is one way that this water mold is dispersed to cause new infections. To isolate *Phytophthora* from a diseased plant root or the soil surrounding the diseased plant, the root and soil are submerged in water. Pieces (leaf disks) of eucalyptus (Figure 30.3), camellia, or rhododendron leaves can be floated on the water surface. The zoospores of *Phytophthora* will be attracted to the leaf disks. After several days, the leaf disks are carefully removed from the water, blotted dry, and placed onto a suitable semiselective medium. If *Phytophthora* was present in the diseased root or in the soil, one should see *Phytophthora* mycelium growing from the leaf disk.

FIGURE 30.3 Baiting for *Phytophthora* spp. with eucalyptus leaf disks.

PATHOGEN DETECTION AND IDENTIFICATION

In recent years, detection assays and identification tools have become more robust, readily available, and economically feasible for use in the field and/or the laboratory. Detection assays are highly useful to detect known pathogens. However, it still takes a practitioner with experience to be able to figure out which detection assay to use. There are commercial companies that manufacture test kits for use in field and in the laboratory. It should be noted that the typical detection assay is used to detect only one known pathogen and is not available for all pathogens.

Detection and Identification with Serological Methods

Serology is the study of immunological reaction—the reaction of antigens (substance foreign to the body) and antibodies (specific molecules produced by a mammal in response to the foreign substance). Common products used for detection assay are the **lateral flow devices (LFDs)** and **enzyme-linked immunosorbent assay (ELISA)** kits. Either of these detection instruments is developed to detect one pathogen or one type of pathogen. Lateral flow devices (Figure 30.4) are field-ready kits used to rapidly detect a pathogen. Typically, plant tissue with a suspected pathogen is macerated in a specific buffer. This slurry is place at one end of the LFD where the liquid of the slurry will absorb onto the absorbent strip and travel by capillary action to the other end. In specific areas of the absorbent strip, the antibodies are embedded. Usually a dark line appears if the particular pathogen is detected. All LFDs include a control line to allow for quality control to ensure that the LFD is working correctly when used. The LFD technology is similar to those used on most home pregnancy tests that are available over the counter.

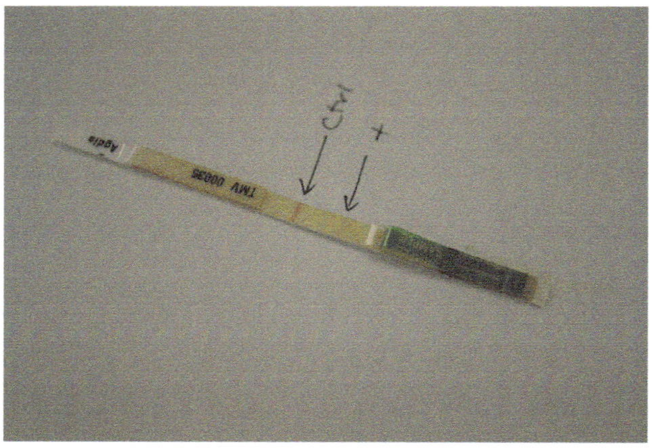

FIGURE 30.4 Lateral flow device for serological testing. Ctrl = positive control *Tobacco mosaic virus* (TMV) and "+" = plant sample is positive for virus.

ELISAs are not as rapid as LFDs, but they are considered to be more sensitive. There are several types of ELISA tests. All rely on an enzyme-mediated colorimetric reaction to detect the presence of the pathogen. ELISA kits are available commercially for the detection of one pathogen or a group (of similar type) of pathogens. For example, there are commercial kits for the detection of *Plum pox virus* (one virus detection) and the potyviruses (detection of a group of viruses). These tests, introduced in the 1970s for use in plant disease diagnoses, were originally used to detect plant viruses only. Today, kits are available for the detection of fungi and bacteria as well.

Many ELISA kits for plant pathogen detection utilize a double-antibody sandwich-ELISA (DAS-ELISA) approach. The DAS-ELISA kit typically comes with an antibody-coated plate or materials to manually coat the bottom of a multiwell plate. These bound antibodies are then exposed to the unknown sample. If the corresponding antigen is present, it would bind to the embedded antibodies. The wells are rinsed to remove any sample and unbound antigens. Next, antibodies conjugated with an enzyme (peroxidase or alkaline phosphatase) are added to the wells. The conjugated antibodies attach only to the corresponding antigens. Finally, a substrate is added, tetra-methyl benzidine (TMB) or p-nitrophenyl phosphate (PNP) for peroxidase or alkaline phosphatase, respectively. When the target pathogen is present, the enzyme acts on the substrate, which results in a color change that can be visually observed or quantified on a spectrophotometer (plate-reader). When the target pathogen is not present, there is no binding of the target and the enzyme-conjugated antibodies, hence no reaction and no color development.

Identification Using Biochemical Methods

Kits utilizing biochemical methods were developed to identify bacteria in the early 1970s. These kits typically employ a number of substrates and an identification index. An example is the Analytical Profile Index (API®) kit, which utilizes 16–32 different substrates for identification purposes. The substrates are exposed to the unknown bacteria and incubated for a defined period of time and conditions as specified by the manufacturer. As the bacteria multiply and utilize substrates, pH changes may occur resulting in color changes in the substrate. Liquefaction of the substrate may occur due to enzyme activity. The results are scored and compared against the index provided by the manufacturer to identify the unknown bacterium. Essentially, this approach relies on the documentation of previously known microorganisms and their reaction to the substrates. The combination of substrates is selected to produce profiles that can be used to differentiate the unknown bacterium from multiple choices. This method utilizes the phenotype of the microorganism for identification.

In the early 1980s, The **Biolog system** was developed, utilizing the similar approach of metabolic profiling. This method makes use of 94 growth substrates on a microtiter plate. After adding an active culture from a single isolate, the microtiter plate is incubated. Utilization of the substrate is scored and compared against an identification database produced by the manufacturers. This system allowed for manual scoring and also automated determination using proprietary computer software. Essentially, the system captures a metabolic "fingerprint" from which bacteria can be identified. Again, the limitation is the database where only previously known bacteria can be identified. This technology has progressed to where such profiling tests can be completed in shorter periods, and the database has grown to include many different bacteria (plant pathogens are a tiny portion of the database). Now it also includes some yeast and fungi.

A different system utilizing fatty acid profiles, Sherlock® microbial identification system (MIS), was developed about the same time as the Biolog system, in response to the need for a rapid phenotypic method to identify plant pathogenic bacteria. This method requires a pure culture of the unknown bacterium to be grown under specified conditions which will then be harvested. The cellular fatty acids are chemically extracted. A gas chromatograph is utilized to separate fatty acids and result in a fatty acid fingerprint profile, which is compared with the database that is provided by the manufacturer. This is accomplished using the proprietary computer software supplied when purchasing the system. Similar limitations exist for the MIS system as with the Biolog system such as the fact that identification is limited to the known bacteria on the MIS database. Recent advances of this system reduces the procedural time to as little as 15 min for a bacterial identification from a pure culture. The range of microorganisms identified by MIS has also expanded to include yeasts, mycoplasmas, and fungi.

Detection and Identification Using Molecular Analytical Methods

Good disease management of plants requires rapid identification of the causal agent. In recent years, the need to rapidly and accurately identify a causal agent has become critical in preventing the introduction and establishment of plant pathogens into new areas. The discovery of **polymerase chain reaction** (**PCR**), which can greatly amplify nucleic acid sequences, has resulted in the development of nucleic acid-based detection tools. As with any detection tool, PCR-based detection assays (Chapter 31) target specific pathogens and rely on the expertise of the practitioner to have a preliminary diagnosis to determine which detection tool, if any, to use.

The components needed to perform a basic PCR include the target DNA (template), two primers, DNA polymerase, deoxynucleoside triphosphates (dNTPs), $MgCl_2$, and buffer solutions. The primers are short sequences that are complementary to the target DNA. These primers serve as the starting point for the amplification process that would amplify a fragment (gene of interest) within the target DNA. DNA polymerase, typically a Taq polymerase, is the enzyme that builds the new DNA strand. This enzyme requires Mg^{+2} as a cofactor and will not work without it. The dNTPs are the basic building blocks of the DNA strand. The buffer solution creates a suitable environment for the PCR to occur. It contains materials that help to maintain the stability and activity of the DNA polymerase.

There are three basic steps in PCR: Denaturation—where high temperature (92°C–99°C) is used to separate DNA into single-stranded DNA. This is followed by the "Annealing" step that lowers the temperature so that the primers can begin to "anneal" to the single-stranded DNA template. The last step is the "Elongation" step where the temperature is typically increased to optimum for the DNA polymerase used. The DNA polymerase builds the complementary strand of DNA to the single-stranded template. At the end of this process, there are two copied fragments of the original target DNA (also known as amplicons). PCR is usually run for 20–40 cycles. Each of these cycle essentially doubles the amount of the DNA target, hence an exponential increase in the amount of the DNA target fragment. The resulting product is visualized using gel electrophoresis. By utilizing known primers or primers that have been developed to identify specific pathogen, the PCR can be utilized as a highly sensitive detection tool.

There have been many variations developed on the conventional PCR method as tools to identify the pathogen. Some of these methods are now economically feasible to be used routinely in a diagnostic clinical setting. For example, PCR will amplify the target DNA regardless of whether the pathogen is alive or dead. A procedural variation to capture information about live microorganism is known as BIO-PCR. BIO-PCR requires an additional step, whereby the microorganisms are cultured prior to DNA extraction and the PCR process. In other words, a microorganism enrichment step is added to the PCR procedure to capture data on living cells only (Schaad et al., 1995).

Reverse transcriptase-PCR (RT-PCR) is a method used to detect pathogens with RNA genome, such as many plant viruses. Since conventional PCR amplifies double-stranded DNA, the RNA must first be converted to DNA prior to being used in a PCR. The modification to conventional PCR include a preceding step where the enzyme, reverse transcriptase, is used to transcribe RNA into its DNA complement. This complementary DNA (cDNA) is amplified in the PCR process.

Real-time PCR is a newer procedure where the PCR products are monitored in "real time" as they accumulate at each reaction cycle. This means that results can be obtained more quickly than conventional PCR. Real-time PCR also allows for the quantification of the template DNA, which can be translated into an estimation of the number of pathogens that may be present. The monitoring of the amplicon (PCR product) is accomplished through fluorescence via an amplicon non-specific or amplicon-specific binding. An example of amplicon non-specific binding is the use of intercalating dyes such as SYBR Green®, which binds to all double-stranded DNA. This will monitor all products that are made in the PCR process. For disease detection purposes, using an amplicon non-specific approach would require highly specific primers for the pathogen. The more common applications for pathogen detection utilized TaqMan® probes, an amplicon-specific method. These specific probes are short, single-stranded oligonucleotides that are labeled with a fluorescent molecule and a corresponding fluorogenic quencher. If a target is present for probe hybridization, the fluorescent molecule is released via the activity of the DNA polymerase resulting in a fluorescent signal. This is a highly specific approach to detect a pathogen.

Multiplex PCR is an approach designed to detect multiple pathogens simultaneously. If successful, this method saves time. Multiplex PCR uses several different primers in the same reaction to detect multiple pathogens. A major challenge to developing a multiplex PCR is the work needed to optimize the PCR conditions to ensure that the primers and DNA polymerases are functioning correctly to amplify any of the target DNA. In addition, the resulting amplicons for the target DNA for each primer set must be of different sizes so that the different pathogens can be easily discriminated when visualized via gel electrophoresis. There are only a few multiplex PCR assays currently used for pathogen detection and diagnosis.

An alternate approach to address the detection of multiple pathogens is using array technology. This strategy embeds probes for different pathogen of a given crop in an array on a glass slide (microarray) or at a larger scale on nitrocellulose or nylon membrane (macroarray). Here, a part of the genome is amplified (typically the ribosomal DNA—ITS region) of the suspect eukaryotic pathogen. The resulting product is exposed to the microarray that contains probes for various pathogens. This technology is used by commercial diagnostic services as a cost-effective approach to screen a particular crop for certain diseases. One such example is the DNA Multiscan®, which screens for multiple pathogens including oomycetes, fungi, and bacteria.

Loop-mediated isothermal amplification (LAMP) PCR is another new technology. This method does not rely on different temperature regimes for DNA amplification. This procedure relies on multiple primers, four or more, to amplify the target sequence. The high specificity of the primer combination can result in higher DNA amplification than conventional PCR. This method is currently used by some commercial diagnostic services. However, LAMP detection assays have been developed for only a few pathogens. It is anticipated that this method will continue to be developed for the detection of more pathogens. It is also anticipated that this technology has the potential to be developed into a tool for pathogen identification useful in the field.

RESEARCH: DEMONSTRATING PROOF OF PATHOGENICITY

The ultimate test to show that a particular organism is the pathogen is to demonstrate proof of pathogenicity. Similarly, if damage was caused by an abiotic agent—the definitive diagnosis would be to reproduce the damage using the agent.

With biological (pathogenic) agents, the proof of pathogenicity is demonstrated using **Koch's postulates**. This procedure was developed by Robert Koch in the late 1800s and was first used to demonstrate the pathogenicity of anthrax and tuberculosis, two human pathogens. It has since been applied to many different pathogens as a method to demonstrate pathogenicity. The four steps to satisfy Koch's postulates, as adapted for plant pathogens are the following:

1. The microorganism must be found in all diseased plant, but not in healthy ones.
2. The microorganism must be isolated from the disease host and grown in pure culture. Characteristics of this microorganism are noted.
3. The cultured microorganism should cause disease symptoms when inoculated into a healthy host plant. These symptoms should be similar to those that were observed on the original diseased plants.
4. The microorganism must be re-isolated from the inoculated diseased experimental plant. The re-isolated microorganism should have characteristics that are identical to the original isolated microorganism (as in Step 2).

In most current diagnostic situations, Koch's postulates are performed infrequently due to time constraints. However, Koch's postulates are performed when a new or previously undescribed pathogen is suspected (Chapter 31). A typical first disease report requires more than one identification method and proof of pathogenicity whenever possible (Agrios, 2005).

Even with the advances that have been made in diagnostics, especially in pathogen detection and identification, knowledge of the plant, the basic skills of observation, and the ability to associate those observations with corresponding descriptions of disease are still first and foremost. As good as the detection technology may be it still requires the practitioner to use clues that are present to be able to direct the diagnosis process in the right direction and to know which technology to use to confirm their suspicions. At the other end of the spectrum, the most convincing demonstration of a disease is the proof of pathogenicity of the pathogen. Newer and more technological advances will be made in the field of molecular diagnostics. Current and new technologies will undoubtedly continue to become cheaper, more readily available, and easier to use in the laboratory or in the field. But with all the new technology to detect and identify a pathogen, the exposure of the pathogenic agent to the plant to recreate the disease or damage is still the gold standard for diagnosis confirmation.

LABORATORY EXERCISES

Experiment 1: The Power of Observation

Basic skills for disease diagnosis are important. "Practice makes perfect" is a great adage for these skills. Good observational skills rely on all of our senses. Our eyes can be trained to notice what is out of the ordinary. Our noses can be trained to smell what is different from normal. Our touch can feel textures that may be a deviation from that of a normal plant. Our ears can be trained to listen to the grower, farmer, or homeowner who is having problems with their plants. The goal is to be able to do all of these things well. Our intellect is needed to combine all the data from these sources and compare with current known knowledge about a particular disease. Experience contributes greatly to this knowledge and the ability to synthesize the collected information into working theories or guesses. One important skill is the ability to describe the symptoms: for self, when researching reference books for a suspect disease causal agent OR when communicating with others, such as a plant pathologist, to discuss case situations.

The following exercise requires a minimal participation of two people. Four or more participants would be more desirable to make this activity fun.

Materials

Each student or group of students will need the following items:

- Collection of photographs of disease plants or several physical samples of symptomatic plants
- Hand lens or loupes (10–20×)
- 4 × 6 note cards
- Timers

Follow the instructions in Procedure 30.1 to complete the exercise.

Anticipated Results

There will be some diseased specimens or their photos that can be easily described, whereas others may be difficult. The varied abilities of the participants may also be revealed. It is typical to find participants with differing abilities to describe their observations. Some problems will be noticed in the details of the description. Issues such as "not clear" or "lack sufficient information" will come up. This will provide an opportunity for participants to critique each other and to assist each other in seeing the finer details. Some will see some things that other will not.

Questions

- What or which descriptor(s) was key in helping to identify the correct corresponding sample?
- What was the biggest challenge when describing the observed problem?
- How would you optimize the way you would collect observational information?

Procedure 30.1
The Power of Observation

Step	Instructions and Comments
1	Each person is assigned a "confidential" sample. This can be a set of photographs of a particular diseased plant or a physical sample of a diseased plant.
2	Set a timer to 3 min. Each person is given 5 min to write a detailed description of the symptoms on the plant sample.
3	All samples are returned to the sample pool.
4	Each person will draw a description of the diseased material on a note card. Set a timer to 3 min.
5	Each person will select a photograph or specimen for the sample that is described on the note card.

Experiment 2: Morphological Identification of a Fungal Pathogen that Causes Leaf Spots

Fungal leaf spots can be identified by associating the fungus with the symptoms on the host's foliage. Typically, fungi are identified by their spore morphology, arrangements of the spores, and/or fruiting bodies that may be present on the leaf. If spores are present in/on the symptomatic plant tissue, then the identification of the fungus may be completed with light microscopy. However, if spores are not present, the symptomatic plant parts should be place in an enclosed, highly humid chamber (a moist chamber), to encourage fungal growth, spore development, and sporulation. After 1–7 days, plant tissue in the moist chamber should be checked for signs of fungal sporulation. If no spores are present, further analysis with culturing or other analytical methods may be needed to identify the pathogen.

Materials

Each student or group of students will need the following items:

- Plant samples with leaf spot symptoms. Some common leaf spot diseases are desirable for this experiment including early blight of tomato, Septoria leaf spot on tomato, blackspot or Cercospora leaf spot on roses, and leaf spot on photinia.
- Microscopes: dissecting and compound microscopes. Microscope slides and coverslips.
- Stains (lactophenol cotton blue).
- Plastic bag or plastic container and paper towel.

Follow the protocols listed in Procedure 30.2 to complete this experiment.

Anticipated Results

Fungal structures (shape of fruiting bodies, conidiophores, or spores) should be observed at 100× and 450× magnification with the compound microscope. The moist chamber should result in sporulation of the leaf spot pathogen. Descriptions, drawings, and/or photographs documenting the observations will be helpful when using references to identify the fungus or to associate the fungus with the disease. Students should gain the experience of identifying the fungus and utilizing references to compare their information against known leaf spot issues on the plant.

Procedure 30.2

Morphological Identification of a Pathogen that Causes Fungal Leaf Spots

Step	Instructions and Comments
1	Assess the sample provided. Record observations of symptoms present. Describe the leaf spot color, shape, size, texture, and so on.
2	Observe leaf spot using a dissecting microscope. Look for evidence of fruiting bodies (bumps or little protrusions) on the leaf spot. Note observations.
3	If fruiting bodies are present, make a wet mount of dead plant tissue and fruiting bodies. Use lactophenol cotton blue to stain the fungal structures for better visualization. Describe or record observation of fruiting bodies and spores.
4	If no fruiting bodies are observed, prepare tape mounts. A tape mount is made by placing the sticky side of cellophane tape on top of a leaf spot. It is then removed and placed on a microscope slide. A tape mount is observed using a compound microscope, typically at 100× and 450×, to observe the presence of spores in greater details. Describe and record observations.
5	Place leaf samples (or a portion of a sample) in a moist chamber (plastic bag or container with a wet or damp paper towel—try to keep leaves from direct contact with wet paper towel). The increased humidity should encourage fungal sporulation. Incubate for 3–7 days at room temperature. Check leaves for additional fruiting bodies and spores. If present, record observations of fungal structures (see Steps 3 and 4).
6	Identify the fungus. Use references such as the *Illustrated Genera of Imperfect Fungi*, 4th edition (Barnett and Hunter, 1998).
7	Use a compendium, such as one from the APS Press Compendia Series for disease identification, to determine if the identified fungus has been shown to cause the symptoms and damage observed on the plant host.

Questions

- How many types of spores were observed (1) before the moist chamber incubation and (2) after the moist chamber incubation? If there is a difference, why?
- When would a tape mount be more effective than a wet mount when preparing sample for examination under the microscope?
- Why should a portion of the sample be used in the moist chamber procedure? Why not place the whole sample in the chamber?

Experiment 3: Inoculation Study for Proof of Pathogenicity

The most convincing diagnosis would be one where proof of pathogenicity can be demonstrated. Unfortunately, this cannot always be completed in a timely fashion. In addition, there are some pathogens that cannot be cultured, which makes it difficult to perform an inoculation study. One of the challenges in doing an inoculation study is the deliberate action required to reproduce the environment that favors disease development. For example, high humidity and optimum growth temperature are two important factors in the development of disease. The method of inoculation can also impact if and how the disease will develop. Some pathogens may require a wound for entry into the plant host. Information gleaned from observations made on diseased plant material may provide clues and ideas to the parameters that are needed to successful onset of disease. This exercise may take 2 to 3 weeks to complete.

Material

Each student or team of students will need the following materials:

- Diseased plant sample. Plant diseases that can be used to demonstrate the process of satisfying Koch's postulate include bitter rot of apples (*Colletotrichum gloeosporioides*), peach brown rot (*Monilinia fructicola*), or early blight of tomato (*Alternaria solani*). Use diseased fruit.
- Corresponding healthy plant material.
- Microscopes: dissecting and compound microscopes.
- Plastic bag or plastic container and paper towel.
- Basic laboratory implements: forceps or tweezers, scalpel or razor blade, probe or needle.
- Petri dishes containing PDA.

Follow the instructions listed in Procedure 30.3 to complete this exercise.

Procedure 30.3

Inoculation Study for Proof of Pathogenicity

Step	Instructions and Comments
1	Surface-sterilize the fruit by spraying the surface, including the disease lesion, with alcohol and blot dry carefully.
2	Aseptically excise (cut-out) sections of fruit that are diseased. Place small sections (2- to 5-mm-diameter) on PDA plates. Incubate at room temperature for 2–7 days.
3	Evaluate fungal growth from sectioned tissue. Record your observations of the fungus on the PDA plate and under the dissecting and compound microscopes.
4	Inoculate the healthy fruit with the fungus. Using forceps or a probe, pick up some mycelium of the fungus. Then, use the tip of the forceps or probe to create a wound on the fruit—use tips of the forceps to pierce the skin of the fruit. Place fruit in a moist chamber. Check every 2–3 days for symptom development. Incubate for up to 10 days, as needed.
5	When symptoms are clearly observed, remove fruit from moist chamber. Record disease symptoms. Compare if symptoms are similar to that found on the original diseased fruit.
6	Surface-sterilize the inoculated fruit as in Step 1.
7	Aseptically excise sections of fruit that are diseased. Place small sections (2- to 5-mm-diameter) on PDA plates. Incubate at room temperature for 2–7 days.
8	Evaluate fungal growth from sectioned tissue. Record observations of the fungus on the PDA plate and under the microscope.
9	Compare observation records of fungal structures (fruiting bodies and/or spores) from re-isolated fungus with the original isolated fungus.
10	Koch's postulates are satisfied when the symptoms on fruit are similar for both before and after inoculation, and the fungal structures observed were similar for both before and after isolation from diseased fruit.

Anticipated Results

Under ideal situations, student should recover the pathogen and use it to cause new disease. The pathogen should be re-isolated from the inoculated fruit. Disease observed on the re-inoculated fruit may not be as severe as found on the original diseased fruit. Fungal structures should be visible at 100× and 450× magnification.

Questions

- What would you conclude if no fungal growth is observed on your initial isolation procedure?
- What would you conclude if more than one fungal type is observed on your initial isolation procedure?
- If more than one fungal type is found on your initial isolation procedure, how would you proceed with this experiment?
- What would you conclude if the re-inoculation did not result in similar disease symptoms and/or interfered with your ability to recover the pathogen?

REFERENCES

Agrios, G. N. 2005. *Plant Pathology*. 5th edition. Burlington, MA: Elsevier Academic Press.

Anon. 2015. On-site testing of growing media and irrigation water. Floriculture Factsheet—British Columbia Ministry of Agriculture and Food. 12 pp. http://www2.gov.bc.ca/assets/gov/farming-natural-resources-and-industry/agriculture-and-seafood/animal-and-crops/crop-production/on-site_testing_of_growing_media_and_irrigation_water_2015.pdf [accessed July 16, 2016].

Barnett, H. L., and B. B. Hunter. 1998. *Illustrated Genera of Imperfect Fungi*, 4th edition. St. Paul, MN: APS Press.

Christie, R. G., and J. R. Edwardson. 1986. Light microscopy techniques for detection of plant virus inclusion. *Plant Diseases* 70:273–279.

Schaad, N., S. S. Cheong, S. Tamaki, E. Hatziloukas, and N. J. Panopoulos. 1995. A combined biological and enzymatic amplification (BIO-PCR) technique to detect *Pseudomonas syringae* pv. *phaseolicola* in bean seed extracts. *Phytopathology* 85:243–248.

Shurtleff, M. C., and C. W. Averre, III. 1997. *The Plant Disease Clinic and Field Diagnosis of Abiotic Diseases*. St. Paul, MN: APS Press.

SUGGESTED READING

Narayanasamy, P. 1997. *Plant Pathogen Detection and Disease Diagnosis*. New York, NY: Marcel Dekker Inc.

Riley, M. B., M. R. Williamson, and O. Maloy. 2002. Plant disease diagnosis. *The Plant Health Instructor*. doi:10.1094/PHI-I-2002-1021-01.

Skoglund, L. G., and T. Blunt 2012. The plant diagnostic lab experience. *APSnet Feature*. http://www.apsnet.org/publications/apsnetfeatures/Pages/diagnostician.aspx [accessed July 16, 2016].

Waller, J. M., J. M. Lenné, and S. J. Waller. 2001. *Plant Pathologist's Pocketbook*. 3rd edition. New York, NY: CABI Publishing.

31 Identifying Obligate, Biotrophic Fungi (and Hosts) Using the Sequence of the Internal Transcribed Spacer (ITS) Region

Robert N. Trigiano and Bonnie H. Ownley

Concept Box

- Koch's postulates are designed to determine whether or not an organism is pathogenic on a specific host plant.
- Not all of Koch's postulates can be performed on obligate, biotrophic pathogens, which typically cannot be isolated and grown on artificial growth medium.
- Obligate pathogens are identified using physical traits such as the morphology of sexual and asexual spores, hyphal characteristics, and so on as well as molecular methods, such as sequencing of a gene or a portion of a gene and comparing these sequence to accessions in various public databases.
- Caution must be exercised when accepting a match of the unknown pathogen to an accession identity because of nomenclatural changes not reflected in the reporting of the accession and/or entry of incorrect identities and details of accessions by contributors.

As plant pathologists, we observe plants in agricultural fields, in protected growing areas, such as glasshouses or high tunnels, and in natural settings for signs and symptoms of disease. Plant pathologists are often called upon by farmers, greenhouse producers, and nursery growers to identify potential disease problems and which organism, if any, cause the disease. Most of the time when plants with signs of a potential pathogenic organism or disease symptoms are observed, our experience allows us to make a diagnosis—we know which organism is inciting the disease as well as the identity of the host plant species. There are occasions when a "new" disease is found on a well-known host, and there are other times when the host species as well as the suspected disease-causing organism is not recognized or known.

Plants are examined for any symptoms including leaf spots, wilting, root rots, abnormal plant morphology, etc., and for any signs of the pathogen such as mycelium, sexual and asexual reproductive bodies, spore types, bacterial ooze, etc. From the macroscopic symptoms of the diseased tissue and signs of the suspected pathogen, the pathogen can usually be assigned to a broad group of organisms, such as a powdery mildews, anthracnose-causing fungi, rusts, etc. Furthermore, microscopic examination of the spores and fruiting structures (if present) provides wonderful clues as to the identity of the pathogen. In many situations, disease compendia or disease indices can be used to learn the probable identity of the pathogen. The offending organism is isolated (axenic culture), used to inoculate healthy specimens; disease symptom development and signs of pathogen are observed, and, at last, the pathogen is re-isolated from the inoculated plants. We have learned in previous chapters (1 and 30) that this is fulfillment of Koch's postulates.

A common practice is to isolate genomic DNA from the pathogen, amplify the internal transcribed spacer (ITS) region of ribosomal DNA, and sequence it for molecular identity of the organism. The ITS sequence of the pathogen is entered (the query) into any number of databases, which are capable of finding sequences that closely match the sequence of the query. The report from this search lists the most likely matches.

However, there are also cases when the identity of the disease organism is not known, and in some instances, we might not be able to accurately identify the host plant. Additionally, the disease-causing organism may be an obligate, biotrophic pathogen, which cannot be cultivated on artificial medium. Parts of Koch's postulates (re-inoculating hosts and observing disease development) can be completed, but isolation and re-isolation of the pathogen are not possible. Therefore, by convention, two independent forms or techniques for identifying the pathogens in lieu of completing Koch's postulates are required. The information most often takes the form of morphological data (spore morphology and dimensions, conidiophore and mycelium characteristics, sexual structures and spores, etc.) and some means of molecular identification, usually sequencing a portion of the ITS region.

This laboratory exercise focuses on identifying obligate, biotrophic pathogens (fungi) and its host plant using the ITS region sequences of both organisms. We suggest that a powdery mildew disease be used for the laboratory exercise because the diseases are abundant on many field crops, in home gardens, and on ornamental plants throughout much of the growing season. Symptomatic host tissues (Figure 31.1) may be used either as freshly collected specimens or from materials stored at −80°C.

FIGURE 31.1 Powdery mildew of purple nettle (*Lamium purpureum*). Note the white "powdery" signs of powdery mildew on the leaves. (Courtesy of R. N. Trigiano.)

Many of the details of the techniques needed for this laboratory can be found in Chapter 33. We suggest that students read the introductory remarks and materials needed for Experiment 1 in Chapter 33 before beginning this laboratory exercise. To make this exercise more realistic, students may be interested in writing a paper in the "Disease Notes" format used by the American Phytopathological Society in the journal *Plant Disease* (http://apsjournals.apsnet.org/userimages/ContentEditor/1236780011229/pd_author_instructions.pdf). This additional exercise will provide valuable logical and organizational writing experiences for students and could possibly lead to a publication if the disease situation has not been reported for their host or in their region.

LABORATORY EXERCISES

ISOLATION OF DNA FROM THE HOST AND PATHOGEN

There is no need to treat the diseased tissue any differently than if only DNA from the host were being isolated. Any DNA isolation kit will yield sufficient DNA of both the organisms to complete this portion of the experiment. Follow the manufacturer's instructions to complete DNA isolation. After DNA isolation, quantify genomic DNA in ng/μL and make a stock of 4 ng/μL by diluting the DNA with sterile, pure water. Store either frozen (−20°C) or in a refrigerator (2°C–4°C) for short periods. This stock will contain DNA from both the plant and fungus.

AMPLIFICATION OF THE ITS REGION OF THE HOST AND FUNGUS

The materials listed in Chapter 33, Experiment 1, will be used in this experiment. We suggest using ITS1 and ITS4 (White et al., 1990) as primers, but other sets of primers will work equally well. The sequences (5'–3') for these two primers are the following, and should be used as 3 μM solutions:

ITS1: TCCGTAGGTGAACCTGCGG
ITS4: TCCTCCGCTTATTGATATGC

We suggest a 30-μL amplification reaction for this exercise. The amplification products from the host and fungus are mixed together in the tube. The ITS regions (as defined by ITS1 and ITS4 primers) for the host plant are about 700–800 bp and the fungus between 500 and 600 bp and must be separated before they can be sequenced. The easiest way to separate them is by agarose or acrylamide gel electrophoresis, and this experiment will include both techniques.

Follow the instructions provided in Procedure 31.1 to complete the amplification of the ITS region of fungus and host.

Procedure 31.1

Amplification of the ITS Regions of Plant Host and Fungus Using Primers ITS1 and ITS4

Step	Instructions and Comments
1	Pipette 11.8 μL of pure, sterile water into a sterile, 0.65 mL Eppendorf tube.
2	Pipette 3.0 μL of 2 mM nucleotides into the tube.
3	Pipette 3.0 μL of 10X Buffer* II (without magnesium chloride) into the tube.
4	Pipette 3.0 μL of 25 mM magnesium chloride* into the tube.
5	Pipette 16 μL (8 × 2 μL) of each ITS1 and ITS4 primers.
6	Pipette 2.0 μL of AmpliTaq Gold DNA polymerase.*
7	Pipette 3.0 μL of fungal DNA template 4.0 ng/μL. Vortex and centrifuge briefly.
8	Place tubes in a thermalcycler and run according to the program provided below.
	Step 1: 95°C for 9 min; 95°C activates the DNA polymerase enzyme. This step is also termed a "hot start."
	Step 2: 96°C for 1 min—denatures DNA
	Step 3: 56°C for 1 min—primers anneal to complementary sites
	Step 4: 72°C for 1 min—extension
	Steps 2 through 4 should be repeated for 35 cycles
	Step 5: 72°C for 7 min final extension
	Step 6: 4°C.

* These reagents are provided in the AmpliTaq Gold DNA polymerase kit (Applied Biosystems). There are several other kits available from different manufacturers that will perform satisfactorily.

Agarose Gel Electrophoresis

Agarose gel electrophoresis is easy to complete and has nontoxic components. It presents a quick way of separating ITS region amplicons, but does have the detractor of having to dissolve the agarose matrix to release the DNA.

Follow Procedure 31.2 to complete agarose gel electrophoresis.

Anticipated Results

Two bands should be present when illuminated with UV light. Be sure to wear protective eyewear as well as cover all exposed skin. UV light is very harmful to your eyes and will cause sunburn.

Isolation of DNA from Agarose Gels

Follow procedure 31.3 to excise the bands from the agarose gel.

Materials

Each student or team of students will require the following materials:

- QIAquick® Gel Extraction Kit (QIAGEN, Valencia, CA USA)
- Two small plastic or glass test tubes
- Centrifuge
- Isopropanol
- 1.5-mL Eppendorf centrifuge tubes
- Vortex
- Scalpel handle and #10 or #11 blade
- Eye and skin protection from UV light

Anticipated Results

The isolation should yield sufficient DNA for sequencing the ITS region from both the plant and fungus. If it does not, use 3 μL of the isolated plant DNA in one Eppendorf tube and 3 μL of fungal DNA in a different tube as templates and reamplify both using the reaction mixture described in Procedure 31.1. If this procedure is used, DNA from both the reamplifications should be purified using Procedure 31.8 beginning with Step 4.

Acrylamide Gel Electrophoresis

The ITS-amplified regions from the plant and fungus can be separated on a 10% acrylamide gel using passive denaturing with urea. The advantage of using acrylamide is that it produces very clean separation of the bands, and the bands can be excised and used as a template for reamplifying the target DNA without dissolving the acrylamide. The disadvantages are that acrylamide in the unpolymerized form is a neurotoxin and requires more complicated procedures to separate DNA than

Procedure 31.2
Electrophoresis of Amplified ITS Products

Step	Instructions and Comments
1	Make a 1% gel by placing 40 mL of 1X TE buffer in a 250-mL flask and add 0.4 g of low-melting-point agarose. Heat in a microwave to melt the agarose.
2	Remove from the microwave, allow the solution to cool to 60°C (comfortably held in the hand) and add 4 µL of SYBR green (used to stain DNA). Mix thoroughly.
3	Pour the molten agarose into the bottom of the gel-casting tray. Try not to introduce air bubbles in the gel. Set the plastic combs (10 teeth) close to one end of the casting unit. Allow the gel to solidify—the preparation will become cloudy. Fill the reservoirs with 1X TE buffer and pour a thin layer (1 cm) over the gel. Carefully remove the plastic comb.
4	Combine 4 µL of loading buffer with 20 µL of amplification products. Skip a well and load the molecular weight (bp) ladder (3 µL of ladder plus 1.5 µL of loading buffer) into the next well.
5	Place the top of electrophoresis unit onto the base (be sure that the negative [black] electrode is closest to the sample wells) and connect the unit to the power supply. Set the power supply to 100 V and allow it to run for about 45 min. The blue tracking dye should migrate through the gel toward the red electrode.
6	Wear gloves. Turn off the power supply and remove the gel tray from the base of the unit. Transport the gel to the transilluminator. Wear goggles that block UV wavelengths and protective clothing that covers exposed skin. Turn off the laboratory lights, and turn on the transilluminator. Note the bands and approximate bp weights of the products. Most fungal ITS1-4 amplicons are between 500 and 600 bp and most amplicons for plants are 700–800 bp. If you see more than two bands per lane, there is probably a contaminant. Try isolating DNA again.
7	Clean the surface of the transilluminator with water and a paper towel, and dispose of the gel in a designated container.

Procedure 31.3
Excision and Recovery of DNA from Agarose Gels

Step	Instructions and Comments
1	Place the gel on a UV transilluminator. Be sure to wear eye protection and cover exposed skin with clothes. With a clean, sterile scalpel blade, carefully excise the heavier plant band (should be a rectangle), including as little of the agarose as possible, and place in one of the small test tubes. Label this tube as Plant ITS.
2	Complete the same process for the lighter fungal band, and label this test tube as Fungal ITS.
3	Follow the instructions provided by the manufacturer of the QIAquick® Gel Extraction Kit except for the final elution of DNA from the column, use sterile distilled water instead of 0.1 TE buffer.
4	Measure the DNA concentration of the plant and fungus samples. Most sequencing centers require about 10 ng/100 bp of sequence. Therefore, the total amount of DNA for the plant will be about 80 ng and about 60 ng for the fungus. Be sure to send 20 µL of 5 µM solutions of both primers to the sequencing center.

agarose gel electrophoresis. The following experimental protocols are modified from exercises that appeared in Trigiano et al. (2008).

Assembling the Gel Apparatus

While wearing acetonitrile gloves, assemble two Protean II (or III) Electrophoresis Cells (Bio-Rad, Hercules, CA) a day before the ITS-amplified products are to be separated electrophoretically. We recommend using 0.5-mm spacers that can be purchased separately for the Protean II apparatus and 0.75-mm spacers for the Protean III. GelBond flexible backing supports (sheets) can be purchased from BioWhittaker Molecular Applications (Rockland, ME). Here are a few helpful hints in assembling the rigs. Meticulously clean the glass plates with running distilled water to remove any dust and polymerized acrylamide from previous experiments. Assemble the rig under running distilled water. Place the hydrophobic surface (the side on which that water beads) of the backing film on and toward the large glass plate and rub the hydrophilic surface until all trapped air is evacuated. If using a Protean II rig, place the spacers on top of the backing film and the small plate on the spacers. If using a Protean III rig, the spacers are built into the large glass backing plate. All gel rig components should be flush at the bottom. Lastly, do not over tighten the knobs (Protean II; Protean III is self-tightening)—the glass plates will bow and produce a thickened center portion of the gel, which will not stain properly. The assembled apparatus should be examined carefully to ascertain that the glass plates, spacers, and backing film are flushed with each other at the bottom. Run a fingernail across the bottom of the apparatus. If it does not feel smooth, or if the fingernail gets "hung-up," the level of the glass plates, spacers, and/or support film need to be adjusted. The gel rigs should be allowed to dry overnight in a place that is dark and dust-free.

Casting the Gels

Follow the protocols listed in Procedures 31.4 and 31.5 to make running buffer and acrylamide stock solution and to assemble and pour gels. Always wear acetonitrile gloves when working with acrylamide and TEMED.

Materials

Each team of students will need the following items:

- Acrylamide stock solution—toxic, wear gloves (Procedure 31.4)
- A 0.22- or 0.45-µm filter and 10-mL syringe
- TEMED—toxic, wear gloves
- 10% ammonium persulfate (100 mg/mL pure water) solution—may be made in bulk, dispensed into 1.5-mL centrifuge tubes and frozen at –20°C for up to one year.
- Two assembled Protean II or III gel rigs, casting stand, and two 0.5-mm combs
- One 10-mL disposable pipette and pipette pump
- One 25- or 50-mL beaker, stir plate and stir bar
- Aluminum foil
- Acetonitrile gloves

Preparing Samples for Electrophoresis

Materials

Each team of students will need the following items:

- A microtiter plate or strips of Parafilm
- Loading buffer (0.25% bromophenol blue, 0.25 xylene cyanol, and 15% type 400 Ficoll in water)
- P10 pipette and flat tips for Protean II and regular tips for Protean III
- Amplification products

Amplification products can be prepared for electrophoresis while the acrylamide is polymerizing. First, carefully pipette two 3-µL volumes of loading buffer into wells in a 6 × 10 microtiter plate. Parafilm can be used to support the samples, but wait to prepare until just before loading—the small volumes that are used evaporate

Procedure 31.4
Composition of 10X TBE Buffer and 10% Polyacrylamide Stock

Step	Instructions or Comments
	10X TBE Buffer
1	Dissolve 121.1 g Tris base, 51.4 g boric acid, and 3.7 g $Na_2EDTA.2H_2O$ in 800 mL of pure water. Bring the final volume to 1 L with pure water; pH = 8.3. Store at room temperature.
2	Note: If room is cool, salts may not remain dissolved. Try making a 5X buffer by dividing the "ingredients" in half and adding 1 L of water.
	10% Polyacrylamide Stock
1	Dissolve 19.6 g acrylamide, 0.4 g PDA (piperazine diacrylamide), and 20.0 g urea in 130 mL of pure water. (Caution: wear protective particle mask, gloves, eyeglasses and clothing—unpolymerized acrylamide is a potent neurotoxin; skin contact and accidental inhalation of the compound should be avoided.) Do not substitute BIS (N, N'-methylene bis-acrylamide for PDA as it adversely affects staining quality of the amplified products.
2	Add 20.0 mL of 10X TBE buffer and 10.0 mL glycerol to the acrylamide solution.
3	Bring the final volume to 200 mL with pure water. Store at 4°C in a brown bottle and discard unused portion after 14 weeks.

Procedure 31.5
Casting Gels

Step	Instructions and Comments
1	Wear gloves! Mount the gel rigs onto the casting stand using the gray rubber gaskets on the bottom. We usually place several equal length and width strips of Parafilm wrap under the gasket to ensure a good seal with the glass plates. A very distinct snap should be heard, as the rigs are set into place on the casting stand. Place the casting stand onto a large piece of aluminum foil on which a 10-mL syringe and a 0.22- or 0.45-μm filter can be placed.
2	Pipette 10 mL of 10% polyacrylamide stock (Procedure 31.4) into a 20-mL beaker containing a small magnetic stir bar. Place the pipette tips containing 14 μL of TEMED and 140 μL of 10% ammonium persulfate solutions into the stirring acrylamide solution and dispense. Dispose of the tips in a safe location. Stir for about 10 s.
3	The following steps in casting the gel should be completed as quickly as possible (usually less than 2 min). Carefully draw the gel solution into a syringe avoiding introduction of air into the barrel. If air bubbles are present, hold the syringe angled upward at 70° away from the body and other people in the laboratory. The air should rise to the top. Slowly depress the plunger until the air is expelled. Mount a nonsterile filter on the open end of the syringe. Slowly express a small amount of acrylamide to wet the filter and release any trapped air.
4	Place the filter tip in the middle of the ledge formed by the small (short) plate and quickly dispense the acrylamide solution into the space between the glass plates. Rotate the casting stand 180° and fill the second gel rig with acrylamide. If there are bubbles trapped in the gel, gently tap the inner (short) glass plate, and with luck, they will rise to the top.
5	Position the 10- or 15-well combs about half way (level) in each of the rigs and examine for small bubbles residing on the bottom surface of the teeth. Protean III combs are made to fit the open area exactly. If bubbles are present, remove and reposition the combs.
6	Allow the acrylamide to polymerize for at least 20 min. If desired, the gels may be cast the day before the laboratory exercise and stored overnight lying flat on the bottom of a plastic container that is lined with wet paper towels. Be careful not to disturb the combs and store in the dark.

quickly. Next, pipette 3 μL of DNA into each drop of loading buffer and mix by repipetting the solution several times. Change tips between samples.

Prerunning Gels and Preparing Stain and Developer Solutions

Materials

Each team of students will need the following items:

- 5 or 10X TBE running buffer
- 1-L graduated cylinder
- Protean II or III reservoir and central stand
- Tuberculin syringe with 25-gauge needle
- Power supply (two or three teams can share this item)

Wear gloves. Make 1 L of 1X TBE buffer by mixing 100 mL of 10X TBE (Procedure 31.4) and 900 mL of water in a 1-L graduated cylinder and mix thoroughly. Dismount the two gel rigs from the casting stand and gently remove any polymerized acrylamide from the bottom of the plates with a laboratory tissue. Rotate the rigs 180° and snap into the central stand. Be careful not to touch or disturb the combs. When both the rigs are mounted, the small plates of the rigs will be toward the interior, facing each other, and the outer plates will form the top buffer reservoir. Fill inner and outer reservoirs with about 800 mL of 1X TBE—do not allow the inner and outer reservoirs to mix. Carefully remove the combs from the gels by gently pulling straight up with equal pressure on both sides; do not damage the wells. Fill a 1-mL (cc) tuberculin syringe, equipped with a 4-cm, 25-gauge needle, with buffer from the central well. Gently insert the needle tip about one-quarter of the way into the top portion of a well and gently force the buffer into the well. This will flush accumulated urea and errant bits of acrylamide from the wells. Repeat the process so that all wells of both gels have been cleaned. Connect the apparatus to the power supply and set to a constant 180–200 V for 15–20 min.

While the gel is prerunning, there will be time to prepare both the silver stain and carbonate developer solutions. Both the solutions may be prepared in bulk including every constituent except formaldehyde and sodium thiosulfate. Silver nitrate solution is light sensitive and should be stored in a brown bottle. The sodium thiosulfate solution should be prepared weekly and stored in the refrigerator. If developing and staining solutions are prepared for daily use, then plan on 75 mL for each gel.

Follow the protocol in Procedure 31.6 for preparing silver stain and developer solutions.

Loading Samples and Running the Gel

After prerunning the gels, clean the wells in one gel as described previously. With a P10 (or equivalent) pipette adjusted to deliver 6.5 μL, load the samples into the Protean II wells using flat tips (Midwest Scientific, Valley Park, MO) or regular 10-μL tips if working with Protean III gels. Use the two middle wells to load the DNA samples. Keep the flat tip parallel to the glass plate, guide it partially into the well and gently dispense the sample into the well. Be careful not to damage the well. Load the next sample with a new flat tip. Reconnect the power supply and run at a constant 180–200 V for about 45–60 min, or until the blue tracking dye reaches the level of the bottom platinum electrode.

Staining and Developing Gels

Turn off the power and disconnect the gel apparatus from the power supply. Wearing gloves, disassemble the gels under distilled water by first loosening the four knobs on the Protean II and gently removing the glass plate sandwich from the apparatus. For the Protean III, the gel assemble will slide out of the holder easily. Holding the "sandwich" with the large glass plate contacting the palm of the left hand and in a stream of or a pan of distilled water, insert the fingernail of the right index finger under the top corner of the small glass plate and gently pry it upward. Let the water do most of the work. The backing film and the gel may now be separated from the large plate and placed in a clear, plastic staining tray or in lids from pipette boxes.

Follow the staining and developing procedures outlined in Procedure 31.7. Remember to add formaldehyde

Procedure 31.6

Composition of Silver Stain and Developer Solutions

Step	Instructions and Comments
	Silver Stain
1	For two gels, dissolve 0.15 g of ACS-certified silver nitrate in 150 mL of pure water.
2	A few minutes before use, add either 750 μL of 16% or 325 μL of 37% formaldehyde in a fume hood.
	Developer
1	For two gels, dissolve 4.5 g of ACS-certified sodium carbonate (Na_2CO_3) in 150 mL of pure water and chill to 8°C–10°C.
2	Add 75 μL of sodium thiosulfate solution (0.2 g/50 mL)
3	Add either 600 μL of 16% or 260 μL of 37% formaldehyde (open and use formaldehyde in a fume hood).

Procedure 31.7

Synopsis of Fixing, Staining, and Developing Gels

Step	Instructions and Comments
1	Fix gels in 7.5% (v/v) acetic acid for 10 min on a rotary shaker (60 rpm).
2	Rinse gels with pure water 3 times each for 2 min on a rotary shaker (60 rpm).
3	Soak gels in silver stain for 20–30 min on a rotary shaker (30 rpm) in a fume hood.
4	Rinse gels in pure water for 5–10 s. Be sure to remove all of the silver nitrate stain.
5	Soak gels in developer for 5–8 min (or until bands are dark) on a rotary shaker (30–40 rpm) in the fume hood.
6	Fix gels in cold (4°C) 7.5% (v/v) acetic acid for 5 min on a rotary shaker (60 rpm).
7	Soak gels in pure, distilled water 2 times each for 5 min on a rotary shaker (60 rpm).
8	Soak gels in anticracking solution for 5 min on a rotary shaker (60 rpm).
9	Hang gels overnight to dry. and allow acetic acid to evaporate.

to silver stain and developer solutions (Procedure 31.6) just prior to use. After silver staining is completed, quickly and completely rinse the gels with pure water to remove all excess silver nitrate solution. Do not pour the cold developing solution directly on the gels; instead introduce the solution onto the bottom of the staining dish and immediately place on a rotary shaker set at about 30–40 rpm. Continue shaking until the bands in sample lanes are dark and sharp. Stop with cold 7.5% (v/v) acetic acid. The gels may be "hung to dry" in a dust-free environment after they are treated with anti-cracking solution (under a fume hood, add 100 mL of glacial acetic acid, 10 mL of glycerol and 370 mL of 95% ethanol to 520 mL of pure water, this solution may be reused many times).

Anticipated Results

There should be two bands present in the gel. The top band is the ITS region of the host plant (about 700–800 bp) and the bottom band is the fungus ITS region (about 500–600 bp). These bands will be used to reamplify the ITS regions to produce sufficient product to sequence.

Selecting Bands for Reamplification

The bands are ready to be excised after drying overnight. If the gels have curled, place them in a moist (not wet) chamber and allow the gels to "relax." The bands will serve as template DNA for reamplification. A more complete accounting of this process is available in Caetano-Anollés and Trigiano (1996). Follow the instructions in Procedure 31.8 to complete this portion of the exercise.

Measure the DNA content of the amplified products. Most sequencing centers require about 10 ng DNA/100 bp of sequence or in this case about 80 ng DNA for the plant and 60 ng DNA for the fungus to complete the Sanger sequencing process. Note that if the target sequence is less than 400 bp, it may be necessary to clone the amplified region using an *Escherichia coli* vector. It is always best to include extra DNA. The sequencing center may also require you to bring the primers used in the original reactions.

FASTA Files and SEQUENCHER

The sequencing center will send an email containing the sequences of the plant and fungus ITS regions in a few days. The email file form is not usable as is and must be converted to a FASTA file using SEQUENCHER (Gene Codes Corporation, Ann Arbor, MI, USA; https://www.genecodes.com). SEQUENCHER software is not free, but the sequencing center should have a copy for your institution. However, there is a free 15-day trial available. There are two YouTube videos with instructions on how to use SEQUENCHER (https://www.youtube.com/watch?v=mPvYTdrPN4M and https://www.youtube.com/

Procedure 31.8

Excising Bands from Acrylamide Gels and PCR Clean-up

Step	Instructions and Comments
1	While working in a sterile flow hood, wipe the surface of the gel with 70% ethanol and allow to dry. This will remove any dust that might have landed on the gel overnight.
2	Using a sterile, #11 scalpel blade, score the outline of the band, and add about 1 uL of sterile, distilled water to rehydrate the plant ITS band. Try not to let the water flow outside the outline of the band. Using the tip of blade, transfer the acrylamide to a 0.65-μL Eppendorf tube or well in a 96-well plate. This may require some scraping. Repeat Steps 1 and 2 for the ITS region of the fungus (Figure 31.2).
3	Add master mix and place in a thermalcycler as described in Procedure 31.1. Note: because template DNA does not have a volume in this procedure, add 3 μL extra of sterile, distilled water to bring to the proper volume. The volume of the reaction mixture may be decreased to 20 μL; adjust the volume of components accordingly.
4	Follow the instructions in the QIAquick® PCR Purification Kit to remove excess reaction components. In Step 7 of this procedure, elute with sterile, distilled water, at pH 7.0–8.5. DO NOT use elution buffers containing EDTA. Most sequencing centers require that the sample be in water.
5	Measure the DNA content of the amplified products. Most sequencing centers require about 10 ng/100 bp or in this case about 80 ng for the plant and 60 ng for the fungus. It is always best to include extra DNA. The sequencing center may also require you to bring the primers, typically 5 or 10 μM used in the original reactions.

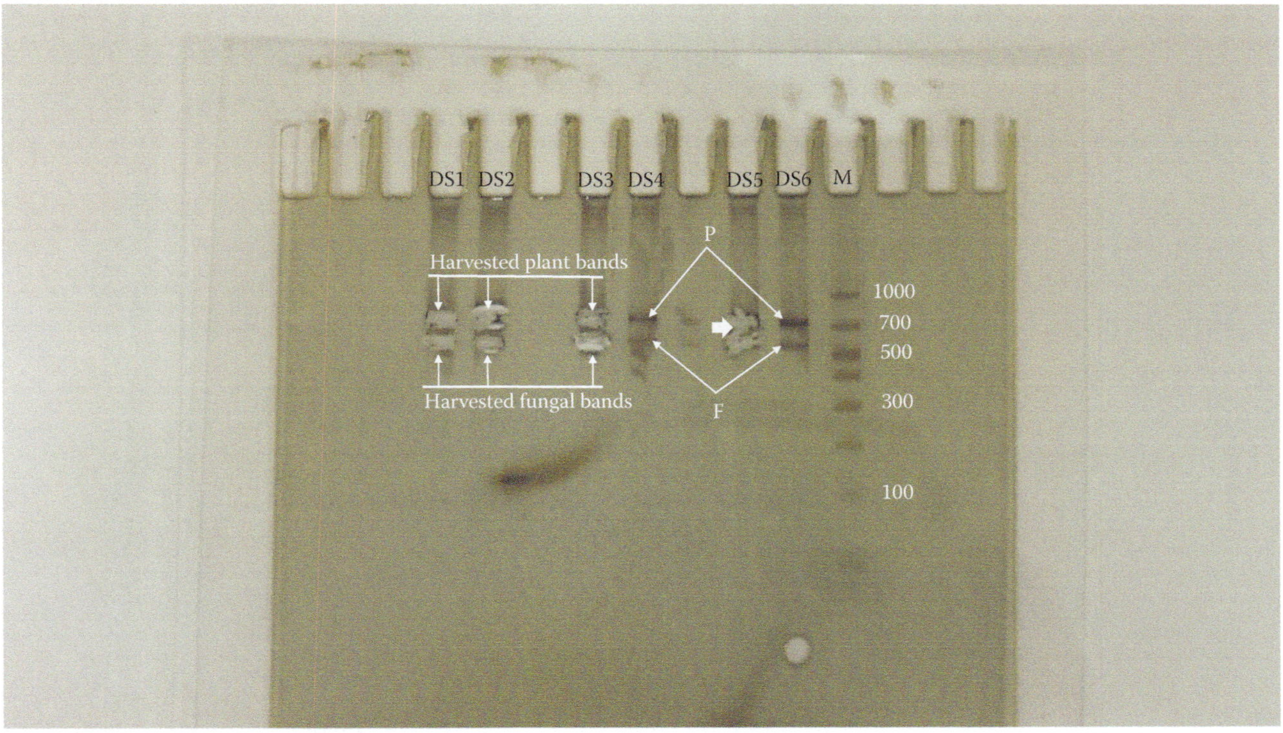

FIGURE 31.2 Selecting bands. After the gels have dried overnight, and the surface has been wiped lightly with 70% ethanol, the amplification of the ITS for the plant (P) and fungus (F) in lanes DS4 and DS6 can be moistened with 2–3 μL of sterile, distilled water. A sterile scalpel is used to remove the band for each of the samples (lanes DS1, DS2, DS3, and DS5). The acrylamide removed from the gels is used as template to reamplify the ITS region of the plant and fungus, respectively. The broad arrow in lane DS5 indicates that excision of the fungal and plant bands was not independent—there is the possibility that the sample removed contains both ITS regions. M = marker and numbers equal base pairs. (Courtesy of R. N. Trigiano.)

watch?v=Lc9QiLFJ3-8). You do not need to trim the primer sequences from the files, and most of the time, the primers cannot be identified.

BLAST

Now that the sequence data has been converted to a FASTA file (Figure 31.3), it can be entered into BLAST (Basic Local Alignment Search Tool). The program is free and may be accessed at the following website: https://blast.ncbi.nlm.nih.gov/Blast.cgi?PAGE_TYPE=BlastSearch.

The program is very easy to use. Follow the instructions in Procedure 31.9 to complete the BLAST search.

The host plant and fungus are now identified. Words of caution, the sequences that are entered in the database are identified to species by the researcher, and there is no editorial check on the accuracy or validity of this identification and/or information. Therefore, couple the morphological data, disease signs and symptoms, and indices with the species provided by BLAST search for positive identification. Also, note that recent name changes may not match the species identified by the BLAST search.

Questions

- After receiving sequences of the ITS region, the primer sequences cannot be found or only partially observed. Why is this?
- If you were to submit the amplified ITS regions of both the host and pathogen in one solution to be sequenced, what information would you receive back and why?
- Why should caution be followed when identifying organisms via blast searches? Are the accessions made by contributors checked and confirmed? Could there be name changes of hosts and pathogens since the information was added to the database?

```
>Pathogen
AAGGATCATTACAGAGCGTGAGATCTGCCCGGGCTTGCCCCGCGCGCAGA
GTTGACCCTCCACCCGTGTTGACTTATCTCATGTTGCTTTGGCGGGCCAG
GTGCCTCGCGCGCCGGCCGGCTCTGTGCTGGCTCGTGTCCGCCAAAGACC
CAACCTAACTCGTGTTGTCGTGTAGTCTGAGGAAAACTATTTGAATTGTT
AAAACTTTCAACAACGGATCTCTTGGCTCTGGCATCGATGAAGAACGCAG
CGAAATGCGATACGTAATGTGAATTGCAGAATTTAGTGAATCATCGAATC
TTTGAACGCACATTGCGCCCCTTGGCATTCCGAGGGGCATGCCTGTTCGA
GCGTCGTCACACCCCCTCAAGCCGCGCCGTGTGTGTGGTTTGGTGTTGGG
GCTCGCCCGTCGGGCGGCCCTTAAAGACAGTGGCGGTGCCGTGGTGGTCT
CTACGCGTAGTACGATTCTCGCGACAGAGCTGCTGTGGCCGCTTGCCAAT
CAATCCATCATCTCAAGGTTGACCTCGA

>Host Plant
TGCGGAAGGATCATTGTCGAACCCTGCATAGCAGAATGACCCGTGAACAA
GTTAACACATCTGGCCTTGCCGGGACCGAAGCATTTGTTTCGGCCCTTGT
GAGTCCTTGTCGACGTGTGTTCATGCATGGACCATACCTTTGGTTTGTCA
TGGATGTCATGTTGACAAAATAACAAACCCCCGGCACGAGATGTGCCAAG
GAAAACCAAAATTAAAGAACCCGTGCTGTTGCGCCCCGTTCGCGGTGTGC
GCGCTGTTCGTGGCGTCTTTGTAAACTTAAAACGACTCTCGGCAACGGAT
ATCTCGGCTCACGCATCGATGAAGAACGTAGCAAAATGCGATACTTGGTG
TGAATTGCAGAATCCCGTGAACCATCGAGTTTTTGAACGCAAGTTGCGCC
CGAAGCCATCCGGTTGAGGGCACGTCTGCCTGGGCGTCACGCATCACGTC
GCCCCCACCAGGCATCCCCTATAGGGCTGTCTTTTGTTGGGGCGGAGATT
GGTCTCCCATGCCCATGGCGTGGTTGGCCTAAATAGGAGTCTCCTCACGA
GGGACGCACGGCTAGTGGTGGTTGATAAGACAGTCGTCTCGTGTCGTGCG
TTTACTTTCTTGAGAGTAGATGCTCTTAAAGTACCCCGATGTGTTGTCTT
ATGACGATGCTTCGATCGCGACCCCAGGTCAGGCGGGACTACCCGCTGAG
TTTAAGCATATCTA
```

FIGURE 31.3 FASTA file. Must have the ">" and any name may be applied to file. (Courtesy of R. N. Trigiano.)

Procedure 31.9

Conducting a BLAST Search to Identify the Host Plant and Fungus

Step	Instructions and Comments
1	Access the BLAST program at https://blast.ncbi.nlm.nih.gov/Blast.cgi?PAGE_TYPE=BlastSearch.
2	Enter FASTA files (Figure 31.3) into the query box. Sequences may be copied and pasted or may be uploaded from a file. More than one file may be entered at each session.
3	Under "Choose Search Set," use the default settings of "Others (nr etc.)" and in the pull-down menu, "Nucleotide collection (nr/nt)." Under "Program Selection," the default is "Highly similar sequences (megablast)." Click on the "BLAST" button.
4	The search will be completed in seconds and return an illustration (Figure 31.4) depicting the similarity (matching) of an organism based on alignment with known (entries in GenBank) sequences to the sequences of the host plant and fungus. The more "red" alignment color present, the better the alignment match.
5	Scroll down. There will be a long list of candidate organisms (Figure 31.5). Look for an "E-value" of close to zero, and an "Ident" (Identity) percentage. There is also an "accession" number and description, which corresponds to the GenBank entry. This will be shown for both the fungus and host plant.

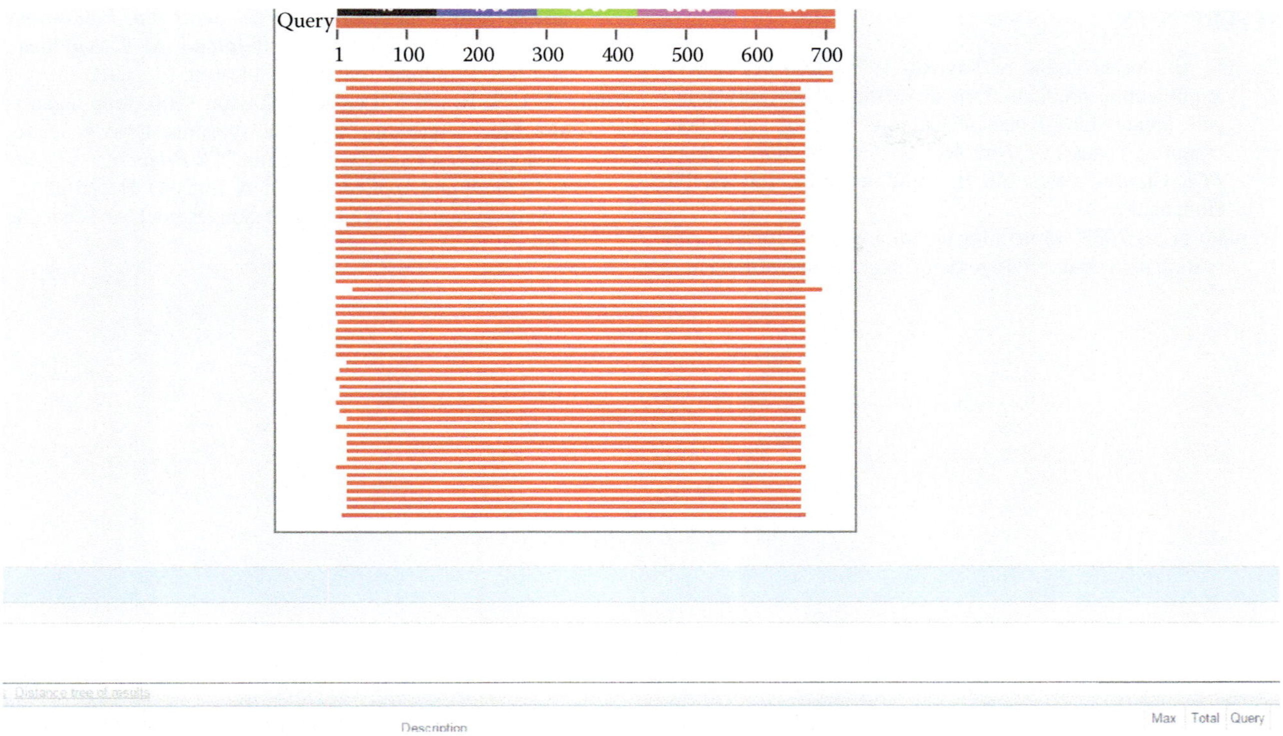

FIGURE 31.4 Screenshot of alignment scores of unknown with BLAST search. Because the bars are all red, sequence alignment is near 100%.

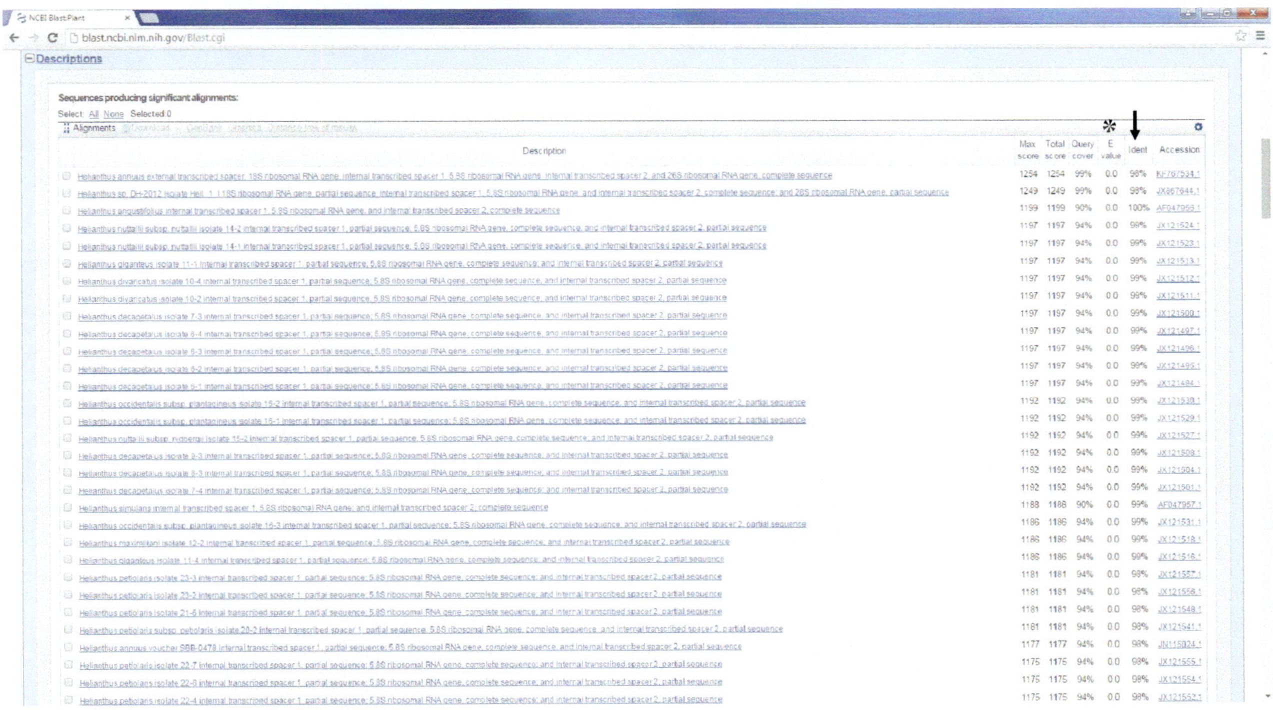

FIGURE 31.5 Screenshot of possible identities of the host plant (in this case sunflower or *Helianthus*). Note the E values of "0" (*) and identity > 98% (arrow). This is a good match.

REFERENCES

Caetano-Anollés, G. and R. N. Trigiano. 1996. Recovery of DNA amplification products from silver stained polyacrylamide gels: applications in nucleic acid fingerprinting and genetic mapping. Pages 11–27 in: *Methods in Molecular Biology: PCR Cloning Protocols*. B. A. White (ed.). Totowa, NJ: Humana Press.

Trigiano et al. 2008. Molecular techniques used for studying systematics and phylogeny of plant pathogens. Pages 269–278 in: *Plant Pathology Concepts and Laboratory Exercises*, 2nd edition. R. N. Trigiano, M.T. Windham, and A.S. Windham (eds.). Boca Raton, FL: CRC Press.

White, T. J., T. Bruns, S. Lee and J. Taylor. 1990. Amplification and direct sequencing of fungal ribosomal RNA genes for phylogenetics. Pages 315–322 in: *PCR Protocols: A Guide to Methods and Applications*. M. A. Innis, D. H. Gelfand, J. J. Sninsky and T. J. White (eds.). San Diego, CA: Academic Press.

32 Extracellular Enzymes Produced by Fungi and Bacteria

Robert N. Trigiano and Laura E. Poplawski

Concept Box

- Production of extracellular enzymes may be either constitutive (no stimulus needed) or inducible (stimulus needed).
- Many extracellular enzymes are either hydrolytic or oxidative.
- Hydrolytic extracellular enzymes "hydrolyze" or add water to complex polymers, such as cellulose and pectin, to form simple compounds such as glucose or galacturonic acid.
- (Poly) phenol oxidases are thought to be involved in lignin degradation and detoxification of phenolic compounds by polymerization.
- Extracellular enzyme activity may be assessed by measuring product liberation, substrate depletion, cofactor usage, and changes in viscosity or pH.
- Most extracellular enzymes, such as pectinase or cellulases, are really complexes or batteries of enzymes and may occur as several isoenzymatic forms.
- Exo-hydrolytic enzymes act on the nonreducing terminus of a polymer liberating a simple molecule. In the case of carbohydrate polymers, reducing sugars are formed and detected using colorimetric methods.
- Endo-hydrolytic enzymes act on bonds in the interior of the polymer and yield shorter, but still complex chain polymers. In the case of carbohydrate polymers, detection of enzyme activity is based on change of viscosity or changes in staining properties of the substrate, or by electrophoretic methods.
- Extracellular enzymes produced by pathogenic fungi may be involved in direct penetration of the host or in haustorium formation by dissolution of the host cell wall.

Fungi, fungilike organisms, and bacteria do not "eat" or obtain nutrition in the ways that many animals do—they lack organized digestive systems in the traditional sense. Instead, they absorb all of the essentials for life directly from the environment. However, many of the simple molecules, such as sugars that are easily transported into the organism, are present only as complex polymers such as cellulose, pectin, hemicelluloses, proteins, and starch in the environment. Unmodified complex carbohydrates and other classes of polymers cannot be used by the microorganisms. Therefore, these organisms must have a means to degrade the carbohydrate polymers, proteins, and so on into their constituent smaller and simple molecules.

Regardless of their relationship to substratum (pathogenic or saprophytic), these classes of organisms produce **extracellular enzymes**, which interact with the environment outside of the cell or hypha. Enzymes are complex proteins that are manufactured in the bacterial cell or fungal hypha, transported across the plasmalemma (cell membrane) and cell wall to the outside environment. These are typically hydrolytic or oxidative and systematically degrade or break down very complex plant (and animal) polymers into simple molecules. In turn, the simple molecules such as sugars, amino acids, fatty acids, and so on resulting from enzymatic actions are absorbed through the cell wall, transported across the plasmalemma, and used for growth, energy, reproduction, and other life processes. Some enzymes that degrade plant cell walls may also be involved in plant pathogenesis. For example, the fungus *Rhizopus stolonifera* and the bacterium *Erwinia amylovora* both produce a battery of **pectinolytic** enzymes (pectinases) that degrade the middle lamella, which is composed chiefly of pectin and lies between plant cells. These organisms cause soft rots via the actions of pectinases. Many obligate parasites, such as *Peronospora tabacina* (blue mold of tobacco), utilize cell wall degrading enzymes (e.g., cellulases) to establish contact between haustoria and plant cell membranes.

The laboratory exercises in this chapter will be concerned with the detection of the activities of extracellular enzymes produced by microorganisms. There are three basic methods used to measure enzyme activities and include the following: detection of products of the enzymatic reaction (e.g., **reducing sugars** [RSs], acids, and change in pH and viscosity); depletion of enzyme substrate (phenolic compounds, etc.); and exhaustion of a cofactor (adenosine triphosphate [ATP], [nicotinamide adenine dinucleotide] [NAD], and others) in the reaction. The enzymes discussed in the exercises generally fall into the following two broad classes: constitutive and inducible. Constitutive enzymes are produced by the organisms at all times, albeit perhaps, at a low level. Inducible enzymes, in contrast, are produced only when the microorganism is grown in the presence of the enzyme substrate. The time for the enzyme to be induced and produced to levels of detection will depend on the organism and the specific enzyme of interest.

In the first exercise designed for undergraduate students, **amylase** (AMY), **lipase** (LIP), and **polyphenol oxidase(s)** (PPO) activities will be qualitatively determined using solid agar media in which the enzyme substrates have been incorporated. These enzymes are usually considered to be constitutive, but the quantity of enzyme produced may be influenced by the presence or absence of the substrate. The second exercise devised for graduate students or special projects for advanced undergraduates will quantitatively determine **cellulolytic** (cellulase) activity of fungi grown in liquid medium. Cellulases are usually considered inducible, and the test organisms will require contact with a cellulose or modified cellulose substrate, such as carboxymethylcellulose (CMC) before any appreciable enzyme activity can be detected. The third exercise is designed for use by either undergraduate or graduate students, and involves detection of endopectinolytic (pectinase) isozymes using **polyacrylamide gel electrophoresis** (PAGE). Pectinases are generally inducible enzymes and will require a specialized liquid medium containing pectin for enzyme production.

LABORATORY EXERCISES

Experiment 1. Qualitative Determination of Some Enzyme Activities

Often the only information desired from an experiment is whether or not the fungus produces a specific enzyme. For example, the fungi that cause dogwood anthracnose, *Discula destructiva* and an undescribed species of *Discula*, can be distinguished from each other by their ability to produce PPOs, which is a presumptive test for the ability to degrade lignin. Qualitative techniques are ideally suited to achieve this goal of detection only when quantification is neither needed nor desired. Typically, evaluation for the ability to produce an enzyme or battery of enzymes can be accomplished using an agar medium in which the substrate for the enzyme(s) has been incorporated. The fungus is allowed to grow on the agar medium containing the substrate for a period of time, and positive enzyme activity is indicated by a color change, including clearing, in the medium or precipitation of a product. For some other qualitative methods, such as some of the pectinolytic enzymes, staining for the original substrate can be used. In this experiment, we will qualitatively evaluate different fungi for the ability to produce AMY, LIP, and PPO. The substrates and enzymatic products for these three enzymes are shown in Figure 32.1.

Follow the protocol outlined in Procedure 32.1 to complete the experiment.

Materials

The following items are needed for each student or team of students:

- Duplicates of Petri dishes (10-cm-diameter) of five to eight cultures of different fungi grown on the appropriate culture medium (Table 32.1). We suggest NOT to use slow-growing fungi such as *Geotrichum* or *Acremonium* or prolific spore- or sporangium-producing fungi, such as *Penicillium* or *Rhizopus*, respectively.
- One dish of uninoculated (without fungi) NA and YEME to use as controls (Table 32.1).
- Three Petri dishes of 60-mm-diameter containing enzyme assay medium for each fungus (Table 32.2).
- A supply of autoclaved plastic drinking straws of 5- to 7-mm-diameter.
- Laminar flow hood (optional).
- Alcohol burner.
- Scalpel and # 11 blade or stainless steel spatula
- Growth room or incubator.
- Iodine reagent consisting of 15 g KI and 3 g I_2 per 1 L of distilled water.

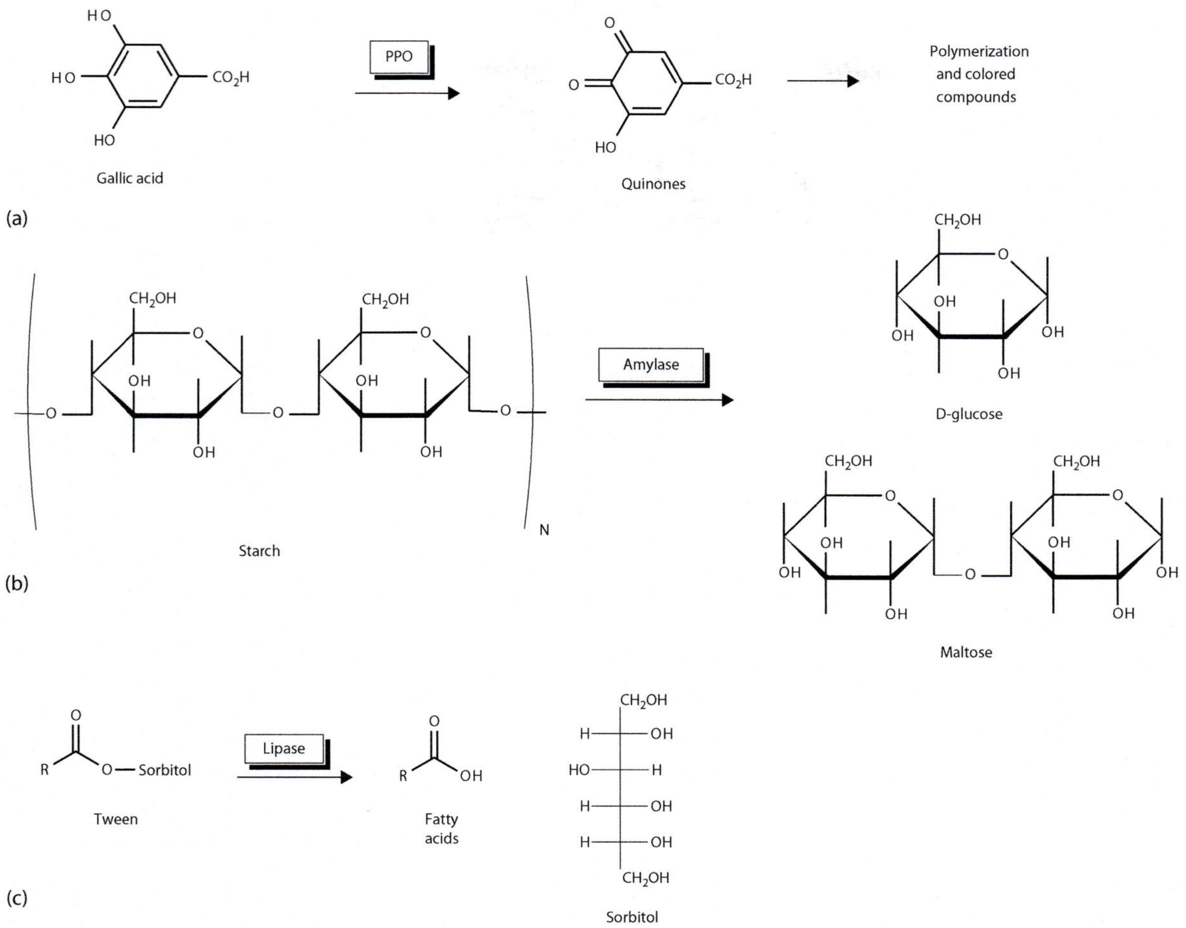

FIGURE 32.1 Substrates and products of polyphenol oxidase (PPO), amylase, and lipase. (a) PPO oxidizes adjacent hydroxyl groups on gallic acid (3,4,5-trihydroxybenzoic acid) and creates quinones, which are unstable and spontaneously polymerize to form pigmented (colored) products. (b) Starch is an α-1-4 polymer of D-glucose resides. Alpha-amylase hydrolyzes the bond between adjacent glucose units to form a random mixture of D-glucose and maltose (two D-glucose molecules linked α-1-4) residues. (c) Tweens are synthetic fats with a sorbitol (a sugar alcohol with five carbons instead of glycerol, which has three carbons) backbone esterified to various fatty acids (R: e.g., lauric or oleic acid). Note that only carbon 1 is shown. Lipase hydrolyzes the ester bond between the carbon in sorbitol and the carbonyl carbon of the fatty acid to form sorbitol and a free fatty acid. Changes in pH and calcium bonding with the free fatty acids combine to produce the white, flocculent precipitate suspended in the medium. (Drawing courtesy of Dr. James Green, University of Tennessee.)

TABLE 32.1
Media for Growing Inoculum

Amylases and Lipases—Nutrient Agar (NA)	Polyphenol Oxidase—Malt Extract—Yeast Extract Agar(YEME)
Nutrient broth (Difco Lab, Detroit, MI), 8 g	Malt extract (Difco Lab, Detroit, MI), 20 g
Agar, 20 g	Yeast extract (Difco Lab, Detroit, MI), 1 g
Distilled water, 1 L	Agar, 20 g
	Distilled water, 1 L

Combine all the ingredients and autoclave at 121°C for 20 min. Dispense media into sterile dishes of 9-cm-diameter when cooled, but not hardened.

- Parafilm®.
- Large forceps.
- Aluminum foil.

Anticipated Results

Almost all species of fungi produce both AMY (Figure 32.3a) and LIP (Figure 32.3b), and the enzymes are easily detectable after a short period (5–10 days) in culture. AMY-positive cultures are indicated by clear zones in the agar after staining with iodine solution. A positive test for LIP is a white flocculent precipitate in the medium. It is possible that not all species classified in the same genus will produce these enzymes.

PPO activity should be expressed as either darkened inoculum plugs and/or darkened assay medium within

Procedure 32.1

Detection of Extracellular Amylase, Lipase, and Polyphenol Oxidase Activity

Step	Instructions and Comments
1	Choose any number of fungi (we suggest five to eight species or include some isolates of the same species) and grow cultures for inoculum in Petri dishes of 9-cm-diameter using the appropriate medium for each of the enzymes (see Table 32.1). Two cultures of each fungus will be adequate for each student or team of students. Also, provide an "blank" Petri dish with medium (without fungus growth) for each enzyme.
2	Cut 10–12 equal diameter plugs of inoculum from the perimeter of the mycelium with a sterile plastic straw for each fungus (see Figure 32.2). If the plugs should become stuck in the straw, squeeze them out into the dish with sterile forceps. Repeat the process for each fungus and the "blank" agar using different straws.
3	Examine the AMY cultures periodically for mycelial growth. After the mycelium has covered 50%–75% of the agar surface, flood one of the cultures with iodine reagent. There is no need to maintain axenic cultures at this time. Using a scalpel, make numerous cuts through the mycelium in one dish and observe any color changes in the assay medium. Cleared zones underneath or in advance of the mycelium indicates AMY activity, whereas blue coloration denotes intact starch polymers and lack of AMY activity (Figure 32.3a). If the test is negative for AMY at this time, allow the fungus in the other dishes to grow for an additional week and retest using the iodine reagent. Record observations.
4	Observe the LIP cultures for white flocculent inclusions either beneath and/or in advance of the mycelial mat (Figure 32.3b). Observation of the medium with a dissecting microscope may help to see the inclusions. Precipitation is a positive indication of LIP activity. As with the AMY assay, if the cultures are negative for LIP at this time, allow another week for growth and look for precipitation. Note where the precipitation occurs in the medium and record the observations.
5	Examine the PPO cultures for darkened plugs and/or assay medium after 24 h and again after 48 h (Figure 32.4). Be sure to compare the "blanks" or control cultures with the inoculated cultures. Dark coloration indicates PPO activity and polymerization of the oxidized gallic acid. Measure the diameter of the discolored area and record observations.

TABLE 32.2
Enzyme Assay Media

Amylase (AMY)	Lipase (LIP)[a]	Polyphenol Oxidase (PPO)[b]
Soluble starch, 2 g	Peptone, 8 g	Solution A:
Nutrient broth (Difco Lab, Detroit, MI), 8 g	$CaCl_2 \cdot H_2O$, 0.1 g	Gallic acid (3,4,5-trihydroxybenzoic acid), 5 g
Agar, 20 g	Agar, 20 g	Distilled water, 250 mL
Distilled water, 1 L	Distilled water, 990 mL	Solution B:
(Society of American Bacteriologists, 1957)	Tween 20 or 80, 10 mL	Malt extract (Difco Lab, Detroit, MI), 5 g
	(Sierra, 1957)	Agar, 20 g
		Distilled water, 750 mL
		(Davidson et al., 1938)

[a] Autoclave Tween (polyoxyethylene sorbitan monolaurate [20] or monooleate [80]) and base medium separately at 121°C; then add to base medium after cooled, but not hardened. Swirl to mix and dispense into sterile, 60-mm-diameter plastic Petri dishes.

[b] Autoclave Solutions A and B separately at 121°C; then combine after cooled, but not hardened. Swirl to mix and dispense into sterile, 60-mm-diameter plastic Petri dishes. This medium should be stored in the dark or wrapped in aluminum foil to exclude light.

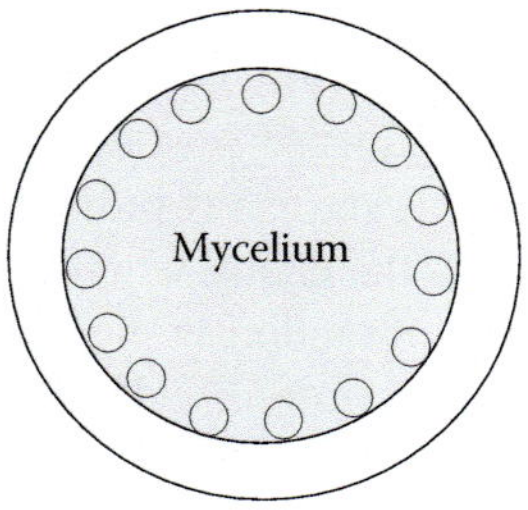

FIGURE 32.2 Diagrammatic representation of harvesting mycelial plugs from cultures. Push the open end of a sterile plastic straw through the periphery of the colony as designated by the circles. Transfer the plugs to the center of the assay medium with the mycelium side of the plug in contact with the medium.

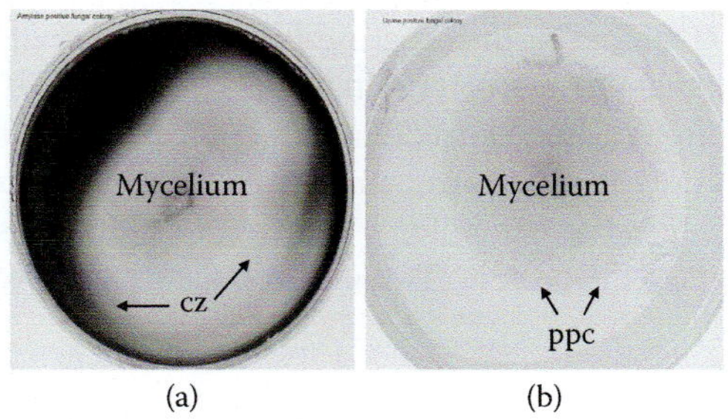

FIGURE 32.3 Assay media that are positive for enzymatic activity. (a) Amylase positive dish. Cleared zone (cz) indicates the lack of starch (amylase positive), whereas the dark (blue) zone has intact starch polymers. (b) Lipase positive dish. Note the flocculent (white) precipitation (ppc) in the medium that indicates lipase activity.

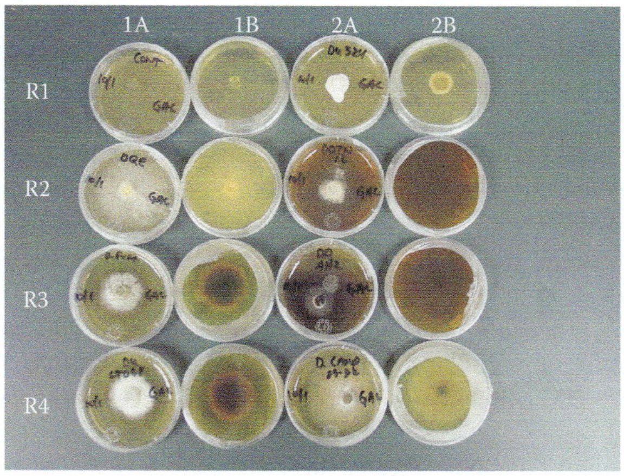

FIGURE 32.4 Solid medium test for polyphenol oxidase (PPO) using gallic acid as the substrate. All fungi were *Discula* spp. and cultures were 10 days post inoculation. Columns 1A and 2A are views of the top surface of the Petri dish and columns 1B and 2B are the corresponding bottom views of the medium. Row 1 (R1) columns 1A and 1B represent the medium inoculated with an agar plug only. Note that there is no discoloration of the medium indicating no PPO activity. The dishes labeled DQE (R2, C1A, C1B), DU324 (R1, C2A, C2B) and D. Camp (R4, C2A, C2B) also show no discoloration of the medium, which indicates no PPO activity. In contrast, all other test organisms produced PPO and corresponding discoloration of the growth medium. Note that most of the fungi grew on the medium, and of those that were positive, the discoloration was limited directly beneath the mycelium (D. frax: R3, C1A, C1B, and LT068: R4, C1A, C1B) or the discoloration extended well beyond the mycelium margin (DDTN12: R2, 2A, 2B and DDAH2: R3, 2A, 2B). The area of discoloration probably indicates the relative abundance of the PPO enzyme, although it is very difficult to quantify.

24–48 h. Generally, the darker and more widespread the discoloration of the medium, the greater the amount of PPO that has been produced (Figure 32.4). In some cultures, the plugs may be very slightly discolored, and it may be difficult to determine if the fungal isolate produces PPO. Many fungi produce PPO, and generally, isolates of the same species are all capable of producing the enzyme. However, there may be great variance within species of the same genus, and this information may be useful in taxonomic considerations. Some fungi grow on PPO assay medium, whereas others do not. Growth or lack of growth of individual species is probably influenced by the low pH of the medium and the toxicity of gallic acid more than the ability to oxidize the substrate.

Questions

- What is an enzyme?
- If a fungus does not grow on PPO assay medium, then why does the assay medium become dark?
- Besides participating in lignin degradation, what other advantage might PPO confer on a plant pathogenic fungus?
- How do enzymes move through the assay medium and what does it mean if enzyme activity is apparent well beyond the perimeter of the fungal colony?
- What role does calcium chloride play in the LIP assay medium? (Hint: what is precipitated in the medium?)

Experiment 2. Quantitative Measurement of Cellulolytic Activity

Quantitative measurement allows more accurate assessment of enzyme activities of individual isolates and comparison of activities between organisms. The procedures described in this experiment are designed to detect products of cellulolytic activity colorimetrically. However, this experiment may be adapted to most hydrolytic enzyme systems, such as pectinases and xylanases, where **reducing sugars** (or other compounds with **anomeric** carbons) are produced.

Crystalline cellulose, 1,4-β linkages of D-glucose residues, is highly insoluble in water at neutral or slightly acidic pH typically found in growth media. An individual polymer consists of 500–15,000 D-glucose units and has a molecular weight in the range of about 5×10^4 to 2.5×10^6 daltons. Therefore, CMCs, which are far more soluble in water, are used as a substrate in most media. CMC has methyl groups esterified through the carboxyl group on carbon 6. Solubility in water is achieved through the partial positive charge imparted by the methyl group(s) to the molecule. Not all of the available carboxyl groups are methylated, and CMC is available in a wide range of percentage methyl substitutions.

Cellulases are inducible enzymes, and in order for organisms to produce them, cellulose or substituted cellulose compounds must be present in the growth medium. Cellulases are hydrolytic (adds a water molecule between individual units) and actually consist of a number or battery of enzymes. The C_1 enzyme or 1,4-β-glucan cellobiohydrolase enzymatically hydrolyzes crystalline cellulose to form cellobiose, a disaccharide composed of two D-glucose residues. The C_{x1} enzyme (1,4-β-exoglucanase), cleaves off individual glucose residues from the end of the chain, whereas the C_{x2} enzyme (1,4-β-endoglucanase) randomly hydrolyzes the cellulose polymer into a mixture of cellulodexans of various lengths. The C_{x3} enzyme is a 1,4-β-glucosidase (cellobiase) that hydrolyzes cellobiose into two D-glucose residues (Figure 32.5).

Experiment 2 is divided into the following three parts: fungal growth and protein isolation; developing standard curves and measuring enzyme activity; and calculating cellulolytic activity. Fungal cultures will take about 2–3 weeks to grow. Two 4- to 6-h laboratory periods should be scheduled to complete the remaining tasks. If this experiment is used as a class activity, we suggest that the instructor initiate the cultures and if time is a limiting factor, isolate total protein from the cultures. Portions of the experiment may also be assigned to teams of students. For example, a team of four students can help isolate the protein. Two of the students can work on

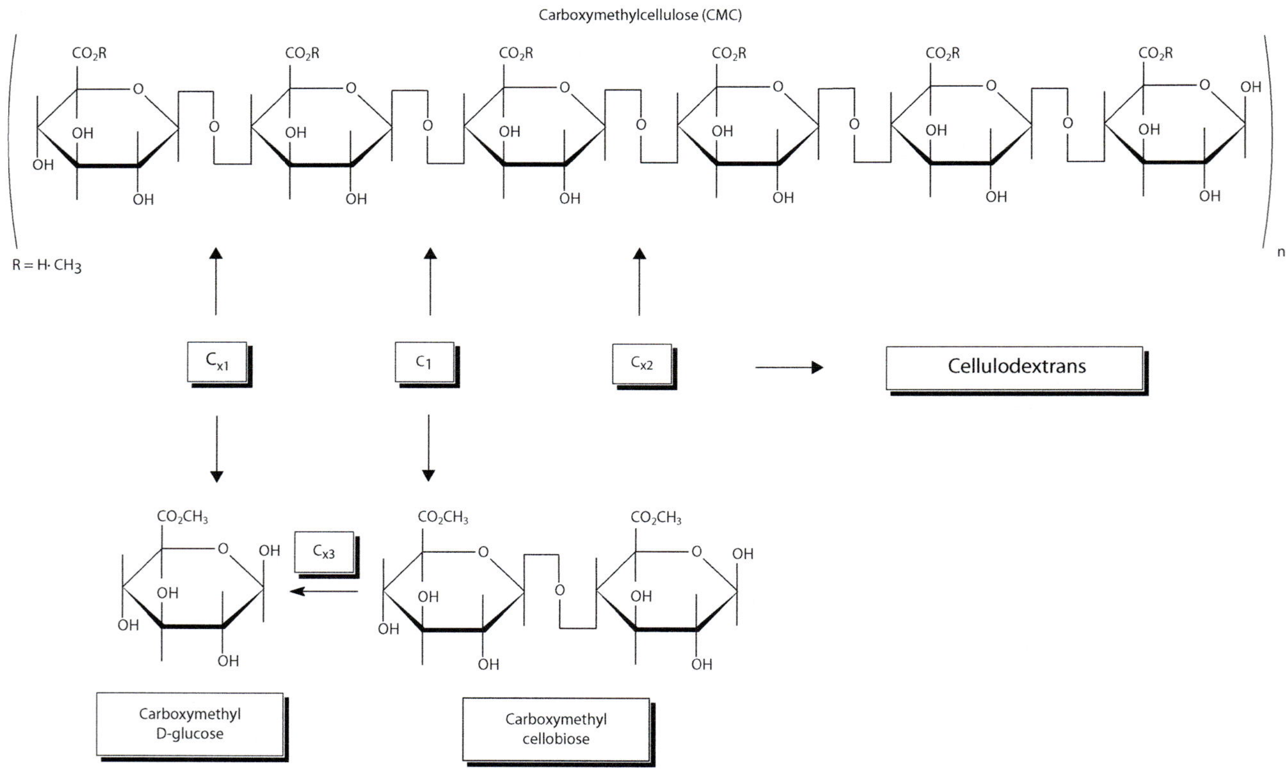

FIGURE 32.5 Diagrammatic representation of the action of cellulolytic enzymes on the substrate carboxymethylcellulose (CMC–OH and C = O groups not shown). The C_1 enzyme (β-1,4-glucan cellobiohydrolase) hydrolyzes the 1-4-β bond on the nonreducing end of the polymer to yield carboxymethyl cellobiose molecules. C_{x3} or cellobiase (1,4-β-glucosidase) hydrolyzes cellobiose to yield carboxymethyl D-glucose residues, which is a reducing sugar. C_{x1} is exocellulase (1,4-β exoglucanase) that cleaves one carboxymethyl D-glucose residue from the nonreducing end of the CMC molecule, whereas C_{x2} is an endocellulase (1,4-β endoglucanase) that internally hydrolyzes bonds to produce various length cellulodextrans. All of the cellulodextrans can be substrates for C_1 or C_{x1} enzymes, and thus, C_{x2} may increase the apparent activity of the other enzymes by providing more substrate (if substrate is a limiting factor in the reaction mixture). (Drawing provided by Dr. James Green, University of Tennessee.)

determining protein concentration, while the other two measure RSs. Note that this experiment essentially measures both endo- and exo-glucanase activities. We have added a simple Petri dish test that measures (for the most part) only endocellulase activity and another that measures endopectinase activity.

Follow the protocol outlined in Procedure 32.2 to complete this part of the exercise.

Procedure 32.2

Growth of Fungi and Isolation of Protein

Step	Instructions and Comments
1	Grow five to eight isolates of fungi (can be different isolates of the same species) in Petri dishes containing nutrient agar as described in Table 32.1.
2	Prepare sufficient liquid cellulase induction medium (about 150 mL per isolate) to dispense 35 mL into each of four capped 125-mL flasks for all isolates included in the experiment. After autoclaving the medium in the flask, some of the CMC may be precipitated as fine white threads at the bottom of the flask, but this will not affect the induction of cellulase. Prepare inoculum plugs using sterile straws as shown in Figure 32.2 and dissect the plugs into quarters using a scalpel and # 11 blade. Transfer four quarters to each of three flasks per isolate. Prepare a control (cut from an uninoculated agar dish—no mycelium) to inoculate the fourth flask. Label the flask with the species name, isolate, and date. Repeat the inoculation step for each species using a different straw each time. Place all flasks in a 25°C incubator for 14–21 days depending on the growth rate(s) of the fungi selected for the experiment.
3	Examine all flasks for contamination, especially for bacteria. The medium should be clear; cloudiness usually indicates bacterial contamination. Pour the liquid contents of a flask into a beaker and repeat this procedure for the remaining two inoculated flasks for the same species or isolate using the same beaker. Pour the contents of the beaker into a 250-mL beaker and record the volume. Repeat the entire procedure for the remaining fungal isolates and the uninoculated flasks.
3a	Alternatively, the liquid from the cultures may be poured into 50-mL centrifuge tubes, taking care not to transfer the mycelial mat to the tube. Balance tubes (two tubes of the same weight), place opposite of each other in the centrifuge, and centrifuge for 10 min at 10,000 g. Pour the liquid into a graduate cylinder and record the volume. Discard the pellet in a biohazard bag and autoclave before disposal.
4	Add 66 g of ammonium sulfate for every 100 mL of induction medium in the 250-mL beakers. For example, if 110 mL of medium was collected, then add 66 g × (110/100) = 66 g × 1.1 or 72.6 g of ammonium sulfate. Stir until **all** of the ammonium sulfate is dissolved and place beakers on ice for at least 30 min.
5	Stir for a few seconds to suspend precipitated proteins and dispense the liquid into 50-mL centrifuge tubes (do not add more than about 45 mL to each tube) and label. Be sure to balance opposite tubes by weighing on a balance (within 1 g is okay) and centrifuge for 10 min at about 3000–5000 g. Decant and discard the supernatant and invert the tube on a paper towel. A pellet may or may not be present at the bottom of the centrifuge tubes. After the excess liquid has drained, dissolve the total protein by adding 0.5 mL of distilled water per tube and shake or vortex. Allow to stand for 10 min. Combine the contents of the tubes of each species or isolate.
6	Cut 15 cm lengths of dialysis tubing, soak in distilled water, and autoclave for 3 min. Tubing should be rinsed several times in distilled water before use. Express all of the water from the tube, and clamp one end. Load the total protein from one species into each of the tubes. Clamp the other end of the tube taking care not to introduce air into the lumen. Immerse the loaded dialysis tubing into chilled, distilled water and place in the refrigerator at 2°–4°C overnight.
7	The next morning, aliquot 0.5 mL from the dialysis tubes to several labeled 1.5-mL Eppendorf tubes. The number of tubes needed per isolate will vary. The Eppendorf tubes may be stored at –70°C for future use, or placed on ice for use later the same day. Note that the samples can only be thawed once from –70°C without permanently denaturing the enzymes or significantly decreasing activity.

Fungal Growth and Protein Isolation

Materials

Each student or team of students will require the following materials and cultures:

- Five-to-eight fungal isolates grown on nutrient medium described in Table 32.1
- Liquid cellulolytic induction medium (Table 32.3; Reese and Mandels, 1963)
- Three 125-mL Erlenmeyer flasks of the above medium for each fungal isolate
- Alcohol lamp
- Sterile straws of 6-to 8-mm-diameter
- Scalpels fitted with # 11 blades
- Incubator at 25°C (other temperatures may be required for specific organisms)
- 250-mL graduated cylinder
- Ammonium sulfate ($(NH_4)_2SO_4$), about 750–1000 g
- 250-mL beakers, magnetic stir bars, and stir plates
- Dialysis tubing (Spectrum Laboratories, Inc., Rancho Dominguez, CA; 800-445-7330) 3500 exclusion and clamps (suture string will also work)
- Ice bath
- Table top centrifuge (3,000–10,000 g; does not need to be refrigerated)
- 50-mL centrifuge tubes (do not need to be sterile)
- Top loading balance
- Squeeze bottle of distilled water
- Distilled water
- Refrigerator
- Water bath set at 40°C

Anticipated Results

Most fungi will grow well in cellulase-inducing medium, and the enzyme(s) should be present in the medium between 1 and 3 weeks after inoculation. Ammonium sulfate yields three osmotically active particles when dissolved in water and will salt-out most proteins (e.g., cellulases) at 90% saturation. It is most important that any undissolved salt crystals are not present when starting the centrifugation step (salt crystals will cause excess water to enter the dialysis tube and dilute the protein). A mat of other insoluble material, mostly carbohydrates, may be present at the top of the centrifuge tubes after centrifugation. This is normal and the mat should be discarded. After overnight dialysis, the contents of the tubes should be increased—some tubes may be very swollen and turgid with water. The quantity of water in the tube is dependent on the amount of salt that was carried over after centrifugation. In many experiments, each tube will yield about 4 mL of an aqueous solution of various proteins and other compounds. Some pigments may have coprecipitated with the proteins, and this is not a cause for concern. Salting out and dialysis concentrate the protein from 15 to 25 times that is found in the inoculated medium.

Developing Standard Curves and Measuring Enzyme Activity

In order to determine enzyme activity, it is necessary to have solutions of known concentration of total protein and enzyme product (in this case RS) to compare with unknowns. Standard solutions are prepared in several concentrations and equations calculated to predict intermediate values of unknowns. The equations are formulated by regressing absorbance on concentration, and these calculations usually follow Beer's law, especially

TABLE 32.3
Composition of Liquid Cellulase Induction and Assay Medium

Cellulase Induction Medium	Amount	Cellulase Assay Medium	Amount
KH_2PO_4	2.0 g	Citric acid H_2O	10.5 g
$(NH_4)_2SO_4$	2.5 g	K_2HPO_4	17.4 g
Urea	0.3 g	Water	1 L
$MgSO_4 \cdot 7H_2O$	0.3 g	CMC[a]*	5.5 g
$CaCl_2 \cdot 2H_2O$	0.5 g	pH = 5.0–5.2	
Peptone	1.0 g	Do not autoclave, store at 4°C	
Water	1 L		
CMC[a]	5.0 g		
pH = 5.5			
Autoclave at 121°C for 15 min[b]			

[a] Adjust medium pH before adding CMC; extensive stirring may be required to dissolve CMC.
[b] Carboxymethylcellulose (CMC) may precipitate as “threads” after autoclaving.

in the mid-range of the standards. That is, the relationship between known concentrations and absorbance is linear, except at very low or high concentration of product, in this case protein or RSs. In this part of the experiment, students will develop standard curves for total protein using the method developed by Lowry et al. (1951) and for RS with dinitrosalicylic acid (DNSA) reagent (Miller, 1959) Note that only the instructor should prepare this reagent, and it should only be used under a fume hood.

Follow the protocol provided in Procedures 32.3 through 32.5 to complete this section of the experiment.

Materials

The following materials will be required for the completion of this part of the experiment:

- D-Glucose and bovine serum albumin (BSA)
- Several 1-, 5-, and 10-mL pipettes
- Visible light spectrophotometer
- Centrifuge that can accommodate 50-mL plastic tubes and develop 3,000–10,000 g
- Fume hood
- Glass test tubes (greater than 10-mL capacity) and racks

Procedure 32.3

Developing a Standard Curve for Protein Concentration

Step	Instructions or Comments
1	Dissolve 100 mg of BSA (protein) in 100 mL of distilled water. This makes a 1 mg BSA/mL solution. Make the following standard solutions using dilutions: 0.1, 0.2, 0.3, and 0.5 mg BSA/mL. Use the following formula: $$C1 \times V1 = C2 \times V2$$ where C = concentration of protein and V = volume. For example, to make a 0.5 mg $$\mathbf{1.0\ mg\ BSA / mL \times 10\ mL = 0.5\ mg\ BSA / mL \times V2}$$ Solve the equation for V2, which is the total volume of the desired solution. $$\mathbf{10\ mg / 0.5\ mg / mL\ or\ 20\ mL = V2}$$ Since a 10-mL volume was initially used: $$\mathbf{V2 - V1 = 10\ mL}$$ Add 10 mL of distilled water to the original 10 mL of 1.0 mg BSA/mL. There should be 20 mL of 0.5 mg BSA/mL. The following amounts of distilled water should be added to 10 mL of 1.0 mg BSA/mL to make the appropriate standards: 90 mL water for 0.1 mg BSA/mL; 40 mL of water for 0.2 mg BSA/mL; and 23.4 mL of water for 0.3 mg BSA/mL. Mix all solutions well, transfer to labeled 50-mL centrifuge tubes, and if not used within the day, store at –20°C.
2	Prepare 200 mL of a 0.1 N NaOH solution by dissolving 0.8 g of NaOH in 200 mL of distilled water. NaOH can be difficult to weigh accurately since it is provided as pellets. Alternatively, dissolve 4 g NaOH in 100 mL (1 N) and dilute 20 mL of this solution with 180 mL of distilled water to make a 0.1 N solution of NaOH. Dissolving NaOH in water is exothermic (generates heat), but is not hazardous. To 200 mL of 0.1 N NaOH, add 4 g of sodium carbonate. Stir until the sodium carbonate is completely dissolved and pipette 1.0 mL of each 1.0% (w/v) cupric sulfate and 2.0% (w/v) sodium tartrate into the solution and mix thoroughly. At this time, dilute 20 mL of Folin–Ciocalteau reagent with an equal amount of distilled water. Wear gloves and eye protection when making this solution.

3	Carefully pipette 1 mL of each of the BSA standard concentrations including a water control (0 mg BSA/mL) into glass test tubes. Prepare three replicates of each BSA concentration. For each fungal species or isolate, pipette 1 mL of the total protein isolated from induction medium into a glass test tube. Add 5 mL of the cupric sulfate solution (Step 2) to each of the test tubes and mix well with a vortex mixer at low speed. After 10 min at room temperature, quickly add 0.5 mL of the diluted Folin–Ciocalteau reagent (Step 2) to each of the tubes and mix. Set aside at room temperature for 30 min.
4	After 30 min, pipette the standards and unknown samples into disposable plastic cuvettes. Determine the absorbance values in a spectrophotometer set at 500 nm and record. Establish the baseline using the "0" BSA samples as a reference. After reading, dispose of the liquid (and cuvettes) in an approved, labeled, hazardous waste container.
5	Calculation of standard curves using linear regression and determination of protein in unknowns are discussed in Procedure 32.6.

Procedure 32.4

Developing a Standard Curve for Reducing Sugars—DNSA Reagent

Step	Instructions and Comments
1	Follow the instructions for the preparation of standard BSA solutions in Procedure 32.3, Step 1, but instead use D-glucose. Prepare the following standard concentrations: 0.1, 0.2, 0.3, 0.4, and 0.5 mg D-glucose/mL.
2	Pipette 1 mL of each standard D-glucose solution (including 0 mg D-glucose/mL) into each of three glass test tubes. RSs in the cellulolytic assay medium samples can be assayed at this time if Procedure 32.5 has been completed. Wearing gloves and eye protection, pipette 3 mL of DNSA reagent (Table 32.4) into each of the tubes.
3	In a fume hood, place the test tube rack with tubes into a water bath containing 100°C water for 15 min. If a water bath is unavailable, the rack can be placed in a plastic dishwashing tub containing freshly autoclaved water.
4	Remove the test tubes and rack from the heated water and allow to cool to room temperature (about 30 min).
5	Transfer D-glucose standards and unknowns to disposable plastic cuvettes, read on a spectrophotometer set at 550 nm and record the absorbance values. If a dual beam spectrophotometer is available, establish the baseline using the "0 mg/mL" D-glucose sample as a reference. After reading, dispose of the liquid and cuvettes in an approved and labeled hazardous waste container.
6	Calculation of standard curves using linear regression and determination of unknowns are discussed in Procedure 32.7. However, for illustrative purposes, the equation $Y = 2.34\,X + (-0.27)$ for RS was calculated from actual data (not shown).

Procedure 32.5

Assaying Isolated Proteins for Cellulase Activity

Step	Instructions and Comments
1	Before starting the procedure, set a water bath at 40°C. Select two 1.5-mL Eppendorf tubes containing 0.5 mL of protein for each fungal species or isolate. If the samples were frozen at –70°C, allow them to come to room temperature—do not heat to thaw. Be sure the tube is labeled and securely closed. Place one of the tubes into boiling water for 15 min to irreversibly denature any enzymes. This will serve as a control.
2	Prepare cellulase liquid assay medium the day before as described in Table 32.3 and store in the refrigerator (4°C). Allow the assay medium to come to room temperature and pipette 4.5 mL into test tubes. Depending on the volume of the protein isolation, additional replicates can be prepared for unknown samples. Dispense 0.5 mL of nondenatured isolated protein (not boiled) into test tubes for each of the unknowns. Dispense

0.5 mL of boiled preparations from Step 1 into separate tubes. Also, prepare at least one tube to which 0.5 mL of sterile water has been added to reaction mixture. Mix the contents of all tubes well and cap with aluminum foil to prevent water loss.

3 Incubate the test tubes in a 40°C water bath for 1 h. Note that if after the RS assay some samples are negative or very low, incubate the samples for an additional hour.

4 Complete the assay for RSs using the DNSA procedure described in Procedure 32.4. Note that the unknowns may be assayed for RS at the same time that the standard concentrations of D-glucose are determined.

- 50-mL centrifuge tubes (not necessarily sterile)
- Disposable, plastic cuvettes, and transfer pipettes
- Folin–Ciocalteau reagent (Fisher Scientific Company)—caution: this reagent is caustic, wear hand and eye protection
- 2% (w/v) sodium carbonate in 0.1 N sodium hydroxide solution
- 2% (w/v) aqueous sodium tartrate solution
- 1% (w/v) aqueous cupric sulfate
- Nitrile gloves and safety glasses
- DNSA reagent (Table 32.4)—caution: these reagents are caustic, wear hand and eye protection; only prepare and heat under fume hood
- 100°C water bath or hot, freshly autoclaved water
- Ice
- Vortex
- Computer with programming capable of completing linear regression

TABLE 32.4

Dinitrosalicylic Acid Reagent for Detecting Reducing Sugars

Dinitrosalicylic (DNSA) Method[a]

Chemical	Amount
3, 5-Dinitrosalicylic acid	8.0 g
Crystalline phenol	6.9 g
Sodium bisulfite	6.9 g
Sodium-potassium tartrate	2.55 g
Sodium hydroxide	15.0 g
Water	1 L

Source: Miller, G. L., *Anal. Chem.* 31:426–428, 1959.

Notes: Prepare at least 24 h before use and store in brown bottle at 4°C. Caution: Wear gloves and eye protection. Prepare solution under fume hood with the supervision of the laboratory instructor.

[a] To begin making the DNSA solution, add 15 g of sodium hydroxide to 900 mL of water and stir until dissolved. Sodium hydroxide generates heat (exothermic) as it dissolves in water. After the sodium hydroxide has dissolved, add all other chemicals and bring the final volume of the solution to 1 L with water.

Anticipated Results

Measurement of protein concentration is quick, easy, and accurate using Lowry et al.'s (1951) method. The blank or "0" protein mixture will develop a very slight blue tint and with increasing protein concentrations, the mixture will be progressively darker. The highest standard concentration of protein will be almost opaque (Figure 32.6). Absorbance measurements of the unknown concentrations should lie within the range of absorbance values of the standards. A general rule is not to extrapolate the line beyond the range of standards. Most unknown preparations will fall between 0.1 and 0.3 mg total protein/mL. However, if any unknown preparation is not within the range, additional standard concentrations, especially slightly above 0.6 mg/mL, may be prepared and the regression line recalculated. Alternatively, the unknown can be diluted 1:1 or 1:2 with water and reassessed. If the concentration is low, the sample can be concentrated by evaporation of water and then remeasured. Two reminders—always use the equation of the line to calculate the protein concentration of the unknown and dispose of all reagents properly.

The color of the DNSA becomes increasingly more red to dark red with higher concentrations of RSs (Figure 32.7). The RS content from unknown assays will usually fall between 0.3 and 0.5 mg D-glucose/mL. The same considerations for determining proteins apply to RSs. If the absorbance of the unknown sample is beyond that of the most concentrated standard, dilute the original sample containing RS 1:1 or 1:2 with assay mixture and reassess the concentration with DNSA. Once again, use the regression equation to compute RSs in the unknown samples. Also, heat DNSA and samples only under the fume hood and dispose of all reagents properly.

Questions

- Why are three replicates of each standard concentration used to estimate the regression line?
- Is the total protein isolated composed entirely of cellulases? What other types of enzymes and/or proteins may be present?
- The activities of which cellulases are measured with the D-glucose (RS) assay procedure. How can the activity of C_{x2} be measured?
- Why is it necessary to include a boiled enzyme in the assay?

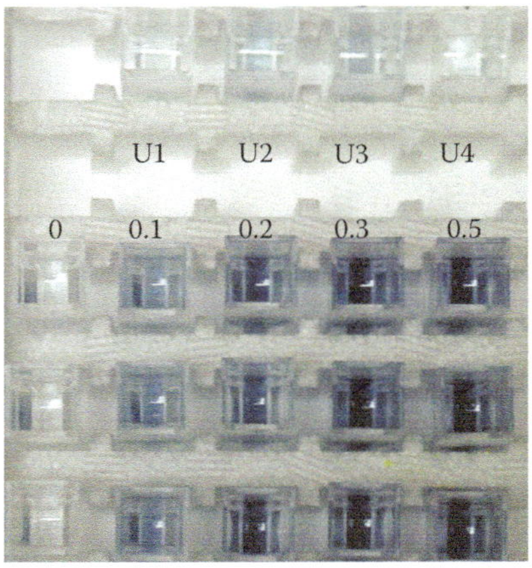

FIGURE 32.6 Protein standards and determination of protein in isolations. The amount of protein in the standards and "unknowns" is linearly related to the amount of blue color developed in the tubes. Three replications of known concentrations of protein (0, 0.1, 0.2, 0.3, and 0.5 mg BSA/mL) are regressed on absorbance, and a standard curve is developed. Protein concentrations from isolations (U1–U4) can then be determined.

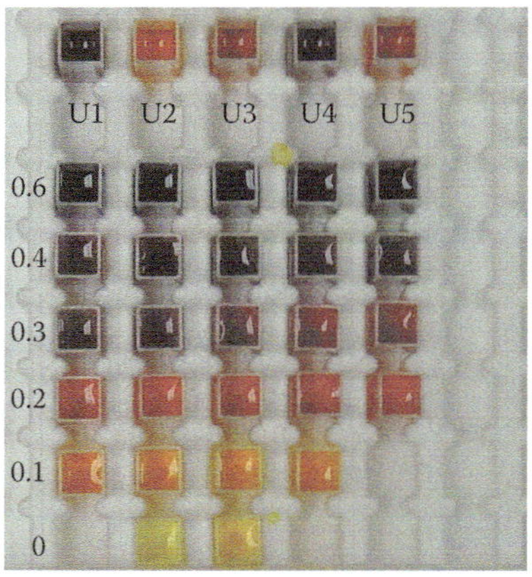

FIGURE 32.7 Reducing sugar standards and determination of reducing sugars produced by the action of cellulases. Using the dinitrosalicylic acid test, the absorbance of three replicates of each standard concentration (0.0, 0.1, 0.2. 0.3, 0.4, and 0.6 mg D-glucose/mL) are determined. Absorbance is regressed (linearly) on D-glucose concentration and a line calculated. The amount of reducing sugar from the action of cellulases in the "unknown tubes" (U1–U5) can then be determined and the activity of the enzyme assessed.

Calculation of Enzyme Activity

All of the essential information for calculating enzyme activity is now available. The hypothetical example will use the standard curves for the data as calculated by Excel. The final expression of data will be in μmoles of RSs (as D-glucose equivalents)/mg of crude protein/hour. Follow the steps as outlined in Procedure 32.6 to calculate the regression equations for protein and RSs and Procedure 32.7 to determine enzyme activity of the unknown samples.

Questions

- Why is it important to calculate a rate of enzyme activity for each fungus? Why not compare only the RS produced in each enzyme assay mixture?
- Would you expect the enzyme activity rate to remain constant over time (i.e., after 1, 2, and 4 weeks of culture)?
- How will the enzyme activity change if the incubation temperature is decreased to 30°C or 20°C or increased to 50°C or 60°C?

Experiment 3. Detection of Endocellulases (C_{x2}) and Endopectinase Activities Using CMC and Pectin Incorporated into an Agarose Gel

The activities of endoacting hydrolytic enzymes, such as C_{x2}, are difficult to measure. Often, their activities are measured as a lessening of viscosity of a solution containing a substrate over time or via gel electrophoresis. In this experiment, endocellulase activity can be visualized in a stained agarose gel containing CMC; intact polymers of CMC are stained with Congo red, whereas endocellulases destroy the structure of CMC by cleaving the polymer into shorter segments, which will not be stained by the dye. Thus, endocellulase activity is expressed as a cleared zone in the agarose medium (protocols adapted from personal communication with Dr. Juan Jurat-Fuentes, University of Tennessee). This is a very qualitative measure of endocellulase activity. Follow the protocol in Procedure 32.2 to grow fungi and isolate extracellular protein. After protein isolation, determine the amount of crude protein in the samples using the protocol described in Procedure 32.3, then follow Procedure 32.8 to complete the experiment with agarose gels.

The activities of endoacting pectinases can also be detected by a similar method. In this alternative experiment, pectin instead of CMC is combined with agarose. The agarose is stained with ruthenium red, and like the endocellulase test, a cleared zone represents the activity of endopectinases. Follow the protocol described in Procedure 32.8 to complete this alternate experiment.

Students should work in groups of two to four.

Materials

The class will require the following materials to complete the laboratory:

- CMC and/or citrus pectin
- Microwave or autoclave
- Agarose
- Hot/stirring plate with stir bar and rotary shaker

Procedure 32.6

Construction of Standard Curves and Determination of Unknowns

Step	Instructions and Comments
1	For calculating protein and RSs standard curves, data should be entered in columns, as shown in Table 32.5. Column A contains the independent variable (concentration of the standards), and Column B the dependent variable (absorbance values recorded from the spectrophotometer). In order to build a data set that can be properly analyzed, you should have multiple samples and corresponding readings for each independent variable. The first measurement taken was to zero the spectrometer. Since there were not many samples to read, the zero measurement was not repeated. For each of the protein concentrations (e.g., 0, 0.1, 0.2, 0.3, and 0.5 mg BSA/mL), three samples were analyzed and their absorbance measurements were recorded. Note that data generated by the class may be similar, but not exactly the same as sample data presented in Table 32.5. The data may be graphed, but it is not necessary in order to calculate either protein or RS concentration. There are other programs available that furnish more control over the appearance of the graph than that provided by either Microsoft Excel or Corel Quattro Pro. In order to use the Regression tool in Excel, the version must have the Analysis Tool Pack installed.
2	For linear regression, enter your data using Microsoft Excel, click Tools/Data Analysis/Regression/OK. In the appropriate box, select the range of Y values. These values are the dependent variables (absorbance values) in the data set. Select the range of X values in the next box. These are independent variables (concentration). Select an area in the spreadsheet in which Excel can place the output of the regression calculations. Click OK. In the results displayed in Table 32.6, the X coefficient (or slope of the line, *m*) is labeled "0" under the "coefficients" heading, and the Y-axis intercept (or b) is labeled "intercept" under the "coefficients" heading. The R^2 value is found at "R Square" under "Regression Statistics" and is a measure of how well the data agrees with the estimated line. From these calculated values, you can construct the equation of the regression line, $y = mx + b$; where m is the slope of the line and b is the Y-intercept. The equation from this analysis is: $Y = 0.99\ X + 0.03$. Use this equation to determine the unknown concentrations. Remember Y is the absorbance value for the unknown; solve for X, the concentration. Microsoft Excel provides a tool to graph the data and apply a trend line to complete regression analysis, but that method is not addressed here. The regression equation is the most accurate way to determine concentration. Trying to determine concentration by using the graph only provides an estimate of the value.
3	Corel Quattro Pro can also be used to calculate a linear regression. Data should be typed in the format shown in Table 32.5. Click Tools/Numeric Tools/Regression. In the appropriate boxes, select the independent (concentration) and dependent (absorbance) values and an area on the spreadsheet in which the output of the regression analysis can appear. Be sure that the Y-intercept radio button is set to "Compute" and click OK. The results from analysis of the data listed in Table 32.5 are shown in Table 32.6 and the calculated regression equation is $Y = 1.01X + 0.02$. There are minor differences in the values of m and b depending on which program (Excel or Quattro Pro) was used to estimate the regression line. Discrepancies are due to variations in the algorithms used in the programs. Very slight differences in the regression equations will not cause major differences in the calculations in either protein or RS concentrations. Choose one spreadsheet program and consistently use it to avoid introducing minor errors into the statistical analysis.

- Petri dishes (60-mm-diameter)
- Sharpie or other marking pen
- Incubator at 30°C
- Commercial cellulase and/or pectinase
- 100- and 200-µL pipettes and tips
- 1 M NaCl solution (58.5 g in 1 L of distilled water)
- 0.1% Congo Red (1 g in 1 L of distilled water)
- 0.1 M malic acid solution (dissolve 13.4 g of malic acid in 700 mL of water and bring to a final volume of 1 L). Store at room temperature.
- 0.01% (100 mg in 1 L of water) ruthenium red aqueous solution. Store at room temperature.
- Micropipette (2–20 µL and 200 µL or equivalent) and disposable tips
- Glacial acetic acid

TABLE 32.5
Sample Data for Protein Standard Curve Entered in Excel Spreadsheet

Column A	Column B
Protein Concentration (mg/mL)	**Absorbance Units (a.u.)**
0	0.000
0.1	0.115
0.1	0.116
0.1	0.125
0.2	0.239
0.2	0.233
0.2	0.235
0.3	0.340
0.3	0.335
0.3	0.334
0.5	0.518
0.5	0.521
0.5	0.519

Anticipated Results

The agarose in control dishes (boiled commercial enzyme) will be stained purple or violet without a cleared zone at the point of inoculation. The agarose in control dishes (inoculated with active commercial enzyme) will have a large cleared (nonviolet) zone indicating endocellulase activity. Agarose inoculated with fungal protein solutions will have variable diameter cleared zones if endocellulases are present; those inoculated without the enzyme will not have a cleared zone. Specifically, the diameter of the cleared zone is in some way proportional to the absolute amount of endocellulases present in the crude protein fraction.

Questions

- How would an increase in incubation temperature affect the diameter of the cleared zone?
- The zone of clearing is also affected by how well the enzymes migrate or diffuse in the agarose gel. How would you investigate this parameter of the assay?

Procedure 32.7
Calculation of Enzyme Activity

Step	Instructions and Comments
1	Calculate total protein in unknown samples. The estimated equation of the regression line given in Procedure 32.6 step 3 is: $$Y = 0.99\,X + 0.03$$ where $\boldsymbol{Y}$ = absorbance measurement in arbitrary units (a.u.) and $\boldsymbol{X}$ = concentration of protein in mg/mL. If the absorbance of an unknown sample (e.g., #1) is 0.33 a.u., the following substitution into the above equation should be made: $$0.33 = 0.99\,X + 0.03$$ Solving for $\boldsymbol{X}$, the concentration yields: $$(0.33 - 0.03)/0.99 = X \text{ or } 0.30 \text{ mg total protein / mL}$$ Note: not all of the protein in the sample are cellulases.
2	Calculate RS/mL in the reaction mixture. The example of estimated equation for the regression line given in Procedure 32.4, Step 6 is: $$Y = 2.34\,X + (-0.27)$$ where $\boldsymbol{Y}$ = absorbance measurement in a.u. and $\boldsymbol{X}$ = concentration of RS in mg/mL. If the absorbance of an unknown RS sample (e.g., #1) is 0.52 a.u., the following substitution into the above equation should be made: $$0.52 = 2.34\,X + (-0.27)$$

Solving for X, the concentration yields:

$$(0.52 + 0.27) / 2.34 = X \text{ or } 0.34 \text{ mg RS / mL of reaction assay medium}$$

3 Calculate total RS in the 5 mL of reaction assay medium using the following equation:

$$\text{Total RS} = \text{mg RS / mL} \times 5 \text{ mL (total volume of the reaction mixture)}$$

Substituting in the equation provides:

$$0.34 \text{ mg RS / mL} \times 5 \text{ mL reaction mixture} = 1.7 \text{ mg RS}$$

Remember to incorporate any dilutions of the assay volume (e.g., if the assay was diluted 1:1 with assay medium, then the total volume would equal 10 mL).

4 Calculate mg of RS/mg of crude protein in the assay. Remember that the assay protocol used only 0.5 mL of protein extract (divide concentration by 2 to determine the amount of protein incubated with the assay mixture). From Step 1, the sample contained 0.30 mg total protein/mL.

$$0.30 \text{ mg total protein / mL} \times 0.5 \text{ mL} = 0.15 \text{ mg of total protein}$$

From Step 3, the total RS in the reaction mixture was 1.7 mg. Divide the RS by the mg of total protein used:

$$1.7 \text{ mg RS / } 0.15 \text{ mg total protein} = 11.33 \text{ mg RS / mg total protein}$$

This calculation provides a measure of the amount of RS that each milligram of crude protein can produce.

5 Calculate mg RS/mg crude protein/h that the assay medium was incubated at 40°C.

$$11.33 \text{ mg RS / mg total protein / 1 h} = 11.33 \text{ mg RS / mg total protein / h.}$$

Most incubation times are 1 h, but occasionally 2 h. This calculation provides a rate for formation of the product.

6 Calculate micromoles of RS using the molecular weight of D-glucose (180.2 g). First, convert g to mg.

$$180.2 \text{ g D} - \text{glucose / mole} \times 1{,}000 \text{ mg / g} = 180{,}200 \text{ mg D} - \text{glucose / mole.}$$

Now, calculate the number of moles of RS produce in the reaction:

$$11.33 \text{ mg RS / } 180{,}200 \text{ mg / mole} = 6.29 \times 10^{-5} \text{ moles RS.}$$

Lastly, convert moles to µmoles:

$$0.0000629 \text{ moles} \times 1{,}000{,}000 \text{ µmoles / mole} = 62.9 \text{ µmoles.}$$

The final expression of enzyme activity becomes:

$$62.9 \text{ µmoles RS / mg total protein / h.}$$

TABLE 32.6

Regression Output for Protein Data Using Excel and Quattro Pro

Output	Excel	Quattro Pro
Constant (Y-intercept or b)	0.029	0.022
Standard error of Y estimate	0.006	0.012
R^2	0.996	0.995
Number of observations	13	13
Degrees of freedom	11	11
X coefficient (slope or m)	0.992	1.015
Standard error of coefficient	0.019	0.021
Estimated equation	$Y = 0.99\,X + 0.029$	$Y = 1.02\,X + 0.022$

Procedure 32.8

Determination of Endocellulase or Endopectinase Activities Using Agarose Augmented with Either Carboxymethylcellulose or Pectin

Step	Instructions and Comments
1a	To determine endocellulase activity, dissolve 0.2 g of CMC in 500 mL of distilled water. Slowly add the CMC to the water with constant stirring; you may apply some mild heat (50°C) to the water to facilitate dissolving the CMC. Be sure that all of the CMC dissolves—no clumps.
1b	Alternatively, add 0.2 g of citrus pectin to 500 mL of cold (4°C) distilled water and stir until completely dissolved.
2	Add 10 g agarose (2%) to the CMC or pectin solution and microwave to melt the agarose. Pour about 15 mL of medium per 60-mm-diameter Petri dish and allow to harden.
3	Draw a circle with a permanent marker at the bottom of the Petri dishes to indicate where the agarose will be inoculated (see Figure 30.8a and b). Add 20 µL of boiled (15 min) commercial enzyme to the agarose surface within the border of the circle. Repeat this for the unboiled control and all fungal isolations. Complete three replications for all controls and samples.
4	Place the endocellulase dishes, agarose side up, in an incubator set at 30°C or 40°C for 2 h; for the endopectinases, go to Step 5b.
5a	For endocellulases, add enough 0.1% Congo red solution to cover the surface of the agarose and place on a rotary shaker (50 rpm). After 20 min, pour off any excess Congo red solution into a waste beaker, and add enough 1 M NaCl to cover the surface of the agarose. Clear zones underneath the inoculated area should appear after 10–20 min if endocellulases are present and have degraded the CMC sufficiently to not stain with Congo red. Pour off excess NaCl. Add 100 µL of glacial acetic acid to the surface—the use of a fume hood is advised as acetic acid can irritate eyes and other mucus membranes. A deep violet color in the agarose will develop quickly where CMC is present (Figure 32.8a).
5b	After the enzyme preparation has been absorbed, cover the dishes with 0.1 M malic acid solution and incubate at room temperature for 1 h. Rinse briefly with distilled water and pour ruthenium red solution over the agarose. After an overnight incubation at room temperature, pour off the stain and rinse briefly with distilled water. Note that the stain solution may be reused several times and stored for a month or more at room temperature in the dark. A clear zone (not staining red) will indicate endopectinase activity (Figure 32.8b).
6	Measure the diameter (mm) of the clear zone (usually circular) and plot the mean diameter against mg protein/mL of solution. Dishes may be photographed by placing on a light box that provides illumination from the bottom (see Figure 30.8a and b).

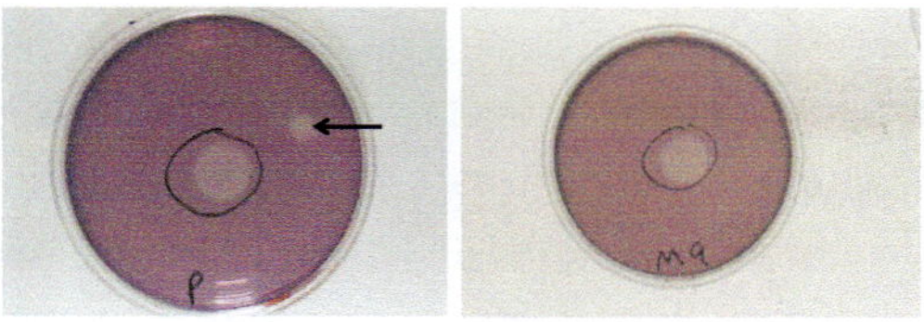

FIGURE 32.8 Detection of endohydrolase activity. (a) Endocellulase activity in agarose dishes. The cleared zone (loss of polymer structure for carboxymethylcellulose) after staining with Congo red indicates endocellulase activity present in crude protein isolated from sample *Pestalotia* species (P). During preparation of the sample, a few microliter of protein solution was dropped outside the target zone (arrow) and demonstrates activity. (b) Endopectinase activity in agarose dishes. The clear zoned (loss of polymer structure of citrus pectin) after staining with ruthenium red indicates endopectinase activity of a *Colletotrichum gloeosporiodes* isolate (M9).

- Activity is expressed as the diameter (mm) of cleared zone/mg crude protein—does this accurately account for differences in activities between various fungi? Hint: is all of the crude protein composed of endocellulases?
- Can you devise an experimental procedure to adapt this protocol for detecting endopectinases? Hint: see Procedure 32.9.

Experiment 4: Detection of Endopectinase Isozymes Using Polyacrylamide Gel Electrophoresis

Pectin polymers are composed of galacturonic acid residues (a hexose sugar acid) linked α-1-4. In some forms of pectin, methyl groups (CH_3) may be esterified to the carboxyl group on carbon 6. There are several classes of pectin-degrading enzymes, some that demethylate

Procedure 32.9

Electrophoresis of Proteins and Detection of Endopectinolytic Isozymes

Step	Instructions and Comments
1	Assemble the horizontal electrophoresis units (Protean III) (2) onto the casting stand. Be sure that the plates are free of dust and dried acrylamide—the gel will stick to the glass surfaces if contaminants on the glass plates are present. Also, it is helpful if a few pieces of Parafilm are cut to fit under the gray gaskets at the bottom of the casting stand. This ensures a good seal between the plates and gasket and will prevent leaks.
2	Wear gloves and lab coat. Weigh 8–10 mg of pectin and pour into a 25-mL beaker with a small stir bar. Pipette 4.0 mL of distilled water, 2.5 mL of resolving gel 4X buffer and 3.3 mL of 30% acrylamide solution into the beaker—this solution should be cool (4°C). Stir until all of the pectin dissolves. Essentially, this is a 10% acrylamide–pectin gel.
3	Pipette 100 μL of 10% ammonium persulfate and 10 μL of TEMED into the acrylamide solution while constantly stirring. Stir for 10 s, draw the solution into a 10-mL syringe and place a 0.22-μm filter at the end of the syringe barrel. Dispense the acrylamide solution between the sets of glass plates. Each gel unit will require about 3.5–4.0 mL of the acrylamide solution—fill to the top of the 0.75-mm gel rig; about 4.5 mL is needed for the 1.0-mm gel rig. Tap the outside glass of the gel units to remove any air bubbles. Insert the 10-tooth combs into the top slots of each gel unit—be certain that there are no air bubbles adhering to the teeth—reset the comb if necessary. Allow to stand for about 30 min or until the acrylamide is gelled. You can determine when the gel has set by observing the remaining solution in the beaker.
4	Clean excess acrylamide from the bottom of the unit with a laboratory tissue. Mount the gels on the central stand and place in the running buffer reservoir (clear plastic box). Fill the inner and outer spaces with running buffer taking care not to overfill the central reservoir. Remember to dilute the 300 mL running buffer stock to 600 mL with distilled water. Gently, remove the combs by lifting directly upward and rinse them in water. Clean the individual wells with a tuberculin syringe (without a needle) using the running buffer. Place the top on the apparatus and run at 100 V/cm^2 for 10 min.
5	Remove protein samples, including the commercial pectinase (CP) preparation, from the freezer and allow to thaw on ice. Mix well. Each sample should contain 25 μL of protein solution and 25 μL of loading buffer. Turn off the power to the electrophoresis unit, remove the top, and clean the wells again using running buffer and the tuberculin syringe. Load 2.5–10 μL of each sample into the wells—you may also "skip" wells between

the samples. The solution should sink to the bottom. There is really no need to change tips between samples. Remember that you are actually loading "backwards." So if you are planning to load five samples + CP, you should begin with sample CP and load it on the left hand side, continue with #5, then #4, and so on to #1. This will ensure that when the electrophoresis is completed, and when the gel is removed and rotated 180°, that the #1 sample will be on the left side and the CP sample on the right side as it faces you. Try not to use the outer lanes of the gels.

5a Note that you may also complete the experiment using a commercial preparation of pectinase (<0.1 mg/mL). Dispense 100 µL into two 0.65-mL Eppendorf tubes. Boil the first tube for 15 min to denature the pectinases and allow to cool to room temperature. Add 100 µL of loading buffer to both the tubes and vortex to mix. Add 5 µL of the boiled preparation to one lane. Change the tip, skip a well (lane) and add 10, 8, 6, 4, and 2 µL of the undenatured (not boiled) sample to the five wells to the right. The order should be as follows: blank, boiled sample, blank, 10 µL, 8 µL, 6 µL, 4 µL, and 2 µL.

6 Replace the top and set the power pack for constant 100 V/cm^2. Allow the tracking dye to migrate to within 1 cm or less of the bottom of the gel. Turn off power, remove the inner glass plate (the smaller one) with the aid of running water and cut a small piece of acrylamide at 45°C to mark the right side of the gel. This will help you keep track of the sample order. Remove the gels from the outer or large glass plates with running water and place in a staining tray (one gel per tray). Pour off any water and cover with 0.1 M malic acid solution at room temperature for 1 h.

7 Rinse briefly with distilled water, and pour ruthenium red solution over gels and allow to stand at room temperature overnight, and then rinse with distilled water. The stain solution may be reused several times. The stain may be stored for a month or more at room temperature in the dark.

8 Note cleared zones (no stain), which may be easier to see with transmitted white light. These are areas where endopectinases migrated during electrophoresis. Draw the patterns for each lane or organism.

pectin, and others that only act at either low (5.0) or high (9.0) pH. Some of the enzymes remove single galacturonic acid units from the end of the pectin polymer (exoenzymes), whereas other enzymes will act at random along the interior length of the pectin polymer, reducing the polymer to a series of smaller chains of variable lengths (endoenzymes). In the previous experiment with cellulase, activity of exocellulases was measured primarily by detecting RSs with DNSA reagent. In contrast, this experiment is designed to detect the activities of individual **isozymes** of endo-pectin methyl galacturonase (endo-PMG) and/or endo-polygalacturonase (endo-PG). Both of these enzymes typically are highly active at pH 5.0.

Isozymes

Isozymes (iso = the same or equal; zyme = enzyme) can be conveniently defined as molecular weight variants of the same functional enzyme that can occur in the same isolate or between different isolates of fungi, bacteria, or other organisms, including plants and animals. Although these enzymes are of different weights, they essentially perform the same task (e.g., degradation of pectin), albeit the specific activities (e.g., how "fast" they work under specified conditions) may be quite different when compared with one another. Isozymes are usually detected by gel electrophoresis, a technique used to separate different molecular weight compounds, such as protein or isoenzymes. A gel is a porous support medium composed of either agarose or polyacrylamide (potato starch may also be used) that permits the migration of charged molecules in an electric (DC) field. Migration of proteins (in our case) through the gel is dependent on the strength of the electrical field (V/cm^2), the composition of the gel (% agarose or acrylamide), which dictates the pore size, and the net charge on and size of the isozymes. The charge on the isozymes is controlled by the pH of the buffer and when all other parameters of electrophoresis remain constant, migration through the gel is dependent on the size of the molecules. Typically, most of the proteins have a net negative charge and migrate toward the cathode or positive pole. However, one must be cognizant of the fact that some proteins, and potentially some of interest, might have a net positive charge in the buffer, move in the opposite direction or toward the anode, and eventually be lost from the gel. The smaller (lower molecular weight) proteins migrate more rapidly and thus end up closer to the cathode (positive pole) at the bottom of the gel compared with large (higher molecular weight) proteins. For a more complete but relatively simple explanation of electrophoresis, refer Trigiano et al. (2008).

The activities of many of the endoenzymes are typically measured by noting the change in viscosity of a standard solution containing the appropriate polymer (or modified polymer) over time or in a substrate containing agar medium. The viscosity of the solution decreases as the endopectinases cleave the polymer into smaller units. However, this methodology determines the total activity of all the endoenzymes, and nothing can be said of how many isozymes are present in the mixture. Another way of detecting activity of endopectinases is to take advantage of the fact that intact pectin or modified pectin polymers incorporated into acrylamide gels stain red-pink with ruthenium red. Electrophoresis of the crude protein will deposit the pectinolytic isozymes at various locations in the gel. If the gel is incubated under the proper conditions (pH, temperature, etc.), then the endoenzymes will degrade the pectin incorporated into the gel so that it will no longer stain with ruthenium red. Each "cleared" or "non-stained" band in each of the lanes represents an isozyme or comigrating enzymes of various endopectinases. Isozyme production from various isolates of the same species or species of different genera can be compared and physiological relatedness, for example, estimated.

In this experiment, endopectinases found in 5- and 10-day-old cultural filtrates of various fungi or bacteria will be isolated with total proteins using the same methods that were used to isolate cellulolytic enzymes. Instructors may opt to complete the experiment up to and including isolating the protein from the cultures and/or pouring the gels.

Follow the instructions in Procedure 32.2 to grow the fungi and isolate protein, except substitute pectin for cellulase. Also in Step 30.7, pipette 25 μL of the isolated protein into several 0.6-mL Eppendorf tubes and 25 μL of loading buffer. Place the samples to be used during the laboratory period on ice, and freeze the remaining samples at –70°C. These samples can be thawed only once before the enzymes are denatured and lose substantial activity. Now, complete Procedure 32.9 to finish this experiment.

Materials

The class will require the following materials to complete the laboratory:

- Five to eight fungal isolates grown on nutrient medium described in Table 32.1.
- Liquid pectinolytic induction medium (Table 32.7).
- Three 125-mL Erlenmeyer flasks of the above medium for each fungal isolate.
- Alcohol lamp.
- Sterile plastic straws of 6- to 8-mm-diameter.
- Scalpels fitted with # 11 blades.
- Incubator at 25°C (other temperatures may be required for specific organisms).

TABLE 32.7

Pectinase Induction Medium[a]

Component	Amount
KCl	1.0 g
$(NH_4)_2SO_4$	2.5 g
$MgSO_4 \cdot 7H_2O$	0.2 g
$CaCl_2 \cdot 2H_2O$	0.5 g
Water	1 L
Citrus pectin	5.0 g

[a] Add all the components of the medium except sodium pectate to 900 mL of water and adjust pH to 5.3. Chill to 4°C in refrigerator, then with constant vigorous stirring, slowly add citrus pectin. Break clumps with a glass rod. Stir until dissolved, autoclave, and dispense 35 mL of medium into 125-mL screw top flasks.

- 250-mL graduated cylinder
- Ammonium sulfate—about 750–1000 g.
- 250-mL beakers, magnetic stir bars, and stir plates.
- 25-mL beakers and magnetic stir bars.
- Dialysis tubing (Spectrum Laboratories, Inc. Rancho Dominguez, CA; 800-445-7330) 3500 exclusion and clamps (suture string or binder clips will also work).
- Ice bath.
- Table top centrifuge (3,000–10,000 g—does not need to be refrigerated).
- 50-mL centrifuge tubes (do not need to be sterile).
- Top loading balance.
- Squeeze bottle of distilled water.
- Large beaker (4 L) with distilled water at 4°C
- Refrigerator.
- Pectin, glycerol (37°C), and 0.01% (10 mg/100 mL) aqueous solution of bromophenol blue tracking dye.
- Acetonitrile gloves, goggles, and lab coat.
- 2X loading buffer: 80 mL distilled water; 40 mL 0.5 M Tris–HCl (7.85 g Tris-HCl in 70 mL distilled water; adjust the pH to 6.8 with NaOH, and bring to 100 mL with distilled water); 32 mL glycerol (much easier to measure if at 37°C), and 1–4 mL 0.01% bromophenol blue. Mix thoroughly and store at room temperature.
- 4X resolving gel (1.5 M Tris) buffer. Dissolve 18.2 g Tris Base in 70 mL of water. Adjust pH with 6 N HCl to 8.8. Bring to 100 mL with water. Store at room temperature.
- 30% acrylamide solution (29.2 acrylamide dissolved in 60 mL of water; add 0.8 g BIS (N,N'-methylene-bis-acrylamide) and bring to 100 mL in a graduated cylinder—store in refrigerator at

4°C). CAUTION: Wear gloves, particle mask and laboratory coat. Clean up any spills immediately.

- Pectinase (commercial preparation) <0.1 mg/1 mL water. Suggest: Pectolase Y-23 Seishin Corporation, Tokyo, Japan and/or Macerozyme R-10 (contains both pectinases and cellulases), Sigma. Dispense 25 µL of the protein and 25 µL of loading buffer into each of several of 0.6-mL Eppendorf tubes.
- Running buffer (dissolve 9.8 g Tris-base and 43.2 g glycine in 700 mL of water and bring to 1 L). To use, dilute 300 mL of this stock with 600 mL distilled water. Store at room temperature.
- Horizontal electrophoresis unit (Protean III) with either 0.75 or 1.0 mm combs and power supply.
- TEMED and 10% ammonium persulfate (1 g in 10 mL of water: aliquot 1 mL into each of 10, 1.5-mL Eppendorf tubes and freeze at −20°C until use).
- 10-mL disposable plastic pipette with pump or bulb.
- 0.1 M malic acid solution (dissolve 13.4 g of malic acid in 700 mL of water and bring to a final volume of 1 L). Store at room temperature.
- 0.01% (100 mg in 1 L of water) ruthenium red aqueous solution. Store at room temperature.
- Micropipette (2–20 µL and 200 µL or equivalent) and disposable tips.
- Plastic staining trays.
- Hot plate with boiling water.

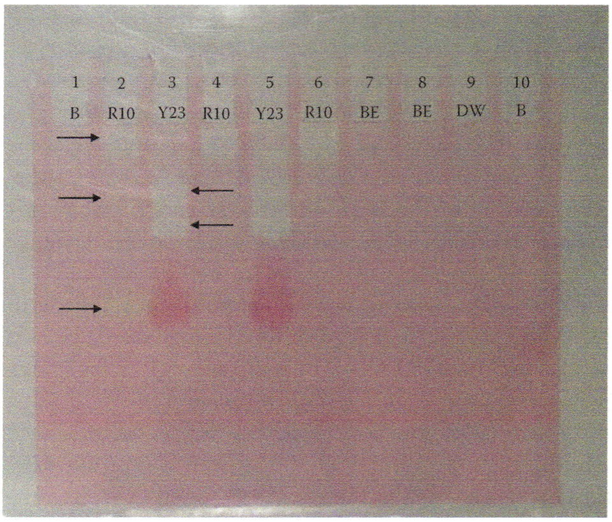

FIGURE 32.9 Polyacrylamide gel electrophoresis of endopectinolytic isozymes from Y-23 and Macrozyme R-10 pectinases. Proteins were electrophoresed for 1 h then incubated in 0.1 M malic acid. Gels were stained in 0.01% ruthenium red overnight at 4°C and then washed with water for 1–2 h. Gels were then viewed on a light box. Arrows indicate cleared (nonstained) areas in the gel and correspond to endopectinase activity. B = Blank (lane 1), R10 (lane 2) = 5 µL of R10, Y23 (lane 3) = 5 µL of Y23, R10 (lane 4) = 2.5 µL of R10, Y23 (lane 5) = 2.5 µL of Y23, R10 (lane 6) = 1 µL of R10, BE (lane 7) = boiled R10 (5 µL), BE (lane 8) = boiled Y23 (5 µL), DW (lane 9) 5 µL of distilled water, and B (lane 10) = blank.

Anticipated Results

See the anticipated results for growth of organisms and isolation of protein with cellulases in Experiment 2. Often the medium containing pectin is cloudy after autoclaving, but will become crystal clear after 7–10 days of fungal growth (the uninoculated medium will remain cloudy). Clearing of the medium probably signifies that endopectinases were produced by the fungi.

Electrophoresis of the proteins isolated from the pectinase induction liquid medium generally takes about an hour. After soaking the gels in 0.1 M malic acid and then staining overnight, cleared or colorless areas in the gel should be apparent (Figure 32.9). Some of these areas will appear as discrete lines or bands, whereas other will be diffused and the entire lane cleared. The "cure" for this condition is to load less protein (less volume) and/or heat the crude protein to 60°C for 1–5 min (determine empirically) before electrophoresis, which lessens enzyme activity during electrophoresis and allows detection of specific bands or isozymes (Willis et al., 2010). Some areas will appear to be extended outside of the lane, which generally indicates that the specific enzymes are relatively abundant or the lane was "overloaded." The "cure" for this condition is to load less protein or less volume. However, if less protein is loaded there is a danger that less abundant or rare isozymes will not be detected. A compromise is to load the gel as explained in the exercise, but additionally, load and run other gels. The other gels could be loaded with a series of less volumes (e.g., 5, 2.5, and 1 µL) or the protein content of all of the protein preparations could be determined after dialysis. Generally when using the smaller gels in the Protean III, 1–2 µg of protein per lane would be sufficient to detect most isozymes. If a larger gel is used, then each lane should be loaded with 3–5 µg of protein.

Questions

- What effect would changing the pH of the incubation solution have on endopectinase activity?
- Do you think that the isozyme "profile" would change over time—for example, if you harvested proteins after 1 week and after 3 weeks—would the "profiles" be the same or different?
- What are the advantages (or disadvantages) of producing isozymes?
- How could you adapt the procedure for detecting endopectinases; to detect endoamylases?

- If the liquid medium remains cloudy throughout the incubation period, do you think there is any endopectinase activity? Why might the medium appear cloudy despite enzyme activity?

REFERENCES

Cruickshank, R. H., and J. I. Pitt. 1987. Identification of species in *Penicillium* subgenus *Penicillium* by enzyme electrophoresis. *Mycologia* 79:614–620.

Davidson, R. W., W. A. Campbell, and D. J. Blaisdell. 1938. Differentiation of wood decaying fungi by their reaction on gallic or tannic acid medium. *J. Agric. Res.* 57:682–695.

Lowry, O. H., N. J. Rosenbrough, A. L. Farr, and R. J. Randell. 1951. Protein measurement with the Folin phenol reagent. *J. Biol. Chem.* 193:265–275.

Miller, G. L. 1959. Use of dinitrosalicylic acid reagent for determination of reducing sugar. *Anal. Chem.* 31:426–428.

Reese, E. T., and M. Mandels. 1963. Enzymic hydrolysis of cellulose and its derivatives. In: D. L. Whisler (ed). *Methods in Carbohydrate Chemistry*. New York, NY: Academic Press, pp. 139–143.

Sierra, G. 1957. A simple method for the detection of lipolytic activity of microorganisms and some observations on the influence of the contact between cells and fatty substrates. *Antonie Van Leeuwenhoek Ned. Tijdschr. Hyg*. 23:15–22.

Society of American Bacteriologists. 1957. *Manual of Microbiological Methods*. New York, NY: McGraw-Hill Book Company.

Trigiano, R. N., B. H. Ownley, A. N. Trigiano, J. Coley, K. D. Gwinn, and J. K. Moulton. 2008. Two simple and inexpensive laboratory exercises for teaching agarose gel electrophoresis and DNA fingerprinting. *HortTechnology* 18:177–187.

Willis, J. D., W. E. Klingeman, C. Oppert, B. Oppret, and J. L. Jurat-Fuentes. 2010. Characterization of cellulolytic activity from digestive fluids of *Dissosteira carolina* (Orthoptera: Acrididae). *Comp. Biochem. Physiol.* 157:267–272.

33 Molecular Tools for Studying Genetic Diversity in Plant Pathogens

Timothy A. Rinehart, Denita Hadziabdic, Phillip A. Wadl, and Robert N. Trigiano

Concept Box

- All living organisms contain DNA that can be assayed by a variety of methods to answer important questions about plant and pathogen diversity.
- DNA fingerprinting techniques such as DAF, RAPD, and AFLP utilize arbitrary priming and do not require prior knowledge of the organism's genome.
- SSR and SNP methods rely on DNA sequence data and provide more detailed information.
- High-throughput or whole-genome DNA sequencing is rapidly becoming an affordable and useful tool for studying genetic diversity, including pathogens.
- There are a variety of free, web-based statistical programs for analyses of population genetics.

The ubiquitous nature of DNA is a central theme for all biology. The nucleus of each cell that makes up an organism contains **genomic DNA**, which is the blueprint for life. The differential expression of genes within each cell gives rise to different tissues, organs, and, ultimately, different organisms. Changes in genomic DNA give rise to functional advantages that make some individuals more successful than others. Success, as measured by the ability to reproduce, dictates that organisms that accumulate useful mutations in their genomic DNA will be more likely to pass those changes on to future generations. Heritable mutations are the **genetic variation** that is visualized by molecular tools. In this chapter, we discuss the many different forms of genetic variation, current molecular methods, programs for characterizing **genetic variation**, and possible questions concerning genetic diversity of plants and pathogens that can be answered with molecular tools.

The methods that we discuss rely heavily on understanding DNA, so it is worthwhile to review its structure. DNA is made up of the following four nucleotide bases: adenine (A), cytosine (C), guanine (G), and thymine (T), which are covalently linked together by a sugar (deoxyribose)-phosphate backbone into a long polymer. Strands of DNA are linear and directional, that is, they are read from 3' to 5' along the sugar-phosphate backbone (Figure 33.1). Within the nucleus, two strands of DNA intertwine to form a double-helix structure where G bonds with C and A bonds with T to form **base pairs (bp)**. Complementary strands encode the same information, a redundancy that ensures fidelity during DNA replication. Under most conditions, complementary strands are annealed to each other via hydrogen bonding. During cell division, complementary DNA strands are split apart by DNA polymerase enzymes and new DNA is synthesized from each of the template strands (Figure 33.1).

Genetic information is determined by the sequence of nucleotides. Regions of DNA that encode functional products are called **genes** and consist of trinucleotide units called **codons**, each of which codes for 1 of 22 possible amino acids. Genes are transcribed by cellular enzymes into a temporary nucleotide monomer called **messenger RNA (mRNA)**. Proteins are produced by machinery that translates the codon sequence of the mRNA and assembles the corresponding amino acids into linear chains. As amino acids are linked together via peptide bonds, they often fold into complex structures that can be combined with other protein subunits, transported, embedded in cellular compartments, or modified to become functional units. Functional proteins are the cellular machinery that makes tissues, organs, and organisms what they are. The unidirectional flow of genetic information from DNA to mRNA to protein, often called the **central paradigm of molecular biology**, is critical for understanding how

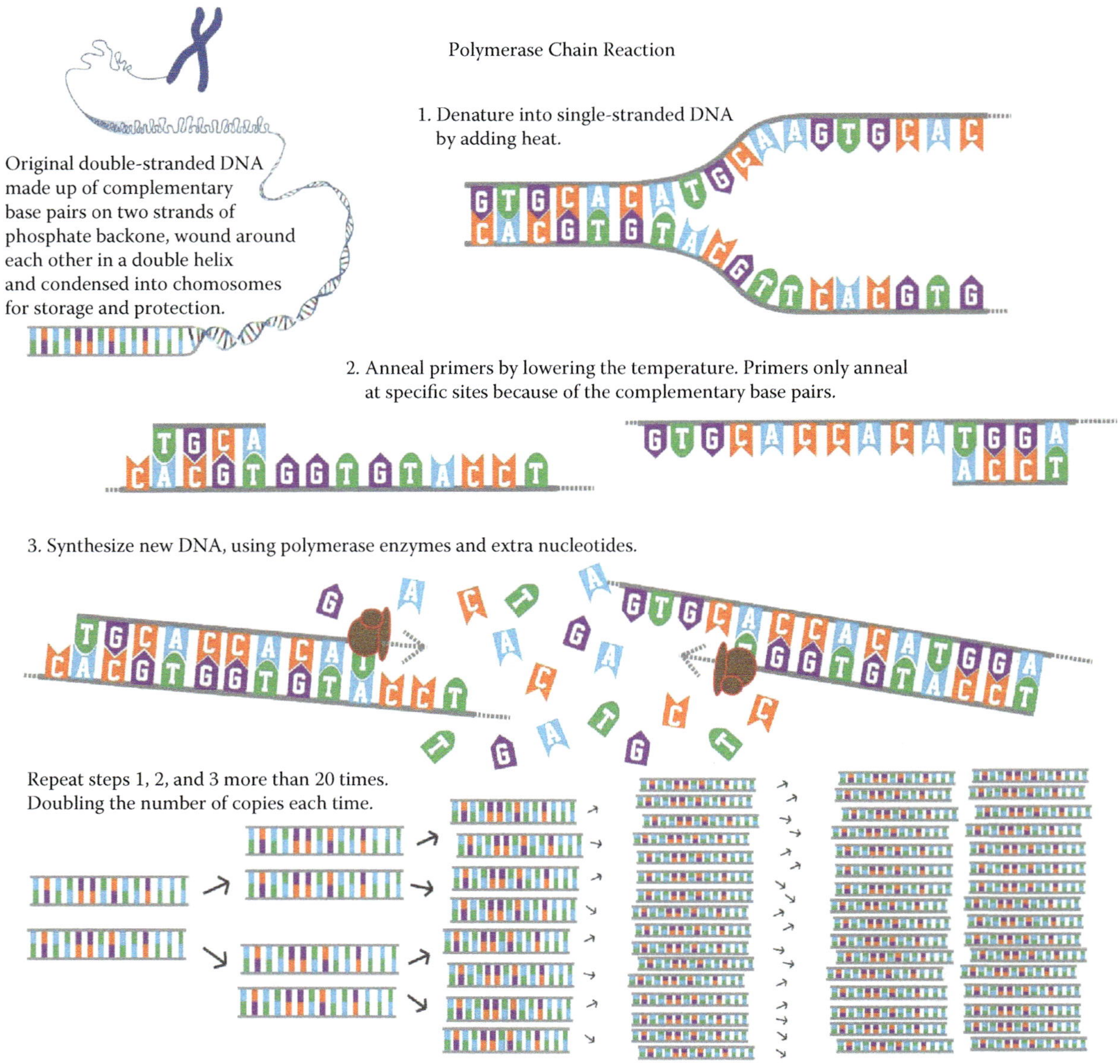

FIGURE 33.1 The complementary structure of DNA and polymerase chain reaction (PCR) amplification of DNA fragments. Nucleotide bases are paired with each to form base pairs (bps) with A and T, and G and C as partners. Primers are short DNA fragments that anneal to specific regions of DNA according to the complementary bp, which serves as a starting point for DNA polymerase to synthesize more DNA. Synthesis proceeds directionally. When PCR primers are designed for both DNA strands facing each other, the resulting cycles of template DNA denaturation, primer annealing, and synthesis of new DNA result in exponential amplification of discreet regions of DNA defined by the primers.

variation in genomic DNA can produce phenotypes that are subjected to natural selection.

The type of genetic variation observed between individuals is highly dependent on where you look within the genome. Genes contain 64 possible combinations of three nucleotide codons that only correspond to 22 possible amino acids. Thus, there is considerable redundancy in codon usage. Altering, or mutating, one nucleotide changes the genomic DNA sequence, but not necessarily the amino acid sequence of the protein. These types of mutations are called **silent mutations** because the phenotypic effect is not apparent and, therefore, not subject to natural selection. Mutations that alter the codon so that a different amino acid is incorporated into the peptide sequence are called **missense mutations**. Mutations can also consist of nucleotide **insertions** or **deletions**. One or two nucleotide insertions or deletions, known as **frameshift mutations**, can disrupt the trinucleotide codon sequence. These mutations typically disrupt protein synthesis and are not observed as often as other mutations. This does not necessarily mean that frameshift mutations occur at a lower rate, just that they are less likely to be passed on due to their deleterious effects. Different regions of the genome demonstrate distinct patterns in the type of mutations they accumulate and the rate at which mutations are observed. Where to look for genetic variation and the types of variation observed plays an important role in how genetic data are analyzed and the conclusions can be made about plant diversity.

Not all genomic DNA codes for cellular machinery. Many DNA sequences are associated with packaging the genome into **chromosomes** and serve as recognition sites for DNA-binding proteins. Other DNA is only useful as spacing between genes or as a buffer against DNA loss. Noncoding regions, especially those with no apparent function at all, typically accumulate mutations at a higher rate. The types of mutations observed vary widely among noncoding sites but often include large insertions and deletions that would not be tolerated in coding regions. Eukaryotic genomes also contain large amounts of mobile DNA, which appear to encode proteins solely for the purpose of copying and inserting additional copies of the DNA, the phenotypic effects of which may not be readily apparent. Regardless, all types of DNA play a vital role when assaying genetic diversity between individuals or populations. In this first part of the chapter, we focus on methods that can be used to genetically identify and describe diversity of plant pathogens. In the second half, we focus on the types of questions that can be answered by a molecular approach to understanding genetic diversity of plant pathogens.

GENETIC IDENTIFICATION AND CHARACTERIZATION OF PLANT PATHOGENS

DNA variation can be used to distinguish between all taxonomic levels of plant pathogens including individuals, populations, strains, species, genera, and families. DNA fingerprinting has significantly accelerated the important task of identifying plant pathogen. If a pathogen has been previously described, new isolates can often be identified from field samples without laboratory culturing, which can be critical for some fungal, microbial, and viral pathogens. Plant pathogen systematics, or the study of pathogen diversity and relationship among pathogens over time, utilizes many different DNA fingerprinting methods to identify and classify plant pathogens or evaluate relationships between pathogens and groups of pathogens. Genetic characterization of plant pathogens can also answer questions regarding population structure, pathogen movement, modes of reproduction, modes of dispersal, and efficacy of chemical and cultural controls.

Some of the most popular molecular tools (Table 33.1) to identify and genetically characterize plant pathogens are arbitrarily primed techniques such as DNA amplification fingerprinting (DAF), random amplified polymorphic DNA (RAPD), and amplified fragment length polymorphism (AFLP). These protocols do not require prior knowledge of the organism's genome. All the three methods make use of **polymerase chain reaction (PCR)** to amplify large amounts of specific DNA from small amounts of total genomic DNA (Mullis and Faloona, 1987). PCR primers are short segments of human-made single-strand DNA that anneal to specific regions in the genome based on the complementary DNA sequence. DNA polymerase synthesizes new DNA using the genomic DNA as a template. When this reaction is repeated several times, the DNA fragment between the PCR primers is exponentially amplified (Figure 33.1). Thermal-stable DNA polymerase is used so that the template DNA can be melted apart (passive denaturation) with heat without destroying the function of the enzyme. A single PCR cycle consists of several seconds at high temperature to denature the DNA, followed by a low temperature to anneal the primers, and finally a period of optimum temperature for the DNA synthesis, usually 68°C–72°C. PCR amplification of short regions, typically less than one kilobase (1 kb = 1000 nucleotides), is more robust than longer amplifications, which require specialized amplification protocols.

DAF and RAPD utilize arbitrary primers of short length, often 12 bases or less, that anneal throughout the genome (Welsh and McClelland, 1990; Williams et al., 1990). The sequence of the primers is random and the probability of two primers annealing within 1.5 kb of each other can be calculated based on the size of the genome being assayed. However, genomes are not composed of random DNA sequences, so primer pairs must be empirically tested. After PCR using arbitrary primers, the DNA fragments can be separated based on length using acrylamide or agarose gels or other inexpensive equipment. Optimization of DAF or RAPD is affordable, especially since hundreds of discreet loci may be amplified during a single PCR amplification. The banding patterns produced indicate genetic similarities and differences between samples (Figure 33.2). Same-sized PCR fragments produced under identical cycling conditions indicate genetic similarity, or shared DNA. Fragments that are different between samples, or polymorphic, suggest changes in one organism when compared with other. Missing bands can be due to nucleotide changes in the primer annealing site, which would eliminate the production of the polymorphic fragment, or an insertion/deletion mutation in the region to be amplified, which would change the size of the fragment. PCR fragments that are unique to a group of samples suggest inheritance of specific DNA and can be used to reconstruct the genetic

TABLE 33.1
Comparison of Molecular Techniques

Technique	Prior Knowledge	Type of Information	Use/Comparison at Level	Difficulty/Expense
DAF	None, PCR uses arbitrary primers	Anonymous bands or loci, dominant data	DNA profiling of individual isolates from the same population or same species	Easy and inexpensive
RAPD	None, PCR uses arbitrary primers	Anonymous bands or loci, dominant data	DNA profiling of individual isolates from the same population or same species	Easy and inexpensive
AFLP	None, PCR uses known primers annealing to linker sites after restriction digestion	Anonymous bands or loci, dominant and codominant data	DNA profiling of individual isolates from the same population, species, or genera	Moderately easy and inexpensive
SSR	DNA sequence of random SSR loci	Allele size variation of known loci, codominant data	DNA profiling of isolates from the same species, genera, or higher order taxa	Moderately difficult, expensive to develop
DNA sequence	DNA sequence data of specific gene(s)	DNA base mutations at known loci, codominant data	DNA profiling of isolates from the same species, genera, or higher order taxa	Moderately difficult, expensive to develop
SNP	DNA sequence data of multiple genes	DNA base mutations at multiple loci, codominant data	DNA profiling of isolates from the same species, genera, or higher order taxa	Moderately difficult, moderately expensive
Whole genome sequencing	DNA sequence data of large genomic regions and/or chromosomes	DNA base mutations at multiple loci, codominant data	DNA profiling at all taxonomic levels	Computationally intensive and expensive

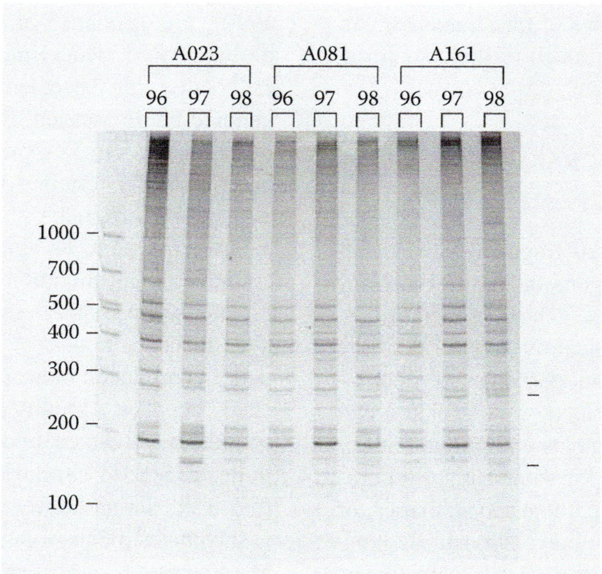

FIGURE 33.2 DNA amplification fingerprinting (DAF) analysis of the fairy ring fungus (*Marasmius oreades*). DNA extracted from the fruiting bodies of several isolates was amplified by using the oligonucleotide primer GTATCGCC; generated fingerprints separated using polyacrylamide gels and fragments stained with silver. Isolates are labeled according to fairy ring clone and the year they were collected. Molecular markers are indicated by base pairs, and selected polymorphic bands are indicated. (From Abesha, E. and Caetano-Anolles, G., Studying the ecology, systematics, and evolution of plant pathogens at the molecular level. In Trigiano, R. N., M.T. Windham. and A.S. Windham [eds.], *Plant Pathology Concepts and Laboratory Exercises*, 2nd edition, Boca Raton, FL: CRC Press, 2008.)

relatedness of individuals and estimate genetic diversity within and between populations.

Reproducibility is often cited as a disadvantage of DAF and RAPD analyses. This is understandable since slight differences in temperature or reagents might bias the amplification process toward or away from certain PCR fragments. When this bias is multiplied by the enormous number of fragments that can be produced, there is considerable justification for being cautious about DAF and RAPD results. However, the overall conclusions from DAF and RAPD analyses have proven reliable as long as the necessary controls are observed (Brown, 1996). Because of the low cost and virtually infinite combinations of primers and amplification conditions, researchers with enough dedication and time can compare an exhaustive number of loci between individuals. Randomly sampling genomic DNA includes comparisons between coding, noncoding, and mobile DNA, which increases the chances of finding a difference since some regions of the genome are more prone to mutation than others. Thus, RAPD and DAF potentially offer greater resolving power between highly related individuals, even when evaluating clonally reproduced pathogens such as the causal agent for Dutch elm disease (*Ophiostoma novo-ulmi*) (Temple et al., 2006) or *Discula destructiva*, the cause of dogwood anthracnose, isolates in the United States (Trigiano et al., 1995).

AFLP is based on restriction fragment length polymorphism (RFLP) where genomic DNA is cut into small segments by restriction endonuclease enzymes. The resulting DNA fragments can be visualized by radiolabeled probes made from known genes (Vos et al., 1995). AFLP employs the same restriction endonucleases to digest genomic DNA, but then utilizes PCR to selectively amplify hundreds of discrete fragments. Unlike DAF and RAPD, which uses random primers, AFLP primers are specific sequences designed to anneal to human-made linker DNA that is attached to the ends of the pieces of fragmented genome. Linkers are annealed to the overhangs left behind by the restriction endonuclease enzymes and attached by ligating them to the sugar-phosphate backbone. The resulting pool of DNA contains small fragments, typically 0.5–2.0 kb, which can be PCR-amplified using primers designed from the linker DNA sequence. Primers are radioactive or fluorescently labeled so that amplified fragments can be visualized after being separated by length, which can differ by as little as a single nucleotide. Since there may be hundreds to thousands of amplifiable and detectable DNA, additional bases are sometimes included in the primer at the 3' end to further reduce the complexity of the amplified DNA. The resulting DNA fingerprint consists of size-separated fragments for each sample that can be compared side by side. Much like RAPD results, fragments that are present or absent in one sample but not the other suggest a genetic difference. Same-sized fragments produced by both organisms indicate genetic similarity. Researchers can choose from a number of restriction endonucleases and primer combinations, which potentially produce hundreds of fragments during a single PCR amplification. Thus, AFLP generates robust DNA fingerprints from loci across the entire genome, but uses specific PCR primers, which increase reliability and repeatability. AFLP has been adapted to run on automated capillary array sequencing instruments using fluorescent labels for greater throughput and automated data analysis (Figure 33.3).

AFLP, DAF, and RAPD produce dominant markers. Binary data are tabulated by identifying DNA fragments that are present (1) or absent (0). The exact nature of the genetic mutation creating differences in the DNA fingerprints is unknown. Genetic similarity between samples and phylogenetic inference regarding shared ancestry are based solely on the mathematical frequency of DNA fragments, not biological models for genome evolution. Amplified DNA fragments that are unique to a specific individual or population can be purified, ligated into plasmid vectors, and cloned into *Escherichia coli*. The plasmid DNA can then be sequenced to identify the underlying nature of the polymorphism. These sequenced regions are referred to as sequence characterized regions (SCARs), which, once described, can be exploited as codominant markers. PCR primers are designed flanking the mutation such that the amplified DNA fragments can be visualized, either by size variation or DNA sequence differences depending on the nature of the mutation, and recorded as alleles specific to each sample. In a sexually reproducing diploid organism, researchers expect to see two alleles per sample, one corresponding to the maternal chromosome and another allele from the paternally contributed chromosome. Conversely, haploid organisms, like many fungi, are visualized as a single allele per sample. Codominant data are not scored as a binary (absence or presence) but as an allele variation. Individuals who contain identical alleles for a SCAR have shared ancestry, whereas different alleles indicate genetic divergence. Codominant data are considered more informative because every PCR amplification produces DNA fragments and a technical failure during PCR amplification cannot be mistakenly scored as absence of a fragment (0).

One technique that generates codominant data is simple sequence repeat (SSR) markers, also known as microsatellite loci (Tautz, 1989). Eukaryotic genomes contain large amounts of repeated DNA, some complex like transposable elements and some simple. Simple repeats generally consist of motifs that are 1–4 bps long and repeated in tandem 10–100 or more times. These repeats have a tendency to change in number when DNA is replicated due to a phenomenon known as DNA polymerase slippage. PCR primers adjacent to an SSR region can amplify the repeat. Size differences in the repeat length

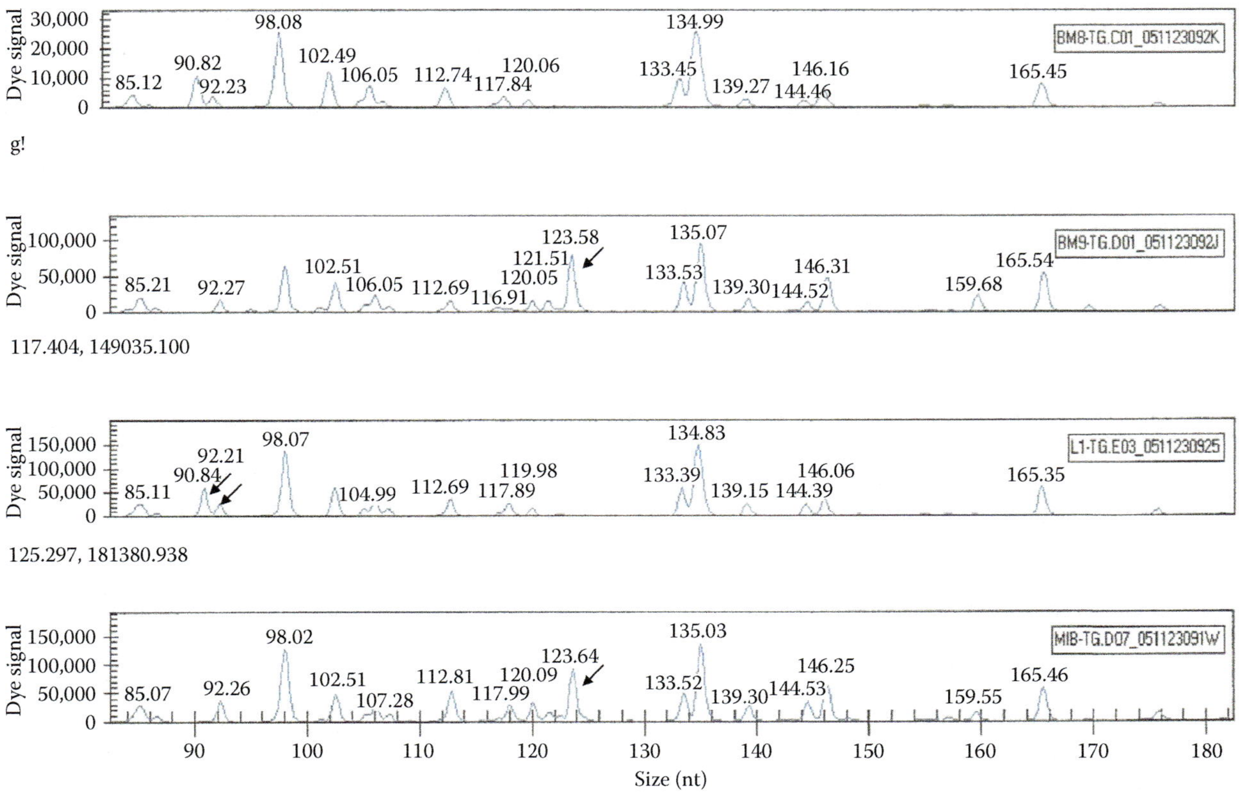

FIGURE 33.3 AFLP electropherogram produced for *Fusarium oxysporum* isolates. Arrows indicate examples of polymorphisms. (From Trigiano, R. N., M.T. Windham. and A.S. Windham [eds.], *Plant Pathology Concepts and Laboratory Exercises*, 2nd edition, Boca Raton, FL: CRC Press, 2008.)

can be visualized by radiolabel or fluorescent molecules incorporated into the PCR products during amplification. Because they are uniformly spread around the genome and some mutate faster than others do, SSRs are robust molecular markers. SSRs are codominant markers since each copy of the repeat is amplified and the maternal and paternal contributions in a diploid organism can be visualized as separate alleles if they differ in repeat number. In this way, SSR markers are particularly suited for diploid organisms (2*n*), such as species of the Oomycota. For haploid organisms (*n*) including many fungi, amplicons are visualized as a single allele, since they contain only one set of chromosomes.

Because SSR markers are specific types of DNA, they are not present in all pathogen genomes. Therefore, viral and bacterial plant pathogens are less likely to produce results. The main disadvantage to the SSR technique is cost. In order to develop SSR markers, researchers must locate and sequence SSR regions before they can develop specific primers for the organism they want to DNA fingerprint. Cross-amplification of SSR markers between genera has been documented for related plant and plant pathogen species, making SSR development more affordable (Dutech et al., 2007; Peakall et al., 1998; Perumal et al., 2008; Trigiano et al., 2012). Despite the extra requirements, SSRs are widely used because data can be easily reproduced and verified by sequencing the amplified products. There are also specific models for the evolution of SSR regions that can be used during data analysis for more accurate conclusions about relatedness and shared ancestry. For example, a trinucleotide repeat consists of 3 bp units. Changes in allele size can be weighted during analysis such that a 15-bp change is weighted more than a 3-bp change because it is likely that multiple slippage events, or more than one mutation, contributed to the 15-bp variant.

The ultimate molecular tool for comparing genetic variation between pathogens is DNA sequencing of the entire genome for each sample. Such an experiment would have been cost prohibitive a decade ago, but recent advances in high-throughput sequencing have lowered costs significantly. However, most DNA sequencing experiments focus on only a few genes. DNA sequence variation can range from highly conserved to highly variable, and researchers often use different regions of the genome to answer different questions. Conserved DNA sequences are more appropriate for evaluating higher-level relationships such as comparing genera or families, whereas regions that are more variable are appropriate for comparing individuals and populations. Public databases such as GenBank (www.ncbi.nih.nlm.gov) contain more than 100 gigabases of DNA sequence data and computational tools to search for analogous DNA sequences, or sequences that share a high level of similarity (Benson et al., 2006). Gene sequences, particularly conserved gene sequences, from related taxa can be used to design PCR primers to amplify the same DNA regions in known and unknown plant pathogens. DNA sequences from species of interest can then be compared with related organisms in order to estimate genetic diversity and relatedness. Studies using conserved loci for species identification are also cataloged in GenBank, making it possible to classify unknown organisms based solely on DNA sequence compared with previously identified plant pathogens (Rinehart et al., 2006). This work builds upon the collective research of others with the expectation that researchers will add their own DNA sequences once studies are published. Genes commonly sequenced include elongation factor genes, tubulin genes, and other universally conserved eukaryotic sequences.

When more variation is desired, which may be necessary for the identification of populations or species, noncoding regions of the genome are more useful since they accumulate mutations rapidly. The optimum scenario is a short hypervariable region sandwiched between conserved DNA sequences such that PCR primers can be designed to anneal to the conserved regions and amplify the more variable internal DNA. For example, internal transcribed spacer (ITS) regions of ribosomal DNA (rDNA; Figure 33.4) are short sections of hypervariable DNA located adjacent to conserved 5.8S, Small Subunit (SSU), and Large Subunit (LSU) regions (White et al., 1990). PCR amplification is robust using universal primers, in part because the primers anneal to the conserved regions and because rDNA is repeated in tandem and found in high copy number per cell. ITS variants containing base changes, nucleotide insertions, and deletions are usually prevalent, sometimes even among individual samples in a population. ITS sequence data for many fungi are available publicly for comparison, which increases the confidence in new results and reduces the labor involved. While DNA sequencing costs more than other molecular methods, sequence data can be easily shared and verified, and objectively appraised to make sure the quality of the data is high.

There is a wide array of genes that could be amplified with PCR and sequenced. Each sample analyzed requires a single genomic DNA extraction, but multiple PCR amplifications and DNA sequencing depending on the number and size of the genes to be analyzed. For this reason, most experiments only utilize a few regions of the genome. Moreover, valid conclusions often require DNA sequencing of related organisms or previously characterized reference plant pathogens in order to compare DNA sequences from unknown pathogens to reference samples. Unless these data already exist in databases, researchers are obligated to repeat the DNA sequencing on many other related pathogens in order to align the DNA sequences and compare genetic variation. This can rapidly increase the number of samples and escalate costs. The greatest advantage of DNA sequencing, aside from the detail inherent to DNA sequence variation, is that sophisticated models for DNA evolution can be incorporated into computer analyses of genetic diversity making conclusions based on DNA sequence data statistically testable. Decisions to use DNA sequencing data to answer question about pathogen diversity generally comes down to how much is already known and how much effort is justified in acquiring DNA sequence data.

If large amounts of DNA sequence data are known or can be generated, then single base differences between samples can be measured. These single nucleotide polymorphisms (SNPs) can be tabulated for several different loci across the genome to generate high-resolution genetic characterization and organism identification. Once an

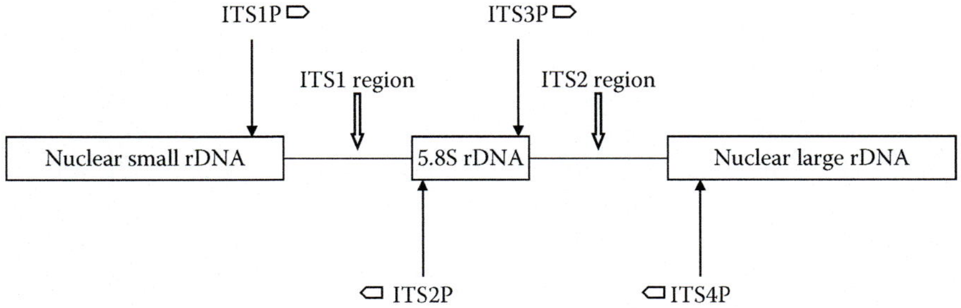

FIGURE 33.4 Diagrammatic representation of internal transcribed spacer (ITS) and ribosomal regions. (Modified from White, T.J.T. et al., Amplification and direct sequencing of fungal ribosomal RNA genes for phylogenetics. In M.A. Innis, D.H. Gelfand, J.J. Sninsky, and T.J. White [eds.], *PCR Protocols: A Guide to Methods and Applications*, Pages 315–322, San Diego, CA: Academic Press, 1990.) ITS1 and ITS2 are primers for ITS-1 region and ITS3 and ITS4 are primers for ITS-2 region. (From Trigiano R. N., M.T. Windham, and A.S. Windham [eds.], *Plant Pathology Concepts and Laboratory Exercises*, 2nd edition, Boca Raton, FL: CRC Press, 2008.)

SNP site has been identified as informative or unique to a particular population or individual, it can be assayed in new or unknown samples, much as SCAR markers generate codominant data. SNP databases are powerful tools since they approximate the strengths of entire genome sequencing by focusing only on polymorphic bp in genes and ignoring noninformative DNA.

Recent advances in high-throughput DNA sequencing technology and analytic methods have lowered costs significantly. Instrumentation for DNA sequences has moved from capillary-based fragment separation to massively parallel technologies such as ion semiconductor sequencing or sequencing by synthesis that process millions of reactions at the same time. Rather than generating data for discrete PCR-amplified regions of the genome, these sequencing methods typically rely on randomly fragmenting the entire genome and processing a size-selected aliquot of the fragments on nano-beads or in micro-wells on a chip using different nucleotide detection strategies. The result is millions of small sequences, generally 100–400 bp in length, which must be assessed for quality and assembled into **contigs** using sophisticated algorithms via computer software(s). These enormous data sets represent random DNA regions from across the entire genome, effectively sampling all types of genetic regions including genes, noncoding, repetitive, and structural parts of all chromosomes in nuclear and organelle genomes. Because the coverage is extensive, assumptions that once had to be considered about sequencing a particular gene or region can be verified by empirically testing data from one part of the genome against other regions. In this way, analyses using large numbers of DNA sequences across the entire genome are much more robust than the other methods previously described. Results often include SNPs, SSRs, and DNA sequences from conserved genes—effectively incorporating multiple marker systems into one approach.

Individual reads are short and may be of low quality depending on the high-throughput DNA sequencing instrument and technology. However, the depth of coverage, that is the number of overlapping or duplicate reads for a region, is much greater in next generation DNA sequencing and can be increased or decreased to mitigate individual read quality concerns. For example, a typical protocol might generate sequence data for the same gene 40–60 times. Increasing depth of coverage increases confidence in the accuracy of the data and allows the discovery of rare variants. On the other hand, methods to barcode samples, or pool multiple samples for one experiment and separate the data after sequencing, can be used to split the coverage between samples and significantly lowered costs. The ability to balance costs per sample with depth of coverage allows researchers to scale sequencing protocols to meet specific needs in data generation. In some cases, costs are low enough to support genome-wide sequencing of each individual in the study, even for nonmodel pathogens.

New methods and statistical analyses have been developed to analyze data across genomes from related populations for studies on phylogeny, taxonomy, and ecology. Additionally, experiments can use RNA as the starting template for next generation sequencing, called transcriptome sequencing, to analyze variation in expressed genes, presumably representing functional differences between organisms. As the cost per sample decreases, high-throughput sequencing is expected to become the standard for markers for all experimental pathogens including systems that are not model organisms. The possibility of looking across the entire genome for all samples in a study has fueled a radical departure in experimental designs, which should significantly enhance our ability to answer questions about genetic diversity of different organisms.

ANSWERING QUESTIONS ABOUT PATHOGEN GENETIC DIVERSITY

Each individual pathogen or isolate of a pathogen contains a different mix of genes that are reshuffled every time a pathogen produces progeny through sexual reproduction. Of course, there are some pathogens that do not reproduce sexually and they employ other methods to increase the mix of genes, albeit less efficiently than sexual reproduction. The “gene pool,” or total number and variety of genes and alleles in a sexually reproducing population, determines what alleles are available for transmission to the next generation. This genetic diversity, or variation between individuals within species, is what allows populations to adapt to changes in local environmental conditions including changes in the host. Without genetic variation, there is little chance to evolve into new forms, populations, or species. Measures of genetic diversity allow researchers to quantify how many potential new forms, populations, or species exist for taxonomic analysis, conservation, breeding manipulation, or commercial exploitation. Estimating genetic diversity is an important first step in understanding pathogen populations.

Allele richness, or the number of genotypes present in a group of samples, is especially influenced by sampling errors. Rare alleles that occur in less than 5% of the genotypes are likely to be missed unless a sufficient number of random individuals are analyzed (Nei, 1987). Thus, using estimates of the total number of alleles as a measure of genetic diversity can be significantly impaired by sampling errors including low number of samples or biased sample collection. Some of the variance due to small sample size can be overcome by using a larger set of molecular markers because the percentage of polymorphic loci can also be considered an estimate of genetic variation. Increasing sample size beyond what is necessary can lead to diminishing returns since the probability of observing a rare allele by adding an individual sample decreases immensely as sample size increases. The general rule of thumb is that sample sizes should be large enough to differentiate between groups of samples to accurately describe genetic variation within a group of samples. To that end, most sampling strategies include analyzing more than one group (i.e., subpopulations, populations, species, or genera). Refined models for determining samples size can be found in Baverstock and Moritz (1996), Crossa et al. (1993), Marshall and Brown (1975), Warburton et al. (2002), Weir (1996), and many other publications.

Genetic distance is the "difference between two entities that can be described by allelic variation" (Nei, 1973). Entities can include individuals, populations, or species, and the differences can be quantified for any of the molecular markers outlined in the first half of this chapter. There are many ways to calculate distance measures. Some statistical calculations are specific to the molecular marker being used, whereas other calculations can be used for any marker system or even combined data sets from multiple marker systems and morphological data. The most common measures of genetic distance include Nei and Li (1979), Jaccard's coefficient (1908), and simple matching (Sokal and Michenener, 1958). All of these measures can be broadly applied to any marker data recorded in binary format. Genetic distances are based on the number of markers present in both individuals, the number absent in both individuals, and the number of markers present in one individual, but not the other. Although similar, all three measures use marker data differently. Jaccard's coefficient does not take into account markers that are absent between individuals, and calculations are based solely on shared markers between individuals. Nei and Li estimates of genetic distance are calculated from the proportion of shared markers divided by the average of the proportion of markers present in each individual. Simple matching calculations use both shared and nonshared markers equally when calculating genetic distance. When molecular marker data can be interpreted as codominant alleles, such as for SSRs, allele frequencies can be calculated directly, much like genetic distance estimates between two individuals is calculated from heterozygosity measures.

Genetic distance measures should be selected in accordance with the molecular marker system being used. For example, dominant and codominant markers are weighted differently when calculating distances with Nei and Li or Jaccard. Higher levels of missing data may also affect distance measures, especially if the presence of matching markers is equally weighted as the absence of matching markers (0-0), which also represents missing or failed marker generation. Jaccard explicitly ignores the absence of matching markers (0-0), so missing data would have minimal effect on genetic distance measures. Likewise, several genetic distance measures explicitly incorporate models for the molecular evolution of the marker system. For example, the observed DNA sequence variation in genes would be significantly different from DNA sequence variation in noncoding regions. Models for the accumulation of mutations in coding DNA take into account the 3-base codon structure, the biochemical bias in the occurrence of mutations (transitions versus transversion), and even the organelle where the DNA originated (mitochondrial, chloroplast, or nuclear genome). The level of genetic diversity being analyzed (individual, population, species) may also influence the selection of a particular genetic distance measure.

Genetic distances, often referred to as pairwise distances, are recorded in a matrix format with each individual compared with all other individuals in the sample (Table 33.2). Oftentimes, researchers will employ more than one measure of genetic distance and look for correspondence between the resulting distance matrixes.

TABLE 33.2
Pairwise Population Φ_{PT} (Analogous to Standardized F_{ST} fozr Haploid Data) Across Seven *Geosmithia morbida* Populations

	Anderson	Blount	Knox	Loudon	Sevier	OR	NC/GRSM
Anderson	0.000						
Blount	0.046	0.000					
Knox	0.083	0.061*	0.000				
Loudon	0.298*	0.187*	0.110*	0.000			
Sevier	0.144	0.047	0.071*	0.231*	0.000		
OR	0.407	0.302*	0.345*	0.508*	0.390*	0.000	
NC/GRSM	0.536*	0.407*	0.463*	0.586*	0.500*	0.369*	0.000

Source: Reprinted from Hadziabdic et al., *Curr. Genet* 60:75–8, 2014. With permission.
*$p < 0.05$.

Correspondence can be statistically tested using the Mantel test (Mantel, 1967) and may add validity to results when there is agreement between different distance measures such as correlation between genetic and geographic distances, also known as isolation by distance (IBD) (Figure 33.5). Similarly, Mantel tests can be applied to genetic distance matrices derived from different molecular marker system that have been used on the same set of samples, such as AFLP and SSRs data from the same samples. Other statistical tests can be used to add credibility to genetic distance measures, such as analysis of molecular variance (AMOVA), but there is not sufficient space to describe them in detail.

Pathogen and specifically fungal systematics use genetic diversity measures to resolve taxonomic issues and accurately classify samples. Most of this work is based within a cladistic framework (also known as phylogenetics), which specifies a hierarchy based on shared ancestors. For example, genetic diversity may be used to support the relatedness between two groups of fungi suggesting a recent common ancestor. The relatedness between these groups is directly proportional to the genetic distances between a most recent common ancestor (MRCA) and each group. Similarly, genetic distances may also be used to differentiate between species and assign taxonomic labels to unknown samples. Cladistic uses for molecular markers often involved cluster analysis of genetic distance measures to group-related samples with relatively high genetic similarity and separate group with high genetic divergence between clusters. Much like the statistical approach to differentiation, clustering analysis uses a genetic distance matrix created from one of the many distance measures noted above, which was derived from molecular marker data generated by one of the marker systems described earlier in the chapter. Computer

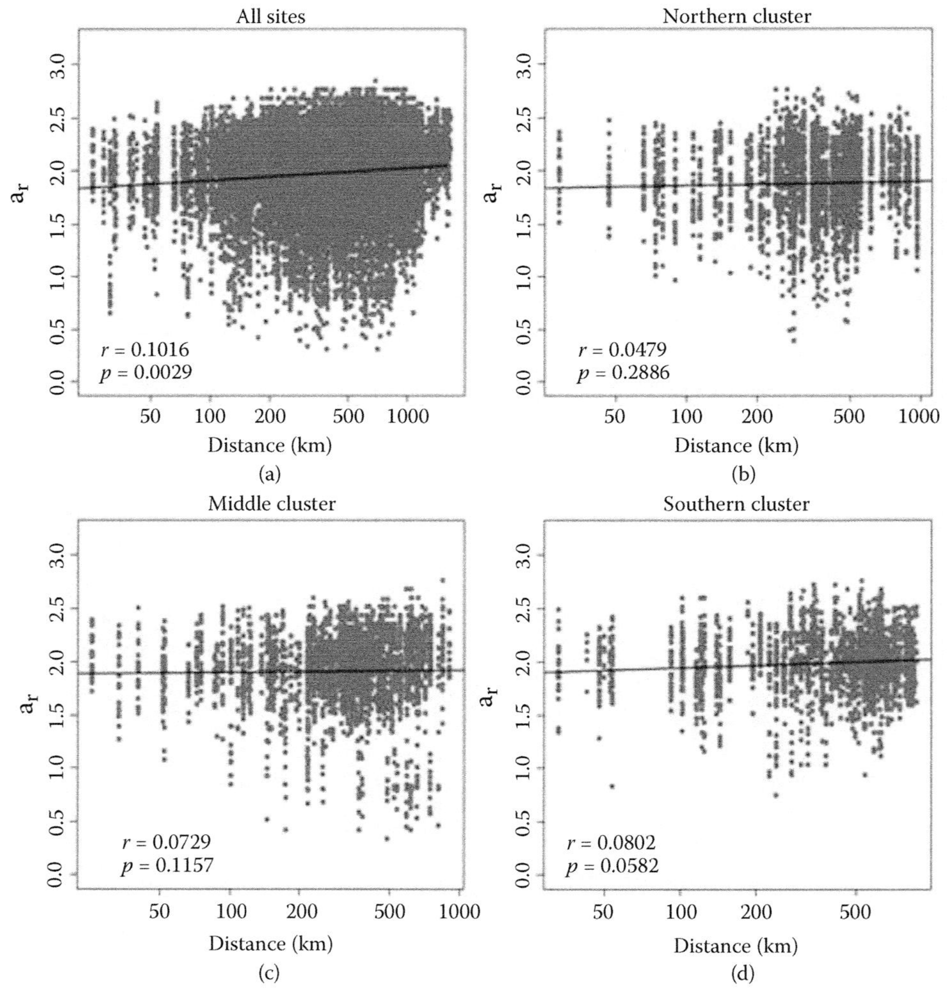

FIGURE 33.5 Isolation by distance (scatterplots of pairwise genetic distances vs. geographical distances) across different flowering dogwood (*Cornus florida*) populations impacted by dogwood anthracnose (*Discula destructiva*). All sampled flowering dogwood individuals (a) are partitioned into sites with predominant membership in the northern cluster (b), middle cluster (c), and southern cluster (d). *p*-Values are from Mantel tests with 10,000 randomizations. Values on the *x*-axis are in log scale. Lines indicate the best fit of least-squares regression of a_r (allelic richness) on log distance. (From Hadziabdic, D. et al., *Genetica*, 138, 2010).

software is necessary to complete the extensive analysis, and the resulting phylogenetic tree, or dendrogram, should be considered a hypothesis of the evolutionary ancestry of the samples in the analysis. Despite advances in software and computational power, the resulting trees are unlikely to be a perfect evolutionary tree, or historical ancestry, of all samples. The use of genetic distance methods has certainly improved the resolution of phylogenetic analysis and several algorithms, such as **U**nweighted **P**air **G**roup **M**ethod with **A**rithmetic Mean (UPGMA) (Figure 33.6) and neighbor-joining (NJ), are commonly accepted as robust measures of cluster-based relatedness.

It is worth noting that phylogenetic analyses can also be conducted on some molecular marker data without generating a genetic distance matrix. Maximum parsimony analysis of DNA sequence data is based on finding the smallest number of evolutionary steps to explain the differences between all samples. Samples are arbitrarily arranged in phylogenetic trees and "scored" based on DNA differences between samples using specific mutational constraints for DNA sequence evolution. The lowest scoring trees are saved for further comparison. and trees with higher scores are discounted. Depending on the number of samples, the number of trees to be compared can be overwhelming, and finding a single, most parsimonious tree is usually impossible. In an effort to reduce the amount of computation, maximum likelihood methods use probability statistics to infer the likelihood of phylogenetic trees and reduce the number of trees to be compared. Likelihood methods were recently improved by including Bayesian statistics to reduce the field of trees by assuming a starting distribution of possible trees. All of these methods are highly computational and require extensive data modeling, but the resulting trees can provide powerful insight into the genetic diversity within a group of samples.

Nonhierarchical clustering analyses do not produce trees. Results are typically viewed as scatter plots with genetic distances among individuals displayed

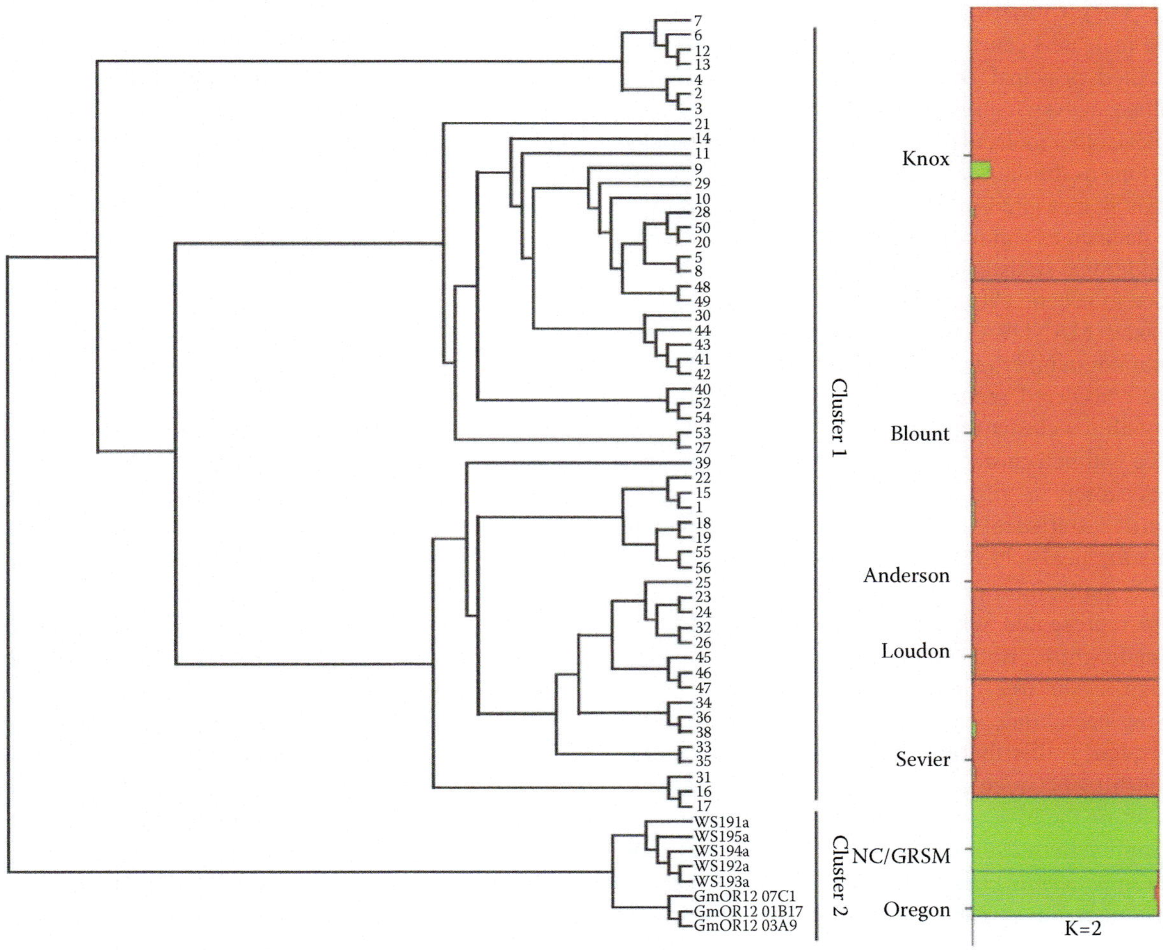

FIGURE 33.6 UPGMA dendrogram of 62 *Geosmithia morbida* isolates was constructed using Nei's genetic distance D. Program STRUCTURE results revealed two distinct clusters among *G. morbida* populations representing five subpopulations from Tennessee (Anderson, Blount, Knox, Loudon, and Sevier) as one cluster (red) and Oregon and North Carolina/Great Smoky Mountains National Park (NC/GRSM) as the second cluster (green). (From Hadziabdic, D. et al., *Curr. Genet.*, 60, 2014.)

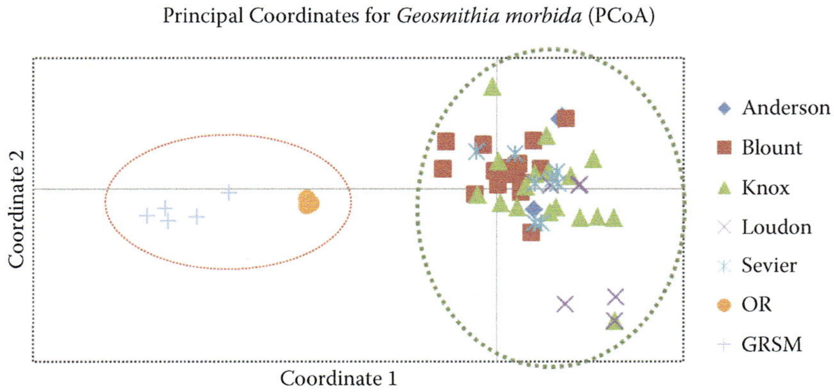

FIGURE 33.7 Principal coordinate analysis (PCoA) of *Geosmithia morbida* populations. Two different clusters were identified and color coded—a red circle representing Oregon and North Carolina/Great Smoky Mountains National Park (GRSM) as one cluster and a green circle representing five populations from Tennessee (Anderson, Blount, Knox, Loudon, and Sevier). (From Hadziabdic, D, et al., *Curr. Genet.*, 60, 2014.)

as proportional distances between dots within the plot (Figure 33.7). Clustering of individuals within the plot indicate increased genetic similarity. Data for these analyses, called principal component analysis or principal coordinate analysis, can also include other measures (morphological, other molecular markers, physiological ratings, etc.) as long as the data are scaled proportionally. Because each axis is measured in genetic diversity units, clusters can be deemed as significant based on user-specified cutoffs. Additional statistic tools are available to ascribe significance to genetic clusters such as multivariant analysis of variance (MANOVA), which is often used to determine the optimum number of clusters. Under MANOVA, clusters are considered separate treatments with each individual within a cluster representing a replicate treatment. Variance within a cluster should be less than the variance between clusters in order to be significant.

Similar statistical tools have been developed to test hierarchical clustering analyses. Bootstrap and Jackknife replication provide support for genetic relationship represented within and between branches on tree dendrogram. Both statistics rely on resampling, or rerunning the tree-building analysis multiple times, and using each "independent observation" to calculate the frequency distribution of the sample hierarchy. Relationships between samples, as shown by branches and nodes on the tree, can be assigned bootstrap or jackknife support based on the number of times the same branching pattern is observed during resampling. Thus, a node connecting two specific samples that is observed in 90 out of 100 reiterations is assigned a bootstrap value of 90%. Bootstrap values less than 50% are generally ignored.

As biotechnological processes become less expensive and more widely available, molecular tools will likely become even more important techniques in understanding pathogen diversity, particularly since these methods require only small amounts of tissue. Advances in molecular genetics rely heavily on public sharing of genetic marker information and computer software. In this chapter, we covered common molecular markers and analyses being used to rapidly quantify genetic diversity, classify pathogens, and increase taxonomic understanding of the relationships between different pathogenic species.

CHARACTERIZING GENE EXPRESSION IN PLANT PATHOGENS

Looking for the genetic underpinning of a trait, such as pathogenicity, typically requires linking the genomic DNA information with gene expression, protein expression, and, finally, protein function. While the central paradigm goes from DNA to RNA to protein, studying this functional pathway does not necessarily have to start with the analysis of genomic DNA. **Expressed sequence tag (EST)** libraries provide snapshots of the genes expressed in a pathogen at a particular time, such as initial infection, and can yield a wealth of genetic data regarding the mechanism of plant–pathogen interactions. EST libraries are created from mRNA isolated from the pathogen, so they reflect only those genes being expressed. A typical mRNA library consists of 10,000 or more clones, each corresponding to an mRNA transcript. Once the mRNA has been converted to cDNA by reverse transcriptase-PCR (rt-PCR), it is usually cloned into a plasmid vector for propagation inside *E. coli*. These bacterial colonies are easily stored, and the plasmid DNA can be extracted and sequenced. Generally, several EST libraries are made from different life stages of the pathogen or infection so that the genes expressed in one library can be compared with the transcripts contained in the other.

Much like SSR and SNP DNA fingerprinting, the cost of DNA sequencing can be a critical factor when deciding to use EST libraries to understand gene expression in plant pathogens. Libraries are often redundant since any mRNA transcript in high copy number will be cloned multiple times. There are molecular methods for removing high copy number DNA and increasing the complexity of the cDNA before cloning. A normalized EST library is one where redundant cDNAs have been reduced by a hybridization step. Similarly, EST libraries can be subtracted from one another to enrich for genes that are unique to each mRNA extraction. DNA sequence data from EST libraries are typically deposited in GenBank or other public databases to accelerate research in multiple labs. Oftentimes, researchers working on the same plant pathogen will pool resources to produce and release EST information. This synergistic approach allows researchers to assemble smaller bits of gene expression information into a coherent model of the genes expressed in a plant pathogen. EST sequence data can also be mined for SNP and SSR marker development.

Differential display is a molecular tool that allows researchers to visualize a large number of expressed genes and uncover those genes that are up or down regulated at a specific time in the plant pathogens life (Liang and Pardee, 1992). Differential display compares mRNA transcripts from two different time points side by side in a manner similar to AFLP fingerprinting. mRNA transcripts are reverse-transcribed into DNA using a mixture of primers anchored to the polyA tail of mRNA and arbitrary primers. The resulting short, labeled fragments are separated and visualized either by acrylamide gel or by capillary array electrophoresis to resolve size differences as small as a single nucleotide. When compared side by side, DNA fragments specific to particular mRNA extraction can be identified as missing from other samples. Results look very similar to DNA fingerprinting using RAPD or AFLP except that the starting template is mRNA, not genomic DNA. Thus, differences do not indicate genetic variation between isolates but rather spatial or temporal differences in gene expression. Unique DNA fragments can be excised from the gel, cloned, and the DNA sequenced.

DNA sequence data for differentially expressed genes can be compared with DNA databases to locate analogous sequences. For example, BLAST searches (Basic Local Alignment Search Tool) of GenBank can uncover previously characterized genes, possibly from other organisms. Even novel gene sequences can be analyzed for possible protein function by theoretical reconstruction of the protein itself. Using computer software, triplet codons are translated into amino acids, which can be strung together and folded into a three dimensional (3D) model of the protein itself. X-ray crystallography has produced a wealth of information regarding the structural components of proteins and their cellular function. For example, membrane-spanning proteins generally alternate characteristic hydrophilic and hydrophobic regions (Seshadri et al, 1998). These models produce reasonable hypotheses about novel protein function, especially when coupled with any analogous DNA or protein sequence information in public databases. Differentially expressed gene sequences can also be used to design specific PCR primers for real-time PCR. In this way, gene expression patterns visualized by differential display can be quantified, verifying that mRNA transcripts in one sample are increased or decreased in another sample. There are several alternative protocols for visualizing differences in gene expression between two samples including subtractive hybridization; RNA-arbitrarily primed PCR (RAP-PCR), representational difference analysis (RDA), and serial analysis of gene expression (SAGE).

Biotechnology and database resources can be leveraged in other ways to understand more about what makes a plant pathogen pathogenic. **DNA microarrays** utilize EST information to create multiple probes for each of the gene transcripts, even when full-length mRNA sequence is not available or gene function has not been characterized. Thousands to tens of thousands of these probes are affixed to a small surface, typically a glass slide, and may represent an equal number of expressed genes. Plant pathogen mRNA is extracted and converted to fluorescent-labeled cDNA. These labeled fragments are then hybridized to the DNA probes that are fixed to the microarray slide. Nonbinding transcripts, or those cDNA that do not share high levels of similarity, are washed off, and the resulting intensity of fluorescence for each microarray probe reflects the number of copies of mRNA transcript present in the sample at that time point. In this manner, the expression level of thousands of genes can be simultaneously compared in a single experiment (Schena et al., 1995). Because microarrays are reusable, additional mRNA extractions and experiments can uncover a coordinated, quantitative picture of the genes that are up- and down-regulated (Figure 33.8).

DNA microarrays require advanced knowledge of expressed genes; however, DNA sequence databases may contain sufficient analogous gene sequences from related genera that the creation and sequencing of new EST libraries is not necessary. For example, the rust pathogen that attacks wheat may be a different species than the rust pathogen that attacks daylily, but the microarray for wheat rust may produce comparable data for daylily rust. Microarrays that are universal to a range of plant pathogens are possible because the DNA sequence variation between genomes, at least the expressed portion of the genome, may be minimal among related pathogens, and multiple probes are

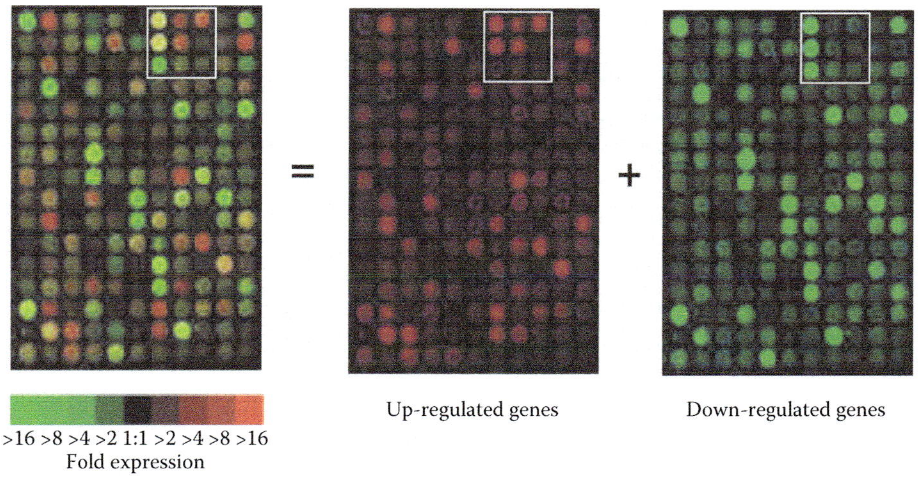

FIGURE 33.8 DNA microarray results demonstrating up- and down-regulated gene expression. (From Trigiano R. N., M.T. Windham. and A.S. Windham [eds.], *Plant Pathology Concepts and Laboratory Exercises*, 2nd edition, Boca Raton, FL: CRC Press, 2008.)

generally included for each expressed gene increasing the chances that hybridization between probes and cDNA will occur for at least some regions of the gene. Thus, investment in DNA microarrays can be spread across several crops or disciplines.

The expanding field of **proteomics** uses computer-assisted analysis of 2D gel-based protein profiles to compare protein fingerprints from different stages or tissues involved in plant–pathogen attacks and then picks unique proteins and analyzes their amino acid composition using mass spectrometry, specifically matrix-assisted laser desorption-ionization time-of-flight (MALDI-TOF) mass spectrometry system. This molecular tool skips genomic DNA and mRNA information and goes straight to the differences in functional protein products. Once proteins are characterized, they can be modeled in three dimensions using computer software. Theoretical mRNA transcripts can also be reverse engineered, and real-time PCR can be used to verify differences in gene expression.

Characterizing gene expression in a plant pathogen does not necessarily rely on theoretical interactions or computational modeling. Targeting induced local lesions in genomes (TILLING) is the process of systematically mutating a specific gene and observing the mutant phenotypes to better understand protein function (Till et al., 2003). Large numbers of identical plant pathogens, or clones, are exposed to a chemical mutagen called ethyl methanesulfonate (EMS), which generally causes single-base mutations in the genomic DNA. Thousands of these mutagenized clones are produced and stored. Genomic sequence data must be available because the randomly mutated pathogens are screened using PCR primers corresponding to expressed genes. The net result is a gigantic collection of mutants that can be characterized at the DNA level for changes in amino acid composition of the protein, possible truncation, or structural changes in the protein folding. Researchers usually place orders for mutations in specific regions of well-characterized genes based on gene function, protein modeling, or gene sequence homology. The collection of mutants is then screened by PCR for base changes in those regions. Any isolate with mutations in the desired region are sent out to the researcher for characterization.

It is also possible to generate specific mutations in well-characterized plant pathogen systems that have transformation systems. The ability to insert genomic DNA into an organism gave rise to the controversial field of genetically modified organisms. For research purposes, the ability to integrate foreign genetic material into the host genome is generally used to turn off specific genes. RNA-mediated interference, abbreviated RNAi, is a molecular tool based on a cellular defense system that is found in most eukaryotic organisms, which shuts down over expressed genes (Fire, et al., 1998). The host system targets RNA transcripts via a double-stranded RNA (dsRNA) intermediate that is complementary to the gene being shut down. Once the dsRNA is detected, ribonuclease enzymes cleave the complementary mRNA transcript into useless chunks. Gene expression is effectively stopped before mRNA can be translated into protein. Because RNAi occurs after transcription, synthetic dsRNA that is added to cells can induce RNAi and selectively reduce the production of a specific protein without having to know where in the genome the mRNA is being expressed. In fact, dsRNA based on analogous genes from other organisms can shut down host genes if there is enough similarity between gene sequences.

The resulting mutant has reduced production of a single protein, which can yield insights into the protein's role and function. RNAi and other tools described in this chapter are universal to plant pathology since they can be applied to the host plant as well. Because pathogens are involved in an arms race with plants, natural selection acts in concert on genes for disease resistance found in the plant and genes for pathogenicity in the plant pathogen.

Increases in biotechnology drive modern biology and have even impacted the role of science in popular culture from television shows and court cases. It is not surprising that molecular tools have become important techniques in plant pathology research. In this chapter we covered molecular tools to rapidly and unambiguously identify pathogenic organisms using DNA markers or DNA sequence data, the classification and increased taxonomic understanding of the relationships between different plant pathogens, and the rapid quantification of how many and what types of pathogens are present. These techniques typically require only small amounts of tissue and new or unknown pathogens may be detected. Advances in molecular genetics rely heavily on public sharing of DNA sequence information and computer software and offers the possibility of understanding the genetic basis for plant pathogen phenotypes.

LABORATORY EXERCISES

Experiment 1. Amplifying the ITS Region in Eukaryotic Pathogens Such as Fungi

The primary objective of this laboratory exercise is to familiarize undergraduate and graduate students (and instructors) with a very powerful molecular technique that is used to characterize **DNA** of plant pathogens and other organisms and/or define relationships between organisms. All of the techniques are similar in that they utilize the **PCR** to amplify or increase copies of **genomic DNA**.

The technique described in this exercise is the selective amplification of the ITS regions that flank the 5.8S nuclear ribosomal unit (rRNA). The PCR is completed with 18–22 bp primer pairs and upon amplification produces a single band or product. The products from individuals are then sequenced (omitted in these exercises), and the sequences compared and relationships among the individuals inferred. **Eukaryotic** ribosomal genes are arranged in tandem repeats with the 5.8S coding region flanked by internal transcribed spacers (ITS; Figure 33.4). Although these regions are important for maturation of nuclear ribosomal RNA (rRNA), ITS regions are usually considered to be under low evolutionary pressure and therefore typically treated as nonfunctional sequences. The sequences of the ITS regions have been used in many phylogenetic studies and molecular systematics of fungi. This experiment will allow students to amplify one or both of the ITS regions (1 and 2) and is largely based on Caetano-Anollés et al. (2001) and White et al. (1990). The experiment will not describe DNA sequencing (usually completed at a university center) or the analysis of sequence data. Instructors and students are encouraged to seek expert help on their campus. This exercise has been successfully completed on the first attempt by several classes of novice undergraduate and graduate students and other researchers.

Before beginning the exercise, a few essential generalities are listed below. First, all pipette tips, Eppendorf centrifuge tubes, water, and reagents used to assemble the amplification reactions should be either autoclaved or filter-sterilized (0.22 μM) and made with sterile, high-quality water. Second, participants should wear either latex or acetonitrile gloves to avoid hazardous materials (e.g., ethidium bromide) and to protect samples from **DNases** found on the skin (Dragon, 1993). Third, where possible, use only the American Chemical Society (ACS)-certified pure chemicals and double-distilled or nanopure water (<16 M ohm/cm: Barnstead/Thermolyne Corp.), hereafter referred to as "pure." It is not necessary to use HPLC-grade water.

Materials

Each team or group of students will require the following items:

- AmpliTaq Gold DNA polymerase kit with buffer II (Applied Biosystems)
- All materials under Exercise 1: experiment 2—DNA amplification except primers
- The following primers (5'–3') in 3 μM concentrations:
  - ITS-1 region use: TCCGTAGGTGAACCTG CGG (ITS1) and
    - GCTGCGTTCTTCATCGATGC (ITS2)
  - ITS-2 region use: GCATCGATGAAG AACG CAGC (ITS-3) and
    - TCCTCCGCTTATTGATATGC (ITS-4), or
    - combine ITS-1 and ITS-2 regions with ITS1 and ITS4 as primers for amplification
- Seven stock fungal DNA samples from various isolates or species diluted to 0.5 ng/ μL
- QiaQuick PCR Purification Kit (Qiagen)
- Low melting point agarose and TE buffer (20 mL of 50X TE combined with 980 mL of distilled water)

- Graduated cylinder, distilled water, and a 250-mL flask
- UV Transilluminator and Cyber green stain
- Small gel casting apparatus including combs and power supply
- Molecular weight ladder (100–1000 bp) and loading buffer with tracking dye (40 mL of water, 80 µL of 0.5 M EDTA, 16 g glycerol and 0.1% of bromophenol blue and xylene cyanol)

The thermalcycler should be programmed as follows:

Step 1: 95°C for 9 min: 95°C activates the DNA polymerase enzyme. This step is also termed a "hot start."
Step 2: 96°C for 1 min: Denatures DNA
Step 3: 56°C for 1 min: Primers anneal to complementary sites
Step 4: 72°C for 1 min: Extension
Steps 2–4 should be repeated for 35 cycles
Step 5: 72°C for 7 min: Final extension
Step 6: 4°C

Follow the protocol as outlined in Procedure 33.1 to amplify the ITS region.

Check for amplification products using electrophoresis with a 2% agarose gel.

Follow the instructions in Procedure 33.2 to complete detection of products.

After the amplified product has been confirmed by gel electrophoresis and before the product can be sequenced, other remaining components of the amplification reaction mixture must be removed. We use a QiaQuick PCR Purification Kit (Qiagen). The instructions in the kit are excellent and easy to follow. Inquire at your sequencing center whether the DNA should be eluted in pure, sterile water or buffer. It is imperative to use sterile water that is between pH 7.0 and 8.5 for the QiaQuick kit. If you use distilled, pure water (pH = ~5.5), the DNA bound to the membrane will not be released. Most sequencing centers require the DNA concentration in the samples to be about 10 ng/µL and at least 10 ng/100 bp of gene to sequence. So, if the ITS region is 500 bp, 50 ng of product will be needed. Typically, you need to supply some of the primers used in the amplification reaction, although some centers have these common primers in stock.

Questions

- What are insertions and deletions?
- Why would you not send DNA in TE (Tris–EDTA) buffer to the sequencing center?
- Would any product be formed using only one primer in the amplification cocktail?

Experiment 2. Analyses of Molecular Data

Using the provided data set, we will evaluate the basic statistical procedures for population genetic analyses including both frequency and distance based analyses. The program GenAlEx 6.5 (Peakall and Smouse, 2006, 2012) offers user-friendly interface, rich graphical outputs using Microsoft Excel platform, and easy navigation for both students and teachers. This procedure will allow us to provide biological interpretation of the statistics described for some real data sets. Additionally, both haploid and diploid data sets can be calculated using program GenAlEx. More

Procedure 33.1
Amplification of ITS-1 and ITS-2 Regions

Step	Instructions and Comments
1	Pipette 46.4 µL (8 × 5.8 µL) of pure water into a sterile, 0.65-mL Eppendorf tube.
2	Pipette 16.0 µL (8 × 2 µL) of 2 mM nucleotides into the tube.
3	Pipette 16.0 µL (8 × 2 µL) of 10X Buffer II (without magnesium chloride) into the tube.
4	Pipette 9.6 µL (8 × 1.2 µL) of 25 mM magnesium chloride into the tube.
5	Pipette 16 µL (8 × 2 µL) of each of either ITS1 and ITS2, or ITS3 and ITS4 primers (note that as an alternative, use ITS1 and ITS4 primers and amplify both ITS-1 and ITS-2 regions plus the gene).
6	Pipette 8 µL (8 × 1 µL) of AmpliTaq Gold DNA polymerase.
7	Vortex and centrifuge briefly. Dispense 16 µL of the reaction mixture into seven, sterile 0.65-mL Eppendorf tubes. Add 4 µL of fungal DNA template (0.5 ng/µL) to each of the seven tubes. Be sure to change pipette tips between different isolates. Vortex and centrifuge briefly.
8	Place tubes in a thermalcycler and run according to the parameters provided.

Procedure 33.2
Electrophoresis of Amplified ITS Products

Step	Instructions and Comments
1	Make a 2% gel by placing 50 ml of 1X TE buffer in the 250-mL flask and add 1 g of low melting point agarose. Place in a microwave to melt the agarose.
2	Remove from the microwave, allow the solution to cool to 60°C (comfortably held in the hand), and add 5 µL of Cyber green (used to stain DNA). Mix thoroughly.
3	Pour the molten agarose into bottom of the gel-casting tray. Try not to introduce air bubbles in the gel. Set the plastic combs (10 teeth) close to one end of the casting unit. Allow the gel to solidify—the preparation will become cloudy. Fill the reservoirs with 1X TE buffer and a thin layer (1 cm) over the gel. Carefully remove the plastic comb.
4	Combine 2 µL of loading buffer with 5 µL of amplification products—repeat for each sample. Pipette 7 µL of sample/loading buffer into the wells. Skip a well and load molecular weight (bp) ladder (3 µL of ladder plus 1.5 µL of loading buffer) into the well.
5	Place the top of electrophoresis unit onto the base (be sure that the negative [black] electrode is closest to the sample wells) and connect the unit to the power supply. Set the power supply to 100 V and allow it to run for about 45 min. The blue tracking dyes should migrate through the gel toward the red electrode.
6	Wear gloves. Turn off the power supply and remove the gel tray from the base of the unit. Transport the gel to the transilluminator. Wear goggles that block UV wavelengths and be sure to wear protective clothing that covers exposed skin. Turn off the lights, and turn on the transilluminator. Note the bands and approximate bp weights of the products. Most fungal ITS1-4 amplicons are between 500 and 750 bp. If you see more than one band per lane, there is a contaminant and the DNA should be isolated from a pure culture, and the process repeated.
7	Clean the surface of the transilluminator and dispose of the gel in a designated container.

information and modules with additional tutorials can be found at GenAlEx website: http://biology.anu.edu.au/GenAlEx/Welcome.html.

Materials

All students or team of students should have the following equipment:

- Computer with access to the Internet and printer

Follow the procedure in Procedure 33.3 to get started.

Questions

- What is the estimated expected heterozygosity (also known as gene diversity [He]) for the total sample, and across each examined population?
- Were you able to obtain Ho calculations? If not, provide an explanation.
- What conclusions concerning overall degree of genetic differentiation (Fst) can be made?
- What are the main conclusions regarding genetic diversity of *Grosmmania clavigera* based on your calculations?
- Can we calculate the Hardy–Weinberg equilibrium (HWE) for a haploid data set? Explain why.

Follow Procedure 33.4 to continue the exercise.

Questions

- What is the estimated expected and observed heterozygosity across eucalyptus (n = 70) and guava (n = 63) populations? Why are other populations excluded from these analyses?
- The results indicated strongly negative fixation indices for both eucalyptus (n = 70) and guava (n = 63) populations. Can you provide an explanation?
- How many private alleles are detected in each population? What conclusions can be drawn from that finding?

To continue with the exercise, follow Procedure 33.5 to complete distance-based analyses—principal coordinate analysis (PCoA).

Procedure 33.3

Haploid Data Analysis—Frequency-Based Analysis

Step	Instructions and Comments
1	Go to the following website to download GenAlEx 6.5 (Peakall and Smouse, 2006; 2012) that offers a wide range of population genetic analyses: http://biology.anu.edu.au/GenAlEx/Welcome.html. The program will be provided as an Excel add-in.
2	Go to DRYAD digital repository and download the following excel file: **GC293-data.xlsx** (http://datadryad.org/resource/doi:10.5061/dryad.48jg20v0; obtained from Tsui et al., 2011, 2012). This is a publically available data set of eight microsatellite loci used for population structure and migration pattern of a conifer fungal pathogen, *G. clavigera*. The ophiostomatoid fungus *G. clavigera* is a haploid filamentous ascomycetes and a symbiont of the mountain pine beetle *Dendroctonus ponderosae*.
3	Note the following insert from **GC293-data.xlsx** (Table 33.3). Column A contains the isolate code (e.g., H12), column B represents the sampling location (e.g., Houston), and columns C–G have the raw data of fragment size (e.g., 216/216 bp). Row 1 contains the following information A1 (8) = total number of microsatellite loci used in the study; B1 (293) = total number of individuals used for data analysis; C1 (19) = total number of geographical populations used for data analysis; D1 (16) = total number of individuals in Houston population; E1 (23) = total number of individuals in Fort St. James population; F1 (13) = total number of individuals in Tumbler Ridge, BC population; G1 (11) = total number of individuals in Fairview, BC population.
4	To start the program, launch *MS Excel*, choose *Open* from the *File* menu to locate GenAlEx *Add-Ins*. Also, the program can be launched by choosing *Add-Ins* tab and selecting *GenAlEx* tab.
5	To start with data analysis, the main data worksheet should be opened, as seen in Table 33.3. Select *Add-Ins/GenAlEx/Frequency* to estimate allele frequency data parameters. Data will be automatically placed into following boxes: #Loci (A1) [8]; #Samples (B1) [293]; #Pops (C1) [19]. User will have an option of choosing data format—one column/locus or two columns/loci. Although *G. clavigera* is haploid pathogen, data were placed in two columns/loci, and therefore, codominant option should be selected.
6	Under *Allele Frequency and Heterozygosity*, select *Frequency, Het, Fstat &Poly by locus* options. Compare your results with Tables 1 and 2 found in Tsui et al. (2012).

TABLE 33.3

Insert of Data Used for Population Structure and Migration Pattern of a Haploid Fungal Pathogen, *Grosmannia clavigera*

	A	B	C	D	E	F	G
1	8	293	19	16	23	13	11
2				Houston	Fort St. James	Tumbler Ridge, BC	Fairview, BC
3							
4	H12	Houston	216	216	169	169	138
5	h17	Houston	216	216	169	169	138
6	h18	Houston	216	216	166	166	138
7	h26	Houston	216	216	169	169	138
8	H27	Houston	216	216	169	169	138
9	h37	Houston	216	216	169	169	138

Sources: Tsui et al. *Dryad Digital Repository*, 2011; Tsui, C.K. et al., *Mol. Ecol.*, 21, 2012.

Procedure 33.4

Diploid Data Analysis—Converting Input Files and Frequency-Based Analyses

Step	Instructions and Comments
1	Numerous population genetics programs are freely available, but they do require special format input. GenAlEx can convert a number of commonly used formats such as Genepop. Go to DRYAD digital repository and download the following excel file: **DRYAD.xlsx** (http://datadryad.org/resource/doi:10.5061/dryad.rk7n8; obtained from Graça et al. 2013b). Data were deposited as CONVERT file (Table 33.4). In order to use that data set in GenAlEx software package, we have to create an input file that the program can read. To do so, download the program CONVERT at: http://www.agriculture.purdue.edu/fnr/html/faculty/rhodes/students%20and%20staff/glaubitz/software.htm. CONVERT is a user-friendly Windows-based platform that facilitates transfer of codominant, diploid genotypic data for commonly used population genetic software packages. Note that only *diploid* data sets can be used with CONVERT program.
2	In DRYAD.xlsx file, two worksheets are available: *Geographic data* and *Allele Score*. Open *Allele Score* and save it as txt file.
3	Open CONVERT and select file tab. The user will be asked about the format of the existing data file. Select CONVERT, load your txt file and choose the desired format for file conversion (note that numerous options are available). Choose Genepop and save the file under a different name (Table 33.5).
4	Open the Excel file and go to the *Add-Ins* tab to select the GenAlEx option. Select *GenAlEx/Import data/Genepop* option. You will be prompted to select newly converted Genepop file. Save workbook and start with data analyses.
5	Select *Frequency option* to estimate allele frequency data parameters. Data will be automatically placed into following boxes: #Loci (A1) [10]; #Samples (B1) [148]; #Pops (C1). Select codominant data under two columns/loci and click ok.
6	Under Allele Frequency and Heterozygosity, select Frequency by Pop and Locus, Het, *Fstat &Poly by Locus* and *Pop* options.
7	Under Allelic Patterns Options select Allelic Patterns, Allele List, and Private Allele List.
8	Under *Multiple Pop Options* select *Nei Distance* and *Pairwise Fst*. Click ok and compare your results with Table 1 from Graça et al. (2013a).

TABLE 33.4

Insert of Data Used for Population Structure, Genetic Diversity, and Evolutionary Origin of Rust Fungus, *Puccinia psidii*

	A	B	C	D	E	F	G	H	I	J
1	PpsidiiALLBR									
2	npops = 7									
3	nloci = 10									
4		501		502		503		504		507
5	pop = EU									
6	SUZ5_EU	233	239	208	212	168	168	153	157	154
7	SUZ6_EU	233	239	208	212	168	168	153	157	154
8	SUZ7_EU	233	239	208	212	168	168	153	157	154
9	SUZ8_EU	233	239	208	212	168	168	153	157	154
10	SUZ9_EU	233	239	208	212	168	168	153	157	154
11	SUZ10_EU	233	239	208	212	168	168	153	157	154
12	SUZ11_EU	233	239	208	212	168	168	153	157	154
13	SUZ12_EU	233	239	208	212	168	168	153	157	154
14	SUZ13_EU	233	239	208	212	168	168	153	157	154
15	SP1_EU	233	239	208	212	168	168	153	157	154

Sources: Graça et al., *Mol. Ecol.*, 22, 2013a,; Graça, R.N. et al., *Dryad Digital Repository*, 2013b.
Note: Submitted data are in CONVERT format.

TABLE 33.5

Insert of Data Used for Population Structure, Genetic Diversity, and Evolutionary Origin of the Rust Fungus, *Puccinia psidii*

	A	B	C	D	E	F	G	H	I	J
1	10	148	7	70	4	63	2	4	3	2
2	PpsidiiALLBR		Import of DRYAD2	Pop1	Pop2	Pop3	Pop4	Pop5	Pop6	Pop7
3	Sample	Pop	501		502		503		504	
4	SUZ5_EU	Pop1	233	239	208	212	168	168	153	157
5	SUZ6_EU	Pop1	233	239	208	212	168	168	153	157
6	SUZ7_EU	Pop1	233	239	208	212	168	168	153	157
7	SUZ8_EU	Pop1	233	239	208	212	168	168	153	157
8	SUZ9_EU	Pop1	233	239	208	212	168	168	153	157
9	SUZ10_EU	Pop1	233	239	208	212	168	168	153	157
10	SUZ11_EU	Pop1	233	239	208	212	168	168	153	157
11	SUZ12_EU	Pop1	233	239	208	212	168	168	153	157
12	SUZ13_EU	Pop1	233	239	208	212	168	168	153	157
13	SP1_EU	Pop1	233	239	208	212	168	168	153	157

Sources: Graça, R.N. et al., *Mol. Ecol.*, 22, 2013a,; Graça, R.N. et al., *Dryad Digital Repository*, 2013b.
Note: Submitted data were converted using the program CONVERT to Genepop format and imported into GenAlEx for data analyses.

Procedure 33.5

Distance-Based Analyses—Principal Coordinate Analysis

Step	Instructions and Comments
1	To visualize the major patterns within a multivariate data set, PCoA option under *PCA* can be used. The mathematical explanation regarding PCoA calculations are beyond the scope of this exercise. Briefly, PCoA is a process in which the major axes of variation are located within a multidimensional data set. The first principal component has the largest possible variance, followed by the second or third axes that typically reveal most of the variations among them.
2	First, we need to create a matrix of genetic distances. Go to *Distance/Genetic* and select *Codom-Genotypic* option for *Distance Calculation*. Select *Liner Genetic* and *Output Total Distance Only* under *Distance Output Options*. Under *Output*, select To Worksheet, as *Tri Matrix*.
3	GenAlEx will create a new worksheet labeled GD. Select GD worksheet and go to *PCA* tab under GenAlEx Add-Ins tab. Select default options for *PCA Parameters*—Input data type as *Tri Distance Matrix*, *Covariance-Standardized* PCA Method, and *Color Code Pops* under Graph Options.

Questions

- What percentage of variation can be explained by first, second, and third axes?
- How well does this PCoA plot represent *Puccinia psidii* populations presented in the manuscript?
- Does the PCoA plot suggest genetic structure among *P. psidii* populations? If so, how many clusters are identified?
- Given the outcome of the PCoA results, what would you expect about the possibility for IBD for this data set?

Follow Procedure 33.6 to conduct the Mantel test.

Questions

- What conclusions can be drawn from the Mantel test? Are the results significant, and what is the value of *Rxy?*
- Is genetic similarity between population pairs associated with the distance between them? In other words, does this data set support the hypothesis of isolation-by-distance?
- How would you explain a correlation between populations and their host associations?

- Explain levels of genetic differentiation among examined populations. Are they high or low? Are those values statistically significant?
- What percentage of variation is based on differences among individuals and what percentage among populations?
- What would happen if we select genetic distance matrix (GD) as the input file for an AMOVA analysis? The resulting worksheet is labeled PhiPT. How do you explain those values?
- Based on your findings, can you draw main conclusions about population structure and genetic diversity of *P. psidii*?
- As a molecular biologist, how would you expand *P. psidii* research?

Procedure 33.6

Mantel Test and Analysis of Molecular Variance

Step	Instructions and Comments
1	To obtain geographic matrix, copy Geographic Data into main data sheet as shown in Table 33.6. Select *Distance/Geographic* and use default options to compute the pairwise geographic distance matrix (our X matrix). Our coordinates are in *Decimal Lat/Long* units so choose that option for *Data*. Select *Output To Worksheet* and *As Tri Matrix*. You can choose to transform your data as well. GenAlEx will create geographic distance matrix (GGD) worksheet that contains geographic matrix.
2	Select GGD worksheet and select *Mantel* option. The Mantel test is used to examine correlation between two matrices, in our case geographic and genetic distance. Select *Input Data Type* as *Tri Distance Matrix* and *999 permutations* for number of permutations. Make sure your *X Matrix* is the GGD and *Y Matrix* is GD.
3	AMOVA is a widely used statistical model for the estimation of molecular variation. It allows hierarchical partitioning of genetic variation among populations as well as estimation of *F*-statistics and/or their analogues. Open the main datasheet as shown in Table 33.5.
4	Select *GenAlEx/AMOVA*. Use default parameters and *Input Data Type as Raw Data. Select Codom-Microsat, Analysis for Total Only*, and *Output as Tri Matrix, Labeled Matrix* and *Sample.* Click ok, and select 999 permutations for both *Total Data Options* and *Pairwise Population Options.* Keep other options at default setting. GenAlEx will create worksheet labeled Rst, RstP, and LinRst.

TABLE 33.6

Insert of Data Used for Population Structure, Genetic Diversity, and Evolutionary Origin of Rust Fungus, *Puccinia psidii*

	A	B	C	D	E	F	G	H	I	J	U	V	W	X	Y
1	10	148	7	70	4	63	2	4	3	2					
2				Pop1	Pop2	Pop3	Pop4	Pop5	Pop6	Pop7					
3	Sample	Pop	501		502		503		504		513			x	y
4	SUZ5_EU	Pop1	233	239	208	212	168	168	153	157	212	212		−17.985	−39.976
5	SUZ6_EU	Pop1	233	239	208	212	168	168	153	157	212	212		−18.335	−39.703
6	SUZ7_EU	Pop1	233	239	208	212	168	168	153	157	212	212		−18.336	−39.702
7	SUZ8_EU	Pop1	233	239	208	212	168	168	153	157	212	212		−18.487	−39.985
8	SUZ9_EU	Pop1	233	239	208	212	168	168	153	157	212	212		−16.174	−39.615
9	SUZ10_EU	Pop1	233	239	208	212	168	168	153	157	212	212		−16.177	−39.632
10	SUZ11_EU	Pop1	233	239	208	212	168	168	153	157	212	212		−16.172	−39.649
11	SUZ12_EU	Pop1	233	239	208	212	168	168	153	157	212	212		−16.038	−39.14
12	SUZ13_EU	Pop1	233	239	208	212	168	168	153	157	212	212		−16.113	−39.316
13	SP1_EU	Pop1	233	239	208	212	168	168	153	157	212	212		−23.868	−47.906

Sources: Graça, R.N. et al., *Mol. Ecol.*, 22, 2013a,; Graça, R.N. et al., *Dryad Digital Repository*, 2013b.

Notes: Data included geographic coordinates for 148 isolate used in this study. Note that coordinates were placed under X and Y columns with one extra (column W) between data and coordinates. Coordinates were labeled X for latitude (decimal degrees, or DD) and Y for longitude (DD).

REFERENCES AND SUGGESTED READINGS

Abesha, E., and G. Caetano-Anolles. 2004. Studying the ecology, systematics, and evolution of plant pathogens at the molecular level. in: *Plant Pathology Concepts and Laboratory Exercises*, 2nd edition. R. N. Trigiano, M. T. Windham. and A. S. Windham (eds). Boca Raton, FL: CRC Press.

Baverstock, P. R., and C. Moritz. 1996. Project design. Pages 17–27 in: *Molecular Systematics*, 2nd edition. D. M. Hillis et al. (ed). Sunderland, MA: Sinauer Associates.

Benali, S., B. Mohamed, H. J. Eddine, and C. Neema. 2011. Advances of molecular markers application in plant pathology research. *Eur. J. Sci. Res.* 50:110–123.

Benson, D. A., I. Karsch-Mizrachi, D. J. Lipman, J. Ostell, and D. L. Wheeler. 2006. GenBank. *Nucleic Acids Res.* 34:16–20.

Brown, J. K. M. 1996. The choice of molecular marker methods for population genetic studies of plant pathogens. *New Phytologist* 133:183–195.

Caetano-Anolles, G. , R. N. Trigiano, and M. T. Windham. 2001. Patterns of evolution in Discula fungi and the origin of dogwood anthracnose in North America, studied using arbitrarily amplified and ribosomal DNA. *Curr. Genet.* 39(5):346-354.

Crossa, J., C. M. Hernandez, P. Bretting, S. A. Eberhart, and S. Taba. 1993. Statistical genetic considerations for maintaining germplasm collections. *Theor. Appl. Genet.* 86:673–678.

Dutech, C., J. Enjalbert J. E. Fournier, F. Delmotte, B. Barres, J. Carlier, D. Tharreau, and T. Giraud. 2007. Challenges of microsatellite isolation in fungi. *Fungal Genetics Biol.* 44:933–949, doi:10.1016/j.fgb.2007.05.003.

Fire, A., S. Xu, M. K. Montgomery, S. A. Kostas, S. E. Driver, and C. C. Mello. 1998. Potent and specific genetic interference by double-stranded RNA in *Caenorhabditis elegans. Nature* 391:806-811.

Graça, R. N., A. L. Ross-Davis, N. B. Klopfenstein, M. Kim M, T. L. Peever, P. G. Cannon, C. P. Aun, E. S. G. Mizubuti, and A. C. Alfenas. 2013a. Rust disease of eucalypts, caused by *Puccinia psidii*, did not originate via host jump from guava in Brazil. *Mol. Ecol.* 22:6033–6047. doi:10.1111/mec.12545.

Graça, R. N., A. L. Ross-Davis, N. B. Klopfenstein, M. Kim, T. L. Peever, P. G. Cannon, C. P. Aun, E. S. G. Mizubuti, and A. C. Alfenas. 2013b. Data from: Rust disease of eucalypts, caused by *Puccinia psidii*, did not originate via host jump from guava in Brazil. *Dryad Digital Repository.* doi:10.5061/dryad.rk7n8

Hadziabdic, D., B. M. Fitzpatrick, X. Wang, P. A. Wadl, T. A. Rinehart, B. H. Ownley, M. T. Windham, and R. N. Trigiano. 2010. Analysis of genetic diversity in flowering dogwood natural stands using microsatellites: the effects of dogwood anthracnose. *Genetica* 138:1047–1057.

Hadziabdic, D., L. Vito, M. Windham, J. Pscheidt, R. Trigiano, and M. Kolarik. 2014. Genetic differentiation and spatial structure of *Geosmithia morbida*, the causal agent of Thousand Cankers Disease in black walnut (*Juglans nigra*). *Curr Genet* 60:75–87. doi:10.1007/s00294-013-0414-x.

Jaccard, P. 1908. Nouvelles researches sur la distribution florale. *Bull. Soc. Vaudoise Sci. Natl.* 44:223–270.

Liang, P., and A. B. Pardee. 1992. Differential display of eukaryotic messenger RNA by means of the polymerase chain reaction. *Science* 257(5072):967-971.

Mantel, N. 1967. The detection of disease clustering and a generalized regression approach. *Cancer Res.* 27:209–220.

Marshall, D. R., and A. H. D. Brown. 1975. Optimum sampling strategies in genetic conservation. Pages 53–80 in: *Crop Genetic Resources for Today and Tomorrow.* O. H. Frankel and J. G. Hawkes (ed). Cambridge, UK: Cambridge Univ. Press.

Mullis, K. B., and F. A. Faloona. 1987. Specific synthesis of DNA in vitro via a polymerase-catalyzed chain reaction. *Meth. Enzymol.* 155:335–50.

Nei, M. 1973. Analysis of gene diversity in subdivided populations. *Proc. Natl. Acad. Sci. (USA)* 70:3321–3323.

Nei, M. and W. Li. 1979. Mathematical model for studying genetic variation in terms of restriction endonucleases. *Proc. Natl. Acad. Sci.* 76:5269–5273.

Nei, M. 1987. *Molecular Evolutionary Genetics.* New York, NY: Columbia University Press.

Peakall, R., S. Gilmore, W. Keys, M. Morgante, and A. Rafalski. 1998. Cross-species amplification of soybean (*Glycine max*) simple sequence repeats (SSRs) within the genus and other legume genera: Implications for the transferability of SSRs in plants. *Mol. Biol. Evol.* 15:1275–1287.

Peakall, R., and P. E. Smouse. 2012. GenAlEx 6.5: Genetic analysis in Excel. Population genetic software for teaching and research-an update. *Bioinformatics* 28:2537–2539.

Peakall, R., and P. E. Smouse. 2006. GENALEX 6: Genetic analysis in Excel. Population genetic software for teaching and research. *Mol. Ecol. Notes* 6:288–295.

Perumal, R., P. Nimmakayala, S. R. Erattaimuthu, E-G. No, U. K. Reddy, L. K. Prom, G. N. Odvody, D. G. Luster, and C. W. Magill. 2008. Simple sequence repeat markers useful for sorghum downy mildew (*Peronosclerospora sorghi*) and related species. *BMC Genetics* 9:77–90. doi:10.1186/1471-2156-9-77.

Rinehart, T. A., C. Copes, T. Toda, and M. Cubeta. 2006. Genetic characterization of binucleate *Rhizoctonia* species causing web blight on azalea in Mississippi and Alabama. *Plant Dis.* 91:616–623.

Schena, M., D. Shalon, R. W. Davis, and P. O. Brown. 1995. Quantitative monitoring of gene expression patterns with a complementary DNA microarray. *Science* 270(5235): 467-470.

Seshadri, K., R. Garemyr, E. Wallin, G.Von Heihne, and A. Elofsson. 1998. Architecture of β-barrel membrane proteins: Analysis of trimeric porins. *Protein Sci.* 7(9):2026-2032.

Shaad, N., and D. Frederick. 2002. Real-time and its application for rapid plant disease diagnostics. *Plant Pathol.* 24:250–258.

Singh, M., J. Saroop, and B. Dhiman. 2004. Detection of intraclonal genetic variability in vegetatively propagated tea using RAPD markers. *Biologia Plantarum* 48:113–115.

Sokal, R. R., and C. D. Michener. 1958. A statistical method for evaluating systematic relationships. *Univ. Kansas Sci. Bull.* 38:1409–1438.

Tautz, D. 1989. Hypervariability of simple sequences as a general source for polymorphic DNA. *Nucleic Acids Res.* 17:6463–6471.

Temple, B., P. A. Pines, and W. E. Hintz. 2006. A nine-year genetic survey of the causal agent of Dutch elm disease, *Ophiostoma novo-ulmi* in Winnipeg, Canada. *Mycol. Res.* 110:594–600.

Till, B. J., S. H. Reynolds, E. A. Greene, C. A. Codomo, L. C. Enns, J. E. Johnson, C. Burtner, A. R. Odden, K. Young, N. E. Taylor, J. G. Henikoff, L. Comai and S. Henikoff. 2003. Large-scale discovery of induced point mutations with high-throughput TILLING. *Genome Res.* 13:524–530.

Trigiano, R. N., G. Caetano-Anollés, B. J. Bassam, and M. T. Windham. 1995. DNA amplification fingerprinting provides evidence that *Discula destructiva*, the cause of dogwood anthracnose in North America, is an introduced pathogen. *Mycologia* 87:490–500.

Trigiano, R. N., P. A. Wadl, D. Dean, D. Hadziabdic, B. E. Scheffler, F. Runge, S. Telle, M. Thines, J. Ristaino, and O. Spring. 2012. Ten polymorphic microsatellite loci identified from a small insert genomic library for *Peronospora tabacina*. *Mycologia* 104:633–640. doi:10.3852/11-288.

Tsui, C. K., A. D. Roe, Y. A. El-Kassaby, A. V. Rice, S. M. Alamouti, F. A. Sperling, J. E. Cooke, J. Bohlmann, and R. C. Hamelin. 2011. Data from: Population structure and migration pattern of a conifer pathogen, *Grosmannia clavigera*, as influenced by its symbiont the mountain pine beetle. *Dryad Digital Repository*. doi:10.5061/dryad.48jg20v0.

Tsui C. K., A. D. Roe, Y. A. El-Kassaby, A. V. Rice, S. M. Alamouti, F. A. Sperling, J. E. Cooke, J. Bohlmann, and R. C. Hamelin. 2012. Population structure and migration pattern of a conifer pathogen, *Grosmannia clavigera*, as influenced by its symbiont, the mountain pine beetle. *Mol. Ecol.* 21:71–86.

Vos, P., R. Hogers, M. Bleeker, M. Reijans, T. van de Lee, M. Hornes, A. Frijters, J. Pot, J. Peleman, and M. Kuiper. 1995. AFLP: A new technique for DNA fingerprinting. *Nucleic Acids Res.* 23:4407–4414.

Warburton, M. L., X. Xianchun, J. Crossa, J. Franco, A. E. Melchinger, M. Frisch, M. Bohn, and D. Hoisington. 2002. Genetic characterization of CIMMYT inbred maize lines and open pollinated populations using large scale fingerprinting methods. *Crop Sci.* 42:1832–1840.

Weinberg, W. 1908. Über den Nachweis der Vererbung beim Menschen. Jahresh Verein f vaterl Naturk Württem 64:368–382. On the demonstration of heredity in man. Pages 4–15 in: *Papers on Human Genetics*. S. H. Boyer (ed). 1963. Englewood Cliffs, NJ: Prentice-Hall.

Weir, B. S. 1996. Intraspecific differentiation. Pages 385–403 in: *Molecular Systematics*. 2nd edition, D. M. Hillis et al. (ed). Sunderland, MA: Sinauer Associates.

Welsh, J., and M. McClelland. 1990. Fingerprinting genomes using PCR with arbitrary primers. *Nucleic Acids Res.* 18:7213–7218.

White, T. J., T. Bruns, S. Lee, and J. Taylor. 1990. Amplification and direct sequencing of fungal ribosomal RNA genes for phylogenetics. Pages 315–322 in: *PCR Protocols: A Guide to Methods and Applications*, M. A. Innis, D. H. Gelfand, J. J. Sninsky, and T. J. White (eds). San Diego, CA: Academic Press.

Williams, J. G. K., A. R. Kubelik, K. J. Livak, J. A. Rafalski, and S. V. Tingey. 1990. DNA polymorphisms amplified by arbitrary primers are useful as genetic markers. *Nucleic Acids Res.* 18:6531–6535.

Wright, S. 1951. The genetical structure of populations. *Ann. Eugen.* 15:323–354.

Glossary

ΔTm. Average melting temperature at which 50% of the DNA molecules are single- versus double-stranded; as temperature increases, hydrogen bonds break and double-stranded DNA separates to single-stranded.

Aberrant ratio. Change in the normal segregation of phenotypic traits during the crossing of two plants; *Barley stripe mosaic virus* (BSMV) causes aberrant ratios in corn that it infects.

Abiotic. Not caused by a biological agent such as an insect, mite, or pathogen. Abiotic disorders are associated with nonliving causal factors.

Abscission. Shedding of leaves or other plant parts that results from a physical weakness in a specialized layer of cells (the abscission layer) that develops at the base of the structure.

Abscission zone. Specialized cells at the base of the structure; referred to as the abscission layer or zone.

Absorption. Movement of water and solutes from the exterior environment into the various locations of the plant, including cells.

Acervulus (pl. acervuli). A flat, saucer-shaped bed of short conidiophores that grow side-by-side within the host tissue beneath the epidermis or cuticle of the host plant.

Action threshold. The pest level at which control measures should be deployed to avoid economic losses. See also **economic threshold**.

Active defense. Defense mechanisms that are present after pathogen recognition.

Additive gene action. When referring to the relationship between different alleles at the same locus, additive action indicates that dominance is largely absent and that both alleles contribute to the same phenotype. When referring to a set of different genes or loci all affecting the same trait, additive action implies that the effect of each allele at each locus on the trait is constant and independent of which alleles are present at all other loci affecting that trait.

Adult-plant resistance (APR). Resistance that becomes apparent during the adult stage of plant growth.

Aeciospore. A dikaryotic spore of a rust fungus that is formed in an aecium.

Aecium (pl. aecia). A dikaryotic stage in the life cycle of some rust fungi that is formed after fertilization; contains chains of aeciospores.

Aerobic. Requiring oxygen for respiration.

Agarose. Highly purified fraction of agar.

Aggressiveness. Amount of disease caused by an isolate of a pathogen.

Agonomycetes. See **Mycelia Sterilia**.

Agroecology. Application of ecological principles, or the relationships among and between living organisms and their environment, to the management of agricultural systems.

Agroecosystem. The totality of all of the influences on agricultural endeavors including, for examples, pollution control, native species preservation, watershed management among others.

Aleuriospore. Asexually produced fungal spore that originates by swelling of a terminal or lateral cell of a hypha.

Alkaloid. Nitrogenous organic compound with alkaline properties, usually bitter in taste, poisonous to animals, and produced by certain plants.

Alleles. One of a pair of genes at a single locus on chromosomes or plasmids.

Alloinfection. Pathogen infection of a host plant by inoculum from another individual plant.

Alternate host. One of the two different plants in the life cycle of a heteroecious rust fungus. Commonly refers to the host plant on which the spermagonial and aecial stages are formed, but sometimes used to refer to the less economically important host in the rust fungus life cycle.

Aminopyrrolizidine. An alkaloid known as a toxic principle in plants poisonous to animals and dangerous to man.

Amphid. Sensory organs located on the head region of nematodes.

Amphimitic. Reproduction by male and female nematodes that results in the formation of eggs.

Amplicon. A portion of DNA amplified or copied during the polymerase chain reaction.

Amplified fragment length polymorphism (AFLP). A method used for detecting polymorphisms in DNA by restriction enzyme digestion of DNA, amplification of restriction fragments, and visualization.

Amylase (amylolytic). Enzymes that degrade starch to maltose and D-glucose.

Anaerobic. Does not require oxygen for respiration (fermentative).

Anaerobic soil disinfestation. Soil incorporation of a labile carbon source to stimulate soil microbial growth and respiration, mulching with plastic to limit gas exchange, and irrigation of the topsoil to flush soil pore space with limited resupply of oxygen.

Analysis. The resolution of problems by representing component processes through mathematical equations.

Anamorph. Conceptually, the asexual aspect of a fungus; contrast to **holomorph** and **teleomorph**.

Anastomosis. Among fungi, the fusion of two hyphae; may sometimes result in genetic recombination.

Anastomosis groups (AG). Refers to *Rhizoctonia* isolates having three or more nuclei per hyphal cell (multinucleate).

Angiosperms. Flowering plants.

Annealing temperature. Temperature at which rejoining of single strands of DNA into the double helix occurs or temperature at which primers pair with DNA.

Annulations. In nematodes, circular grooves around the outside of the body, providing flexibility to the nematode during movement. May be very shallow, appearing as just markings, or very deep, giving the appearance of distinct segmentations, or rings.

Annulus. The ring around the stipe that is a remnant of the inner veil in the developing mushroom that once enveloped only the gills or pores.

Anomeric. A carbon atom existing in two isomeric forms, α and β forms of carbon 4 in a glucose molecule. The α and β forms are sometimes termed "up" and "down," respectively, and refer to the hydroxyl group. A molecule such as D-glucose can be a **reducing sugar** because of the anomeric carbon.

Antagonism. Association between organisms in which one or more of the participants is harmed or has its activities limited by the other through antibiosis, competition, parasitism, or predation.

Antagonist. Organism that harms another organism through its activities.

Antheridium (pl. antheridia). Specialized cell in which nuclei are formed, which function as male gametes, and is therefore a male gametangium; contrast to gynecium.

Anthracnose. Sunken lesions that occur on leaves, stems, and fruits.

Antibiosis. Inhibition or destruction of one organism by a metabolite produced by another organism.

Antibiotic. Metabolite produced by one organism that is harmful to another organism.

Antibody. New or modified protein produced by an animal lymphatic system in response to a foreign substance (called an antigen); the antibody binds with the antigen, and this binding causes the antigen to become inactive.

Antigen. Usually a foreign protein (occasionally a lipid, polysaccharide, or nucleic acid) that is introduced into animal tissue and reacts by producing antibodies that specifically bind and inactivate the protein.

Antisense RNA. RNA strand that is complementary or antisense to a positive sense RNA strand.

Antiserum (pl. antisera). The serum fraction of the blood from an animal that has been immunized with an antigenic substance; the serum contains a number of different immunoglobulin types and reactivities.

Apophysate. A natural swelling, projection, or outgrowth of an organ or part.

Apoplast. A space, outside the plasma membrane, composed of cell walls and air spaces, where the transport of water is facilitated.

Apoptosis. Biochemical events that lead to characteristic morphological cell changes and cell death.

Apothecium (pl. apothecia). Ascomycete fruiting structure that is cup-shaped or based on a cup shape in which exposed asci are formed; the structure develops as a result of nuclear exchange, typically involving an ascogonium and antheridium.

Appressorium (pl. appressoria). Terminal or intercalary cell of a germ tube of a fungus used for attachment to the host.

Arbitrarily amplified DNA (AAD). Collective term that describes those nucleic acid amplification techniques that use arbitrary primers. AAD includes several well-established techniques that provide genetic information from a plurality of sites (loci) in a genome or nucleic acid molecule, such as random amplified polymorphic DNA (RAPD), arbitrarily primed polymerase chain reaction (AP-PCR), DNA amplification fingerprinting (DAF), and AFLP.

Arbitrary primers. Generally, short (7–12 base pairs) segments of DNA that do not correspond to known genomic sequences and thus may pair many sites (arbitrary). They are used in PCR methods such as RAPD and DAF.

Arbuscule. Special, dichotomously branched haustorium that develops intracellularly and serves as the site of nutrient and carbon exchange in glomalean endomycorrhizae.

Archaea. The domain that includes unicellular microorganisms that are genetically distinct from bacteria and eukaryotes. These organisms often inhabit extreme environments and include methanogens, halophiles, and thermophiles.

Area under the curve. The region of a graph bounded by axes and a line, where the line depicts disease levels over time.

Argonaute protein family. A group of proteins that regulate RNAi by binding noncoding regions.

Arthropods. Any member of the Phylum Arthropoda; arthropods have distinct body segments, exoskeleton, jointed appendages, and bilateral symmetry.

Arthrospore. An asexual spore formed when hyphal fragments break into unicellular sections.

Ascocarp. Fungal tissue in or on which asci are formed; a fruiting body. Synonym: ascoma.

Ascogenous hyphae. Dikaryotic hyphae that will give rise to asci.

Ascogonium (pl. ascogonia). Specialized cell that functions to accept nuclei from an antheridium and is therefore a female gametangium.

Ascomycetes. Member of the Ascomycota.

Ascomycota. A diverse group (phylum) of fungi that are characterized by the production of asci usually containing eight ascospores.

Ascospores. Fungal spores produced in a saclike structure called an ascus, most often following sexual recombination.

Ascus (pl. asci). Specialized cell in which nuclei fuse, undergo meiosis, and are subsequently incorporated into ascospores.

Aseptic. Free of all living organisms.

Asexual reproduction. Vegetative reproduction of an organism accompanied by mitotic divisions of nuclei in which the chromosome number is not reduced by meiosis.

AUDPC. Area under the disease progress curve; see also **Area under the curve**.

Autoclave. A chamber that sterilizes materials using pressurized steam; a temperature of 121°C for 15 min is required to kill most microbes.

Autoecious. A rust fungus that completes its life cycle on a single host species.

Autoinfection. Pathogen infection of a host plant by inoculum arising from that same individual plant.

Autotrophs. Organisms that produce their own food by photosynthesis.

Auxins. Plant growth regulators (hormones) affecting shoot elongation and other functions.

Avirulence. Inability of a pathogen to cause a compatible (susceptible) reaction on a host cultivar with genetic resistance.

Avirulence (Avr) gene. A gene whose translated product interacts with a host resistance gene product, resulting in a resistant (incompatible) interaction.

Avoidance. Creating a situation or environment to reduce the likelihood of disease development.

Axenic. Literally, without a stranger. Free from all microbial contamination (describing a pure culture).

Bacilliform viruses. Viruses that are short thick rods which are rounded on both ends; these viruses are constructed uniquely and often have a glycoprotein also in their capsid.

Bactericide. A chemical compound that kills or inhibits the growth of bacteria.

Bacteriocin. Antibiotics that inhibit specific groups of microorganism that are closely related to the microorganism that produced the chemical.

Bacteriophage. A phage "virus" that infects bacteria.

Bacterium. Any number of prokaryotic organisms that reproduce by binary fission and may or may not produce spores.

Baiting. The use of a living organism, instead of artificial culture media, to promote the growth of a particular microorganism.

Bark splitting. Localized death of the cambium layer of tree trunk that leads to bark pulling away from the sapwood.

Base pairs. A pair of complementary nitrogenous bases in a DNA molecule that form a "rung of the DNA ladder." Also, the unit of measurement for DNA sequences.

Basidiocarp. Fruiting structure in the Basidiomycota.

Basidiomycetes. Member of the Basidiomycota.

Basidiomycota. Diverse group (phylum) of fungi that are characterized by the production of basidia and typically four basidiospores.

Basidiospore. Ephemeral, thin-walled, haploid (N) spore of basidiomycetes formed following nuclear fusion and meiosis, formed on a **basidium**. In rusts and smuts formed upon germination of teliospores.

Basidium (pl. basidia). A structure in the life cycle of a member of Basidiomycota in which karyogamy and meiosis occurs, typically producing a finite number of haploid **basidiospores**.

Binary fission. Type of asexual reproduction in unicellular organisms (e.g., bacteria) where the parent cell grows and splits into two approximately equal daughter cells.

Binomial. Genus and species designation for an organism.

Bioassay. A technique using a plant cell, tissue, or organ as a source to test their response against another organism or chemicals under controlled conditions. May also be used to detect chemical agents.

Biocontrol agents (BCAs). Natural or genetically modified pathogens, parasites, or predators that are deliberately introduced in the environment or, if already present, are encouraged to multiply with the purpose of reducing the survival and/or activity of pest organisms.

Biofilms. Bacterial cells embedded in extracellular polysaccharides. Zoogloea are examples of biofilms occurring among plant cells.

Biofumigation. A process in which amendments to soil that suppresses the number and growth of pathogens. One example is the use of brassica crops (e.g., mustard or broccoli) incorporated into the soil. See **glucosinolates** and **isothiocyanates**.

Biolog system. A method that employs the ability to utilize different growth substrates to identify bacteria.

Biological control. Use of natural or modified organisms, genes or gene products, to reduce the effects of undesirable organisms such as plant pathogens, and to favor desirable organisms such as crop plants.

Biopesticides. Certain types of pesticides derived from such natural materials as animals, plants, bacteria, and certain minerals. They include naturally occurring metabolites, live organisms, or pesticidal substances produced by genetically modified plants.

Biotic disease. Disease caused by living pathogens including viruses.

Biotroph (biotrophic). Organism that can live and multiply only on or within another living organism (i.e., obligate parasite).

Birefringence (double refraction). Division of a light ray into two rays when it passes through certain types of material, such as crystals, starch grains, and thick cell walls.

Bitunicate ascus. Ascus with a double wall.

Blastospores. Spores produced by budding.

Bleaching. Loss of color.

Blight. Rapid blackening of host tissue that may include leaves, stems, and flowers.

Bordeaux mixture. Copper sulfate and lime; one of the first fungicides.

Callus (pl. calli). Tissue mass (usually not uniform) arising from disorganized proliferation of cells and tissues.

Canker. Localized infection of the bark on the stem or trunk.

Capillary water. Soil water that is held cohesively as a continuous layer around soil particles and in pores between soil particles.

Capsid. The protein coat or shell that encapsulates or encloses a viral genome.

Cellulase (cellulolytic). A collection of hydrolytic enzymes capable of degrading native and modified celluloses to cellobiose and D-glucose.

Cellulose. β-(1-4) linkages of D-glucose; major component of the cell wall of the Oomycota and plants.

Central paradigm of molecular biology. The unidirectional flow of genetic information from DNA to **mRNA** to protein.

Central vacuole. The large membrane-bound fluid-filled body found in plant cells. It functions in storage of chemicals, enzymes, and maintenance of osmotic potential.

Ceratobasidium anastomosis group (CAG). *Rhizoctonia* isolates having two nuclei per hyphal cell (binucleate).

Character. An observable feature of an organism that can distinguish it from another.

Chasmothecium. A powdery mildew ascocarp.

Chelating agent. A chemical with the ability to bind divalent ions such as calcium or magnesium.

Chemical control. The act of killing or suppressing the growth and reproduction of a pathogen with the use of various compounds such as a fungicide.

Chemotaxis. Swimming orientation of a zoospore in response to an external chemical stimulus.

Chilling. Low-temperature damage to plants associated with temperatures above freezing, but low enough to cause injury. Chilling injury is due to a sudden drop in temperature during an active period of plant growth or development.

Chimera. A single plant with two genetically different tissue types, for example, leaf variegation.

Chitin. β-(1-4) linkages of *n*-acetylglucosamine; chief structural component of the cell wall of higher fungi (kingdom Fungi).

Chitinases. An enzyme that degrades chitin, a primary polymer of cell walls of fungi, which can elicit plant defense responses.

Chlamydospore. Thick wall resting spore characteristic of some fungi.

Chloroplast. Specialized membrane-bound cellular organelle containing the chlorophyll pigments that are involved in the photosynthetic processes of the plant.

Chlorosis. Abnormal yellowing or fading of the normal green color of leaves due to loss of chlorophyll production or increase in chlorophyll degradation.

Chlorotic. The absence or decrease in the pigment chlorophyll causing distinctive yellowing or whitening.

Chromosome(s). One of the threadlike "packages" that contains genetic information such as genes and other DNA in the nucleus of a cell. Chromosomes generally occur in pairs: one obtained from the mother; the other from the father.

Chytridiomycota. Fungal phylum where a true mycelium is lacking, uniflagellate zoospores are present and where asexual reproduction occurs when the mature vegetative body transforms into thick-walled resting spores; in the kingdom Fungi.

Cibarial pump. The muscularized chamber in Homopterans that serves to intake fluid during feeding.

Circumnutation. Also called revolving nutation, is the spiral movement of seedlings and twining stems caused by unequal growth at the meristem or turgor pressure differences in cells on one side of the stem versus the other.

Cirrus (pl. cirri). A mass of spores, usually in a gelatinous matrix, that forms a droplet or ribbon as it is forced from the fungal fruiting body.

Clade. A monophyletic (single lineage or origin inferred through cladistic analysis) group that are each other's closest relatives.

Clamp connection. A structure connecting two hyphal cells and involved in continuation of the dikaryotic condition for some basidiomycetous fungi.

Classification. A systematic arrangement in groups or categories according to established criteria.

Clavate. Club-shaped.

Clavicipitaceous fastidious fungi. An **endophyte** of cool season grasses.

Cleistothecium (pl. cleistothecia). Sexual fruiting body of an ascomycete that lacks a natural opening for spore discharge. Spores are usually dispersed following breakage by freezing, thawing, or wear and tear.

Cluster analysis. Clustering is grouping a set of objects in such a way that objects in the same group (called a cluster) are more similar (in some sense or another) to each other than to those in other groups (clusters).

Cockles. Wrinkled areas formed on seeds.

Codon. Set of three adjoined nucleotides, or triplet, that specifies a particular amino acid or signal sequence for a polypeptide chain during protein synthesis.

Coelomycetes. An old informal class of mitosporic fungi where spores are produced in conidiomata.

Coenocytic. Lacking, or mostly lacking, cross walls or aseptate.

Colonization. The establishment of a pathogen on a host or an inoculated substrate.

Colony. Discrete or diffuse, circular visible cluster of fungal hyphae and/or spores, or bacterial cells growing on the surface of or within a solid medium, presumably arising from one cell or one grouping of cells.

Colony forming units (CFU). Number of colonies formed per unit of volume or mass of a cell or spore suspension

Color break. A mosaic pattern that develops in the floral parts of the plants due to changes in the color pigments.

Columella. Swollen tip of a conidiophore that bears asexual sporangiospores (or mitospores); in the Zygomycota.

Commensalism. Two organisms living in close association and neither having an obvious advantage.

Competition. A process where two or more organisms try to utilize the same food or mineral source or occupy the same niche or infection site.

Competitive saprophytic ability. An intrinsic characteristic of a fungal species that results in successful competitive colonization of organic substrates.

Competitors. Chemical agents that are utilized to show a reaction by competing with the substrate of interest for enzymatic binding site or reactive chemicals.

Competitors. Chemical compounds that are utilized to slow down a reaction by competing with the substrate of interest for enzymatic binding sites or for reactive chemicals.

Concatenate. Things, such as spores, that are linked together in a chain or series.

Conidiogenous cell. The hyphal cell from which a conidium is formed.

Conidiomata. Asexual fruiting structure (e.g., acervulus, pycnidium, and synnema).

Conidiophore. Simple or branched hyphae where mitosis occurs and conidia are produced.

Conidium (pl. conidia). Asexually produced spore borne on a conidiophore.

Conjugation. The one-way transfer of DNA between bacteria in cellular contact.

Conservation tillage. Method of soil cultivation that allows **crop residue** to remain on top of the soil to reduce soil erosion and runoff.

Contact fungicides. Chemicals applied as foliar sprays to protect above ground plant parts from infection; they may also be used as seed treatments.

Contig. Assemblage of several or many small (100–400 bp) DNA segments into larger units.

Contrast. The ratio of light and dark. To produce a good image, you must have good contrast.

Convergent evolution. Two plant lineages that are currently very similar, but where each lineage derived from phylogenetically distinct ancestors. The term can also be used to describe characters (traits) within such lineages.

Cover crops. Plants that are grown primarily for improvement of the agroecosystem (e.g., biological nitrogen fixation, increased soil organic matter, erosion prevention, and pest suppression) rather than for harvest and market.

Crop residue. The vegetative portions of the crop that remains after harvest.

Crop rotation. Growing various plants that differ in disease susceptibility in a given area.

Cross protection. A form of competition in which an avirulent or weakly virulent strain of a pathogen (e.g., virus) is used to protect against infection from a more virulent strain of the same or closely related pathogen (e.g., virus).

Cross protection. The utilization of an avirulent or weakly virulent strain of a virus to provide protection from a more virulent strain of the same or closely related pathogen.

Cultural control of plant diseases. Any practice other than use of chemical fungicides or bactericides used to prevent disease incidence or spread. May include site selection, managing fertility, etc.

Cultural practice(s). Methods used in planting and maintaining crops and ultimately suppressing plant pathogens.

Culture. To grow a microorganism or other live cells or tissue on an artificially prepared nutritional medium.

Culture medium. An artificially prepared solid or liquid medium on which microorganisms or other cells are grown.

Cuticle. The waxy covering of plant epidermal cells whose main function is to prevent desiccation.

Cytokinins. Plant growth regulators (hormones) affecting cell division and other functions.

Damage threshold. The amount of crop damage that is greater than the cost of management measures.

Damping-off. Rotting of seeds in the soil or the death of seedlings soon after germination.

Demicyclic. Rust fungus that lacks a uredinial (repeating) stage in the life cycle.

Deletions. Losing a nucleotide base from the sequence of a gene causing a reading **frameshift**.

Deleterious rhizosphere microorganisms. Organisms that do not penetrate plant vascular tissue but that have a negative impact on crop growth and associated with **yield decline**.

Depth of field. Distance along the optical path throughout that the specimen can be seen with clarity.

Diagnosis. Process or procedure for determining the cause of a plant disease or other problem; the process of disease identification.

Diazotrophs. Beneficial bacteria that can fix atmospheric nitrogen via the nitrogenase enzyme complex. The largest groups are rhizobia, which form symbiotic associations with leguminous plants.

Dicots. Plants having two cotyledons (as beans).

Dieback. Necrosis that begins at the top of a plant and progresses downward.

Dikaryotic (Dikaryon). Having two haploid nuclei per cell (N + N), a condition found in most basidiomycete cells and the ascogenous hyphae of ascomycetes.

Diploid. Having a single nucleus containing two sets of chromosomes (2N), one from each parent.

Disease. Condition detrimental to the normal development of a plant resulting from the continuous interaction between the plant and a causal agent leading to the production of symptoms.

Disease complexes. Diseases caused by a combination of pathogens such as fungi and nematodes.

Disease cycle. Series of events, processes, and structures involving both host and pathogen in the development of a disease, repeated occurrence of the disease in a population, and injurious effects of the disease on the host.

Disease diagram. Pictorial representation of a disease level on a leaf or plant.

Disease incidence. Number or proportion of plants in a population that are affected by disease.

Disease progress. Development of disease, usually in time.

Disease resistance. Ability of a plant to block establishment or reproduction of the pathogen.

Disease suppressive soils. Soils in which the pathogen does not establish or persist, establishes, but causes little or no damage, or establishes and causes disease for a while, but thereafter the disease is less important.

Disease-free (pathogen-free). Loosely used phrase to denote that specific plant pathogens are not present on or in selected plants. Other pathogens may or may not be present.

Disinfectant. Chemical agent that is used to eliminate plant pathogens from the surface of plants, seeds, or inanimate objects such as greenhouse benches, tools, pots, and flats.

Disinfested. Process of removing or killing unwanted microorganisms. Synonym: **surface sterilization**.

Dispersal. To separate or move in different directions.

Disperse. Distribute or spread a disease over an area.

Distortion. Changes in the lamina of the host leaves or in host fruits resulting in areas that are twisted, deformed, or distorted; Deformations may be described as blisters, bubbles, rumpling, rugosity, or twisting. *Bean pod mottle virus* (BPMV) produces distortions in infected soybean leaves.

DNA amplification fingerprinting (DAF). Arbitrarily amplified DNA technique that uses one or more synthetic oligonucleotide primers of typically 5–10 nucleotides in length and high primer-to-template ratios. Amplification products are generally visualized by polyacrylamide gel electrophoresis and silver staining, and fingerprints are highly complex. Synonym: DNA profile.

DNAases. Enzymes that degrade DNA.

DNA homology. Degree or percentage of hybridization between the DNA of different organisms. Nucleic acid sequences are similar because they have a common evolutionary origin.

DNA microarrays. DNA microarray is a collection of microscopic DNA spots attached to a solid surface, such as glass, plastic, or silicon chip forming an array.

DNA polymerase. Enzyme that makes the DNA polymer.

DNA polymorphism. A variation in a DNA sequence detected through DNA sequencing or DNA analysis (fingerprinting, profiling, etc.).

DNAse. Enzyme that degrades or destroys DNA.

Dolipore septum. Complex septum found in the hyphae of basidiomycetes that prevents the movement of organelles between cells.

Dominant. A relationship between alleles of a gene in which one allele masks the expression (phenotype) of another allele at the same locus; only one copy of a dominant allele is needed to produce the corresponding phenotype.

Downy mildews. Diseases caused by obligate fungal-like parasites classified in the phylum Oomycota and the family Peronosporaceae.

Durable resistance. Resistance that remains effective during a prolonged and widespread use in an environment favorable for disease.

Dwarfing. Reduction in the size of a plant; dwarfing and stunting are closely related terms.

Ebb and flow irrigation (flood and drain). Potted plants are placed in a tray or nonporous floor, and water is used to fill the tray or floor area. After the potting medium reaches the desired moisture level, the excess water is drained away. Often, the excess water is returned to a reservoir and used again at the next watering.

Economic injury level. Pathogen or disease level that causes loss of production.

Economic threshold. The costs of controlling the pest or disease costs more than the return in production and cash value. See also **action threshold**.

Ectomycorrhizae. Symbiotic, thought to be mutualistic, association between a plant root and fungus. The fungus forms a layer or mantle around the root and grows inward between the outer cortical root cells. The complex of hyphae around and in the root is referred to as a "Hartig net." Fungus receives carbohydrates from the plant, and the plant gains greater access to water and mineral nutrients and some protection from root pathogens.

Ectoparasite. In nematology, parasitic nematodes that feed on roots from the outside.

Ectotrophic hyphae. Root-infecting hyphae that develop mainly on the root surface.

Effector(s). A protein or small molecule produced by the pathogen that facilitates infection by altering host-cell structure or function. Many effectors suppress host defense responses and kill or weaken the host cell to make it more susceptible to pathogenesis.

Effector triggered immunity (ETI). Plant immune response triggered by pathogen effectors. The ETI response is reliant on **R genes** and is activated by specific pathogen strains. An ETI response often causes **apoptosis** or the **hypersensitive response**.

Effector(s). A pathogen molecule, usually a protein, that is translocated into host cells where it may act to directly manipulate host innate immunity, also known as **effector triggered immunity (ETI)**.

Egestion. The expulsion of food materials through the aphid stylet; aphids alternate between ingestion (the intake of food materials) and egestion during feeding. It may serve to ensure that the stylet is not blocked during feeding.

Electrophoresis (gel). Movement of charged molecules, such as DNA and proteins, through a supporting medium (a gel) that is usually composed of starch, agarose, or acrylamide. Molecules are usually separated by weight.

ELISA. Enzyme-linked immunosorbent assay; a serological technique that employs binding antibodies or antigens to a polystyrene microtiter plate and utilizes antibodies linked to an enzyme by glutaraldehyde for detection.

Enations. Small overgrowths occurring on the leaf are called enations; Enations are typically produced on the underside of the leaf following the vein patterns. *Pea enation mosaic virus* (PEMV) is named for the enations that it produces.

Encapsidate. Process of packaging the nucleic acid of a virus within its protein coat or capsid.

Endemic. Prevalent in or peculiar to a particular locality or region.

Endobiotic. Living entirely in the interior of a host; in the Plasmodiophoromycota.

Endoconidium (pl. endoconidia). A conidium produced in the interior of a conidiogenous cell.

Endonuclease. Enzymes that cleave the phosphodiester bonds within a polynucleotide chain at a specific sequence.

Endoparasites. In nematology, nematodes that penetrate the root and feed on tissue inside the root.

Endophyte. Plant, fungus, or bacteria developing and living inside another plant. Some are pathogens or parasites and others are apparently nonpathogenic inhabitants. The degree of endophytic intimacy varies; some endophytes live within cells and others colonize the surface of living cells within the plant (pseudomonads and xanthomonads).

Endosphere. The internal transport system of plants (vascular system composed of xylem, phloem, and associated tissues).

Endopolygalacturonases (EPGs). See **pectinases**.

Endoxylanases. Degrades cross-linking glycans (hemicelluloses) found in the cell wall of plants. See **xylanases** and **pectinases** for explanation of endo and exo.

Environment. The conditions, influences, or forces that influence living forms.

Enzyme. A protein that is a catalyst for chemical reactions without being altered or degraded during the process; in fungi and bacteria.

Enzyme-linked immunosorbent assay (ELISA). A biochemical technique used to detect the presence of an antibody or an antigen in a sample.

Epidemic. Any increase in the occurrence of disease in a population.

Epidemiology. The study of factors that lead to a change in disease.

Epidermis. A tissue composed of the outer layer of cells on a plant.

Epigenetic. Environmentally induced changes in appearance or response of an organism that are not permanent.

Epiillumination. Light is projected onto the specimen from above. This is typically used with opaque samples.

Epinasty. A downward curling of plant leaves or parts due to the overgrowth of the upper surface.

Epiphytic. Growing on the plant surface without harming it.

Eradicant fungicides. Chemical compounds that halt fungal growth if applied shortly after infection.

Eradicate. To eliminate a pathogen from a plant or field.

Eradication. Eliminating a pathogen from an area by destruction of infected materials.

Ergot. A sclerotium produced by *Claviceps purpurea* and related species in the floret of infected grains.

Etiolation. Excessive elongation of internodes in plants usually in response to low light or disease.

Etiology. The study of all the causes of a disease or abnormal condition.

Eukaryote (eukaryotic). Organisms that have a nucleus as well as other double membrane organelles.

Exoenzymes. Enzyme that acts outside the cell that produces it or an enzyme that acts on the terminal end of a polymer. Synonym: **extracellular enzyme**.

Exogenous. Not produced within the organism.

Exoglycome. Oligosaccharide chains on the outer surface of the plasmalemma.

Exopolysaccharide(s). See **Extracellular polysaccharide(s) (EPS)**.

Exotic host. A host that is introduced into a new geographic area.

Exotic pathogen. A pathogen that is introduced into a new geographical area.

Exponential model. A mathematical equation used to represent disease development in time; this model allows infinite increase in disease.

Expressed sequence tag (EST). A short strand of DNA (approximately 200 base pairs long) that is part of a cDNA. Because an EST is usually unique to a particular cDNA, and because cDNAs correspond to a particular gene in the genome, ESTs can be used to help identify unknown genes and to map their position in the genome.

Extension temperature. Temperature in PCR cycle that is optimum for the activity of DNA polymerase (cf. 72°C).

Extracellular enzyme. A protein (enzyme) produced with the cytoplasm of fungi and bacteria that is transported across plasmalemma to act on a substrate in the environment, for example, cellulases. May be involved in pathogenesis or nutrition of the organism or both.

Extracellular polysaccharide(s) (EPS). Polysaccharides excreted by bacterial cells. See also **Exopolysaccharides**.

Extracorporeal digestion. Digestive enzymes produced in the salivary glands of nematodes that are injected through the stylet into a plant cell.

Facultative fermenter. An organism that can derive energy using oxygen (aerobic) or fermentative (anaerobic) processes.

Facultative parasite. An organism that normally lives as a saprophyte but under certain conditions can become a parasite.

Facultative saprophyte. An organism that normally lives as a parasite but under certain conditions can become a saprophyte.

Fallow. Tilled land that during the growing season has not been seeded with a crop. This is a tactic used in **crop rotation**.

Fastidious prokaryotes. Organisms that have very rigid nutritional requirements for growth and reproduction.

Fertility. The ability of a soil to sustain plant growth and development, and yield.

Field. The diameter of the viewing area. As magnification is increased, the field of view is decreased.

Fingerprints. A bar-code-like occurrence which can distinguish the uniqueness of one individual from another.

Flagellum (pl. flagella). A motile appendage projecting from a zoospore. Some flagella have hairs (e.g., stramenipilous flagellum), whereas others are hairless.

Flexuous hypha. A hypha emerging from a spermagonium that fuses with a spermatium (pycniospore) of a compatible mating type and establishes the dikaryotic nuclear state in rusts. Also called **receptive hypha**.

Flexuous rod virus. One of the elongated rod virus types; flexuous rods tend to be longer and narrower than rigid rods. They are also capable of bending and twisting into many formations.

Fluorometer. A single wavelength spectrophotometer used to measure DNA concentration.

Forecast. To estimate in advance; predict.

Forma specialis (pl. formae speciales). Special forms of a pathogenic fungal species, separation based on morphology or host.

Formulation. Pesticides consist of an active ingredient and inert ingredients in a dry or liquid form such as granules or emulsifiable concentrates.

Frameshift mutations. Disruption by deletions and insertions into a codon that cause the codons to be misread and disrupts the proper amino acid sequence of proteins. The protein is nonfunctional.

Free-living nematodes. Generally soil-inhabiting nematodes that are not plant-parasitic. Important links in the soil food web and nutrient recycling.

Functional genetic variation. Variation in genetic sequence that results in phenotypic differences.

Fungicide. A chemical compound that kills fungi.

Fungus (pl. fungi). Eukaryotic, heterotrophic, absorptive organism that develops a microscopic, diffuse, branched, tubular thread called a hypha.

Fusiform. In nematodes, cigar-shaped body, relatively short and wide, descriptive of ring nematodes.

Galacturonases. An array of hydrolytic enzymes that degrade pectin compounds found in plant cell walls and middle lamellae. See **pectinase**.

Gall. An abnormal, overgrowth caused by hyperplasia and hypertrophy of infected plant cells that are produced in response to some insects, nematodes, fungi, or bacteria.

Gametangial contact. Sexual reproduction in the Oomycota involving antheridia and oogonia in contact, but not fusing, resulting in a diploid oospore. Contrast to Gametangial copulation.

Gametangial copulation. Sexual reproduction in the Zygomycota involving the fusion of gametangia and resulting in the formation of a diploid zygospore.

Gametangium (pl. gametangia). Sex organ that contains gametes.

Gametes. Sexual cells, for example, sperm and eggs or their functional equivalents.

Gas chromatography. Movement of molecules through a gas medium so that molecules of different molecular weight will be separated; used to identify bacteria by profiles of bacterial fatty acids.

Gel electrophoresis. Technique utilizing the migration of electrically charged nucleic acids or proteins in a matrix of agarose or polyacrylamide and an electric field. Also, **polyacrylamide gel electrophoresis**.

Gene. The fundamental physical and functional unit of heredity or an ordered sequence of nucleotides located in a particular position on a particular chromosome that encodes a specific functional product (i.e., a protein or RNA molecule).

Gene-for-gene. A host–pathogen interaction in which resistance in the host and avirulence in the pathogen is in both cases conditioned by a single dominant gene.

Gene pool. The "gene pool" is the total of all genes, or genetic information, in any population. This can be divided up into the primary gene pool consisting mainly of members of the same species; the secondary gene pool, consisting of members of closely related species that can still cross and produce fertile hybrids with the species of interest; and the tertiary gene pool, which consists of more distantly related species that can only cross with the species in question by means of artificial interventions such as embryo rescue or induced polyploidy. (The term quaternary gene pool is sometimes used to refer to genes available for transgenic introduction into the target species; i.e., all the genes available in all living organisms.)

Gene pyramid(s). Multiple genes for resistance to a specific disease in a single line.

Generation time. The time interval needed for the cell(s) to divide. In the case of bacteria, the time to undergo binary fission (20 -30 minutes for some). Compare to human generation time that averages between 22 and 32 years.

Genetic adaptation. The change in the genome occurring via some selection pressure. Also natural selection.

Genetic locus/loci. A specific location of a gene or DNA sequence on a chromosome.

General resistance. Resistance effective against all biotypes of a pathogen.

General secretory pathway. Protein secretory pathway involving the *sec*-pathway.

Genetic engineering. Addition or alteration of genetic material to the genome of the organism.

Genetic transformation. The genetic alteration of a cell or organism resulting from the uptake and incorporation of a portion of foreign DNA.

Genetic variation. Variation in DNA from one individual to the next within a species which allows for selection, natural or guided, to have a foothold for implementing change.

Genome. Nucleic acid sequences that contain the coding required for the function and reproduction of an organism.

Genomic DNA. The entire genome of an organism that includes nuclear, mitochondrial, and sometimes chloroplast DNA.

Genomic selection (GS). A breeding approach that assigns breeding values to multiple molecular markers spaced throughout the entire genome and that uses all these values to predict phenotypes and improve breeding selections.

Genotype. The genetic makeup of an organism.

Genus (pl. genera). Taxonomic classification below that of family that includes related species.

Giant cells. Enlarged, specialized feeding cells formed in plant roots by root-knot nematodes.

Gill. Plate on the undersurface of the pileus that supports the production of basidia and basidiospores.

Glandular trichomes. Plant hairs that are capable of secreting chemicals, usually from the base of the trichome.

Glucans. Long chains of glucosyl residues; with chitin, a structural component of the cell wall of higher fungi (kingdom Fungi).

Glucanases. Any number of enzymes that hydrolyze compounds composed of glucose subunits. See **cellulases** or **cellulolytic**.

Glucanase inhibitor proteins (GIPs). Proteins produced by fungi that prevent degradation of their cell walls. See **chitinase**.

Glycanases. An array of cell wall-degrading, hydrolytic enzymes that fragment plant cell wall polymers. See **galacturonases**, **xylanases**, and **glucanases**.

Glycocalyx. Cell wall; composed chiefly of polysaccharide that is slimy as in slime molds or firm as in most other fungi.

Glycoconjugate. Carbohydrates chemically bonded to other classes of molecules such as proteins and lipids.

Glycoforms. Any number of proteins with attached sugar moieties in various forms.

Glycogen. Storage polysaccharide used as a major carbon reserve compound in fungal cells.

Glycolysis. Enzymatic breakdown of glucose and other carbohydrates to form lactic acid or pyruvic acid and the production of energy in the form of adenosine triphosphate (ATP).

Glucosinolates. Sulfur-containing compounds produced by plant species in the Brassicaceae that are hydrolyzed to **isothiocyanates**, which are biologically active in controlling disease causing organisms.

Glyomics. The comprehensive study of entire complement of sugars in an organism, including genetic, physiologic, pathologic, and other aspects.

Gradient. A slope or the change in disease or inoculum concentration with distance from a source.

Gravitational potential. Gravitational forces acting on soil water.

Gravitational water. Free water that moves through the soil due to the force of gravity.

H-body. Two conjugated, basidiospores of *Tilletia* spp.; the filiform shape of the basidiospores and the short, connecting conjugation tube gives the structure the shape of a capital letter H.

Halo. Chlorosis surrounding a necrotic spot.

Haploid. Having one set of chromosomes in the nucleus.

Harpin(s). Glycine-rich proteins secreted via a type III secretion system of plant pathogenic bacteria. These proteins can elicit hypersensitive responses and defense systems of plants.

Hartig net. Intercellular network of fungal hyphae in the root cortex that does not penetrate cortical cells and that serves as an exchange site for nutrients in certain groups of mycorrhizae; connected to the mantle.

Haustorium (pl. haustoria). A specialized fungal structure formed by obligate fungal parasites inside a host plant cell that is surrounded by the host plasma membrane; facilitates absorption of nutrients from the host. (2) Morphologically modified root that physically connects a parasitic plant to its plant host.

Helper virus. An independently replicating virus that assists a satellite virus or subviral agent with replication.

Hemacytometer. A type of counting chamber, originally designed for counting blood cells that can be used to determine the concentration of fungal spores in a liquid suspension.

Hemibiotroph (hemibiotrophic). A pathogen that infects and colonizes initially without killing host cells and later begins killing host cells to continue its life cycle on dead tissues.

Hemiendophytic. Living partly within plant tissue.

Hemiparasite. A parasite that is photosynthetic and that obtains water and nutrients from the host xylem. Some hemiparasites are only photosynthetic during particular life cycle stages.

Hemolymph. The circulatory fluid of insects.

Herbicide injury. Plant injury caused by the misapplication, drift, or residue of a herbicide on nontarget plants.

Heteroecious. A rust fungus that requires two, unrelated host plants for completion of the life cycle.

Heterokaryon. Hyphal cells have two or more genetically different nuclei.

Heterothallic. Self-sterile; a complementary mating type is needed for sexual reproduction.

Heterkont. An organism that possess two flagella of unequal length.

Heterotroph (heterotrophic). Organism living only on organic food substances as primary sources of energy.

Holocarpic. Entire thallus matures to form thick-walled resting spores (Chytridiomycota).

Holomorph. Collective concept that refers to both the sexual and asexual aspects of a fungus; contrast with anamorph and teleomorph.

Holoparasite. A nonphotosynthetic parasite that obtains water and nutrients from the host xylem and photosynthates from the host phloem.

Homoeologous. Having similar gene content, but sufficiently different in content and gene order to inhibit or completely prevent chromosomal pairing at meiosis.

Homologous. Identical in terms of gene content and linear ordering; homologous chromosomes pair and recombine with one another at meiosis.

Homothallic. Self-fertile; only one mating type required for sexual reproduction.

Horizontal gene transfer. Transfer of genes between organisms in a manner other than traditional reproduction from the parental generation to offspring via sexual or asexual reproduction.

Horizontal Resistance. See **Quantitative Resistance**.

Hormone. Organic compound produced by organisms at minute levels and are generally transported to other parts of the organism to control various growth and physiological processes of the organism; in plants also referred to as plant growth regulators or phytohormones.

Host growth stage. Particular physical or physiological point in plant host development.

Host index. A list of all the hosts on which a pathogen has been identified.

Host plant resistance. An approach to disease control that highlights the development of disease or insect resistant plant varieties.

Host range. Variety of plant genera and species that a pathogen can infect.

Housekeeping genes. Constitutive genes that are required for maintenance of basic cellular function and are expressed in all cells of an organism under normal and pathophysiological conditions.

***hrp* genes.** Genes of plant pathogenic bacteria responsible for eliciting hypersensitive responses in nonhosts or in resistant hosts and for causing disease in susceptible hosts.

Hydrolysis. Chemical process in which water molecules are added to a substance, resulting in the breakdown of the substance.

Hydrophobin gene. Essential for pathogenicity as it acts as a cellular relay for adhesion and trigger development of the **appressorium**.

Hydroponic systems. A system in which crops are grown in a nutrient water solution without soil.

Hygroscopic water. Component of soil water that is held so tightly on the surface of soil particles that it is not available to plant roots.

Hydroxyproline-rich proteins. Proteins found in the cell walls of plants.

Hymenium (pl. hymenia). Layer of hyphae where sexual reproductive cells, such as asci and basidia, and other sterile cells occur.

Hyperauxiny. Increased levels of indole acetic acid in host tissues that result in tumorous tissue development and hypocotyl elongation.

Hyperparasitism. Parasitism of parasites by other organisms.

Hyperplasia. Abnormal tissue growth caused by an increase in the number of cells.

Hypersensitive response (HR). Rapid, programmed cell death in plant tissue that results from an incompatible interaction with pathogens or from the recognition of various elicitors.

Hypertrophy. Abnormal tissue growth caused by the enlargement of individual cells.

Hypha (pl. hyphae). Thread-like filaments that form the mycelium or vegetative body of a fungus that may or may not have cross walls.

Hyphal cells. In reference to parasitic plants, this is the tissue that differentiates inside the *Cuscuta* prehaustorium and grows through this tissue and into the host; it is also called the inner haustorium.

Hyphomycetes. An informal class of mitosporic fungi where spores are produced on separate conidiophores.

Hyphopodium (pl. hyphopodia). Stalked, thick-walled, lobed cells that stick to plant surfaces; the term is also used to describe the infection structures produced by ectotrophic hyphae of certain root-infecting fungi such as *Gaeumannomyces* (take-all pathogen).

Hypovirulence. Reduced level of virulence in a strain of pathogen resulting from genetic changes in the pathogen or from the effects of an infectious agent on the pathogen.

Icosahedral virus. A 20-faceted shape; viruses that appear to be spherical are actually icosahedral in shape.

Immunobinding. Technique of binding proteins to membrane sheets and utilizing antibodies for their detection.

Immunogenic. Relating to or denoting substances able to produce an immune response.

Immunoglobulins. Glycoproteins with the ability to recognize and attach to antigenic regions or epitopes of proteins or other substances.

Immunomodulation. A molecular technique that allows interference with cellular metabolism, signal transduction, or pathogen infectivity.

Incidence. The rate of occurrence of disease; for example, the proportion of plants in a crop that are symptomatic.

Incubation period. Time between inoculation and symptom development.

Indexing. Examination and testing of plants in order to detect the presence of certain specific pathogens (see **Disease-free**).

Indolediterpenes. Mycotoxins of a diverse range of chemical structures that share a common biosynthetic origin.

Induced systemic resistance (ISR). Active defense mechanism in which a root-colonizing bacterium causes distant induction of general defense mechanisms that serve to protect the plant. In ISR, jasmonic acid and ethylene signal of distal portions of the plant and antimicrobial proteins are not produced.

Infection court. Site of entry and establishment of a pathogen in or on a host plant.

Infection. Establishment of a food relationship between the host and pathogen and is the second stage of the disease cycle.

Infectivity. Ability of a pathogen (virus) to establish infection.

Infectivity curve. The comparison of diluted virus inoculum and the number of infections caused by the dilution.

Inflorescences. Flowering parts of plants variously arranged.

Ingestion. Process by which some insects intake food materials.

Inhibitors. Compounds with the ability to slow or stop a specified reaction.

Initial inoculum. Parts or forms of a pathogen that are able to infect after survival through time periods when a host is absent.

Inoculation. Placement of the propagule of a pathogen on or near a host cell or at a site where it can infect the host.

Inoculum. Part of the pathogen that can infect the plant. This includes sexual or asexual spores, vegetative mycelium or other infectious bodies that serve to initiate infection in a host organism (being restricted to only specific spore stages in rusts and smuts).

Inoculum density. Number of pathogen propagules per unit volume of soil or other medium.

Inoperculate. Refers to asci that lack a hinged lid or **operculum**.

Insertions. Adding an additional nucleotide base into a gene sequence causing a **frameshift**.

Integrated pest management (IPM). A system for controlling disease, mite, and insect problems using multiple strategies (cultural, biological, and chemical).

Intensity. The level or degree of occurrence; in plant pathology, this is often used when disease severity and incidence are combined.

Interactome. The whole set of molecular interactions in cells.

Intercalary. A structure formed as an interruption of a continuous strand, as with certain fungal spores (e.g., chlamydospores).

Introgression. The movement of a gene or genetic locus from one line or individual into another line or individual; in plant breeding, the deliberate transfer of a desirable gene.

In vitro selection. Process by which cells are selected by screening with an appropriate selection agent in a container or under laboratory conditions.

Irrigation. Various methods of providing water for crop growth and development.

IPM labeling. Designating that crops and food have been grown according to the best practices of IPM.

Isolate. Any propagated culture of a virus, bacterium, or fungus with a unique origin or history.

Isolines. Lines of a crop species that differ at one genetic locus, but are otherwise uniform.

Isometric. Geometric three-dimensional form that is equal on all sides or facets.

Isothiocyanates. See **glucosinolates**.

Isozymes. Enzymes that catalyze the same reaction, but may differ in weight and possibly substrate specificity.

Jasmonic acid. See **Induced systemic resistance (ISR)**.

Juveniles. In nematodes, the immature life stages before mature adults are formed. Delineated by molts, but not referred to a larvae since there is no true metamorphosis.

Karyogamy. Fusion of two haploid nuclei to form a diploid nucleus; final step in fertilization.

Koch's postulates. Guidelines for establishing proof of pathogenicity, that is, proof that a specific pathogen is the cause of a specific disease.

Koehler illumination. A highly effective illumination designed by August Köhler that involves focusing and centering and all the elements in the light path.

Landrace. A local variety of a cultivated crop species that developed outside of a formal breeding process.

Larva (pl. larvae). Juvenile organisms that undergo several growth stages prior to reaching adult development.

Latent period. Time between inoculation and reproduction.

Lateral flow devices (LFD). Field ready kits used to rapidly detect a pathogen.

Leaf rolling. Pronounced curling of plant leaves beginning at the edge; leaf rolling typically curls upward, but in some cases, it may bend the leaf downward.

Leaf spot. Localized area of necrosis on a leaf.

Lectinomics. The study of various proteins (lectins) that can bind to membranes.

Lectins. Receptor are proteins of nonimmune origin that bind in a stable manner to carbohydrates.

Lesion (local). Abnormal appearance in a localized area due to a wound or disease.

Life cycle. All the successive stages, morphological and cytological, in an organism that occur between the first appearance of a stage and the next appearance of that same stage, and including all the intervening stages.

Lignin. Complex organic material derived from phenylpropane that imparts rigidity and strength to cells in many plant tissues.

Line pattern. Symptomatic patterns of varied pigmentation found near the edge of the leaf or tracing the outline of the leaf. Line patterns are related to ringspots and are often formed as extensions of ringspots. *Rose mosaic virus* produces a striking yellow line pattern on rose.

Linnaeus. Carl Linnaeus, a Swedish botanist, physician, and zoologist, laid the foundations for the modern biological naming scheme of binomial nomenclature.

Lipase (lipolytic). Enzyme that hydrolyzes ester bonds between fatty acids and glycerol (or sorbitol in Tween) backbone found in fats.

Lipidomics. The study of metabolic pathways and structures involving lipids in organisms.

Lipopolysaccharides(LPS). Large molecules consisting of a lipid and a polysaccharide composed of O-antigen, outer core, and inner core joined by a covalent bond; also referred to as lipoglycans and endotoxins. They are found in the outer membrane of Gram-negative bacteria, and elicit strong immune responses in animals.

Lipoxygenase (LOX) products. Volatile compounds, such as perioxides, formed by the action of lipoxygenases on host lipid-containing compounds (membranes).

Local lesion. Small disease lesion that forms by the cell-to-cell movement of the pathogen; local lesions may be formed by the host's defense reaction limiting the pathogen movement or by the inability of the pathogen to move systemically. Local lesions may be either chlorotic or necrotic.

Locule. Small cavity.

Lodge. Falling over of a plant generally at the soil line.

Logistic model. A mathematical equation used to represent disease development in time; this model is symmetrical, with slower start and end, and a limit to disease increase.

Loline. Fungal secondary metabolite present in grasses symbiotic with endophytes in the genera *Epichloë* and *Neotyphodium*.

Lolitrems. Indolediterpenes that are potent neurotoxins; cause tremors in mammals and deter insect feeding.

Lyse. To induce lysis, or to cause dissolution or destruction of a cell membrane with lysin.

Maceration. The blending, grinding, or pulverizing of tissue until a smooth homogenous mixture results; may also refer to sot rots caused by enzymatic degradation of host cell walls.

Macrocyclic. A rust fungus life cycle that includes the aecial, uredinial and telial stages; the spermagonial stage may be present or absent in a macrocyclic rust.

MAMP/PAMP. Synonymous acronyms for microbe/pathogen-associated molecular patterns. These are molecular patterns (or sets of molecules) specifically associated with microbes that are used by a host to detect their presence and can influence innate immunity of the host.

Mantle. Fungal sheath enveloping the outside of infected feeder roots; in ectomycorrhizae and some ericaceous mycorrhizae.

Marcescence. Retention of dead leaves by deciduous plant species.

Marginal. Along the edge (e.g., leaf).

Marker-assisted selection. Use of a molecular marker to indirectly select individuals bearing a trait such as disease resistance that is associated with that marker.

Marker effect. The effect on the trait of interest associated with substituting one copy of a particular marker allele for its alternative allele.

Mass flow. Mechanism that describes the movement of carbohydrates in the phloem of plants. Requires active transport of sugar into phloem cells at the source and out of the phloem cells at the sink. This sets up a diffusion and turgor pressure gradient that moves water and sugar through the phloem. Also called pressure flow or bulk flow.

Material data sheet (MSDS). Contains important information about pesticides, such as adverse effects of overexposure, first aid measures, firefighting measures, how to handle a spill or leak, proper storage conditions and information about worker protection standards including required personal protection equipment.

Mating types. "Sexes"; complementary mating types are needed for sexual reproduction in **heterothallic** fungi.

Matric potential. Forces independent of gravity that act on soil water and are due to the adsorptive force by which soil particles attract water and capillary forces which results from the surface tension of water and its contact angles with soil particles.

Mechanical inoculation. The entry of a plant virus into a plant cell through a nonlethal wound in the epidermis, cell wall, and cell membrane; these wounds are typically formed by an abrasive when plants are intentionally infected.

Mechanical transmission. Transmission of a virus through a wound generated during mechanical inoculation.

Meiosis. Sequence of cellular events in sexual cell division, leading to the division of the nucleus into four nuclei that are genetically dissimilar due to chromosome recombination and sorting.

Melanin. Dark brown or black pigment found in fungal cell walls.

Mesophile (mesophilic). Organism that grows well between 10°C and 40°C.

Metabolomics. The study of all metabolites within an organism.

Microbiome. Microorganisms inhabiting the human body.

Microbivorous nematode. Nematodes that feed on fungi and bacteria.

Microcyclic. A rust fungus life cycle that lacks aecial and uredinial stages.

Microorganism. Microscopic organism, particularly a bacterium, virus, or fungus.

Microsclerotium (pl. microsclerotia). Small group of dark-colored cells with thickened walls that can form a germ tube or hypha. Microsclerotia are characteristic of fungi such as *Verticillium*.

Microtome. Mechanical device used to cut biological specimens into very thin sections for microscopic examination.

Migratory ectoparasite. Nematodes that move along roots and graze on epidermal or cortical cells, but rarely enter the root.

Migratory endoparasite. Nematode that enters the root and then moves through and feeds on cortical cells.

Migratory. Nematodes that move from one host plant to another or an individual nematode that feeds from many sites along a root.

Mildew. A thin coating of mycelial growth and spores on the surfaces of infected plant parts

Millardet. Pierre-Marie-Alexis Millardet, a professor of botany at Bordeaux University, who studied and promoted the use of a mixture of copper sulfate and lime (Bordeaux mixture) to manage downy mildew of grape and other diseases.

Missense mutations. Mutations that alter the codon so that a different amino acid is incorporated into the peptide sequence.

Mitochondrion (pl. mitochondria). Cytoplasmic double membrane bound organelle that generates energy via respiration in the form of ATP.

Mitosis. Sequence of cellular events in asexual cell division, leading to the division of the nucleus into two genetically similar nuclei.

Mitosporic fungi. Fungi that reproduce asexually via conidia or mitospores.

Model. Any physical, mathematical, or theoretical representation of some process.

Moldboard plowing. Method used to reduce soil compaction and improve aeration in the layer being disturbed, 6–12 in deep, and inverted the soil to bury **crop residue**.

Molecular marker. Distinguishing molecular feature that can be used to identify a segment or region of a genome, chromosome, or genetic linkage group.

Mollicutes. Cell wall-less prokaryotes thriving in high osmoticum habitats protecting the integrity of their plasma membranes from lysis. Mollicutes include the polymorphic phytoplasmas and helical spiroplasmas. These organisms are nutritionally fastidious, growing only in phloem sap in their plant and hemolymph in their insect vectors.

Monoclonal antibodies. Single-type antibodies that are produced by fusing a spleen cell from an immunized mouse to a murine myeloma cell.

Monocots. Plants having a single cotyledon (as grasses).

Monoculture. Production of the same crop or plant species over a large area and usually for an extended period of time.

Monocyclic. Relative to plant disease, a process of pathogen infection, growth, production of symptoms, and reproduction without repetition.

Monogenic resistance. Resistance conditioned by a single gene.

Monokaryotic. Cell having genetically identical haploid nuclei.

Monomolecular model. Mathematical equation used to represent disease development in time; this model has a limit to disease increase.

Monophyletic. Descendants from a single common ancestor.

Mosaic. Pattern of lighter and darker pigmentation in an infected plant; most mosaics are formed by effects on the plant's chlorophyll or chloroplasts. They are often patterns of darker green with light green yellow or white areas.

Mottle. Pattern of lighter and darker pigmentation in an infected plant; mottles and mosaics are very similar symptoms and are often produced by viruses.

MTI/PTI. Synonymous terms standing for MAMP/PAMP-triggered immunity; disease resistance resulting from a low-level defense response triggered by the recognition of a MAMP (PAMP).

Messenger RNA (mRNA). Single-stranded molecule of "messenger" ribonucleic acid that directs protein production (translation).

Multigene resistance. Resistance encoded by several to many genes; may be similar to horizontal resistance.

Multilines. Mixtures of isolines (lines of a crop species differing at only one or two loci).

Multilocus sequence analysis (MLSA) or multilocus sequence typing (MLST). A technique in molecular biology for the typing of multiple loci. The procedure characterizes isolates of microbial species using the DNA sequences of internal fragments of multiple housekeeping genes.

Multipartite genome. The complete genome of an organism is contained on multiple strands of nucleic acid.

Mutualism. A form of symbiosis in which two or more organisms of different species are living together for the benefit of both or all.

Mycelia Sterilia. A group of fungi that do not produce spores; archaic taxa term, no longer in use.

Mycelial fan. Fan-shaped aggregation of hyphae found under the bark of trees that invades and kills the cambial tissues of the host; common in *Armillaria* species.

Mycelium (pl. mycelia). Mass of highly branched threadlike filaments (hyphae) that constitute the vegetative body of a fungus.

Mycoheterotroph. A plant, either photosynthetic or not, that obtains nutrients via a mycorrhizal fungus that is attached to a tree root.

Mycology. The study of fungi and traditional allies.

Mycoparasitism. One fungus being parasitic on another fungus.

Mycopathogenic. Use of nonpathogenic fungi and bacteria to control pathogens.

Mycorrhizae. "Fungus root"; a mutualistic relationship between certain fungi and the roots of plants, also mycorrhizal fungi.

Mycotoxin. Toxin produced by a fungus in infected grain or legumes that is poisonous to humans or animals when consumed.

Naked ascus. Ascus not produced in an ascocarp.

Necrosis. Abnormal death of living tissue.

Necrotic. A cell, tissue, or organism that is displaying necrosis.

Necrotroph (necrotrophic). A plant pathogenic fungus that kills its host cells for nutrition.

Nematicide(s). Chemicals used to kill or interrupt the life cycle of plant-parasitic nematodes.

Nematology. The study of nematodes.

Nodulation. Nitrogen-fixing swellings on the roots of legumes, not to be confused with galls caused by root-knot nematodes.

Nomenclature. A list of names in a taxonomic systems; the process of assigning an organism to a taxonomic group.

Nonculturable bacteria. Bacteria in which there is no known artificial medium that will support growth.

Nonhost resistance. Resistance exhibited by plants against the majority of potentially pathogenic microorganisms, in which no member of a pathogen species can parasitize the host species.

Nonparametric statistics. Statistics not based on numbers or distribution of numbers, but includes analyses of descriptive and/or inferential data.

Nonseptate. Not divided by a septum or not having a septum.

Nucleic acid. A single- or double-stranded polynucleotide containing either deoxyribonucleotides (e.g., DNA) or ribonucleotides (e.g., RNA) linked by 3'-5'-phosphodiester bonds.

Numerical Aperture (N.A.). N.A. indicates the resolving power of a lens. A lens with a larger numerical aperture will be able to visualize finer details than a lens with a smaller numerical aperture. Lenses with larger numerical apertures also collect more light and will generally provide a brighter image.

Obligate aerobes (anaerobes). Require oxygen for respiration and metabolism. Obligate anaerobes are poisoned by oxygen; they use compounds other than oxygen as electron acceptors for respiration and metabolism.

Obligate biotroph. A parasite that requires a living host for growth and reproduction; obligate biotrophic fungi generally cannot be grown in axenic culture.

Obligate parasite. (1) In reference to parasitic plants, a plant that must attach to a host to complete its life cycle. All holoparasites are obligate, whereas only some hemiparasites are. (2) Microbial parasite that requires a living host to feed; cannot survive on dead or decaying plant material.

Ocular micrometer. A precisely etched circular glass element that is placed inside an ocular. It contains a scale that enables the observer to take accurate measurements. It is calibrated with a stage micrometer.

Odontophore. Flanges on which the odontostylet is attached.

Odontostylet. A stylet with or without flanges and found in dagger and needle nematodes.

Oidium (pl. oidia). Asexual spore that splits off in succession from the tip of a short branch or oidiophore; for example, powdery mildews.

Oligogalacturonides (OGAs). Degradation product (9–15 degree of polymerization) of **endopolygalacturonases (EPGs)** acting on pectin substrates that can elicit plant defenses.

Oligonucleotide. A DNA molecule consisting of a few to many nucleotides.

Omnivorous nematode. Nematodes that feed on several sources of food or switching between food sources depending on developmental stage.

Onchiostylet. A toothlike stylet found in stubby-root nematodes.

Onset. The beginning or early stages; point in time when disease is initially seen in a population of plants.

Oogonium (pl. oogonia). Female gametangium of the fungi-like species of the Oomycota.

Oomycota. "Water molds"; a funguslike phylum where sexual reproduction results in production of oospores; in the kingdom Stramenopila.

Oospore. A thick-walled resting spore that develops from the fertilization of an oogonium by one or more antheridium.

Operculum. A structure that closes or covers an aperture.

Ordinal scale. A scale in which individuals are classified by rank. In an ordinal scale, the degree of difference between consecutive classes is not necessarily constant or even measurable.

Organic Foods Production Act. Adopted to "develop, implement, and administer" standards for the production of foods labeled as organic.

Organic production. A system that is managed to respond to site-specific conditions by integrating cultural, biological, and mechanical practices that foster cycling of resources, promote ecological balance, and conserve biodiversity.

Organic system plan. Conditions for use of limited number of synthetic substances approved for use in organic crop production and must be clearly documented.

Osmotic potential. Forces that result from ions and molecules dissolved in soil solution. These forces predominate when water is moved across a semipermeable membrane.

Ostiole. An opening or pore in a pycnidium or perithecium.

Oxidative phosphorylation. Metabolic processes where energy in the form of ATP is stored during the transfer of electrons through the electron transport pathway.

PAMPs. Pathogen-associated molecular patterns, which include ion fluxes, signaling molecules, and mitogen-activated protein kinases. that are recognized by plant or animal receptors and can influence innate immunity of the host.

PAMP-triggered immunity (PTI). Immunity resulting from recognition of pathogen-associated molecular patterns **(PAMPs)** by transmembrane pattern recognition receptors encoded by **Resistance genes (R genes)**, resulting in immunity or no disease.

Papilla (pl. papillae). Deposit of callose and other material (lignin) on the inside of a plant cell wall that are produced in response to pathogen ingress.

Parametric statistics. Sample data is obtained from a population and that data follows a probability distribution (several types e.g. normally distributed). Typically used with "hard" data, which includes measurements.

Parasite (parasitic or parasitism). An organism, virus or viroid living with, in or on another living organism and obtaining food from it.

Parfocal. A microscope designed so that specimens that are in focus at one magnification will remain in focus when the objective is changed.

Parthenogenesis. Female nematodes produce viable eggs without mating.

Partial resistance. Resistant reactions in which some symptoms develop or pathogen reproduction occurs, but the amount is less than on a fully susceptible reaction. Partial resistance implies nothing about inheritance, gene expression or race specificity of the resistance.

Pasteurization. A process by which pathogens are eliminated from a medium (e.g., soil) with heat (55°C–60°C).

Passive defense. Defense mechanisms that are present before pathogen recognition.

Parasexual recombination. A form of genetic recombination observed in fungi that does not require meiosis.

Pathogen (phytopathogen). Organism or agent capable of causing a disease on a host plant.

Pathogen fitness. The ability of the pathogen to survive or compete in the environment where the host is grown.

Pathogen race. Population of pathogen isolates that have the same virulences and avirulences.

Pathogenesis-related proteins (PR). Synthesis and accumulation of defense-related proteins following pathogen infection of plants.

Pathogenesis. Development of disease.

Pathogenicity factor. Component of a pathogen that is essential for pathogenesis.

Pathogenicity. Ability of a pathogen to cause disease on a specific host species.

Pathosystem. The organisms, host and pathogen, involved in a disease.

Pathotype strain. Group of strains that cause the same disease because they share a set of virulence genes.

Pathovar. Bacterial strain or set of strains with the same or similar characteristics, that is differentiated at infra-sub-specific level from other strains of the same species or subspecies on the basis of pathogenicity to one or more plant hosts.

Pectin. A methylated polymer of galacturonic acid found in the middle lamella and the primary cell wall of plants; enzymatic degradation of pectin by pathogens causes soft rots.

Pectic acid. Polygalacturonic acid.

Pectinase (pectinolytic). Complex battery of enzymes capable of degrading pectin and substituted pectin to a number of simpler products including galacturonic acid. Enzymes may be endotype (**endopolygalacturonases**), which essentially hydrolyze bonds in the interior of the polymer or exotype (**exopolygalacturonases**), which cleave individual molecules of the repeating unit from the nonreducing terminus.

Pectin lyases. See **pectinases**.

Pectin methyl esterases. See **pectinases**.

Pedicle. Stalk that hold each flower in an inflorescence that contains more than one flower.

Peloton. An intra- or inter-cellular hyphal coil in the roots of some endomycorrhizae.

Penetration peg. Specialized, slender fungal hypha that penetrates the cell wall of a host plant through a combination of enzymatic and mechanical force. Usually produced from the undersurface of an appressorium.

Percent DNA/DNA hybridization. Process of combining two complementary single-stranded DNA molecules and allowing them to form a single double-stranded molecule through base pairing. The percentage of hybridization that occurs is used by bacterial taxonomists to determine relatedness between bacterial strains.

Pericarp. Outermost wall of a ripened ovary or fruit; the "seed" coat, that is, the outer layer of the caryopsis in cereals and other grasses.

Periplasm. Space between the inner and outer membranes of a Gram-negative bacterium.

Peridium (pl. peridia). The outer wall or delimiting membrane of a fungal structure.

Perithecium (pl. perithecia). A flask-shaped structure with an opening (ostiole) at the top of the flask in which asci form in a single layer (hymenium); the structure develops as a result of nuclear exchange, typically involving an ascogonium and antheridium.

Permanent wilt. Wilt from which plants do not recover.

Permeability. The ability of a membrane to control the passage of materials through it.

Pest. Organism that injures or has a harmful effect on a plant.

Petiole. One of three possible parts of a leaf; the part the joins the blade or stamina of the leaf to the stem.

pH. The measure of acidity or alkalinity within a range of 0–14 where pH 7 is neutral. The pH scale is logarithmic so a change of one unit is equal to a 10-fold change of hydrogen ion concentration.

Phenol (phenolic). Compound having one or more hydroxyl groups substituted in single or multiple benzene rings.

Phenotype (phenotypic). The physical appearance or physiological attributes of an organism; in the case of disease resistance, the observable response to a pathogen.

Phenotypic plasticity. The capacity of an organism, via controlled changes in the transcriptome, proteome, and metabolome, to originate multiple phenotypes from a single genotype, depending on environmental conditions.

Phialide. A bottle-shaped type of conidiogenous cell that produces conidia in basipetal succession. In basipetal succession, conidia mature from the apex toward the base, that is, the conidium at the base is the youngest and most recently formed.

Phialoconidium (pl. phialoconidia). Asexual reproductive fungal spore formed by abstriction from the top of a **phialide**.

Phialospore. Synonymous with **phialoconidium** (pl. phialoconidia).

Phloome. Proteins transported in the phloem and a subset of the **proteome**.

Photorespiration. Metabolic process in plants in which oxygen, instead of carbon dioxide, is added to ribulose-1,5-bisphosphate in chlorophylls.

Photosynthesis. Metabolic process that obtains energy from light to produce glucose and oxygen from carbon dioxide and water.

Phyllody. The growth of leaflike structures on other plant organs such as flowers or on fruits.

Phyllosphere. The aerial surfaces of plants (leaves, stems, fruit, and flowers) have habitats for phytopathogenic prokaryotes during the resident phase or pathogenesis.

Phylogenetic analysis. The use of hypothesis of character transformation to group taxa hierarchically into nested sets and then interpreting these relationships as a phylogenetic tree. Phylogenetic relationships are "reconstructed" using distance, parsimony, and maximum likelihood methods.

Phylogenetic tree. A hypothesis of genealogical or evolutionary relationship among a group of taxa.

Phylotype. Similarities that classify a group of organisms by their evolutionary relationships.

Phylum (pl. phyla). A taxonomic rank between kingdom and class (formerly referred to as division).

Physiology. Study of basic activities conducted within the cells and tissues of a living organism as they are related to the chemical and physical processes of the organism.

Phytoalexins. Active defense, antimicrobial small molecular weight compounds that are produced after infection or elicitation by abiotic agents.

Phytoanticipins. Passive defense antimicrobial small molecular weight compounds that are stored in the plant cell or are released from a glucoside.

Photoassimilates. Molecules produced by plants involving photosynthesis such as sucrose and asparagine.

Phytobacteriology. The study of bacteria living in association with plants including pathogenic and beneficial species.

Phytogenic volatile organic compounds (PVOCs). Volatile metabolites emitted into the atmosphere by plants that are involved in chemical communication between organisms.

Phytopharmaceuticals. Chemicals used to treat or prevent diseases of plants.

Phytotoxic. Causing injury to plants, as a pesticide or a **phytotoxin**.

Phytotoxin. A chemical produced by a plant pathogenic microbe in the process of disease development.

Pileus. The cap of a mushroom basidiocarp.

Plant-associated and environmental microbes database (PAMDB). Multilocus sequence typing and analysis website and database (http://genome.ppws.vt.edu/cgi-bin/MLST/home.pl) specifically designed for the identification of plant-associated and environmental microbes and for the study of their epidemiology, population genetics, and molecular evolution.

Plant disease diagnosis. The act or process of determining and identifying a disease problem.

Plantibodies. Antibodies or antibody fragments produced in plants by the expression of corresponding encoding genes or gene fragments.

Planting date. When crops are sown. Consideration should be given to environmental factors such as temperature and soil moisture.

Plasmalemma. Cell membrane that bounds the protoplast; found inside the cell wall in eukaryotic organisms including fungi and plants.

Plasmid. A self-replicating piece of circular DNA that is present in some bacteria and is not a part of the chromosomal DNA. Its replication is not associated with cell division.

Plasmodiophoromycota. A phylum within the kingdom Protista characterized by organisms that have cruciform nuclear division and produce plasmodia.

Plasmodium (pl. plasmodia). A cell without a wall (e.g., naked protoplast of a slime mold), usually consisting of a mass of protoplasm with many nuclei.

Plasmogamy. Fusion of protoplasts to form a dikaryotic cell; first step in fertilization.

Polarizing filter. A filter that produces polarized light in which the light waves are aligned in one plane.

Pollution. Contamination of air, water, or soil by particulates or chemical substances.

Polyculture. Mixtures of crop species in the same field or land.

Polycyclic. Relative to plant disease, the repeating process of pathogen infection, growth, production of symptoms, and reproduction.

Polyetic. Taking place over more than 1 year.

Polygenic resistance. Resistance conditioned by several genes.

Polygalacturonases. See **pectinases**.

Polymerase chain reaction (PCR). An in vitro nucleic acid amplification technique capable of increasing the mass (number) of a specific DNA region (fragment). The method uses two synthetic oligonucleotide primers (15- to 30-nucleotide long) that specifically hybridize to opposite DNA strands flanking the region to be amplified. A series of temperature cycles involving

DNA denaturation, primer annealing, and the extension of annealed primers by the activity of a thermostable DNA polymerase amplify the target region many million-fold.

Polymorphism. More than one form. In DNA fingerprinting, the presence or absence of a specific band or PCR product at a specific locus.

Polyphasic classification. A consensus approach to bacterial systematics that takes into account all available phenotypic and genotypic data and integrates them in a consensus type of classification, framed in a general phylogeny derived from 16S rRNA sequence analysis.

Polyphenol oxidase. An enzyme that oxidizes adjacent hydroxyl groups on phenolic compounds.

Pour-thru method. Distilled-deionized water is poured into the pot, and the leachate collected. **Total soluble salts** are determined with a conductivity meter.

Powdery mildew. A plant disease caused by a fungus in the order Erysiphales, generally causing white, powdery patches on the surface of leaves, stems, or flower parts. See **Mildew**.

Predacious (predatory) nematode. Nematodes that feed on live prey including other nematodes.

Predispose. To make more susceptible (to disease).

Prehaustorium. In regard to parasitic plants, a bumplike structure formed on the stem of *Cuscuta* that first attaches to the host surface by specialized epidermal cells. Also called the upper haustorium or adhesive disk.

Primary gene pool. Consists of individuals in the same or a very closely related species with which crossing (sexual reproduction) is easiest.

Primary host. A plant species on which the telial stage is formed in a heteroecious rust fungus life cycle; may also be used for the more economically important host in the life cycle.

Primary inoculum. Inoculum that begins a disease cycle, also known as initial inoculum or overwintering inoculum.

Primary literature. Original or the first report of research; not derived from other works.

Programmed cell death (PCD) or apoptosis. An active defense mechanism commonly associated with reproductive and xylem tissue development. In PCD the attacked cell and several plant cells around it die in response to chemical signals. The sacrifice of these cells isolates the pathogen.

Prokaryote. Bacteria and mollicutes that differ from eukaryotes (higher plants, fungi, and animals) in that they lack organelles (membrane-bound nuclei, mitochondria, and chloroplasts), and their proteins are translated from mRNA exclusively on 70S ribosomes.

Promycelium (pl. promycelia). A filamentous basidium formed upon germination of teliospores in smut fungi, particularly species of *Tilletia*.

Proprioceptor. Sensory receptor located deep in the tissues such as skeletal muscles, tendons, etc.

Protectant fungicides. Chemicals applied to plant surfaces prior to the arrival of any pathogen inoculum that prevent infections and disease from occurring.

Protein subunits. The individual proteins composing the viral **capsid**.

Proteome (proteomics). The study of the structure and function of proteins, including the way they work and interact with each other inside cells.

Pseudothecium (pl. pseudothecia). A flask-shaped structure with an opening (ostiole) at the top of the flask in which asci form in a single layer (hymenium); the structure develops before any nuclear exchange involving an ascogonium and antheridium.

Psychrophile (psychrophilic). An organism that grows well at colder temperatures (<10°C).

Puccinia path. The path, some 600–900 miles (1000–1300 km) wide, extending from northern Mexico and southern Texas to the Prairie Provinces of Canada, over which urediniospores of stem and leaf rust fungi annually blow northward and successively reproduce, extending those diseases from Mexico and Texas to Canada.

Pure culture. See **axenic culture**.

Pycnidium (pl. pycnidia). Conidiomata that are flask-shaped and contain exposed conidia and conidiophores.

Pycnium (pl. pycnia). A globose or flask-shaped haploid (n) fruiting body of a rust fungus that contains the **receptive hyphae** and **pycniospores.**

Pycniospore. Thin-walled, monokaryotic (N) gamete produced in a pycnium (spermatium), the nuclei of which pair, but not fuse, with compatible nuclei in receptive, flexuous hyphae to establish the dikaryotic (N + N) condition of the rust fungus life cycle. See **spermatia**.

Pyramiding genes. A breeding strategy in which multiple genes that confer a similar phenotype, for example, resistance to a particular disease, are incorporated into the same plant genotype.

Pyriform. In nematodes, largely swollen or flask-shaped adult female body, found in root-knot, cyst, and several other nematode species.

qPCR. Short for quantitative PCR; a form of PCR that allows for quantification of the original amount of template.

Qualitative resistance. Resistance that can be placed into distinct categories that form the basis of Mendelian ratios.

Quantitative resistance. Resistant host reactions that are not differentiated into distinct classes because there is continuous variation from resistant to susceptible phenotypes.

Quarantine. Exclusion of pathogen from a designated area where they have not been identified; quarantines are a form of legal control for plant diseases.

Quorum-sensing. Bacterial cell-to-cell communication system based on population density.

R-gene. Dominant plant gene responsible for resistance to disease.

Race. A population of pathogen isolates that have the same virulence.

Race-specific resistance. Resistance that is effective against certain races (virulence types) of a pathogen, but not others.

Randomly amplified polymorphic DNA (RAPD). Amplified DNA polymorphism uncovered by the use of arbitrarily amplified DNA.

Rating scale. A progressive classification, as of size, amount, importance, or disease grades.

Reaction mixture or cocktail (slang). The components of a PCR; usually includes buffer, primer, nucleotides, template DNA, magnesium ion, DNA polymerase, and water.

Receptive hyphae. Portion of the rust fungus **pycnium (spermagonium)** that receives the nucleus of a **pycniospore (spermatium).**

Recessive. Refers to an allele causing a phenotype that is only apparent in a homozygote.

Recovery. A decrease in severity or disappearance of symptoms that was previously expressed.

Reducing agent. A chemical that donates electrons to reduce another chemical reaction.

Reducing sugar (compound). Any sugar (or sugar-acid) that has an anomeric carbon; one in which the hydroxyl group around a carbonyl carbon can exist in different isomeric forms (e.g., D-glucose).

Refuges. A small group of plants susceptible to a disease or pest located near a much larger planting of resistant plants.

Regurgitant. The fluid mixture produced by beetles during feeding; it contains high levels of ribonucleases, deoxyribonucleases, and proteases.

Reisolates. In Koch's postulates, the suspected causal organism of the disease isolated from the inoculated host plant.

Repeating spore. A spore stage that gives rise to the same type of mycelium and subsequent type of spore production as that from which it developed.

Repeating stage. The uredinial stage in rust fungi; in heteroecious rusts, this is the only stage that is able to infect the host on which the spore stage is formed.

Rep-PCR. Repetitive extragenic palindromic sequence PCR.

Resident phase. Epiphytic colonization of plants by phytopathogenic prokaryotes without apparent disease production. Among phytopathogenic prokaryotes lacking resistant spores, resident phases are important for the survival of the pathogen until conditions favor pathogenesis.

Resistance. In the host, the ability of the host to reduce the growth, reproduction, and/or disease-producing activities of the pathogen. In pathogens, lack of sensitivity to a pesticide, acquired through a genetic change.

Resistance (R-) gene. A gene conferring a high level of disease resistance.

Resistant. Describes a condition where plants are not hosts for a specific pathogen and the pathogen cannot cause a disease on that plant. Often used to describe varieties of a plant species on which the pathogen cannot cause disease.

Respiration. Metabolic processes where carbohydrates and lipids are converted into energy with the uptake of oxygen from the environment.

Resting spore (or cyst). A spore that functions as a survival structure under conditions of extreme temperature and moisture.

Restriction fragment length polymorphisms (RFLP). The variation in the length of a DNA fragment produced by a specific restriction endonuclease and generally detected by Southern hybridization with a nucleic acid probe.

Rhizocompetent. The ability of microorganisms such as bacteria and fungi to colonize the rhizosphere or roots of plants.

Rhizoid. A branched fungal hypha that resembles a root and penetrates a substrate; in the Zygomycota.

Rhizomorph. Rootlike structure composed of thick strands of somatic hyphae; facilitates the dispersal of some fungi to new substrates.

Rhizoplane. The plant surface–soil interface on roots.

Rhizosphere. The volume of soil that is physically, chemically, or biologically affected by the presence of living roots and that adheres to the plant root after loose soil has been removed by shaking.

Rhizoxin. A cyclopeptide toxin produced by species in the Zygomycota that act as a pathogenicity factor and can damage mammalian livers.

Ribosomal DNA ITS. Internal transcribed spacer (ITS) refers to noncoding DNA between the small-subunit *ribosomal RNA* and large-subunit *rRNA genes.*

Ribozyme. Self-cleaving RNA strand that can specifically bind and cleave other RNA strands.

Rigid rod virus. One of the two major elongated or rod virus types; rigid rods tend to be shorter and slightly wider than other rod viruses with a more defined central canal.

Ringspot. Plant disease symptom characterized by chlorotic or necrotic circular lesions.

RNAse. Enzyme that destroys or degrades RNA.

RNA-interference (RNAi). Mechanism found in most organisms whereby short (usually 20–24 base pairs) fragments of RNA can reduce the expression of genes to which they are homologous.

Rogue (rouging). Removal and destruction of undesired plants.

Root cap. A mass of cells that covers and protects the growing root tip.

Root exudates. A mixture of compounds, mainly sugars and amino acids, that leak from growing and expanding sections of roots, and from broken cells at exit points of lateral roots, and diffuse into soil.

Rosette. A type of stunting caused by shortening of the internodes so that the nodes are literally stacked on top of each other; *Groundnut rosette virus* produces rosette symptoms in peanuts.

Rot. A generalized (not localized) type of necrosis.

Rotation diversity. The number of different crop species and/or botanical families within a crop rotation sequence.

Rotation length. The number of years within a rotation sequence before a crop is repeated.

Rust diseases. Any number of diseases caused by fungi classified in the Basidiomycetes (Pucciniales, previously known as Uredinales). Signs of the disease typically resemble various shades of "iron rust."

Saline. Containing a salt or salts.

Salivary gland. Organs responsible for the production of salivary fluids.

Sanitation. Control methods used to prevent the introduction or spread of plant pathogens.

Saprophyte (adj. saprotrophic, adv. saprophytically). An organism that obtains its food from a nonliving (dead) host.

Satellite. A subviral agent that cannot function independently without the assistance of a **helper virus**.

Sclerotium (pl. sclerotia). Spherical resting structure 1 mm- to- 1 cm in diameter with a thick-walled rind and a central core of thin-walled cells with abundant lipid and glycogen reserves; individual hyphae have lost identity.

Scouting. Comprehensive, systematic check of fields at regular intervals to gather information on the crop's progress and pest level.

Searching hypha. In regard to parasitic plants, differentiated from hyphal cells, searching hyphae are elongated cells whose growth is directed toward host xylem and phloem. The invasion of host tissues is both inter- and intra-cellular and involves mechanical as well as enzymatic degradation.

Second-stage juvenile. A nematode that has undergone a first molting, sometimes within the egg.

Secondary gene pool. Primarily consists of species closely related to the crop of interest. See **Primary gene pool** for comparison.

Secondary inoculum. Inoculum that is produced on infected tissue and is capable of infecting a new host immediately.

Secondary metabolite. A compound not directly associated with the processes that support growth but that presumably plays some other role (e.g., defense compounds).

Secondary pests. Some pesticide applications may lead to the development of a different pest that would not ordinarily be a problem.

Secretins. A family of proteins that are integral pore-forming outer membrane proteins.

Secreton. Collective term for the components of a type II secretion machinery.

Sedentary. A nematode that is nonmigratory or remains at a single location.

Seed certification. Practice of providing disease-free seed by certified clean production practices and verifying that the seeds are free from the pathogen of interest.

Segregating genes. Allele pairs or genes separate or segregate from each other into separate gametes.

Selection. The process of isolating and preserving individuals or characters from a group of individuals or characters.

Selection pressure. An abstract force that shapes the evolution of a population by differentially advantaging some traits and disadvantaging others; the driving force of natural selection.

Sensillum (pl. sensilla). An epithelial sense organ composed of one or a few cells with a nerve connection and taking the form of a spine, plate, rod, cone, or peg.

Septation. A structure that delimits cells.

Sequevar. Strain that has a particular sequence.

Serology. The study of antigen and antibody reactions and of their application to identification and purification of antigens.

Seta (pl. setae). A sterile, bristle-like hair.

Severity. The proportion of an individual (leaf, plant, or single field) affected by disease.

Sexual dimorphism. Distinct morphological forms of male and female nematodes.

Sexual reproduction. Reproduction of an organism requiring the union of compatible nuclei, forming the diploid chromosomal number, followed by a meiotic division that reduces the chromosomal number to the haploid state, followed by the subsequent formation of gametic cells that unite, reestablishing the diploid condition.

Shoe stringing. A severe thinning of plant structures such as leaves; when shoe stringing is very severe, the plant leaves may be almost vine like in appearance.

Short cycle rusts. Alternative term used for rust fungi that lack one or more spore stages in the life cycle.

Shot holes. Small holes in bark, leaves, fruit, or another plant part caused by an insect or a plant pathogen.

Siderophore. Low-molecular-weight compounds released by an antagonist that inhibit the growth of other organism by chelating or sequestering iron.

Sieve tube element. Found in phloem, this elongated cell is connected end to end to others, thereby transporting carbohydrate in the plant. The sieve tube element is cytoplasmic but lacks a nucleus. A nucleated companion cell is often adjacent to the sieve tube element.

Sign. The presence of the pathogen or its parts on a host plant.

Signal recognition/transduction. Active defense mechanism by which the plant recognizes the pathogen and begins the production of induced defenses. Ion fluxes, oxidative bursts, protein phosphorylation, and signal molecules are involved in signal recognition and transduction.

Signal words. Words such as Caution, Warning, or Danger appear on pesticide labels depending on the toxicity of the product.

Silent mutations. Altering, or mutating, one nucleotide changes the genomic DNA sequence, but not necessarily the amino acid sequence of the protein.

Simple sequence repeats (SSR). One of many DNA sequences dispersed throughout fungal, plant, and animal genomes, composed of short (2–10 base pairs) sequences that are repeated in tandem and are usually highly variable. Also known as microsatellites.

Single gene resistance. Resistance conditioned by a single gene; See **monogenic**.

Sink cells. Cells in a plant that accumulate carbohydrates, such as those that require energy (e.g., actively dividing meristems), or those functioning in storage such as roots and fruits.

Site selection. Careful consideration of many factors that influence where to plant crops or ornamental landscapes to lessen the chance of disease.

Smut. A basidiomycete fungus (Ustilaginales) that typically produces brown teliospores in **sori**.

Sod-based rotations. Perennial forage grasses and legumes that reduce the incidence of some diseases and function by either introducing a perennial crop that does not serve as a host for a given pathogen or creating a cropping system environment that is unfavorable to disease organisms.

Soft rots. Typically a postharvest disease caused by some bacteria or fungi that produce pectinases, which destroys the middle lamella between the plant cells.

Soil fumigant. A gas or volatile substance applied to soil to kill or inhibit the growth of microorganisms, nematodes, or other pests.

Soil fungistasis. Phenomenon in natural soils whereby fungal growth or spore germination is inhibited. The effect of fungistasis is often overcome by root or seed exudates.

Soil inhabitant. Organism that survives in soil as propagules and through saprophytic colonization of dead organic matter. Soil inhabitants can maintain their population density or increase in number in the absence of a host plant.

Soil invader. Organism that survives in soil as propagules, but does not have an active phase in soil, such as saprophytic colonization. Over time, in the absence of a host plant, the population of a soil invader declines.

Soil water potential. The difference in potential energy per unit quantity of water between soil water and pure, free water (reference state set at 0 water potential). The components of soil water potential include gravitational, matric, osmotic, and pressure potential.

Soilborne plant pathogen. Diverse group of plant pathogens that can survive in soil for an extended period of time in the absence of host plants. Group includes fungi, bacteria, viruses, nematodes, and parasitic higher plants.

Soilless media. Material used in pots, flats, or other containers in which plants are grown. Components of soilless media may include peat moss, sand, vermiculite, perlite, rockwool, styrofoam, compost, bark, or other organic materials. Soil test laboratories usually consider a potting medium "soilless" if it is composed of less than 25% field soil.

Solarization. Use of sunlight to heat soil to temperatures lethal to plant pathogens and other biological contaminants.

Sorus (pl. sori). Fruiting structure in rust and smut fungi containing a mass of spores.

Source cells. Cells in a plant that generate carbohydrates via photosynthesis, such as the leaf mesophyll (palisade parenchyma).

Spatial. Of or pertaining to existing in space.

Species complex. A collection of very closely related organisms that cannot be resolved into distinct species.

Species name. Specific epithet of an organism.

Species. A group of individuals that are genetically or morphologically distinct from other organisms.

Specific epithet. The species name of a member of a genus.

Specific resistance (synonym race-specific resistance). Resistance effective against certain biotypes or races of a pathogen but ineffective against other biotypes or races.

Spectrophotometer. An instrument used to measure the intensity of light of a specific wavelength transmitted by a substance or a solution. This measurement indicates the amount of material in the solution absorbing the light.

Spermagonium (pl. spermagonia). In rust fungi, a globose or flask-shaped enclosed haploid fruiting body composed of **receptive hyphae** and **spermatium (pl. spermatia** or **pycniospores.** Synonym **Pycnium (pl. pycnia).**

Spermatium (pl. spermatia). Spores that function as male gametes in the Ascomycota and Basidiomycota.

Spermosphere. The volume of soil that is physically, chemically, or biologically affected by the presence of living seeds.

Spontaneous generation. Life arising directly from inert matter.

Sporangiole. A small sporangium that lacks a columella and that contains only a few spores (Zygomycetes).

Sporangiophore. The hyphal stalk that bears a sporangium.

Sporangiospore. Motile or nonmotile asexual spore produced in a sporangium.

Sporangium (pl. sporangia). A saclike structure that is internally converted into spores or a single spore (downy mildews).

Spore. The cellular reproductive unit of many organisms (similar in function to seeds of plants).

Sporidium (pl. sporidia). Basidiospores, or spore other than a teliospore formed in smut fungus life cycle.

Sporocarp. Fruiting structure that bears spores.

Sporodochium (pl. sporodochia). Conidiomata with short conidiophores that are clustered into a rosette and produced on a superficial mycelial layer.

Sporophore. Stalk that supports a spore (see **conidiophore**; **sporangiophore**).

Stabilizing selection. A type of natural selection in which genetic diversity (such as a virulence mutation) decreases as the population stabilizes.

Stage micrometer. A glass slide that contains a precisely etched scale. The largest unit is typically a millimeter (1000 µm) and the smallest unit is usually 0.01 mm (10 µm). It is used to calibrate an ocular micrometer. It is also used to determine the magnification of images recorded with a microscope.

Steam disinfestation. Use of heat to reduce numbers of soilborne plant pathogens and plant-parasitic nematodes prior to planting a susceptible crop.

Sterile fungi. Fungi that do not produce spores. See **Mycelia sterilia**.

Sterility. An inability of a plant to produce or set viable seed.

Sterilization. The process for killing all forms of life.

Stipe. The stalk of a mushroom basidiocarp that supports the pileus.

Stolon. Branch of hyphae that skips over the substrate – a runner; in the Zygomycota.

Strain. A virus isolate that differs from the type isolate of the species in a definable character, but does not differ enough to be a new species.

Streak. A mosaic pattern that occurs on a monocot host; the leaf lamina in monocots is bounded by the parallel veins found in their leaves. Thus, a mosaic on a monocot host forms between the parallel veins and is seen as a streak of altered pigmentation. Streaks may tend to be shorter and more broken in early stages than stripes.

Stripe. A mosaic pattern that occurs on a monocot leaf when the mosaic forms in the leaf lamina between the parallel veins of a monocot host; Stripes and streaks are both formed in this manner and are often used interchangeably.

Stroma (pl. stromata). A mass of vegetative hyphae with or without tissue of the host or substrate; may bear spores or fruiting structures.

Stunting. A reduction in plant size compared with an uninfected plant growing under the same conditions. Can occur through the reduction in the entire plant or through reduction of a particular plant structure such as internodes.

Stylet. A needlelike structure found in the anterior end of nematodes and used to penetrate cell walls for feeding and movement inside roots. Also, a type of piercing-sucking mouthpart found in Homoptera.

Subcuticular. Beneath the waxy cuticle.

Surface sterilization. Removing or killing microorganisms from a surface, such as a leaf or root. See **disinfected**.

Sun scald. Injury that occurs to fruits or leaves when exposed to a sudden increase in light intensity.

Survival. Dormant stage or period of inactivity for a pathogen.

Susceptibility (susceptible). Inability of the host to reduce the growth, reproduction and/or disease-producing activities of the pathogen when environmental conditions are favorable (also **susceptible**).

Symplast. A space within the bounds of the plant plasma membrane. Composed of cell spaces linked together by their plasmodesmata, and has cytoplasmic continuity.

Symptom suppression. A decline or disappearance of symptoms that had previously been expressed during an infection.

Symptom (symptomatic). The response of the plant to the presence of a pathogen or adverse conditions.

Symptomless. An infection that is marked by the absence of observable symptoms in the host.

Synascus. A compound ascus found in the Protomycetaceae in which meiosis occurs and multiple ascospores are formed by mitosis.

Syncytium. A specialized, multinucleate feeding cell induced by cyst nematodes inside plant roots.

Syndrome. A series of symptoms that are characteristic for a given disease.

Synnema (pl. synnemata). Conidiomata with compact or aggregated clusters of erect conidiophores with conidia formed at or near the apex.

Systemic acquired resistance (SAR). An active defense mechanism in which a necrotizing pathogen causes distant induction of general defense mechanisms that serve to protect the plant. Salicylic acid or methyl salicylate are produced as a primary and secondary responses; antimicrobial proteins are also formed.

Systemic fungicides. Fungicides that have one or more of the following characteristics: the ability to enter a plant through roots or leaves, water solubility to enhance movement in the plant's vascular system, and stability within the plant. Systemic fungicides are often more vulnerable to the development of resistance in the target fungus population.

TAL effectors. Acronym standing for transcription activator-like effectors (sometimes also referred to ta TALEs). Proteins secreted by *Xanthomonas* bacteria into their host's cells, which can bind promoter sequences in the host plant and activate the expression of plant genes that aid bacterial infection. They recognize plant DNA sequences through a predictable amino acid motif.

Tannins. Polyphenolic compounds occurring in plants that have the ability to bind and precipitate proteins.

Target pest resurgence. Chemical pesticide applications also reduce the number of natural enemies of the target pest, and the populations of the target pest rebound.

Taxon (pl. taxa). The name applied to a taxonomic group (e.g., genus, species) in a formal system of nomenclature.

Taxonomist. A person who studies taxonomy.

Taxonomy. The study of the general principles of scientific classification. Synonym: systematics.

Teleomorph. Conceptually, the sexual aspect of a fungus; contrast with **holomorph** and **anamorph**.

Teliospore. Thick-walled, over-wintering or resting spore of rusts and smuts, in which karyogamy (nuclear fusion) occurs and from which the basidium arises.

Telium (pl. telia). Binucleated cell(s) that produce teliospores.

Temporal. Of or limited by time.

Tendril. A specialized stem in twining plants that attaches them to their support. In parasitic plants such as *Cuscuta*, these also form the haustorial connections to the host plant.

Terpenes. A diverse class of organic molecules that are odiferous.

Terroirs. The set of environmental factors that influences epigenic aspects of plants when grown in a specific location.

Tertiary gene pool. Species that are more distantly related to the species in the breeding program. See **Primary gene pool** and **Secondary gene pool** for comparison.

Thallus. The vegetative (nonreproductive) body or soma of a fungus.

Thermalcycler. The machine that is programmed for automatic temperature changes (cycles) required for the PCR. Typically has set points for denaturing, annealing, and extension temperatures.

Thermophile (thermophilic). An organism that grows well and reproduces at warmer temperatures (> 45°C).

Thigmotropism. Contact response to a solid or rigid surface that results in orientation of an organism or one of its parts.

Tillage. One of the primary methods of site preparation for many crops.

Tolerance. The ability of a cultivar to perform well under adverse conditions. "True tolerance" occurs if one cultivar sustains less damage than another cultivar when the amount of infection is the same for both cultivars. Tolerance is also used synonymously with partial or **general resistance** in which case the improved performance under adverse conditions is due to a lesser amount of disease on the "tolerant" cultivar (i.e., partial resistance).

Total soluble salts. The total amount of ions in the soil usually measured as electrical conductivity in mSiemens per cm (mS/cm).

Toxin(s). Substances that disturb normal metabolic pathways and lead to the localized or systemic death of an organism.

Tracheid. Elongated cells that transport water and minerals in vascular plants. One of two cell types in plant xylem, the other being vessel elements.

Transcription. The transfer of genetic information associated with deoxyribonucleic acid (DNA) to messenger ribonucleic acid (mRNA).

Transcriptomic. The study of the entire complement of RNA transcripts in an organism.

Transduction. Process by which DNA is transferred from one bacterium to another by a viral vector.

Transelimination (β-elimination). Chemical reaction catalyzed by lyases that results in the production of unsaturated products.

Transfer cell. A specialized cell in the host plant that facilitates transfer of food to a sedentary endoparasitic nematode.

Transformation. Stable genetic change of one organism that occurs through uptake of DNA without associated proteins and organelles.

Transgenic. Containing a gene from another species that was introduced by artificial means (genetic transformation).

Transillumination. The inspection of a specimen by passing a light through it. This is typically used with compound microscopes and can be used with stereoscopes that have transilluminator bases.

Translation. The transfer of genetic information associated with the messenger ribonucleic acid (mRNA) into a sequence of amino acids forming a polypeptide chain ultimately forming a protein.

Translatome. The entirety of the proteins translated from mRNA. See **Translation**.

Translocation. Movement of materials, including water, minerals, and organic materials through the vascular system of the plant.

Transpiration. Loss of water in the vapor phase from leaves and other aboveground plant parts.

Transportable. The ability to be moved or transported in an organism or system.

Tree of life. A project examining the diversity of organisms on Earth and how they might be related.

Trehalose. Disaccharide of glucose used as a carbon reserve compound in fungal cells.

Trichome. Plant hair.

Trophic groups. The division of nematodes into groups by preferred food, ecological niche, and taxonomic relationships.

True nucleus. Nuclei of eukaryotic organisms in which chromosomes are surrounded with a nuclear membrane.

Tumor. Large overgrowths caused by hyperplasia and hypertrophy of infected cells; *Wound tumor virus* (WTV) is an example of a virus that produces tumors in its host.

Tylosis (pl. tyloses). Occurs in xylem tissue; outgrowth of a parenchyma (ray) cell that partially or completely blocks the lumen of the vessel. Often associated with **vascular wilts diseases**.

Type III effector(s). Proteins secreted via a type III secretion system of Gram-negative plant pathogenic bacteria.

Type strain. A living culture that serves as a fixed reference point for the assignment of bacterial and archaeal names, thus often also denoted as a reference strain.

Unculturable bacterium. A species of bacteria that cannot be grown or cultured on artificial medium.

Unipartite genome. The complete genome of an organism, such as a virus, is contained on a single strand of nucleic acid.

Unitunicate ascus. An ascus having a single wall.

Urediniospore. Dikaryotic (N + N), repeating, "summer" spore stage of rust fungi; the only spore stage in the **heteroecious** rust life cycle that infects the same host on which it is produced.

Uredinium (pl. uredinia). The dikaryotic, repeating stage in rust fungus life cycle; a structure in which **urediniospores** are formed.

Vacuole. A membrane-bound vessel found in fungi and other eukaryotic organisms that is filled with various substances including water, enzymes, salts, nutrients, wastes, or secondary metabolites, among others.

Variety mixtures. Mixed plantings of crop varieties that complement each other by having different disease resistances, yield potential, and/or other traits.

Vascular wilt diseases. Diseases caused by pathogens that infect the xylem.

Vector. A nematode, insect, or other organism that can transmit a virus or other agent into or onto a plant.

Vector. An organism that moves the inoculum for a disease to a new plant or even a new area; plant viruses are strongly associated with the vectors that transmit them.

Vegetative compatibility. Vegetative hyphae that are able to fuse and maintain the heterokaryotic genetic state.

Vein banding. Areas of intense pigmentation form bordering the veins of the leaves; Vein banding is typically seen as dark green bands bordering major leaf veins. Vein banding may be narrow or several millimeters wide. *Bean common mosaic virus* (BCMV) is an example of a virus that causes vein banding.

Vein clearing. A loss of pigmentation (clearing or translucence of tissue) in the veins; It can be best observed by allowing light to shine through the leaf. This clearing is associated with enlargement of the cells near the vein in some viruses. Vein clearing sometimes precedes the formation of a mosaic.

Veraison. Period of grape ripening during which berry growth slows down and coloration develops.

Vermiform. In nematodes, worm-shaped body, much longer than it is wide. Threadlike. Common shape of most plant-parasitic nematodes.

Vertical resistance. See **Qualitative resistance.**

Vesicle. Intra- or inter-cellular, ovate to spherical structure that contains storage lipids and may also serve as a propagule in some endomycorrhizae. Also, the thin-walled, balloon-like structure in which zoospores of *Pythium* species (Oomycota) are differentiated.

Virion. Individual virus particle composed of a DNA or RNA core surrounded by a protein coat or capsid.

Viroid. Subcellular pathogens simply composed of small "naked" RNA genomes; viroids do not have a protein coat nor do they produce proteins during pathogenesis.

Virulence. The ability of a pathogen to cause a compatible (susceptible) reaction on a host cultivar with genetic resistance.

Virulence factor. Component of a pathogen that contributes to pathogenesis.

Virulence formula. Identification of isolates of pathogens based on effective and ineffective host genes.

Viruliferous. A vector that has acquired virus and is capable of transmitting it.

Virus. A pathogen that comprises either RNA or DNA and is enclosed in a protective protein coat.

Viral inclusion bodies. Aggregations of virus particles in the plant cell, which are visible using a compound microscope, and useful for classifying individual virus groups.

Viscin. A sticky mucilaginous pulp that coats the seed of dwarf mistletoes.

Volva. At the base of the stipe, a remnant of the universal veil that once enveloped the entire developing mushroom.

White rusts. Diseases caused by species of *Albugo*, *Pustula*, and *Wilsoniana* (Oomycota, Albuginaceae).

White smut. A basidiomycete fungus producing spores in white sori.

Whole Genome Sequencing. The entire DNA sequence of an organism's genome determine in a single process.

Wilt (Wilting). General response to loss of water brought about by diseases that impeded or degrade the vascular system; may also be caused by lack of available water.

Wilting. A disease symptom marked by a visible loss of turgor and often leading to plant's collapse.

Witches' brooms. Distortions of normal shoot growth in which a loss of apical dominance creates a bunched growth of shoots.

Wood decay. Enzymatic and chemical process by which microorganisms, such as basidiomycetes, degrade plant cell walls to obtain nutrients for growth and reproduction. Generally white and brown rots are recognized.

Yield decline. A common problem in monoculture crops grown without rotation or in very simple rotations.

Xylanases. A number or battery of enzymes that degrade polymers containing D-xylose, a five-carbon sugar component of the plant cell wall polymer, hemicellulose. Also known as hemicellulases.

Xylome. Proteins transported in xylem and a subset of the **proteome**.

Yellows. A generalized yellowing or lack of chlorophyll affecting the majority of the plant; often striking in appearance due to their bright yellow color.

Yield loss. The difference in yield between a host infected with a disease and another host of the same species that is not infected with a disease. Yield loss comparison must be made with hosts grown in as close to the same environment as possible.

Zone of root elongation. The region of a root tip, just behind the root apical meristem, where cells undergo elongation. Cellular elongation pushes the root cap and apical meristem through the soil.

Zoospore. A flagellated spore produced during reproduction capable of moving in water (i.e., "swimming spore").

Zygomycota. Fungal phylum where sexual reproduction results in the production of zygospores; includes common bread mold and decay fungi; in the kingdom Fungi.

Zygospore. A diploid, thick-walled spore that results from sexual reproduction; in the **Zygomycota**.

Index

C

D

E

P

T

U

V

W

X

Y

Z

For Product Safety Concerns and Information please contact our EU representative GPSR@taylorandfrancis.com Taylor & Francis Verlag GmbH, Kaufingerstraße 24, 80331 München, Germany

Batch number: 10733561

Printed by Printforce, the Netherlands